tenth edition

Human Genetics

Concepts and Applications

Ricki Lewis

Genetic Counselor
CareNet Medical Group
Schenectady, New York

Adjunct Assistant Professor of
Medical Education
Alden March Bioethics Institute
Albany Medical College

HUMAN GENETICS: CONCEPTS AND APPLICATIONS, TENTH EDITION

Published by McGraw-Hill, a business unit of The McGraw-Hill Companies, Inc., 1221 Avenue of the Americas, New York, NY 10020. Copyright © 2012 by The McGraw-Hill Companies, Inc. All rights reserved. Printed in the United States of America. Previous editions © 2010, 2008, and 2007. No part of this publication may be reproduced or distributed in any form or by any means, or stored in a database or retrieval system, without the prior written consent of The McGraw-Hill Companies, Inc., including, but not limited to, in any network or other electronic storage or transmission, or broadcast for distance learning.

Some ancillaries, including electronic and print components, may not be available to customers outside the United States.

This book is printed on acid-free paper.

2 3 4 5 6 7 8 9 0 DOW/DOW 1 0 9 8 7 6 5 4 3 2

ISBN 978–0–07–131583–8
MHID 0–07–131583–7

All credits appearing on page or at the end of the book are considered to be an extension of the copyright page.

www.mhhe.com

About the Author

Ricki Lewis has built an eclectic career in communicating the excitement of genetics. She earned her Ph.D. in genetics in 1980 from Indiana University. It was the dawn of the modern biotechnology era, which Ricki chronicled in many magazines and journals. She published one of the first articles on DNA fingerprinting in *Discover* magazine in 1988, and a decade later one of the first articles on human stem cells in *The Scientist*.

Ricki has taught a variety of life science courses at Miami University, the University at Albany, Empire State College, and community colleges. She has authored or co-authored several university-level textbooks and is the author of *The Forever Fix: Gene Therapy and the Boy Who Saved It*, as well as an essay collection and a novel. On the clinical front, Ricki has been a genetic counselor for a private medical practice since 1984 and has been a hospice volunteer since 2005. She is a frequent public speaker.

Ricki presently teaches an online course on "Genethics" for the Alden March Bioethics Institute of Albany Medical College. She lives in upstate New York and sometimes Martha's Vineyard, with husband Larry, many cats, and a tortoise. She can be reached at rickilewis54@gmail.com and writes a blog, Genetic Linkage, at www.rickilewis.com.

Dedicated to
Benzena Tucker and
Glenn Nichols, who
taught me the value
of optimism.

Brief Contents

Contents

CHAPTER 21

Reproductive Technologies 406

CHAPTER 22

Genomics 423

Preface

Human Genetics Comes Full Circle

I like to peek at the ending to a novel. I just can't wait to know what will happen to the characters. Human genetics has become the same way, and so I invite you to read the very last section of this tenth edition now, at the beginning. It is about the personal genome sequencing of one man, and how he is using the information. As the cost of such sequencing plummets, we all may be able to look to our genomes for echoes of our pasts and hints of our futures—if we so choose. We may also learn what we can do to counter our inherited tendencies and susceptibilities. Genetic knowledge is informative and empowering. This book shows you how and why this is true.

Ricki Lewis

Today, human genetics is for everyone. It is about our variation more than about our illnesses, and increasingly about the common rather than the rare. Once an obscure science or an explanation for an odd collection of symptoms, human genetics is now part of everyday conversation. At the same time, it is finally being recognized as the basis of medical science. Despite the popular tendency to talk of "a gene for" this or that, we now know that for most traits and illnesses, several genes interact with each other and environmental influences.

What Sets This Book Apart

Current Content

The exciting narrative writing style, with clear explanations of concepts and mechanisms propelled by stories, reflects Dr. Lewis's eclectic experience as a science writer, instructor, and genetic counselor, coupled with her expertise in genetics. Updates to this landmark edition include

- Genetic tests, from preconception to old age
- From Mendel to molecules: family genome analysis
- Allelic diseases: one gene, two diseases
- Denisova hominins, coexisting with us and Neanderthals
- Disease-in-a-dish stem cell technology
- The reemergence of gene therapy
- Personal genome sequencing: promises and limitations

This edition continues the shift in focus from rare single-gene inheritance to more common multifactorial traits and diseases.

The Human Touch

Human genetics is about people, and their voices echo throughout these pages.

Compelling Stories and Cases When the parents of children with visual loss stood up at a conference to meet other families with the same very rare inherited disease, Dr. Lewis was there, already composing the opening essay to chapter 5. She met the little girl in Reading 2.2, who is 1 in 51 in the world with her disease, as well as the magazine editor in Reading 18.1, who was one of the first to benefit from a wonder drug for her leukemia. In the photo that starts chapter 20, Dr. Lewis reads birthday cards to Max, who had gene therapy for Canavan disease. Readers of earlier editions of this book made the cards for Max, who has appeared here since age 3. The experiences of real people flesh out the science of genetics.

Practical Application of Human Genetics A working knowledge of the principles and applications of human genetics is critical to being an informed citizen and health care consumer. Broad topics of particular interest include

- The roles that genes play in disease risk, physical characteristics, and behavior, with an eye toward the dangers of genetic determinism
- Biotechnologies, including next-generation DNA sequencing, genetic testing, stem cell technology, gene expression profiling, genome-wide association studies, gene therapy, familial DNA searches, whole exome sequencing, cell-free fetal DNA testing, and personal genome sequencing
- Ethical concerns that arise from the interface of genetic information and privacy.

The Lewis Guided Learning System

Each chapter begins with two views of the content. "*Learning Outcomes*" embedded in the table of contents guide the student in mastering content. "*The Big Picture*" encapsulates the overall theme of the chapter. Content flows logically through three to five major sections per chapter that are peppered with high-interest boxed readings ("*In Their Own Words,*" "*Readings,*" "*Bioethics: Choices for the Future,*" and "*Technology Timelines*"). End-of-chapter pedagogy progresses from straight recall to applied and creative questions and challenges. Overall the pedagogical features of each chapter reinforce key ideas and prompt students to think more deeply about the application of the content.

Dynamic Art

Outstanding photographs and dimensional illustrations, vibrantly colored, are featured throughout *Human Genetics*. Figure types include process figures with numbered steps, micro to macro representations, and the combination of art and photos to relate stylized drawings to real-life structures.

New to This Edition!

New and updated information is integrated throughout the chapters. Highlights from the revision are included here.

Chapter 1 *Overview of Genetics*
- A university encourages genetic testing of incoming freshmen—until stopped.

Chapter 2 *Cells*
- A "disease-in-a-dish" for a rare condition opens the chapter.

Chapter 3 *Meiosis and Development*
- Can sirtuins slow aging?

Chapter 4 *Single-Gene Inheritance*
- Family genome analysis bridges Mendelian and molecular genetics.

Chapter 5 *Beyond Mendel's Laws*
- Genetic heterogeneity is seen through the eyes of families at a conference for hereditary blindness.

Chapter 6 *Matters of Sex*
- Wiskott-Aldrich syndrome opens the chapter.
- Hemophilia B affected the royal families, not hemophilia A.

Chapter 7 *Multifactorial Traits*
- "Genetics of Athletics" opens the chapter.
- More emphasis on epigenetics.
  Faces: a compelling example of normal variation revealed with genome-wide association studies.

Chapter 8 *Genetics of Behavior*
- Behavioral disorders may not be distinct at the genetic level.
- How are autism and schizophrenia related?

Chapter 10 *Gene Action: From DNA to Protein*
- "Whole exome sequencing" opens the chapter.

Chapter 11 *Gene Expression and Epigenetics*
- How famine sowed the seeds of schizophrenia, thanks to epigenetics.

Chapter 12 *Gene Mutation*
- Restoring protein folding to treat cystic fibrosis opens the chapter.
- Allelic diseases explain old mysteries.

Chapter 14 *Constant Allele Frequencies*
- Forensics update: touch DNA, familial DNA searches, fish fraud.

Chapter 15 *Changing Allele Frequencies*
- Update on dog and cat genetics.
- "Mutiny on the Bounty" and migraine: a profound founder effect.

Chapter 16 *Human Ancestry*
- Exploiting the Havasupai in *Bioethics: Choices for the Future.*
- Neanderthal genome reveals we interbred.
- Denisova hominins shared Europe with our ancestors and Neanderthals.
- DNA of the Khoisan is the most diverse.
- A newly discovered paleo-Eskimo from Greenland reveals an ancient migration route.

Chapter 17 *Genetics of Immunity*
- The Berlin patient: curing leukemia and HIV.
- A new view of allergy, due to mutation.

Chapter 19 *Genetic Technologies: Amplifying, Modifying, and Monitoring DNA*
- Experimental evolution and the Gulf oil spill.

Chapter 20 *Genetic Testing and Treatment*
- Max Randell turns 13, thanks to gene therapy for Canavan disease.
- Preconception comprehensive carrier testing for 400+ diseases.
- Cell-free fetal DNA reveals entire fetal genomes.
- Repurposed drugs treat common genetic diseases.
- Gene therapy returns with three successes.

Chapter 21 *Reproductive Technologies*
- "The Twiblings" open the chapter.
- The ethics of postmortem gamete retrieval, male and female.

Chapter 22 *Genomics*
- Updates on genome projects.
- Sequencing the *Cacao* genome (chocolate).
- The human microbiome.
- How useful is a personal genome sequence?

ACKNOWLEDGMENTS

Human Genetics: Concepts and Applications, Tenth Edition, would not have been possible without the editorial and production dream team: Developmental Editor Wendy Langerud, Project Manager Kelly Heinrichs, copyeditor Beatrice Sussman, and photo editor extraordinaire, Toni Michaels. Special thanks to Rita Ryan at Cor Jesu Academy in St. Louis; Don Watson, dedicated learner, who provides meticulous feedback to every edition; and John Davis, who filled two notebooks with helpful suggestions and criticism. I would also like to thank the people with inherited disease in their families who shared their stories, and Stephen Quake for permission to write about his genome. Renad Zhdanov and Shelly Queneau Bosworth clarified certain concepts. As always, many thanks to my wonderful husband Larry for his support and encouragement and to my three daughters, who grew up with this book.

Tenth Edition Reviewers

Andy Andres
  Boston University
Marne Bailey
  Lewis University
Judy Bluemer
  Morton College
Dean Bratis
  Villanova University
Peter E. Busher
  Boston University

Dan A. Dixon
  University of South Carolina
Kim R. Finer
  Kent State University
Mike Ganger
  Gannon University
Bradley Isler
  Ferris State University
Bridget Joubert
  Northwestern State University
Anthony Jay Julis
  Washington University
Brenda Knotts
  Eastern Illinois University
Nicholas J. LoCascio
  Niagara County Community College
Bernard Possidente
  Skidmore College
Cheryl Wistrom
  Saint Joseph's College
Kathy Zoghby
  University of Richmond

This book continually evolves thanks to input from instructors and students. Please let me know your thoughts and suggestions for improvement. (rickilewis54@gmail.com)

Applying Human Genetics

Chapter Openers

The Human Touch

Readings

In Their Own Words

Bioethics: Choices for the Future

The Lewis Guided Learning System

Learning Outcomes enable students to practice and apply learning in the real world.

Numbered Sections help organize the content.

Chapter Openers show how the content relates to real life.

Key Concepts boxes summarize what a student should know before leaving each numbered section.

The Big Picture encapsulates the chapter content in a sentence or two.

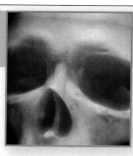

Comparing skulls among modern humans, our modern primate cousins, and fossilized hominins can reveal much about our ancestors and our evolution.

CHAPTER 16

Human Ancestry

Learning Outcomes

16.1 Human Origins

1. Distinguish between hominoids and hominins.
2. Explain why more than one species of *Australopithecus* coexisted.
3. Distinguish between *Australopithecus* and *Homo*.
4. Explain what genome sequencing has revealed about the ancestry of Neanderthals and us.

16.2 Molecular Evolution

5. Explain how DNA information can be used to shed light on evolution.
6. List genes that were important in our evolution.
7. Explain how chromosome banding patterns and protein sequences reveal evolutionary trends.
8. Explain what mitochondrial DNA and Y chromosome sequences reveal about human ancestry.

16.3 The Peopling of the Planet

9. Explain what mitochondrial Eve represents.
10. Describe how people expanded out of Africa and then Eurasia, populating the world.

The Big Picture: Our genes and genomes are informational molecules, and their sequences hold clues to our deep past as well as our present diversity.

The Hobbits

It's odd to be the only ones of our kind, which may be why a dual humanity theme persists in science fiction. *The Time Machine* looked at two battling breeds of people. In *Darwin's Children*, a virus scrambles the genomes of a group of newborns, starting a new species. In other stories, a Neanderthal lives in modern-day Tajikistan and a caveman in Kenya.

Fossils indicate that from 2 to 6 million years ago, humans and prehumans overlapped, in time if not place. The discovery of preserved bones of several ancient humans on the island of Flores in Indonesia in 2004 suggested a recent coexistence of two types of people. A female skeleton found 17 feet beneath a cave floor with pieces of others nearby was named *Homo floresiensis*, popularly called the Hobbit. She was about half as tall as a modern human, with a brain about a third of the size. She lived about 18,000 years ago.

The Hobbits exhibited "island dwarfism," an effect of natural selection on small, isolated, island populations. With limited resources, those who need less food are more likely to reproduce. Who were the Hobbits? At first, researchers thought that Hobbits were direct descendants of *Homo erectus*, who lived before us. Then, analysis of limb bones revealed feet and proportions like those of an ape, despite a more humanlike skull. Therefore, the Hobbits may have been direct descendants of a primate older than *Homo erectus*, who evolved in a different direction on their isolated island.

---

tall plants of unknown genotype with short (*tt*) plants. If a tall plant crossed with a *tt* plant produced both tall and short progeny, it was genotype *Tt*; if it produced only tall plants, it must be *TT*.

Crossing an individual of unknown genotype with a homozygous recessive individual is called a test cross. The logic is that the homozygous recessive is the only genotype that can be identified by its phenotype—that is, a short plant is always *tt*. The homozygous recessive is a "known" that can reveal the unknown genotype of another individual to which it is crossed.

Key Concepts

1. Mendel deduced that "elementen" for height segregate, then combine at random with those from the opposite gamete at fertilization.
2. A homozygote has two identical alleles, and a heterozygote has two different alleles. The allele expressed in a heterozygote is dominant; the allele not expressed is recessive.
3. A monohybrid cross yields a genotypic ratio of 1:2:1 and a phenotypic ratio of 3:1.
4. Punnett squares display expected genotypic and phenotypic ratios among progeny.
5. A test cross uses a homozygous recessive individual to reveal an unknown genotype.

4.2 Single-Gene Inheritance Is Rare

Mendel's first law addresses traits determined by single genes. **Reading 4.1** describes a few unusual single-gene traits. Inheritance of single genes is also called Mendelian, or monofactorial, inheritance. Single-gene disorders, such as sickle cell disease and muscular dystrophy, are rare compared to infectious diseases, cancer, and multifactorial disorders, affecting 1 in 10,000 or fewer individuals. Because of the rarity of single-gene diseases, getting an accurate diagnosis can be difficult if physicians are unfamiliar with the phenotype.

Single-gene inheritance is much more complicated than it might appear from considering such obvious traits as green or yellow pea color. Sequencing the human genome and using SNPs (points in the genome where people vary) to catalog inherited variation in genome-wide association studies have

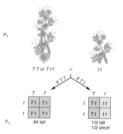

P_1

T T or *T t*?   ×   *tt*

F_1 All tall   1/2 tall 1/2 short

Figure 4.5 Test cross. Breeding a tall pea plant with homozygous recessive short plants reveals whether the tall plant is ...

Technology Timeline

PATENTING LIFE AND GENES

1790 U.S. patent act enacted. A patented invention must be new, useful, and not obvious.

1873 Louis Pasteur is awarded first patent on a life form, for yeast used in industrial processes.

1930 New plant variants can be patented.

1980 First patent awarded on a genetically modified organism, a bacterium given four DNA rings that enable it to metabolize components of crude oil.

1988 First patent awarded for a transgenic organism, a mouse that manufactures human protein in its milk. Harvard University granted patent for "OncoMouse" transgenic for human cancer.

1992 Biotechnology company awarded patent for all forms of transgenic cotton. Groups concerned that this will limit the rights of subsistence farmers contest the patent several times.

1996–1999 Companies patent partial gene sequences and certain

In-Chapter Review Tools, such as chapter glossaries and timelines of major discoveries, are handy tools for reference and study.

Bioethics: Choices for the Future boxes include Questions for Discussion.

Bioethics: Choices for the Future

Banking Stem Cells

The parents-to-be were very excited by the DVD that came in the mail shortly after they began seeing an obstetrician:

Bank your baby's cord blood stem cells and benefit from breakthroughs. Be prepared for the unknowns in life.

The short film profiled children who were saved from certain deadly diseases because their parents had stored their umbilical cord blood. The statistics quoted were persuasive: More than 70 diseases are currently treatable with cord blood transplants, and 10,000 procedures have already been done.

With testimonials like that, it is little wonder that parents collectively spend more than $100 million per year to store cord blood. The ads and statistics are accurate but misleading, because of what they *don't* say. Most people never actually use the umbilical cord blood stem cells that they store. The scientific reasons go beyond the fact that treatable diseases are very rare. In addition, cord blood stem cells are not nearly as pluripotent as some other stem cells, limiting their applicability. Perhaps the most compelling reason that stem cell banks are rarely used is based on logic: For a person with an inherited disease, *healthy* stem cells are required—not his or her own, which could cause the disease all over again. The patient needs a donor.

Commercial cord blood banks may charge more than $1,000 for the initial collection plus an annual fee. However, the U.S. National Institutes of Health and organizations in many other nations have supported not-for-profit banks for years, and do not charge fees. Donations of cord blood to these facilities are not to help the donors directly, but to help whoever can use the cells.

As stem cell science has leaped forward, both commercial cell banks and anecdotal reports of successes have captured much media attention. This was the case for an 18-month-old boy whose cerebral palsy greatly improved after he was treated with his own cord blood cells. Whether he would have improved without the treatment isn't known.

Commercial stem cell banks are not just for newborns. One company, for example, offers to bank "very small embryonic-like stem cells" for an initial charge of $7,500 and a $750 annual fee, "enabling people to donate and store their own stem cells when they are young and healthy for their personal use in times of future medical need." The cells come from a person's blood and, in fact, one day may be very useful, but the research has yet to be done supporting use of the cells in treatments.

Questions for Discussion

1. Storing stem cells is not regulated by the U.S. government the way that a drug or a surgical procedure is because it is a service that will be helpful for treatments not yet invented. Do you think such banks should be regulated, and if so, by whom and how?
2. What information do you think that companies offering to store stem cells should present on their websites?
3. Do you think that advertisements for cord blood storage services that have quotes and anecdotal reports, but do not mention that most people who receive stem cell transplants do not in fact receive their own cells, are deceptive? Or do you think it is the responsibility of the consumer to research and discover this information?
4. How can medical consumers become aware that the government funds facilities to store stem cells?
5. It is likely that in the future, stem cell–based treatments will be possible, following large-scale clinical trials. What is the fairest way to prepare for this type of future medical treatment?

Each chapter ends with a point-by-point **Chapter Summary.**

Review Questions assess content knowledge.

Applied Questions help students develop problem-solving skills.

Summary

11.1 Gene Expression Through Time and Tissue

1. Changes in gene expression occur over time at the molecular level (globin switching), at the tissue level (blood plasma), and at the organ/gland level (pancreas development).
2. **Proteomics** catalogs the types of proteins in particular cells, tissues, organs, or entire organisms under specified conditions.

11.2 Control of Gene Expression

3. Acetylation of certain histones enables the transcription of associated genes. Phosphorylation and methylation are also important in **chromatin remodeling**.
4. **MicroRNAs** bind to certain mRNAs, blocking translation.

11.3 Maximizing Genetic Information

5. A small part of the genome encodes protein, but these genes specify a much greater number of proteins.
6. Alternate splicing, use of introns, and cutting proteins translated from a single gene contribute to protein diversity.

11.4 Most of the Human Genome Does *Not* Encode Protein

7. The nonprotein-encoding part of the genome includes viral sequences, noncoding RNAs, **pseudogenes,** introns, promoters and other controls, and repeats.

www.mhhe.com/lewisgenetics10

Answers to all end-of-chapter questions can be found at **www.mhhe.com/lewisgenetics10.** You will also find additional practice quizzes, animations, videos, and vocabulary flashcards to help you master the material in this chapter.

Review Questions

1. Why is control of gene expression necessary?
2. Define *epigenetics.*
3. Distinguish between the type of information that epigenetics provides and the information in the DNA sequence of a protein-encoding gene.
4. Describe three types of cells and how they differ in gene expression from each other.
5. Explain how a mutation in a promoter can affect gene expression.
6. What is the environmental signal that stimulates globin switching?
7. How does development of the pancreas illustrate differential gene expression?
8. How do histones control gene expression, yet genes also control histones?
9. Name a mechanism that silences transcription of a gene and a mechanism that blocks translation of an mRNA.
10. What controls whether histones enable DNA wrapped around them to be transcribed?
11. What are two ways that microRNA functioning is complex?
12. Describe three ways that the number of proteins exceeds the number of protein-encoding genes in the human genome.
13. How can alternate splicing generate more than one type of protein from the information in a gene?
14. In the 1960s, a gene was defined as a continuous sequence of DNA, located permanently at one place on a chromosome, that specifies a sequence of amino acids from one strand. List three ways this definition has changed.
15. Give an example of a discovery mentioned in the chapter that changed the way we think about the genome.

Applied Questions

1. Several new drugs inhibit the enzymes that either put acetyl groups on histones or take them off. Would a drug that combats a cancer caused by too little expression of a gene that normally suppresses cell division add or remove acetyl groups?
2. Chromosome 7 has 863 protein-encoding genes, but many more proteins. The average gene is 69,877 bases, but the average mRNA is 2,639 bases. Explain both of these observations.

Web Activities

1. Gene expression profiling tests began to be marketed just a few years ago. Google "Oncotype DX," "MammaPrint," or simply "gene expression profiling in cancer" and describe how classifying a particular cancer based on gene expression profiling can improve diagnosis and/or treatment. (Or apply this question to a different type of disease.)

Forensics Focus

1. Establishing time of death is critical information in a murder investigation. Forensic entomologists can estimate the "postmortem interval" (PMI), or the time at which insects began to deposit eggs on the corpse, by sampling larvae of specific insect species and consulting developmental charts to determine the stage. The investigators then count the hours backwards to estimate the PMI. Blowflies are often used for this purpose, but their three larval stages look remarkably alike in shape and color, and development rate varies with environmental conditions. With luck, researchers can count back 6 hours from the developmental time for the largest larvae to estimate the time of death.

   In many cases, a window of 6 hours is not precise enough to narrow down suspects when the victim visited several places and interacted with many people in the hours before death. Suggest a way that gene expression profiling might be used to more precisely define the PMI and extrapolate a probable time of death.

Case Studies and Research Results

1. Jerrold is 38 years old. His body produces too much of the hormone estrogen, which has enlarged his breasts. He had a growth spurt and developed pubic hair by age 5, and then his growth dramatically slowed so that his adult height is well below normal. He has a very high-pitched voice and no facial hair, which reflect the excess estrogen. Jerrold's son, Timmy, is 8 years old and has the same symptoms.

   Jerrold and Timmy have an overactive gene for aromatase, an enzyme required to synthesize estrogen. Five promoters control expression of the gene in different tissues, and each promoter is activated by a different combination of hormonal signals. The five promoters lead to estrogen production in skin, fat, brain, gonads (ovaries and testes) and placenta. In premenopausal women, the ovary-specific promoter is highly active, and estrogen is abundant. In men and postmenopausal women, however, only small amounts of estrogen are normally produced, in skin and fat. The father and son have a wild type aromatase gene, but high levels of estrogen in several tissues, particularly fat, skin, and blood. They do, however, have a mutation that turns around an adjacent gene so that the aromatase gene falls under the control of a different promoter. Suggest how this phenotype arises.

Web Activities encourage students to use the latest tools and databases in genetic analysis.

Capitalizing on students' interest in forensic science, new **Forensics Focus** questions make students think about the genetic principles involved in the collection and use of genetic information in criminal investigations.

Cases and Research Results use stories based on accounts in medical and scientific journals; real clinical cases; posters and reports from professional meetings; and fiction to ask students to analyze data and predict results.

Dynamic Art Program

Multilevel Perspective

Illustrations depicting complex structures show macroscopic and microscopic views to help students see the relationship between increasingly detailed drawings.

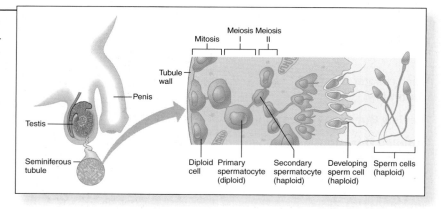

Combination Art

Drawings of structures are paired with micrographs to give the student the best of both perspectives: the realism of photos and the explanatory clarity of line drawings.

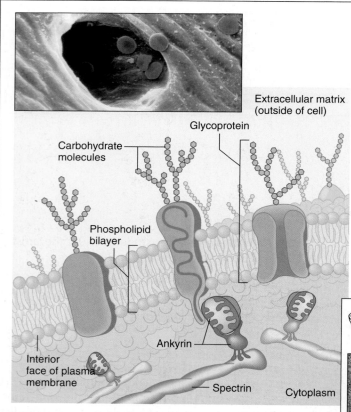

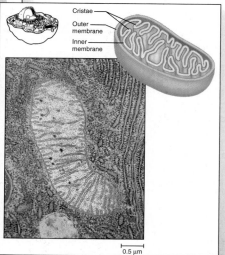

Complex Content in Context

Molecular and cellular information is put into a familiar context to help students make connections.

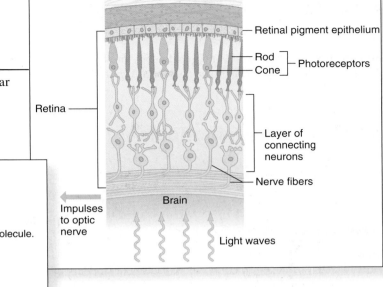

Retinal pigment epithelium

Rod
Cone — Photoreceptors

Retina

Layer of connecting neurons

Nerve fibers

Brain

Impulses to optic nerve

Light waves

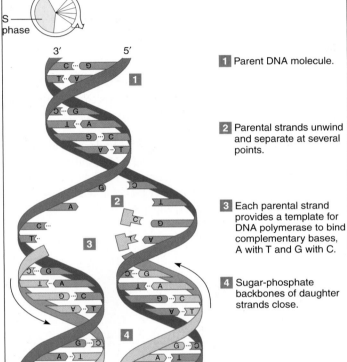

S phase

3′   5′

1 Parent DNA molecule.

2 Parental strands unwind and separate at several points.

3 Each parental strand provides a template for DNA polymerase to bind complementary bases, A with T and G with C.

4 Sugar-phosphate backbones of daughter strands close.

5′   3′   5′   3′

Process Figures

Complex processes are broken down into a series of smaller steps that are easy to follow. Here, organelles interact to produce and secrete a familiar substance—milk.

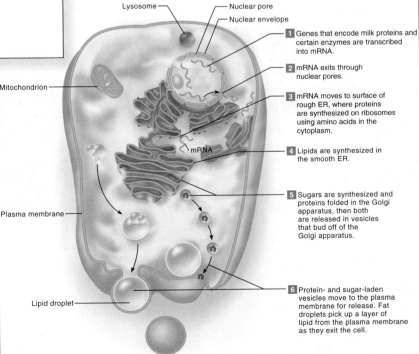

Lysosome

Nuclear pore

Nuclear envelope

1 Genes that encode milk proteins and certain enzymes are transcribed into mRNA.

2 mRNA exits through nuclear pores.

3 mRNA moves to surface of rough ER, where proteins are synthesized on ribosomes using amino acids in the cytoplasm.

Mitochondrion

mRNA

4 Lipids are synthesized in the smooth ER.

5 Sugars are synthesized and proteins folded in the Golgi apparatus, then both are released in vesicles that bud off of the Golgi apparatus.

Plasma membrane

6 Protein- and sugar-laden vesicles move to the plasma membrane for release. Fat droplets pick up a layer of lipid from the plasma membrane as they exit the cell.

Lipid droplet

Supplements

McGraw-Hill offers various tools and technology products to support Lewis, *Human Genetics: Concepts and Applications,* Tenth Edition.

Teaching and Learning Tools

www.mhhe.com/lewisgenetics10

McGraw-Hill Connect® Genetics

McGraw-Hill Connect Genetics provides online presentation, assignment, and assessment solutions. It connects your students with the tools and resources they'll need to achieve success.

With Connect Genetics you can deliver assignments, quizzes, and tests online. A robust set of questions and activities are presented and aligned with the textbook's learning outcomes. As an instructor, you can edit existing questions and author entirely new problems. Track individual student performance—by question, assignment, or in relation to the class overall—with detailed grade reports. Integrate grade reports easily with Learning Management Systems (LMS), such as WebCT an Blackboard®. And much more.

ConnectPlus® Genetics provides students with all the advantages of Connect Genetics, plus 24/7 online access to an eBook. This media-rich version of the book is available through the McGraw-Hill Connect platform and allows seamless integration of text, media, and assessments.

To learn more, visit
www.mcgrawhillconnect.com

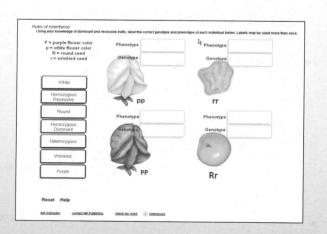

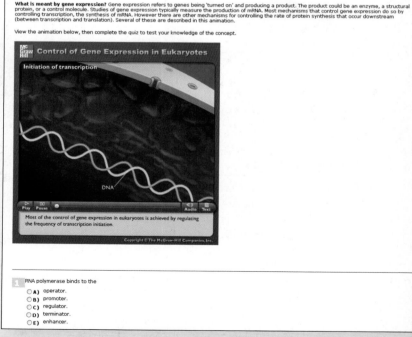

Presentation Tools

Everything you need for outstanding presentations in one place! This easy-to-use table of assets includes

- **Image PowerPoint® Files**—Including every piece of art, nearly every photo, all tables, as well as unlabeled art pieces.
- **Animation PowerPoint Files**—Numerous full-color animations illustrating important processes are also provided. Harness the visual impact of concepts in motion by importing these files into classroom presentations or online course materials.

- **Lecture PowerPoint Files**—with animations fully embedded!
- **Labeled and Unlabeled JPEG Images**—Full-color digital files of all illustrations ready to incorporate into presentations, exams, or custom-made classroom materials.
- **Student Response System PowerPoint Files**—Each chapter has a short set of "clicker" questions that can be used with a student response system to gauge student comprehension of core concepts and increase interactivity.

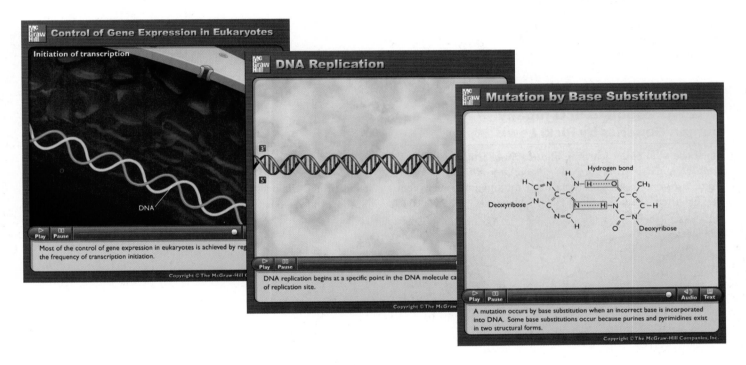

Presentation Center

Accessed from the website for *Human Genetics*, Presentation Center is an online digital library containing assets such as photos, artwork, animations, PowerPoints, and other types of media that can be used to create customized lectures, visually enhanced tests and quizzes, compelling course websites, or attractive printed support materials.

Computerized Test Bank written by Ricki Lewis!

The author has rewritten and expanded the test bank to include many more cases and problems. Terms match those used in the text, and the questions follow the order of topics within the chapters. This comprehensive bank of questions is provided within a computerized test bank powered by McGraw-Hill's flexible electronic testing program EZ Test Online. EZ Test Online allows you to create paper and online tests or quizzes in this easy-to-use program!

Imagine being able to create and access your test or quiz anywhere, at any time without installing the testing software. Now, with EZ Test Online, instructors can select questions from multiple McGraw-Hill test banks or author their own, and then either print the test for paper distribution or give it online.

Access the computerized test bank for Lewis, *Human Genetics* by going to www.mhhe.com/lewisgenetics10 and clicking on Instructor Resources.

Companion Website

www.mhhe.com/lewisgenetics10

The companion website to Lewis, *Human Genetics: Concepts and Applications,* provides students easy access to a variety of digital learning tools, including the following.

- Chapter-level quizzing
- Animations for viewing, the same animations available for classroom presentation
- Vocabulary flashcards to help students learn new terminology
- Answers to end-of-chapter questions

Visit the author's website (**www.rickilewis.com**) to read and comment on genetics updates on her blog Genetic Linkage, or to communicate with her directly.

Case Workbook to Accompany Human Genetics by Ricki Lewis

For those who enjoy learning and teaching from cases, *In the Family: A Case Workbook to Accompany Human Genetics, Tenth Edition,* bases questions on a multigenerational blending of three core families. Each chapter in the workbook

My Lectures-Tegrity

McGraw-Hill Tegrity Campus™ records and ditributes your class lecture, with just a click of a button. Students can view anytime/anywhere via computer, iPod, or mobile device. It indexes as it records your PowerPoint presentations and anything shown on your computer so students can use keywords to find exactly what they want to study.

corresponds to a textbook chapter and highlights a section of the overall connected pedigree. The casebook is a fun, highly innovative way to apply genetics concepts. Through the narrative and dialog style of the workbook, readers will come to know the various family members, while learning genetics.

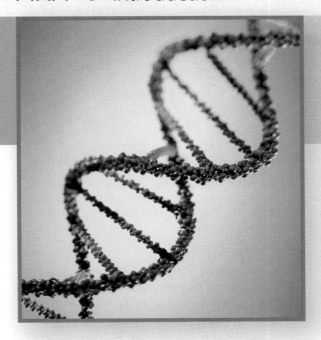

Personal genetic information is now readily available. People use genetic information to learn about their health risks and trace their ancestry.

Overview of Genetics

Learning Outcomes

1.1 Introducing Genes

1. Explain what genetics is, and what it is not.
2. Distinguish between gene and genome.
3. Define *bioethics*.

1.2 Levels of Genetics

4. Describe the levels of genetics, from nucleic acids to chromosomes, to cells, body parts, families, and populations.
5. Explain how genetics underlies evolution.

1.3 Genes and Their Environment

6. Discuss how genes and environmental factors interact to sculpt traits.
7. Define *genetic determinism*.

1.4 Applications of Genetics

8. Provide examples of how genetics is used in identification of people, in health care, in agriculture, and in ecology.

The Big Picture: Genes affect nearly all aspects of our lives, from our identities, to our health, to what we eat, and how we interact with others.

Direct-to-Consumer Genetic Testing

Genetic tests were once used solely to diagnose conditions so rare that doctors could not often match a patient's symptoms to a recognized illness. Today, taking a genetic test is as simple as ordering a kit on the Internet, swishing a plastic swab inside the mouth, and mailing the collected cell sample to a testing company or research project. The returned information can reach back to the past to chart a person's ancestry, or into the future to estimate disease risk.

Some "direct-to-consumer" (DTC) genetic tests identify well-studied mutations that cause certain diseases. Yet other tests are based on "associations" of patterns of genetic variation that appear much more often in people who share certain traits or illnesses. Because these new types of tests are drawn from population studies, they might not apply to a particular person. If interpreted carefully, information from genetic tests can be used to promote health or identify whether individuals are related.

Eve is curious about her ancestry and future health, so she takes tests that provide clues to both, rather than test all of her DNA. Her DNA sample is scanned for genes inherited from her mother and compared to a database of patterns from 20 nations and 200 ethnic groups in and near Africa. Eve learns that her family on her mother's side came from Gambia. The testing company will notify her of others who share this part of her deep ancestral roots. She already knows about her father's background from family lore.

The health tests require more thought. Eve dismisses tests for traits she considers frivolous, such as ear wax consistency and eye color. She skips tests for cancer and Alzheimer disease, for now.

Eve selects her health tests based on her family history. Because she, a sister, and her father often have respiratory infections, she asks for her DNA to be tested for gene variants that affect breathing, causing or contributing to such conditions as cystic fibrosis, asthma, emphysema, nicotine dependence, and lung cancer. She checks the boxes for heart and blood vessel diseases, too. Her reasoning: She can do something proactive to prevent or delay these conditions, such as breathing clean air, exercising, not smoking, and following a healthy diet.

Is genetic testing something that you would do?

Figure 1.1 **Inherited traits.** This young lady owes her red hair, fair skin, and freckles to a variant of a gene that encodes a protein (the melanocortin 1 receptor) that controls the balance of pigments in her skin.

1.1 Introducing Genes

Genetics is the study of inherited traits and their variation. Sometimes people confuse genetics with genealogy, which considers relationships but not traits. Because some genetic tests can predict illness, genetics has even been compared to fortune-telling! However, genetics is neither genealogy nor fortune-telling—it is a life science. Heredity is the transmission of traits between generations, and genetics is the study of how that happens.

Inherited traits range from obvious physical characteristics, such as the freckles and red hair of the girl in **figure 1.1,** to many aspects of health, including disease. Talents, quirks, behaviors, and other difficult-to-define characteristics might appear to be inherited if they affect several family members, but may reflect a combination of genetic and environmental influences. Some traits attributed to genetics border on the silly—such as sense of humor, fondness for sports, and whether or not one votes or joins a gang.

Until the 1990s, genetics was more an academic than a clinical science, except for rare diseases inherited in clear patterns in families. As the century drew to a close, researchers completed the human genome project, which deciphered the complete set of our genetic instructions. The next step—surveying our genetic variability—was already under way. Today, genetics has emerged as an informational as well as a life science that is having a big societal impact. Genetic information is accessible to anyone, and we are learning the contribution of genes to the most common traits and disorders.

Like all sciences, genetics has its own vocabulary. Many terms may be familiar, but actually have precise technical definitions. All of the terms and concepts in this chapter are merely introductions that set the stage for the detail in subsequent chapters.

Genes are the units of heredity. Genes are biochemical instructions that tell **cells,** the basic units of life, how to manufacture certain proteins. These proteins, in turn, impart or control the characteristics that create much of our individuality. A gene is the long molecule **deoxyribonucleic acid (DNA).** The DNA transmits information in its sequence of four types of building blocks.

The complete set of genetic instructions characteristic of an organism, including protein-encoding genes and other DNA sequences, constitutes a **genome.** Nearly all of our cells contain two copies of the genome. Researchers are still analyzing what all of our genes do, and how genes interact and respond to environmental stimuli. Only a tiny fraction of the 3.2 billion building blocks of our genetic instructions determines the most interesting parts of ourselves—our differences. Comparing and analyzing genomes, which constitute the field of **genomics,** reveals how closely related we are to each other and to other species.

Genetics directly affects our lives, as well as those of our relatives, including our descendants. Principles of genetics also touch history, politics, economics, sociology, anthropology, art, and psychology. Genetic questions force us to wrestle with concepts of benefit and risk, even tapping our deepest feelings about right and wrong. A field of study called **bioethics** was founded in the 1970s to address moral issues and controversies that arise in applying medical technology. Bioethicists today confront concerns that arise from new genetic technology, such as privacy and discrimination. Essays throughout this book address bioethical issues.

Many of the basic principles of genetics were discovered before DNA was recognized as the genetic material, from experiments and observations on patterns of trait transmission in families. For many years, genetics textbooks presented concepts in the order that they were understood, discussing pea plant experiments before DNA structure. Now, since even gradeschoolers know what DNA is, a "sneak preview" of DNA structure and function (**Reading 1.1**) is appropriate to consider the early discoveries in genetics (see chapter 4) from a modern perspective.

Introducing DNA

We have probably wondered about heredity since our beginnings, when our distant ancestors noticed family traits such as a flat nose or an ability to run fast. Awareness of heredity appears in ancient Jewish law that excuses a boy from circumcision if his brothers or cousins bled to death following the ritual. Nineteenth-century biologists thought that body parts controlled traits, and they gave the hypothetical units of inheritance such colorful names as "pangens," "ideoblasts," "gemules," and simply "characters."

In the late nineteenth century, when Gregor Mendel bred pea plants to follow trait transmission, establishing the basic laws of inheritance, he inferred that units of inheritance were at play. He had no knowledge of cells, chromosomes, or DNA. This short reading explains, very briefly, what Mendel did not know—how DNA confers inherited traits. Chapter 9 examines DNA in detail, including how many researchers provided the clues that James Watson and Francis Crick assembled to deduce the three-dimensional structure of the molecule.

DNA resembles a spiral staircase or double helix in which the "rails," or backbone of alternating sugars and phosphates, is the same from molecule to molecule, but the "steps" are pairs of four types of building blocks, or DNA bases, whose sequence varies (**figure 1**). The bases are adenine (A) and thymine (T), which attract, and cytosine (C) and guanine (G), which attract. DNA holds information in the sequences of A, T, C, and G. The two strands are oriented in opposite directions.

DNA uses its information in two ways. If the sides of the helix part, each half can reassemble its other side by pulling in free building blocks—A and T attracting and G and C attracting. This process, called DNA replication, maintains the information when the cell divides. DNA also directs the production of specific proteins. In a process called transcription, the sequence of part of one strand of a DNA molecule is copied into a related molecule, messenger RNA. Each three RNA bases in a row attract another type of RNA that functions as a connector, bringing with it a particular amino acid, which is a building block of protein. The synthesis of a protein is called translation. As the two types of RNA temporarily bond, the amino acids align and join, forming a protein that is then released. DNA, RNA, and proteins can be thought of as three related languages of life (**figure 2**).

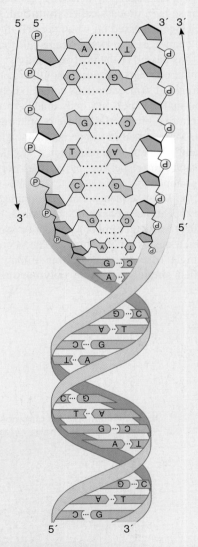

Figure 1 **The DNA double helix.** The 5′ and 3′ labels indicate the head-to-tail organization of the DNA double helix. A, C, T, and G are bases. S stands for sugar and P for phosphate.

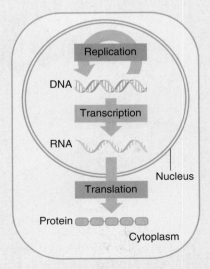

Figure 2 **The language of life:** DNA to RNA to protein.

1.2 Levels of Genetics

Genetics considers the transmission of information at several levels. It begins with the molecular level and broadens through cells, tissues and organs, individuals, families, and finally to populations and the evolution of species (**figure 1.2**).

The Instructions: DNA, Genes, Chromosomes, and Genomes

Genes consist of sequences of the four types of DNA building blocks, or bases—adenine (A), guanine (G), cytosine (C), and thymine (T). Each base bonds to a sugar and a phosphate group to form a unit called a nucleotide, and nucleotides are linked into long DNA molecules. In genes, DNA bases provide an alphabet of sorts. Each consecutive three DNA bases is a code for a particular amino acid, and amino acids are the building blocks of proteins. Another type of molecule, **ribonucleic acid (RNA),** uses the information in certain DNA sequences to construct specific proteins. Messenger RNA (mRNA) carries the gene's base sequence, whereas two other major types of RNA assemble the protein's building blocks. The protein confers the trait. DNA remains in the part of the cell called the nucleus, and is passed on when a cell divides. A genome's worth of DNA is like a database that is accessed to run the cell.

Proteomics is a field that considers the types of proteins made in a particular type of cell. A muscle cell, for example, requires abundant contractile proteins, whereas a skin cell contains mostly scaly proteins called keratins. A cell's proteomic profile changes as conditions change. A cell lining the stomach, for example, would produce more protein-based digestive enzymes after a meal than when a person hasn't eaten in several hours.

The human genome has about 20,325 protein-encoding genes. The few thousand known to cause disorders or traits are described in a database called Mendelian Inheritance in Man (MIM) (www.omim.org). Throughout this text, the first mention of a disease includes its MIM number. Reading 4.1 describes some of the more colorful traits in MIM.

Despite knowing the sequence of DNA bases of the human genome, there is much we still do not know. For example, only about 1.5 percent of our DNA encodes protein. The rest includes many DNA sequences that assist in protein synthesis or turn protein-encoding genes on or off. The functions of many parts of the human genome are yet to be discovered.

The same protein-encoding gene may vary slightly in base sequence from person to person. These variants of a gene are called **alleles**. The changes in DNA sequence that distinguish alleles arise by a process called **mutation**. Once a gene mutates, the change is passed on when the cell that contains it divides. If the change is in a sperm or egg cell that becomes a fertilized egg, it is passed to the next generation.

Some mutations cause disease, and others provide variation, such as freckled skin. Mutations can also help. For example, a mutation makes a person's cells unable to manufacture a surface protein that binds HIV. These people are resistant to HIV infection. Many mutations have no visible effect because they do not change the encoded protein in a way that affects its function, just as a minor spelling *error* does not obscure the meaning of a sentence.

Parts of the DNA sequence can vary among individuals, yet not change appearance or health. Such a variant in sequence that is present in at least 1 percent of a population is called a polymorphism, which means "many forms." The genome includes millions of single base sites that differ among individuals. These are called **single nucleotide polymorphisms (SNPs,**

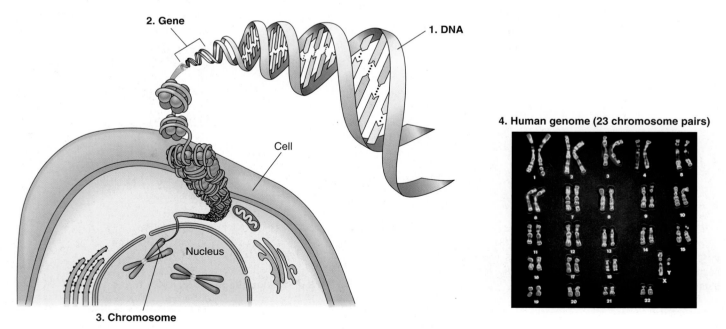

2. Gene

1. DNA

Cell

Nucleus

4. Human genome (23 chromosome pairs)

3. Chromosome

Figure 1.2 **Levels of genetics.** Genetics can be considered at several levels, from DNA, to genes, to chromosomes, to genomes, to the more familiar individuals, families, and populations. (A gene is actually several hundred or thousand DNA bases long.)

pronounced "snips"). SNPs can be associated with diseases or just mark places in the genome where people differ.

SNPs can yield interesting information when compared among a large enough group of individuals to associate the pattern of variation with a specific trait or disease. Many research groups are conducting **genome-wide association studies** that look at a million or more SNPs, in thousands of people. These SNP patterns can then be used to estimate risk of the disease in people who are not yet sick but have inherited the same DNA variants. Such a genome-wide association study lies in between the sequencing of a gene and a genome in scope. If a genome is like a detailed Google map of the entire United States and a gene is like a Google map showing the streets of a specific neighborhood, then SNPs that speckle a genome are like a map of the United States with only the names of states and interstate highways indicated—just clues.

Sequences of DNA bases, whether for single genes or entire genomes, provide a structural view of genetic material. Another way to look at DNA, called **gene expression profiling,** highlights a cell's activities by measuring the abundance of different RNA molecules, which reflects protein production. The power of the approach is in comparisons. A muscle cell from a bedridden person, for example, would have different levels of contractile proteins than a muscle cell from an athlete. **Table 1.1** summarizes types of information that DNA sequences provide.

DNA molecules are very long. They wrap around proteins and wind tightly, forming structures called **chromosomes**. A human somatic (non-sex) cell has 23 pairs of chromosomes. Twenty-two pairs are **autosomes,** which do not differ between the sexes. The autosomes are numbered from 1 to 22, with 1 the largest. The other two chromosomes, the X and the Y, are **sex chromosomes**. The Y chromosome bears genes that determine maleness. In humans, a female has two X chromosomes and a male has one X and one Y. Charts called **karyotypes** display the chromosome pairs from largest to smallest.

A human cell has two complete sets of genetic information. The 20,325 or more protein-encoding genes are scattered among 3.2 billion DNA bases in each set of 23 chromosomes.

Table 1.1	Types of Information in DNA Sequences
Level	**Description**
Single gene	Hundreds to thousands of DNA bases that encode a protein or parts of a protein
Genome	The entire 3.2-billion base sequence of the genetic material in a human cell
Genome-wide association study	Patterns of single-base variants (SNPs) associated with traits or medical conditions
Gene expression profiling	Levels of mRNAs in specific cells under specific conditions that reflect physiology and reveal malfunction

The Body: Cells, Tissues, and Organs

A human body consists of approximately 50 to 100 trillion cells. All cells except red blood cells contain the entire genome, but cells differ in appearance and activities because they use only some of their genes—and which ones they access at any given time depends upon environmental conditions both inside and outside the body.

The genome is like the Internet in that it contains a wealth of information, but only some of it need be accessed. The expression of different subsets of genes drives the **differentiation,** or specialization, of distinctive cell types. An adipose cell is filled with fat, but not the collagen and elastin proteins of connective tissue cells. Both cell types, however, have complete genomes. Groups of differentiated cells assemble and interact with each other and the nonliving material that they secrete to form aggregates called tissues.

The body has four basic tissue types, composed of more than 260 types of cells. Tissues intertwine and layer to form the organs of the body, which in turn connect into organ systems. The stomach shown at the center of **figure 1.3**, for example, is a sac made of muscle that also has a lining of epithelial tissue, nervous tissue, and a supply of blood, which is a type of connective tissue. **Table 1.2** describes tissue types.

Many organs include rare, unspecialized **stem cells**. A stem cell can divide to yield another stem cell and a cell that differentiates. Thanks to stem cells, organs can maintain a reserve supply of cells to grow and repair damage.

Relationships: From Individuals to Families

Two terms distinguish the alleles that are *present* in an individual from the alleles that are *expressed*. The **genotype** refers to the underlying instructions (alleles present), whereas the **phenotype** is the visible trait, biochemical change, or effect on health (alleles expressed). Alleles are further distinguished by how many copies it takes to affect the phenotype. A **dominant** allele has an effect when present in just one copy (on one chromosome), whereas a **recessive** allele must be present on both chromosomes to be expressed.

Individuals are genetically connected into families. A person has half of his or her genes in common with each parent and each sibling, and one-quarter with each grandparent. First cousins share one-eighth of their genes.

For many years, transmission (or Mendelian) genetics dealt with single genes in families. The scope of transmission genetics has greatly broadened in recent years. Family genetic studies today often trace more than one gene at a time, or traits that have substantial environmental components. Molecular genetics, which considers DNA, RNA, and proteins, often begins with transmission genetics, when an interesting family trait or illness comes to a researcher's attention. Charts called **pedigrees** represent the members of a family and indicate which individuals have particular inherited traits. Sometimes understanding a rare single-gene condition in a family leads to treatments for the greater number of people with similar disorders that are not inherited. This is the case for the statin drugs widely used to lower cholesterol.

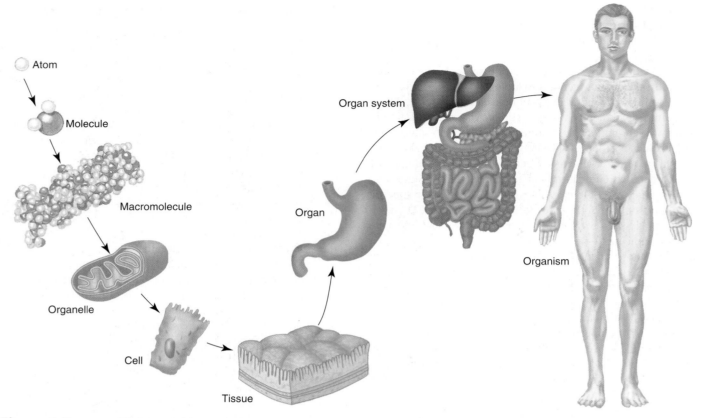

Figure 1.3 Levels of biological organization.

Table 1.2	Tissue Types
Tissue	**Function/Location/Description**
Connective tissues	A variety of cell types and materials around them that protect, support, bind to cells, and fill spaces throughout the body; include cartilage, bone, blood, and fat
Epithelium	Tight cell layers that form linings that protect, secrete, absorb, and excrete
Muscle	Cells that contract, providing movement
Nervous	Neurons transmit information as electrochemical impulses that coordinate movement and sense and respond to environmental stimuli; neuroglia are cells that support and nourish neurons

The Bigger Picture: From Populations to Evolution

Above the family level of genetic organization is the population. In a strict biological sense, a population is a group of interbreeding individuals. In a genetic sense, a population is a large collection of alleles, distinguished by their frequencies. People from a Swedish population, for example, would have a greater frequency of alleles that specify light hair and skin than people from a population in Ethiopia, who tend to have dark hair and skin. The fact that groups of people look different and may suffer from different health problems reflects the frequencies of their distinctive sets of alleles. All the alleles in a population constitute the **gene pool**. (An individual does not have a gene pool.)

Population genetics is applied in health care, forensics, and other fields. It is also the basis of evolution, which is defined as changing allele frequencies in populations. These small-scale genetic changes foster the more obvious species distinctions we most often associate with evolution.

Comparing DNA sequences for individual genes, or the amino acid sequences of the proteins that the genes encode, can reveal how closely related different types of organisms are (**figure 1.4**). The underlying assumption is that the more similar the sequences are, the more recently two species diverged from a shared ancestor. This is a more plausible explanation than two species having evolved similar or identical gene sequences by chance.

Both the evolution of species and family patterns of inherited traits show divergence from shared ancestors. This is based on logic. It is more likely that a brother and sister share approximately half of their gene variants because they have the same parents than that half of their genetic material is identical by chance.

Genome sequence comparisons reveal more about evolutionary relationships than comparing single genes, simply because there are more data. Humans, for example, share more than 98 percent of the DNA sequence with chimpanzees. Our genomes differ from theirs more in gene organization and in the number of copies of genes than in the overall sequence. Learning the functions of the human-specific genes may explain the differences between us

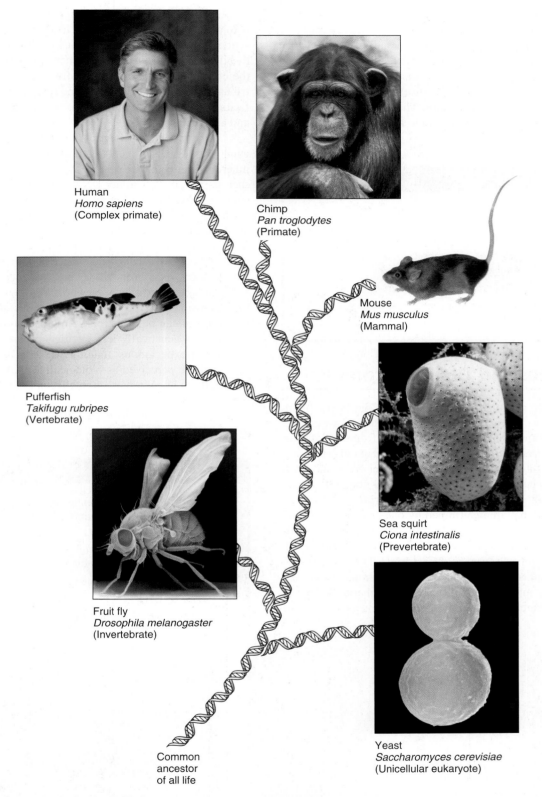

Human
Homo sapiens
(Complex primate)

Chimp
Pan troglodytes
(Primate)

Mouse
Mus musculus
(Mammal)

Pufferfish
Takifugu rubripes
(Vertebrate)

Sea squirt
Ciona intestinalis
(Prevertebrate)

Fruit fly
Drosophila melanogaster
(Invertebrate)

Yeast
Saccharomyces cerevisiae
(Unicellular eukaryote)

Common
ancestor
of all life

Figure 1.4 Genes and genomes reveal our place in the world. All life is related, and different species share a basic set of genes that makes life possible. The more closely related we are to another species, the more genes we have in common. This illustration depicts how humans are related to certain contemporaries whose genomes have been sequenced.

During evolution, species diverged from shared ancestors. For example, humans diverged more recently from chimps, our closest relative, than from mice, pufferfish, sea squirts, flies, or yeast.

and them—such as our lack of hair and use of spoken language. Reading 16.1 highlights some of our distinctively human traits.

Comparisons of people at the genome level reveal that we are much more like each other genetically than are other mammals. Chimpanzees are more distinct from each other than we are! The most genetically diverse modern people are from Africa, where humanity arose. The gene variants among different modern ethnic groups include subsets of our ancestral African gene pool.

Key Concepts

1. Genetics is the study of inherited traits and their variation.
2. Genetics can be considered at the levels of DNA, genes, chromosomes, genomes, cells, tissues, organs, individuals, families, and populations.
3. A gene can exist in more than one form, or allele.
4. Comparing genomes among species reveals evolutionary relatedness.

1.3 Genes and Their Environment

Most, if not all, genes do not function alone but interact with other genes as well as with environmental factors. For example, genes control how many calories we extract from food, but so do the numbers and types of bacteria that live in our intestines. Because gene variants and gut bacteria differ from person to person, some people can eat a great deal and not gain weight, yet others gain weight easily. **Multifactorial,** or complex, **traits** are those that are determined by one or more genes and the environment (**figure 1.5**). (The term *complex traits* has different meanings in a scientific and a popular sense, so this book uses the more precise term *multifactorial*.) The same symptoms may be inherited or not, and if inherited, may be caused by one gene or more than one. Usually the inherited forms of an illness are rarer. The more factors that contribute to a trait

or illness—inherited or environmental—the more difficult it is to predict the risk of occurrence in a particular family member. Osteoporosis illustrates the various factors that can contribute to a disease. It mostly affects women past menopause, thinning the bones and increasing risk of fractures. Several genes contribute to susceptibility to the condition, as well as do lifestyle factors, including smoking, lack of weight-bearing exercise, and a calcium-poor diet.

The modifying effect of the environment on gene action counters the idea of **genetic determinism,** which is that an inherited trait is inevitable. The idea that "we are our genes," or such phrases as "it's in her DNA," ignore environmental influences. In predictive testing for inherited disease, which detects a disease-causing genotype in a person without symptoms, results are presented as risks, rather than foregone conclusions, because the environment can modify gene expression. A woman might be told "You have a 45 percent chance of developing this form of breast cancer," not, "You will get breast cancer."

Genetic determinism may be harmful or helpful, depending upon how we apply it. As part of social policy, genetic determinism can be disastrous. An assumption that one group of people is genetically less intelligent than another can lead to lowered expectations and/or fewer educational opportunities for those perceived as biologically inferior. Environment, in fact, has a huge impact on intellectual development. On the other hand, knowing the genetic contribution to a trait can be helpful when it gives us more control over health by guiding us in influencing noninherited factors, such as diet.

Key Concepts

1. Inherited traits are determined by one gene (Mendelian) or by one or more genes and the environment (multifactorial).
2. Even the expression of single genes is affected to some extent by the actions of other genes.
3. Genetic determinism is the idea that an inherited trait cannot be modified.

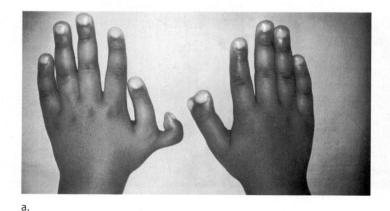
a.

b.

Figure 1.5 Mendelian versus multifactorial traits. (a) Polydactyly—extra fingers and/or toes—is a Mendelian trait (single-gene). **(b)** Hair color is multifactorial, controlled by at least three genes plus environmental factors, such as the bleaching effects of sun exposure.

1.4 Applications of Genetics

Genetics is in the news and blogosphere on a daily basis. It is impacting many areas of our lives, from health care choices, to what we eat and wear, to unraveling our pasts and guiding our futures. Thinking about genetics evokes fear, hope, anger, and wonder, depending upon context and circumstance. Following are glimpses of applications of genetics that we will explore more fully in subsequent chapters.

Establishing Identity

A technique called DNA profiling compares DNA sequences to establish or rule out identity, relationships, or ancestry. DNA profiling today looks at SNPs and short, repeated DNA sequences, but as the cost of sequencing entire genomes continues to fall, it may compare more information and become even more powerful in distinguishing among individuals.

Forensics

Before September 11, 2001, the media reported on DNA profiling (then known as DNA fingerprinting) to identify plane crash victims or to provide evidence in high-profile criminal cases. After the 2001 terrorist attacks in the United States, investigators compared DNA sequences in bones and teeth collected from the attack scenes to hair and skin samples from hairbrushes, toothbrushes, and clothing of missing people, and to DNA samples from relatives. It was a massive undertaking, but small compared to use of the technique to identify victims of natural disasters, such as tsunamis, hurricanes, and earthquakes.

In law enforcement, a rare DNA sequence in tissue left at a crime scene is compared to DNA from a suspect. A match is statistically strong evidence that the accused person was at the crime scene (or that someone planted evidence). DNA databases of convicted felons often provide "cold hits" when DNA at a crime scene matches a criminal's DNA in the database. This is especially helpful when there is no suspect.

DNA profiling is used to overturn convictions, too. Illinois led the way in 1996, when DNA tests exonerated the Ford Heights Four—men convicted of a gang rape and double murder who had spent 18 years in prison, 2 of them on death row. In 1999, the men received $36 million for their wrongful convictions. A journalism class at Northwestern University initiated the investigation, leading to new state laws granting death row inmates new DNA tests if their convictions could have arisen from mistaken identity, or if DNA tests were performed when they were far less accurate. The Innocence Project is an organization that has used DNA profiling to exonerate more than 250 death row prisoners. One of them is introduced in the opening essay to chapter 14.

DNA profiling helps adopted individuals locate blood relatives. The Kinsearch Registry maintains a database of DNA information on people adopted in the United States from China, Russia, Guatemala, and South Korea, which are the sources of most foreign adoptions. Adopted individuals can provide a DNA sample and search the database by country of origin to find siblings. Websites allow children of sperm donors to find their biological fathers, if the men wish to be contacted.

History and Ancestry

DNA analysis can reveal or clarify details of history. A famous case concerns the offspring of Thomas Jefferson and his slave Sally Hemings (**figure 1.6**). Abundant evidence existed that the pair had children together. The president was near Hemings nine months before each of her seven children was born, some of the children looked just like him, and the children themselves claimed to be his. A Y chromosome analysis revealed that Thomas Jefferson could have fathered Hemings' youngest son, Eston—but so could any of 26 other Jefferson family members. The Y chromosome, because it is only in males, passes from father to son. Researchers identified very unusual DNA sequences on the Y chromosomes of descendants of Thomas Jefferson's paternal uncle, Field Jefferson. (These men were checked because the president's only son with wife Martha died in infancy, so Thomas Jefferson had no direct descendants.) The Jefferson family's unusual Y chromosome matched that of descendants of Eston Hemings, supporting the talk of the time.

Reaching farther back, DNA profiling can clarify relationships from Biblical times. Consider a small group of Jewish people, the cohanim, who share distinctive Y chromosome DNA sequences and enjoy special status as priests. By considering the number of DNA differences between cohanim and other Jewish people, how long it takes DNA to mutate, and the average generation time of 25 years, researchers extrapolated that five similar Y chromosome patterns are in the cohanim and not others, and originated about 3,200 years ago. This includes the time when Moses lived. According to religious documents, Moses' brother Aaron was the first priest.

Figure 1.6 **DNA reveals and clarifies history.** After DNA evidence showed that Thomas Jefferson likely fathered a son of his slave, descendants of both sides of the family met.

The Jewish priest DNA signature also appears today among the Lemba, a population of South Africans with black skin. Researchers looked at them for the telltale Jewish gene variants because the Lemba customs suggest a Jewish origin—they do not eat pork (or hippopotamus), they circumcise their newborn sons, and they celebrate a weekly day of rest (**figure 1.7**).

Studies that compare many parts of the genome in modern Jewish to non-Jewish communities reveal that Jewish people (except those from Ethiopia and India) descend from people in the Middle East and that they are more closely related to each other than thought before. Most of today's Jewish people are fourth or fifth cousins.

DNA testing can provide glimpses into past bouts with infectious disease. For example, analysis of DNA in the mummy of the Egyptian king Tutankhamun revealed the presence of the microorganism that causes malaria. The child king likely died from complications of malaria following a leg fracture from weakened bones rather than from intricate murder plots, a kick from a horse, or fall from a chariot, as had been thought. His tomb included a cane and drugs, supporting the diagnosis based on DNA evidence.

Health Care

Looking at diseases from a genetic point of view is changing health care. Many diseases, not just inherited ones, are now viewed as the consequence of complex interactions among genes and environmental factors. Even the classic single-gene diseases are sensitive to the environment. A child with cystic fibrosis (MIM 219700), for example, is more likely to suffer frequent respiratory infections if she regularly breathes second-hand smoke. A genetic approach to health is as much common sense as it is technological.

Diseases can result from altered proteins or too little or too much of a protein, or proteins made at the wrong place or time. Gene expression profiling studies are revealing the sets of genes that are turned on and off in specific cells and tissues as health declines. Genes also affect how people respond to particular drugs. For example, inheriting certain gene variants can make a person's body very slow at breaking down an anti-clotting drug, or extra sensitive to the drug. Such an individual could experience dangerous bleeding at the same dose that most patients

Figure 1.7 Y chromosome DNA sequences reveal origins. The Lemba, a modern people with dark skin, have the same Y chromosome DNA sequences as the cohanim, a group of Jewish priests. The Lemba practiced Judaism long before DNA analysis became available.

Table 1.3	Pharmacogenomic Tests
Antidepressants	
Chemotherapies	
HIV drugs	
Smoking cessation drugs	
Statins (cholesterol-lowering drugs)	
Warfarin (anti-clotting)	

tolerate. Identifying individual drug reactions based on genetics is a growing field called pharmacogenomics. **Table 1.3** lists some examples.

Single-Gene Diseases

Inherited illness caused by a single gene differs from other types of illnesses in several ways (**table 1.4**). In families, we can predict inheritance of a single-gene disease by knowing exactly how a person is related to an affected relative. In contrast, an infectious disease requires that a pathogen pass from one person to another, which is much less predictable.

A second distinction of single-gene disorders is that tests can sometimes predict the risk of developing symptoms. This is possible because all cells harbor the mutation. A person with a family history of Huntington disease (HD; MIM 143100), for example, can have a blood test that detects the mutation at any age, even though symptoms typically do not occur until near age 40. Inheriting the HD mutation predicts illness with near certainty. For many conditions, predictive power is much lower. For example, inheriting one copy of a particular variant of a gene called *APOE* raises risk of developing Alzheimer disease by three-fold, and inheriting two copies raises it 15-fold. But without absolute risk estimates and no treatments for this disease, would you want to know?

A third feature of single-gene diseases is that they may be much more common in some populations than others. Genes do not like or dislike certain types of people; rather, mutations stay in certain populations because we marry people like ourselves. While it might not seem politically correct to offer a "Jewish genetic disease" screen, it makes biological and economic

Table 1.4	How Single-Gene Diseases Differ from Other Diseases
1. Risk can be predicted for family members.	
2. Predictive (presymptomatic) testing may be possible.	
3. Different populations may have different characteristic disease frequencies.	
4. Correction of the underlying genetic abnormality may be possible.	

sense—two dozen disorders are much more common in this population. A fourth characteristic of a genetic disease is that it may be "fixable" by altering the abnormal instructions.

Redefining Disease to Reflect Gene Expression

Diseases are increasingly being described in terms of gene expression patterns, which is not the same as detecting mutations. Gene expression refers to whether a gene is "turned on" or "turned off" from being transcribed and translated into protein (see Reading 1.1). Mutations are changes in the gene's structure.

Tracking gene expression can reveal new information about diseases and show how diseases are related to each other. Diseases with different symptoms might be variations of the same underlying defect, or conditions with similar symptoms might be distinct at the molecular level.

Figure 1.8 shows part of a huge disease map called the "**diseasome.**" It connects diseases that share genes that have altered expression. The diseasome reveals relationships

among diseases that were not obvious from traditional medical science, which is based on observing symptoms, detecting pathogens or parasites, or measuring changes in body fluid composition. Some of the links and clusters are well known, such as obesity, hypertension, and diabetes. Others are surprises, such as Duchenne muscular dystrophy (DMD; see figure 2.1) and heart attacks. The muscle disorder has no treatment, but heart attack does—researchers are now testing cardiac drugs on boys with DMD. In other cases, the association of a disease with genes whose expression goes up or down can suggest targets for new drugs.

Comparing gene expression profiles is a new way of looking at the body's function. An interesting illustration is cigarette smoking. It has been known for half a century that smoking causes lung disease. However, recent studies show that smokers overexpress more than 300 genes that nonsmokers do not overexpress. These genes affect the immune system, cell death, cancer, and response to poisons. Identifying and studying these genes can suggest new drug targets to help smokers quit or more effectively treat smoking-related illnesses.

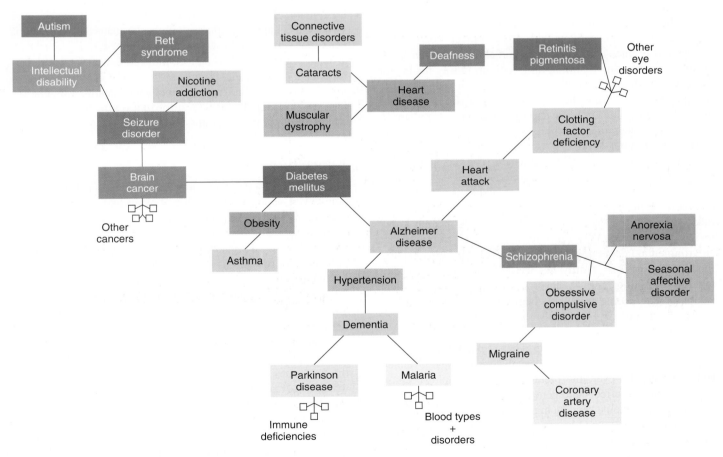

Figure 1.8 **Part of the diseasome.** This tool links diseases by shared gene expression. That is, a particular gene may be consistently overexpressed or underexpressed in two diseases, compared to the healthy condition. The lines refer to at least one gene connecting the disorders depicted in the squares. The conditions are not necessarily inherited because gene expression changes in all situations. For example, Alzheimer disease is linked to heart attack because both conditions entail a buildup of cholesterol. Finding diseasome links can suggest existing drugs that can be "repurposed" to treat different illnesses. The cholesterol-lowering statin drugs are being tested on Alzheimer patients, for example. (Based on the work of A-L Barabási and colleagues.)

Genetic Testing

Tests to identify about 1,200 single-gene disorders, most of them very rare, have been available for years. Direct-to-consumer (DTC) genetic testing, via websites and cheek cell samples, is bringing many kinds of DNA-based tests to many more people. Before passage of the Genetic Information Nondiscrimination Act (GINA) in the United States in 2008, it was common for people to avoid genetic testing for fear of the misuse of genetic information or to take tests under false names so the result would not appear in their medical records. Some people refused to participate in clinical trials of new treatments if genetic information could be traced to them. Today the opposite is true. Many people take genetic tests.

Under GINA, employers cannot use genetic information to hire, fire, or promote an employee, or require genetic testing. Similarly, health insurers cannot require genetic tests nor use the results to deny coverage. GINA also clearly defines a genetic test: It is an analysis of human DNA, RNA, chromosomes, proteins, or metabolites, to detect genotypes, mutations, or chromosomal changes. The law defines "genetic information" as tests or phenotypes (traits or symptoms) in individuals and/or families.

The long-awaited GINA legislation, however, raises new issues. Consider two patients with breast cancer—one with a strong family history and a known mutation, the other diagnosed after a routine mammogram, with no family history or identified mutation. A health insurer could refuse to cover the second woman, but not the first. Other limitations of GINA are that it does not apply to companies with fewer than 15 employees, it does not overrule state law, it does not protect privacy, it does not cover life insurance, and it does not spell out how discrimination will be punished. These concerns will be addressed as the law is put into practice.

In the long term, genetic tests, whether for single-gene disorders or the more common ones with associated genetic risks, may actually lower health care costs. If people know their inherited risks, they can forestall or ease symptoms that environmental factors might trigger—such as by eating healthy foods suited to their family history, not smoking, exercising regularly, avoiding risky behaviors, having frequent medical exams and screening tests, and beginning treatments earlier. The protection of GINA will also help recruit participants for clinical trials.

The day is fast approaching when it becomes routine to sequence a person's genome, and analyze it for health-related gene variants. However, interpretation will be complex because of environmental effects and gene interactions. Even if we can achieve cheap genome sequencing, the knowledge and skills of clinicians will continue to be important in extracting and interpreting actionable information.

Treatments

Only a few single-gene diseases can be treated. Supplying a missing protein directly can prevent some symptoms, such as giving a clotting factor to a person with a bleeding disorder. Some inborn errors of metabolism (see Reading 2.1) in which an enzyme deficiency leads to the buildup of a biochemical in cells, can be counteracted by tweaking diet to minimize the accumulation or by supplying the enzyme. Treatment at the DNA level—gene therapy—replaces the faulty instructions for producing the protein in cells that are affected in the illness.

For some genetic diseases, better understanding of how mutations cause the symptoms suggests that an existing drug for another condition might work. For example, experiments in mice with tuberous sclerosis complex, a disease that causes autism, memory deficits, and intellectual disability in humans (MIM 191100), led to clinical trials of a drug, rapamycin, already in use to lessen transplant rejection. Tuberous sclerosis affects the same enzyme that the drug targets. Chapter 20 discusses various approaches to treating genetic disease.

Genome information is useful for treating infectious diseases, because the microorganisms and viruses that make us sick also have genetic material that can be sequenced and detected. In one interesting case, three patients died from infection 6 weeks after receiving organs from the same donor. All tests for known viruses and bacteria were negative, so medical researchers sampled DNA from the infected organs, removed human DNA sequences and those of known pathogens, and examined the remainder for sequences that resemble those of bacteria and viruses. This approach picked up genetic material from pathogens that cannot be grown in the laboratory. Using the DNA sequence information to deduce and reconstruct physical features of the pathogens, the researchers were able to identify a virus that caused the transplant recipients' deaths. Researchers then developed a diagnostic test for future transplant recipients who have the same symptoms.

Agriculture

The field of genetics arose from agriculture. Traditional agriculture is the controlled breeding of plants and animals to select individuals with certain combinations of inherited traits that are useful to us, such as seedless fruits or lean meat. **Biotechnology,** which is the use of organisms to produce goods (including foods and drugs) or services, is an outgrowth of agriculture.

One ancient example of biotechnology is using microorganisms to ferment fruits to manufacture alcoholic beverages, a technique the Babylonians used by 6000 B.C. Beer brewers in those days experimented with different yeast strains cultured under different conditions to control aroma, flavor, and color. Today, researchers have sequenced the genomes of the two types of yeast that are crossed to ferment lager beer, which requires lower temperatures than does ale. The work has shown that beers from different breweries around the world have unique patterns of gene expression, suggesting ways to brew new types of beer.

Traditional agriculture is imprecise because it shuffles many genes—and, therefore, many traits—at a time, judging them by taste or appearance. In contrast, DNA-based techniques manipulate one gene at a time. Organisms altered to have new genes or to over- or underexpress their own genes are termed "genetically modified" (GM). If the organism has genes from another species, it is termed transgenic. Golden rice, for example, manufactures twenty-three times as much beta carotene (a vitamin A precursor) as unaltered rice, thanks to "transgenes" from corn and bacteria. Golden rice also stores twice as much iron as unaltered rice because one of its own genes is overexpressed.

People in the United States have been safely eating GM foods for more than a decade, but some people object to them, for several reasons. Officials in France and Austria have called such crops "not natural," "corrupt," and "heretical." Food labels in some nations indicate whether a product is "GM-free," which can prevent allergic reaction to an ingredient in a food that wouldn't naturally be there, such as a peanut protein in corn.

Field tests may not adequately predict the effects of GM crops on ecosystems. GM plants have been found far beyond where they were planted, thanks to wind pollination. For example, a sampling of 400 canola plants growing along roads in North Dakota found that 86 percent of them had a gene indicating that they descended from GM plants. Planting GM crops may also lead to genetic uniformity, which could be disastrous. Some GM organisms, such as fish that grow to twice normal size or can survive at temperature extremes, may disrupt natural ecosystems. **Figure 1.9** shows an artist's rendition of these fears.

Agriculture also benefits from genome-level analysis, sometimes called "landscape genomics." Researchers are probing the genomes of domesticated animal species that make them able to survive, or even thrive, under extreme environmental conditions. Perhaps these genes can be swapped into the genomes of less-hardy relatives.

Ecology

We share the planet with many thousands of other species. We aren't familiar with many of Earth's residents because we can't observe their habitats, or we can't grow them in laboratories.

Figure 1.9 An artist's view of biotechnology. Artist Alexis Rockman vividly captures some fears of biotechnology, including a pig used to incubate spare parts for sick humans, a muscle-boosted boxy cow, a featherless chicken with extra wings, a mini-warthog, and a mouse with a human ear growing out of its back.

"Metagenomics" is a field that is revealing and describing much of the invisible living world by sequencing all of the DNA in a particular habitat. Such areas range from soil, to an insect's gut, to garbage. Metagenomics studies are revealing how species interact, and may yield new drugs and reveal novel energy sources.

Metagenomics researchers collect and sequence DNA, then consult databases of known genomes to imagine what the organisms might be like. The first metagenomics project described life in the Sargasso Sea. This 2-million-square-mile oval area off the coast of Bermuda has long been thought to lack life beneath its thick cover of seaweed, which is so abundant that Christopher Columbus thought he'd reached land when his ships came upon it. Many a vessel has been lost in the Sargasso Sea, which includes the area known as the Bermuda Triangle. Researchers collected more than a billion DNA bases from the depths, representing about 1,800 microbial species and including more than a million previously unknown genes.

A favorite site for metagenomics analysis is the human body. The human microbiome project is exploring the other forms of life within us. Genome profiling on various parts of our anatomy reveals that 90 percent of the cells in a human body are not actually human! A human body is, in fact, a vast ecosystem. This is possible because bacterial cells are so much smaller than ours. Humans have a "core microbiome" of bacterial species that everyone has, but also many others that reflect our differing environments, habits, ages, diets, and health.

About 10 trillion bacteria live in our digestive tracts. The human mouth is home to about 500 different species of bacteria, only about 150 of which can grow in the laboratory. Analysis of their genomes yields practical information. For example, the genome of one bacterium, *Treponema denticola,* showed how it survives amid the films other bacteria form in the mouth, and how it causes gum disease. The other end of the digestive tract is easy to study too, because feces are very accessible research materials that are full of bacteria from the intestines. One study examined soiled diapers from babies regularly during their first year, chronicling the establishment of the gut bacterial community. Newborns start out with blank slates—clean intestines—and after various bacteria come and go, very similar species remain by their first birthdays.

In parallel to metagenomics, several projects are exploring biodiversity with DNA tags to "bar-code" species, rather than sequencing entire genomes. DNA sequences that vary reveal more about ancestries, because they are informational, than do comparisons of physical features, such as body shape or size, which formed the basis of traditional taxonomy (biological classification).

A Global Perspective

Genetics is more than a branch of life science, because it affects us intimately. Equal access to genetic testing, misuse of genetic information, and abuse of genetics to intentionally cause harm are compelling issues that parallel scientific progress.

Genetics and genomics are spawning technologies that may vastly improve quality of life. But at first, tests and treatments will be costly and not widely available. While advantaged people in economically and politically stable nations may

Genetic Testing and Privacy

The field of bioethics began in the 1950s to address issues raised by medical experimentation during World War II. Bioethics initially centered on informed consent, paternalism, autonomy, allocation of scarce medical resources, justice, and definitions of life and death. Today, the field covers medical and biotechnologies and the dilemmas they present. Genetic testing is a key issue in current bioethics because its informational nature affects privacy. Consider these situations.

Testing Incoming Freshmen

When incoming freshmen received information from the University of California, Berkeley, in the summer of 2010, they got more than a class schedule and dorm assignments. They also received kits to send in DNA samples to test for three genes that control three supposedly harmless traits. Participation was voluntary, and because the intent was to gather data, informed consent was not required. However, after genetics groups, bioethicists, policy analysts, and consumer groups protested, the Department of Public Health ruled that the tests provided personal medical information, and should be conducted by licensed medical labs. Since this quintupled the cost, the university changed the program to collect aggregate data, rather than individually identified results.

The three genetic tests were to detect lactose intolerance, alcohol metabolism, and folic acid metabolism. The alcohol test detects variants of a gene that cause a facial flush, nausea, and heart palpitations after drinking, particularly in East Asians—who make up a significant part of the freshman class. Certain mutations in this gene raise the risk of developing esophageal cancer, and so test results may be useful, but they could also encourage drinking.

Testing Tissue from Deceased Children

When parents approve genetic testing for a sick child, they usually assume that their consent applies only when the child is still living, but research may continue after the child is gone. If a newly discovered gene function explains the condition of a child who had never received an accurate diagnosis, should the parents be informed?

The consensus of medical and scientific organizations is that posthumous genetic test information should be disclosed only if the results have been validated (confirmed), the results can lead to testing or treatment for other individuals, and if the parents have not indicated that they do not want to know. For example, several years after a 7-year-old girl died of then-mysterious symptoms, her mother read an article about Rett syndrome (MIM 312750 and Reading 6.2), and thought it described her daughter's small head, hands, and feet; poor socialization skills; cognitive impairment; and hand-wringing. Researchers tested the girl's DNA from a baby tooth the mother had

saved. It was Rett syndrome. The other children in the family were tested and found not to be affected or able to pass it on.

The Military

A new recruit hopes that the DNA sample given when military service begins is never used—it is stored to identify remains. Until now, genetic tests have only been performed for two specific illnesses that could endanger soldiers under certain environmental conditions. Carriers of sickle cell disease (MIM 603903) can develop painful blocked circulation at high altitudes, and carriers of G6PD deficiency (MIM 305900) react badly to anti-malaria medication. Carriers wear red bands on their arms to alert officers to keep the soldiers from harmful situations. In the future, the military may use genetic information to identify soldiers at risk for such conditions as depression and post-traumatic stress disorder. Deployments can be tailored to personal risks, minimizing suffering.

Genome-Wide Association Studies and Disappearing Privacy

The first genome-wide association studies typed people for only a few hundred SNPs. This limited analysis ensured privacy because there were many more people than genotypes, so that it was highly unlikely that an individual could be the only one to have a particular genotype. That is no longer true. Because studies now probe a million or more SNPs, an individual's genotype can be traced to a particular group being investigated—revealing, for example, that a person has a particular disease. That is, the more ways that we can detect that people vary, the easier it is to identify any one of them. It is a little like adding four digits to a zip code to increase the pool of identifiers.

Questions for Discussion

1. If a genetic test reveals a mutation that could harm a blood relative, should the first person's privacy be sacrificed to inform the second person?

2. How might an informed consent document be worded to ask parents if they would like to receive research updates on their child's inherited disease after the child has died?

3. What measures can physicians, the military, and researchers take to maintain the privacy of genetic information?

4. Some student athletes have died of complications from being carriers of sickle cell disease. What are the risks and benefits of testing student athletes for sickle cell disease carrier status?

5. Do you think that passenger screening at airports should include quick DNA scans, as at least one company is offering?

look forward to genome-based individualized health care, poor people in other nations just try to survive, often lacking basic vaccines and medicines. In an African nation where two out of five children suffer from AIDS and many die from other infectious diseases, newborn screening for rare single-gene defects hardly seems practical. However, genetic disorders weaken people so that they become more susceptible to infectious diseases, which they can pass to others.

Human genome information can ultimately benefit everyone. Genome information from humans and our pathogens and parasites is revealing new drug targets. Global organizations, including the United Nations, World Health Organization, and the World Bank, are discussing how nations can share new diagnostic tests and therapeutics that arise from genome information.

Individual nations are adopting guidelines for how to use genetic information to suit their particular strengths (**table 1.5**).

India, for example, has many highly inbred populations with excellent genealogical records, and is home to one-fifth of the world's population. Studies of genetic variation in East Africa are especially important because this region is the cradle of humanity—home of our forebears. The human genome belongs to us all, but efforts from around the world will tell us what our differences are and how they arose. *Bioethics: Choices for the Future* discusses instances when genetic testing can be intrusive.

Table 1.5	Nations Plan for Genomic Medicine
Nation	**Program**
China	The genomes of 100 people are being sequenced.
Gambia	A DNA databank has samples from 57,000 people.
India	A national databank stores DNA from 15,000 people. A company is genotyping the entire Parsi population of 69,000. Other efforts are examining why many drugs only help some people. Laws prevent foreign researchers from sampling tissue from Indians without permission.
Mexico	The National Institute for Genomic Medicine has genotyped 1,200$^+$ people to look for correlations to common diseases. "Safari research" legislation requires approval for foreign researchers to sample DNA from Mexicans.
South Africa	Studies of human genetic diversity among indigenous tribes and susceptibility to HIV and tuberculosis among many populations are under way.
Thailand	A database stores information on genetic susceptibility to dengue fever, malaria, other infectious diseases, and posttraumatic stress disorder from the 2004 tsunami.

Key Concepts

1. Genetics has diverse applications. Matching DNA sequences can clarify relationships, which is useful in forensics, establishing identity, and understanding historical events.
2. Inherited disease differs from other disorders in its predictability; characteristic frequencies in different populations; and the potential of gene therapy.
3. Agriculture and biotechnology apply genetic principles.
4. Collecting DNA from habitats and identifying the sequences in databases is used to analyze ecosystems.
5. Human genome information has tremendous potential but must be carefully managed.

Summary

1.1 Introducing Genes

1. **Genes** are the instructions to manufacture proteins, which determine inherited traits.
2. A **genome** is a complete set of genetic information. A **cell,** the unit of life, contains two genomes of **DNA. Genomics** is the study of many genes and their interactions.

1.2 Levels of Genetics

3. Genes encode proteins and the **RNA** molecules that synthesize proteins. RNA carries the gene sequence information so that it can be utilized, while the DNA is transmitted when the cell divides. Much of the genome does not encode protein.

4. Variants of a gene, called **alleles,** arise by **mutation.** Alleles may differ slightly from one another, but encode the same product. A polymorphism is a site or sequence of DNA that varies in 1 percent or more of a population.

5. **Genome-wide association studies** compare landmarks across the genomes among individuals who share a trait. **Gene expression profiling** examines which genes are more or less active in particular cell types.

6. **Chromosomes** consist of DNA and protein. The 22 types of **autosomes** do not include genes that specify sex. The X and Y **sex chromosomes** bear genes that determine sex.

7. Cells **differentiate** by expressing subsets of genes. **Stem cells** divide to yield other stem cells and cells that differentiate.

8. The **phenotype** is the gene's expression. An allele combination constitutes the **genotype.** Alleles may be **dominant** (exerting an effect in a single copy) or **recessive** (requiring two copies for expression).

9. Pedigrees are diagrams used to study traits in families.

10. Genetic populations are defined by their collections of alleles, termed the **gene pool**. Genome comparisons among species reveal evolutionary relationships.

1.3 Genes and Their Environment

11. Single genes determine Mendelian traits. **Multifactorial traits** reflect the influence of one or more genes and the environment. Recurrence of a Mendelian trait is predicted based on Mendel's laws; predicting the recurrence of a multifactorial trait is more difficult.

12. **Genetic determinism** is the idea that the expression of an inherited trait cannot be changed.

1.4 Applications of Genetics

13. DNA profiling can establish identity, relationships, and origins.

14. In health care, single-gene diseases are more predictable than other diseases, but gene expression profiling is revealing how many types of diseases are related.

15. Agriculture is selective breeding. **Biotechnology** is the use of organisms or their parts for human purposes. A transgenic organism harbors a gene or genes from a different species.

16. In metagenomics, DNA collected from habitats, including the human body, is used to reconstruct ecosystems.

www.mhhe.com/lewisgenetics10

Answers to all end-of-chapter questions can be found at **www.mhhe.com/lewisgenetics10.** You will also find additional practice quizzes, animations, videos, and vocabulary flashcards to help you master the material in this chapter.

Review Questions

1. Place the following terms in size order, from largest to smallest, based on the structures or concepts they represent:
   a. chromosome
   b. gene pool
   c. gene
   d. DNA
   e. genome

2. Distinguish between
   a. an autosome and a sex chromosome.
   b. genotype and phenotype.
   c. DNA and RNA.
   d. recessive and dominant traits.
   e. pedigrees and karyotypes.
   f. gene and genome.

3. Explain how DNA encodes information.

4. Explain how all humans have the same genes, but vary genetically.

5. Explain how a genome-wide association study, gene expression profiling, and DNA sequencing of a gene or genome differ.

6. Explain how all cells in a person's body have the same genome, but are of hundreds of different types that look and function differently.

7. Suggest a practical example of gene expression profiling.

8. Explain the protections under the Genetic Information Nondiscrimination Act, and the limitations.

9. Explain what an application of a "diseasome" type of map, such as in figure 1.8, might provide.

10. Cite an example of a phrase that illustrates genetic determinism.

11. Give an example of a genome that is in a human body, but is not human.

Applied Questions

1. If you were ordering a genetic test panel, which traits and health risks would you like to know about, and why?

2. Two roommates go grocery shopping and purchase several packages of cookies that supposedly each provide 100 calories. After a semester of eating the snacks, one roommate has gained 6 pounds, but the other hasn't. Assuming that other dietary and exercise habits are similar, explain the roommates' different response to the cookies.

3. A study comparing feces of vegetarians, people who eat mostly meat (carnivores), and people who eat a variety of foods (omnivores) found that the microbiome of the vegetarians is much more diverse than that of the other types of diners. Explain why this might be so.

4. One variant in the DNA sequence for the gene that encodes part of the oxygen-carrying blood protein hemoglobin differs in people who have sickle cell disease. Newborns are tested for this mutation. Is this a single-gene test, a genome sequencing, a genome-wide association study, or a gene expression profile?

5. Consider the following two studies:

   ■ Gout is a form of arthritis that often begins with pain in the big toe. In one study, researchers looked at 500,000 SNPs in 100 people with gout and 100 who do not have gout, and found a very distinctive pattern in the people with painful toes.

   ■ About 1 percent of people who take cholesterol-lowering drugs (statins) experience muscle pain. Researchers discovered that their muscle cells have different numbers and types of mRNA molecules than most people who tolerate the drugs well.

   Which description is of a genome-wide association study and which a gene expression study?

6. A 54-year-old man is turned down for life insurance because testing following a heart attack revealed that he had inherited cardiac myopathy, and this had most likely caused the attack. He cites GINA, but the insurer says that the law does not apply to his case. Who is correct?

7. How does GINA benefit
   a. health care consumers?
   b. employers?
   c. insurers?
   d. researchers?

8. An ad for a skin cream proclaims it will "boost genes' activity and stimulate the production of youth proteins." Which technology described in the chapter could be used to test the ad's claim?

9. What are the possible benefits and risks of a government requiring that all citizens have genetic profiles or DNA test results on file?

10. What body part would you like to explore metagenomically, and why?

Web Activities

1. Consult a website for a direct-to-consumer genetic testing company, such as 23andMe, Navigenics, or deCODE Genetics. Choose three tests, and explain why you would want to take them. Also discuss a genetic test that you would not wish to take, and explain why not.

2. Many organizations are using DNA bar codes to classify species. Consult the websites for one of the following organizations and describe an example of how they are using DNA sequences:

   Consortium for the Barcode of Life (International)
   Canadian Barcode of Life Network
   Species 2000 (UK)
   Encyclopedia of Life (Wikipedia)

3. Human microbiome projects have different goals. Consult the websites for two of the following projects and compare their approaches:

   The Human Microbiome Project (NIH)
   Meta-Gut (China)
   Metagenomics of the Human Intestinal Tract (European Commission)
   Human Gastric Microbiome (Singapore)

   Australian Urogenital Microbiome Consortium
   Human MetaGenome Consortium (Japan)
   Canadian Microbiome Initiative

4. Look at the website for the McLaughlin-Rotman Centre for Global Health (www.mrcglobal.org). Describe a nation's plan to embrace genomic medicine.

5. The Gopher Kids Study collects DNA from children aged 1 to 11 at the Minnesota State Fair, in a saliva sample. The goal is to identify genes that contribute to normal health and development. The website is http://www.peds.umn.edu/gopherkids/why/index.htm. If you were a parent of a child at the fair, what questions would you ask before donating his or her DNA?

6. The Multiplex Initiative (http://www.genome.gov/25521052) asked 2,000 young adults (a) whether genetics or lifestyle habits affect health more, and (b) which of the following conditions they would want to take a genetic test for: osteoporosis, type 2 diabetes, high cholesterol, high blood pressure, lung cancer, and others. Most participants said that lifestyle factors are more important. Do you agree or disagree? State a reason for your answer.

Forensics Focus

1. Consult the websites for a television program that uses, or is based on, forensics (CSI or Law and Order, for example), and find an episode in which species other than humans are critical to the case. Explain how DNA bar coding could help to solve the crime.

2. On an episode of the television program House, the main character, Dr. House, knew from age 12 that his biological father was a family friend, not the man who raised him. At

his supposed father's funeral, the good doctor knelt over the body in the casket and sneakily snipped a bit of skin from the corpse's earlobe—for a DNA test.

   a. Do you think that this action was an invasion of anyone's privacy? Was Dr. House justified?

   b. Dr. House often orders treatments for patients based on observing symptoms. Suggest a way that he can use DNA testing to refine his diagnoses.

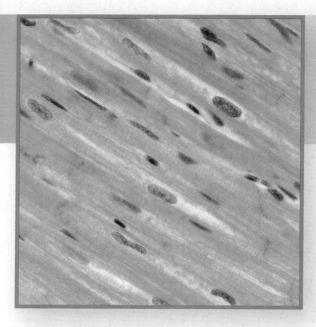

Heart muscle cells are larger and more rigid in LEOPARD syndrome than in these normal cells. Researchers reprogrammed patients' skin cells to become heart muscle cells in order to study how this rare inherited disease begins, suggesting new ways to treat it.

Cells

Learning Outcomes

2.1. Introducing Cells

1. Explain why it is important to know the cellular basis of a disease.

2. Define *differentiated cell.*

2.2 Cell Components

3. List the four major chemicals in cells.

4. Describe how organelles interact.

5. Describe the structure and function of a biological membrane.

6. List the components of the cytoskeleton.

2.3 Cell Division and Death

7. Distinguish between mitosis and apoptosis.

8. Describe the events and control of the cell cycle.

2.4 Cell-Cell Interactions

9. Explain how chemical signals enter a cell.

10. List the steps of cellular adhesion.

2.5 Stem Cells

11. List the characteristics of a stem cell.

12. Define stem and progenitor cell.

The Big Picture: Our bodies are built of trillions of cells that interact in complex ways to keep us alive. All cells in a body use the same genome, but have different structures and functions because they access different parts of the genome.

A Disease in a Dish

A new way to learn about a disease is to watch it arise in a dish. However, this isn't always possible because some cells, such as heart muscle and nerve cells, do not easily grow outside a body. A way around this limitation is to use a technology called cellular reprogramming to derive one cell type from another. The approach takes cells back to a state similar to stem cells, then coaxes them to specialize in a different way.

Researchers used cellular reprogramming to investigate LEOPARD syndrome (MIM 151100). The name is an acronym for the symptoms, which include an enlarged heart with blocked valves to the lungs, freckles, abnormal genitals, and deafness. Treatment is symptomatic. The researchers took skin cells from two patients and added a "cocktail" of genes that reprogrammed the cells so that they could give rise to almost any cell type. Then a different cocktail stimulated the cells to form pulsating heart cells that had the mutation. Sure enough, compared to healthy heart muscle cells, the LEOPARD cells were too big, with abnormally rigid muscle filaments, and with a different set of genes being expressed, indicating that the diseased cells are tuned into a different set of signals than healthy cells.

By discovering what is wrong in a disease by studying the right types of cells nurtured from patients, researchers can "repurpose" existing drugs, or invent new ones that will treat the disease at its source. Reprogrammed cells are giving researchers unprecedented peeks into the genesis of disease.

2.1 Introducing Cells

Our inherited traits, quirks, and illnesses arise from the activities of cells. Understanding cell function reveals how a healthy body works, and how it develops from one cell to trillions. Understanding what goes wrong in certain cells to cause pain or other symptoms can suggest ways to treat the condition, because we learn what must be repaired or replaced. For example, genes tell cells how to make the proteins that align to form the contractile apparatus of muscles. In Duchenne muscular dystrophy (MIM 310200), one type of muscle protein is missing, and as a result muscle cells collapse under forceful contraction. Certain muscles become very weak in early childhood. The little boy in **figure 2.1** has overdeveloped calf muscles, and he cannot stand normally. Identifying the protein revealed exactly what must be replaced—but doing so has been difficult because many muscle cells must be corrected.

Our bodies include more than 260 specialized, or differentiated, cell types. These include bone, nerve, and muscle, and subtypes of those. They are **somatic cells,** also called body cells. Somatic cells have two copies of the genome and are said to be **diploid.** In contrast, the rarer sperm and egg have one copy of the genome and are **haploid.** The meeting of sperm and egg restores the diploid state. Especially important in many-celled organisms are **stem cells,** which are diploid cells that both give rise to differentiated cells and replicate themselves in a process called self-renewal. Stem cells enable a body to develop, grow, and repair damage.

Cells interact. They send, receive, and respond to information. Some cells aggregate with others of like function, forming tissues, which in turn interact to form organs and organ systems. Other cells move about the body. Cell numbers are important, too—they are critical to development, growth, and healing. Staying healthy reflects a precise balance between cell division, which adds cells, and cell death, which takes them away.

2.2 Cell Components

All cells share certain features that enable them to perform the basic life functions of reproduction, growth, response to stimuli, and energy use. Specialized features emerge as cells express different subsets of the thousands of protein-encoding genes. Many other genes control which protein-encoding genes a cell expresses.

All multicellular organisms, including other animals, fungi, and plants, have differentiated cells. Some single-celled organisms, such as the familiar paramecium and amoeba, have very distinctive cells as complex as our own. The most abundant organisms on the planet, however, are simpler and single-celled, such as bacteria. These microorganisms are nonetheless successful life forms because they have occupied Earth much longer than we have, and even live in our bodies.

Biologists recognize three broad varieties of cells that define three major "domains" of life: the Archaea, the Bacteria, and the Eukarya. A domain is a broader classification than the familiar kingdom.

Members of the Archaea and Bacteria are single-celled, but they differ from each other in the sequences of many of their genes and in the types of molecules in their membranes. Archaea and Bacteria are both **prokaryotes.** This means that they lack a **nucleus,** the structure that contains DNA in the cells of other types of organisms, which comprise the third domain of life, the Eukarya. Also known as **eukaryotes,** this group includes single-celled organisms that have nuclei, as well as all multicellular organisms (**figure 2.2**). Eukaryotic cells are also distinguished from prokaryotic cells in that they have structures called **organelles,** which perform specific functions. The cells of all three domains contain globular assemblies of RNA and protein called **ribosomes** that are essential

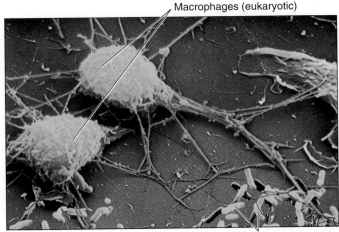

Macrophages (eukaryotic)

Bacteria (prokaryotic)

Figure 2.2 Eukaryotic and prokaryotic cells. A human cell is eukaryotic and much more complex than a bacterial cell, while an archaean cell looks much like a bacterial cell. Here, human macrophages (blue) capture bacteria (yellow). Note how much larger the human cells are. (A few types of giant bacteria are larger than some of the smaller human cell types.)

 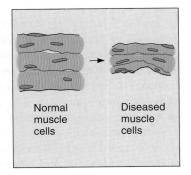

Normal muscle cells

Diseased muscle cells

Figure 2.1 Genetic disease at the whole-person and cellular levels. An early sign of the boy on the right's Duchenne muscular dystrophy is overdeveloped calf muscles that result from his inability to rise from a sitting position the usual way. Lack of the protein dystrophin causes his skeletal muscle cells to collapse when they contract.

for protein synthesis. The eukaryotes may have arisen from an ancient fusion of a bacterium with an archaean.

Chemical Constituents

Cells are composed of molecules. Some of the chemicals of life (biochemicals) are so large that they are called macromolecules.

The major macromolecules that make up cells and that cells use as fuel are **carbohydrates** (sugars and starches), **lipids** (fats and oils), **proteins,** and **nucleic acids** (DNA and RNA). Cells require vitamins and minerals in much smaller amounts.

Carbohydrates provide energy and contribute to cell structure. Lipids form the basis of several types of hormones, form membranes, provide insulation, and store energy. Proteins have many diverse functions in the human body. They participate in blood clotting, nerve transmission, and muscle contraction, and form the bulk of the body's connective tissue. Antibodies that fight bacterial infection are proteins. **Enzymes** are especially important proteins because they facilitate, or catalyze, biochemical reactions so that they occur swiftly enough to sustain life. Most important to the study of genetics are the nucleic acids DNA and RNA, which translate information from past generations into specific collections of proteins that give a cell its individual characteristics.

Macromolecules often combine in cells, forming larger structures. For example, the membranes that surround cells and compartmentalize their interiors consist of double layers (bilayers) of lipids embedded with carbohydrates, proteins, and other lipids.

Life is based on the chemical principles that govern all matter; genetics is based on a highly organized subset of the chemical reactions of life. **Reading 2.1** describes diseases that affect these major types of biological molecules.

Organelles

A typical eukaryotic cell holds a thousand times the volume of a bacterial or archaeal cell. To carry out the activities of life in such a large cell, organelles divide the labor by partitioning off certain areas or serving specific functions. The coordinated functioning of the organelles in a eukaryotic cell is much like the organization of departments in a big-box store, compared to the prokaryote-like simplicity of a small grocery store. In general, organelles keep related biochemicals and structures close enough to one another to interact efficiently. This eliminates the need to maintain a high concentration of a particular biochemical throughout the cell.

Organelles have a variety of functions. They enable a cell to retain as well as to use its genetic instructions, acquire energy, secrete substances, and dismantle debris. Saclike organelles sequester biochemicals that might harm other cellular constituents. Some organelles consist of membranes studded with enzymes embedded in the order in which they participate in the chemical reactions that produce a particular molecule. **Figure 2.3** depicts organelles.

The most prominent organelle of most cells is the nucleus. It is enclosed in a layer called the nuclear envelope.

Biochemicals can exit or enter the nucleus through nuclear pores, which are rings of proteins like portholes in a ship's side (**figure 2.4**).

On the inner face of the nuclear membrane is a layer of fibrous material called the nuclear lamina. This layer has several important functions. The DNA in the nucleus touches the nuclear lamina as the cell divides. The nuclear lamina also provides mechanical support and holds in place the nuclear pores. Chapter 3 discusses very rare, accelerated aging disorders that result from an abnormal nuclear lamina.

Inside the nucleus is an area that appears darkened under a microscope, called the nucleolus ("little nucleus"). Here, ribosomes are produced. The nucleus is filled with DNA complexed with many proteins to form chromosomes. Other proteins form fibers that fill out the nucleus, giving it a roughly spherical shape. RNA is abundant too, as are enzymes and proteins required to synthesize RNA from DNA. The fluid in the nucleus, minus these contents, is called nucleoplasm.

The remainder of the cell—that is, everything but the nucleus, organelles, and the outer boundary, or **plasma membrane**—is **cytoplasm.** Other cellular components include stored proteins, carbohydrates, and lipids; pigment molecules; and various other small chemicals. We now take a closer look at three cellular functions: secretion, digestion inside cells, and energy production.

Secretion—The Eukaryotic Production Line

Organelles interact in ways that coordinate basic life functions and sculpt the characteristics of specialized cell types. Secretion, which is the release of a substance from a cell, illustrates one way that organelles function together.

Secretion begins when the body sends a biochemical message to a cell to begin producing a particular substance. For example, when a newborn first suckles the mother's breast, her brain releases hormones that signal cells in her breast to rapidly increase the production of the complex mixture that makes up milk (**figure 2.5**). In response, information in certain genes is copied into molecules of **messenger RNA (mRNA)**, which then exit the nucleus (see steps 1 and 2 in figure 2.5). In the cytoplasm, the mRNAs, with the help of ribosomes and another type of RNA called **transfer RNA,** direct the manufacture of milk proteins. These include nutritive proteins called caseins, antibodies that protect against infection, and enzymes.

Most protein synthesis occurs on a maze of interconnected membranous tubules and sacs called the **endoplasmic reticulum (ER)** (see step 3 in figure 2.5). The ER winds from the nuclear envelope outward to the plasma membrane. The section of ER nearest the nucleus, which is flattened and studded with ribosomes, is called rough ER, because the ribosomes make it appear fuzzy when viewed under an electron microscope. Messenger RNA attaches to the ribosomes on the rough ER. Amino acids from the cytoplasm are then linked, following the instructions in the mRNA's sequence, to form particular proteins that will either exit the cell or become part of membranes (step 3, figure 2.5). Proteins are also synthesized on ribosomes not associated with the ER. These proteins remain in the cytoplasm.

Inborn Errors of Metabolism Affect the Major Biomolecules

Enzymes, by speeding specific chemical reactions, control a cell's production of all types of macromolecules. When an enzyme is not produced or cannot function, too much or too little of the product of the biochemical reaction may be made. These biochemical buildups and breakdowns may cause symptoms. Genetic disorders that result from deficient or absent enzymes are called "inborn errors of metabolism." Following are some examples.

Carbohydrates

The newborn yelled and pulled up her chubby legs in pain a few hours after each feeding. She developed watery diarrhea, even though she was breastfed. Finally, a doctor diagnosed *lactase deficiency* (MIM 223000). This is lack of the enzyme lactase, which enables the digestive system to break down the carbohydrate lactose. Bacteria multiplied in the undigested lactose in the child's intestines, producing gas, cramps, and bloating. Switching to a soybean-based, lactose-free infant formula helped. A different condition with milder symptoms is lactose intolerance (MIM 150200), common in adults (see the opening essay to chapter 15).

Lipids

A sudden sharp pain began in the man's arm and spread to his chest. At age 36, he was younger than most people who suffer heart attacks, but he had inherited a gene variant that halved the number of protein receptors for cholesterol on his liver cells. Because cholesterol could not enter liver cells efficiently, it built up in his arteries, constricting blood flow in his heart and causing a mild heart attack. A fatty diet and lack of exercise had accelerated his *familial hypercholesterolemia*. He hopes that taking a cholesterol-lowering drug and starting an exercise program will lower his risk of suffering future heart attacks.

Proteins

Newborn Tim slept most of the time, and he vomited so often that he hardly grew. A blood test revealed *maple syrup urine disease* (MIM 248600), which makes urine smell like maple syrup. Tim could not digest three types of amino acids (protein building blocks), which accumulated in his bloodstream. A diet very low in these amino acids controlled the symptoms. Today this inborn error is one of many dozen that are detected with blood tests shortly after birth. Newborn screening is discussed in chapter 20.

Nucleic Acids

From birth, Troy's wet diapers contained orange, sandlike particles, but otherwise he seemed healthy. By 6 months of age, he was in pain when urinating. A physician noted that Troy's writhing movements were involuntary rather than normal crawling.

The orange particles in Troy's diaper indicated *Lesch-Nyhan syndrome* (MIM 300322), caused by the deficiency of an enzyme called HGPRT. Troy's body could not recycle two of the four types of DNA building blocks, instead converting them into uric acid, which crystallizes in urine. Other symptoms that began later were not as easy to explain—severe intellectual disability, seizures, and aggressive and self-destructive behavior. By age 3, he responded to stress by uncontrollably biting his fingers and lips. On doctors' advice, his parents had his teeth removed to keep him from harming himself, and he was kept in restraints. Troy would probably die before the age of 30 of kidney failure or infection.

Vitamins

Vitamins enable the body to use the carbohydrates, lipids, and proteins we eat. Julie inherited *biotinidase deficiency* (MIM 253260), which greatly slows her body's use of the vitamin biotin. If Julie hadn't been diagnosed as a newborn and quickly started on biotin supplements, by early childhood she would have shown biotin deficiency symptoms: intellectual disability, seizures, skin rash, and loss of hearing, vision, and hair. Her slow growth, caused by her body's inability to extract energy from nutrients, would have eventually proved lethal.

Minerals

Ingrid, in her thirties, lived in the geriatric ward of a mental hospital, unable to talk or walk. She grinned and drooled, but she was alert and communicated using a computer. When she was a healthy high-school senior, symptoms of *Wilson disease* (MIM 277900) began as her weakened liver could no longer control the excess copper her digestive tract absorbed from food. The initial symptoms were stomachaches, headaches, and an inflamed liver (hepatitis). Then other changes began—slurred speech; loss of balance; a gravelly, low-pitched voice; and altered handwriting. A psychiatrist noted the telltale greenish rings around her irises, caused by copper buildup, and diagnosed Wilson disease (**figure 1**). Finally Ingrid received penicillamine, which enabled her to excrete the excess copper in her urine. The treatment halted the course of the illness, saving her life. She now lives with a relative.

Figure 1 **Wilson disease.** A greenish ring around the brownish iris is one sign of the copper buildup of Wilson disease.

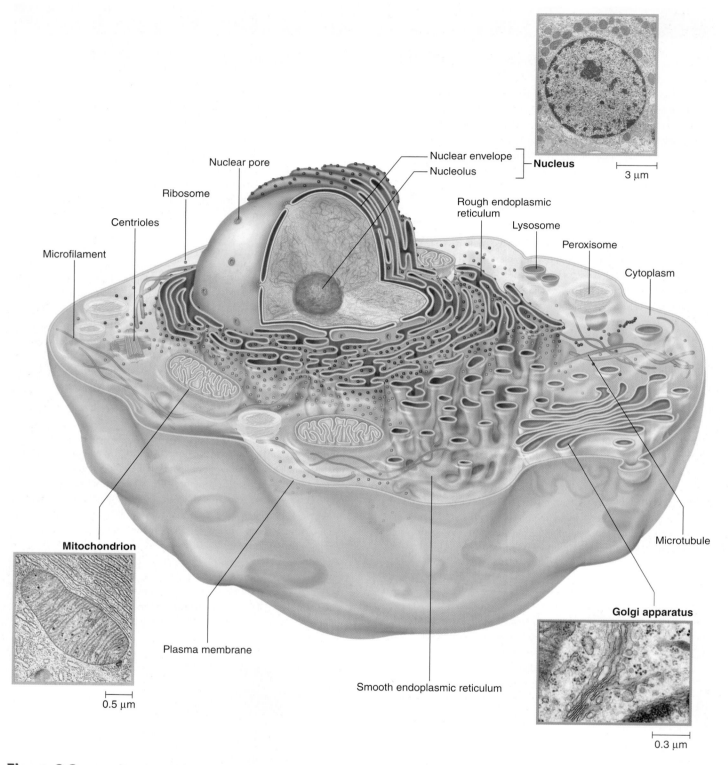

Nucleus

3 μm

Nuclear pore

Ribosome

Centrioles

Microfilament

Nuclear envelope

Nucleolus

Rough endoplasmic reticulum

Lysosome

Peroxisome

Cytoplasm

Microtubule

Mitochondrion

0.5 μm

Plasma membrane

Smooth endoplasmic reticulum

Golgi apparatus

0.3 μm

Figure 2.3 **Generalized animal cell.** Organelles provide specialized functions for the cell. Most of these structures are transparent; colors are used here to distinguish them. Different cell types have different numbers of organelles. All cell types have a single nucleus, except for red blood cells, which expel their nuclei as they mature.

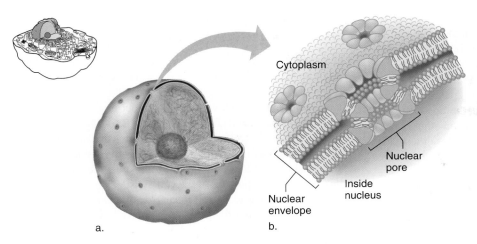

Figure 2.4 **The nucleus is the genetic headquarters.** (a) The largest structure in a typical human cell, the nucleus lies within two membrane layers that make up the nuclear envelope. (b) Nuclear pores allow specific molecules to move in and out of the nucleus through the envelope.

The ER acts as a quality control center for the cell. Its chemical environment enables the forming protein to start folding into the three-dimensional shape necessary for its specific function. Misfolded proteins are pulled out of the ER and degraded, much as an obviously defective toy might be pulled from an assembly line at a toy factory and discarded. Misfolded proteins can cause disease, as discussed further in chapter 10.

As the rough ER winds out toward the plasma membrane, the ribosomes become fewer, and the tubules widen, forming a section called smooth ER. Here, lipids are made and added to the proteins arriving from the rough ER (step 4, figure 2.5). The lipids and proteins are transported until the tubules of the smooth ER narrow and end.

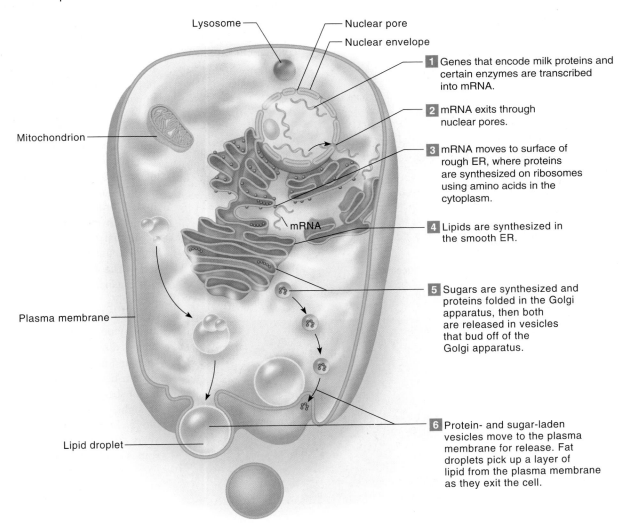

Figure 2.5 **Secretion: making milk.** Milk production and secretion illustrate organelle functions and interactions in a cell from a mammary gland: (1) through (6) indicate the order in which organelles participate in this process. Lipids are secreted in separate droplets from proteins and their attached sugars. This cell is highly simplified.

Then the proteins exit the ER in membrane-bounded, saclike organelles called **vesicles,** which pinch off from the tubular endings of the membrane. Lipids are exported without a vesicle, because a vesicle is itself made of lipid.

A loaded vesicle takes its contents to the next stop in the secretory production line, a **Golgi apparatus** (step 5, figure 2.5). This processing center is a stack of flat, membrane-enclosed sacs. Here, the milk sugar lactose is synthesized and other sugars are made that attach to proteins to form glycoproteins or to lipids to form glycolipids, which become parts of plasma membranes. Proteins finish folding in the Golgi apparatus.

The components of complex secretions, such as milk, are temporarily stored in the Golgi apparatus. Droplets of proteins and sugars then bud off in vesicles that move outward to the plasma membrane, fleetingly becoming part of it until they are secreted to the cell's exterior. Lipids exit the plasma membrane directly, taking bits of it with them (step 6, figure 2.5).

In the breast, epithelial (lining) cells called lactocytes form tubules, into which they secrete the components of milk. When the baby suckles, contractile cells squeeze the milk through the tubules and out of holes in the nipples. This "ejection reflex" is so powerful that the milk can actually shoot across a room!

Intracellular Digestion—Lysosomes and Peroxisomes

Just as clutter and garbage accumulate in an apartment, debris builds up in cells. Organelles called **lysosomes** handle the garbage. Lysosomes are membrane-bounded sacs that contain enzymes that dismantle bacterial remnants, worn-out organelles, and other material such as excess cholesterol (**figure 2.6**). The enzymes also break down some digested nutrients into forms that the cell can use.

Lysosomes fuse with vesicles carrying debris from outside or within the cell, and the lysosomal enzymes then degrade the contents. This process of the cell's disposing of its own trash is called "autophagy," which means "eating self." For example, a type of vesicle that forms from the plasma membrane, called an endosome, ferries extra low-density lipoprotein (LDL) cholesterol to lysosomes. A loaded lysosome moves toward the plasma membrane and fuses with it, releasing its contents to the outside. The word *lysosome* means "body that lyses"; *lyse* means "to cut." Lysosomes maintain the very acidic environment that their enzymes require to function, without harming other cellular constituents that could be destroyed by acid.

Cells differ in number of lysosomes. Certain white blood cells and macrophages that move about and engulf bacteria are loaded with lysosomes. Liver cells require many lysosomes to break down cholesterol, toxins, and drugs.

All lysosomes contain more than 40 types of digestive enzymes, which must be maintained in a correct balance. Absence or malfunction of an enzyme causes a "lysosomal storage disease." In these inherited disorders, which are a type of inborn error of metabolism, the molecule that the missing or abnormal enzyme normally degrades accumulates. The lysosome swells, crowding organelles and interfering with the cell's functions. In Tay-Sachs disease (MIM 272800), for example, an enzyme is deficient that normally breaks down lipids in the cells that surround nerve cells. As the nervous system becomes buried in lipid, the infant begins to lose skills, such as sight, hearing, and the ability to move. Death is typically within 3 years. Even before birth, the lysosomes of affected cells swell.

Peroxisomes are sacs with outer membranes that are studded with several types of enzymes. These enzymes perform a variety of functions, including breaking down certain lipids and rare biochemicals, synthesizing bile acids used in fat digestion, and detoxifying

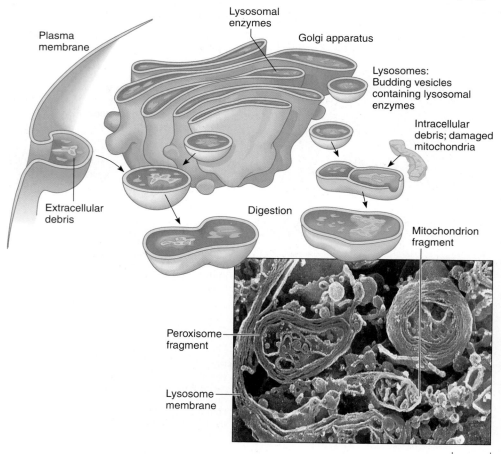

Figure 2.6 **Lysosomes are trash centers.** Lysosomes fuse with vesicles or damaged organelles, activating the enzymes within to recycle the molecules. Lysosomal enzymes also dismantle bacterial remnants. These enzymes require a very acidic environment to function.

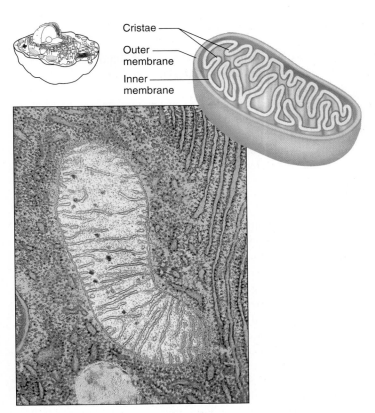

Cristae

Outer membrane

Inner membrane

Figure 2.7 **A mitochondrion extracts energy.** Cristae, infoldings of the inner membrane, increase the available surface area containing enzymes for energy reactions in a mitochondrion.

compounds that result from exposure to oxygen free radicals. Peroxisomes are large and abundant in liver and kidney cells, which handle toxins.

The 1992 film *Lorenzo's Oil* recounted the true story of a child with an inborn error of metabolism caused by an absent peroxisomal enzyme. Lorenzo had adrenoleukodystrophy (MIM 202370), in which a type of lipid called a very-long-chain fatty acid builds up in the brain and spinal cord. Early symptoms include low blood sugar, skin darkening, muscle weakness, altered behavior, and irregular heartbeat. The patient eventually loses control over the limbs and usually dies within a few years. Lorenzo's parents devised a combination of edible oils that theoretically diverts another enzyme to compensate for the missing one. Lorenzo died a day after his thirtieth birthday, but even his father admits to not knowing whether the oil or Lorenzo's excellent care prolonged his life. Stem cell transplants and gene therapy can

halt progression of the disease, preventing symptoms if started early enough.

Energy Production—Mitochondria

The activities of secretion and the many chemical reactions taking place in the cytoplasm require continual energy. Organelles called **mitochondria** provide energy by breaking the chemical bonds that hold together the nutrient molecules in food.

A mitochondrion has an outer membrane similar to those in the ER and Golgi apparatus and an inner membrane that forms folds called cristae (**figure 2.7**). These folds hold enzymes that catalyze the biochemical reactions that release energy from nutrient molecules. The freed energy is captured and stored in the bonds that hold together a molecule called adenosine triphosphate (ATP). In this way, ATP functions a little like an energy debit card.

The number of mitochondria in a cell varies from a few hundred to tens of thousands, depending upon the cell's activity level. A typical liver cell, for example, has about 1,700 mitochondria, but a muscle cell, with its very high energy requirements, has many more. Mitochondria contain a small amount of DNA different from the DNA in the nucleus (see figure 5.8). Chapter 5 discusses mitochondrial inheritance, and chapter 16 describes how mitochondrial genes provide insights into early human migrations.

Table 2.1 summarizes the structures and functions of organelles.

The Plasma Membrane

Just as the character of a community is molded by the people who enter and leave it, the special characteristics of different cell types are shaped in part by the substances that enter and leave.

Table 2.1	Structures and Functions of Organelles	
Organelle	**Structure**	**Function**
Endoplasmic reticulum	Membrane network; rough ER has ribosomes, smooth ER does not	Site of protein synthesis and folding; lipid synthesis
Golgi apparatus	Stacks of membrane-enclosed sacs	Site where sugars are made and linked into starches or joined to lipids or proteins; proteins finish folding; secretions stored
Lysosome	Sac containing digestive enzymes	Degrades debris; recycles cell contents
Mitochondrion	Two membranes; inner membrane enzyme-studded	Releases energy from nutrients, participates in cell death
Nucleus	Porous sac containing DNA	Separates DNA within cell
Peroxisome	Sac containing enzymes	Breaks down and detoxifies various molecules
Ribosome	Two associated globular subunits of RNA and protein	Scaffold and catalyst for protein synthesis
Vesicle	Membrane-bounded sac	Temporarily stores or transports substances

The plasma membrane controls this process. It forms a selective barrier that completely surrounds the cell and monitors the movements of molecules in and out. How the chemicals that comprise the plasma membrane associate with each other determines which substances can enter or leave the cell. Membranes similar to the plasma membrane form the outer boundaries of several organelles, such as the mitochondrion, and some organelles consist entirely of membranes, such as the endoplasmic reticulum and Golgi apparatus. A cell's membranes are more than mere coverings, because some of their molecules have specific functions.

A biological membrane has a distinctive structure. It is a double layer (bilayer) of molecules called phospholipids. A phospholipid is a fat molecule with attached phosphate groups. A phosphate group [PO_4] is a phosphorus atom bonded to four oxygen atoms. A phospholipid is often depicted as a head with two parallel tails.

Membranes form because phospholipid molecules self-assemble into sheets **figure 2.8**. The molecules do this because their ends react oppositely to water: The phosphate end of a phospholipid is attracted to water, and thus is hydrophilic ("water-loving"); the other end, which consists of two chains of fatty acids, moves away from water, and is therefore hydrophobic ("water-fearing"). Because of these forces, phospholipid molecules in water spontaneously form bilayers. Their hydrophilic surfaces are exposed to the watery exterior and interior of the cell, and their hydrophobic surfaces face each other on the inside of the bilayer, away from water.

A phospholipid bilayer forms the structural backbone of a biological membrane. Proteins are embedded in the bilayer. Some traverse the entire structure, while others extend from one or both faces (**figure 2.9**).

The proteins, glycoproteins, and glycolipids that extend from a plasma membrane create surface topographies that are important in a cell's interactions with other cells. The surfaces of your cells indicate not only that they are part of your body, but also that they are part of a particular organ and a particular tissue type.

Many molecules that extend from the plasma membrane are **receptors,** which are structures that have indentations or other shapes that fit and hold molecules outside the cell. The molecule that binds to the receptor, called the **ligand,** may set into motion a cascade of chemical reactions inside the cell that carries out a particular activity, such as dividing.

The phospholipid bilayer is oily, and some proteins move within it like ships on a sea. Proteins with related functions may

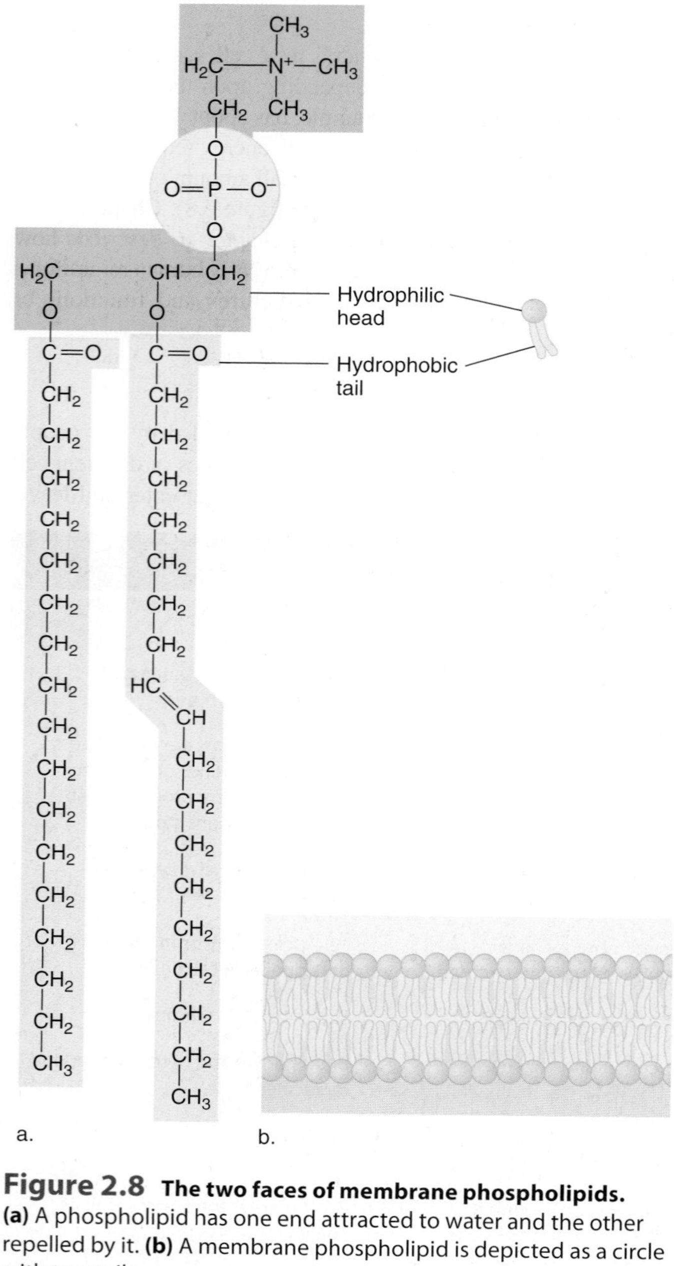

Figure 2.8 **The two faces of membrane phospholipids.** (a) A phospholipid has one end attracted to water and the other repelled by it. (b) A membrane phospholipid is depicted as a circle with two tails.

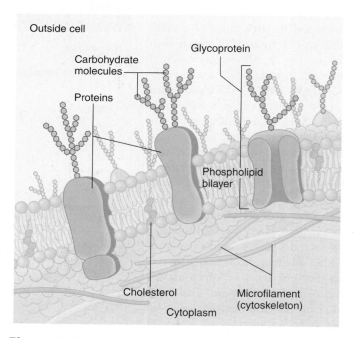

Figure 2.9 **Anatomy of a plasma membrane.** Mobile proteins are embedded throughout a phospholipid bilayer. Carbohydrates jut from the membrane's outer face.

cluster on "lipid rafts" that float on the phospholipid bilayer. The rafts are rich in cholesterol and other types of lipids. This clustering of proteins eases their interaction.

Proteins aboard lipid rafts have several functions. They contribute to the cell's identity; act as transport shuttles into the cell; serve as gatekeepers; and can let in certain toxins and pathogens. HIV, for example, enters a cell by breaking a lipid raft.

The inner hydrophobic region of the phospholipid bilayer blocks entry and exit to most substances that dissolve in water. However, certain molecules can cross the membrane through proteins that form passageways, or when they are escorted by a "carrier" protein. Some membrane proteins form channels for ions, which are atoms or molecules with an electrical charge. **Reading 2.2** describes "channelopathies"—diseases that stem from faulty ion channels.

Faulty Ion Channels Cause Inherited Disease

What do abnormal pain intensity, irregular heartbeats, and cystic fibrosis have in common? All result from abnormal ion channels in plasma membranes.

Ion channels are protein-lined tunnels in the phospholipid bilayer of a biological membrane. These passageways permit electrical signals in the form of ions (charged particles) to pass through membranes.

Ion channels are specific for calcium (Ca^{2+}), sodium (Na^+), potassium (K^+), or chloride (Cl^-) ions. A plasma membrane may have a few thousand ion channels for each of these ions. In order for a cell to function normally, it must allow certain types of ions in and out at specific rates. Ion channels control these movements. Ten million ions can pass through an ion channel in one second! The following "channelopathies" result from abnormal ion channels.

Unusual Pain and Sodium Channels

A 10-year-old boy who lived in a northern Pakistani town was completely unable to feel pain. He became a performer, stabbing knives through his arms and walking on hot coals to entertain crowds. Several other people in this community where relatives often married relatives were also unable to feel pain. Researchers studied the connected families and discovered a mutation that alters sodium channels on certain nerve cells so that the message to feel pain cannot be sent. The boy died at age 13 from jumping off a roof.

A different mutation affecting the same sodium channel causes very different symptoms. In "burning man syndrome," the channels become extra sensitive, opening and flooding the body with pain easily, in response to exercise, an increase in room temperature, or just putting on socks. In another condition, "paroxysmal extreme pain disorder," the sodium channels stay open too long, causing excruciating pain in the rectum, jaw, and eyes. Researchers are using the information from studies of these genetic disorders to develop new painkillers.

Long-QT Syndrome and Potassium Channels

Four children in a Norwegian family were born deaf, and three of them died at ages 4, 5, and 9. All of the children had inherited from unaffected carrier parents "long-QT syndrome associated with deafness" (MIM 176261). ("QT" refers to part of a normal heart rhythm.) These children had abnormal potassium channels in the cells of the heart muscle and in the inner ear. In the heart cells, the malfunctioning ion channels disrupted electrical activity, fatally disturbing heart rhythm. In the cells of the inner ear, the abnormal ion channels increased the extracellular concentration of potassium ions, impairing hearing.

Cystic Fibrosis and Chloride Channels

A seventeenth-century English saying, "A child that is salty to taste will die shortly after birth," described the consequence of abnormal chloride channels in cystic fibrosis (CF). The chloride channel is called CFTR, for cystic fibrosis transductance regulator. In most cases, CFTR protein remains in the cytoplasm, unable to reach the plasma membrane, where it would normally function (**figure 1**).

CF is inherited from carrier parents. The major symptoms of difficulty breathing, frequent severe respiratory infections, and a clogged pancreas that disrupts digestion all result from a buildup of extremely thick mucous secretions.

Abnormal chloride channels in cells lining the lung passageways and ducts of the pancreas cause the symptoms of CF. The primary defect in the chloride channels also disrupts sodium channels. The result: Salt trapped inside cells draws moisture in and thickens surrounding mucus. Reading 4.2 discusses CF further.

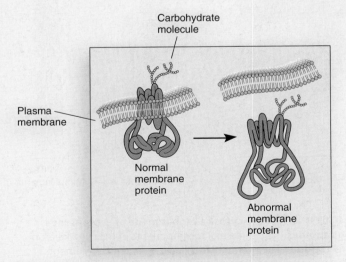

Figure 1 In cystic fibrosis, CFTR protein remains in the cytoplasm, rather than anchoring in the plasma membrane. This prevents normal chloride channel function.

The Cytoskeleton

The **cytoskeleton** is a meshwork of protein rods and tubules that molds the distinctive structures of a cell. It positions organelles and provides three-dimensional shapes. The proteins of the cytoskeleton are continually broken down and built up as a cell performs specific activities. Some cytoskeletal elements function as rails that transport cellular contents; other parts, called motor molecules, power the movement of organelles along these rails by converting chemical energy to mechanical energy.

The cytoskeleton includes three major types of elements —**microtubules, microfilaments,** and **intermediate filaments (figure 2.10).** They are distinguished by protein type, diameter, and how they aggregate into larger structures. Other proteins connect these components, creating the framework that provides the cell's strength and ability to resist force and maintain shape.

Long, hollow microtubules provide many cellular movements. A microtubule is composed of pairs (dimers) of a protein, called tubulin, assembled into a hollow tube. The cell can change the length of the tubule by adding or removing tubulin molecules.

Cells contain both formed microtubules and individual tubulin molecules. When the cell requires microtubules to carry out a specific function—cell division, for example—free tubulin dimers self-assemble into more tubules. After the cell divides, some of the microtubules fall apart into individual tubulin dimers, replenishing the cell's supply of building blocks. Cells are perpetually building up and breaking down microtubules.

Microtubules also form cilia, which are hairlike structures (**figure 2.11**). Cilia were the first organelles seen under a microscope, and their name means "cells' eyelashes" in Latin. Coordinated movement of cilia generates a wave that moves the cell or propels substances along its surface. Cilia beat particles up and out of respiratory tubules and move egg cells in the female reproductive tract. Because cilia are so widespread in the body, defects in them affect health. One such "ciliopathy" is Bardet-Biedl syndrome (MIM 209900), which causes obesity, visual loss, diabetes, cognitive impairment, and extra fingers and/or toes.

Cilia that do not move are also found in several types of cells. These cilia serve as antennae, receiving biochemical signals and passing them to particular locations within the cell. Nonmoving cilia are important in stimulating cells to move, such as in forming organs in an embryo and wound healing. Absence of nonmoving cilia can cause disease, such as polycystic kidney disease (MIM 173900).

Microfilaments are long, thin rods composed of many molecules of the protein actin. They are solid and narrower than microtubules, enable cells to withstand stretching and compression, and help anchor one cell to another. Microfilaments provide many other functions in the cell through proteins that interact with actin. When any of these proteins is absent or abnormal, a genetic disease results.

Intermediate filaments have diameters intermediate between those of microtubules and microfilaments, and are made of different proteins in different cell types. However, all

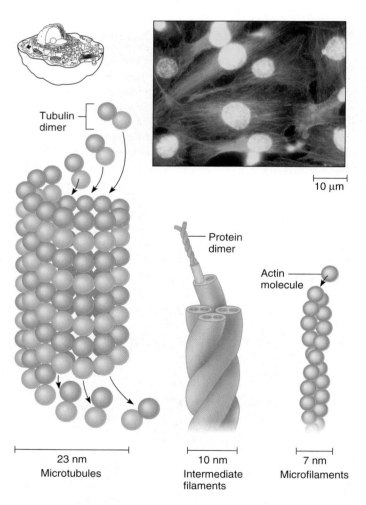

| 23 nm | 10 nm | 7 nm |
| Microtubules | Intermediate filaments | Microfilaments |

Figure 2.10 The cytoskeleton is made of protein rods and tubules. The three major components of the cytoskeleton are microtubules, intermediate filaments, and microfilaments. Through special staining, the cytoskeletons in these cells appear orange under the microscope. (The abbreviation nm stands for nanometer, which is a billionth of a meter.)

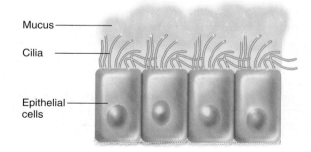

Figure 2.11 Microtubules form cilia, which are hairlike structures that wave, moving secretions such as mucus on the cell surfaces.

intermediate filaments consist of paired proteins entwined into nested coiled rods. Intermediate filaments are scarce in many cell types but are very abundant in nerve cells and skin cells. In a group of inherited conditions called epidermolysis bullosa (MIM 226500, 226650, 131750), abnormal intermediate filaments cause the skin to blister easily as tissue layers separate. The *In Their Own Words* essay describes how abnormal intermediate filaments affect a little girl, who has a different condition, giant axonal neuropathy (MIM 256850).

Disruption of how the cytoskeleton interacts with other cell components can be devastating. Consider hereditary spherocytosis (MIM 182900), which disturbs the interface between the plasma membrane and the cytoskeleton in red blood cells.

The doughnut shape of normal red blood cells enables them to squeeze through the narrowest blood vessels. Their cytoskeletons provide the ability to deform. Rods of a protein called spectrin form a meshwork beneath the plasma membrane, strengthening the cell. Proteins called ankyrins attach the spectrin rods to the plasma membrane (**figure 2.12**). Spectrin molecules also attach to microfilaments and microtubules. Spectrin molecules are like steel girders, and ankyrins are like nuts and bolts. If either molecule is absent, the red blood cell collapses.

In hereditary spherocytosis, the ankyrins are abnormal, and parts of the red blood cell plasma membrane disintegrate. The cell balloons out, obstructing narrow blood

A Little Girl with Giant Axons

A nerve cell (neuron) communicates by receiving electrochemical signals at one highly branched end, and sending signals from the other end, which is a single branch called an axon. Intermediate filaments, called neurofilaments, control the axon's shape. In giant axonal neuropathy (GAN), a key neurofilament protein, gigaxonin, is not dismantled and recycled as it normally is, and instead builds up in axons, distending them. The giant axons stifle nerve transmission, affecting the ability to move. Proteins in hair are affected too, resulting in characteristically kinky, pale hair. An affected individual is wheelchair-bound by adolescence, and does not survive his or her twenties. Lori Sames tells about her daughter, Hannah, who has GAN.

Hannah Sarah Sames is a beautiful little girl who was born on March 5, 2004. She has extremely curly blonde hair, a slight build, a precocious smile, and a charming personality. She loves to sing and dance, and play outdoors. Hannah is a beaming light of love.

When Hannah was 2 years, 5 months old, her grandmother noticed her left arch seemed to be rolling inward. I took Hannah to an orthopedist and a podiatrist, and was told Hannah would be fine. But by her third birthday, we suspected something was wrong—both arches were now involved, and her gait had become awkward. Her pediatrician gave her a rigorous physical exam and agreed she had an awkward gait, but felt that was just how Hannah walks.

Two months later, I took Hannah to another orthopedist, who told me to just let her live her life, she would be fine. Convinced otherwise, my sister showed cell phone video of Hannah walking to a physical therapist she works with, who thought Hannah's gait was like that of a child with muscular dystrophy. Our pediatrician referred us to a pediatric neurologist and a pediatric geneticist, and 6 months of testing for various diseases began. Results: all normal. During another visit with the pediatric neurologist, he took out a huge textbook and showed us a photo of a skinny little boy with kinky hair and a high forehead and braces that went just below the knee—he had GAN. He looked exactly like Hannah. So off we went to a children's hospital in New York City for more tests, and the diagnosis of GAN was confirmed.

Meeting with a genetic counselor 3 days later brought devastation. Matt and I are each carriers, and we passed the disease to Hannah. Each of our two other daughters has a 2 in 3 chance of being a carrier. We learned GAN is a rare "orphan genetic disorder" for which there is no cure, no treatment, no clinical trial and no ongoing research. "So you are telling us this is a death sentence?" I asked. And, we were told, "Yes."

Matt and I walked around in a state of shock, anger, disbelief, and grief for two days. Then, we realized, as with any disease, someone has to be the first to be cured. Some family has to be the first to raise funds and awareness and pull the medical community together to find treatment. This is how Hannah's Hope Foundation was born! As a result, we held the world's first symposium for GAN, where clinicians and scientists brainstormed. Our foundation is now funding a number of projects aimed at treating GAN.
Lori Sames
http://www.hannahshopefund.org/

Hannah Sames has giant axonal neuropathy, a disorder that affects intermediate filaments in nerve cells. Her beautiful curls are one of the symptoms.

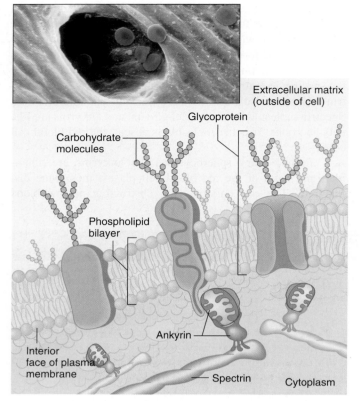

Carbohydrate molecules

Extracellular matrix (outside of cell)

Glycoprotein

Phospholipid bilayer

Ankyrin

Interior face of plasma membrane

Spectrin

Cytoplasm

Figure 2.12 **The red blood cell plasma membrane.** The cytoskeleton that supports the plasma membrane of a red blood cell withstands the turbulence of circulation. Proteins called ankyrins bind molecules of spectrin from the cytoskeleton to the inner membrane surface. On its other end, ankyrin binds proteins that help ferry molecules across the plasma membrane. In hereditary spherocytosis, abnormal ankyrin collapses the plasma membrane. The cell balloons—a problem for a cell whose function depends upon its shape. The inset shows normal red blood cells.

vessels—especially in the spleen, the organ that normally disposes of aged red blood cells. Anemia develops as the spleen destroys red blood cells more rapidly than the bone marrow can replace them. The result is great fatigue and weakness. Removing the spleen can treat the condition.

Key Concepts

1. Cells are the units of life. They consist mostly of carbohydrates, lipids, proteins, and nucleic acids.

2. Organelles subdivide specific cell functions. They include the nucleus, the endoplasmic reticulum (ER), Golgi apparatus, mitochondria, lysosomes, and peroxisomes.

3. The plasma membrane is a flexible, selective phospholipid bilayer with embedded proteins and lipid rafts.

4. The cytoskeleton is an inner framework made of protein rods and tubules, connectors, and motor molecules.

2.3 Cell Division and Death

In a human body, new cells form as old ones die, at different rates in different tissues. Growth, development, maintaining health, and healing from disease or injury require an intricate interplay between the rates of **mitosis** and **cytokinesis**, which divide the DNA and the rest of the cell, respectively, and **apoptosis,** a form of cell death (**figure 2.13**).

About 10 trillion of a human body's 100 or so trillion cells are replaced daily. Yet, cell death must happen to mold certain organs, just as a sculptor must remove some clay to shape the desired object. Apoptosis carves toes, for example, from weblike structures that telescope out from an embryo's developing form. "Apoptosis" is Greek for "leaves falling from a tree," and it is a precise, genetically programmed sequence of events that is a normal part of development.

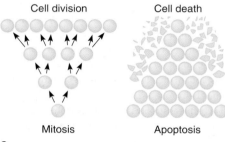

Cell division

Cell death

Mitosis

Apoptosis

a.

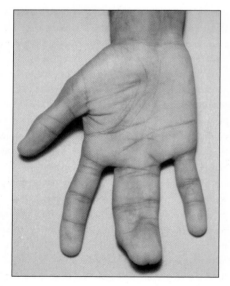

b.

Figure 2.13 **Mitosis and apoptosis mold a body.** **(a)** Cell numbers increase from mitosis and decrease from apoptosis. **(b)** In the embryo, fingers and toes are carved from webbed structures. In syndactyly, normal apoptosis fails to carve digits, and webbing persists.

The Cell Cycle

Many cell divisions transform a fertilized egg into a many-trillion-celled person. A series of events called the **cell cycle** describes the sequence of activities as a cell prepares to divide and then does.

Cell cycle rate varies in different tissues at different times. A cell lining the small intestine's inner wall may divide throughout life, whereas a neuron in the brain may never divide. A cell in the deepest skin layer of a 90-year-old may divide as long as the person lives. Frequent mitosis enables the embryo and fetus to grow rapidly. By birth, the mitotic rate slows dramatically. Later, mitosis maintains the numbers and positions of specialized cells in tissues and organs.

The cell cycle is continual, but we divide it into stages based on what we observe. The two major stages are **interphase** (not dividing) and mitosis (dividing) (**figure 2.14**). In mitosis, a cell duplicates its chromosomes, then in cytokinesis it apportions one set into each of two resulting cells, called daughter cells. This division maintains the set of 23 chromosome pairs characteristic of a human somatic cell. Another form of cell division, meiosis, produces sperm or eggs, which have half the amount of genetic material in somatic cells, or 23 single chromosomes. Chapter 3 discusses meiosis.

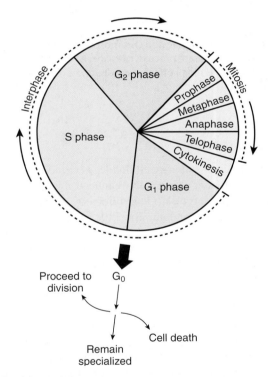

Figure 2.14 **The cell cycle.** The cell cycle is divided into interphase, when cellular components are replicated, and mitosis, when the cell distributes its contents into two daughter cells. Interphase is divided into G_1 and G_2, when the cell duplicates specific molecules and structures, and S phase, when it replicates DNA. Mitosis is divided into four stages plus cytokinesis, when the cells separate. G_0 is a "time-out" when a cell "decides" which course of action to follow.

Interphase—A Time of Great Activity

Interphase is a very active time. The cell continues the basic biochemical functions of life and also replicates its DNA and other subcellular structures. Interphase is divided into two gap (G_1 and G_2) **phases** and one synthesis (**S**) **phase.** In addition, a cell can exit the cell cycle at G_1 to enter a quiet phase called G_0. A cell in G_0 maintains its specialized characteristics but does not replicate its DNA or divide. From G_0, a cell may also proceed to mitosis and divide, or die. Apoptosis may ensue if the cell's DNA is so damaged that cancer might result. G_0, then, is when a cell's fate is either decided or put on hold.

During G_1, which follows mitosis, the cell resumes synthesis of proteins, lipids, and carbohydrates. These molecules will contribute to building the extra plasma membrane required to surround the two new cells that form from the original one. G_1 is the period of the cell cycle that varies the most in duration among different cell types. Slowly dividing cells, such as those in the liver, may exit at G_1 and enter G_0, where they remain for years. In contrast, the rapidly dividing cells in bone marrow speed through G_1 in 16 to 24 hours. Cells of the early embryo may skip G_1 entirely.

During S phase, the cell replicates its entire genome. As a result, each chromosome then consists of two copies joined at an area called the **centromere.** In most human cells, S phase takes 8 to 10 hours. Many proteins are also synthesized during this phase, including those that form the mitotic **spindle** that will pull the chromosomes apart. Microtubules form structures called **centrioles** near the nucleus. Centriole microtubules join with other proteins and are oriented at right angles to each other, forming paired, oblong structures called **centrosomes** that organize other microtubules into the spindle.

G_2 occurs after the DNA has been replicated but before mitosis begins. More proteins are synthesized during this phase. Membranes are assembled from molecules made during G_1 and are stored as small, empty vesicles beneath the plasma membrane. These vesicles will merge with the plasma membrane to enclose the two daughter cells.

Mitosis—The Cell Divides

As mitosis begins, the replicated chromosomes are condensed enough to be visible, when stained, under a microscope. The two long strands of identical chromosomal material in a replicated chromosome are called **chromatids** (**figure 2.15**). At a certain point during mitosis, a replicated chromosome's centromere splits. This allows its chromatid pair to separate into two individual chromosomes. (Although the centromere of a replicated chromosome appears as a constriction, its DNA is replicated.)

During **prophase,** the first stage of mitosis, DNA coils tightly. This shortens and thickens the chromosomes, which enables them to more easily separate (**figure 2.16**). Microtubules assemble from tubulin building blocks in the cytoplasm to form the spindles. Toward the end of prophase, the nuclear membrane breaks down. The nucleolus is no longer visible.

Metaphase follows prophase. Chromosomes attach to the spindle at their centromeres and align along the center of the cell, which is called the equator. Metaphase chromosomes

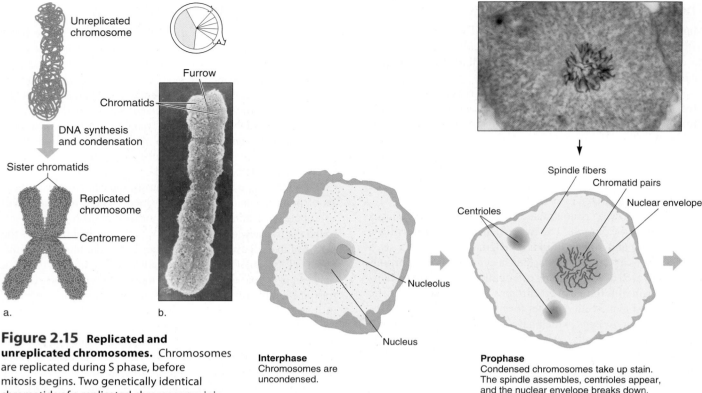

Figure 2.15 **Replicated and unreplicated chromosomes.** Chromosomes are replicated during S phase, before mitosis begins. Two genetically identical chromatids of a replicated chromosome join at the centromere **(a).** In the photograph **(b),** a human chromosome is forming two chromatids.

Interphase
Chromosomes are uncondensed.

Prophase
Condensed chromosomes take up stain. The spindle assembles, centrioles appear, and the nuclear envelope breaks down.

Figure 2.16 **Mitosis in a human cell.** Replicated chromosomes separate and are distributed into two cells from one. In a separate process, cytokinesis, the cytoplasm and other cellular structures distribute and pinch off into two daughter cells. (Not all chromosome pairs are depicted.)

are under great tension, but they appear motionless because they are pulled with equal force on both sides, like a tug-of-war rope pulled taut.

Next, during **anaphase,** the plasma membrane indents at the center, where the metaphase chromosomes line up. A band of microfilaments forms on the inside face of the plasma membrane, constricting the cell down the middle. Then the centromeres part, which relieves the tension and releases one chromatid from each pair to move to opposite ends of the cell—like a tug-of-war rope breaking in the middle and the participants falling into two groups. Microtubule movements stretch the dividing cell. During the very brief anaphase stage, a cell temporarily contains twice the normal number of chromosomes because each chromatid becomes an independently moving chromosome, but the cell has not yet physically divided.

In **telophase,** the final stage of mitosis, the cell looks like a dumbbell with a set of chromosomes at each end. The spindle falls apart, and nucleoli and the membranes around the nuclei re-form at each end of the elongated cell. Division of the genetic material is now complete. Next, during **cytokinesis,** organelles and macromolecules are distributed between the two daughter cells. Finally, the microfilament band contracts like a drawstring, separating the newly formed cells.

Control of the Cell Cycle

When and where a somatic cell divides is crucial to health. Illness can result from abnormally regulated mitosis. Control of mitosis is a daunting task. Quadrillions of mitoses occur in a lifetime, and not at random. Too little mitosis, and an injury goes unrepaired; too much, and an abnormal growth forms.

Groups of interacting proteins function at times in the cell cycle called checkpoints to ensure that chromosomes are faithfully replicated and apportioned into daughter cells (**figure 2.17**). A "DNA damage checkpoint," for example, temporarily pauses the cell cycle while special proteins repair damaged DNA. An "apoptosis checkpoint" turns on as mitosis begins. During this checkpoint, proteins called survivins override signals telling the cell to die, ensuring that mitosis (division) rather than apoptosis (death) occurs. Later during mitosis, the "spindle assembly checkpoint" oversees construction of the spindle and the binding of chromosomes to it.

Cells obey an internal "clock" that tells them approximately how many times to divide. Mammalian cells grown (cultured) in a dish divide about 40 to 60 times. The mitotic clock ticks down with time. A connective tissue cell from a fetus, for example, will divide about 50 more times. A similar cell from an adult divides only 14 to 29 more times.

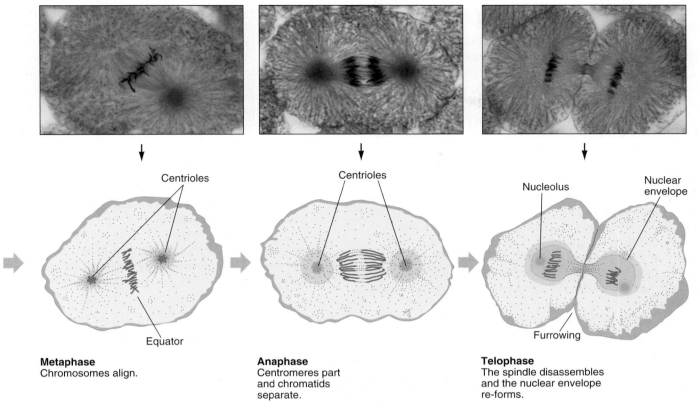

Metaphase
Chromosomes align.

Anaphase
Centromeres part and chromatids separate.

Telophase
The spindle disassembles and the nuclear envelope re-forms.

How can a cell "know" how many divisions remain? The answer lies in the chromosome tips, called **telomeres** (**figure 2.18**). Telomeres function like cellular fuses that burn down as pieces are lost from the ends. Telomeres consist of hundreds to thousands of repeats of a specific six DNA-base sequence. At each mitosis, the telomeres lose 50 to 200 endmost bases, gradually shortening the chromosome. After about 50 divisions, a critical length of telomere DNA is lost, which signals mitosis to stop. The cell may remain alive but not divide again, or it may die.

Not all cells have shortening telomeres. In eggs and sperm, in cancer cells, and in a few types of normal cells that must continually supply new cells (such as bone marrow cells), an enzyme called telomerase keeps chromosome tips

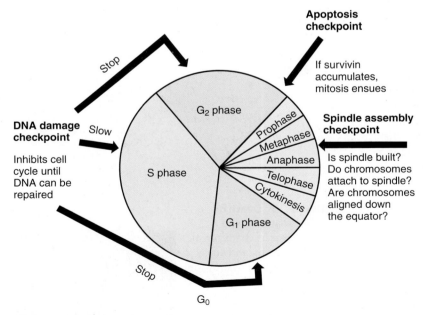

Figure 2.17 **Cell cycle checkpoints.** Checkpoints ensure that mitotic events occur in the correct sequence. Many types of cancer result from faulty checkpoints.

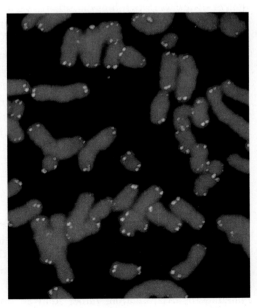

Figure 2.18 **Telomeres.** Fluorescent tags mark the telomeres in this human cell.

long. However, most cells do not produce telomerase, and their chromosomes gradually shrink.

The rate of telomere shortening provides a "clock" for a cell's existence. The telomere clock may not only count down a cell's remaining lifespan, but may also sense environmental stimuli. For example, chronic stress, obesity, and elevated blood sugar are associated with accelerated telomere shortening. It may be possible to alter the environment to keep telomeres longer. One study compared healthy identical twins, one of whom was physically active and the other sedentary. The twins who exercised had on average 200 more DNA bases on their chromosome tips than did the sedentary twins.

Factors from outside the cell can also affect a cell's mitotic clock. Crowding can slow or halt mitosis. Normal cells growing in culture stop dividing when they form a one-cell-thick layer lining the container. This limitation to division is called contact inhibition. If the layer tears, the cells that border the tear grow and divide, filling in the gap. They stop dividing once the space is filled. Perhaps a similar mechanism in the body limits mitosis.

Chemical signals control the cell cycle from outside. Hormones and growth factors are biochemicals from outside the cell that influence mitotic rate. A **hormone** is a substance synthesized in a gland and transported in the bloodstream to another part of the body, where it exerts a specific effect. Hormones secreted in the brain, for example, signal the cells lining a woman's uterus to build up each month by mitosis in preparation for possible pregnancy. A growth factor acts more locally. Epidermal growth factor (EGF), for example, stimulates cell division in the skin beneath a scab. Certain cancer drugs work by plugging growth factor receptors on cancer cells, blocking the signals to divide.

Two types of proteins, cyclins and kinases, interact inside cells to activate the genes whose products carry out mitosis. The two types of proteins form pairs. Cyclin levels fluctuate regularly throughout the cell cycle, while kinase levels stay the same. A certain number of cyclin-kinase pairs turn on the genes that trigger mitosis. Then, as mitosis begins, enzymes degrade the cyclin. The cycle starts again as cyclin begins to build up during the next interphase.

Apoptosis

Apoptosis rapidly and neatly dismantles a cell into membrane-enclosed pieces that a phagocyte (a cell that engulfs and destroys another) can mop up. It is a little like packing the contents of a messy room into garbage bags, then disposing of it all. In contrast is necrosis, a form of cell death associated with inflammation, rather than an orderly, contained destruction.

Like mitosis, apoptosis is a continuous process. It begins when a "death receptor" on the cell's plasma membrane receives a signal to die. Within seconds, enzymes called caspases are activated inside the doomed cell, stimulating each other and snipping apart various cell components. These killer enzymes:

- destroy enzymes that replicate and repair DNA;
- activate enzymes that cut DNA into similarly sized small pieces;
- tear apart the cytoskeleton, including threads that support the nucleus, which collapses, condensing the DNA within;
- cause mitochondria to release molecules that trigger further caspase activity, end the cell's energy supply, and destroy these organelles;
- abolish the cell's ability to adhere to other cells; and
- send a certain phospholipid from the plasma membrane's inner face to its outer surface, where it attracts phagocytes that dismantle the cell remnants.

A dying cell has a characteristic appearance (**figure 2.19**). It rounds up as contacts with other cells are cut off, and the plasma membrane undulates, forming bulges called blebs. The

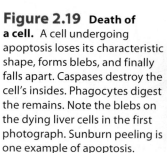

Death receptor on doomed cell binds signal molecule. Caspases are activated within.

Caspases destroy various proteins and other cell components. Cell undulates.

Blebs

Cell fragments

Phagocyte attacks and engulfs cell remnants. Cell components are degraded.

Figure 2.19 Death of a cell. A cell undergoing apoptosis loses its characteristic shape, forms blebs, and finally falls apart. Caspases destroy the cell's insides. Phagocytes digest the remains. Note the blebs on the dying liver cells in the first photograph. Sunburn peeling is one example of apoptosis.

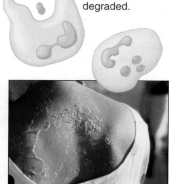

nucleus bursts, releasing same-sized DNA pieces. Mitochondria decompose. Then the cell shatters. Almost instantly, pieces of membrane encapsulate the cell fragments, which prevents inflammation. Within an hour, the cell is gone.

From the embryo onward through development, mitosis and apoptosis are synchronized, so that tissue neither overgrows nor shrinks. In this way, a child's liver retains much the same shape as she grows into adulthood, yet enlarges. During early development, mitosis and apoptosis orchestrate the ebb and flow of cell number as new structures form. Later, these processes protect—mitosis produces new skin to heal a scraped knee; apoptosis peels away sunburnt skin cells that might otherwise become cancerous. Cancer is a profound derangement of the balance between cell division and death. In cancer, mitosis occurs too frequently or too many times, or apoptosis happens too infrequently. Chapter 18 discusses cancer in detail.

Key Concepts

1. Mitosis and apoptosis regulate cell numbers during development, growth, and repair.

2. The cell cycle includes interphase and mitosis. During G_0, the cell "decides" to divide, die, or stay differentiated. Interphase includes two gap (G) phases and a synthesis (S) phase that prepares the cell for mitosis. During S phase, DNA is replicated. Proteins, carbohydrates, and lipids are synthesized during G_1 and more proteins are synthesized in G_2. During mitosis, replicated chromosomes condense, align, split, and distribute into daughter cells.

3. The cell cycle is controlled by checkpoints; telomeres; hormones and growth factors from outside the cell; and cyclins and kinases from within.

4. During apoptosis, cells receive a death signal, activate caspases, and break apart in an orderly fashion.

2.4 Cell-Cell Interactions

Precisely coordinated biochemical steps orchestrate the cell-cell interactions that make multicellular life possible. Defects in cell communication and interaction cause certain inherited illnesses. Two broad types of interactions among cells are signal transduction and cellular adhesion.

Signal Transduction

In **signal transduction,** molecules on the plasma membrane assess, transmit, and amplify incoming messages to the cell's interior. *Transduce* means to change one form of something (such as energy or information) into another. In signal transduction, the cell changes various types of stimuli into specific biochemical reactions. A cell's existence may depend upon particular signal molecules binding receptors on the cell surface. Yet other signals must be ignored for cell survival, such

as a signal to divide when cell division is not warranted. A cell's response to the many signals it receives is very complex.

Some proteins that carry out signal transduction are in the cytoplasm and some are embedded in the plasma membrane, from which they extend from one or both faces. The proteins act in a sequence, beginning at the cell surface. First, a receptor binds an incoming molecule, called the "first messenger." The receptor then contorts, touching a nearby protein called a regulator (**figure 2.20**). Next, the regulator activates a nearby enzyme, which catalyzes (speeds) a specific chemical reaction. The product of this reaction, called the "second messenger," is the key part of the entire process because it elicits the cell's response. This is usually activation of certain enzymes.

A single stimulus can trigger the production of many second messenger molecules. This is how signal transduction amplifies incoming information. Because cascades of proteins carry out signal transduction, it is a genetically controlled process.

Defects in signal transduction underlie many inherited disorders. In neurofibromatosis type 1 (NF1) (MIM 162200), for example, tumors (usually benign) grow in nervous tissue, particularly under the skin. At the cellular level, NF1 occurs when cells fail to block transmission of a growth factor signal that triggers cell division. Affected cells misinterpret the signal and divide when it is inappropriate.

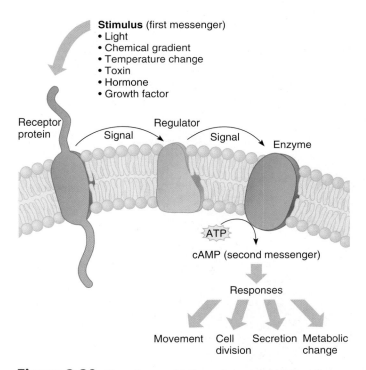

Figure 2.20 Signal transduction. A receptor binds a first messenger, triggering a cascade of biochemical activity at the cell's surface. An enzyme catalyzes a reaction inside the cell that circularizes ATP to cyclic AMP, the second messenger. cAMP then stimulates responses, such as cell division, metabolic changes, and muscle contraction. Splitting ATP also releases energy.

Cellular Adhesion

Cellular adhesion is a precise sequence of interactions among the proteins that connect cells. In one type of cellular adhesion, inflammation, white blood cells (leukocytes) move in the circulation to the injured or infected body part. There they squeeze between cells of the blood vessel walls to exit the circulation and reach the injury site, where they cause painful, red swelling. Cellular adhesion molecules, or CAMs, help guide white blood cells to the injured area.

Three types of CAMs carry out the inflammatory response: selectins, integrins, and adhesion receptor proteins (**figure 2.21**). First, selectins attach to the white blood cells and slow them to a roll by also binding to carbohydrates on the capillary wall. (This is a little like putting out your arms to slow your ride down a slide.) Next, clotting blood, bacteria, or decaying tissues release chemical attractants that signal white blood cells to stop moving. The chemical attractants activate CAMs called integrins, which latch onto the white blood cells, and CAMs called adhesion receptor proteins, which extend from the capillary wall at the injury site. The integrins and adhesion receptor proteins then guide the white blood cells between the tile-like lining cells to the injury site.

If the signals that direct white blood cells to injury sites fail, a condition called leukocyte-adhesion deficiency (MIM 116920) results. The first symptom is often teething sores that do not heal. These and other small wounds never accumulate the pus (bacteria, cellular debris, and white blood cells) that indicates the body is fighting infection. The person lacks the CAMs that enable white blood cells to stick to blood vessel walls, and so blood cells travel right past wounds. An affected individual must avoid injury and infection, and receive anti-infective treatments for even the slightest wound.

More common disorders may also reflect abnormal cellular adhesion. Cancer cells journey easily from one part of the body to another thanks to impaired cellular adhesion. Arthritis may occur when the wrong adhesion molecules pull in white blood cells, inflaming a joint where no injury exists.

Cellular adhesion is critical to many other functions. CAMs guide cells surrounding an embryo to grow toward maternal cells and form the placenta, the supportive organ linking a pregnant woman to the fetus. Sequences of CAMs also help establish connections among nerve cells in the brain that underlie learning and memory.

Key Concepts

1. In signal transduction, cell surface receptors receive information from first messengers (stimuli) and pass them to second messengers, which then trigger a cellular response.

2. Cellular adhesion molecules (CAMs) guide white blood cells to injury sites using a sequence of cell-protein interactions.

2.5 Stem Cells

Bodies grow and heal thanks to cells that retain the ability to divide, generating both new cells like themselves and cells that go on to specialize. **Stem cells** and **progenitor cells** renew tissues so that as the body grows, or loses cells to apoptosis, injury, and disease, other cells are produced that take their places.

Cell Lineages

A stem cell divides by mitosis to yield either two daughter cells that are stem cells like itself, or one that is a stem cell and one that is a partially specialized progenitor cell (**figure 2.22**). The characteristic of **self-renewal** is what makes a stem cell a stem cell—its ability to continue the lineage of cells that can divide to give rise to another cell like itself. A progenitor cell cannot self-renew, and its daughters specialize as any of a restricted number of cell types. A fully differentiated cell, such as a mature blood cell, descends from a sequence of increasingly specialized progenitor cell intermediates, each one less like a

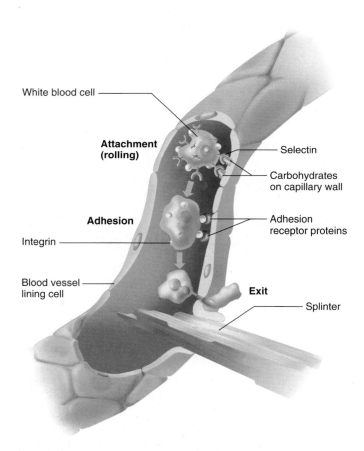

Figure 2.21 Cellular adhesion. Cellular adhesion molecules (CAMs), including selectins, integrins, and adhesion receptor proteins, direct white blood cells to injury sites.

White blood cell

Attachment (rolling)

Selectin

Carbohydrates on capillary wall

Adhesion

Integrin

Adhesion receptor proteins

Blood vessel lining cell

Exit

Splinter

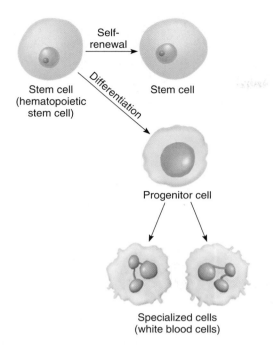

Figure 2.22 **Stem cells and progenitor cells.** A stem cell is less specialized than the progenitor cell that descends from it by mitosis. Various types of stem cells provide the raw material for producing the specialized cells that comprise tissues, while retaining the ability to generate new cells. A hematopoietic stem cell resides in the bone marrow and can produce progenitors whose daughter cells may specialize as certain blood cell types.

stem cell and more like a blood cell. Our more than 260 differentiated cell types develop from sequences, called lineages, of stem and progenitor cells. **Figure 2.23** shows parts of a few lineages.

Stem cells and progenitor cells are described in terms of developmental potential—that is, according to the number of possible fates of their daughter cells. A fertilized ovum is the ultimate stem cell. It is totipotent, which means that it can give rise to every cell type, including the cells of the membranes that support the embryo. Other stem cells and progenitor cells are pluripotent: Their daughter cells have fewer possible fates. Some are multipotent: Their daughter cells have only a few developmental "choices." This is a little like a college freshman's consideration of many majors, compared to a junior's more narrowed focus in selecting courses.

As stem cell descendants specialize, they express some genes and ignore others. An immature bone cell forms from a progenitor cell by manufacturing mineral-binding proteins and enzymes. In contrast, an immature muscle cell forms from a muscle progenitor cell that accumulates contractile proteins. The bone cell does not produce muscle proteins, nor does the muscle cell produce bone proteins. All cells, however, synthesize proteins for basic "housekeeping" functions, such as energy acquisition and protein synthesis.

Many, if not all, of the organs in an adult human body have stem or progenitor cells. These cells can divide when injury or illness occurs and generate new cells to replace damaged ones. Stem cells in the adult may have been set aside in the embryo or fetus in particular organs as repositories of future healing. Evidence suggests that some stem cells, such as those from bone marrow, can travel to and replace damaged or dead cells elsewhere in the body, in response to signals that are released in injury or disease. Because every cell contains all of an individual's genetic material, any cell type, given appropriate signals, can in theory become any other. This concept is the basis of much of stem cell technology.

Researchers are investigating stem cells to learn more about basic biology and to develop treatments for a great variety of diseases and injuries—not just inherited conditions. Clinical trials are currently testing stem cell–based treatments. These cells come from donors as well as from patients' own bodies, as **figure 2.24** illustrates. The cells can be mass-produced in laboratory glassware, and if they originate with a patient's cell or nucleus, they are a genetic match.

Stem Cell Sources

There are three general sources of human stem cells.

Embryonic stem (ES) cells are not actually cells from an embryo, but are created in a laboratory dish using certain cells from a region of a very early embryo called the inner cell mass (ICM) (see figure 3.14). Some ICM cells, under certain conditions, become pluripotent and can self-renew—they are stem cells. The ICM cells used to derive ES cells can come from two sources: "left-over" embryos from fertility clinics that would otherwise be destroyed, and from nuclear transfer, in which a nucleus from a person's somatic cell (such as a skin fibroblast) is transferred to an egg cell that has had its own nucleus removed. Nuclear transfer is popularly called "cloning" because it copies the nucleus donor's genome.

Since 2007, nuclear transfer has been largely replaced by another source of stem cells: "reprogramming" somatic cells to differentiate into any of several cell types. The cells that result from reprogramming are called **induced pluripotent stem (iPS) cells.** Reprogramming a cell may require a journey back through developmental time to an ES cell–like state, and then the cell divides and gives rise to cells that specialize as a different, desired cell type. Or, cells can be reprogrammed directly into another cell type. Reprogramming instructions consist of a few genes that can alter a cell's fate, certain RNA molecules, or other chemical factors. Deriving iPS cells does not require the use of any cells from an embryo.

For now, iPS cells are a research tool and not used to treat disease. This is because researchers do not yet know whether iPS cells function exactly as ES cells, and whether the immune system ignores them. For research purposes, human ES cells remain the "gold standard" because they more closely approximate normal development than do iPS cells. Once we learn more about how accurately iPS cells mimic normal development, human ES cells might no longer be required in research, and iPS cells might be used in treatment.

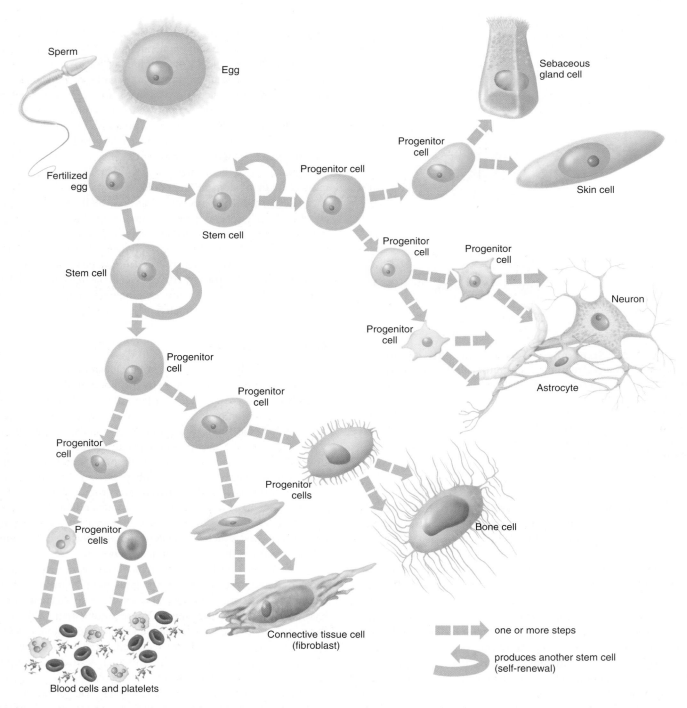

Figure 2.23 **Pathways to cell specialization.** All cells in the human body descend from stem cells, through the processes of mitosis and differentiation. The differentiated cells on the lower left are all connective tissues (blood, connective tissue, and bone), but the blood cells are more closely related to each other than they are to the other two cell types. On the upper right, the skin and sebaceous gland cells share a recent progenitor, and both share a more distant progenitor with neurons and supportive astrocytes. Imagine how complex the illustration would be if it embraced all 260-plus types of cells in a human body!

To return to the college major analogy, the idea of reprogramming a cell is like a senior in college deciding to change his or her major. If, for example, a French major wanted to become a mechanical engineer, he'd have to start over, taking very different courses. But if a biology major wanted to become a chemistry major, she would not need to start from scratch because many of the same courses apply to both majors. So it is for stem cells. Taking a skin cell from a man with heart disease and turning it

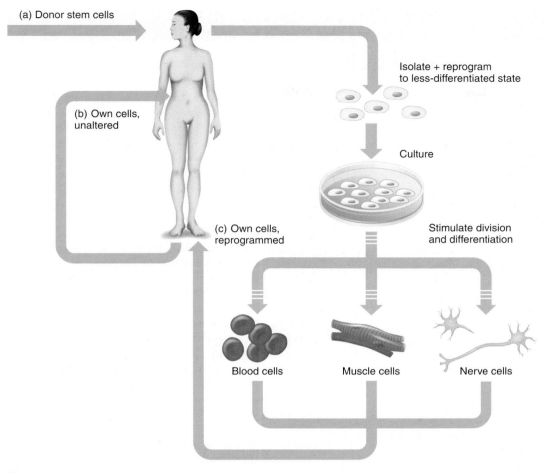

(a) Donor stem cells

(b) Own cells, unaltered

Isolate + reprogram to less-differentiated state

Culture

(c) Own cells, reprogrammed

Stimulate division and differentiation

Blood cells

Muscle cells

Nerve cells

Figure 2.24 **Using stem cells to heal.** **(a)** Stem cells from donors (bone marrow or umbilical cord blood) are already in use. **(b)** A person's cells may be used, unaltered, to replace damaged tissue, such as bone marrow. **(c)** It is possible to "reprogram" a person's cells in culture, taking them back to a less specialized state and then nurturing them to differentiate as a needed cell type.

impossible to culture. Drugs are tested on these cells. Liver and heart cells derived from stem cells are particularly useful in testing drugs for side effects, because the liver detoxifies many drugs, and the heart is easily harmed by drugs. Using stem cells in drug development can minimize the need to experiment on animals and can weed out drugs with adverse effects before they are tested on people.

A second application of stem cells growing in culture is to observe the earliest signs of a disease, which may begin long before symptoms appear in a person. The first disease for which human iPS cells were derived was amyotrophic lateral sclerosis (ALS), also known as Lou Gehrig's disease. In ALS, motor neurons that enable a person to move gradually fail. A few years after the first signs of weakness or stumbling, death comes from failure of the respiratory muscles. ALS had been difficult to study because motor neurons cannot be cultured in the laboratory because they do not divide. However, iPS cells derived from fibroblasts in patients' skin are reprogrammed in culture to become ALS motor neurons. Thanks to these and other iPS cells, researchers are now observing the beginnings of hundreds of diseases, many of them inherited—and discovering new ways to treat them. The chapter opener describes use of iPS cells from patients to study a heart condition.

The third application of stem cells is to create tissues and organs, for use in implants and transplants, or to study. This approach is not new—the oldest such treatment, a bone marrow transplant, has been around for more than half a century. Many other uses of adult stem cells, delivered as implants, transplants, or simply infusions into the bloodstream, are being tested. A patient's own bone marrow cells, for example, can be removed, bathed in selected factors, and reinfused. They follow signals that lead them to damaged tissues. Researchers are also using stem cells to study how organs form. For example, ES cells are used to build a simple version of a retina, the part of the eye that provides vision.

into a healthy heart muscle cell might require taking that initial cell back to an ES or iPS state, because these cells come from very different lineages. But turning a skeletal muscle cell into a smooth muscle cell requires fewer steps backwards because the two cell types differentiate from the same type of progenitor cell.

Another source of stem cells are those that naturally are part of the body, called "adult" stem cells. They are more accurately called tissue-specific or somatic stem cells because they are found in the tissues of fetuses, embryos, and children, and not just in adult bodies. Adult stem cells self-renew, but most are multipotent, giving rise to a few types of specialized daughter cells. Researchers are still discovering niches of adult stem cells in the body. Many potentially valuable adult stem cells are routinely discarded as medical waste.

Stem Cell Applications

Stem cells are being used in four basic ways. In drug discovery and development, stem cell cultures supply the human cells that are affected in a particular disease, which may be difficult or

Banking Stem Cells

The parents-to-be were very excited by the DVD that came in the mail shortly after they began seeing an obstetrician:

> Bank your baby's cord blood stem cells and benefit from breakthroughs. Be prepared for the unknowns in life.

The short film profiled children who were saved from certain deadly diseases because their parents had stored their umbilical cord blood. The statistics quoted were persuasive: More than 70 diseases are currently treatable with cord blood transplants, and 10,000 procedures have already been done.

With testimonials like that, it is little wonder that parents collectively spend more than $100 million per year to store cord blood. The ads and statistics are accurate but misleading, because of what they *don't* say. Most people never actually use the umbilical cord blood stem cells that they store. The scientific reasons go beyond the fact that treatable diseases are very rare. In addition, cord blood stem cells are not nearly as pluripotent as some other stem cells, limiting their applicability. Perhaps the most compelling reason that stem cell banks are rarely used is based on logic: For a person with an inherited disease, *healthy* stem cells are required—not his or her own, which could cause the disease all over again. The patient needs a donor.

Commercial cord blood banks may charge more than $1,000 for the initial collection plus an annual fee. However, the U.S. National Institutes of Health and organizations in many other nations have supported not-for-profit banks for years, and do not charge fees. Donations of cord blood to these facilities are not to help the donors directly, but to help whoever can use the cells.

As stem cell science has leaped forward, both commercial cell banks and anecdotal reports of successes have captured much media attention. This was the case for an 18-month-old boy whose cerebral palsy greatly improved after he was treated with his own cord blood cells. Whether he would have improved without the treatment isn't known.

Commercial stem cell banks are not just for newborns. One company, for example, offers to bank "very small embryonic-like stem cells" for an initial charge of $7,500 and a $750 annual fee, "enabling people to donate and store their own stem cells when they are young and healthy for their personal use in times of future medical need." The cells come from a person's blood and, in fact, one day may be very useful, but the research has yet to be done supporting use of the cells in treatments.

Questions for Discussion

1. Storing stem cells is not regulated by the U.S. government the way that a drug or a surgical procedure is because it is a service that will be helpful for treatments not yet invented. Do you think such banks should be regulated, and if so, by whom and how?

2. What information do you think that companies offering to store stem cells should present on their websites?

3. Do you think that advertisements for cord blood storage services that have quotes and anecdotal reports, but do not mention that most people who receive stem cell transplants do not in fact receive their own cells, are deceptive? Or do you think it is the responsibility of the consumer to research and discover this information?

4. How can medical consumers become aware that the government funds facilities to store stem cells?

5. It is likely that in the future, stem cell–based treatments will be possible, following large-scale clinical trials. What is the fairest way to prepare for this type of future medical treatment?

The fourth application of stem cells became clear with the creation of iPS cells. It might be possible to introduce the reprogramming proteins directly into the body to stimulate stem cells in their natural niches. Once we understand the signals, we might not need the cells. The applications of stem cells seem limited only by our imaginations. *Bioethics: Choices for the Future* discusses stem cell banking.

Key Concepts

1. A stem cell self-renews and gives rise to any of several differentiated cell types. All cells descend from stem cells and progenitor cells, which do not self-renew.
2. Cells differentiate down cell lineages by differential gene expression.
3. Stem cells are present throughout life and provide growth and repair.
4. Stem cells in health care include embryonic stem cells, induced pluripotent stem cells, and adult stem cells. ES and reprogrammed cells give researchers views of disease origins.

Summary

2.1 Introducing cells

1. Cells are the fundamental units of life and comprise the human body. Inherited traits and illnesses can be understood at the cellular and molecular levels.

2. All cells share certain features, but they are also specialized because they express different subsets of genes.

3. **Somatic** (body) **cells** are **diploid** and sperm and egg cells are **haploid. Stem cells** produce new cells.

2.2 Cell Components

4. Prokaryotic cells are small and lack **nuclei** and other **organelles. Eukaryotic** cells have organelles, and their DNA is in a **nucleus**.

5. Cells consist primarily of water and several types of macromolecules: **carbohydrates, lipids, proteins,** and **nucleic acids.**

6. Organelles sequester related biochemical reactions, improving the efficiency of life functions and protecting the cell. The cell also consists of **cytoplasm** and other chemicals.

7. The nucleus contains DNA and a nucleolus, which is a site of ribosome synthesis. **Ribosomes** provide scaffolds for protein synthesis; they exist free in the cytoplasm or complexed with the rough **endoplasmic reticulum** (**ER**).

8. In secretion, the rough ER is the site of protein synthesis and folding, the smooth ER is the site of lipid synthesis, transport, and packaging, and the **Golgi apparatus** packages secretions into vesicles, which exit through the **plasma membrane. Lysosomes** contain enzymes that dismantle debris, and **peroxisomes** house enzymes that perform a variety of functions. Enzymes in **mitochondria** extract energy from nutrients.

9. The plasma membrane is a protein-studded phospholipid bilayer. It controls which substances exit and enter the cell, and how the cell interacts with other cells.

10. The **cytoskeleton** is a protein framework of hollow microtubules, made of tubulin, and solid microfilaments, which consist of actin. Intermediate filaments are made of more than one protein type and are abundant in skin. The cytoskeleton and the plasma membrane distinguish different types of cells.

2.3 Cell Division and Death

11. Coordination of cell division (**mitosis**) and cell death (**apoptosis**) maintains cell numbers, enabling structures to enlarge during growth and development but preventing abnormal growth.

12. The **cell cycle** describes whether a cell is dividing (mitosis) or not (**interphase**). Interphase consists of two gap phases, when proteins and lipids are produced, and a synthesis phase, when DNA is replicated.

13. Mitosis proceeds in four stages. In **prophase,** replicated chromosomes consisting of two **chromatids** condense, the **spindle** assembles, the nuclear membrane breaks down, and the nucleolus is no longer visible. In **metaphase,** replicated chromosomes align along the center of the cell. In **anaphase,** the **centromeres** part, equally dividing the now unreplicated chromosomes into two daughter cells. In **telophase,** the new cells separate. Cytokinesis apportions other components into daughter cells.

14. Internal and external factors control the cell cycle. Checkpoints are times when proteins regulate the cell cycle. **Telomere** (chromosome tip) length determines how many more mitoses will occur. Crowding, hormones, and growth factors signal cells from the outside; the interactions of cyclins and kinases trigger mitosis from inside.

15. In apoptosis, a receptor on the plasma membrane receives a death signal, which activates caspases that tear apart the cell in an orderly fashion. Membrane surrounds the pieces, preventing inflammation.

2.4 Cell-Cell Interactions

16. In **signal transduction,** a stimulus (first messenger) activates a cascade of action among membrane proteins, culminating in the production of a second messenger that turns on enzymes that provide the response.

17. **Cellular adhesion** molecules enable cells to interact. Selectins slow the movement of leukocytes, and integrins and adhesion receptor proteins guide the blood cell through a capillary wall to an injury site.

2.5 Stem Cells

18. **Stem cells** produce daughter cells that retain the ability to divide and daughter cells that specialize. Progenitor cells give rise to more specialized daughter cells but do not self-renew.

19. A fertilized ovum is totipotent. Some stem cells are pluripotent, and some are multipotent. Cells are connected in lineages.

20. The three sources of stem cells are **embryonic stem (ES) cells, induced pluripotent stem (iPS) cells,** and adult stem cells.

21. Stem cell technology enables researchers to observe the origins of diseases and to devise new types of treatments.

www.mhhe.com/lewisgenetics10

Answers to all end-of-chapter questions can be found at **www.mhhe.com/lewisgenetics10.** You will also find additional practice quizzes, animations, videos, and vocabulary flashcards to help you master the material in this chapter.

Review Questions

1. Match each organelle to its function.

   Organelle
   a. lysosome
   b. rough ER
   c. nucleus
   d. smooth ER
   e. Golgi apparatus
   f. mitochondrion
   g. peroxisome

   Function
   1. lipid synthesis
   2. houses DNA
   3. energy extraction
   4. dismantles debris
   5. detoxification
   6. protein synthesis
   7. processes secretions

2. What advantage does compartmentalization provide to a large and complex cell?

3. Explain the functions of the following proteins:

   a. tubulin and actin
   b. caspases
   c. cyclins and kinases
   d. checkpoint proteins
   e. cellular adhesion molecules

4. List four types of controls on cell cycle rate.

5. How can all of a person's cells contain exactly the same genetic material, yet specialize as bone cells, nerve cells, muscle cells, and connective tissue cells?

6. Distinguish between

   a. a bacterial cell and a eukaryotic cell.
   b. interphase and mitosis.
   c. mitosis and apoptosis.
   d. rough ER and smooth ER.
   e. microtubules and microfilaments.
   f. a stem cell and a progenitor cell.
   g. totipotent and pluripotent.

7. How are intermediate filaments similar to microtubules and microfilaments, and how are they different?

8. What role does the plasma membrane play in signal transduction?

9. Distinguish among ES cells, iPS cells, and adult stem cells, and state the pros and cons of working with each to develop a therapy.

Applied Questions

1. How might abnormalities in each of the following contribute to cancer?

   a. cellular adhesion
   b. signal transduction
   c. balance between mitosis and apoptosis
   d. cell cycle control
   e. telomerase activity

2. How do stem cells maintain their populations within tissues that consist of mostly differentiated cells?

3. Why wouldn't a cell in an embryo likely be in phase G_0?

4. In which organelle would a defect cause fatigue?

5. If you wanted to create a synthetic organelle to test new drugs for toxicity, which natural organelle's function would you try to replicate?

6. An inherited form of migraine is caused by a mutation in a gene (SCN1A) that encodes a sodium channel in neurons. What is a sodium channel, and in which cell structure is it located?

7. How can signal transduction, the plasma membrane, and the cytoskeleton function together?

8. A single stem cell in skin gives rise to skin cells, hair follicle cells, and sebaceous (oil) gland cells. Suggest a treatment that might use these cells.

9. Invent a stem cell therapy. Choose a disease, identify the affected cell types, describe how they are abnormal, and explain the type of stem cell (ES, iPS, or adult) you would use. Explain how the treatment would work, and what the dangers might be.

10. Ads proclaiming "*Stem Cells: The Future of Skin Rejuvenation*" have appeared in many magazines. The product is actually secretions from stressed, cultured skin cells from behind the ears of healthy young people. The secretions are dried out and mixed with conventional skin cream. The ad promises that the goop contains "signals delivered by skin stem cells," because skin has stem cells.

    a. Is this product worth $155 for a 1.7-ounce jar? Why or why not?
    b. What evidence might convince you to buy such a product?

Web Activities

1. Several companies offer expensive "banking" of stem cells, even if treatments using the cells have not yet been invented. Find a website offering to bank the cells, and discuss whether or not the company provides enough information for you to make an informed choice as to whether or not to use the service.

2. Nations vary greatly in which stem cell technologies they consider ethical. Go to www.hinxtongroup.org/wp.html and use the interactive map feature to find examples of

   a. a nation that allows use of human embryos from clinics and nuclear transfer.
   b. a nation that allows use of clinic embryos but not nuclear transfer.
   c. a nation that bans derivation of any new human ES cell lines.

3. Consult Online Mendelian Inheritance in Man for the following disorders, and state what they have in common at the cellular level: Meckel syndrome (MIM 249000), Joubert syndrome (MIM 213300), Ellis van Creveld disease (MIM 225500), and Senior-Loken disease (MIM 266900).

Forensics Focus

1. Michael Mastromarino was recently sentenced to serve many years in prison for trafficking in body parts. As the owner of Biomedical Tissue Services in Fort Lee, NJ, Mastromarino and his cohorts dismembered corpses taken, without any consent, from funeral homes in Pennsylvania, New Jersey, and New York. Thousands of parts from hundreds of bodies were used in surgical procedures, including hip replacements and dental implants. The most commonly used product was a paste made of bone. Many family members testified at the trials. What is the basis of the fact that cells from bone tissue can be matched to blood or cheek lining cells from blood relatives?

Case Studies and Research Results

1. Nadine has a form of neuronal ceroid lipofuscinosis (MIM 610127). She suffers from seizures, loss of vision, and lack of coordination, and will likely not survive beyond 10 years of age. Her cells lack an enzyme that normally breaks down certain proteins, causing them to accumulate and destroy the nervous system. Name two organelles that are affected in this illness.

2. Studies show that women experiencing chronic stress, such as from caring for a severely disabled child, have telomeres that shorten at an accelerated rate. Suggest a study that would address the question of whether men have a similar reaction to chronic stress.

Meiosis and Development

Learning Outcomes

3.1 The Reproductive System

1. Describe the structures of the male and female reproductive systems.

3.2 Meiosis

2. Explain why meiosis is necessary to reproduce.
3. Summarize the events of meiosis.

3.3 Gamete Maturation

4. List the steps in sperm and oocyte formation.

3.4 Prenatal Development

5. Describe early prenatal development.
6. Explain how the embryo differs from the fetus.

3.5 Birth Defects

7. Define *critical period*.
8. List some teratogens.

3.6 Maturation and Aging

9. Describe common disorders that begin in adulthood.
10. Explain how rapid aging disorders occur.

The Big Picture: Our reproductive systems enable us to start a new generation. First our genetic material must be halved, so it can combine with that of a partner to reconstitute a full diploid genome. Then genetic programs unfold as the initial cell divides and its daughter cells specialize. The forming tissues fold into organs and the organs interact, slowly building a new human body.

Selling Eggs: Vanessa's Story

"I couldn't believe the ad in the student newspaper—a semester's tuition for a few weeks of discomfort! So I applied. I was 18, on the volleyball team, healthy except for some acne, and had a 3.8 GPA. Since I didn't plan on having children at the time, or at all, I thought why not?

"I passed the physical and psychological screens, and my family history seemed OK. Then three weeks later, I got the call. A young couple who couldn't have a child because the woman had had cancer wanted to use donor eggs, to be fertilized *in vitro* by the man's sperm. They'd seen my photo and read my file, and thought I'd be a good match. I was thrilled, but the warnings scared me: bleeding, infection, cramping, mood swings, and scarred ovaries.

"For the first 10 days, I gave myself shots in the thigh of a drug to suppress my ovaries. Then for the next 12 days, I injected myself with two other drugs in the back of the hip, to mature my egg cells. Frequent ultrasounds showed that my ovaries looked like grape clusters, with the maturing eggs popping to the surface. Towards the end of the regimen I felt a dull aching in my belly.

"The egg retrieval wasn't bad. I was sedated, had anesthesia, and the doctor removed 20 eggs using a needle passed through the wall of my vagina. My abdomen ached at night and the next day, and I felt bloated for a few days. But they got a dozen eggs! Two were implanted, four were frozen, and the rest were donated for stem cell research.

"That was 4 years ago. One of my eggs may now be a preschooler!"

3.1 The Reproductive System

Genes orchestrate our physiology from a few days after conception through adulthood. Expression of specific sets of genes sculpts the differentiated cells that interact, aggregate, and fold, forming the organs of the body. Abnormal gene functioning can affect health at all stages of development. Certain single-gene mutations act before birth, causing broken bones, dwarfism, or cancer. Many other mutant genes exert their effects during childhood, appearing as early developmental delays that might not seem unusual to new parents. Inherited forms of heart disease and breast cancer can appear in early or middle adulthood, which is earlier than multifactorial forms of these conditions. Pattern baldness is a very common single-gene trait that may not become obvious until well into adulthood. This chapter explores the stages of the human life cycle that form the backdrop against which genes function.

The first cell that leads to development of a new individual forms when a **sperm** from a male and an **oocyte** (also called an egg) from a female join. Sperm and oocytes are **gametes,** or sex cells. They provide a way to mix genetic material from past generations. Because we have thousands of genes, some with many variants, each person (except for identical twins) has a unique combination of inherited traits.

Sperm and oocytes are produced in the reproductive system. The reproductive organs are organized similarly in the male and female. Each system has

- paired structures, called **gonads,** where the sperm and oocytes are manufactured;
- tubular structures that transport these cells; and
- hormones and secretions that control reproduction.

The Male

Sperm cells develop within a 125-meter-long network of seminiferous tubules, which are packed into paired, oval organs called **testes** (testicles) (**figure 3.1**). The testes are the male gonads. They lie outside the abdomen within a sac, the scrotum. This location keeps the testes cooler than the rest of the body, which is necessary for sperm to develop. Leading from each testis is a tightly coiled tube, the epididymis, in which sperm cells mature and are stored. Each epididymis continues into another tube, the ductus deferens. Each ductus deferens bends behind the bladder and joins the urethra, which is the tube that carries sperm and urine out through the penis.

Along the sperm's path, three glands add secretions. The ductus deferentia pass through the prostate gland, which produces a thin, milky, alkaline fluid that activates the sperm to swim. Opening into the ductus deferens is a duct from the seminal vesicles, which secrete fructose (an energy-rich sugar) and hormonelike prostaglandins, which may stimulate contractions in the female that help sperm and oocyte meet. The bulbourethral glands, each about the size of a pea, join the urethra where it passes through the body wall. They secrete an alkaline mucus that coats the urethra before sperm are released. All of these secretions combine to form the seminal fluid that carries sperm.

During sexual arousal, the penis becomes erect so that it can penetrate and deposit sperm in the female reproductive tract. At the peak of sexual stimulation, a pleasurable sensation called orgasm occurs, accompanied by rhythmic muscular contractions that eject the sperm from each ductus deferens through the urethra and out the penis. The discharge of sperm from the penis, called ejaculation, delivers about 200 to 600 million sperm cells.

The Female

The female sex cells develop in paired organs in the abdomen called **ovaries (figure 3.2),** which are the female gonads. Within each ovary of a newborn girl are about a million immature oocytes. Each individual oocyte nestles within nourishing follicle cells, and each ovary houses oocytes in different stages of development. After puberty, about once a month, one ovary releases the most mature oocyte. Beating cilia sweep the mature oocyte into the fingerlike projections of one of two uterine (also called fallopian) tubes. The tube carries the oocyte into a muscular, saclike organ called the uterus, or womb.

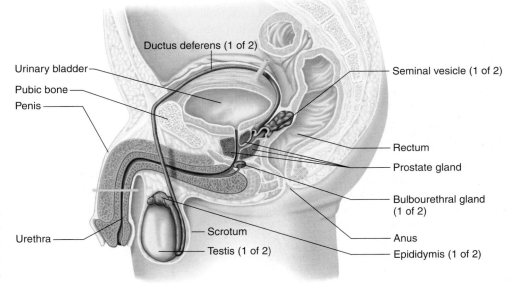

Figure 3.1 The human male reproductive system. Sperm cells are manufactured in the seminiferous tubules, which wind tightly within the testes, which descend into the scrotum. The prostate gland, seminal vesicles, and bulbourethral glands add secretions to the sperm cells to form seminal fluid. Sperm mature and are stored in the epididymis and exit through the ductus deferens. The paired ductus deferentia join in the urethra, which transports seminal fluid from the body.

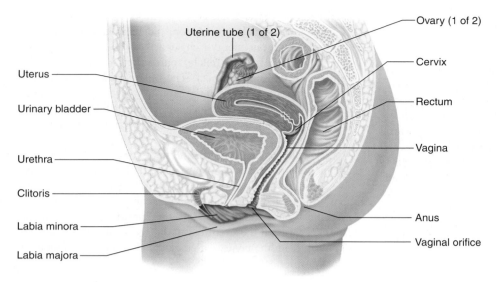

Figure 3.2 **The human female reproductive system.** Oocytes mature in the paired ovaries. Once a month after puberty, an ovary releases one oocyte, which is drawn into a nearby uterine tube. If a sperm fertilizes the oocyte in the uterine tube, the fertilized ovum continues into the uterus, where for 9 months it divides and develops. If the oocyte is not fertilized, the body expels it, along with the built-up uterine lining. This is the menstrual flow.

The released oocyte may encounter a sperm. This usually occurs in a uterine tube. If the sperm enters the oocyte and the DNA of the two gametes merges into a new nucleus, the result is a fertilized ovum. After about a day, this first cell divides while moving through the uterine tube. It then settles into the lining of the uterus, where it may continue to divide and an embryo develops. If fertilization does not occur, the oocyte, along with much of the uterine lining, is shed as the menstrual flow. Hormones coordinate the monthly menstrual cycle.

The lower end of the uterus narrows and leads to the cervix, which opens into the tubelike vagina. The vaginal opening is protected on the outside by two pairs of fleshy folds. At the upper juncture of both pairs is the 2-centimeter-long clitoris, which is anatomically similar to the penis. Rubbing the clitoris triggers female orgasm. Hormones control the cycle of oocyte maturation and the preparation of the uterus to nurture a fertilized ovum.

Key Concepts

1. Sperm develop in the seminiferous tubules, mature and collect in each epididymis, enter the ductus deferentia, and move through the urethra in the penis. The prostate gland adds an alkaline fluid, seminal vesicles add fructose and prostaglandins, and bulbourethral glands secrete mucus to form seminal fluid.

2. In the female, ovaries contain oocytes. Each month, an ovary releases an oocyte, which enters a uterine tube leading to the uterus. If the oocyte is fertilized, it begins rapid cell division and nestles into the uterine lining to divide and develop. Otherwise, the oocyte exits the body. Hormones control the cycle of oocyte development.

3.2 Meiosis

Gametes form from special cells, called germline cells, in a type of cell division called **meiosis** that halves the chromosome number. A further process, maturation, sculpts the distinctive characteristics of sperm and oocyte. The organelle-packed oocyte has 90,000 times the volume of the streamlined sperm.

Meiosis halves the amount of genetic material, so that the full amount is restored when sperm meets egg. Gametes contain 23 different chromosomes, constituting one copy of the genome. Somatic cells contain 23 *pairs*, or 46 chromosomes. One member of each pair comes from the person's mother and one comes from the father. The chromosome pairs are called **homologous pairs,** or *homologs* for short. Homologs have the same genes in the same order but may carry different alleles, or variants, of the same gene. Recall from chapter 2 that gametes are **haploid** (1*n*), which means that they have only one of each type of chromosome, and somatic cells are **diploid** (2*n*), with two copies of each chromosome type.

Without meiosis, the sperm and oocyte would each contain 46 chromosomes, and the fertilized ovum would have twice the normal number of chromosomes, or 92. Such a genetically overloaded cell, called a polyploid, usually does not develop far enough to be born. Other chromosome abnormalities, such as a missing or extra chromosome, are more common than polyploidy.

In addition to producing gametes, meiosis mixes up trait combinations. For example, a person might produce one gamete containing alleles encoding green eyes and freckles, yet another gamete with alleles encoding brown eyes and no freckles. Meiosis explains why siblings differ genetically from each other and from their parents.

In a much broader sense, meiosis, as the mechanism of sexual reproduction, provides genetic diversity. A population of sexually reproducing organisms is made up of individuals with different genotypes and phenotypes. Genetic diversity may enable a population to survive an environmental challenge. A population of asexually reproducing organisms, such as bacteria or genetically identical crops, consists of individuals with the same genome sequence. Should a new threat arise, such as an infectious disease that kills only organisms with a certain genotype, then the entire asexual population could be wiped out. However, in a sexually reproducing population, individuals that inherited a certain combination of genes might survive. This differential survival of certain genotypes is the basis of evolution by natural selection, discussed in chapter 15.

Meiosis entails two divisions of the genetic material. The first division is called a **reduction division** (or meiosis I) because it reduces the number of replicated chromosomes from 46 to 23. The second division, called an **equational division** (or meiosis II), produces four cells from the two cells formed in the first division by splitting the replicated chromosomes. **Figure 3.3** shows an overview of the process, and **figure 3.4** depicts the major events of each stage.

As in mitosis, meiosis occurs after an interphase period when DNA is replicated (doubled) (**table 3.1**). For each chromosome pair in the cell undergoing meiosis, one homolog comes from the person's mother, and one from the father. In figures 3.3 and 3.4, the colors represent the contributions of the two parents, whereas size indicates different chromosomes.

After interphase, prophase I (so called because it is the prophase of meiosis I) begins as the replicated chromosomes condense and become visible when stained. A spindle forms. Toward the middle of prophase I, the homologs line up next to one another, gene by gene, in an event called synapsis. A mixture of RNA and protein holds the chromosome pairs together. At this time, the homologs exchange parts, or **cross over** (**figure 3.5**). All four chromatids that comprise each homologous

Table 3.1	Comparison of Mitosis and Meiosis
Mitosis	**Meiosis**
One division	Two divisions
Two daughter cells per cycle	Four daughter cells per cycle
Daughter cells genetically identical	Daughter cells genetically different
Chromosome number of daughter cells same as that of parent cell (2n)	Chromosome number of daughter cells half that of parent cell (1n)
Occurs in somatic cells	Occurs in germline cells
Occurs throughout life cycle	In humans, completes after sexual maturity
Used for growth, repair, and asexual reproduction	Used for sexual reproduction, producing new gene combinations

chromosome pair are pressed together as exchanges occur. After crossing over, each homolog bears some genes from each parent. Prior to this, all of the genes on a homolog were derived from one parent. New gene combinations arise from crossing over when the parents carry different alleles. Toward the end of prophase I, the synapsed chromosomes separate but remain attached at a few points along their lengths.

To understand how crossing over mixes trait combinations, consider a simplified example. Suppose that homologs carry genes for hair color, eye color, and finger length. One of

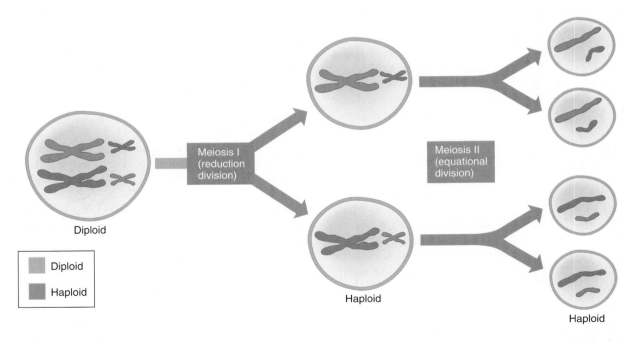

Diploid

Meiosis I (reduction division)

Meiosis II (equational division)

Diploid
Haploid

Haploid

Haploid

Figure 3.3 Overview of meiosis. Meiosis is a form of cell division in which certain cells are set aside and give rise to haploid gametes. This simplified illustration follows the fate of two chromosome pairs rather than the true 23 pairs. In actuality, the first meiotic division reduces the number of chromosomes to 23, all in the replicated form. In the second meiotic division, the cells essentially undergo mitosis. The result of the two meiotic divisions (in this illustration and in reality) is four haploid cells. In this illustration, homologous pairs of chromosomes are indicated by size, and parental origin of chromosomes by color.

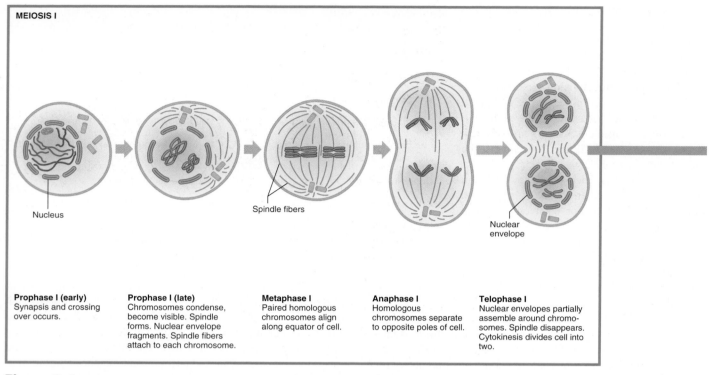

Prophase I (early)	**Prophase I (late)**	**Metaphase I**	**Anaphase I**	**Telophase I**
Synapsis and crossing over occurs.	Chromosomes condense, become visible. Spindle forms. Nuclear envelope fragments. Spindle fibers attach to each chromosome.	Paired homologous chromosomes align along equator of cell.	Homologous chromosomes separate to opposite poles of cell.	Nuclear envelopes partially assemble around chromosomes. Spindle disappears. Cytokinesis divides cell into two.

Figure 3.4 Meiosis. An actual human cell undergoing meiosis has 23 chromosome pairs.

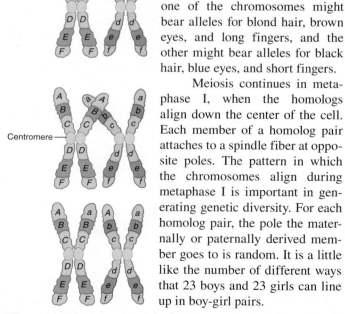

Figure 3.5 Crossing over recombines genes. Crossing over generates genetic diversity by recombining genes, mixing parental traits. The capital and lowercase forms of the same letter represent different variants (alleles) of the same gene. A chromosome has hundreds to thousands of genes.

the chromosomes carries alleles for blond hair, blue eyes, and short fingers. Its homolog carries alleles for black hair, brown eyes, and long fingers. After crossing over, one of the chromosomes might bear alleles for blond hair, brown eyes, and long fingers, and the other might bear alleles for black hair, blue eyes, and short fingers.

Meiosis continues in metaphase I, when the homologs align down the center of the cell. Each member of a homolog pair attaches to a spindle fiber at opposite poles. The pattern in which the chromosomes align during metaphase I is important in generating genetic diversity. For each homolog pair, the pole the maternally or paternally derived member goes to is random. It is a little like the number of different ways that 23 boys and 23 girls can line up in boy-girl pairs.

The greater the number of chromosomes, the greater the genetic diversity generated at this stage. For two pairs of homologs, four (2^2) different metaphase alignments are possible. For three pairs of homologs, eight (2^3) different alignments can occur. Our 23 chromosome pairs can line up in 8,388,608 (2^{23}) different ways. This random alignment of chromosomes causes **independent assortment** of the genes that they carry. Independent assortment means that the fate of a gene on one chromosome is not influenced by a gene on a different chromosome (**figure 3.6**). Independent assortment accounts for a basic law of inheritance discussed in chapter 4.

Homologs separate in anaphase I and finish moving to opposite poles by telophase I. These movements establish a haploid set of still-replicated chromosomes at each end of the stretched-out cell. Unlike in mitosis, the centromeres of each homolog in meiosis I remain together. During a second interphase, chromosomes unfold into very thin threads. Proteins are manufactured, but DNA is not replicated a second time. The single DNA replication, followed by the double division of meiosis, halves the chromosome number.

Prophase II marks the start of the second meiotic division. The chromosomes are again condensed and visible. In metaphase II, the replicated chromosomes align down the center of the cell. In anaphase II, the centromeres part, and the newly formed chromosomes, each now in the unreplicated form, move to opposite poles. In telophase II, nuclear envelopes form around the four nuclei, which then separate into individual cells. The net result of meiosis is four haploid cells, each carrying a new assortment of genes and chromosomes that represent a single copy of the genome.

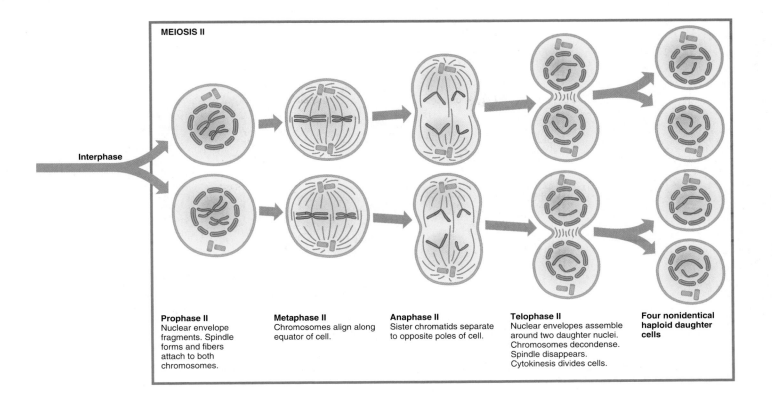

MEIOSIS II

Interphase

Prophase II
Nuclear envelope fragments. Spindle forms and fibers attach to both chromosomes.

Metaphase II
Chromosomes align along equator of cell.

Anaphase II
Sister chromatids separate to opposite poles of cell.

Telophase II
Nuclear envelopes assemble around two daughter nuclei. Chromosomes decondense. Spindle disappears. Cytokinesis divides cells.

Four nonidentical haploid daughter cells

Meiosis generates astounding genetic variety. Any one of a person's more than 8 million possible combinations of chromosomes can meet with any one of the more than 8 million combinations of a partner, raising potential variability to more than 70 trillion $(8,388,608^2)$ genetically unique individuals! Crossing over contributes almost limitless genetic variability.

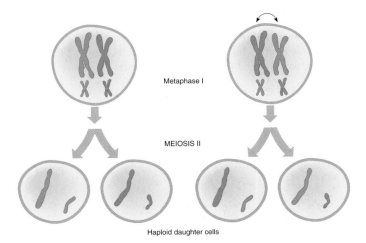

Metaphase I

MEIOSIS II

Haploid daughter cells

Figure 3.6 Independent assortment. The pattern in which homologs randomly align during metaphase I determines the combination of maternally and paternally derived chromosomes in the daughter cells. Two pairs of chromosomes can align in two ways to produce four possibilities in the daughter cells. The potential variability that meiosis generates skyrockets when one considers all 23 chromosome pairs and the effects of crossing over.

Key Concepts

1. The haploid sperm and oocyte are derived from diploid germline cells by meiosis and maturation.
2. Meiosis maintains the chromosome number over generations and mixes gene combinations.
3. In the first meiotic (or reduction) division, the number of replicated chromosomes is halved.
4. In the second meiotic (or equational) division, each of two cells from the first division divides again, yielding four cells from the original one.
5. Chromosome number is halved because the DNA replicates once, but the cell divides twice.
6. Crossing over and independent assortment generate further genotypic diversity by creating new combinations of alleles.

3.3 Gamete Maturation

Meiosis occurs in both sexes, but further steps elaborate the very different-looking sperm and oocyte. Each type of gamete is haploid, but different distributions of other cell components create their distinctions. The cells of the maturing male and female proceed through similar stages, but with sex-specific terminology and different timetables. A male begins manufacturing sperm at puberty and continues throughout life, whereas a female begins meiosis when she is a fetus. Meiosis in the female completes only if a sperm fertilizes an oocyte.

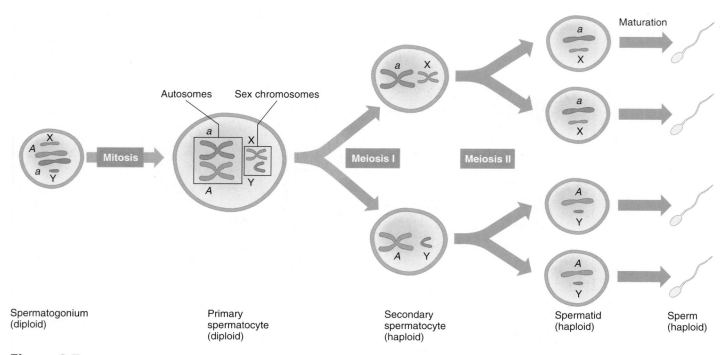

Figure 3.7 Sperm formation (spermatogenesis). Primary spermatocytes have the normal diploid number of 23 chromosome pairs. The large pair of chromosomes represents autosomes (non-sex chromosomes). The X and Y chromosomes are sex chromosomes.

Sperm Formation

Spermatogenesis, the formation of sperm cells, begins in a diploid stem cell called a **spermatogonium (figure 3.7)**. This cell divides mitotically, yielding two daughter cells. One continues to specialize into a mature sperm. The other daughter cell remains a stem cell, able to self-renew and continually produce more sperm.

Bridges of cytoplasm attach several spermatogonia, and their daughter cells enter meiosis together. As these spermatogonia mature, they accumulate cytoplasm and replicate their DNA, becoming primary spermatocytes.

During reduction division (meiosis I), each primary spermatocyte divides, forming two equal-sized haploid cells called secondary spermatocytes. In meiosis II, each secondary spermatocyte divides to yield two equal-sized spermatids. Each spermatid then develops the characteristic sperm tail, or flagellum. The base of the tail has many mitochondria, which will split ATP molecules to release energy that will propel the sperm inside the female reproductive tract. After spermatid differentiation, some of the cytoplasm connecting the cells falls away, leaving mature, tadpole-shaped spermatozoa (singular, *spermatozoon*), or sperm. **Figure 3.8** presents an anatomical view showing the stages of spermatogenesis within the seminiferous tubules.

A sperm, which is a mere 0.006 centimeter (0.0023 inch) long, must travel about 18 centimeters (7 inches) to reach an oocyte. Each sperm cell consists of a tail, body or midpiece, and a head region **(figure 3.9)**. A membrane-covered area on the front end, the acrosome, contains enzymes that help the sperm cell penetrate the protective layers around the oocyte. Within the large sperm head, DNA is wrapped around proteins. The sperm's DNA at this time is genetically inactive. A male manufactures trillions of sperm in his lifetime. Although many of these will come close to an oocyte, very few will actually touch one.

Meiosis in the male has built-in protections that help prevent sperm from causing some birth defects. Spermatogonia that are exposed to toxins tend to be so damaged that they never mature into sperm. More mature sperm cells exposed to toxins are often so damaged that they cannot swim.

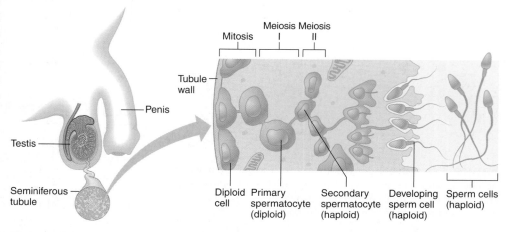

Figure 3.8 Meiosis produces sperm cells. Diploid cells divide through mitosis in the linings of the seminiferous tubules. Some of the daughter cells then undergo meiosis, producing haploid spermatocytes, which differentiate into mature sperm cells.

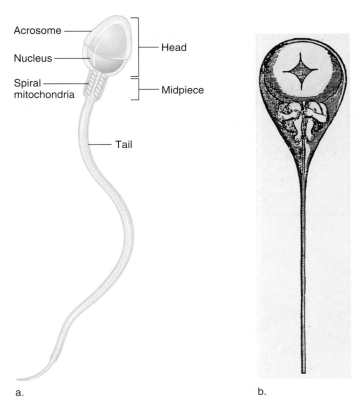

a.   b.

Figure 3.9 **Sperm.** **(a)** A sperm has distinct regions that assist in delivering DNA to an oocyte. **(b)** This 1694 illustration by Dutch histologist Niklass Hartsoeker presents a once-popular hypothesis that a sperm carries a preformed human called a homunculus.

Oocyte Formation

Meiosis in the female, called **oogenesis** (egg making), begins with a diploid cell, an **oogonium.** Unlike male cells, oogonia are not attached. Instead, follicle cells surround each oogonium. As each oogonium grows, cytoplasm accumulates, DNA replicates, and the cell becomes a primary oocyte. The ensuing meiotic division in oogenesis, unlike the male pathway, produces cells of different sizes.

In meiosis I, the primary oocyte divides into two cells: a small cell with very little cytoplasm, called a first **polar body,** and a much larger cell called a secondary oocyte (**figure 3.10**). Each cell is haploid, with the chromosomes in replicated form. In meiosis II, the tiny first polar body may divide to yield two polar bodies of equal size, with unreplicated chromosomes; or the first polar body may decompose. The secondary oocyte, however, divides unequally in meiosis II to produce another small polar body, with unreplicated chromosomes, and the mature egg cell, or ovum, which contains a large volume of cytoplasm. **Figure 3.11** summarizes meiosis in the female, and **figure 3.12** provides an anatomical view of the process.

Most of the cytoplasm among the four meiotic products in the female ends up in only one cell, the ovum. The woman's body absorbs the polar bodies, which normally play no further role in development. Rarely, a sperm fertilizes a polar body. When this happens, the woman's hormones respond as if she is pregnant, but a disorganized clump of cells that is not an embryo grows for a few weeks, and then leaves the woman's body. This event is a type of miscarriage called a "blighted ovum."

Before birth, a female's million or so oocytes arrest in prophase I. (This means that when your grandmother was pregnant with your mother, the oocyte that would be fertilized and eventually become you was already there.) By puberty, about 400,000 oocytes remain. After puberty, meiosis I continues in one or several oocytes each month, but halts again at metaphase II. In response to specific hormonal cues each month, one ovary releases

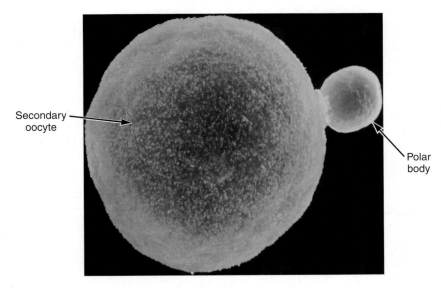

Figure 3.10 **Meiosis in a female produces a secondary oocyte and a polar body.** Unequal division and apportioning of cell parts enable the cell destined to become a fertilized ovum to accumulate most of the cytoplasm and organelles from the primary oocyte, but with only one genome copy. The oocyte accumulates abundant cytoplasm that would have gone into the meiotic product that became the polar body if the division had been equal.

a secondary oocyte; this event is ovulation. The oocyte drops into a uterine tube, where waving cilia move it toward the uterus. Along the way, if a sperm penetrates the oocyte membrane, then female meiosis completes, and a fertilized ovum forms. If the secondary oocyte is not fertilized, it degenerates and leaves the body in the menstrual flow, meiosis never completed.

A female ovulates about 400 oocytes between puberty and menopause. However, experiments in mice suggest that stem cells may produce oocytes even past menopause. Most oocytes degrade, because fertilization is so rare. Furthermore, only one in three of the oocytes that do meet and merge with a sperm cell will continue to grow, divide, and specialize to eventually form a new individual.

Key Concepts

1. An oogonium accumulates cytoplasm and replicates its DNA, becoming a primary oocyte.

2. In meiosis I, the primary oocyte divides, forming a small polar body and a large, haploid secondary oocyte.

3. In meiosis II, the secondary oocyte divides, yielding another small polar body and a mature haploid ovum.

4. Oocytes arrest at prophase I until puberty, after which one or several oocytes complete the first meiotic division each month. The second meiotic division completes at fertilization.

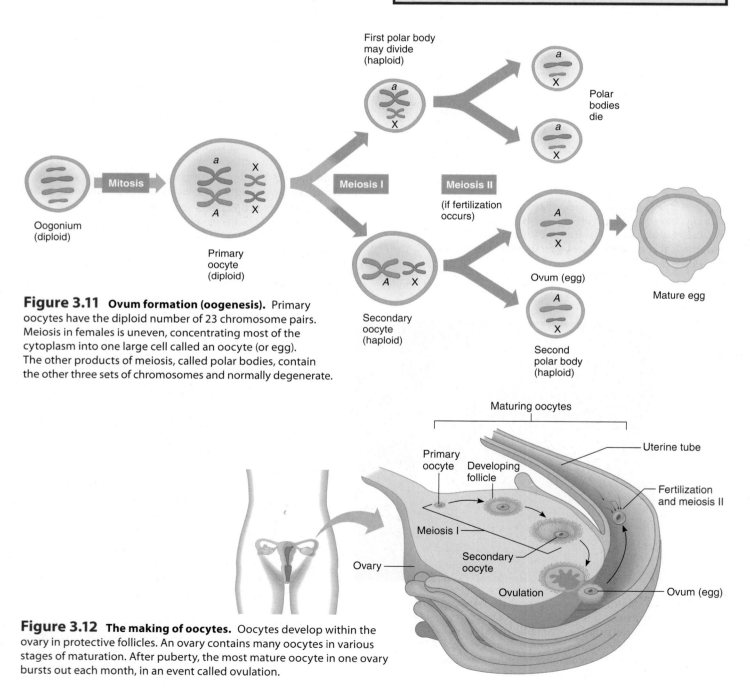

Figure 3.11 Ovum formation (oogenesis). Primary oocytes have the diploid number of 23 chromosome pairs. Meiosis in females is uneven, concentrating most of the cytoplasm into one large cell called an oocyte (or egg). The other products of meiosis, called polar bodies, contain the other three sets of chromosomes and normally degenerate.

Figure 3.12 The making of oocytes. Oocytes develop within the ovary in protective follicles. An ovary contains many oocytes in various stages of maturation. After puberty, the most mature oocyte in one ovary bursts out each month, in an event called ovulation.

3.4 Prenatal Development

A prenatal human is considered an **embryo** for the first 8 weeks. During this time, rudiments of all body parts form. The embryo in the first week is considered to be in a "preimplantation" stage because it has not yet settled into the uterine lining. Some biologists do not consider a prenatal human an embryo until it begins to develop tissue layers, at about 2 weeks.

Prenatal development after the eighth week is the fetal period, when structures grow and specialize. From the start of the ninth week until birth, the prenatal human organism is a **fetus.**

Fertilization

Hundreds of millions of sperm cells are deposited in the vagina during sexual intercourse. A sperm cell can survive in the woman's body for up to 3 days, but the oocyte can only be fertilized in the 12 to 24 hours after ovulation.

The woman's body helps sperm reach an oocyte. A process in the female called capacitation chemically activates sperm, and the oocyte secretes a chemical that attracts sperm. Sperm are also assisted by contractions of the female's muscles and by the moving sperm tails. Still, only 200 or so sperm come near the oocyte.

A sperm first contacts a covering of follicle cells, called the corona radiata, that guards a secondary oocyte. The sperm's acrosome then bursts, releasing enzymes that bore through a protective layer of glycoprotein (the zona pellucida) beneath the corona radiata. Fertilization, or conception, begins when the outer membranes of the sperm and secondary oocyte meet (**figure 3.13**). The encounter is dramatic. A wave of electricity spreads physical and chemical changes across the entire oocyte surface. The changes keep other sperm out. More than one sperm can enter an oocyte, but the resulting cell has too much genetic material for development to follow.

Usually only the sperm's head enters the oocyte. Within 12 hours of the sperm's penetration, the ovum's nuclear membrane disappears, and the two sets of chromosomes, called pronuclei, approach one another. Within each pronucleus, DNA replicates. Fertilization completes when the two genetic packages meet and merge, forming the genetic instructions for a new individual. The fertilized ovum is called a **zygote.** The *Bioethics: Choices for the Future* reading describes cloning, which is a way to start development without a fertilized egg.

Cleavage and Implantation

About a day after fertilization, the zygote divides by mitosis, beginning a period of frequent cell division called **cleavage** (**figure 3.14**). The resulting early cells are called **blastomeres.** When the blastomeres form a solid ball of sixteen or more cells, the embryo is called a **morula** (Latin for "mulberry," which it resembles).

During cleavage, organelles and molecules from the secondary oocyte's cytoplasm still control cellular activities, but some of the embryo's genes begin to function. The ball of cells hollows out, and its center fills with fluid, creating a **blastocyst.** "Cyst" refers to the fluid-filled center. Some of the cells form a clump on the inside lining called the **inner**

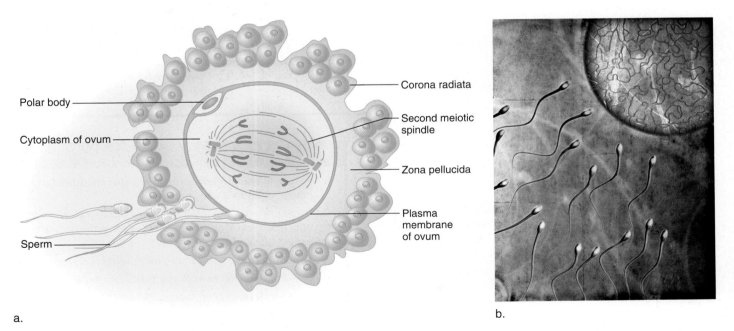

a.

b.

Figure 3.13 **Fertilization.** **(a)** Fertilization by a sperm cell induces the oocyte (arrested in metaphase II) to complete meiosis. Before fertilization occurs, the sperm's acrosome bursts, spilling enzymes that help the sperm's nucleus enter the oocyte. **(b)** Only about 200 sperm come near an oocyte, and only one can enter.

Why a Clone Is Not an Exact Duplicate

Cloning creates a genetic replica of an individual. In contrast, normal reproduction and development combine genetic material from two individuals. In fiction, scientists have cloned Nazis, politicians, dinosaurs, and organ donors. Real scientists have cloned sheep, mice, cats, pigs, dogs, and amphibians.

Cloning transfers a nucleus from a somatic cell into an oocyte whose nucleus has been removed. The technique is more accurately called "somatic cell nuclear transfer" or just "nuclear transfer." It cannot produce an exact replica of a person, for several reasons:

- **Premature cellular aging.** In some species, telomeres of chromosomes in the donor nucleus are shorter than those in the recipient cell (see chapter 2). Premature aging, seen in shortened telomeres, may be why the first cloned mammal, Dolly the cloned sheep, died young of a severe respiratory infection.

- **Altered gene expression.** In normal development, for some genes, one copy is turned off, depending upon which parent transmits it. This phenomenon is called genomic imprinting. In cloning, genes in a donor nucleus skip passing through a sperm or oocyte, and thus are not imprinted. Lack of imprinting may somehow cause cloned animals to be unusually large. Regulation of gene expression is abnormal at many times during prenatal development of a clone (see chapter 6).

- **More mutations.** DNA from a donor cell has had years to accumulate mutations. A mutation might not be noticeable in one of millions of somatic cells, but it could be devastating if that nucleus is used to program development of a new individual (see chapter 12).

- **X inactivation.** At a certain time in early prenatal development in female mammals, one X chromosome is inactivated. Whether that X chromosome is from the mother or the father occurs at random in each cell, creating an overall mosaic pattern of expression for genes on the X chromosome. The pattern of X inactivation of a female clone would probably not match that of her nucleus donor, because X inactivation occurs in the embryo, not the first cell (see chapter 6).

- **Mitochondrial DNA.** Mitochondria contain DNA. A clone's mitochondria descend from the recipient oocyte, not from the donor cell, because mitochondria are in the cytoplasm, not the nucleus.

The environment is another powerful factor in why a clone isn't an identical copy. For example, coat color patterns differ in cloned calves and cats. When the animals were embryos, cells destined to produce pigment moved in a unique way in each individual, producing different color patterns. In humans, experience, nutrition, stress, exposure to infectious disease, and many other factors join our genes in molding who we are.

Cloned cats. A company called "Genetic Savings and Clone" tried to sell cloned cats for $50,000, but lowered the price to only $32,000 when customers were scarce. The company went out of business in 2006.

---

cell mass. Its formation is the first event that distinguishes cells from each other by their relative positions, other than inside and outside the morula. The inner cell mass continues developing, forming the embryo.

A week after conception, the blastocyst nestles into the uterine lining. This event, called implantation, takes about a week. As it starts, the outermost cells of the blastocyst, called the trophoblast, secrete the "pregnancy hormone" human chorionic gonadotropin (hCG), which prevents menstruation. hCG detected in a woman's urine or blood is one sign of pregnancy.

Key Concepts

1. Following sexual intercourse, sperm are capacitated and drawn to the secondary oocyte.
2. Acrosomal enzymes assist the sperm's penetration of the oocyte. Chemical and electrical changes in the oocyte's surface block additional sperm.
3. The two sets of chromosomes meet, forming a zygote.
4. Cleavage cell divisions form a morula and then a blastocyst.
5. The outer layer of cells invades and implants in the uterine lining.
6. The inner cell mass develops into the embryo.
7. Certain blastocyst cells secrete hCG.

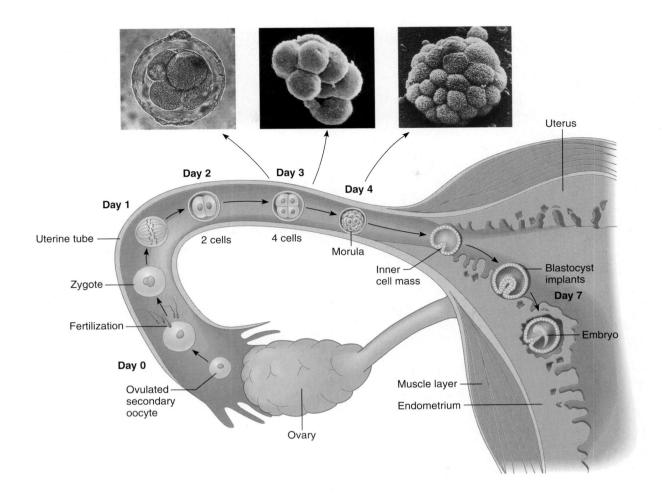

Figure 3.14 **Cleavage: From ovulation to implantation.** The zygote forms in the uterine tube when a sperm nucleus fuses with the nucleus of an oocyte. The first divisions proceed while the zygote moves toward the uterus. By day 7, the zygote, now called a blastocyst, begins to implant in the uterine lining.

The Embryo Forms

During the second week of prenatal development, a space called the amniotic cavity forms between the inner cell mass and the outer cells anchored to the uterine lining. Then the inner cell mass flattens into a two-layered embryonic disc. The layer nearest the amniotic cavity is the **ectoderm;** the inner layer, closer to the blastocyst cavity, is the **endoderm.** Shortly after, a third layer, the **mesoderm,** forms in the middle. This three-layered structure is called the primordial embryo, or the **gastrula (figure 3.15).**

Once these three layers, called **primary germ layers,** form, many cells become "determined" or fated, to develop as a specific cell type. Each primary germ layer gives rise to certain structures. Cells in the ectoderm become skin, nervous tissue, or parts of certain glands. Endoderm cells form parts of the liver and pancreas and the linings of many organs. The middle layer of the embryo, the mesoderm, forms many structures, including muscle, connective tissues, the reproductive organs, and the kidneys.

Genes called homeotics control how the embryo develops parts in the right places. Mutations in these genes cause some very interesting conditions, including forms of intellectual disability, autism, blood cancer, and blindness. The homeotic mutations were originally studied in fruit flies that had legs growing where their antennae should be.

Table 3.2 summarizes the stages of early human prenatal development.

Supportive Structures Form

As an embryo develops, structures form that support and protect it. These include chorionic villi, the placenta, the yolk sac, the allantois, the umbilical cord, and the amniotic sac.

By the third week after conception, fingerlike outgrowths called chorionic villi extend from the area of the embryonic disc close to the uterine wall. The villi project into pools of the woman's blood. Her blood system and the embryo's are separate, but nutrients and oxygen diffuse across the chorionic villi from her circulation to the embryo. Wastes

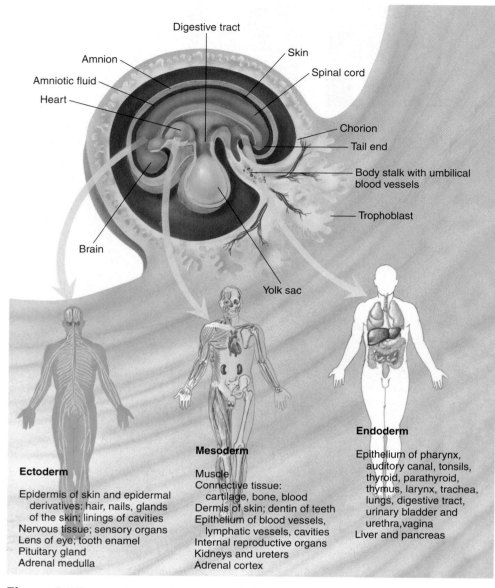

Digestive tract

Amnion

Amniotic fluid

Heart

Skin

Spinal cord

Chorion

Tail end

Body stalk with umbilical blood vessels

Trophoblast

Brain

Yolk sac

Ectoderm

Epidermis of skin and epidermal derivatives: hair, nails, glands of the skin; linings of cavities
Nervous tissue; sensory organs
Lens of eye; tooth enamel
Pituitary gland
Adrenal medulla

Mesoderm

Muscle
Connective tissue: cartilage, bone, blood
Dermis of skin; dentin of teeth
Epithelium of blood vessels, lymphatic vessels, cavities
Internal reproductive organs
Kidneys and ureters
Adrenal cortex

Endoderm

Epithelium of pharynx, auditory canal, tonsils, thyroid, parathyroid, thymus, larynx, trachea, lungs, digestive tract, urinary bladder and urethra, vagina
Liver and pancreas

Figure 3.15 **The primordial embryo.** When the three primary germ layers of the embryo form at gastrulation, many cells become "fated" to follow a specific developmental pathway. Each layer retains stem cells as the organism develops. Under certain conditions, these cells may produce daughter cells that can specialize as many cell types.

Table 3.2	Stages and Events of Early Human Prenatal Development	
Stage	**Time Period**	**Principal Events**
Fertilized ovum	12–24 hours following ovulation	Oocyte fertilized; zygote has 23 pairs of chromosomes and is genetically distinct
Cleavage	30 hours to third day	Mitosis increases cell number
Morula	Third to fourth day	Solid ball of cells
Blastocyst	Fifth day through second week	Hollowed ball forms trophoblast (outside) and inner cell mass, which implants and flattens to form embryonic disc
Gastrula	End of second week	Primary germ layers form

leave the embryo's circulation and enter the woman's circulation to be excreted.

By 10 weeks, the placenta is fully formed. This organ links woman and fetus for the rest of the pregnancy. The placenta secretes hormones that maintain pregnancy and alter the woman's metabolism to send nutrients to the fetus.

Other structures nurture the developing embryo. The yolk sac manufactures blood cells, as does the allantois, a membrane surrounding the embryo that gives rise to the umbilical blood vessels. The umbilical cord forms around these vessels and attaches to the center of the placenta. Toward the end of the embryonic period, the yolk sac shrinks, and the amniotic sac swells with fluid that cushions the embryo and maintains a constant temperature and pressure. The amniotic fluid contains fetal urine and cells.

Two of the supportive structures that develop during pregnancy provide the material for prenatal tests (see figure 13.5), discussed in chapter 13. Chorionic villus sampling examines chromosomes from cells snipped off the chorionic villi at 10 weeks. Because the villi cells and the embryo's cells come from the same fertilized ovum, an abnormal chromosome in villi cells should also be in the embryo. In amniocentesis, a sample of amniotic fluid is taken and fetal cells in it are examined for biochemical, genetic, and chromosomal anomalies.

The umbilical cord is another prenatal structure that has medical applications. Its cells, not only the blood stem cells mentioned in *Bioethics: Choices for the Future* in chapter 2, are valuable. They can be cultured to differentiate as cells from any of the three primary germ layers, including bone, fat, nerve, cartilage, and muscle cells. Stem cells from the cord are used to treat a respiratory disease of newborns that scars and inflames the lungs. The stem cells become two types of needed lung cells: the type that secretes surfactant, which is the chemical that inflates the microscopic air sacs, and the cell type that exchanges oxygen for carbon dioxide. Stem cells from umbilical cords are abundant, easy to obtain and manipulate, and can become almost any cell type.

Multiples

Twins and other multiples arise early in development. Twins are either fraternal or identical. Fraternal, or **dizygotic** (DZ), twins result when two sperm fertilize two oocytes. This can happen if ovulation occurs in two ovaries in the same month, or if two oocytes leave the same ovary and are both fertilized. DZ twins are no more alike than any two siblings, although they share a very early environment in the uterus. The tendency to have DZ twins may run in families if the women sometimes ovulate two oocytes in a month.

Identical, or **monozygotic** (MZ), twins descend from a single fertilized ovum and therefore are genetically identical. They are natural clones. Three types of MZ twins can form, depending upon when the fertilized ovum or very early embryo splits (**figure 3.16**). This difference in timing determines which supportive structures the twins share. About a third of all MZ twins have completely separate chorions and amnions, and about

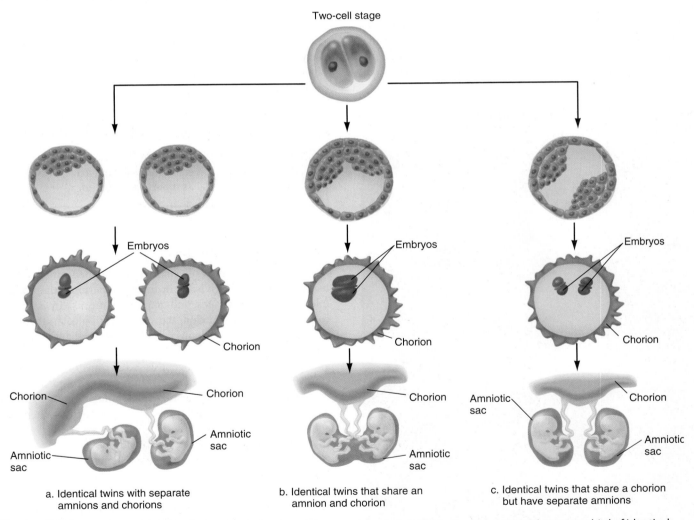

a. Identical twins with separate amnions and chorions

b. Identical twins that share an amnion and chorion

c. Identical twins that share a chorion but have separate amnions

Figure 3.16 **Types of identical twins.** Identical twins originate at three points in development. **(a)** In about one-third of identical twins, separation of cells into two groups occurs before the trophoblast forms on day 5. These twins have separate chorions and amnions. **(b)** About 1 percent of identical twins share a single amnion and chorion, because the tissue splits into two groups after these structures have already formed. **(c)** In about two-thirds of identical twins, the split occurs after day 5 but before day 9. These twins share a chorion but have separate amnions. Fraternal twins result from two sperm fertilizing two secondary oocytes. These twins develop their own amniotic sacs, yolk sacs, allantois, placentas, and umbilical cords.

two-thirds share a chorion but have separate amnions. Slightly fewer than 1 percent of MZ twins share both amnion and chorion. (The amnion is the fluid-filled sac that surrounds the fetus. The chorion develops into the placenta.) These differences may expose the different types of MZ twins to slightly different uterine environments. For example, if one chorion develops more attachment sites to the maternal circulation, one twin may receive more nutrients and gain more weight than the other.

In 1 in 50,000 to 100,000 pregnancies, an embryo divides into twins after the point at which the two groups of cells can develop as two individuals, between days 13 and 15. The result is conjoined or "Siamese" twins. The latter name comes from Chang and Eng Bunker, who were born in Thailand, then called Siam, in 1811. They were joined by a band of tissue from the navel to the breastbone, and could easily have been separated today. Chang and Eng lived for 63 years, attached. They fathered 22 children and divided each week between their wives.

For Abigail and Brittany Hensel, shown in **figure 3.17,** the separation occurred after day 9 of development, but before day 14. Biologists know this because the girls' shared organs have derivatives of ectoderm, mesoderm, and endoderm; that is, when the lump of cells divided incompletely, the three primary germ layers had not yet completely sorted themselves out.

Figure 3.17 **Conjoined twins.** Abby and Brittany Hensel are the result of incomplete twinning during the first 2 weeks of prenatal development.

The Hensel girls are extremely rare "incomplete twins." They are "dicephalic," which means that they have two heads. They are very much individuals.

Each girl has her own neck, head, heart, stomach, gallbladder, and lungs. Each has one leg and one arm, and a third arm between their heads was surgically removed. Each girl also has her own nervous system. The twins share a large liver, a single bloodstream, and all organs below the navel. They have three kidneys. Because at birth Abby and Brittany were strong and healthy, doctors suggested surgery to separate them. But their parents, aware of other cases where only one child survived separation, declined surgery.

As teens, Abby and Brittany are glad their parents did not choose to separate them, because they would have been unable to walk or run, as they can today. They enjoy kickball, volleyball, basketball, and cycling. Like any teen girls, they have distinctive tastes in clothing and in food.

MZ twins occur in 3 to 4 pregnancies per 1,000 births worldwide. In North America, twins occur in about 1 in 81 pregnancies, which means that 1 in 40 of us is a twin. However, not all twins survive to be born. One study of twins detected early in pregnancy showed that up to 70 percent of the eventual births are of a single child. This is called the "vanishing twin" phenomenon.

Key Concepts

1. Dizygotic (fraternal) twins arise from two fertilized ova.
2. Monozygotic (identical) twins arise from a single fertilized ovum and may share supportive structures.

The Embryo Develops

As the days and weeks of prenatal development proceed, different rates of cell division in different parts of the embryo fold the forming tissues into intricate patterns. In a process called embryonic induction, the specialization of one group of cells causes adjacent groups of cells to specialize. Gradually, these changes mold the three primary germ layers into organs and organ systems. Organogenesis is the transformation of the simple three layers of the embryo into distinct organs. During the weeks of organogenesis, the developing embryo is particularly sensitive to environmental influences such as chemicals and viruses.

During the third week of prenatal development, a band called the primitive streak appears along the back of the embryo. The primitive streak gradually elongates to form an axis that other structures organize around as they develop. The primitive streak eventually gives rise to connective tissue progenitor cells and the notochord, which is a structure that forms the basic framework of the skeleton. The notochord induces a sheet of overlying ectoderm to fold into the hollow **neural tube,** which develops into the brain and spinal cord (central nervous system). If the neural tube does not completely zip up by day 20, a birth defect called a neural tube defect (NTD) occurs. Parts of the brain or spinal cord protrude from the open head or spine, and the person cannot move body parts below the defect. Surgery can correct some NTDs (see *Bioethics: Choices for the Future* box in chapter 15). Lack of the B vitamin folic acid can cause NTDs in embryos with a genetic susceptibility. For this reason, the U.S. government adds the vitamin to grains, and pregnant women take supplements. A blood test during the 15th week of pregnancy detects a substance from the fetus's liver called alpha fetoprotein (AFP) that leaks at an abnormally rapid rate into the woman's circulation if there is an open NTD.

Some nations designate day 14 of prenatal development and primitive streak formation as the point beyond which they ban research on the human embryo. The reasoning is that the primitive streak is the first sign of a nervous system and day 14 is when implantation completes.

Appearance of the neural tube marks the beginning of organ development. Shortly after, a reddish bulge containing the heart appears. The heart begins to beat around day 18, and this is easily detectable by day 22. Soon the central nervous system starts to form.

The fourth week of embryonic existence is one of spectacularly rapid growth and differentiation (**figure 3.18**). Arms and legs begin to extend from small buds on the torso. Blood cells form and fill primitive blood vessels. Immature lungs and kidneys begin to develop.

By the fifth and sixth weeks, the embryo's head appears too large for the rest of its body. Limbs end in platelike structures with tiny ridges, and gradually apoptosis sculpts the fingers and toes. The eyes are open, but they do not yet have lids or irises. By the seventh and eighth weeks, a skeleton composed of cartilage forms. The embryo is now about the length and weight of a paper clip. At eight weeks of gestation, the prenatal human has tiny versions of all of the structures that will be present at birth. It is now a fetus.

 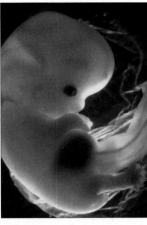

a. 28 days   4–6 mm   b. 49 days   13–22 mm

Figure 3.18 **Human embryos at (a) 28 days, and (b) 49 days.**

Key Concepts

1. During week 3, the primitive streak appears, followed rapidly by the notochord, neural tube, heart, central nervous system, limbs, digits, facial features, and other organ rudiments.

2. By week 8, all of the organs that will be present in the newborn have begun to develop.

The Fetus Grows

During the fetal period, body proportions approach those of a newborn. Initially, the ears lie low, and the eyes are widely spaced. Bone begins to replace the softer cartilage. As nerve and muscle functions become coordinated, the fetus moves.

Sex is determined at conception, when a sperm bearing an X or Y chromosome meets an oocyte, which always carries an X chromosome. An individual with two X chromosomes is a female, and one with an X and a Y is a male. A gene on the Y chromosome, called *SRY* (for "sex-determining region of the Y"), determines maleness.

Differences between the sexes do not appear until week 6, after the *SRY* gene is expressed in males. Male hormones then stimulate male reproductive organs and glands to differentiate from existing, indifferent structures. In a female, the indifferent structures of the early embryo develop as female organs and glands, under the control of other genes. Differences may be noticeable on ultrasound scans by 12 to 15 weeks. Sexual development is discussed further in chapter 6.

By week 12, the fetus sucks its thumb, kicks, makes fists and faces, and has the beginnings of teeth. It breathes amniotic fluid in and out, and urinates and defecates into it. The first trimester (3 months) of pregnancy ends.

By the fourth month, the fetus has hair, eyebrows, lashes, nipples, and nails. By 18 weeks, the vocal cords have formed, but the fetus makes no sound because it doesn't breathe air. By the end of the fifth month, the fetus curls into a head-to-knees position. It weighs about 454 grams (1 pound). During the sixth month, the skin appears wrinkled because there isn't much fat beneath it, and turns pink as capillaries fill with blood. By the end of the second trimester, the woman feels distinct kicks and jabs and may even detect a fetal hiccup. The fetus is now about 23 centimeters (9 inches) long.

In the final trimester, fetal brain cells rapidly link into networks as organs elaborate and grow. A layer of fat forms beneath the skin. The digestive and respiratory systems mature last, which is why infants born prematurely often have difficulty digesting milk and breathing. Approximately 266 days after a single sperm burrowed its way into an oocyte, a baby is ready to be born.

The birth of a healthy baby is against the odds. Of every 100 secondary oocytes exposed to sperm, 84 are fertilized. Of these 84, 69 implant in the uterus, 42 survive one week or longer, 37 survive 6 weeks or longer, and only 31 are born alive. Of the fertilized ova that do not survive, about half have chromosomal abnormalities that cause problems too severe for development to proceed.

3.5 Birth Defects

Certain genetic abnormalities or toxic exposures can affect development in an embryo or fetus, causing birth defects. Only a genetic birth defect can be passed to future generations. Although development can be derailed in many ways, about 97 percent of newborns appear healthy at birth.

The Critical Period

The specific nature of a birth defect reflects the structures developing when the damage occurs. The time when genetic abnormalities, toxic substances, or viruses can alter a specific structure is its **critical period (figure 3.19).** Some body parts, such as fingers and toes, are sensitive for short periods of time. In contrast, the brain is sensitive throughout prenatal development, and connections between nerve cells continue to change throughout life. Because of the brain's continuous critical period, many birth defect syndromes include learning disabilities or intellectual disability.

About two-thirds of all birth defects arise from a disruption during the embryonic period. More subtle defects, such as learning disabilities, that become noticeable only after infancy, are often caused by interventions during the fetal period. A disruption in the first trimester might cause intellectual disability; in the seventh month of pregnancy, it might cause difficulty in learning to read.

Some birth defects can be attributed to an abnormal gene that acts at a specific point in prenatal development. In a rare inherited condition called phocomelia (MIM 276826), for example, a mutation halts limb development from the third to the fifth week of the embryonic period, causing "flippers" to develop in place of arms and legs. The risk that a genetically caused birth defect will affect a particular family member can be calculated.

Many birth defects are caused not by mutant genes but by toxic substances the pregnant woman encounters. These environmentally caused problems will not affect other family members unless they, too, are exposed to the environmental trigger. Chemicals or other agents that cause birth defects are called **teratogens** (Greek for "monster-causing"). While it is best to avoid teratogens while pregnant, some women may need to continue to take potentially teratogenic drugs to maintain their own health.

Teratogens

Most drugs are not teratogens. Whether or not exposure to a particular drug causes birth defects may depend upon a woman's genes. For example, certain variants of a gene that control the body's use of an amino acid called homocysteine affect whether or not the medication valproic acid causes birth defects. Valproic acid is used to prevent seizures and symptoms of bipolar disorder. Rarely, it can cause NTDs, heart defects, hernias, and clubfoot. Women can be tested for this gene variant (*MTHFR C677T,* MIM 607093) and if they have it, switch to a different medication when they try to conceive. Experiments using stem cells are investigating exactly how valproic acid disrupts development.

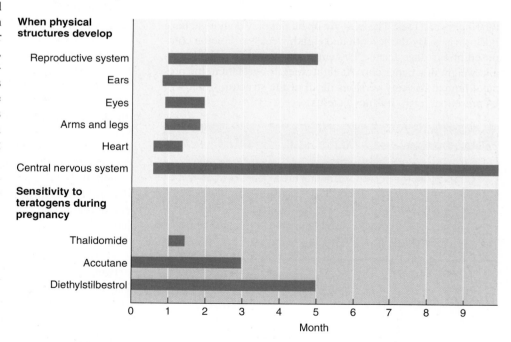

Figure 3.19 Critical periods of development. The nature of a birth defect resulting from drug exposure depends upon which structures were developing at the time of exposure. The time when a particular structure is vulnerable is called its critical period. Isotretinoin (Accutane) is an acne medication that causes cleft palate and eye, brain, and heart defects. Diethylstilbestrol (DES) was used in the 1950s to prevent miscarriage. It caused vaginal cancer in some "DES daughters." Thalidomide was used to prevent morning sickness.

Thalidomide

The idea that the placenta protects the embryo and fetus against harmful substances was tragically disproven between 1957 and 1961, when 10,000 children were born in Europe with what seemed, at first, to be phocomelia. Doctors realized that this genetic disorder is very rare, and therefore couldn't be the cause of the sudden problem. They discovered that the mothers had all taken a mild tranquilizer, thalidomide, to alleviate nausea early in pregnancy, during the critical period for limb formation. Many "thalidomide babies" were born with incomplete or missing legs and arms.

The United States was spared from the thalidomide disaster because an astute government physician noted the drug's adverse effects on laboratory monkeys. Still, several "thalidomide babies" were born in South America in 1994, where pregnant women were given the drug. In spite of its teratogenic effects, thalidomide is used to treat leprosy, AIDS, and certain blood and bone marrow cancers.

Cocaine

When cocaine use soared in the late 1980s and 1990s, physicians predicted a wave of birth defects in children who were affected by their mothers' cocaine use while pregnant. However, analysis of studies on more than 4,000 of these children indicates that effects, if any, are mild, and are less severe than those of prenatal exposure to alcohol or tobacco. Cocaine may slow fetal growth, particularly of the head, but the babies quickly catch up. No effects were seen on either language skills or IQ. Children exposed to cocaine in the uterus may have difficulty concentrating and behave poorly, but it is difficult to tease out whether this is due to the past drug exposure, or to the problems associated with poverty, such as malnutrition and hunger. Researchers are continuing to follow the progress of these young people to see if problems develop that can be linked to prenatal cocaine exposure.

Cigarettes

Chemicals in cigarette smoke stress a fetus. Carbon monoxide crosses the placenta and prevents the fetus's hemoglobin molecules from adequately binding oxygen. Other chemicals in smoke block nutrients. Smoke-exposed placentas lack important growth factors, causing poor growth before and after birth. Cigarette smoking during pregnancy increases the risk of spontaneous abortion, stillbirth, prematurity, and low birth weight.

Alcohol

A pregnant woman who has just one or two alcoholic drinks a day, or perhaps a large amount at a single crucial time, risks fetal alcohol syndrome (FAS) in her unborn child. Tests for gene variants that encode proteins that regulate alcohol metabolism may be able to predict which women and fetuses are at elevated risk for developing FAS, but until these tests are validated, pregnant women are advised to avoid all alcohol.

A child with FAS has a small head and a flat face (**figure 3.20**). Growth is slow before and after birth. Teens and young adults who have FAS are short and have small heads. More than 80 percent of them retain the facial characteristics of a young child with FAS.

The long-term mental effects of prenatal alcohol exposure are more severe than the physical vestiges. Intellectual impairment ranges from minor learning disabilities to intellectual disability. Many adults with FAS function at early grade-school level. They often lack social and communication skills and find it difficult to understand the consequences of actions, form friendships, take initiative, and interpret social cues.

Greek philosopher Aristotle noticed problems in children of alcoholic mothers more than 23 centuries ago. In the United States today, 1 to 3 of every 1,000 infants has the syndrome, meaning 2,000 to 6,000 affected children are born each year. Many more children have milder "alcohol-related effects." A fetus of a woman with active alcoholism has a 30 to 45 percent chance of harm from prenatal alcohol exposure.

Nutrients

Certain nutrients ingested in large amounts, particularly vitamins, act as drugs. The acne medicine isotretinoin (Accutane) is a vitamin A derivative that causes spontaneous abortion and defects of the heart, nervous system, and face in exposed embryos. Physicians first noted the tragic effects of this drug 9 months after dermatologists began prescribing it to young women in the early 1980s. Another vitamin A–based drug, used to treat psoriasis, as well as excesses of vitamin A itself, also cause birth defects. Some forms of vitamin A are stored in body fat for up to 3 years.

Excess vitamin C can harm a fetus that becomes accustomed to the large amounts the woman takes. After birth, when the vitamin supply suddenly plummets, the baby may develop symptoms of vitamin C deficiency (scurvy), such as bruising and becoming infected easily.

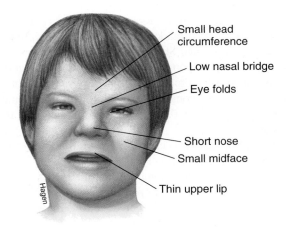

Small head circumference

Low nasal bridge

Eye folds

Short nose

Small midface

Thin upper lip

Hagen

Figure 3.20 Fetal alcohol syndrome. Some children whose mothers drank alcohol during pregnancy have characteristic flat faces.

Malnutrition threatens a fetus. The opening essay to chapter 11 describes the effects of starvation on embryos during the bleak "Dutch Hunger Winter" of 1944–1945. Poor nutrition later in pregnancy affects the development of the placenta and can cause low birth weight, short stature, tooth decay, delayed sexual development, and learning disabilities.

Occupational Hazards

Teratogens are present in some workplaces. Researchers note increased rates of spontaneous abortion and children born with birth defects among women who work with textile dyes, lead, certain photographic chemicals, semiconductor materials, mercury, and cadmium. Men whose jobs expose them to sustained heat, such as smelter workers, glass manufacturers, and bakers, may produce sperm that can fertilize an oocyte and then cause spontaneous abortion or a birth defect. A virus or a toxic chemical carried in semen may also cause a birth defect.

Viral Infection

Viruses are small enough to cross the placenta and reach a fetus. Some viruses that cause mild symptoms in an adult, such as the chickenpox virus, may devastate a fetus. Men can transmit viral infections to an embryo or fetus during sexual intercourse.

HIV can reach a fetus through the placenta or infect a newborn via blood contact during birth. Fifteen to 30 percent of infants born to untreated HIV-positive women are HIV positive. The risk of transmission is significantly reduced if a pregnant woman takes anti-HIV drugs. All fetuses of HIV-infected women are at higher risk for low birth weight, prematurity, and stillbirth if the woman's health is failing.

German measles (rubella) is a well-known viral teratogen. In the United States, in the early 1960s, an epidemic of the usually mild illness caused 20,000 birth defects and 30,000 stillbirths. Children who were exposed during the first trimester of pregnancy could develop cataracts, deafness, and heart defects. Fetuses exposed during the second or third trimester of pregnancy may have as a result developed learning disabilities, speech and hearing problems, and type 1 diabetes mellitus.

The incidence of these problems, called congenital rubella syndrome, has dropped markedly since vaccination eliminated the disease in the United States. However, the syndrome resurfaces in unvaccinated populations. In 1991 among a cluster of unvaccinated Amish women in rural Pennsylvania, 14 of every 1,000 newborns had congenital rubella syndrome, compared to an incidence then of 0.006 per 1,000 in the general U.S. population.

Herpes simplex virus can harm a fetus or newborn whose immune system is immature. Forty percent of babies exposed to active vaginal herpes lesions become infected, and half of them die. Of the survivors, 25 percent sustain severe nervous system damage, and another 25 percent have skin sores. A surgical (caesarean) delivery can protect the child.

Pregnant women are routinely checked for hepatitis B infection, which in adults causes liver inflammation, great fatigue, and other symptoms. Each year in the United States, 22,000 infants are infected with this virus during birth. These babies are healthy, but at high risk for developing serious liver problems as adults. When infected women are identified, a vaccine can be given to their newborns to help prevent complications.

Key Concepts

1. The critical period is the time during prenatal development when a structure is sensitive to damage from a faulty gene or environmental insult.
2. Most birth defects develop during the embryonic period and are more severe than problems that arise during fetal development.
3. Teratogens are agents that cause birth defects.

3.6 Maturation and Aging

"Aging" means moving through the life cycle. In adulthood, as we age, the limited life spans of cells are reflected in the waxing and waning of biological structures and functions. Although some aspects of our anatomy and physiology peak very early—such as the number of brain cells or hearing acuity, which do so in childhood—age 30 seems to be a turning point for decline. Some researchers estimate that, after this age, the human body becomes functionally less efficient by about 0.8 percent each year.

Can we slow aging? In the quest to extend life, people have sampled everything from turtle soup to owl meat to human blood. Recently, attention has turned to a component of red wine called resveratrol (**figure 3.21**). It is a member of a class of enzymes called sirtuins that regulate energy use in cells by altering the expression of certain sets of genes. Through their effect on energy metabolism, the sirtuins seem to prevent or delay several diseases that are more common in the aged, such as heart disease and neurodegenerative disorders. In addition, studies in rodents show that sirtuins are responsible for the life extension that is seen with caloric restriction—eating very little. Drug companies are exploring compounds that are 1,000 times as active as resveratrol to fight particular diseases. However, we do not know nearly enough about the sirtuins to take them to forestall aging—at least not yet.

Many diseases that begin in adulthood, or are associated with aging, have genetic components. Often these disorders are multifactorial, because it takes many years for environmental exposures to alter gene expression in ways that noticeably affect health. Following is a closer look at how genes may impact health throughout life.

Adult-Onset Inherited Disorders

Human prenatal development is a highly regulated program of genetic switches that are turned on in specific body parts at specific times. Environmental factors can affect how certain genes are expressed before birth in ways that create risks that

Figure 3.21 **Sirtuins may extend healthy life.** Chemicals related to a component of the skin of red grapes, resveratrol, may alter gene expression in ways that extend the number of healthy years that a person may enjoy.

appear much later. For example, adaptations that enable a fetus to grow despite near-starvation become risk factors for certain common conditions of adulthood, when conserving calories is no longer needed. Such disorders include coronary artery disease, obesity, stroke, hypertension, and type 2 diabetes mellitus. A fetus that does not receive adequate nutrition has intrauterine growth retardation (IUGR), and though born on time, is very small. Premature infants, in contrast, are small but are born early, and are not predisposed to conditions resulting from IUGR.

More than 100 studies correlate low birth weight due to IUGR with increased incidence of cardiovascular disease and diabetes later in life. Much of the data come from war records because enough time has elapsed to study the effects of prenatal malnutrition as people age. How can poor nutrition before birth cause disease decades later? Perhaps to survive, the starving fetus redirects its circulation to protect vital organs, as muscle mass and hormone production change to conserve energy. Growth-retarded babies have too little muscle tissue, and since muscle is the primary site of insulin action, glucose metabolism changes. Thinness at birth, and the accelerated weight gain in childhood that often compensates, sets the stage for coronary heart disease and type 2 diabetes much later.

In contrast to the delayed effects of fetal malnutrition, symptoms of single-gene disorders can begin at any time. In general, inherited conditions that affect children are recessive. Even a fetus can have symptoms of inherited disease, such as various types of osteogenesis imperfecta ("brittle bone disease").

Most dominantly inherited conditions start to affect health in early to middle adulthood. In polycystic kidney disease (MIM 173900), cysts that may have been present but undetected in the kidneys during one's twenties begin causing bloody urine, high blood pressure, and abdominal pain in the thirties. The joint destruction of osteoarthritis may begin in one's thirties, but not cause pain for 20 years. The uncontrollable movements, unsteady gait, and diminishing mental faculties of Huntington disease typically become obvious near age 40, but subtle changes begin years earlier.

Five to 10 percent of Alzheimer disease cases are inherited and produce initial symptoms in the forties and fifties. German neurologist Alois Alzheimer first identified the condition in 1907 as affecting people in mid-adulthood. Noninherited Alzheimer disease typically begins later in life.

Whatever the age of onset, Alzheimer disease starts gradually. Mental function declines steadily for 3 to 10 years after the first symptoms of depression and short-term memory loss appear. The person gradually loses cognitive skills, even becoming unable to walk because he or she forgets how to put one foot in front of the other. Confused and forgetful, Alzheimer patients often wander away from family and friends. Finally, the patient cannot recognize loved ones and can no longer perform basic functions such as speaking or eating.

The brains of Alzheimer disease patients contain deposits of a protein, called beta amyloid, in learning and memory centers. Alzheimer brains also contain structures called neurofibrillary tangles, which consist of a protein called tau. Tau binds to and disrupts microtubules in nerve cell branches, destroying the shape of the cell, which is essential to its ability to communicate. It isn't clear whether the "plaques and tangles" of the Alzheimer brain are causes or effects. Reading 5.1 discusses Alzheimer disease further.

Disorders That Resemble Accelerated Aging

Genes control aging both passively (as structures break down) and actively (by initiating new activities). A group of "rapid aging" inherited disorders are very rare, but may hold clues to how genes control aging in all of us.

The most severe rapid aging disorders are the segmental progeroid syndromes. Most of these disorders are caused by impairment of cells' ability to repair DNA, which is discussed in chapter 12. Mutations that would ordinarily be corrected persist. Over time, the accumulation of mutations destabilizes the entire genome, and even more mutations occur in somatic cells. The various changes that we associate with aging ensue.

Table 3.3 lists the more common segmental progeroid syndromes. They vary in severity. People with Rothmund-Thomson syndrome, for example, may have a normal life span, but develop gray hair or baldness, cataracts, cancers, and osteoporosis at young ages. Werner syndrome becomes apparent before age 20, causing death before age 50 from diseases associated with aging. Young adults with Werner syndrome develop atherosclerosis, type 2 diabetes mellitus, hair graying and loss, osteoporosis, cataracts, and wrinkled skin. They are short because they skip the growth spurt of adolescence.

Table 3.3	Rapid Aging Syndromes		
Disorder	**Incidence**	**Average Life Span**	**MIM Number**
Ataxia telangiectasia	1/60,000	20	208900
Cockayne syndrome	1/100,000	20	216400
Hutchinson-Gilford syndrome	<1/1,000,000	13	176670
Rothmund-Thomson syndrome	<1/100,000	Normal	268400
Trichothiodystrophy	<1/100,000	10	601675
Werner syndrome	<1/100,000	50	277700

The child in **figure 3.22,** Megan, has Hutchinson-Gilford progeria syndrome, but is unusually beautiful. An affected child typically appears normal at birth but slows in growth by the first birthday. Within just a few years, the child becomes wrinkled and bald, with easily bruised skin and the facial features characteristic of advanced age. The body ages on the inside as well, as arteries clog with fatty deposits. The child usually dies of a heart attack or a stroke by age 13, although some patients live into their twenties.

The child's cells show aging-related changes too. Normal cells growing in culture divide about 50 times before dying. Cells from Hutchinson-Gilford progeria syndrome patients die in culture after only 10 to 30 divisions.

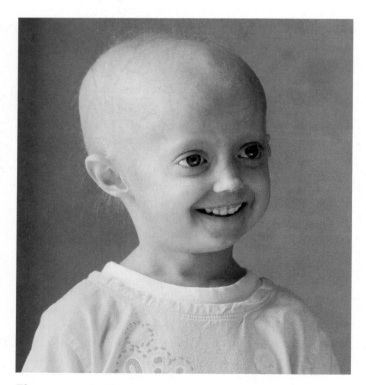

Figure 3.22 Accelerated aging. Megan has Hutchinson-Gilford progeria syndrome, which is extremely rare.

Only 50 people who currently have Hutchinson-Gilford progeria syndrome are known worldwide, and only 100 have ever been reported in the medical literature. A campaign called "Find the Other 150" is searching the globe for unidentified children with the syndrome predicted to exist, both so that they can be helped, and so that they can provide insights into the common health problems that accompany aging. Unlocking the molecular cause of the disease has suggested several uncommon as well as common drugs that are now in clinical trials to help stall the symptoms.

Hutchinson-Gilford progeria is caused by a single DNA base change in the gene that encodes a protein called lamin A. That one base is a site that determines how parts of the protein are cut and joined, and when it is altered, the protein lacks 50 amino acids. The shortened protein is called progerin. It remains stuck to the endoplasmic reticulum, instead of being transported into the nucleus through the nuclear pores, as happens to normal lamin A protein. Instead, progerin diffuses into the tubules of the ER and travels within them to the nuclear membrane, which happens because these two organelles are continuous. This route of entry stresses the nuclear membrane, causing it to bubble or "bleb" inward, altering the way that the nuclear lamina (the layer on the inside face) binds the chromatin (DNA complexed with protein) within. Somehow, disturbing the chromatin hampers DNA repair, allowing mutations associated with the signs of aging to occur. Several drugs being tested block the molecule that holds progerin to the ER.

Studies on stem cells from bone marrow of patients add further evidence: Progerin shifts the activities of certain genes in ways that promote bone formation and suppress fat deposition. This skewed development perhaps explains the failure to thrive and skeletal appearance of affected individuals. The molecular view of progeria suggests that DNA repair is what enables us to live many years. These children lack that protection, and the mutations that age us all accumulate much faster.

Is Longevity Inherited?

Aging reflects gene activity plus a lifetime of environmental influences. Families with many very aged members have a lucky collection of gene variants plus shared environmental influences such as good nutrition, excellent health care, devoted relatives, and other advantages. A genome-level approach to identifying causes of longevity identified a region of chromosome 4 that houses gene variants associated with long life. Genome comparisons among people who've passed their 100th birthdays to those who have died of the common illnesses of older age are revealing other genes that influence longevity (**Reading 3.1**).

Genes and Longevity

The human genome is like a vast library that holds the clues to good health. One way to identify those clues is to probe the genomes of individuals who have lived past 100 years. These fortunate people are called centenarians **(figure 1).** Usually they enjoy excellent health and are socially active, then succumb rapidly to diseases that typically claim people decades earlier.

Figure 1 This woman has enjoyed living for more than a century. Researchers are discovering clues to good health by probing the genomes of centenarians.

Centenarians fall into three broad groups—about 20 percent of them never get the diseases that kill most people; 40 percent get these diseases, but at a much older age than average; and the other 40 percent live with and survive the more common disorders of aging. Researchers hope that learning which gene variants offer this protection will lead to better understanding of the common disorders of later adulthood—heart disease, stroke, cancers, type 2 diabetes mellitus, and dementias.

While the environment seems to play an important role in the deaths of people ages 60 to 85, past that age, genes predominate. That is, someone who dies at age 68 of lung cancer can probably blame a lifetime of cigarette smoking. But a smoker who dies at age 101 of the same disease probably had gene variants that protected against lung cancer. Centenarians have higher levels of large lipoproteins that carry cholesterol (high-density lipoprotein [HDL]) than other people, which researchers estimate adds 20 years of life.

Children and siblings of centenarians tend to be long-lived as well, supporting the idea that longevity is inherited. Brothers of centenarians are 17 times as likely to live past age 100 as the average man, and sisters are 8.5 times as likely. The fact that some people more than 100 years of age have less-than-healthful habits suggests that genes are protecting them. One researcher suggests that the saying, "The older you get, the sicker you get" be replaced with "The older you get, the healthier you've been."

Centenarians have inherited two types of gene variants—those that directly protect them, and wild type alleles of genes that, when mutant, cause disease. Research focuses on individual genes as well as genome-wide scans to identify gene variants that make it more likely that a person will live past age 100. **Table 1** lists some "candidate" gene types that may control longevity. To find other gene variants that promote long life, researchers are conducting genome-wide association studies on centenarians and comparing the patterns to those of people with particular conditions associated with aging.

Several studies are identifying gene variants that contribute to living long and well. The New England Centenarian Study, headed at Boston University, began in 1988 to amass information on families of the oldest citizens in the United States. The researchers are compiling a "healthy standard genome." Investigators from the Coriell Institute in New Jersey are probing the genomes of people over age 90 who live in nursing homes. So far, what these people have in common is never having had heart disease and never having smoked. Several had cancer, indicating that cancers are often survivable.

Researchers at the University of Pittsburgh have identified places in the genome that harbor "successful aging genes" that have variants that preserve cognition. Other studies are looking at known genes implicated in the diseases that kill most of us, while others seek patterns of variation or the alleles that long-lived siblings share. Considered together, perhaps these studies will provide information that will help the majority of us who have not been fortunate enough to have inherited longevity gene variants.

Table 1

Single genes important in aging affect

- control of insulin secretion and glucose metabolism,
- immune system functioning,
- control of the cell cycle,
- lipid (cholesterol) metabolism,
- response to stress, and
- production of antioxidant enzymes.

It is difficult to tease apart inborn from environmental influences on life span. One approach compares adopted individuals to both their biological and adoptive parents. In a study from Denmark, adopted individuals with one biological parent who died of natural causes before age 50 were more than twice as likely to die before age 50 as were adoptees whose biological parents lived beyond this age, suggesting an inherited component to longevity. Interestingly, adopted individuals whose natural parents died early due to infection were more than five times as likely to also die early of infection, perhaps because of inherited immune system deficiencies. The adoptive parents' ages at death had no influence on that of their adopted children. Chapter 7 explores the "nature versus nurture" phenomenon more closely.

Key Concepts

1. Starvation before birth can set the stage for later disease by affecting gene expression in certain ways.

2. Most single-gene disorders are recessive and strike early in life. Single-gene disorders with an adult onset are more likely to be dominant.

3. The segmental progeroid syndromes are single-gene disorders that speed the signs of aging.

4. Families with many aged members can thank their genes as well as the environment. Chromosome 4 houses longevity genes, and genome-wide screens are identifying others.

5. Adoption studies compare the effects of genes versus environmental influences on longevity.

Summary

3.1 The Reproductive System

1. The male and female reproductive systems include paired **gonads** and networks of tubes in which **sperm** and **oocytes** are made.

2. Male **gametes** originate in seminiferous tubules within the testes, then pass through the epididymis and ductus deferentia, where they mature before exiting the body through the urethra during sexual intercourse. The prostate gland, the seminal vesicles, and the bulbourethral glands add secretions.

3. Female gametes originate in the ovaries. Each month after puberty, one ovary releases an oocyte into a uterine tube. The oocyte then moves to the uterus for implantation (if fertilized) or expulsion.

3.2 Meiosis

4. **Meiosis** reduces the chromosome number in gametes from **diploid** to **haploid,** maintaining the chromosome number between generations. Meiosis ensures genetic variability by **independently assorting** combinations of genes into gametes as chromosomes randomly align and **cross over.**

5. Meiosis I, a **reduction division,** halves the number of chromosomes. Meiosis II, an **equational division,** produces four cells from the two that result from meiosis I, without another DNA replication.

6. Crossing over occurs during prophase I. It mixes up paternally and maternally derived genes on **homologous pairs** of chromosomes.

7. Chromosomes segregate and independently assort in metaphase I, which determines the distribution of genes from each parent.

3.3 Gamete Maturation

8. **Spermatogenesis** begins with spermatogonia, which accumulate cytoplasm and replicate their DNA, becoming primary spermatocytes. After meiosis I, the cells become haploid secondary spermatocytes. In meiosis II, the secondary spermatocytes divide, each yielding two spermatids, which then differentiate into spermatozoa.

9. In **oogenesis,** some oogonia grow and replicate their DNA, becoming primary oocytes. In meiosis I, the primary oocyte divides to yield one large secondary oocyte and a small **polar body.** In meiosis II, the secondary oocyte divides to yield the large ovum and another polar body. Female meiosis is completed at fertilization.

3.4 Prenatal Development

10. In the female, sperm are capacitated and drawn toward a secondary oocyte. One sperm burrows through the oocyte's protective layers with acrosomal enzymes. Fertilization occurs when the sperm and oocyte fuse and their DNA combines in one nucleus, forming the **zygote.** Electrochemical changes in the egg surface block additional sperm from entering. **Cleavage** begins and a 16-celled **morula** forms. Between days 3 and 6, the morula arrives at the uterus and hollows, forming a **blastocyst** made up of **blastomeres.** The trophoblast and **inner cell mass** form. Around day 6 or 7, the blastocyst implants, and trophoblast cells secrete hCG, which prevents menstruation.

11. During the second week, the amniotic cavity forms as the inner cell mass flattens. **Ectoderm** and **endoderm** form, and then **mesoderm** appears, establishing the **primary germ layers.** Cells in each germ layer begin to develop into specific organs. During the third week, the placenta, yolk sac, allantois, and umbilical cord begin to form as the amniotic cavity swells with fluid. **Monozygotic** twins result when one fertilized ovum splits. **Dizygotic** twins result from two fertilized ova. Organs form throughout the embryonic period. The primitive streak, notochord and **neural tube,** arm and leg buds, heart, facial features, and skeleton develop.

12. At the eighth week, the **embryo** becomes a **fetus,** with all structures present but not fully grown. Organ rudiments grow and specialize. The developing organism moves and reacts, and its body proportions come to resemble those of a baby. In the last trimester, the brain develops rapidly, and fat is deposited beneath the skin. The digestive and respiratory systems mature last.

3.5 Birth Defects

13. Birth defects can result from a mutation or an environmental intervention.

14. A substance that causes birth defects is a **teratogen.** Environmentally caused birth defects are not transmitted to future generations.

15. The time when a structure is sensitive to damage from an abnormal gene or environmental intervention is its **critical period.**

3.6 Maturation and Aging

16. Genes cause or predispose us to illness throughout life. Single-gene disorders that strike early tend to be recessive; most adult-onset single-gene conditions are dominant.

17. Malnutrition before birth can alter gene expression in ways that cause illness later in life.

18. The segmental progeroid syndromes are single-gene disorders that speed aging-associated changes.

19. Long life is due to genetics and environmental influences.

www.mhhe.com/lewisgenetics10

Answers to all end-of-chapter questions can be found at **www.mhhe.com/lewisgenetics10.** You will also find additional practice quizzes, animations, videos, and vocabulary flashcards to help you master the material in this chapter.

Review Questions

1. How many sets of human chromosomes are in each of the following cell types?
   a. an oogonium
   b. a primary spermatocyte
   c. a spermatid
   d. a cell from either sex during anaphase of meiosis I
   e. a cell from either sex during anaphase of meiosis II
   f. a secondary oocyte
   g. a polar body derived from a primary oocyte

2. List the structures and functions of the male and female reproductive systems.

3. A dog has 39 pairs of chromosomes. Considering only the random alignment of chromosomes, how many genetically different puppies are possible when two dogs mate? Is this number an underestimate or overestimate of the actual total? Why?

4. How does meiosis differ from mitosis?

5. What do oogenesis and spermatogenesis have in common, and how do they differ?

6. How does gamete maturation differ in the male and female?

7. Why is it necessary for spermatogenesis and oogenesis to generate stem cells?

8. Describe the events of fertilization.

9. Write the time sequence in which the following structures begin to develop: notochord, gastrula, inner cell mass, fetus, zygote, morula.

10. Why does exposure to teratogens produce more severe health effects in an embryo than in a fetus?

11. The same birth defect syndrome can be caused by a mutant gene or exposure to a teratogen. How do the consequences of each cause differ for future generations?

12. List four teratogens, and explain how they disrupt prenatal development.

13. Why is an "anti-aging" pill, diet, or device impossible?

14. Cite two pieces of evidence that genes control aging.

Applied Questions

1. Up to what stage, if any, do you think it is ethical to experiment on a prenatal human? Cite reasons for your answer.

2. Under a microscope, a first and second polar body look alike. What structure would distinguish them?

3. Armadillos always give birth to identical quadruplets. Are the offspring clones?

4. Some Vietnam War veterans who were exposed to the herbicide Agent Orange claim that their children—born years after the exposure—have birth defects caused by dioxin, a contaminant in the herbicide. What types of cells would the chemical have to have affected in these men to cause birth defects years later?

5. In about 1 in 200 pregnancies, a sperm fertilizes a polar body instead of an oocyte. A mass of tissue that is not an embryo develops. Why can't a polar body support the development of an entire embryo?

6. Should a woman be held legally responsible if she drinks alcohol, smokes, or abuses drugs during pregnancy and it harms her child? Should liability apply to all substances that can harm a fetus, or only to those that are illegal?

7. Would you want to have your genome scanned to estimate how long you are likely to live? Why or why not?

8. What types of evidence have led researchers to hypothesize that a poor prenatal environment can raise the risk for certain adult illnesses? How are genes part of this picture?

9. Why is an "anti-aging" skin product not possible?

10. Do you think that use of a particular recreational drug should be outlawed or restricted if it is a teratogen? Should any restrictions apply only to illegal drugs? Alcohol and cigarettes are more dangerous to a fetus than cocaine, and their use is legal.

Web Activities

1. Look over the "Living to 100 Life Expectancy Calculator" at www.livingto100.com and list 10 ways that you can change your behavior to possibly live longer. What does this quiz suggest about the relative role of genes and the environment in determining longevity?

2. Go to the Motherisk website at http://www.motherisk.org./ Click on "Women, Partners, Family and Friends." Then click on "drugs in pregnancy" in the left-hand list. Identify three drugs that are safe to take during pregnancy and three that are not safe, and list the associated medical problems.

Case Studies and Research Results

1. Human embryonic stem cells can be derived and cultured from an 8-celled cleavage embryo and from a cell of an inner cell mass. Explain the difference between these stages of human prenatal development.

2. Victor, a 34-year-old artist, was killed in a car accident. He and his wife Emma hadn't started a family yet, but planned to soon. The morning after the accident, Emma asked if some of her husband's sperm could be collected and frozen, for her to use to have a child. Do you think that this "post-mortem sperm retrieval" should be done?

3. Miguel and Maria are carriers of cystic fibrosis (CF), and the condition is severe in their families. They have a procedure called preimplantation genetic diagnosis (see figure 21.4) to ensure that they conceive a child who does not inherit CF. Maria's oocyte is fertilized with Miguel's sperm in a laboratory dish, and it develops to the 8-cell stage. One cell is removed and tested for the mutant allele that Miguel and Maria carry. Only the wild type allele is detected.

   Anna and Peter are also carriers of a genetic disorder that can affect either sex. They cannot get into a preimplantation genetic diagnosis clinical trial, which would be free; their insurance will not cover the procedure; and they cannot afford it. So, they choose chorionic villus sampling (CVS), in which a cell from the developing placenta is tested for the mutant allele, at the tenth week of gestation. Their fetus is found to be a carrier, like them.

   A third couple, Vivian and Max, are not willing to take the higher risk of miscarriage associated with CVS, so they wait until the sixteenth week, and Vivian has amniocentesis. Fetal cells are sampled from the amniotic fluid and the mutation that causes the clotting disorder hemophilia in Vivian's family checked. Vivian may be a carrier for hemophilia A. If she is, a son would face a 50 percent chance of inheriting the disorder. The amniocentesis indicates a daughter.

   a. Why can a genetic test on a cell from an embryo, the placenta, or a fetus predict future health?
   b. At the time of preimplantation genetic diagnosis, is the embryo a cleavage embryo, an inner cell mass, or a gastrula?
   c. What structures are present in Vivian and Max's fetus that have not yet developed in Anna and Peter's at the time of their prenatal tests?

4. Surgical separation of conjoined twins is more likely to succeed if fewer body parts are shared or attached. This was the case for Maria de Jesus and Maria Theresa, born in Guatemala in 2001 and separated before their first birthday. They were joined at the head, but facing opposite directions, so they could not move much. The surgery took 23 hours! Today they are well. The outcome wasn't good for Mandan and Label Bikini, 29-year-old Iranian conjoined twins who could no longer stand being joined along their heads, with their brains fused. They died shortly after 50 hours of surgery in 2003.

   If you had conjoined twins, what would you do? Would you attempt surgical separation?

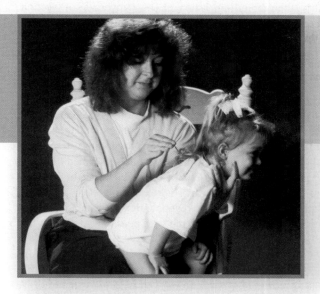

An autosomal recessive disease such as cystic fibrosis, which affects this little girl but not her parents, can skip generations. An autosomal dominant disease, such as Huntington disease, cannot skip generations.

Single-Gene Inheritance

Learning Outcomes

4.1 Following the Inheritance of One Gene

1. Describe how Mendel deduced that recessive traits seem to disappear in hybrids.

2. Define and distinguish *heterozygote* and *homozygote*; *dominant* and *recessive*; *phenotype* and *genotype*.

3. Explain how the law of segregation reflects the events of meiosis.

4. Describe a Punnett square.

4.2 Single-Gene Inheritance Is Rare

5. Explain how a gene alone usually does not solely determine a trait.

6. Distinguish between autosomal recessive and autosomal dominant inheritance.

4.3 Following the Inheritance of More Than One Gene

7. Explain how Mendel's experiments followed the inheritance of more than one gene.

8. Explain how the law of independent assortment reflects the events of meiosis.

4.4 Pedigree Analysis

9. Explain how pedigrees show single-gene transmission.

4.5 Family Genome Analysis

10. Explain how genome sequencing in a family can reveal Mendelian inheritance patterns.

The Big Picture: Gregor Mendel deduced the basis of inheritance patterns. His two laws brilliantly described how chromosomes behave in meiosis, which had not yet been discovered.

A Tale of Two Families

Henry T. is a healthy college student trying to decide whether to take a genetic test to see if he has inherited the mutation that causes Huntington disease (HD). His grandfather Emmett died last year in a nursing home where his younger sister, who also has HD, still lives. When Henry turned 12, his parents told him about his grandfather and great aunt and took him for a visit. His grandfather was restrained in bed by tightly tucked blankets to stop his constant writhing, and his face grimaced and ticked. Now Henry's mother Ann is starting to show symptoms, moving her left cheek repeatedly, and she's irritable. He knows that HD does not skip generations, unless, by chance, no one inherits the mutation. Did he or his sister Sue?

Sean P. has also lived with genetic disease in his family. His parents are healthy, but his younger sister Ellen has cystic fibrosis (CF). As an infant, Ellen couldn't gain weight, and she had frequent respiratory infections. When she had pneumonia at 8 months old, her doctor tested her for CF, and she indeed had two copies of the most common mutation. If Ellen had been born today, newborn screening would have detected her CF, and treatment begun right away—enzymes to help digestion and breathing, antibiotics to prevent infection, inhaled drugs to clear her airways, and twice-daily therapy to shake free the thick mucus that clogs her lungs. She also takes a new drug to correct the defect in the protein that causes the disease.

Ellen's childhood was rough, with frequent hospital stays and constant vigilance to keep her free of infection. One summer, Sean went to a summer camp for CF families, where he realized that he could be a carrier, like his parents. A few years later, he took a CF test at the campus health center and learned that he, too, is a carrier.

4.1 Following the Inheritance of One Gene

Single-gene diseases such as Huntington disease and cystic fibrosis affect families in patterns termed **modes of inheritance**. Knowing these patterns makes it possible to predict risks that particular family members inherit the mutation. HD is **autosomal dominant,** which means that it affects both sexes and appears every generation. In contrast, CF is **autosomal recessive,** which means that the disease affects both sexes and can skip generations through carriers.

This chapter explains how we came to understand the patterns of inheritance. The tale begins with pea plants in a long-ago garden in a small village in what is now the Czech Republic, and continues through today's whole-genome sequencing of families.

Gregor Mendel was the first thinker to probe the underlying rules of logic that make it possible to predict inheritance of specific traits. By breeding pea plants, Mendel described units of inheritance that pass traits from generation to generation. He called these units "elementen." He could not see them, but he inferred their existence from the appearances of his plants. Although Mendel knew nothing of DNA, chromosomes, or cells, his "laws" of inheritance have not only stood the test of time, but explain trait transmission in any diploid species.

Mendel's Experiments

The son of a farmer and grandson of a gardener, Mendel learned early how to tend fruit trees. At age 10 he left home to attend a special school for bright students, supporting himself by tutoring. He eventually became a priest at a monastery where the priests were teachers and did research in natural science. Here, Mendel learned how to artificially pollinate crop plants to control their breeding. Later, at the University of Vienna, courses in the sciences and statistics got him thinking about a question that had confounded other plant breeders: Why did certain traits disappear in one generation, yet reappear in the next? To solve this puzzle, Mendel bred pea plant hybrids and applied statistics.

From 1857 to 1863, Mendel crossed and cataloged traits in 24,034 plants, through several generations. He deduced that consistent ratios of traits in the offspring indicated that the plants transmitted distinct units. He derived two hypotheses to explain how inherited traits are transmitted. Mendel described his work to the Brnö Medical Society in 1865 and published it in the organization's journal the next year. His remarkably clear paper discussed plant hybridization, the reappearance of traits in the third generation, and the joys of working with peas, plus data. But few people read it.

Shortly after Mendel's paper was republished in English in 1901, three botanists independently rediscovered the laws of inheritance. Once they read Mendel's paper, they credited him. Mendel came to be regarded as the "father of genetics."

Peas are ideal for probing heredity because they are easy to grow, develop quickly, and have many traits that take one of two easily distinguishable forms. **Figure 4.1** illustrates the seven traits that Mendel followed through several pea generations. When analyzing genetic crosses, the first generation is the parental generation, or P_1; the second generation is the first filial generation, or F_1; the next generation is the second filial generation, or F_2, and so on.

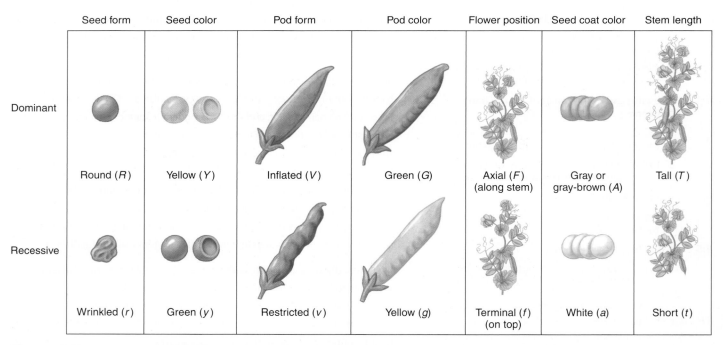

Figure 4.1 **Traits Mendel studied.** Gregor Mendel studied the transmission of seven traits in the pea plant. Each trait has two easily distinguished expressions, or phenotypes.

Mendel's first experiments followed single traits with two expressions, such as "short" and "tall." He set up all combinations of possible artificial pollinations, manipulating fertilizations to cross tall with tall, short with short, and tall with short plants. This last combination, plants with one trait variant crossed to plants with the alternate, produces hybrids, which are offspring that inherit a different gene variant (allele) from each parent.

Mendel noted that short plants crossed to other short plants were "true-breeding," always producing short plants. The crosses of tall plants to each other were more confusing. Some tall plants were true-breeding, but others crossed with each other yielded short plants in about one-quarter of the next generation. In some tall plants, tallness appeared to mask shortness. One trait that masks another is said to be **dominant;** the masked trait is **recessive**.

Mendel conducted up to 70 hybrid crosses for each of the seven traits. This experiment is called a **monohybrid cross** because it follows one trait and the parents are hybrids.

When Mendel allowed the non-true-breeding tall plants—monohybrids—to self-fertilize, the progeny were in the ratio of one-quarter short to three-quarters tall plants (**figure 4.2**). In further crosses, he found that two-thirds of the tall plants from the monohybrid F_1 cross were non-true-breeding, and the remaining third were true-breeding.

In these experiments, Mendel confirmed that hybrids hide one expression of a trait—short, in this case—which reappears when hybrids are self-crossed. He tried to explain how this happened: Gametes distribute "elementen" because these cells physically link generations. Paired sets of elementen separate as gametes form. When gametes join at fertilization, the elementen combine anew. Mendel reasoned that each elementen was packaged in a separate gamete. If opposite-sex gametes combine at random, he could mathematically explain the different ratios of traits produced from his pea plant crosses. Mendel's idea that elementen separate in the gametes would later be called the **law of segregation**.

When Mendel's ratios were demonstrated in several species in the early 1900s, just when chromosomes were being discovered, it became apparent that elementen and chromosomes had much in common. Both paired elementen and pairs of chromosomes separate at each generation and are transmitted—one from each parent—to offspring. Both are inherited in random combinations. Chromosomes provided a physical mechanism for Mendel's hypotheses. In 1909, English embryologist William Bateson renamed Mendel's elementen *genes* (Greek for "give birth to"). In the 1940s, scientists began investigating the gene's chemical basis, discussed in chapter 9.

In the twentieth century, researchers discovered the molecular basis of some of the traits that Mendel studied. "Short" and "tall" plants reflect expression of a gene that enables a plant to produce the hormone gibberellin, which elongates the stem. One tiny change to the DNA, and a short plant results. Likewise, "round" and "wrinkled" peas arise from the *R* gene, whose encoded protein connects sugars into branching polysaccharide molecules. Seeds with a mutant *R* gene cannot attach the sugars. As a result, water exits the cells, and the peas wrinkle.

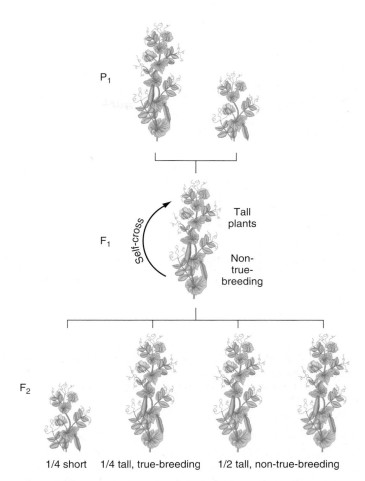

Figure 4.2 A monohybrid cross. When Mendel crossed true-breeding tall plants with short plants, the next generation plants were all tall. When he self-crossed the F_1 plants, one-quarter of the plants in the next generation, the F_2, were short, and three-quarters were tall. Of the tall plants in the F_2, one-third were true-breeding, and the other two-thirds were not true-breeding. He could tell this by conducting further crosses of the tall plants to short plants, to see which bred true.

Terms and Tools to Follow Segregating Genes

The law of segregation reflects the actions of chromosomes and the genes they carry during meiosis. Because a gene is a long sequence of DNA, it can vary in many ways. An individual with two identical alleles for a gene is **homozygous** for that gene. An individual with two different alleles is **heterozygous**—what Mendel called "non-true-breeding" or "hybrid."

When considering a gene with two alleles, the dominant one is shown as a capital letter and the recessive with the corresponding small letter. If both alleles are recessive, the individual is homozygous recessive, shown with two small letters. An individual with two dominant alleles is homozygous dominant, and has two capital letters. One dominant and one recessive allele, such as *Tt* for non-true-breeding tall pea plants, indicates a heterozygote.

An organism's appearance does not always reveal its alleles. Both a *TT* and a *Tt* pea plant are tall, but *TT* is a

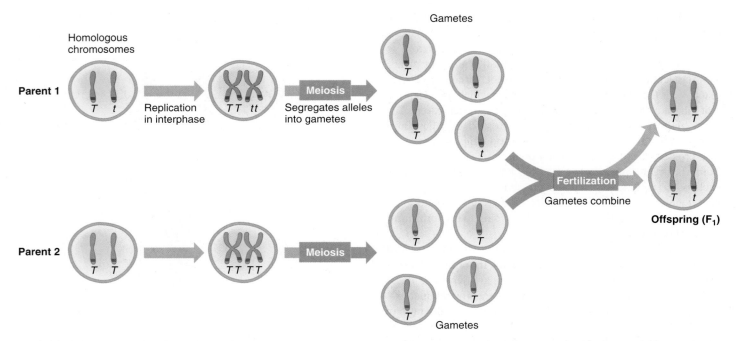

Figure 4.3 **Mendel's first law—gene segregation.** During meiosis, homologous pairs of chromosomes and their genes separate and are packaged into separate gametes. At fertilization, gametes combine at random. Green and blue denote different parental origins of the chromosomes. This cross yields offspring of genotypes *TT* and *Tt*.

homozygote and *Tt* a heterozygote. The **genotype** describes the organism's alleles, and the **phenotype** describes the outward expression of an allele combination. A **wild type** phenotype is the most common expression of a particular allele combination in a population. The wild type allele may be recessive or dominant. A **mutant** phenotype is a variant of a gene's expression that arises when the gene undergoes a change, or **mutation**.

Mendel was observing the events of meiosis. When a gamete is produced, the two copies of a gene separate with the homologs that carry them. In a plant of genotype *Tt*, for example, gametes carrying either *T* or *t* form in equal numbers during anaphase I. Gametes combine at random. A *t*-bearing oocyte is neither more nor less attractive to a sperm than is a *T*-bearing oocyte. These two factors—equal allele distribution into gametes and random combinations of gametes—underlie Mendel's law of segregation (**figure 4.3**).

When Mendel crossed short plants (*tt*) with true-breeding tall plants (*TT*), the seeds grew into F₁ plants that were all tall (genotype *Tt*). Next, he self-crossed the F₁ plants. The progeny were *TT, tt,* and *Tt*. A *TT* individual resulted when a *T* sperm fertilized a *T* oocyte; a *tt* plant resulted when a *t* oocyte met a *t* sperm; and a *Tt* individual resulted when either a *t* sperm fertilized a *T* oocyte, or a *T* sperm fertilized a *t* oocyte.

Because two of the four possible gamete combinations produce a heterozygote, and each of the others produces a homozygote, the genotypic ratio expected of a monohybrid cross is 1*TT*: 2*Tt*: 1*tt*. The corresponding phenotypic ratio is three tall plants to one short plant, a 3:1 ratio. Mendel saw these

results for all seven traits that he studied (**table 4.1**). A diagram called a **Punnett square** shows these ratios (**figure 4.4**). A Punnett square represents how particular genes in gametes join, assuming they are on different chromosomes. Experimental crosses yielded numbers of offspring that approximate these ratios.

Mendel distinguished the two genotypes resulting in tall progeny—*TT* from *Tt*—with more crosses (**figure 4.5**). He bred

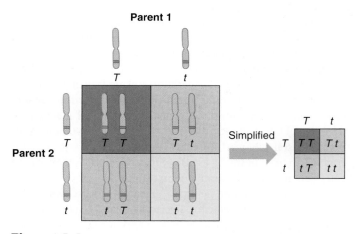

Figure 4.4 **A Punnett square.** A Punnett square illustrates how alleles combine in offspring. The different types of gametes of one parent are listed along the top of the square, with those of the other parent listed on the left-hand side. Each compartment displays the genotype that results when gametes that correspond to that compartment join.

Table 4.1

Table 4.1   Mendel's Law of Segregation

Experiment	Total	Dominant	Recessive	F$_2$ Phenotypic Ratios
1. Seed form	7,324	5,474	1,850	2.96:1
2. Seed color	8,023	6,022	2,001	3.01:1
3. Seed coat color	929	705	224	3.15:1
4. Pod form	1,181	882	299	2.95:1
5. Pod color	580	428	152	2.82:1
6. Flower position	858	651	207	3.14:1
7. Stem length	1,064	787	277	2.84:1
				Average = 2.98:1

tall plants of unknown genotype with short (*tt*) plants. If a tall plant crossed with a *tt* plant produced both tall and short progeny, it was genotype *Tt;* if it produced only tall plants, it must be *TT*.

Crossing an individual of unknown genotype with a homozygous recessive individual is called a test cross. The logic is that the homozygous recessive is the only genotype that can be identified by its phenotype—that is, a short plant is always *tt*. The homozygous recessive is a "known" that can reveal the unknown genotype of another individual to which it is crossed.

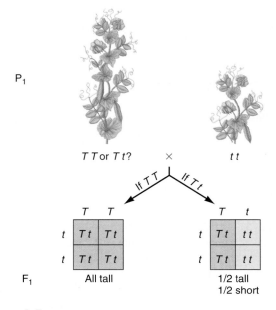

Figure 4.5   **Test cross.** Breeding a tall pea plant with homozygous recessive short plants reveals whether the tall plant is true-breeding (*TT*) or non-true-breeding (*Tt*). Punnett squares usually indicate only the alleles.

Key Concepts

1. Mendel deduced that "elementen" for height segregate, then combine at random with those from the opposite gamete at fertilization.

2. A homozygote has two identical alleles, and a heterozygote has two different alleles. The allele expressed in a heterozygote is dominant; the allele not expressed is recessive.

3. A monohybrid cross yields a genotypic ratio of 1:2:1 and a phenotypic ratio of 3:1.

4. Punnett squares display expected genotypic and phenotypic ratios among progeny.

5. A test cross uses a homozygous recessive individual to reveal an unknown genotype.

4.2 Single-Gene Inheritance Is Rare

Mendel's first law addresses traits determined by single genes. **Reading 4.1** describes a few unusual single-gene traits. Inheritance of single genes is also called Mendelian, or monofactorial, inheritance. Single-gene disorders, such as sickle cell disease and muscular dystrophy, are rare compared to infectious diseases, cancer, and multifactorial disorders, affecting 1 in 10,000 or fewer individuals. Because of the rarity of single-gene diseases, getting an accurate diagnosis can be difficult if physicians are unfamiliar with the phenotype.

Single-gene inheritance is much more complicated than it might appear from considering such obvious traits as green or yellow pea color. Sequencing the human genome and using SNPs (points in the genome where people vary) to catalog inherited variation in genome-wide association studies have revealed that the phenotypes associated with single genes are influenced by other genes as well as by environmental factors.

It's All in the Genes

Do you have uncombable hair, misshapen toes or teeth, or a pigmented tongue tip? Are you unable to smell a squashed skunk, or do you sneeze repeatedly in bright sunlight? Do you lack teeth, eyebrows, eyelashes, nasal bones, thumbnails, or fingerprints? If so, your unusual trait may be one of thousands described in the online database Mendelian Inheritance in Man.

Genes control whether hair is blond, brown, or black, has red highlights, and is straight, curly, or kinky. Widow's peaks, cowlicks, a whorl in the eyebrow, and white forelocks run in families; so do hairs with triangular cross-sections. Some people have multicolored hairs, like cats; others have hair in odd places, such as on the elbows, nose tip, knuckles, palms, or soles. Teeth can be missing or extra, protuberant or fused, present at birth, shovel-shaped, or "snowcapped." A person can have a grooved tongue, duckbill lips, flared ears, egg-shaped pupils, three rows of eyelashes, spotted nails, or "broad thumbs and great toes." Extra breasts are known in humans and guinea pigs, and one family's claim to genetic fame is a double nail on the littlest toe.

Unusual genetic variants can affect metabolism, producing either disease or harmless, yet noticeable, effects. Members of some families experience "urinary excretion of odoriferous component of asparagus" or "urinary excretion of beet pigment," producing a strange odor or dark pink urine after consuming the offending vegetable. In blue diaper syndrome, an infant's urine turns blue on contact with air, thanks to an inherited inability to break down the amino acid tryptophan.

One bizarre inherited illness is the Jumping Frenchmen of Maine syndrome (MIM 244100). This exaggerated startle reflex was first noted among French-Canadian lumberjacks from the Moosehead Lake area of Maine, whose ancestors were from the Beauce region of Quebec. Physicians first reported the condition at a medical conference in 1878. Geneticists videotaped the startle response in 1980, and the condition continues to appear in genetics journals. MIM offers a most vivid description:

> If given a short, sudden, quick command, the affected person would respond with the appropriate action, often echoing the words of command. . . . For example, if one of them was abruptly asked to strike another, he would do so without hesitation, even if it was his mother and he had an ax in his hand.

The Jumping Frenchmen of Maine syndrome may be an extreme variant of the more common Tourette syndrome, which causes tics and other uncontrollable movements. **Figure 1** illustrates some other genetic variants.

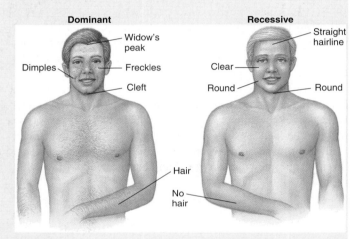

Figure 1 Inheritance of some common traits. Freckles, dimples, hairy arms, widow's peak, and a cleft chin are examples of dominant traits.

That is, the single gene controls trait transmission, but other genes and the environment affect the degree of the trait or severity of the illness. Eye color is a good example of the new view of single-gene traits.

Eye Color

Most people have brown eyes; blue and green eyes are almost exclusively in people of European ancestry. The color of the iris is due to melanin pigments, which come in two forms—the dark brown/black eumelanin, and the red-yellow pheomelanin. In the eye, cells called melanocytes produce melanin, which is stored in structures called melanosomes in the outermost layer of the iris. People differ in the amount of melanin and number of melanosomes, but have about the same number of melanocytes in their eyes.

Nuances of eye color—light versus dark brown, clear blue versus greenish or hazel—arise from the distinctive peaks and valleys at the back of the iris. Thicker regions darken appearance of the pigments, rendering brown eyes nearly black in some parts and blue eyes closer to purple. The bluest eyes have thin irises with very little pigment. The effect of the iris surface on color is a little like the visual effect of a rough-textured canvas on paint.

A single gene on chromosome 15, *OCA2* (MIM 611409), confers eye color by controlling melanin synthesis. If this gene is missing, albinism results, causing very pale skin and red eyes (see figure 4.15). A recessive allele of this gene confers blue color and a dominant allele confers brown. But inheritance of eye color is more complicated than this. Near the *OCA2* gene on chromosome 15 is a second gene, *HERC2*, that controls expression of the *OCA2* gene. A certain SNP in *HERC2* abolishes the control over *OCA2*, and blue eyes result. A person must inherit two copies of this SNP to have blue eyes.

If blue eye color is the disruption of a "normal" function, why has it persisted? A clue comes from evolution. The

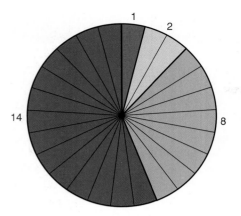

Figure 4.6 **Eye color in humans.** This pie chart shows the distribution of eye colors among the students in a classroom.

HERC2 gene is found in many species, indicating that it is ancient and important, because it has persisted. Researchers hypothesize that mutations in *HERC2* arose long ago among hunter-gatherers in Europe, and the unusual individuals with the pale eyes were, for whatever reason, more desirable as sexual partners. Over time, this sexual selection would have increased the proportion of the population with blue eyes. **Figure 4.6** shows the distribution of eye color in an average classroom in the United States.

Modes of Inheritance

Modes of inheritance are rules that explain the common patterns of single-gene transmission. They are derived from Mendel's laws. Knowing mode of inheritance makes it possible to calculate the probability that a particular couple will have a child who inherits a particular condition. The way that a trait is passed depends on whether the gene that determines it is on an autosome or on a sex chromosome, and whether the particular allele is recessive or dominant.

In autosomal dominant inheritance, a trait can appear in either sex because an autosome carries the gene. If a child has the trait, at least one parent also has it. That is, autosomal dominant traits do not skip generations. If no offspring inherit the trait in one generation, its transmission stops because the offspring can pass on only the recessive form of the gene.

Huntington disease is an autosomal dominant condition. Henry, from the chapter opener, has an affected parent and grandparent, but he has not inherited the mutation. Because his sister decided not to have children after learning she had inherited the mutation, disease transmission in their branch of the family tree stopped. The Punnett square in **figure 4.7** depicts inheritance of an autosomal dominant trait or condition, and **table 4.2** summarizes the criteria. Many autosomal dominant diseases do not cause symptoms until adulthood. The *Bioethics: Choices for the Future* box on page 76 looks at the dilemma that adult-onset inherited diseases presents for reproductive choices.

An autosomal recessive trait can appear in either sex. Affected individuals have a homozygous recessive genotype,

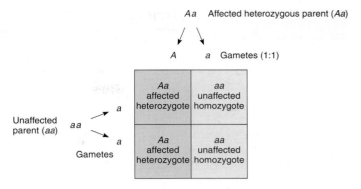

Figure 4.7 **Autosomal dominant inheritance.** When one parent has an autosomal dominant condition and the other does not, each offspring has a 50 percent probability of inheriting the mutant allele and the condition. The affected parent is *Aa* here, and not *AA*, because for many dominant disorders, the homozygous dominant (*AA*) phenotype is either lethal or very rare because both parents of the person with the *AA* genotype would have to have the disorder.

whereas in heterozygotes (carriers) the wild type allele masks expression of the mutant allele. Ellen, the little girl with CF described in the chapter opener, inherited a mutant allele from each of her carrier parents. They were unaffected because they each also had a dominant allele that encodes enough functional protein for health. **Figure 4.8** depicts the autosomal recessive inheritance pattern of a harmless trait, curly hair.

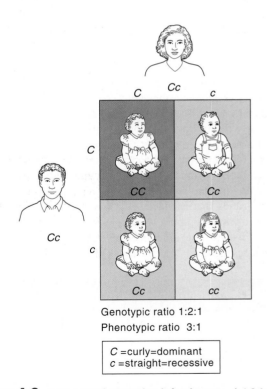

Genotypic ratio 1:2:1
Phenotypic ratio 3:1

C =curly=dominant
c =straight=recessive

Figure 4.8 **Autosomal recessive inheritance.** A 1:2:1 genotypic ratio results from a monohybrid cross, whether in peas or people. Curly hair (*C*) is dominant to straight hair (*c*). This pedigree depicts a monohybrid cross for curly hair.

When Diagnosing a Fetus Also Diagnoses a Parent: Huntington Disease (HD)

When Peter and Martha were 24 and expecting their first child, they learned that Peter's mother, who was adopted, had early signs of HD. Her clumsiness and slurred speech would eventually progress to a near-constant writhing and repetitive, dance-like movements, until she would die, probably of infection, in 15 to 20 years.

A genetic counselor told the couple that because HD is autosomal dominant, Peter had a 50 percent (1 in 2) chance of having inherited the condition. He was considered "at-risk" until (and if) symptoms began. However, he could take a "predictive" genetic test that would reveal whether he had inherited HD. If he had, he would be considered "pre-manifest" before signs and symptoms begin. The mutation extends the wild type gene with repeats of a three-base DNA sequence. The more repeats, the sooner symptoms begin. Peter didn't want that information, but Martha felt differently. She did not want to have a child who would have HD. Could the fetus be checked for the mutation? Then, if the disease had been inherited, the couple could end the pregnancy. But if the fetus had the mutation, then so did Peter, and he didn't want to know. He also questioned ending a pregnancy, considering that symptoms do not begin until later in life, by which time treatment might exist.

The couple researched how others had handled the dilemmas that predictive testing for an untreatable, adult-onset illness raise. They read that only about 10 percent of people offered the test for HD take it, and those that do can handle the results well. After much soul-searching, Martha chose to respect Peter's wishes. He and the fetus were not tested. So far, both are healthy.

Postscript: Although Peter still does not wish to know whether he will get HD, he is participating in a large-scale study. He and 1,000 other at-risk individuals have given a blood sample to researchers. The participants are being followed medically for 10 years to identify the very earliest signs and symptoms of the disease—which might suggest treatments, perhaps even existing drugs.

Questions for Discussion

1. Some geneticists advise that people under 18 not be tested for the HD mutation, unless they are very mature and well-informed. Do you agree with this practice?

2. Choose an autosomal dominant condition from MIM, familyvillage.org, genetests.org, or Wikipedia, and argue for or against testing children for the mutation.

3. Discuss issues of privacy that may arise in genetic counseling for an autosomal dominant condition.

Mendel's first law can be used to calculate the probability that an individual will have either of two phenotypes. The probabilities of each possible genotype are added. For example, the chance that a child whose parents are both carriers of cystic fibrosis will *not* have the condition is the sum of the probability that she has inherited two normal alleles (1/4) plus the chance that she herself is a heterozygote (1/2), or 3/4. Note that this also equals 1 minus the probability that she is homozygous recessive and has the condition.

The ratios that Mendel's first law predicts for autosomal recessive inheritance apply to each offspring anew. If a couple has a child with an autosomal recessive illness, each of their next children faces the same 25 percent risk of inheriting the condition.

Table 4.2	Criteria for an Autosomal Dominant Trait

1. Males and females can be affected. Male-to-male transmission can occur.

2. Males and females transmit the trait with equal frequency.

3. Successive generations are affected.

4. Transmission stops after a generation in which no one is affected.

Most autosomal recessive conditions appear unexpectedly in families. **Reading 4.2** discusses one of them, cystic fibrosis. However, a situation in which an autosomal recessive condition is more likely to recur is when blood relatives have children together. The higher risk of having a child with a particular autosomal recessive condition is because the related parents may carry the same alleles inherited from an ancestor that they have in common, such as a grandparent. Marriage between relatives introduces **consanguinity,** which means "shared blood"—a figurative description, since genes are not passed in blood. Alleles inherited from shared ancestors are said to be "identical by descent."

Consanguinity is part of many cultures. For example, marriage between first cousins is common in some Islam communities. In Qatar, half of all marriages, until very recently, were between first cousins. Today the figure is about 22 percent.

Logic explains why consanguinity raises risk of inheriting autosomal recessive diseases. An unrelated man and woman have eight different grandparents, but first cousins have only six, because they share one pair through their parents, who are siblings (see figure 4.14c). That is, the probability of two relatives inheriting the same disease-causing recessive allele is greater than that of two unrelated people having the same allele by chance. However, genome-wide studies show that cousins

Cystic Fibrosis, Then and Now

Alex Deford was born in 1972, daughter of sportswriter Frank Deford. Alex did not gain much weight as an infant, and as a young child suffered severe lung infections and digestive difficulties before dying of CF at age 8. Wrote her father about his feelings at the time of diagnosis in his book, *Alex, The Life of a Child*:

> I went to the encyclopedia and read about this cystic fibrosis. To me, at that point, it was one of those vague diseases you hear about now and then . . . One out of every 20 whites carries the defective gene, as I do, as Carol does, as perhaps 10 million other Americans do.

In the 1970s, it was common for a child with CF not to survive to see adolescence. Today, thanks to treatments and therapies to minimize symptoms, people with CF are living much longer. Within a decade, more than half of all people with the disease will be over the age of 18, and some are living normal life expectancies. Identification of more than 1,600 mutations has revealed milder guises of the condition, such as chronic sinus infections and/or bronchitis.

Diagnosis

CF is caused by faulty chloride ion channels in cells lining passageways, such as in the lungs and pancreas. The result is thickened secretions that impair breathing and digestion of fats. Diagnosis of CF is based on any one of the following phenotypes:

- Sinus and lung disease—The bronchial tubes may be dilated, inflamed, and collapsed, and nasal polyps and infected sinuses are common. Pneumonia is caused by specific pathogens.
- Pancreatic insufficiency causes large, fatty bowel movements, malnutrition from poor fat absorption, and intestinal blockage.
- Males lack the ductus deferens, blocking sperm.

The course of CF varies greatly, but in general people lose from 1 to 4 percent of lung function a year. It may be episodic, with periods of fatigue, shortness of breath, a productive cough, weight loss, and poor lung function.

People with CF are highly susceptible to pneumonia caused by a few types of bacteria. Children are infected with *Staphylococcus aureus* and *Hemophilus influenzae*, and adults with *Pseudomonas aeruginosa* and *Burkholderia cepacia*. Particularly dangerous is *B. cepacia*, which not only infects rapidly, but easily spreads among patients, such as children at summer camp. However, these bacterial infections may seem predominant only because we can grow the microbes in the lab. When researchers sequenced bacterial DNA in lung fluid from children with CF, they discovered genes representing more than 60 species of bacteria! It's clear that we have more to learn about the role of infection in CF.

Treatment

People with cystic fibrosis are living longer thanks to many ways to prevent and treat symptoms (**figure 1**). Airway clearance exercises or wearing a vibrating vest helps to shake free sticky mucus (**figure 2**). This is done up to four times a day. Inhaling a saline solution and use of a DNA-digesting enzyme that destroys white blood cells at infection sites also thins the mucus. Intravenous antibiotics combat infections and anti-inflammatory drugs lessen pain. Enzymes mixed into soft foods compensate for poor fat digestion due to a blocked pancreas. Newer drugs treat the condition at the source, either correcting a misfolded CFTR protein, or enabling CFTR stalled inside the cell to reach the plasma membrane, where it functions as an ion channel (see Reading 2.2).

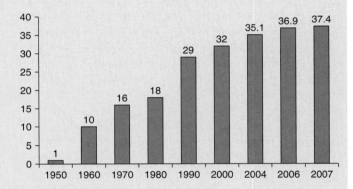

Figure 1 **Living with cystic fibrosis.** Many people who have CF are surviving to adulthood.

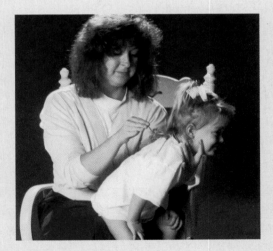

Figure 2 **Treating cystic fibrosis.** In cystic fibrosis, the thick, sticky mucus that clogs airways must be coughed up at least twice every day. Bronchial drainage treatments—tapping hard on the chest and wearing a vibrating vest—help to shake free the mucus.

often inherit different parts of the shared ancestor's genome, which explains why these populations are healthier than might be expected given the high frequency of consanguinity. In addition, the Qatari people descend from three groups: Bedouin tribes, Persians (Iran), and Africans who speak the Bantu language. As a result, modern Qatari genomes are diverse.

The nature of the phenotype is important when evaluating the transmission of single-gene traits. Some diseases are too severe for people to live long enough or feel well enough to have children. For example, each adult sibling of a person who is a known carrier of Tay-Sachs disease has a two-thirds chance of being a carrier. The probability is two-thirds, and not one-half, because only two genotypes are possible for an adult—homozygous for the wild type allele or a carrier who inherits the mutant allele from either mother or father. A homozygous recessive individual would not have survived childhood, due to brain degeneration.

Geneticists who study human traits and illnesses can hardly set up crosses as Mendel did, but they can pool information from families whose members have the same trait or illness. Consider a simplified example of 50 couples in whom both partners are carriers of sickle cell disease. If 100 children are born, about 25 of them would be expected to have sickle cell disease. Of the remaining 75, theoretically 50 would be carriers like their parents, and the remaining 25 would have two wild type alleles. **Table 4.3** lists criteria for an autosomal recessive trait.

Solving a Problem for Single-Gene Inheritance

Using Mendel's laws to predict phenotypes and genotypes requires a careful reading of the problem to identify and organize relevant information. Sometimes common sense is useful, too. The following general steps can help to solve a problem that addresses Mendel's first law, which describes the inheritance of a single-gene trait:

1. List all possible genotypes and phenotypes for the trait.
2. Determine the genotypes of the individuals in the first (P_1) generation. Use information about those individuals' parents.
3. After deducing genotypes, derive the possible alleles in gametes each individual produces.
4. Unite these gametes in all combinations to reveal all possible genotypes. Calculate ratios for the F_1 generation.

Table 4.3	Criteria for an Autosomal Recessive Trait

1. Males and females can be affected.

2. Affected males and females can transmit the gene, unless it causes death before reproductive age.

3. The trait can skip generations.

4. Parents of an affected individual are heterozygous or have the trait.

5. To extend predictions to the F_2 generation, use the genotypes of the specified F_1 individuals and repeat steps 3 and 4.

As an example, consider curly hair, depicted in figure 4.8. If C is the dominant allele, conferring curliness, and c is the recessive allele, then both CC and Cc genotypes result in curly hair. A person with a cc genotype has the straight hair phenotype.

Wendy has beautiful curls, and her husband Rick has straight hair. Wendy's father is bald, but once had curly hair, and her mother has stick-straight hair. What is the probability that Wendy and Rick's child will have straight hair? Steps 1 through 5 solve the problem:

1. State possible genotypes: CC, Cc = curly; cc = straight.
2. Determine genotypes: Rick must be cc, because his hair is straight. Wendy must be Cc, because her mother has straight hair and therefore gave her a c allele.
3. Determine gametes: Rick's sperm carry only c. Half of Wendy's oocytes carry C, and half carry c.
4. Unite the gametes:

Wendy

		C	c
Rick	c	Cc	cc
	c	Cc	cc

5. Conclusion: Each child of Wendy and Rick's has a 50 percent chance of having curly hair (Cc) and a 50 percent chance of having straight hair (cc).

On the Meaning of Dominance and Recessiveness

Knowing whether an allele is dominant or recessive is critical in determining the risk of inheriting a particular condition (phenotype). Dominance and recessiveness arise from the genotype, and reflect the characteristics or abundance of a protein.

Mendel based his definitions of dominance and recessiveness on what he could see—one allele masking the other. Today we can often add a cellular or molecular explanation. Consider inborn errors of metabolism caused by absent enzymes. These disorders are typically recessive because the half normal amount of the enzyme that a carrier produces is usually sufficient to maintain health. The one normal allele, therefore, compensates for the mutant one, to which it is dominant. The situation is similar in pea plants. Short stem length results from deficiency of an enzyme that activates a growth hormone, but the Tt plants produce enough hormone to attain the same height as TT plants.

A recessive trait is said to arise from a "loss of function" because the recessive allele usually prevents the production or activity of the normal protein. In contrast, some dominantly inherited disorders are said to be due to a "gain of function," because they result from the action of an abnormal protein that interferes with the function of the normal protein. Huntington disease is a "gain of function" disorder. In HD, the dominant mutant allele encodes an abnormally long protein that prevents the normal protein from functioning in certain brain cells. Huntington disease is a gain of function because individuals who are missing one copy of the gene do not have the illness. That is, the protein encoded by the mutant HD allele must be abnormal, not absent, to cause the disease.

Recessive disorders tend to be more severe, and produce symptoms earlier than dominant disorders. Disease-causing recessive alleles remain in populations because healthy heterozygotes pass them to future generations. In contrast, if a dominant mutation arises that harms health early in life, people who have the allele are either too ill or do not survive long enough to reproduce. The allele eventually becomes rare in the population unless it arises anew by mutation. Dominant disorders whose symptoms do not appear until adulthood, or that do not severely disrupt health, remain in a population because they do not prevent a person from having children.

Under certain circumstances, for some genes, a heterozygous individual (a carrier) can develop symptoms. This is the case for sickle cell disease. Carriers can develop a life-threatening breakdown of muscle if exposed to the combination of environmental heat, intense physical activity, and dehydration. Several college athletes died from these symptoms, prompting sports authorities to begin testing athletes for sickle cell disease carrier status.

Key Concepts

1. Single genes can determine trait transmission patterns (modes of inheritance), but usually other genes and/or the environment modify phenotypes.

2. Modes of inheritance reveal whether a single-gene trait is dominant or recessive and whether the gene that controls it is carried on an autosome or a sex chromosome.

3. Autosomal dominant traits do not skip generations and can affect both sexes; autosomal recessive traits can skip generations and can affect both sexes.

4. Rare autosomal recessive disorders sometimes recur in families when parents are related.

5. Mendel's first law, which can predict the probability that a child will inherit a single-gene trait, applies anew to each child.

6. Genetic problems are solved with logic and by applying Mendel's laws to follow gametes.

7. Dominance is the ability of a protein encoded by one allele to compensate for a missing or abnormal protein encoded by another allele.

4.3 Following the Inheritance of More Than One Gene

The law of segregation follows the inheritance of two alleles for a single gene. In a second set of experiments, Mendel examined the inheritance of two traits at a time.

Mendel's Second Law

The second law, the **law of independent assortment,** states that for two genes on different chromosomes, the inheritance of one does not influence the chance of inheriting the other. The two genes thus "independently assort" because they are packaged into gametes at random (**figure 4.9**). Two genes that are far apart on the same chromosome also appear to independently assort, because so many crossovers occur between them that it is as if they are carried on separate chromosomes (see figure 3.5.)

Mendel looked at seed shape, which was either round or wrinkled (determined by the R gene), and seed color, which was either yellow or green (determined by the Y gene). When he crossed true-breeding plants that had round, yellow seeds to true-breeding plants that had wrinkled, green seeds, all the progeny had round, yellow seeds. These offspring were double heterozygotes, or dihybrids, of genotype $RrYy$. From their appearance, Mendel deduced that round is dominant to wrinkled, and yellow to green.

Next, he self-crossed the dihybrid plants in a **dihybrid cross,** so named because two genes and traits are followed. Mendel found four types of seeds in the next, third generation: 315 plants with round, yellow seeds; 108 plants with round, green seeds; 101 plants with wrinkled, yellow seeds; and 32 plants with wrinkled, green seeds. These classes occurred in a ratio of 9:3:3:1.

Mendel then crossed each plant from the third generation and to plants with wrinkled, green seeds (genotype $rryy$). These test crosses established whether each plant in the third generation was true-breeding for both genes (genotypes $RRYY$ or $rryy$), true-breeding for one gene but heterozygous for the other (genotypes $RRYy, RrYY, rrYy,$ or $Rryy$), or heterozygous for both genes (genotype $RrYy$). Mendel could explain the 9:3:3:1 proportion of progeny classes only if one gene does not influence transmission of the other. Each parent would produce equal numbers of four different types of gametes: $RY, Ry, rY,$ and ry. Each of these combinations has one gene for each trait. A Punnett square for this cross shows that the four types of seeds:

1. round, yellow ($RRYY, RrYY, RRYy,$ and $RrYy$),
2. round, green ($RRyy$ and $Rryy$),
3. wrinkled, yellow ($rrYY$ and $rrYy$), and
4. wrinkled, green ($rryy$)

are present in the ratio 9:3:3:1, just as Mendel found (**figure 4.10**).

Solving a Problem in Following More Than One Segregating Gene

A Punnett square for three genes has 64 boxes; for four genes, 256 boxes. An easier way to predict genotypes and phenotypes in multigene crosses is to use the mathematical laws of probability on which Punnett squares are based. Probability predicts the likelihood of an event.

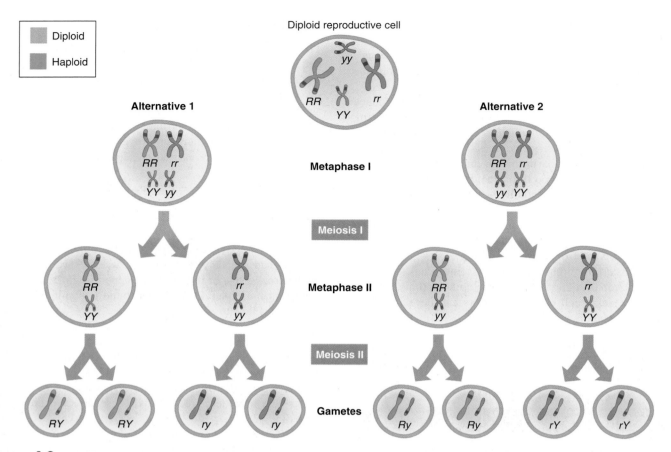

Figure 4.9 **Mendel's second law—independent assortment.** The independent assortment of genes carried on different chromosomes results from the random alignment of chromosome pairs during metaphase of meiosis I. An individual of genotype *RrYy,* for example, manufactures four types of gametes, containing the dominant alleles of both genes (*RY*), the recessive alleles of both genes (*ry*), and a dominant allele of one with a recessive allele of the other (*Ry* or *rY*). The allele combination depends upon which chromosomes are packaged together in a gamete—and this happens at random.

An application of probability theory called the product rule can predict the chance that parents with known genotypes can produce offspring of a particular genotype. The product rule states that the chance that two independent events will both occur equals the product of the chance that either event will occur alone. Consider the probability of obtaining a plant with wrinkled, green peas (genotype *rryy*) from dihybrid (*RrYy*) parents. Do the reasoning for one gene at a time, then multiply the results (**figure 4.11**).

A Punnett square depicting a cross of two *Rr* plants indicates that the probability of producing *rr* progeny is 25 percent, or 1/4. Similarly, the chance of two *Yy* plants producing a *yy* plant is 1/4. Therefore, the chance of dihybrid parents (*RrYy*) producing homozygous recessive (*rryy*) offspring is 1/4 multiplied by 1/4, or 1/16. Now consult the 16-box Punnett square for Mendel's dihybrid cross again (see figure 4.10). Only one of the 16 boxes is *rryy,* just as the product rule predicts. **Figure 4.12** shows how these tools can be used to predict offspring genotypes and phenotypes for three human traits simultaneously.

Mendel followed only a few traits compared to genetics investigations today. It is common now to screen for hundreds or thousands of alleles or expressed genes at once. Although computer analysis of many gene combinations has largely replaced Punnett squares, the diagrams remain useful for easily following traits in a family.

Key Concepts

1. Mendel's law of independent assortment considers genes transmitted on different chromosomes.
2. In a dihybrid cross of heterozygotes for seed color and shape, the phenotypic ratio of 9:3:3:1 revealed that transmission of one gene does not influence that of another.
3. Meiotic events explain independent assortment.
4. Punnett squares and probability can be used to follow independent assortment.
5. Today, computers analyze many genes at a time.

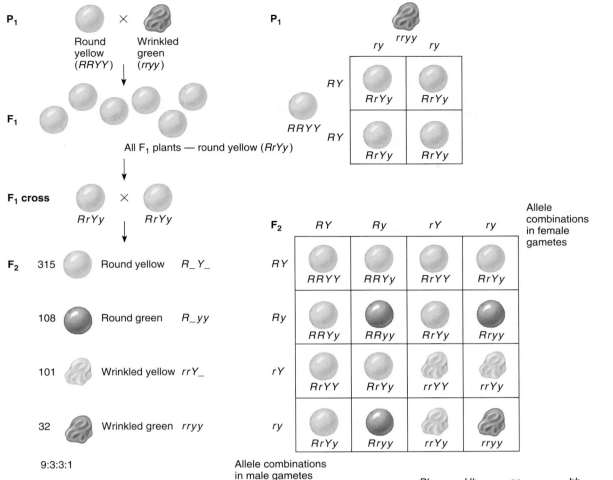

Figure 4.10 Plotting a dihybrid cross. A Punnett square can represent the random combinations of gametes produced by dihybrid individuals. An underline in a genotype (in the F₂ generation) indicates that either a dominant or a recessive allele is possible. The numbers in the F₂ generation are Mendel's experimental data.

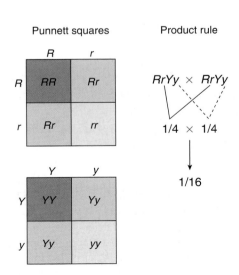

Figure 4.11 The product rule.

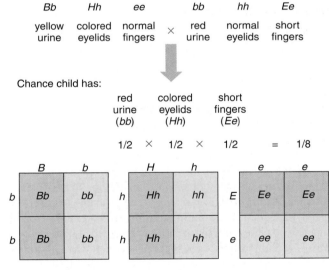

Bb yellow (normal) urine
bb beeturia (red urine after eating beets)
Hh colored eyelids
hh normal eyelids
Ee brachydactyly (short fingers)
ee normal fingers

Figure 4.12 Using probability to track three traits. A man with normal urine, colored eyelids, and normal fingers wants to have children with a woman who has red urine after she eats beets, normal eyelids, and short fingers. The chance that a child of theirs will have red urine after eating beets, colored eyelids, and short fingers is 1/8.

4.4 Pedigree Analysis

For genetics researchers, families are tools, and the bigger the family the better—the more children in a generation, the easier it is to see modes of inheritance. Charts called **pedigrees** display family relationships and depict which relatives have specific phenotypes and, sometimes, genotypes. A human pedigree serves the same purpose as one for purebred dogs or cats or thoroughbred horses—it represents relationships. A pedigree in genetics differs from a family tree in genealogy, and from a genogram in social work, in that it indicates disorders or traits as well as relationships and ancestry. Pedigrees may also include molecular data, test results, haplotypes (genes or SNPs linked in segments on a chromosome), and even genome-wide association study information.

A pedigree consists of lines that connect shapes. Vertical lines represent generations; horizontal lines that connect two shapes at their centers depict partners; shapes connected by vertical lines that are joined horizontally represent siblings. Squares indicate males; circles, females; and diamonds, individuals of unspecified sex. Roman numerals designate generations. Arabic numerals or names indicate individuals. **Figure 4.13** shows these and other commonly used pedigree symbols. Colored or shaded shapes indicate individuals who express a trait, and half-filled shapes are known carriers. A genetic counselor may sketch a pedigree while interviewing a client, then use a computer program and add test results that indicate genotypes.

Pedigrees Then and Now

The earliest pedigrees were strictly genealogical, not indicating traits. **Figure 4.14a** shows such a pedigree for a highly inbred part of the ancient Egyptian royal family. The term *pedigree* arose in the fifteenth century, from the French *pie de grue*, which means "crane's foot." Pedigrees at that time, typically depicting large families, showed parents linked by curved lines to their many offspring. The overall diagram often resembled a bird's foot.

One of the first pedigrees to trace an inherited illness was an extensive family tree of several European royal families, indicating which members had the clotting disorder hemophilia (see figure 6.6). The mutant gene probably originated in Queen Victoria of England in the nineteenth century. In 1845, a genealogist named Pliny Earle constructed a pedigree of a family with colorblindness using musical notation—half notes for unaffected females, quarter notes for colorblind females, and filled-in and squared-off notes to represent the many colorblind males. In the early twentieth century, eugenicists tried to use pedigrees to show that traits such as criminality, feeblemindedness, and promiscuity were the consequence of faulty genes. Figure 15.18 shows one of these pedigrees.

Today, pedigrees are important both for helping families identify the risk of transmitting an inherited illness and as starting points for identifying a gene from the human genome sequence. People who have kept meticulous family records, such as the Mormons and the Amish, are invaluable in helping

Symbols

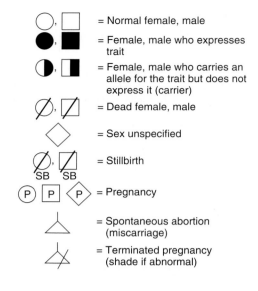

Lines

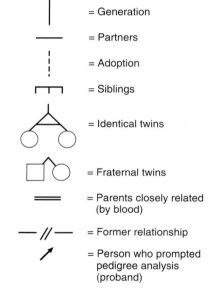

Numbers

Roman numerals = generations

Arabic numerals = individuals in a generation

Figure 4.13 Pedigree components. Symbols representing individuals are connected to form pedigree charts, which display the inheritance patterns of particular traits.

researchers follow the inheritance of particular genes. Very large pedigrees can provide information on many individuals with a particular rare disorder. The researchers can then search affected individuals' DNA to identify a particular sequence they have all inherited that is not found in healthy family members. This is where the causative mutation lies. Discovery of the gene that causes HD, for example, took researchers to a remote village in

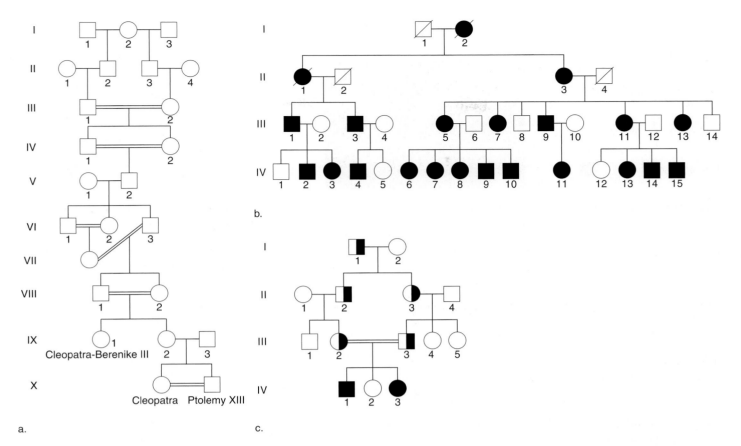

Figure 4.14 **Unusual pedigrees.** **(a)** A partial pedigree of Egypt's Ptolemy dynasty shows only genealogy, not traits. It appears almost ladderlike because of the extensive consanguinity. From 323 B.C. to Cleopatra's death in 30 B.C., the family experienced one pairing between cousins related through half-brothers (generation III), four brother-sister pairings (generations IV, VI, VIII, and X), and an uncle-niece relationship (generations VI and VII). Cleopatra married her brother, Ptolemy XIII, when he was 10 years old! These marriage patterns were an attempt to preserve the royal blood. **(b)** A family with polydactyly (extra fingers and toes) extends laterally, with many children because the phenotype doesn't affect reproduction. **(c)** This pedigree shows marriage of first cousins. They share one set of grandparents, and therefore risk passing on the same recessive alleles to offspring.

Venezuela to study an enormous family. The gene was eventually traced to a Portuguese sailor who introduced the mutation in the nineteenth century.

Pedigrees Display Mendel's Laws

Visual learners can easily "see" a mode of inheritance in a pedigree. Consider a pedigree for an autosomal recessive trait, albinism. Homozygous recessive individuals in the third (F_2) generation lack an enzyme necessary to manufacture the pigment melanin and, as a result, hair and skin are very pale (**figure 4.15**). Their parents are inferred to be heterozygotes (carriers). One partner from each pair of grandparents must also be a carrier. For some disorders, carriers can be identified by detecting half the wild type amount of a key biochemical in a body fluid, such as blood or urine.

An autosomal dominant trait does not skip generations and can affect both sexes. A typical pedigree for an autosomal dominant trait has some squares and circles filled in to indicate affected individuals in each generation. Figure 4.14b

illustrates an autosomal dominant trait, extra fingers and toes (polydactyly).

A pedigree may be inconclusive, which means that either autosomal recessive or autosomal dominant inheritance can explain the pattern of filled-in symbols. **Figure 4.16** shows one such pedigree, for a type of hair loss called alopecia areata (MIM 104000). According to the pedigree, this trait can be passed in an autosomal dominant mode because it affects both males and females and is present in every generation. However, the pedigree can also depict autosomal recessive inheritance if the individuals represented by unfilled symbols are carriers. Inconclusive pedigrees tend to arise when families are small and the trait is not severe enough to impair fertility.

Pedigrees may be difficult to construct or interpret for several reasons. People may hesitate to supply information because they are embarrassed by the symptoms. Family relationships can be complicated by adoption, children born out of wedlock, serial relationships, blended families, and assisted reproductive technologies (see chapter 21). Many people cannot trace their families back far enough to reveal a mode of inheritance.

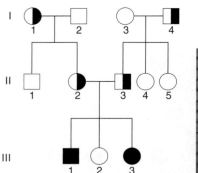

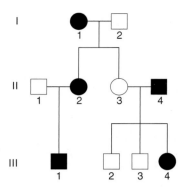

Figure 4.15 **Albinism is autosomal recessive.** Albinism affects males and females and can skip generations, as it does in generations I and II in this pedigree. Homozygous recessive individuals lack an enzyme needed to produce melanin, which colors the eyes, skin, and hair.

Figure 4.16 **An inconclusive pedigree.** This pedigree could account for an autosomal dominant trait or an autosomal recessive trait that does not prevent affected individuals from having children. (Unfilled symbols could represent carriers.)

Solving a Problem in Conditional Probability

Pedigrees and Punnett squares can be useful in tracing a conditional probability, which is when an offspring's genotype depends on the parents' genotypes, which may not be obvious from their phenotypes. For example, if a person has an autosomal recessive trait or condition, his or her parents are inferred to be carriers (unless a new mutation arises in the affected individual). Consider the family represented in **figure 4.17**. Deshawn has sickle cell disease. His unaffected parents, Kizzy and Ike, must each be heterozygotes (carriers). Deshawn's sister, Taneesha, is also healthy, and she is expecting her first child. Taneesha's husband, Antoine, has no family history of sickle cell disease. Taneesha wants to know the probability that her child will inherit the mutant allele from her and be a carrier.

Taneesha raises two questions. First, what is the probability that she is a carrier? Because she is the product of a monohybrid cross, and she is not homozygous recessive (sick), she has a 2 in 3 chance of being a carrier. If so, the chance that

she will transmit the mutant allele is 1 in 2, because she has two copies of the gene, and only one allele goes into each gamete. To calculate the overall risk to her child, multiply the probability that she is a carrier (2/3) by the chance that if she is, she will transmit the mutant allele (1/2). The result is 1/3.

4.5 Family Genome Analysis

Gregor Mendel inferred events that occur inside cells by looking at the appearance of his experimental subjects, peas. He followed one, two, or three traits at a time. Today, that type of analysis is at an entirely different level—but Mendelian inheritance prevails.

Family genome analysis compares the complete genome sequences of the members of a family. Researchers at the University of Washington in Seattle pioneered this tool using a four-person family in which the mother and father are healthy, but the two children each have two autosomal recessive diseases

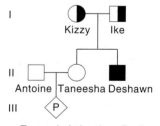

a. Taneesha's brother, Deshawn, has sickle cell disease.

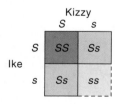

b. Probability that Taneesha is a carrier: $^2/_3$

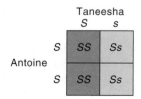

c. If Taneesha is a carrier, chance that fetus is a carrier: $^1/_2$

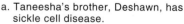

Total probability = $^2/_3 \times ^1/_2 = ^1/_3$

Figure 4.17 **Making predictions.** Taneesha's brother Deshawn has sickle cell disease **(a).** Taneesha wonders if her fetus has inherited the sickle cell allele. First, she must calculate the chance that she is a carrier. The Punnett square in **(b)** shows that this risk is 2 in 3. (She must be genotype SS or Ss, but cannot be ss because she does not have the disease.) The risk that the fetus is a carrier, assuming that the father is not a carrier, is half Taneesha's risk of being a carrier, or 1 in 3 **(c).**

whose causative genes were unknown. The idea was that the genes would be located in parts of the genome that are the same in the siblings—areas where they are essentially twins.

According to the steps of meiosis, siblings share about half of their genes. However, they actually share more than this because the parents may have identical variants for some or even many genes, especially if they are related. We see this sometimes in startling family resemblances (**figure 4.18**). The researchers were indeed able to narrow down the locations of the two disease-causing genes by identifying gene candidates in the "twin" sites in the siblings' genomes. As the cost of genome sequencing falls, consulting the parents' genomes is helping to reveal the genes behind very rare diseases. Genome sequencing, of individuals as well as of families and even populations, is going to both validate Gregor Mendel's brilliant observations and conclusions, and provide valuable personal information.

Figure 4.18 Family genome analysis. Now that the cost of genome sequencing has come down, researchers can identify the mutation behind a rare disease in a family by looking for parts of the genome that affected siblings share. The physical resemblance between this sister and brother is strong, but they share about 50 percent of their inherited health characteristics too.

Key Concepts

1. Pedigrees depict family relationships and the transmission of inherited traits. Squares represent males, and circles, females; horizontal lines link partners, vertical lines show generations, and elevated horizontal lines depict siblings. Heterozygote symbols are half-shaded, and symbols for individuals who express a trait are completely shaded.

2. Pedigrees can reveal modes of inheritance. They and Punnett squares apply Mendel's laws to predict the recurrence risks of inherited disorders or traits.

3. Family genome analysis compares genome sequences to trace genetic disease origins. It also validates and illustrates Mendel's laws.

Summary

4.1 Following the Inheritance of One Gene

1. Mendel's laws, based on pea plant crosses, derive from the actions of chromosomes during meiosis. They apply to all diploid organisms.

2. Mendel used statistics to investigate why some traits vanish in hybrids. The **law of segregation** states that alleles of a gene are distributed into separate gametes during meiosis. Mendel demonstrated this in seven traits in pea plants.

3. A diploid individual with two identical alleles of a gene is **homozygous.** A **heterozygote** has two different alleles of a gene. A gene may have many alleles.

4. A **dominant** allele masks the expression of a **recessive** allele. An individual may be homozygous dominant, homozygous recessive, or heterozygous.

5. When Mendel crossed two true-breeding types, then bred the resulting hybrids to each other, the two variants of the trait appeared in a 3:1 phenotypic ratio. Crossing these progeny further revealed a genotypic ratio of 1:2:1.

6. A **Punnett square** follows the transmission of alleles and is based on probability.

4.2 Single-Gene Inheritance Is Rare

7. Eye color illustrates how a single-gene trait can be affected by other genes.

8. In **autosomal dominant** inheritance, males and females may be affected, and the trait does not skip generations.

9. Inheritance of an **autosomal recessive** trait may affect either males or females and may skip generations. Autosomal recessive conditions are more likely to occur in families with **consanguinity.** Recessive disorders tend to be more severe and cause symptoms earlier than dominant disorders.

10. Genetic problems can be solved by tracing alleles as gametes form and then combine in a new individual.

11. Dominance and recessiveness reflect how alleles affect the abundance or activity of the gene's protein product.

4.3 Following the Inheritance of More Than One Gene

12. Mendel's second law, the **law of independent assortment,** follows the transmission of two or more genes on different chromosomes. A random assortment of maternally and paternally derived chromosomes during meiosis yields gametes with different gene combinations.

13. The chance that two independent genetic events will both occur is equal to the product of the probabilities that each event will occur. This product rule is useful in calculating the risk that individuals will inherit a particular genotype and in following the inheritance of genes on different chromosomes.

4.4 Pedigree Analysis

14. A **pedigree** is a chart that depicts family relationships and patterns of inheritance for particular traits. A pedigree can be inconclusive.

4.5 Family Genome Analysis

15. Comparing genome sequences of family members extends Mendelian analysis.

www.mhhe.com/lewisgenetics10

Answers to all end-of-chapter questions can be found at **www.mhhe.com/lewisgenetics10.** You will also find additional practice quizzes, animations, videos, and vocabulary flashcards to help you master the material in this chapter.

Review Questions

1. State two ways that Huntington disease (HD) and cystic fibrosis (CF) affect families differently.

2. How does meiosis explain Mendel's laws?

3. Discuss how Mendel derived the two laws of inheritance without knowing about chromosomes.

4. Distinguish between
   a. autosomal recessive and autosomal dominant inheritance.
   b. Mendel's first and second laws.
   c. a homozygote and a heterozygote.
   d. a monohybrid and a dihybrid cross.
   e. a Punnett square and a pedigree.

5. Why would Mendel's results for the dihybrid cross have been different if the genes for the traits he followed were near each other on the same chromosome?

6. Why are extremely rare autosomal recessive disorders more likely to appear in families in which blood relatives have children together?

7. How does the pedigree of the ancient Egyptian royal family in figure 4.14a differ from a pedigree a genetic counselor might use today?

8. What are the genotypes of the parents of a person who has two HD mutations?

9. What is the probability that two individuals with an autosomal recessive trait, such as albinism, will have a child with the same genotype and phenotype as they have?

Applied Questions

1. Predict the phenotypic and genotypic ratios for crossing the following pea plants:
   a. short × short
   b. short × true-breeding tall
   c. true-breeding tall × true-breeding tall

2. What are the genotypes of the pea plants that would have to be bred to yield one plant with restricted pods for every three plants with inflated pods?

3. If pea plants with all white seed coats are crossed, what are the possible phenotypes of their progeny?

4. Pea plants with restricted yellow pods are crossed to plants that are true-breeding for inflated green pods and the

F_1 crossed. Derive the phenotypic and genotypic ratios for the F_2 generation.

5. More than 100 genes cause deafness when mutant. What is the most likely mode of inheritance in families in which all children are born deaf?

6. The MacDonalds raise Labrador retrievers. In one litter, two of eight puppies, a male and a female, have a condition called exercise-induced collapse. After about 15 minutes of intense exercise, the dogs wobble about, develop a fever, and their hind legs collapse. The parents are healthy. What is the mode of inheritance?

7. Draw a pedigree to depict the following family: One couple has a son and a daughter with normal skin pigmentation.

Another couple has one son and two daughters with normal skin pigmentation. The daughter from the first couple has three children with the son of the second couple. Their son and one daughter have albinism; their other daughter has normal skin pigmentation.

8. Chands syndrome (MIM 214350) is autosomal recessive and causes very curly hair, underdeveloped nails, and abnormally shaped eyelids. In the following pedigree, which individuals must be carriers?

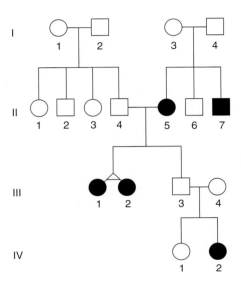

Chands syndrome

9. Lorenzo has a double row of eyelashes (MIM 126300), which he inherited from his mother as a dominant trait. His maternal grandfather is the only other relative to have it. Fatima, who has normal eyelashes, marries Lorenzo. Their first child, Nicola, has normal eyelashes. Now Fatima is pregnant again and hopes for a child with double eyelashes. What chance does the child have of inheriting double eyelashes? Draw a pedigree of this family.

10. Peeling skin syndrome (MIM 270300) causes the outer skin layer to fall off on the upper surfaces of the hands and feet. The pedigrees depict three families with this condition. What do they share that might explain the appearance of this otherwise rare condition?

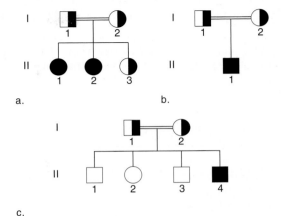

Peeling skin syndrome

11. Sclerosteosis (MIM 269500) (see below) causes overgrowth of the skull and jaws that produces a characteristic face, gigantism, facial paralysis, and hearing loss. The overgrowth of skull bones can cause severe headaches and even sudden death. In the pedigree shown below for a family with sclerosteosis:

    a. What is the relationship between the individuals who are connected by slanted double lines?

    b. Which individuals in the pedigree must be carriers?

12. The child referenced in figure 4.12 who has red urine after eating beets, colored eyelids, and short fingers, is of genotype *bbHhEe*. The genes for these traits are on different chromosomes. If he has children with a woman who is a trihybrid for each of these genes, what are the expected genotypic and phenotypic ratios for their offspring?

13. Sketch a pedigree of your family or of a pet's or favorite animal's family.

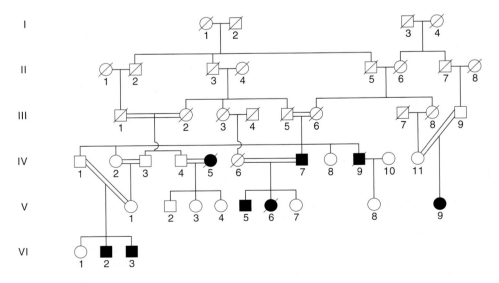

Sclerosteosis

Web Activities

1. Go to the website for the National Organization for Rare Disorders. Identify an autosomal recessive disorder and an autosomal dominant disorder. Create a family for each one, and describe transmission of the disease over three generations.

2. Go to the website for Gene Gateway—Exploring Genes and Genetic Disorders. Select two disorders or traits that would demonstrate independent assortment if present in the same family, and two that would not.

3. Construct your family's health history using websites and talking to your relatives. The U.S. government provides a tool at https://familyhistory.hhs.gov/fhh-web/ home.action. Background information is at http://www.hhs. gov/familyhistory/ and tips can be found at the websites for the Genetic Alliance, the Mayo Clinic, and the University of Utah.

Forensics Focus

1. A woman, desperate to complete her family tree for an upcoming family reunion, cornered a stranger in a fast-food restaurant. Her genealogical research had identified him as a distant cousin, and she needed his DNA. He refused to cooperate, looked scared, and ran off. The woman took his discarded coffee cup and collected DNA from traces of saliva, which she sent on a swab to a DNA ancestry testing company. (This is a true story.)

   a. Do you think the woman was justified in her action? Why or why not?

   b. What are the strengths and limitations of using genealogical information (family records, word-of-mouth) versus DNA testing to construct a pedigree?

2. A young woman walking to her car in a parking lot late at night was attacked and brutally raped. She kept her wits about her and was able to recall striking details of her rapist's attire—a green silk shirt, a belt with a metal buckle that left a bruise on her abdomen, jeans, and running shoes. She also remembered that he had white skin, a shaved head, and startling blue eyes. She recalled footsteps approaching and then someone yanking the man off her, but her head was let go and hit the pavement, knocking her unconscious.

   A few weeks after the rape, a naked male body washed up in a nearby river, and police found a belt buckle farther downstream that matched the pattern of the woman's bruise. The body, however, was headless and the neck ragged, as if cut. How can police determine the eye color of the corpse to help in identifying or ruling him out as the rapist?

Case Studies and Research Results

1. On the daytime drama "The Young and the Restless," several individuals suffer from SORAS, which stands for "soap opera rapid aging syndrome." It is not listed in MIM. In SORAS, a young child is sent off to boarding school and returns three months later an angry teenager. In the Newman family, siblings Nicholas and Victoria SORASed from childhood to teenhood in a few months. Half-sister Abby does not have the condition. Her mother is Ashley. Victoria and Nick's parents, Victor and Nikki, are curiously not affected; in fact, they never seem to age at all. What is the mode of inheritance of SORAS, and how do you know this?

2. More than a dozen recessive illnesses that are very rare in most of the world are fairly common among the Bedouin people who live in the Negev Desert area of Israel. More than 65 percent of Bedouins marry their first or second cousins. This practice helped the group to survive a nomadic existence in the harsh environment in the past. Recently, two physicians and a geneticist set up a service that enables people wishing to marry to take genetic tests to learn if they are carriers for the same diseases. Prenatal testing has also been introduced to provide the option of terminating pregnancies that would otherwise lead to the births of children who would die of a recessive disorder in early childhood. Discuss the pros and cons of introducing genetic testing in this community. Should medical science interfere with a society's long-held cultural practices?

3. The Cleaver family awaited the birth of Claudette's puppies with great anticipation. But shortly after the standard poodle gave birth to six pups, two of them, a male and a female, developed seizures and died. One of the surviving pups, Sylvester, was bred when he was a year old to Minuette, who was an offspring of Sylvester's father, Otis. Alas, one of Minuette's four pups also died of the canine seizure disorder.

   a. Draw a pedigree for the poodle family.

   b. What is the mode of inheritance of the disorder?

   c. What advice would you give to the Cleavers about successfully breeding their poodles in the future?

4. A man named Grady Stiles performed in circuses as "Lobster Boy," a condition called ectrodactyly, or split-hand/split-foot syndrome (MIM 225300). Lobster Boy had only two digits on each hand and foot, but was otherwise healthy, living long enough to see the trait appear in two of his four children. He added them to his traveling act. A granddaughter has the condition too. Grady, who died in 1992, traced the family trait back six generations to William Stiles, born in 1805. Ectrodactyly may be inherited as an autosomal dominant or an autosomal recessive trait. Which one is the more likely explanation for the trait in this family? Cite a reason for your answer.

Beyond Mendel's Laws

Learning Outcomes

5.1 A New View of Mendelian Genetics

1. Explain how single-gene inheritance is not simple.

5.2 When Gene Expression Appears to Alter Mendelian Ratios

2. Discuss phenomena that can appear to alter expected Mendelian ratios.

5.3 Mitochondrial Genes

3. Describe the mode of inheritance of a mitochondrial trait.

4. Explain how mitochondrial DNA differs from nuclear DNA.

5.4 Linkage

5. Explain how linked traits are inherited differently from Mendelian traits.

6. Discuss the basis of linkage in meiosis.

7. Explain how linkage is the basis of genetic maps and genome-wide association studies.

The Big Picture: Patterns of inheritance can be obscured when genes have many variants, interact with each other or the environment, are in mitochondria, or are linked on the same chromosome.

A Gene Search to Explain a Child's Blindness

"Genetic heterogeneity" refers to mutations in different genes that cause the same symptoms. This technical concept came to life on a sunny summer Saturday in Philadelphia, where families with a form of visual loss called Leber congenital amaurosis (LCA) gathered to hear researchers talk about gene therapy to cure one type of the condition. It had already worked in one youngster (see figure 20.10). At least eighteen genes cause LCA.

Jennifer Pletcher was at the conference. Her daughter Finley has a mutation in a gene called *RDH12*. Jennifer had met a few LCA families on Facebook, but they had mutations in different genes. At the conference, one of the speakers asked attendees to stand when their mutation was called. When she said "*RDH12*," "we stood up, and soon three families were around us. It was awesome! Now we are friends and can keep in contact about changes in our kids," said Jennifer.

Not all were able to stand and be counted among those with known mutations. Jennifer and Troy Stevens were there to learn which mutation their 2-year-old son Gavin had. The parents were to meet with the head of the lab testing their DNA, but Gavin's mutation was not found. The young parents were discouraged, because the first step in gene therapy is knowing which gene is mutant. Usually a negative result on a genetic test is good news, but for Gavin's parents, it means their quest must go on. They hope that family genome sequencing (see section 4.5) will identify Gavin's mutation.

5.1 A New View of Mendelian Genetics

Gregor Mendel's analysis of inheritance, in which "either-or" Punnett squares based on simple probabilities represent proportions of peas that are green or yellow, wrinkled or smooth, was brilliant for its time, and correct. That single-gene view is becoming greatly altered through today's genome-wide lens. Unlike Mendel's tallying of what he could see in breeding experiments, the sequencing of the human genome and tracking of how our DNA sequences vary have provided peeks at the tangled complexity of inheritance. Even single-gene inheritance is not simple.

Single genes rarely completely control a phenotype in the way that Mendel's experiments suggested. Genes interact with each other, and with environmental influences, in many ways that we are just beginning to understand. However, when transmission patterns of a visible trait do not exactly fit autosomal recessive or autosomal dominant inheritance, Mendel's laws are still operating. The underlying genotypic ratios persist, but other factors affect the phenotypes. Chapters 7 and 11 explore some of these "outside-the-gene" influences, which include:

- other protein-encoding genes;
- other DNA sequences, such as SNPs in non-protein-encoding regions of the genome;
- RNA sequences that turn groups of genes on and off;
- epigenetic alterations of DNA, such as chemical groups that either activate or shield genes; and
- environmental stimuli.

This chapter considers three general phenomena that seem to be exceptions to Mendel's laws, but are really not: gene expression, mitochondrial inheritance, and linkage.

5.2 When Gene Expression Appears to Alter Mendelian Ratios

In several circumstances, phenotypic ratios appear to contradict Mendel's laws, but do not.

Lethal Allele Combinations

A genotype (allele combination) that causes death is, by strict definition, lethal. Death from genetic disease can occur at any stage of development or life. Tay-Sachs disease is lethal by age 3 or 4, whereas Huntington disease may not be lethal until late middle age. In a population and evolutionary sense, a lethal genotype has a more specific meaning—it causes death before the individual can reproduce, which prevents passage of genes to the next generation.

In organisms used in experiments, such as fruit flies, pea plants, or mice, lethal allele combinations remove an expected progeny class following a specific cross. For example, in a cross of heterozygous flies, homozygous recessive progeny die as embryos, leaving only heterozygous and homozygous dominant adult fly offspring. In humans, early-acting lethal alleles cause spontaneous abortion. When both a man and a woman carry a recessive lethal allele for the same gene, each pregnancy has a 25 percent chance of spontaneously aborting—this is the homozygous recessive class.

A double dose of a dominant allele may be lethal, as is the case for Mexican hairless dogs (**figure 5.1**). Inheriting one dominant allele confers the coveted hairlessness trait, but inheriting two dominant alleles is lethal to the unlucky embryo. Breeders cross hairless to hairy ("powderpuff") dogs, rather than hairless to hairless, to avoid losing the lethal homozygous dominant class—a quarter of the pups. The dogs also have missing or unusually shaped teeth. A mutation in a gene called *forkhead box I3 (FOXI3)* causes the condition, which is called canine ectodermal dysplasia. Humans also have the *FOXI3* gene (MIM 612351), but so far it has not been associated with a phenotype.

An example of a lethal genotype in humans is achondroplastic dwarfism (MIM 100800), which has a very distinct phenotype of a long trunk, very short limbs, and a large head with a flat face. It is inherited as an autosomal dominant trait, but is most often the result of a spontaneous (new) mutation. Each child of two people with achondroplasia has a one in four chance of inheriting both mutant alleles. However, because such homozygotes are not seen, this genotype is presumed to be lethal. Observations of homozygotes for achondroplasia mutations in other species suggest that the homozygotes would be unable to breathe because the lungs do not have room to inflate. The mutation is in the gene that encodes a receptor for a growth factor. Without the receptor, growth is severely stunted.

Multiple Alleles

An individual has two alleles for any autosomal gene, one on each homolog. However, a gene can exist in more than two allelic forms in a population because it can mutate in many ways. That is, the sequence of hundreds of DNA bases that makes up a gene can be altered in many ways, just as mistakes can occur anywhere in a written sentence. Different allele combinations can produce variations in the phenotype. The more alleles, the more variations of the phenotype are possible. An individual with two different mutant alleles for the same gene is called a *compound heterozygote*.

It is useful when knowing the genotype enables a physician to predict the course of an illness. However, this is often difficult because other genes and environmental effects can modify a phenotype. Two disorders for which allele identification *can* predict severity and types of symptoms are phenylketonuria (PKU) and cystic fibrosis.

In PKU, too little or lack of an enzyme causes the amino acid phenylalanine to build up in brain cells. Hundreds of different mutant alleles pair, in compound heterozygotes or as homozygotes, causing four basic phenotypes:

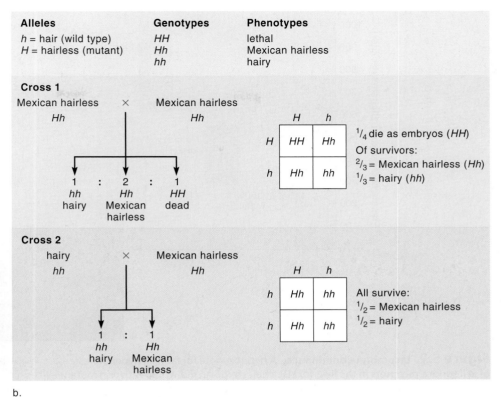

Alleles
h = hair (wild type)
H = hairless (mutant)

Genotypes
HH
Hh
hh

Phenotypes
lethal
Mexican hairless
hairy

Cross 1

Mexican hairless   ×   Mexican hairless
Hh   Hh

	H	h
H	HH	Hh
h	Hh	hh

¹/₄ die as embryos (HH)
Of survivors:
²/₃ = Mexican hairless (Hh)
¹/₃ = hairy (hh)

1 : 2 : 1
hh   Hh   HH
hairy Mexican dead
       hairless

Cross 2

hairy   ×   Mexican hairless
hh   Hh

	H	h
h	Hh	hh
h	Hh	hh

All survive:
¹/₂ = Mexican hairless
¹/₂ = hairy

1 : 1
hh   Hh
hairy Mexican
       hairless

Figure 5.1 Lethal alleles. (a) This Mexican hairless dog has inherited a dominant allele that makes it hairless. Inheriting two such dominant alleles is lethal to embryos. **(b)** Breeders cross Mexican hairless dogs to hairy ("powderpuff") dogs to avoid dead embryos and stillbirths that represent the *HH* genotypic class.

a.

b.

- classic PKU with profound intellectual disability;
- moderate PKU;
- mild PKU; and
- asymptomatic PKU, with excretion of excess phenylalanine in urine.

Eating a special diet extremely low in phenylalanine allows normal brain development. Knowing the allele combination can guide how strict and prolonged the diet should be. Most people stay on it for many years.

Multiple alleles are considered in carrier testing for CF, which is done routinely in early pregnancy (see Reading 4.2). When the *CF* gene was discovered in 1989, researchers identified one mutant allele, Δ*F508,* that causes about 70 percent of cases in many populations when homozygous. The genotype is called a "double delta." As the allele list grew, researchers discovered that not all allele combinations cause the same symptoms. People homozygous for Δ*F508* have frequent, severe respiratory infections, very congested lungs, and poor weight gain. Another genotype increases susceptibility to bronchitis and pneumonia, and another causes only absence of the ductus deferens, causing male infertility. Genetic tests probe panels of *CF* mutations that are the most common in a patient's ethnic group, maximizing the likelihood of detecting carriers and avoiding the cost of testing for 1,500-plus alleles. If a pregnant woman has a disease-causing allele, then the father-to-be is tested, and if he has a mutant allele too, then the fetus may be tested to see if it has inherited the disease.

Different Dominance Relationships

In complete dominance, one allele is expressed, while the other isn't. In **incomplete dominance,** the heterozygous phenotype is intermediate between that of either homozygote.

Enzyme deficiencies in which a threshold level is necessary for health illustrate both complete and incomplete dominance. For example, on a whole-body level, Tay-Sachs disease displays complete dominance because the heterozygote (carrier) is as healthy as a homozygous dominant individual. However, if phenotype is based on enzyme level, then the heterozygote is intermediate between the homozygous dominant (full enzyme level) and homozygous recessive (no enzyme). Half the normal amount of enzyme is sufficient for health, which is why at the whole-person level, the wild type allele is completely dominant.

For many genes, researchers can measure the expression levels associated with various genotypes, demonstrating that even heterozygotes whose phenotypes are the same as those of homozygotes are distinctive at the biochemical level. Often, they produce half the normal amount of a protein, but this is sufficient for health.

Familial hypercholesterolemia (FH) is an example of incomplete dominance in humans that can be observed on both the molecular and whole-body levels. A person with two disease-causing alleles lacks receptors on liver cells that take up the low-density lipoprotein (LDL) form of cholesterol from the bloodstream. A person with one disease-causing allele

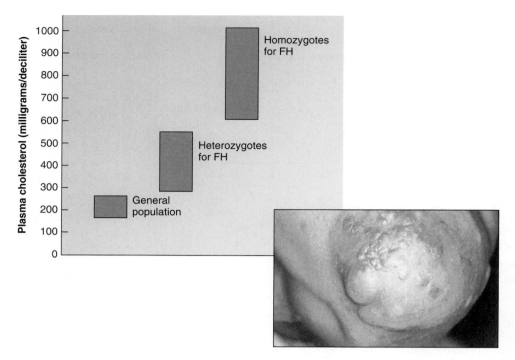

Figure 5.2 Incomplete dominance. A heterozygote for familial hypercholesterolemia (FH) has approximately half the normal number of cell surface receptors in the liver for LDL cholesterol. An individual with two mutant alleles has the severe form of FH, with liver cells that totally lack the receptors. As a result, serum cholesterol level is very high. The photograph shows cholesterol deposits on the elbow of an affected young man. Cholesterol is also deposited in joints and many other body parts.

has half the normal number of receptors. Someone with two wild type (the most common) alleles has the normal number of receptors. **Figure 5.2** shows how measurement of plasma cholesterol reflects these three genotypes. The phenotypes parallel the number of receptors—those with two mutant alleles die in childhood of heart attacks, those with one mutant allele may suffer heart attacks in young adulthood, and those with two wild type alleles do not develop this inherited form of heart disease.

Different alleles that are both expressed in a heterozygote are **codominant**. The ABO blood group is based on the expression of codominant alleles.

Blood types are determined by the patterns of molecules on the surfaces of red blood cells. Most of these molecules are proteins embedded in the plasma membrane with attached sugars that extend from the cell surface. The sugar is the antigen, which is the molecule that the immune system recognizes. People who belong to blood group A have an allele that encodes an enzyme that adds a final piece to a certain sugar to produce antigen A. In people with blood type B, the allele and its encoded enzyme are slightly different, which causes a different piece to attach to the sugar, producing antigen B. People in blood group AB have both antigen types. Blood group O reflects yet a third allele of this gene. It is missing just one DNA nucleotide, but this changes the encoded enzyme in a way that removes the sugar chain from its final piece (**figure 5.3**). Type O red blood cells lack both A and B antigens.

The A and B alleles are codominant, and both are completely dominant to O. Considering the genotypes reveals how

these interactions occur. In the past, ABO blood types have been described as variants of a gene called "I," although MIM now abbreviates the designations. The older I system is easier to understand. ("I" stands for isoagglutinin.) The three alleles are I^A, I^B, and i. People with blood type A have antigen A on the surfaces of their red blood cells, and may be of genotype $I^A I^A$

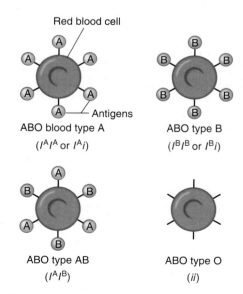

Figure 5.3 ABO blood types illustrate codominance. ABO blood types are based on antigens on red blood cell surfaces. This depiction greatly exaggerates the size of the A and B antigens. Genotypes are in parentheses.

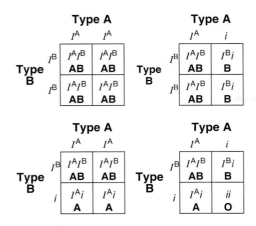

Figure 5.4 Codominance. The I^A and I^B alleles of the I gene are codominant, but they follow Mendel's law of segregation. These Punnett squares follow the genotypes that could result when a person with type A blood has children with a person with type B blood.

or $I^A i$. People with blood type B have antigen B on their red blood cell surfaces, and may be of genotype $I^B I^B$ or $I^B i$. People with the rare blood type AB have both antigens A and B on their cell surfaces, and are genotype $I^A I^B$. People with blood type O have neither antigen, and are genotype ii.

Fiction plots often misuse ABO blood type terminology, assuming that a child's ABO type must match that of one parent. This is not true, because a person with type A or B blood can be heterozygous. A person who is genotype $I^A i$ and a person who is $I^B i$ can jointly produce offspring of any ABO genotype or phenotype, as **figure 5.4** illustrates.

Epistasis

Mendel's laws can appear not to operate when one gene masks or otherwise affects the phenotype of another. This phenomenon is called **epistasis**. It refers to interaction between different genes, not between the alleles of the same gene. A gene that affects expression of another is called a modifier gene.

In epistasis, the blocked gene is expressed normally, but the product of the modifier gene inactivates it, removes a structure needed for it to contribute to the phenotype, or otherwise counteracts its effects. An obvious example of epistasis is the hairless gene in dogs. Genes that color dog hairs have no effect if there are no hairs. An epistatic interaction seen in many species is albinism, in which one gene blocks the action of genes whose products confer color.

A more complex example of epistasis is a blood type called the Bombay phenotype. It results from an interaction between a gene called *H* and the *I* gene that confers ABO blood type. The *H* gene controls the placement of a molecule to which antigens A and B attach on red blood cell surfaces. In a person of genotype *hh*, that molecule isn't made, so the A and B antigens have no way to attach to the red blood cell surface. The A and B antigens fall off—and the person tests as type O blood. However, any ABO genotype is possible.

Epistasis can explain why siblings who inherit the same disorder can suffer to differing degrees. One study examined siblings who both inherited spinal muscular atrophy 1 (MIM 253300), in which nerves cannot signal muscles. The muscles weaken and atrophy, usually proving fatal in early childhood. The mutation encodes an abnormal protein that shortens axons, which are the extensions on nerve cells that send messages. Some siblings who inherited the SMA mutation never developed symptoms. They can thank another gene, *plastin 3* (MIM 300131), which increases production of the cytoskeletal protein actin that extends axons. Because these children inherited the ability to make extra long axons, the axon-shortening effects of SMA were not harmful.

Penetrance and Expressivity

The same allele combination can produce different degrees of a phenotype in different individuals because a gene does not act alone. Every gene that is expressed functions against a backdrop of the expression of other genes, as well as effects of environmental influences such as nutrition, exposure to toxins, and stress. This is why two individuals who have the same *CF* genotype may have different clinical experiences. One person may be much sicker because she also inherited genes predisposing her to develop asthma and respiratory allergies. Even identical twins with the same genetic disease may be affected to different degrees due to environmental influences.

Many single-gene traits and illnesses have distinctive phenotypes, despite all of these influences. The terms *penetrance* and *expressivity* describe degrees of expression of a single gene. **Penetrance** refers to the all-or-none expression of a genotype; **expressivity** refers to severity or extent.

An allele combination that produces a phenotype in everyone who inherits it is completely penetrant. This is very rare. Huntington disease (see *Bioethics: Choices for the Future* in chapter 4) is nearly completely penetrant. Almost all people who inherit the mutant allele will develop symptoms if they live long enough.

A genotype is incompletely penetrant if some individuals do not express the phenotype (have no symptoms). Polydactyly (see figure 1.5) is incompletely penetrant. Some people who inherit the dominant allele have more than five digits on a hand or foot. Yet others who we know have the allele because they have an affected parent and child have ten fingers and ten toes. Penetrance is described numerically. If 80 of 100 people who inherit the dominant polydactyly allele have extra digits, the genotype is 80 percent penetrant.

A phenotype is variably expressive if symptoms vary in intensity among different people. One person with polydactyly might have an extra digit on both hands and a foot, but another might have just one extra fingertip. Polydactyly is both incompletely penetrant and variably expressive.

Pleiotropy

A single-gene disorder with many symptoms, or a gene that controls several functions or has more than one effect, is termed **pleiotropic**. Such conditions can be difficult to trace through families because people with different subsets of symptoms may

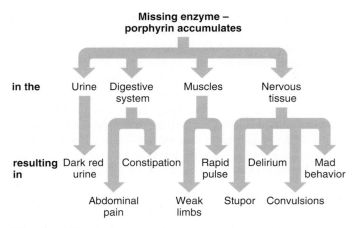

Figure 5.5 Pleiotropy. King George III suffered from the autosomal dominant disorder porphyria variegata, and so did several other family members. Because of pleiotropy, the family's varied illnesses and quirks appeared to be different, unrelated disorders. Symptoms appear every few years in a particular order.

appear to have different disorders. A classic example of pleiotropy is porphyria variegata, an autosomal dominant disease that affected several members of the royal families of Europe (**figure 5.5**).

King George III ruled England during the American Revolution. At age 50, he first experienced abdominal pain and constipation, followed by weak limbs, fever, a fast pulse, hoarseness, and dark red urine. Next, nervous system signs and symptoms began, including insomnia, headaches, visual problems, restlessness, delirium, convulsions, and stupor. His confused and racing thoughts, combined with actions such as ripping off his wig and running about naked while at the peak of a fever, convinced court observers that the king was mad. Just as Parliament was debating his ability to rule, he recovered.

But the king's ordeal was far from over. He relapsed 13 years later, then again 3 years after that. Always the symptoms appeared in the same order, beginning with abdominal pain, fever, and weakness, and progressing to nervous system symptoms. Finally, an attack in 1811 placed George in a prolonged stupor, and the Prince of Wales dethroned him. George III lived for several more years, experiencing further episodes.

In George III's time, doctors were permitted to do very little to the royal body, and their diagnoses were based on what the king told them. Twentieth-century researchers found that porphyria variegata caused George's red urine. It is one of several types of porphyrias, which result from deficiency of any of several enzymes required to manufacture heme. The king's disorder arises from lack of enzyme #7 in the heme pathway shown in **figure 5.6**. Heme is part of hemoglobin, the molecule that carries oxygen in the blood and imparts the red color. In the disease, a part of heme called a porphyrin ring is routed into the urine instead of being broken down and metabolized in cells. Porphyrin also builds up and attacks the nervous system.

Physicians' records of George's royal relatives reported the disorder as several different illnesses. Today, porphyria variegata remains rare, and is often misdiagnosed as a seizure disorder. Unfortunately, some seizure medications and anesthetics

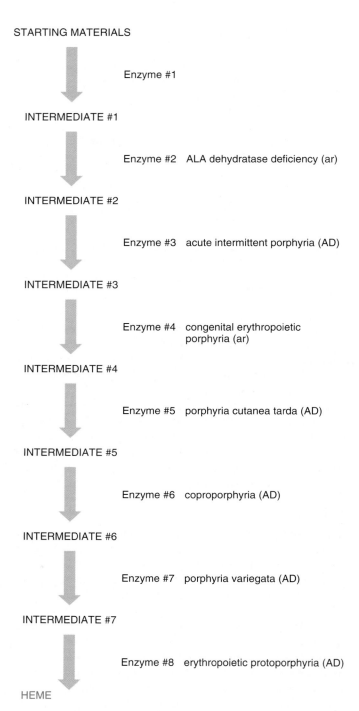

Figure 5.6 The porphyrias. Errors in the heme biosynthetic pathway cause seven related, yet distinct, diseases. In each disorder, the intermediate biochemical that a deficient enzyme would normally affect builds up. The excess is excreted in the urine or accumulates in blood, feces, or inside red blood cells, causing symptoms. People with various porphyria-related symptoms may have inspired the vampire and werewolf legends, including reddish teeth, pink urine, excess hair, and photosensitivity (avoidance of daylight).

worsen symptoms. Ironically, treatment may have worsened King George III's disease. Medical records and hair analysis indicate that a medicine based on the element antimony was forced upon the king in the madhouse. Antimony was often contaminated with arsenic, and arsenic inactivates several of the enzymes in the heme biosynthetic pathway.

On a molecular level, pleiotropy occurs when a single protein affects different body parts, participates in more than one biochemical reaction, or has different effects in different amounts. Consider Marfan syndrome. The most common form of this autosomal dominant condition is a defect in an elastic connective tissue protein called fibrillin (MIM 134797). The protein is abundant in the lens of the eye, in the aorta (the largest artery in the body, leading from the heart), and in the bones of the limbs, fingers, and ribs. The symptoms are lens dislocation, long limbs, spindly fingers, and a caved-in chest (**figure 5.7**). The most serious symptom is a life-threatening weakening in the aorta, which can suddenly burst. If the weakening is detected early, a synthetic graft can replace the section of artery wall.

Certain proteins that form a structure in the eye called lens crystallins beautifully illustrate pleiotropy. If in low abundance as single molecules, these proteins are metabolic enzymes, functioning in many cell types. At higher abundance, however, they join and form crystallins, which aggregate to create a transparent lens whose physical properties enable it to focus incoming light on the retina.

Genetic Heterogeneity

Mutations in different genes that produce the same phenotype is a phenomenon called **genetic heterogeneity**. In these cases, it may appear that Mendel's laws are not operating, even though they are. For example, at least seventeen different genes cause Leber congenital amaurosis, the visual disorder described in the chapter opener. The different forms of the disease arise because there are many ways that a mutation can disrupt the functioning of the rods and cones, the cells that provide vision (**figure 5.8**). If a man who is homozygous recessive for a mutation in one of the genes that causes the condition has a child with a woman who is homozygous recessive for a different gene, then the child would not inherit Leber congenital amaurosis, because he or she would be heterozygous for both genes.

Discovering additional genes that can cause a known disorder is happening more as the human genome is analyzed, and can have practical repercussions. Consider osteogenesis imperfecta, in which abnormal collagen causes children's bones to break easily. Often when a child is brought to the hospital and past fractures are discovered, child abuse is suspected. A test for the most common type of osteogenesis imperfecta (MIM 166210) can rule this out. However, some parents who

Figure 5.7 **Marfan syndrome.** Whether or not Abraham Lincoln had Marfan syndrome is a matter of debate. In 1964, a physician published an article diagnosing the president based on body proportions and a sunken chest. Although a distant Lincoln cousin had Marfan, at least a dozen other conditions are associated with the long limbs and face. It will take a DNA sample from the man himself to diagnose this connective tissue disorder, which displays both pleiotropy and genetic heterogeneity.

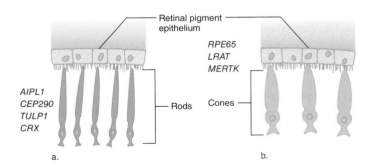

Figure 5.8 **Many routes to blindness.** Mutations in more than 100 genes cause degeneration of the retina, the multilayered structure at the back of the eye that includes the rods and cones, which are the photoreceptor cells that signal incoming light to the brain. This illustration indicates seven of these genes. Their encoded proteins have various functions: activating vitamin A, cleaning up debris, providing energy, and maintaining the functioning of the rods and cones.

insisted there was no abuse did not have the common mutation. They were cleared of charges when researchers discovered a rare mutation in a different gene (MIM 610854). This gene normally encodes an enzyme that adds a small chemical group to collagen, enabling it to function. Since then, even more forms of the disease have been discovered, due to mutations that disrupt the interactions of collagen chains.

Genetic heterogeneity can occur when genes encode enzymes that catalyze the same biochemical pathway, or different proteins that are part of the pathway. **Reading 5.1** describes a common genetically heterogeneic condition—Alzheimer disease.

Phenocopies

An environmentally caused trait that appears to be inherited is a **phenocopy**. Such a trait can either produce symptoms that resemble those of a known single-gene disorder or mimic inheritance patterns by affecting certain relatives. For example, the limb birth defect caused by the drug thalidomide, discussed in chapter 3, is a phenocopy of the inherited illness phocomelia. Physicians recognized the environmental disaster when they began seeing many children born with what looked like the very rare phocomelia. A birth defect caused by exposure to a teratogen was more likely than a sudden increase in incidence of a rare inherited disease.

An infection can be a phenocopy. Children who have AIDS may have parents who also have the disease, but these children acquired AIDS by viral infection, not by inheriting a gene. A phenocopy caused by a highly contagious infection can seem to be inherited if it affects more than one family member.

Sometimes, common symptoms may resemble those of an inherited condition until medical tests rule heredity out. For example, an underweight child who has frequent colds may show some signs of cystic fibrosis, but may instead suffer from malnutrition. Negative test results for several common CF alleles would alert a physician to look for another cause.

The Human Genome Sequence Adds Perspective

As researchers continue to describe the genes sequenced in the human genome project, it is becoming clear that phenomena once considered to only rarely complicate single-gene inheritance may be common. As a result, terms such as *epistasis* and *genetic heterogeneity* are beginning to overlap and blur. Consider Marfan syndrome. Most affected individuals have a mutation in the fibrillin gene. However, some people with the syndrome instead have a mutation in the gene that encodes the transforming growth factor beta receptor (TGFβR) (MIM 190181). Fibrillin and TGFβR are part of the same biochemical pathway. The conditions fit the definition of genetic heterogeneity because mutations in different genes cause identical symptoms. Yet they are also epistatic because a mutation in TGFβR blocks the activity of fibrillin.

Gene interactions also underlie penetrance and expressivity, once thought to be strictly a characteristic of a particular gene. Even genes that do not directly interact, in space or time, can affect each other's expression. This is the case for Huntington disease, described in chapter 4, in which cells in a certain part of the brain die, typically beginning in young adulthood. Siblings who inherit the exact same HD mutation may differ in the number of cells that they have in the affected brain area, thanks to variants of other genes that affected the division rate of neural stem cells in the brain during embryonic development. As a result, an individual who inherits HD, but also extra brain cells, might develop symptoms much later in life than a brother or sister who does not have such a built-in reserve supply. If the delay is long enough that death comes from another cause, HD would then be nonpenetrant.

DNA microarrays that reveal gene expression patterns in different tissues are painting detailed portraits of pleiotropy, showing that inherited disorders may affect more tissues or organs than are obvious as symptoms. Finally, more cases of genetic heterogeneity are being discovered as researchers identify genes with redundant or overlapping functions.

Table 5.1 summarizes phenomena that appear to alter single-gene inheritance. Our definitions and designations are changing as improving technology enables us to describe and differentiate disorders in greater detail. Phenomena such as variable expressivity, incomplete penetrance, epistasis, pleiotropy, and genetic heterogeneity, once considered unusual characteristics of single genes, are turning out to be the norm.

Gregor Mendel derived the two laws of inheritance working with traits conferred by genes located on different chromosomes in the nucleus. When genes do not conform to these conditions, however, the associated traits may not appear in Mendelian ratios. The rest of this chapter considers two types of gene transmission that do not fulfill the requirements for single-gene inheritance.

Key Concepts

1. A lethal genotype does not appear as a progeny class.
2. In incomplete dominance, the heterozygote phenotype is intermediate between those of the homozygotes; in codominance, two different alleles are expressed.
3. In epistasis, one gene influences expression of another.
4. Genotypes vary in penetrance and expressivity of the phenotype.
5. A gene with more than one expression is pleiotropic.
6. In genetic heterogeneity, different genes cause the same phenotype.
7. A trait caused by the environment but resembling a known genetic trait or occurring in certain family members is a phenocopy.

The Genetic Roots of Alzheimer Disease

"What is that thing for, that you put in your ear?" asked 72-year-old Ginny for the fifth time in half an hour.

"Mom, don't you know you've asked me that several times? It gets cell phone calls," answered her son, trying not to become annoyed.

"No, I've never asked you that before." She paused, looking puzzled. "What did you say it is?"

In the months following that conversation, Ginny's short-term memory declined further. She could rarely concentrate long enough to finish reading a newspaper article, or follow a conversation. In the grocery store, she couldn't find items she'd been buying for decades. Aware of her growing deficits, she became depressed. Finally, her son suggested she have a complete neurological exam. By the time she could see a physician, other signs had emerged. Ginny couldn't recall her zip code or the name of the small town where she grew up. Sometimes she couldn't remember where things belonged—she put a cantaloupe in the bathtub, and still had trouble with that contraption her son put in his ear to receive phone calls.

Ginny was showing signs of "mild cognitive impairment"(MCI). It could be the start of Alzheimer disease, which affects 26 million people worldwide. A neurologist started Ginny on a drug to slow breakdown of the neurotransmitter acetylcholine in the brain. She also prescribed an anti-depressant, which revived Ginny enough so that she was more willing to leave her apartment. If the MCI progressed to Alzheimer disease, Ginny would continue to lose thinking, reasoning, learning, and communicating abilities, and would one day no longer recognize her loved ones. Eventually, she would no longer speak and smile, and would cease walking, not because her legs wouldn't function, but because she would forget how to walk. Yet the haze would sometimes seem to lift from her eyes and the old Ginny would return.

In Alzheimer disease, certain brain parts—the amygdala (seat of emotion) and hippocampus (the memory center)—become buried in two types of protein. Amyloid precursor protein normally converts iron into a safe form. In Alzheimer disease, the protein is cut into unusually sized amyloid beta peptides, which aggregate to form "plaques" outside brain cells because cells cannot remove the material fast enough. At the same time, unsafe iron accumulates inside brain neurons, and zinc builds up in the plaques outside them. Ginny's brain scan showed plaques accumulating. The second type of protein, called tau, was building up too, clumping into "tangles" inside brain cells (**figure 1**). The telltale plaques and tangles, present to a lesser extent in everyone, could cause the symptoms or, more likely, result from them. Either way, their abundance in a spinal fluid test, combined with Ginny's cognitive problems, strongly suggested Alzheimer disease.

Fewer than 1 percent of Alzheimer cases are familial (inherited), caused by mutations in any of at least three genes that are involved with amyloid accumulation or clearance (**table 1**). A variant of a fourth gene, *APOE4,* increases the risk of developing a late-onset form three-fold in a heterozygote and 15-fold in a homozygote. Other genes raise the risk, too. A variant of a gene called *CALHM1* raises risk to about a tenth of the extent of *APOE4,* and genome-wide association studies have identified several places in the genome that have risk genes. However, they raise risk to such a small degree that they cannot be used to predict the disease.

Table 1	Genes Associated with Alzheimer Disease		
Causative Gene	**MIM**	**Chromosome**	**Mechanism**
Amyloid precursor protein (*APP*)	104760	21	Unusually sized pieces aggregate outside brain cells.
Presenilin 1	607822 104311	14	Forms part of secretase (enzyme) that cuts APP.
Presenilin 2	606889 600759	1	Forms part of secretase (enzyme) that cuts APP.
Risk Gene			
Apolipoprotein E4 (*APOE4*)	104310 107741	19	Unusually sized pieces add phosphates to tau protein, making it accumulate and impairing microtubule binding.

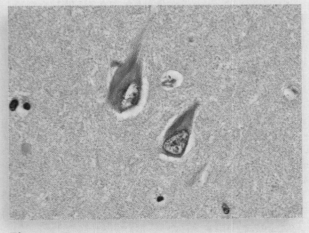

Figure 1 The dark dots in these two brain neurons are "neurofibrillary tangles" of tau protein, characteristic of Alzheimer disease.

Table 5.1	Factors That Alter Single-Gene Phenotypic Ratios	
Phenomenon	**Effect on Phenotype**	**Example**
Lethal alleles	A phenotypic class does not survive to reproduce.	Spontaneous abortion
Multiple alleles	Many variants or degrees of a phenotype occur.	Cystic fibrosis
Incomplete dominance	A heterozygote's phenotype is intermediate between those of two homozygotes.	Familial hypercholesterolemia
Codominance	A heterozygote's phenotype is distinct from and not intermediate between those of the two homozygotes.	ABO blood types
Epistasis	One gene masks or otherwise affects another's phenotype.	Bombay phenotype
Penetrance	Some individuals with a particular genotype do not have the associated phenotype.	Polydactyly
Expressivity	A genotype is associated with a phenotype of varying intensity.	Polydactyly
Pleiotropy	The phenotype includes many symptoms, with different subsets in different individuals.	Porphyria variegata
Phenocopy	An environmentally caused condition has symptoms and a recurrence pattern similar to those of a known inherited trait.	Infection
Genetic heterogeneity	Different genotypes are associated with the same phenotype.	Leber congenital amaurosis

5.3 Mitochondrial Genes

The basis of the law of segregation, that both parents contribute genes equally to offspring, does not apply for genes in mitochondria (see figure 2.7). Each of the hundreds to thousands of mitochondria in each human cell contains several copies of a "mini-chromosome" that carries just 37 genes. Mitochondrial DNA (mtDNA)-encoded genes act in the mitochondrion, but the organelle also requires the activities of certain genes from the nucleus.

The inheritance patterns and mutation rates for mitochondrial genes differ from those for genes in the nucleus. Mitochondrial genes are maternally inherited. They are passed only from an individual's mother because sperm almost never contribute mitochondria when they fertilize an oocyte. In the rare instances when mitochondria from sperm enter an oocyte, they are usually selectively destroyed early in development. Pedigrees that follow mitochondrial genes therefore show a woman passing the trait to all her children, while a male cannot pass the trait to any of his (**figure 5.9**).

DNA in the mitochondria differs functionally from DNA in the nucleus in several ways (**table 5.2** and **figure 5.10**). Mitochondrial DNA does not cross over. It mutates faster than DNA in the nucleus for two reasons: It has fewer ways to repair DNA (discussed in chapter 12), and the mitochondrion is the site of energy reactions that produce oxygen free radicals that damage DNA. Also unlike nuclear DNA, mtDNA is not wrapped

in proteins, nor are genes "interrupted" by DNA sequences that do not encode protein. Finally, a cell has one nucleus but many mitochondria, and each mitochondrion harbors several copies of its chromosome. Mitochondria with different alleles for the same gene can reside in the same cell.

Mitochondrial Disorders

Mitochondrial genes encode proteins that participate in protein synthesis and energy production. Twenty-four of the 37 genes encode RNA molecules (22 transfer RNAs and 2 ribosomal RNAs) that help assemble proteins. The other 13 mitochondrial genes encode proteins that function in cellular respiration, which is the process that uses energy from digested nutrients to synthesize ATP, the biological energy molecule.

Several diseases result from mutations in mitochondrial genes. They are called mitochondrial myopathies and have specific names, but news reports often lump them together as "mitochondrial disease." Symptoms arise from tissues whose

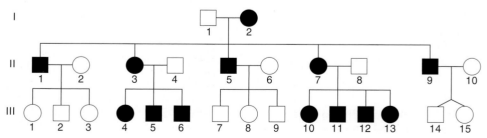

Figure 5.9 Inheritance of mitochondrial genes. Mothers pass mitochondrial genes to all offspring. Fathers do not transmit mitochondrial genes because sperm only very rarely contribute mitochondria to fertilized ova. If mitochondria from a male do enter, they are destroyed.

Table 5.2	Features of Mitochondrial DNA
No crossing over	
Fewer types of DNA repair	
Inherited from the mother only	
Many copies per mitochondrion and per cell	
High exposure to oxygen free radicals	
No histones (DNA-associated proteins)	
Genes not interrupted	

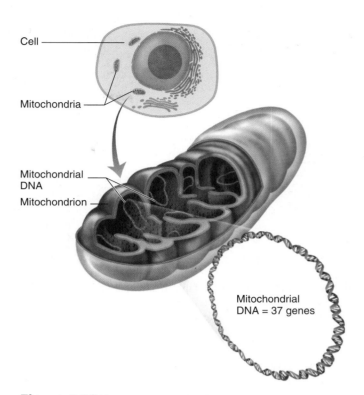

Figure 5.10 Mitochondrial DNA. A mitochondrion contains several rings of DNA. Different alleles can reside on different copies of the mitochondrial chromosome. A cell typically has thousands of mitochondria, each of which has many copies of its "mini-chromosome."

cells have many mitochondria, such as skeletal muscle. It isn't surprising that major symptoms are great fatigue, weak and flaccid muscles, and intolerance to exercise. Skeletal muscle fibers appear "red and ragged" when stained and viewed under a light microscope, their abundant abnormal mitochondria visible beneath the plasma membrane.

A mutation in a mitochondrial gene that encodes a tRNA or rRNA can be devastating because it impairs the organelle's general ability to manufacture proteins. Consider what happened to Lindzy S., a once active and articulate dental hygienist. In her forties, Lindzy gradually began to slow down at work. She heard a buzzing in her ears and developed difficulty talking and walking. Then her memory began to fade in and out, she became lost easily in familiar places, and her conversation made no sense. Her condition worsened, and she developed diabetes, seizures, and pneumonia and became deaf and demented. She was finally diagnosed with MELAS, which stands for "mitochondrial myopathy encephalopathy lactic acidosis syndrome" (MIM 540000). Lindzy died. Her son and daughter will likely develop the condition because they inherited her mitochondria.

A woman with a mitochondrial disorder can avoid transmitting it to her children if mitochondria from a healthy woman's oocyte are injected into her oocytes. Such a bolstered oocyte is fertilized in a laboratory dish by the partner's sperm, and the zygote is implanted in her uterus. Several dozen children, apparently free of mitochondrial disease, have been born from this technique.

About 1 in 200 people has a mutation in a mitochondrial gene that could cause disease. However, mitochondrial diseases are very rare, apparently because of a weeding-out process during egg formation. Such a mutation may disrupt energy acquisition so greatly in an oocyte that it cannot survive.

Heteroplasmy

The fact that a cell contains many mitochondria makes possible a rare condition called **heteroplasmy.** In this state, a particular mutation is in some mitochondrial chromosomes, but not others. At each cell division, the mitochondria are distributed at random into daughter cells. Over time, the chromosomes within a mitochondrion tend to be all wild type or all mutant

for any particular gene. But different mitochondria can have different alleles predominating. As an oocyte matures, the number of mitochondria drops from about 100,000 to 100 or fewer. If the woman is heteroplasmic for a particular mutation, by chance, she can produce an oocyte that has mostly mitochondria that are wild type for that gene, mostly mitochondria that have the mutation, or anything in between (**figure 5.11**). In this way, a woman who does not have a mitochondrial disorder, because the mitochondria bearing the mutation are either rare or not abundant in affected cell types, can nevertheless pass the associated condition to a child.

Heteroplasmy has several consequences for the inheritance of mitochondrial phenotypes. Expressivity may vary widely among siblings, depending upon how many mutation-bearing mitochondria were in the oocyte that became each brother or sister. Severity of symptoms reflects which tissues have cells whose mitochondria bear the mutation. This is the case for a family with Leigh syndrome (MIM 256000), which affects the enzyme that directly produces ATP. Two boys died of the severe form of the disorder because the brain regions that control movement rapidly degenerated. Another sibling was blind and had central nervous system degeneration. Several relatives, however, suffered only mild impairment of their peripheral vision. The more severely affected family members had more brain cells that received the mutation-bearing mitochondria.

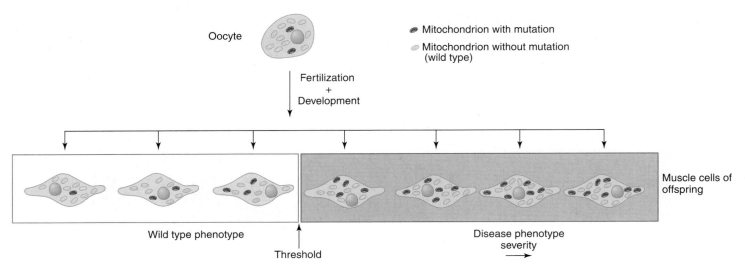

Oocyte

Mitochondrion with mutation

Mitochondrion without mutation
(wild type)

Fertilization
+
Development

Wild type phenotype

Threshold

Disease phenotype
severity

Muscle cells of
offspring

Figure 5.11 Mitochondrial inheritance. Mitochondria and their genes are only passed from the mother. Cells have many mitochondria. If an oocyte is heteroplasmic, differing numbers of copies of a mitochondrial mutation may be transmitted. The phenotype reflects the proportion of mitochondria bearing the mutation.

The most severe mitochondrial illnesses are heteroplasmic. This is presumably because *homo*plasmy—when all mitochondria bear the mutant allele—too severely impairs protein synthesis or energy production for embryonic development to complete. Often, severe heteroplasmic mitochondrial disorders do not produce symptoms until adulthood because it takes many cell divisions, and therefore years, for a cell to receive enough mitochondria bearing mutant alleles to cause symptoms.

Heteroplasmy helped to clarify a compelling question of history. On a July night in 1918, Tsar Nicholas II of Russia and his family, the royal Romanovs, were shot, their bodies damaged with acid and dumped into a shallow grave. In another July, in 1991, two amateur historians found the grave. DNA testing revealed the number of buried people, Y chromosome sequences distinguished males from females, and mitochondrial DNA sequences identified a mother (the Tsarina) and three daughters (**figure 5.12;** see also figure 6.6 for the pedigree). When researchers consulted the DNA of modern-day relatives to link the remains to the royal family, they encountered a problem: The remains of the suspected Tsar and his living great-grandniece Xenia differed at nucleotide position 16169 in the mitochondrial DNA. More puzzling, retesting the remains showed that for this site in the mitochondrial genome, the purported Tsar was in some samples thymine (T), yet in others cytosine (C). Had there been a lab error?

In yet another July, in 1994, researchers dug up the body of Nicholas's brother, the Grand Duke Georgij Romanov, and solved the mystery. Mitochondrial DNA position 16169 in bone cells from the Grand Duke also went both ways, with a T or a C. The family had heteroplasmy.

When heteroplasmy was discovered in the Romanov family, researchers thought it was rare. Since then, with the sequencing of many mitochondrial genomes, it is clear that about one in ten of the DNA bases in the mitochondrial genome are heteroplasmic—that is, they can differ within an individual. The mutations that generate these single-base variations in mtDNA probably occur all the time, but only those that originated early in development have a chance to accumulate enough so that they can be detected. The variability of mtDNA complicates its use in forensic investigations—as the Romanov story showed.

Figure 5.12 Heteroplasmy. DNA analysis identified the remains of the murdered Romanovs—and revealed an interesting genetic quirk, heteroplasmy.

Mitochondrial DNA Reveals the Past

Interest in mtDNA extends beyond the medical. mtDNA provides a powerful forensic tool used to link suspects to crimes, identify war dead, and support or challenge historical records. mtDNA is used in forensics because it is more likely to remain after extensive damage, because cells have many copies of it.

Sequencing mtDNA identified the son of Marie Antoinette and Louis XVI, who supposedly died in prison at age 10. In 1845, the boy was given a royal burial, but some people thought the buried child was an imposter. His heart had been stolen at the autopsy, and through a series of bizarre events, wound up, dried out, in the possession of the royal family. Recently, researchers compared mtDNA sequences from cells in the boy's heart to corresponding sequences in heart and hair cells from Marie Antoinette (her decapitated body identified by her fancy underwear), two of her sisters, and living relatives Queen Anne of Romania and her brother. The mtDNA evidence showed that the buried boy was indeed the prince, Louis XVII. Chapter 16 discusses how researchers consult mtDNA sequences to reconstruct ancient migration patterns.

Key Concepts

1. Mitochondrial genes are maternally inherited and mutate rapidly. A cell contains many mitochondria, which have many copies of the mitochondrial genome.
2. Mitochondrial genes encode RNAs or proteins that function in protein synthesis or energy metabolism.
3. In heteroplasmy, cells contain mitochondria that have different alleles of a gene.

5.4 Linkage

Most of the traits that Mendel studied in pea plants were conferred by genes on different chromosomes. When genes are close to each other on the same chromosome, they usually do not segregate at random during meiosis and therefore do not support Mendel's predictions. Instead, genes close on a chromosome are packaged into the same gametes and are said to be "linked" (**figure 5.13**). Linkage has this very precise meaning in genetics. The term is popularly used in a much more general sense to mean any association between two events or observations.

Linkage refers to the transmission of genes on the same chromosome. Linked genes do *not* assort independently and do *not* produce Mendelian ratios for crosses tracking two or more genes. Understanding and using linkage as a mapping tool has been critical in identifying disease-causing genes, and helped pave the way for genome-wide association studies.

Discovery in Pea Plants

William Bateson and R. C. Punnett first observed the unexpected ratios indicating linkage in the early 1900s, again in pea plants. They crossed true-breeding plants with purple flowers

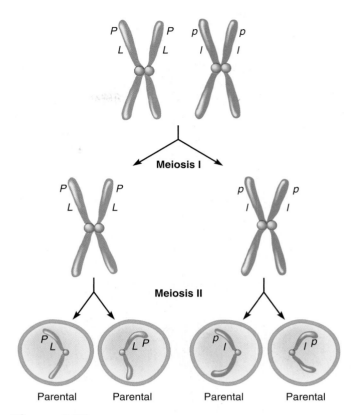

Figure 5.13 **Inheritance of linked genes.** Genes linked closely to one another are usually inherited together when the chromosome is packaged into a gamete.

and long pollen grains (genotype *PPLL*) to true-breeding plants with red flowers and round pollen grains (genotype *ppll*). The plants in the next generation, of genotype *PpLl*, were then self-crossed. But this dihybrid cross did not yield the expected 9:3:3:1 phenotypic ratio that Mendel's second law predicts (**figure 5.14**).

Bateson and Punnett noticed that two types of third-generation peas, those with the parental phenotypes *P_L_* and *ppll*, were more abundant than predicted, while the other two progeny classes, *ppL_* and *P_ll*, were less common (the blank indicates that the allele can be dominant or recessive). The more prevalent parental allele combinations, Bateson and Punnett hypothesized, could reflect genes that are transmitted on the same chromosome and that therefore do not separate during meiosis. The two less common offspring classes could also be explained by a meiotic event, crossing over. Recall that this is an exchange between homologs that mixes up maternal and paternal gene combinations without disturbing the sequence of genes on the chromosome (**figure 5.15**).

Progeny that exhibit this mixing of maternal and paternal alleles on a single chromosome are called **recombinant**. *Parental* and *recombinant* are relative terms. Had the parents in Bateson and Punnett's crosses been of genotypes *ppL_* and *P_ll,* then *P_L_* and *ppll* would be recombinant rather than parental classes.

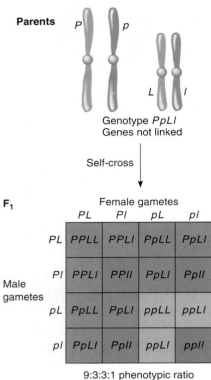

Parents

P   p

L   l

Genotype *PpLl*
Genes not linked

Self-cross ↓

F₁

Female gametes

	PL	Pl	pL	pl
PL	PPLL	PPLl	PpLL	PpLl
Pl	PPLl	PPll	PpLl	Ppll
pL	PpLL	PpLl	ppLL	ppLl
pl	PpLl	Ppll	ppLl	ppll

Male gametes

9:3:3:1 phenotypic ratio

a.

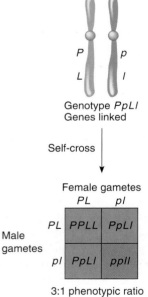

P   p

L   l

Genotype *PpLl*
Genes linked

Self-cross ↓

Female gametes

	PL	pl
PL	PPLL	PpLl
pl	PpLl	ppll

Male gametes

3:1 phenotypic ratio

b.

Figure 5.14 **Expected results of a dihybrid cross. (a)**
Unlinked genes assort independently. The gametes represent all
possible allele combinations. The expected phenotypic ratio of
a dihybrid cross is 9:3:3:1. **(b)** If genes are linked, only two allele
combinations are expected in the gametes. The phenotypic ratio
is 3:1, the same as for a monohybrid cross.

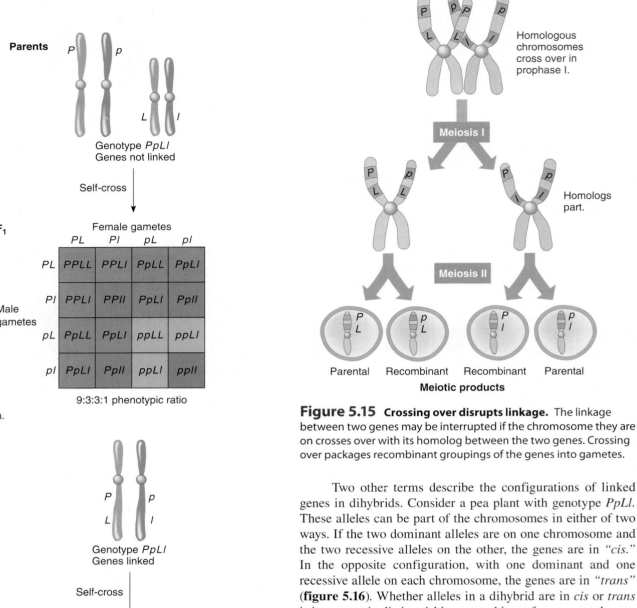

Homologous chromosomes cross over in prophase I.

Meiosis I

Homologs part.

Meiosis II

P   p   P   p
L   L   l   l

Parental   Recombinant   Recombinant   Parental

Meiotic products

Figure 5.15 **Crossing over disrupts linkage.** The linkage
between two genes may be interrupted if the chromosome they are
on crosses over with its homolog between the two genes. Crossing
over packages recombinant groupings of the genes into gametes.

Two other terms describe the configurations of linked
genes in dihybrids. Consider a pea plant with genotype *PpLl*.
These alleles can be part of the chromosomes in either of two
ways. If the two dominant alleles are on one chromosome and
the two recessive alleles on the other, the genes are in *"cis."*
In the opposite configuration, with one dominant and one
recessive allele on each chromosome, the genes are in *"trans"*
(**figure 5.16**). Whether alleles in a dihybrid are in *cis* or *trans*
is important in distinguishing recombinant from parental progeny classes in specific crosses.

Linkage Maps

As Bateson and Punnett were discovering linkage in peas,
geneticist Thomas Hunt Morgan and his coworkers at Columbia University were doing the same using the fruit fly *Drosophila melanogaster*. They assigned genes to relative positions
on chromosomes and compared progeny class sizes to assess
whether traits were linked. They soon saw that the pairs of traits
fell into four groups. Within each group, crossed dihybrids did
not produce offspring classes according to Mendel's second
law. Also, the number of linkage groups—four—matched the
number of chromosome pairs in the fly. Coincidence? No. The
traits fell into four groups because their genes are inherited
together on the same chromosome.

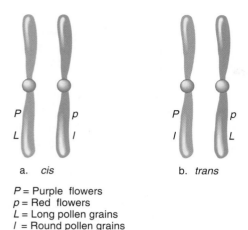

P = Purple flowers
p = Red flowers
L = Long pollen grains
l = Round pollen grains

Figure 5.16 Allele configuration is important. Parental chromosomes can be distinguished from recombinant chromosomes only if the allele configuration of the two genes is known—they are either in *cis* (**a**) or in *trans* (**b**).

The genius of the work on linkage in fruit flies was that the researchers translated their data into actual maps depicting positions of genes on chromosomes. Morgan wondered why the sizes of the recombinant classes varied for different genes. In 1911, Morgan's undergraduate student, Alfred Sturtevant, proposed that the farther apart two genes are on a chromosome, the more likely they are to cross over simply because more physical distance separates them (**figure 5.17**). Sturtevant stayed up very late one night and devised a way to represent the correlation between crossover frequency and the distance between genes as a **linkage map**. These diagrams showed the order of genes on chromosomes and the relative distances between them. The distance was represented using "map units" called centimorgans, where 1 centimorgan equals 1 percent recombination. These units are still used today to construct single nucleotide polymorphism (SNP) maps. They provide an estimate of genetic distance along a chromosome. Sturtevant's approach led to the construction of linkage maps of human genes that were important in the sequencing of the human genome.

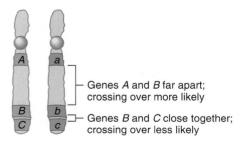

Genes *A* and *B* far apart; crossing over more likely

Genes *B* and *C* close together; crossing over less likely

Figure 5.17 Breaking linkage. Crossing over is more likely between widely spaced linked genes *A* and *B*, or between *A* and *C*, than between more closely spaced linked genes *B* and *C*, because there is more room for an exchange to occur.

The frequency of a crossover between any two linked genes is inferred from the proportion of offspring from a cross that are recombinant. Frequency of recombination is based on the percentage of meiotic divisions that break the linkage between—that is, separate—two parental alleles. Genes at opposite ends of the same chromosome cross over often, generating a large recombinant class. Genes lying very close on the chromosome would only rarely be separated by a crossover. The probability that genes on opposite ends of a chromosome cross over approaches the probability that, if on different chromosomes, they would independently assort—about 50 percent. **Figure 5.18** illustrates this distinction.

The situation with linked genes (or SNPs) can be compared to a street lined with stores on both sides. There are more places to cross the street between stores at opposite ends on opposite sides than between two stores in the middle of the block on opposite sides of the street. Similarly, more crossovers, or progeny with recombinant genotypes, are seen when two genes are farther apart on the same chromosome.

As the twentieth century progressed, geneticists in Columbia University's "fly room" mapped several genes on all four chromosomes, as researchers in other labs assigned many genes to the human X chromosome. Localizing genes on the X chromosome was easier than doing so on the autosomes, because X-linked traits follow an inheritance pattern that is distinct from the one all autosomal genes follow. In human males, with their single X chromosome, recessive alleles on the X are expressed and observable. We return to this point in chapter 6.

By 1950, geneticists began to think about mapping genes on the 22 human autosomes. To start, a gene must be matched to its chromosome. This became possible when geneticists identified people with a particular inherited condition or trait and an unusual chromosome.

In 1968, researcher R. P. Donohue was looking at chromosomes in his own white blood cells when he noticed a dark area consistently located near the centromere of one member of his largest chromosome pair (chromosome 1). He examined chromosomes from several family members for the dark area, noting also whether each family member had a blood type called Duffy. (Recall that blood types refer to the patterns of sugars on red blood cell surfaces.) Donohue found that the Duffy blood type was linked to the chromosome variant. He could predict a relative's Duffy blood type by whether or not the chromosome had the telltale dark area. This was the first assignment of a trait in humans to an autosome.

Finding a chromosomal variation linked to a family trait like Donohue did was unusual. More often, researchers mapped genes in experimental organisms, such as fruit flies, by calculating percent recombination (crossovers) between two genes with known locations on a chromosome. However, because human parents do not have hundreds of offspring, nor do they produce a new generation every 10 days, getting enough information to establish linkage relationships for us requires observing the same traits in many families and pooling the results. Today, even though we know the human genome sequence, linkage remains a powerful tool to track disease-associated genes.

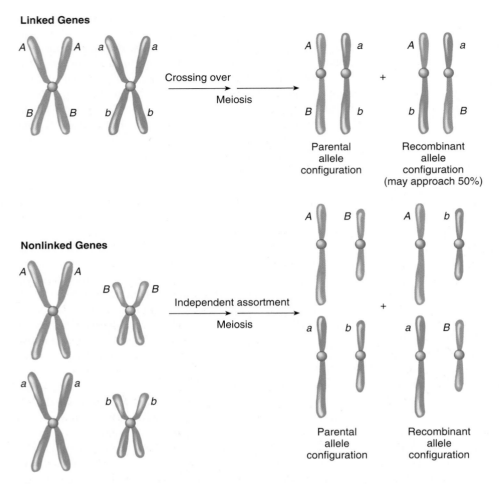

Linked Genes

Crossing over

Meiosis

Parental allele configuration

Recombinant allele configuration (may approach 50%)

Nonlinked Genes

Independent assortment

Meiosis

Parental allele configuration

Recombinant allele configuration

Figure 5.18 **Linkage versus nonlinkage (independent assortment).** When two genes are widely separated on a chromosome, the likelihood of a crossover is so great that the recombinant class may approach 50 percent—which may appear to be the result of independent assortment.

Solving Linkage Problems Uses Logic

As an example of determining the degree of linkage by percent recombination using several families, consider the traits of Rh blood type (MIM 111700) and a form of anemia called elliptocytosis (MIM 130500). An Rh positive phenotype corresponds to genotypes *RR* or *Rr*. (This is simplified.) The anemia corresponds to genotypes *EE* or *Ee*.

In 100 one-child families, one parent is Rh negative with no anemia (*rree*), and the other parent is Rh positive with anemia (*RrEe*), and the *R* and *E* (or *r* and *e*) alleles are in *cis*. Of the 100 offspring, 96 have parental genotypes (*re/re* or *RE/re*) and four are recombinants for these two genes (*Re/re* or *rE/re*). Percent recombination is therefore 4 percent, and the two linked genes are 4 centimorgans apart.

Consider another pair of linked genes in humans. Nail-patella syndrome (MIM 161200) is a rare autosomal dominant trait that causes absent or underdeveloped fingernails and toenails, and painful arthritis in the knee and elbow joints. The gene is 10 map units from the *I* gene that determines

the ABO blood type, on chromosome 9. Geneticists determined the map distance by pooling information from many families. The information can be used to predict genotypes and phenotypes in offspring, as in the following example.

Greg and Susan each have nail-patella syndrome. Greg has type A blood. Susan has type B blood. What is the chance that their child inherits normal nails and knees and type O blood? A genetic counselor deduces their allele configurations using information about the couple's parents (**figure 5.19**).

Greg's mother has nail-patella syndrome and type A blood. His father has normal nails and type O blood. Therefore, Greg must have inherited the dominant nail-patella syndrome allele (*N*) and the *I*^A allele from his mother, on the same chromosome. We know this because Greg has type A blood and his father has type O blood—therefore, he couldn't have gotten the *I*^A allele from his father. Greg's other chromosome 9 must carry the alleles *n* and *i*. His alleles are therefore in *cis*.

Susan's mother has nail-patella syndrome and type O blood, and so Susan inherited *N* and *i* on the same chromosome. Because her father has normal nails and type B blood, her homolog from him bears alleles *n* and *I*^B. Her alleles are in *trans*.

Determining the probability that Susan and Greg's child could have normal nails and knees and type O blood is the easiest question the couple could ask. The only way this genotype can arise from theirs is if an *ni* sperm (which occurs with a frequency of 45 percent, based on pooled data) fertilizes an *ni* oocyte (which occurs 5 percent of the time). The result—according to the product rule—is a 2.25 percent chance of producing a child with the *nnii* genotype.

Calculating other genotypes for their offspring is more complicated, because more combinations of sperm and oocytes could account for them. For example, a child with nail-patella syndrome and type AB blood could arise from all combinations that include *I*^A and *I*^B as well as at least one *N* allele (assuming that *NN* has the same phenotype as *Nn*).

The Rh blood type and elliptocytosis, and nail-patella syndrome and ABO blood type, are examples of linked gene pairs. A linkage map begins to emerge when percent recombination is known between all possible pairs of three or more linked genes, just as a road map with more landmarks provides more information on distance and direction. Consider genes *x*, *y*, and *z* (**figure 5.20**). If the percent recombination between

	Greg	Susan
Phenotype	nail-patella syndrome, type A blood	nail-patella syndrome, type B blood
Genotype	$NnI^A i$	$NnI^B\_\_$
Allele configuration	$\dfrac{N\quad I^A}{n\quad i}$	$\dfrac{N\quad i}{n\quad I^B}$

Gametes:	sperm	frequency	oocytes
Parental	$N\ I^A$	45%	$N\ i$
	$n\ i$	45%	$n\ I^B$
Recombinants	$N\ i$	5%	$N\ I^B$
	$n\ I^A$	5%	$n\ i$

N = nail-patella syndrome
n = normal

Figure 5.19 **Inheritance of nail-patella syndrome.** Greg inherited the N and I^A alleles from his mother; that is why the alleles are on the same chromosome. His n and i alleles must therefore be on the homolog. Susan inherited alleles N and i from her mother, and n and I^B from her father. Population-based probabilities are used to calculate the likelihood of phenotypes in the offspring of this couple. Note that in this figure, map distances are known and are used to predict outcomes.

x and y is 10, between x and z is 4, and between z and y is 6, then the order of the genes on the chromosome is x-z-y. This is the only order of the three genes that accounts for the percent recombination data. It is a little like deriving a geographical map from distances between cities, or calculating the mileage from different legs of a flight.

Genetic maps derived from percent recombination between linked genes accurately reflect the order on the chromosome, but the distances are estimates because crossing over is not equally likely across the genome. Some DNA sequences are nearly always inherited together, like two inseparable friends. This nonrandom association between DNA sequences is called **linkage disequilibrium** (**LD**). The human genome consists of many "LD" blocks where stretches of alleles stick together, interspersed with areas where crossing over is prevalent. Chapter 7 discusses the use of LD blocks, called **haplotypes,** to track genes of interest in populations.

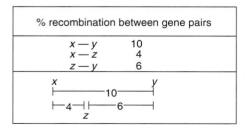

% recombination between gene pairs	
$x - y$	10
$x - z$	4
$z - y$	6

Figure 5.20 **Recombination mapping.** If we know the percent recombination between all possible pairs of three genes, we can determine their relative positions on the chromosome.

From Linkage to Genome-Wide Associations

The first human genes mapped to their chromosomes encoded blood proteins, because these were easy to study. In 1980, researchers began using DNA sequences near genes of interest as landmarks called **genetic markers**. These markers need not encode proteins that cause a phenotype. They might be DNA sequence differences that alter where a DNA cutting enzyme cuts, or differing numbers of short repeated sequences of DNA with no obvious function, or SNPs (see figure 7.10). The term "genetic marker" is used popularly to mean any DNA sequence that is associated with a particular phenotype, usually one affecting health.

The old idea of linkage lies behind the new strategy of **genome-wide association studies**. These investigations scan the genomes of thousands of individuals who have a particular phenotype for DNA sequences that they share, but are much rarer in people who do not have the trait (see figure 7.10). The genetic marker "signposts" are SNPs and repeats. The assumption is that these shared genome regions point the way toward genes that control or contribute to the phenotype. **Table 5.3** reviews types of genetic markers mentioned throughout the book.

Computers tally how often genes and their markers are inherited together. The "tightness" of linkage between a marker and a gene of interest is represented as a LOD score, which stands for "logarithm of the odds." A LOD score indicates the likelihood that particular crossover frequency data indicate linkage, rather than the inheritance of two alleles by chance. The higher the LOD score, the closer the two genes.

A LOD score of 3 or greater signifies linkage. It means that the observed data are 1,000 (10^3) times more likely to have occurred if the two DNA sequences (a disease-causing allele and its marker) are linked than if they reside on different chromosomes and just happen to often be inherited together by chance. It is somewhat like deciding whether two coins tossed together 1,000 times always come up both heads or both tails by chance, or because they are taped together side by side in that position, as are linked genes. If the coins land with the same sides up in all 1,000 trials, it indicates they are very likely taped.

Before sequencing of the human genome, genetic markers were used to predict which individuals in some families were most likely to have inherited a particular disorder, before symptoms began. Such tests are no longer necessary, because tests directly detect disease-causing genes. However, genetic markers are still used to distinguish parts of chromosomes. In pedigrees, designations for markers linked into haplotypes are sometimes placed beneath the symbols to further describe chromosomes. In the family with cystic fibrosis depicted in **figure 5.21,** each set of numbers beneath a symbol represents

Table 5.3	Types of Genetic Markers
Genetic Element	**Description**
Chromosomal	Trait appears with unusual chromosome
RFLPs	DNA cutting site
VNTRs	Repeats
STRs	Repeats
SNPs	Single-base differences
CNVs	Copy number variants

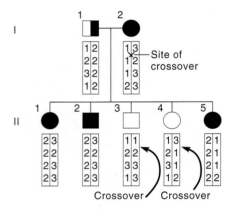

Figure 5.21 Haplotypes. The numbers in bars beneath pedigree symbols enable researchers to track specific chromosome segments with markers. Disruptions of a marker sequence indicate crossover sites.

a haplotype. Knowing the haplotype of individual II–2 reveals which chromosome in parent I–1 contributes the mutant allele. Because Mr. II–2 received haplotype 3233 from his affected mother, his other haplotype, 2222, comes from his father. Since Mr. II–2 is affected and his father is not, the father must be a heterozygote, and 2222 must be the haplotype linked to the mutant *CFTR* allele.

Today, new associations between DNA sequences and traits are reported daily. Genetics has indeed evolved from an obscure science to a robust source of knowledge about ourselves.

Key Concepts

1. Genes on the same chromosome are linked. They are inherited in different patterns than are unlinked genes.
2. Crosses involving linked genes produce a large parental class and a small recombinant class (caused by crossing over).
3. The farther apart two genes are on a chromosome, the more likely they are to cross over. Linkage maps translate crossover frequency into relative distances between genes on a chromosome.
4. Unusual chromosomes revealed the first linkage associations.
5. Linkage disequilibrium is a linkage combination that is stronger than that predicted by gene frequencies in a population. Genetic markers were used in early linkage mapping and today in genome-wide association studies.
6. Linkage maps reflect the percent recombination between linked genes. LOD scores describe the tightness of linkage. Haplotypes indicate linked DNA sequences, and reflect a variety of marker types.

Summary

5.1 A New View of Mendelian Genetics

1. Single genes do not act as independently as Mendel's experiments suggested.
2. Changes in gene expression, inheritance of mitochondrial genes, and linkage may seem to oppose Mendel's laws, but do not.

5.2 When Gene Expression Appears to Alter Mendelian Ratios

3. Homozygosity for lethal recessive alleles stops development before birth, eliminating an offspring class.
4. A gene can have multiple alleles because its sequence can deviate in many ways. Different allele combinations may produce different variations of the phenotype.
5. Heterozygotes of **incompletely dominant** alleles have phenotypes intermediate between those associated with the two homozygotes. **Codominant** alleles are both expressed in the phenotype.
6. In **epistasis,** one gene affects the phenotype of another.
7. An incompletely **penetrant** genotype is not expressed in all individuals who inherit it. Phenotypes that vary in intensity among individuals are variable in **expressivity**.
8. **Pleiotropic** genes have several expressions.
9. In **genetic heterogeneity,** two or more genes specify the same phenotype.
10. A **phenocopy** is a characteristic that appears to be inherited but is environmentally caused.

5.3 Mitochondrial Genes

11. Only females transmit mitochondrial genes; males can inherit such a trait but cannot pass it on.
12. Mitochondrial genes do not cross over, and they mutate more frequently than nuclear DNA.
13. The 37 mitochondrial genes encode tRNA, rRNA, or proteins involved in protein synthesis or energy reactions.

14. Many mitochondrial disorders are **heteroplasmic,** with mitochondria in a single cell harboring different alleles.

5.4 Linkage

15. Genes on the same chromosome are **linked** and, unlike genes that independently assort, produce many individuals with parental genotypes and a few with **recombinant** genotypes.

16. **Linkage maps** depict linked genes. Researchers can examine a group of known linked DNA sequences (a **haplotype**) to follow the inheritance of certain chromosomes.

17. Knowing whether linked alleles are in *cis* or *trans,* and using crossover frequencies from pooled data, can be used to predict the probabilities that certain genotypes will appear in progeny.

18. Genetic linkage maps assign distances to linked genes based on crossover frequencies. Today many genetic markers spanning the genome are used to compare large groups of individuals, one of which has a particular phenotype. These **genome-wide association studies** are used in discovering genetic variants that contribute to multifactorial conditions.

www.mhhe.com/lewisgenetics10

Answers to all end-of-chapter questions can be found at **www.mhhe.com/lewisgenetics10.** You will also find additional practice quizzes, animations, videos, and vocabulary flashcards to help you master the material in this chapter.

Review Questions

1. Explain how each of the following phenomena can disrupt Mendelian phenotypic ratios.
   a. lethal alleles
   b. multiple alleles
   c. incomplete dominance
   d. codominance
   e. epistasis
   f. incomplete penetrance
   g. variable expressivity
   h. pleiotropy
   i. a phenocopy
   j. genetic heterogeneity

2. How does the relationship between dominant and recessive alleles of a gene differ from epistasis?

3. Why can transmission of an autosomal dominant trait with incomplete penetrance look like autosomal recessive inheritance?

4. How does inheritance of ABO blood type exhibit both complete dominance and codominance?

5. How could two people with albinism have a child who has normal skin pigment?

6. How do the porphyrias exhibit variable expressivity, pleiotropy and genetic heterogeneity?

7. How can epistasis explain incomplete penetrance?

8. The lung condition emphysema may be caused by lack of an enzyme, or by smoking. Which cause is a phenocopy?

9. List three ways that mtDNA differs from DNA in a cell's nucleus.

10. Describe why inheritance of mitochondrial DNA and linkage are exceptions to Mendel's laws.

11. How does a pedigree for a maternally inherited trait differ from one for an autosomal dominant trait?

12. If researchers could study pairs of human genes as easily as they can study pairs of genes in fruit flies, how many linkage groups would they detect?

13. Describe three types of genetic markers.

14. The popular media often use words that have precise meanings in genetics, but more general common meanings. Explain the two types of meanings of "linked" and "marker."

Applied Questions

1. For each of the diseases described in situations *a* through *i,* indicate which of the following phenomena (A–H) is at work. More than one may apply.
   A. lethal alleles
   B. multiple alleles
   C. epistasis
   D. incomplete penetrance
   E. variable expressivity
   F. pleiotropy
   G. a phenocopy
   H. genetic heterogeneity
   a. A woman with severe neurofibromatosis type 1 has brown spots on her skin and several large tumors beneath her skin. A genetic test shows that her son has

the disease-causing autosomal dominant allele, but he has no symptoms.

b. A man would have a widow's peak, if he wasn't bald.

c. A man and woman have six children. They also had two stillbirths—fetuses that died shortly before birth.

d. Mutations in any of at least three genes cause familial ALS.

e. A woman with dark brown skin uses a bleaching cream with a chemical that darkens her fingertips and ears, just like the inherited disease alkaptonuria.

f. In Labrador retrievers, the *B* allele confers black coat color and the *b* allele brown coat color. The *E* gene controls the expression of the *B* gene. If a dog inherits the *E* allele, the coat is golden no matter what the *B* genotype is. A dog of genotype *ee* expresses the *B* (black) phenotype.

g. Two parents are heterozygous for genes that cause albinism, but each gene specifies a different enzyme in the biochemical pathway for skin pigment synthesis. Their children thus do not face a 25 percent risk of having albinism.

h. Alagille syndrome (MIM 118450), in its most severe form, prevents the formation of ducts in the gallbladder, damaging the liver. Affected children also usually have heart murmurs, unusual faces, and butterfly-shaped vertebrae. Such children often have one otherwise healthy parent who has a heart murmur, unusual face, and butterfly vertebrae.

i. Two young children in a family have very decayed teeth. Their parents think it is genetic, but the true cause is a babysitter who puts them to sleep with juice bottles in their mouths.

2. If many family studies for a particular autosomal recessive condition reveal fewer affected individuals than Mendel's law predicts, the explanation may be either incomplete penetrance or lethal alleles. How might you use haplotypes to determine which of these two possibilities is the cause?

3. A man who has type O blood has a child with a woman who has type A blood. The woman's mother has AB blood, and her father, type O. What is the probability that the child has the following blood types?

a. O

b. A

c. B

d. AB

4. Enzymes are used in blood banks to remove the A and B antigens from blood types A and B. This makes the blood type O. Does this alter the phenotype or the genotype?

5. Ataxia-oculomotor apraxia syndrome (MIM 208920), which impairs the ability to feel and move the limbs, usually begins in early adulthood. The molecular basis of the disease is impairment of ATP production in mitochondria, but the mutant gene is in the nucleus of the cells. Would this disorder be inherited in a Mendelian fashion? Explain your answer.

6. What is the chance that Greg and Susan, the couple with nail-patella syndrome, could have a child with normal nails and type AB blood?

7. A gene called secretor (MIM 182100) is 1 map unit from the *H* gene that confers the Bombay phenotype on chromosome 19. Secretor is dominant, and a person of either genotype *SeSe* or *Sese* secretes the ABO and H blood type antigens in saliva and other body fluids. This secretion, which the person is unaware of, is the phenotype. A man has the Bombay phenotype and is not a secretor. A woman does not have the Bombay phenotype and is a secretor. She is a dihybrid whose alleles are in *cis*. What is the chance that their child will have the same genotype as the father?

8. In prosopagnosia (MIM 610382), a person has "face blindness"—he or she cannot identify individuals by their faces. It is inherited as an autosomal dominant trait, and affects people to different degrees. Some individuals learn early in life to identify people by voice or style of dress, and so appear not to have the condition. Only a small percentage of cases are inherited; most are the result of stroke or brain injury. (www.faceblind.org offers tests to help you imagine what it is like not to be able to recognize faces). Which of the following does face blindness demonstrate? Explain your choices.

a. incomplete penetrance

b. variable expressivity

c. pleiotropy

d. phenocopy

9. Many people who have the "iron overload" disease hereditary hemochromatosis (MIM 235200; see section 20.1) are homozygous for a variant of the *C282Y* gene. How would you determine the penetrance of this condition?

10. A Martian creature called a gazook has 17 chromosome pairs. On the largest chromosome are genes for three traits—round or square eyeballs (*R* or *r*); a hairy or smooth tail (*H* or *h*); and 9 or 11 toes (*T* or *t*). Round eyeballs, hairy tail, and 9 toes are dominant to square eyeballs, smooth tail, and 11 toes. A trihybrid male has offspring with a female who has square eyeballs, a smooth tail, and 11 toes on each of her three feet. She gives birth to 100 little gazooks, who have the following phenotypes:

- 40 have round eyeballs, a hairy tail, and 9 toes
- 40 have square eyeballs, a smooth tail, and 11 toes
- 6 have round eyeballs, a hairy tail, and 11 toes
- 6 have square eyeballs, a smooth tail, and 9 toes
- 4 have round eyeballs, a smooth tail, and 11 toes
- 4 have square eyeballs, a hairy tail, and 9 toes

a. Draw the allele configurations of the parents.

b. Identify the parental and recombinant progeny classes.

c. What is the crossover frequency between the *R* and *T* genes?

Web Activities

1. Look at the gene descriptions at RetNet: http://www.sph. uth.tmc.edu/retnet/disease.htm. Select three retinal diseases and use Mendelian Inheritance in Man or other sources to describe how they affect vision differently. (Some researchers consider Leber congenital amaurosis to be a subtype of retinitis pigmentosa.)

2. Go to the websites for the Genetic Alliance (www. geneticalliance.org) or the National Institutes of Health's Office of Rare Diseases Research (http://rarediseases.info.nih. gov/). Learn about different inherited diseases, and identify one that exhibits pleiotropy.

3. Go to the United Mitochondrial Disease Foundation website and describe the phenotype of a mitochondrial disorder.

4. Browse the National Center for Biotechnology Information (NCBI) site, and list three sets of linked genes. Consult MIM to describe the trait or disorder that each specifies.

5. Use MIM to identify a genetically heterogeneic condition, and explain why this description applies.

6. For some of the porphyrias, attacks are precipitated by an environmental trigger. Using MIM, describe factors that can trigger an attack of any of the following:

   a. acute intermittent porphyria
   b. porphyria cutanea tarda
   c. coproporphyria
   d. porphyria variegata
   e. erythropoietic protoporphyria

Forensics Focus

1. "Earthquake McGoon" was 32 years old when the plane he was piloting over north Vietnam was hit by groundfire on May 6, 1954. Of the five others aboard, only two survived. McGoon, actually named James B. McGovern, was well known for his flying in World War II, and for his jolliness. Remains of a man about his height and age at death were discovered in late 2002, but could not be identified by dental records. However, DNA sampled from a leg bone enabled forensic scientists to identify him. Describe the type of DNA likely analyzed, and what further information was needed to make the identification.

Case Studies and Research Results

1. Shiloh Winslow is deaf. In early childhood, she began having fainting spells, especially when she became excited. When she fainted while opening Christmas gifts, her parents took her to the hospital, where doctors said, again, that there wasn't a problem. As the spells continued, Shiloh became able to predict the attacks, telling her parents that her head hurt beforehand. Her parents took her to a neurologist, who checked Shiloh's heart and diagnosed long QT syndrome with deafness, a severe form of inherited heartbeat irregularity (see Reading 2.2). Ten different genes can cause long QT syndrome. The doctor told them of a case from 1856: a young girl, called at school to face the headmaster for an infraction, became so agitated that she dropped dead. The parents were not surprised; they had lost two other children to great excitement.

   The Winslows visited a medical geneticist, who discovered that each parent had a mild heartbeat irregularity that did not produce symptoms. Shiloh's parents had normal hearing. Shiloh's younger brother Pax was also hearing-impaired and suffered night terrors, but had so far not fainted during the day. Like Shiloh, he had the full syndrome. Vivienne, still a baby, was also tested. She did not have either form of the family's illness; her heartbeat was normal.

   Today, Shiloh and Pax are treated with beta blocker drugs, and each has an implantable defibrillator to correct a potentially fatal heartbeat. Shiloh's diagnosis may have saved her brother's life.

   a. Which of the following applies to the condition in this family?

      i. genetic heterogeneity
      ii. pleiotropy
      iii. variable expressivity
      iv. incomplete dominance
      v. a phenocopy

   b. How is the inheritance pattern of this form of long QT syndrome similar to that of familial hypercholesterolemia?

   c. How is it possible that Vivienne did not inherit either the serious or asymptomatic form of the illness?

   d. Do the treatments for the condition affect the genotype or the phenotype?

2. Barnabas Collins has congenital erythropoietic porphyria, and his wife Angelique is a carrier of ALA dehydratase deficiency. What is the chance that, if they have a child, he or she will have a porphyria?

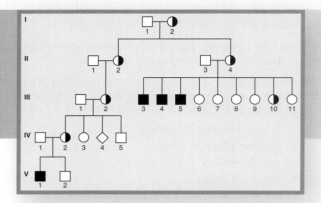

In 1937, Alfred Wiskott described the family depicted in this pedigree, which shows classic X-linked inheritance—males affected, female carriers. Wiskott-Aldrich syndrome is treatable today with a combination of stem cell and gene therapies.

Matters of Sex

Learning Outcomes

6.1. Our Sexual Selves

1. Describe the factors that contribute to whether we are and feel male or female.

2. Distinguish between the X and Y chromosomes.

3. Discuss how manipulating sex ratio can affect societies.

6.2. Traits Inherited on Sex Chromosomes

4. Distinguish between Y linkage and X linkage.

5. Compare and contrast X-linked recessive inheritance and X-linked dominant inheritance.

6.3 Sex-Limited and Sex-Influenced Traits

6. Discuss the inheritance pattern of a trait that appears in only one sex.

7. Define *sex-influenced trait*.

6.4 X Inactivation

8. Explain why X inactivation is necessary.

9. Explain how X inactivation is an epigenetic change.

10. Discuss how X inactivation affects the phenotype in female mammals.

6.5 Genomic Imprinting

11. Explain the chemical basis of silencing the genetic contribution from one parent.

12. Discuss how genomic imprinting is seen in Prader-Willi and Angelman syndromes.

The Big Picture: Sex affects our lives in many ways. Which sex chromosomes we are dealt at conception sets the developmental program for maleness or femaleness, but gene expression before and after birth greatly influences how that program unfolds.

Stem Cell and Gene Therapies Save Boys' Lives

In 1937, Alfred Wiskott, a pediatrician in Germany, saw a family that had six healthy girls, but three baby boys who died of the same illness. They had bruises, a skin condition, bloody diarrhea, ear infections, and pneumonia. All died from bleeding in their digestive tracts and infection in the blood. Dr. Wiskott noted the inherited nature of the condition and that it affected only boys.

In 1954, Robert Aldrich described the disorder in a large Dutch family. Sick boys who survived childhood developed cancers of the blood or autoimmune disorders. The disease was named Wiskott-Aldrich syndrome (MIM 301000), and the mutant gene is on the X chromosome. Only boys are affected because, unlike girls, they lack a second X chromosome to block the effects of the mutation.

In 2006, German researchers contacted living relatives of the original family with the three sick boys. Genetic tests revealed a mutation in the *WAS* gene. An affected little boy, the first cousin twice removed of the three original boys, was successfully treated with a stem cell transplant from a matched, unrelated donor. Today, patients are cured with a combination of stem cell and gene therapy. They receive their own bone marrow stem cells bolstered with functional *WAS* genes.

6.1 Our Sexual Selves

Whether we are male or female is enormously important in our lives. It affects our relationships, how we think and act, and how others perceive us. Gender is, at one level, dictated by genes, but it is also layered with psychological and sociological components.

Maleness or femaleness is determined at conception, when he inherits an X and a Y chromosome, or she inherits two X chromosomes (**figure 6.1**). Another level of sexual identity comes from the control that hormones exert over the development of reproductive structures. Finally, both biological factors and social cues influence sexual feelings, including the strong sense of whether we are male or female.

Sexual Development

Gender differences become apparent around the ninth week of prenatal development. During the fifth week, all embryos develop two unspecialized gonads, which are organs that will develop as either testes or ovaries. Each such "indifferent" gonad forms near two sets of ducts that offer two developmental options. If one set of tubes, called the Müllerian ducts, continues to develop, female sexual structures form. If the other set, the Wolffian ducts, persist, male sexual structures form.

The choice to follow either developmental pathway occurs during the sixth week, depending upon the sex chromosome constitution and actions of certain genes. If a gene on the Y chromosome called *SRY* (for "sex-determining region of the Y") is activated, hormones steer development along a male route. In the absence of *SRY* activation, a female develops. However, being female is not just a "default" option

that acts in the absence of maleness. Sex determination is more accurately described as a fate imposed on ambiguous precursor structures. Several genes besides *SRY* guide early development of reproductive structures. A mutation in a gene called *Wnt4*, for example, causes a female to have high levels of male sex hormones and lack a vaginal canal and uterus. Hence, the gene is essential for development and maturation as a female.

Sex Chromosomes

The sex with two different sex chromosomes is called the **heterogametic sex,** and the other, with two of the same sex chromosome, is the **homogametic sex**. In humans, males are heterogametic (XY) and females are homogametic (XX). Some other species are different. In birds and snakes, for example, males are ZZ (homogametic) and females are ZW (heterogametic).

The sex chromosomes differ in size and capacity. The human X chromosome has more than 1,500 genes and is much larger than the Y chromosome, which has 231 protein-encoding genes. In meiosis in a male, the X and Y chromosomes act as if they are a pair of homologs.

Identifying genes on the human Y chromosome has been difficult. Before the human genome sequence became available, researchers inferred the functions and locations of Y-linked genes by observing how men missing parts of the chromosome differ from normal. Creating linkage maps for the Y was not possible because the Y does not cross over along all of its length.

Analysis of the human genome sequence revealed why mapping the Y chromosome was so hard: It has a very unusual organization. In the 95 percent of the chromosome that harbors male-specific genes, many DNA segments are palindromes. In written languages, palindromes are sequences of letters that read the same in both directions—"Madam, I'm Adam," for example. This symmetry in a DNA sequence, compared to "a hall of mirrors," destabilizes DNA replication so that during meiosis, sections of a Y chromosome attract each other. This can loop out parts in between, which may account for infertility caused by missing parts of the Y. Yet this "hall of mirrors" organization may also provide a way for the chromosome to recombine with itself, essentially sustaining its structure. Two researchers—one an XX, one an XY—take a lighthearted look at the curious structure of the human Y chromosome in *In Their Own Words* on page 114.

The Y chromosome has a very short arm and a long arm (**figure 6.2**). At both tips are pseudoautosomal regions, termed PAR1 and PAR2. They comprise only 5 percent of the chromosome and include 63 pseudoautosomal genes. The term "pseudoautosomal" means that the DNA sequences have counterparts on the X chromosome and can cross over with them. The pseudoautosomal genes encode a variety of proteins that function in both sexes. These genes control bone growth, cell division, immunity, signal transduction, the synthesis of hormones and receptors, fertility, and energy metabolism.

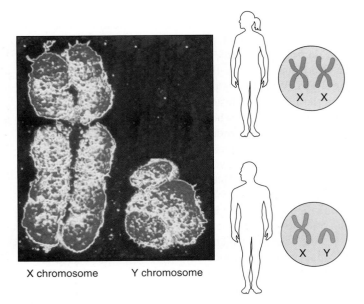

X chromosome   Y chromosome

Figure 6.1 **The sex chromosomes.** A human male has one X chromosome and one Y chromosome. A female has two X chromosomes.

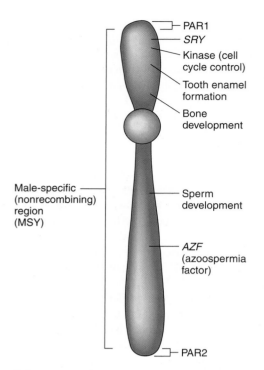

Figure 6.2 **Anatomy of the Y chromosome.** The Y chromosome has two pseudoautosomal regions (PAR1 and PAR2) and a large central area (MSY) that comprises about 95 percent of the chromosome. A few genes are indicated here. *SRY* determines sex. *AZF* encodes a protein essential to producing sperm; mutations in it cause infertility.

Most of the Y chromosome is the male-specific region, or MSY, that lies between the two pseudoautosomal regions. It consists of three classes of DNA sequences. About 10 to 15 percent of the MSY sequence is 99 percent identical to counterparts on the X chromosome. Protein-encoding genes are scarce here. Another 20 percent of the MSY consists of DNA sequences that are somewhat similar to X chromosome sequences and may be remnants of an ancient autosome that long ago gave rise to the X chromosome. The remainder of the MSY includes palindrome-ridden regions. The genes in the MSY include many repeats and specify protein segments that combine in different ways, which is one reason why counting the number of protein-encoding genes on the Y chromosome has been difficult. Many of the genes in the MSY are essential to fertility, including *SRY*.

The Y chromosome was first seen under a light microscope in 1923, and researchers soon recognized its association with maleness. For many years, they sought to identify the gene or genes that determine sex. Important clues came from two very interesting types of people—men who have two X chromosomes (XX male syndrome), and women who have one X and one Y chromosome (XY female syndrome). A close look at their sex chromosomes revealed that the XX males actually had a small piece of a Y chromosome, and the XY females lacked a small part of the Y chromosome. The part of the Y chromosome present in the XX males was the same part that was missing in the XY females. This critical area accounted for half a percent of the Y chromosome. In 1990, researchers isolated and identified the *SRY* gene here.

The Phenotype Forms

The *SRY* gene encodes a very important type of protein called a **transcription factor,** which controls the expression of other genes. The *SRY* transcription factor stimulates male development by sending signals to the indifferent gonads that destroy potential female structures while stimulating development of male structures.

Prenatal sexual development is a multistep process, and mutations can intervene at several points (**figure 6.3**). The result may be an XY individual who looks female because of a block in the development of male structures. For example, in androgen insensitivity syndrome (MIM 300068), caused by a mutation in a gene on the X chromosome, the absence of receptors for androgens (the male sex hormones testosterone and dihydrotestosterone [DHT]) stops cells in early reproductive structures from receiving the signal to develop as male. The person looks female, but is XY.

Several terms are used to describe individuals whose genetic/chromosomal sex and physical structures, both internal and external, are not consistent with one gender. *Hermaphroditism* is an older and more general term for an individual with both male and female sexual structures. *Intersex* refers to individuals whose internal structures are inconsistent with external structures, or whose genitalia are ambiguous. It is the preferred term outside of medical circles. *Pseudohermaphroditism* refers to the presence of both female and male structures but at different life stages.

Living with pseudohermaphroditism can be confusing. Consider 5-alpha reductase deficiency (MIM 264600), which is autosomal recessive. Affected individuals have a normal Y chromosome, a wild type *SRY* gene, and testes. The internal male reproductive tract develops and internal female structures do not, so the male anatomy is there on the inside. But the child is unaware of the insides, and on the outside looks like a girl. Without 5-alpha reductase, which normally catalyzes the reaction of testosterone to form DHT, a penis cannot form. At puberty, when the adrenal glands, which sit atop the kidneys, start to produce testosterone, this XY individual, who thought she was female, starts to experience the signs of maleness—deepening voice, growth of facial hair, and the sculpting of muscles into a masculine physique. Instead of developing breasts and menstruating, the clitoris enlarges into a penis. Usually sperm production is normal. XX individuals with 5-alpha reductase deficiency look female.

The degree to which pseudohermaphroditism disturbs the individual depends as much on society as it does on genetics. In the Dominican Republic in the 1970s, 22 young girls reached the age of puberty and began to transform into boys. They had a form of 5-alpha reductase deficiency that was fairly common in the population due to consanguinity (relatives having children with relatives). The parents were happy that they had had sons after all, and so these special adolescents were given

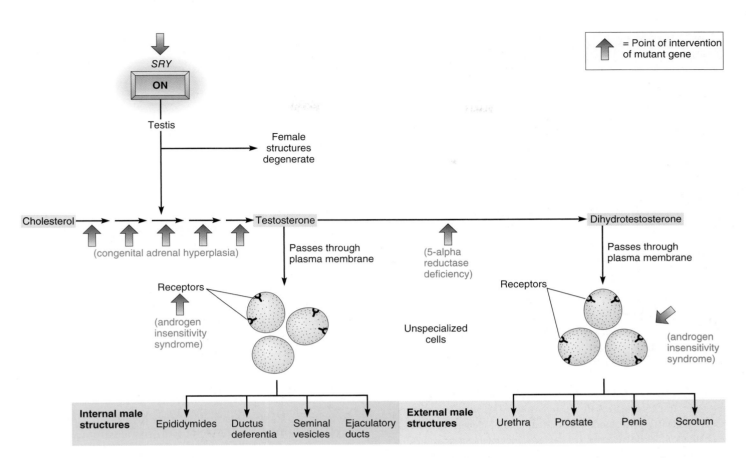

Figure 6.3 Mutations that affect male sexual development. In normal male prenatal development, activation of a set of genes beginning with *SRY* sends signals that destroy female rudiments, while activating the biochemical pathway that produces testosterone and dihydrotestosterone, which promote the development of male structures. The green arrows indicate where mutations disrupt normal sexual development. Diseases are indicated in red.

their own gender name—*guevedoces,* for "penis at age 12." They were fully accepted as whatever they wanted to be. This isn't always the case. A very realistic novel, *Middlesex,* tells the story of a young man with this condition who grew up as a female.

In a more common form of pseudohermaphroditism, congenital adrenal hyperplasia (CAH) due to 21-hydroxylase deficiency (MIM 201910), an enzyme block causes testosterone and DHT to accumulate. It is autosomal recessive, and both males and females are affected. The higher levels of androgens cause precocious puberty in males or male secondary sex characteristics to develop in females. Boys may enter puberty as early as 3 years old, with well-developed musculature, small testes, and enlarged penises. At birth, girls may have swollen clitorises that appear to be small penises. They are female internally, but as they reach puberty, their voices deepen, facial hair grows, and they do not menstruate.

Prenatal tests that detect chromosomal sex have changed the way that pseudohermaphroditism is diagnosed. Before, the condition was detected only after puberty, when a girl began to look like a boy. Today pseudohermaphroditism is suspected when a prenatal chromosome check reveals an X and a Y chromosome, but the newborn looks like a girl.

Transgender is a poorly understood condition related to sexual identity. A transgendered individual has the phenotype and sex chromosomes of one gender, but identifies extremely strongly with the opposite gender. It is a much more profound condition than transvestitism, which refers to a male who prefers women's clothing. The genetic or physical basis of transgender is not known. Some affected individuals have surgery to better match their physical selves with the gender that they feel certain they are.

Is Homosexuality Inherited?

No one really knows why we have feelings of belonging to one gender or the other, or of being attracted to the same or opposite gender, but these feelings are intense. In homosexuality, a person's phenotype and genotype are consistent, and physical attraction is toward members of the same sex. Homosexuality is seen in all human cultures and has been observed for thousands of years. It has been documented in more than 500 animal species.

Homosexuality reflects complex input from genes and the environment. The genetic influence may be seen in the strong feelings that homosexual individuals have as young

The Y Wars

Researcher Jennifer Marshall-Graves predicts that the Y chromosome will "self-destruct" within the next 10 million years. Her comparison of Y chromosomes in a wide variety of mammals indicates that, gradually, important genes are being transferred to other chromosomes. David Page, who has led the mapping of the Y chromosome, has a more optimistic view. Each researcher spoke out, in jest, at two scientific conferences. Here is some of what they had to say:

The Rise and Fall of the Human Y Chromosome

Jennifer A. Marshall-Graves, Australian National University

The Y chromosome is unique in the human genome. It is small, gene-poor, prone to deletion and loss, variable among species, and useless. You can lack a Y and not be dead, just female. It is impossible to understand why this chromosome is so weird without understanding where it came from. It is a sad decline, and I predict its imminent loss.

The X is a decent sort of chromosome. It accounts for 4 percent of the genome, with about 1,500 perfectly normal genes. The Y is a pathetic little chromosome that has few genes interspersed with lots of junk. And those genes are a weird lot. They are particularly concerned with male sexual development. Some are quite bizarre and many inactive. The Y clearly diverged from the X.

There are several models of the Y (**figure 1**). The dominant Y model of a macho Y reflects the fact that the Y contains the male-determining *SRY* gene. The selfish Y model predicts that the Y kidnapped genes from elsewhere. The wimp Y model says that the Y is just a relic of the once glorious X chromosome. This model was first proposed by biologist Susumo Ohno in 1967 in the theory that the X and Y originated as a pair of autosomes. Then the Y acquired the male-determining locus, and other genes that are required for spermatogenesis gathered nearby. This led to suppressed

recombination in this region of the Y, which allowed all sorts of horrible genetic accidents to occur that could not be repaired. Mutations, deletions, and insertions accumulated until almost nothing was left, except bits at the top and bottom that still pair with the X. A few genes survived because they found a useful male-specific function, and many of these have made copies of themselves in a desperate race to stave off disappearing altogether.

The Y is degrading fast, losing genes at the rate of 5 per million years. I predict that it will be completely gone in 5 to 10 million years. Will we have males? The males in the audience can take comfort from the mole vole *Ellobius lutescens* (**figure 2**). It has no Y, but it does have males and females. It has no *SRY*, no Y chromosome at all. Both sexes are XO. How do they do it? We don't know. Clearly another gene takes over and new sex genes start evolving. Will there be new sex chromosome evolution in humans? Maybe it will happen in different ways in different populations, and we will split into two species.

Rethinking the Rotting Y Chromosome

David Page, Massachusetts Institute of Technology and Howard Hughes Medical Institute Investigator

The Y chromosome has had a public relations problem for a long time. For most of the last half of the past century, people thought that the Y chromosome was a junk heap. We can now update that model.

Back 300 million years ago, when we were reptiles, we had no sex chromosomes, only ordinary autosomes. Shortly after our ancestors parted company with the ancestors of birds, a mutation arose on one member of a pair of ordinary autosomes to give rise to *SRY*. The process of shutting down XY crossing over began, first in the

Models of the Human Y

Dominant Y Selfish Y Wimp Y

Figure 1 **Models of the human Y chromosome.** Researcher Jennifer Marshall-Graves offers a tongue-in-cheek look at the Y chromosome, but her research findings are serious—the chromosome is shrinking.

Figure 2 **Life without a Y?** Males of all mammals, except two species of mole voles, have Y chromosomes. Birds and reptiles do not. The Y chromosome probably arose from an X chromosome about 310 million years ago. The X lost many genes and gained a few that set their carriers on the road to maleness. This animal is a Y-less male mole vole—it reproduces just fine.

(Continued)

vicinity of *SRY*, and then in an expanding region. Once a piece of the Y was no longer able to recombine with the X, its genes began to rot. The purpose of sex (recombination in meiosis) is not just to generate new gene combinations, but to allow genes to rid themselves of mildly deleterious mutations that accumulate. Y genes are not protected because they have lots of areas of no crossing over. Genes decayed, except for *SRY* and the tips. It wasn't a very flattering model for the Y.

When Jennifer Marshall-Graves and John Aitken wrote that the Y would self-destruct in 10 million years, it truly frightened the people in my lab. We decided we needed to pick up the pace.

Based on the sequencing of the Y, we've been able to rethink its evolution, and realized that the chromosome may have found a way around its seemingly inevitable problems. We looked closely at

the male-specific region of the Y, reanalyzing sequences in a different way, chopped into smaller bits. And we found that each piece would find a match elsewhere on the Y. So segments on the Y are effectively functioning as alleles—30 percent have a perfect match elsewhere on the chromosome. These are not simple repeats, but highly complex sequences of tens to hundreds of kilobases. The region includes eight palindromes and one inverted repeat (**figure 3**). We propose that there is intense recombination within the palindromes. And so the Y has two forms of productive recombination: conventional routine recombination of crossing over with the X at pseudoautosomal regions, and recombination within the Y. It's not that the Y doesn't recombine, it just does it its own way. The Y does copying that preserves its identity.

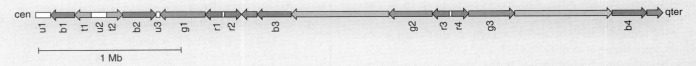

Figure 3 **The Y chromosome is highly repetitive.** A section of the Y chromosome that David Page studies, called *AZFc* (for azoospermia factor c), consists of DNA sequences that read the same in either direction, an organization that can lead to instability as well as provide a mechanism to generate new alleles. Other parts of the chromosome house similar repeats. Matching colors in this depiction represent identical sequences. Same-color arrows that point in opposite directions indicate inverted repeats.

children, long before they know of the existence or meaning of the term. Other evidence comes from identical twins, who are more likely to both be homosexual than are both members of fraternal twin pairs.

Experiments in the 1990s identified genetic markers on the X chromosome that were more often identical among pairs of homosexual brothers than among other pairs of brothers. This finding led to the idea that a single gene, or a few genes, dictates sexual preference but these results could not be confirmed. Further studies on twins indicated what many people have long suspected—the roots of homosexuality are not simple.

Twin studies compare a trait between identical and fraternal twin pairs, to estimate the rough proportion of a trait that can be attributed to genes. Chapter 7 discusses this approach further. Such a study done on all of the adult twins in Sweden found that in males, genetics contributes about 35 percent to homosexuality, whereas among females the genetic contribution is about 18 percent. Clearly, homosexuality reflects the input of many genes and environmental factors, and may in fact arise in a variety of ways.

Table 6.1 summarizes the components of sexual identity.

Sex Ratio

Mendel's law of segregation predicts that populations should have approximately equal numbers of male and female newborns. That is, male meiosis should yield equal numbers of X-bearing and Y-bearing sperm. After birth, societal and environmental factors may favor survival of one gender over the other.

The proportion of males to females in a human population is called the **sex ratio**. Sex ratio is calculated as the number

Table 6.1	Sexual Identity	
Level	**Events**	**Timing**
Chromosomal/genetic	XY = male XX = female	Fertilization
Gonadal sex	Undifferentiated structure begins to develop as testis or ovary	6 weeks after fertilization
Phenotypic sex	Development of external and internal reproductive structures continues as male or female in response to hormones	8 weeks after fertilization, puberty
Gender identity	Strong feelings of being male or female develop	From childhood, possibly earlier
Sexual orientation	Attraction to same or opposite sex	From childhood

of males divided by the number of females multiplied by 1,000, for people of a particular age. (Some organizations describe sex ratio based on 1.0.) A sex ratio of equal numbers of males and females would be designated 1,000. The sex ratio at conception is called the primary sex ratio. In the United States for the past six decades, newborn boys have slightly outnumbered newborn girls, with the primary sex ratio averaging 1,050. The slight excess of boys may reflect the fact that Y-bearing sperm weigh slightly less than X-bearing sperm, and so they may reach the oocyte faster.

Sex ratio at birth is termed secondary and at maturity is called tertiary. Sex ratio can change markedly with age. This reflects medical conditions that affect the sexes differently, as well as environmental factors that affect one sex more than the other, such as participation in combat or engaging in other dangerous behaviors.

It is interesting to see what happens when a society intentionally alters the sex ratio. This has been done in India and China, where researchers have identified great numbers of "missing females." In these societies, prenatal diagnostic techniques were used to identify XX fetuses. Termination of XX fetuses, underreporting of female births, and, rarely, selective infanticide of girl babies all contributed to a very unnaturally skewed sex ratio favoring males.

The effects of altering sex ratio echo for years. In China, by the year 2020, 20 million men will lack female partners as a long-term consequence of that nation's "one-child policy." It began in 1979, with financial incentives to control runaway population growth. If a couple had a second child, the government revoked benefits. Some families, wanting their only child to be a boy, failed to continue or report female pregnancies. The reasoning was societal: A son would care for his aging parents, but a daughter would care for her in-laws.

China's one-child policy prevented hundreds of millions of births. The average number of births per woman fell from 5.4 in 1971 to 1.8 in 2001. By the turn of the century, 117 boys were being born for every 100 girls. But today, many children in China have few siblings, cousins, aunts, or uncles. Young women, now rare, have become valued once again. The Chinese government is now promoting a "Care for Girls" program, which awards housing subsidies and scholarships to families that have girls. Government officials hope that a more natural sex ratio will return.

India is experiencing male bias similar to the situation in China. A survey of 1.1 million families revealed that the ratio of boys to girls was about equal when the first or second child was a boy, but if the first child and especially if the first two children were girls, then the secondary sex ratio fell to about 750 girls to every 1,000 boys. Families were using prenatal diagnosis to detect female pregnancies and were terminating about a fourth of them. Researchers estimate that India has about 100 million "missing females."

At the other end of the human life cycle, sex ratio favors females in most populations. For people over the age of 65 in the United States, for example, the sex ratio is 720, meaning that there are 72 men for every 100 women. The ratio among older people is the result of disorders that are more likely to be fatal in males as well as behaviors that may shorten their life spans compared to women.

Key Concepts

1. The human female has two X chromosomes and a male has one X and one Y.
2. The Y chromosome has few genes. It includes palindromes and sequences similar to sequences on the X chromosome.
3. Activation of *SRY* starts gene action that causes testes to develop. Anti-Müllerian hormone stops the development of female structures and testosterone stimulates development of male internal structures and directs development of external structures. In several disorders, chromosomal, gonadal, and/or phenotypic sex are inconsistent.
4. Genes and the environment contribute to homosexuality.
5. Sex ratio is the number of males divided by the number of females multiplied by 1,000, for people of a particular age.

6.2 Traits Inherited on Sex Chromosomes

Genes on the Y chromosome are **Y-linked** and genes on the X chromosome are **X-linked**. Y-linked traits are rare because the chromosome has few genes, and many have counterparts on the X chromosome. These traits are passed from male to male, because a female does not have a Y chromosome. No other Y-linked traits besides infertility (which obviously can't be passed on) are yet clearly defined, although certain gene products have been identified. Claims that "hairy ears" is a Y-linked trait did not hold up—it turned out that families hid their affected female members!

The X chromosome not only has many more genes than the Y chromosome, but a disproportionate number of X-linked genes make us sick when mutant. The chromosome includes 4 percent of all the genes, but accounts for about 10 percent of Mendelian (single-gene) diseases.

Genes on the X chromosome have different patterns of expression in females and males, because a female has two X chromosomes and a male just one. In females, X-linked traits are passed just like autosomal traits—that is, two copies are required for expression of a recessive allele, and one copy for a dominant allele. In males, however, a single copy of an X-linked allele causes expression of the trait or illness because there is no copy of the gene on a second X chromosome to mask the effect. A man inherits an X-linked trait only from his mother. The human male is considered **hemizygous** for X-linked traits, because he has only one set of X-linked genes.

Understanding how sex chromosomes are inherited is important in predicting phenotypes and genotypes in offspring.

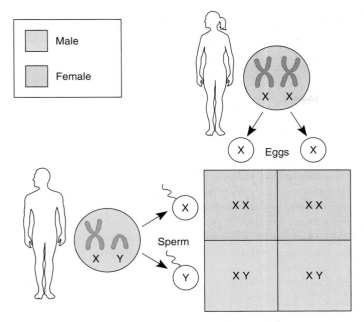

Figure 6.4 Sex determination in humans. An oocyte has a single X chromosome. A sperm cell has either an X or a Y chromosome. If a Y-bearing sperm cell with a functional *SRY* gene fertilizes an oocyte, the zygote is a male (XY). If an X-bearing sperm cell fertilizes an oocyte, then the zygote is a female (XX).

A male inherits his Y chromosome from his father and his X chromosome from his mother (**figure 6.4**). A female inherits one X chromosome from each parent. If a mother is heterozygous for a particular X-linked gene, her son or daughter has a 50 percent chance of inheriting either allele from her. X-linked traits are always passed on the X chromosome from mother to son or from either parent to daughter, but there is no male-to-male transmission of X-linked traits.

X-Linked Recessive Inheritance

An X-linked recessive trait is expressed in females if the causative allele is present in two copies. Many times, an X-linked trait passes from an unaffected heterozygous mother to an affected son. **Table 6.2** summarizes the transmission of an X-linked recessive trait.

Table 6.2	Criteria for an X-Linked Recessive Trait

1. Always expressed in the male.

2. Expressed in a female homozygote but very rarely in a heterozygote.

3. Passed from heterozygote or homozygote mother to affected son.

4. Affected female has an affected father and a mother who is affected or a heterozygote.

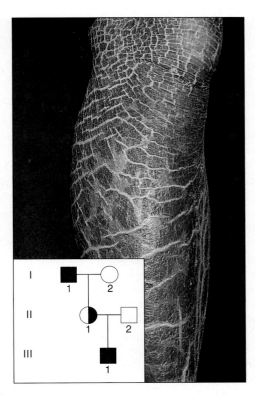

Figure 6.5 An X-linked recessive trait. Ichthyosis is transmitted as an X-linked recessive trait. A grandfather and grandson were affected in this family.

If an X-linked condition is not lethal, a man may be healthy enough to transmit it to offspring. Consider the small family depicted in **figure 6.5**. A middle-aged man who had rough, brown, scaly skin did not realize his condition was inherited until his daughter had a son. By a year of age, the boy's skin resembled his grandfather's. In the condition, called ichthyosis (MIM 308100), an enzyme deficiency blocks removal of cholesterol from skin cells. The upper skin layer cannot peel off as it normally does, and appears brown and scaly. A test of the daughter's skin cells revealed that she produced half the normal amount of the enzyme, indicating that she was a carrier.

Colorblindness is another X-linked recessive trait that does not hamper the ability of a man to have children. About 8 percent of males of European ancestry are colorblind, as are 4 percent of males of African descent. Only 0.4 percent of females in both groups are colorblind. **Reading 6.1,** on page 118, takes a closer look at this interesting trait.

Figure 6.6 shows part of an extensive pedigree for another X-linked recessive trait, the blood-clotting disorder hemophilia B (MIM 306900), also known as Christmas disease. Note the combination of pedigree symbols and a Punnett square to trace transmission of the trait. Dominant and recessive alleles are indicated by superscripts to the X and Y chromosomes. In the royal families of England, Germany, Spain, and Russia, the mutant allele arose in one of Queen Victoria's X chromosomes; it was either a new mutation or she inherited it.

Colorblindness

English chemist John Dalton saw things differently from most people. Sealing wax that appeared red to other people was as green as a leaf to Dalton and his brother. Pink wildflowers were blue. The Dalton brothers had X-linked recessive colorblindness.

Curious about the cause of his colorblindness, Dalton asked his physician, Joseph Ransome, to dissect his eyes after he died. When that happened, Ransome snipped off the back of one eye, removing the retina, where the cone cells that provide color vision are nestled among the more abundant rod cells that impart black-and-white vision. Because Ransome could see red and green normally when he peered through the back of his friend's eyeball, he concluded that it was not an abnormal filter in front of the eye that altered color vision. He stored the eyes in dry air, enabling researchers at the London Institute of Ophthalmology to analyze DNA in Dalton's eyeballs in 1994. Dalton's remaining retina lacked one of the three types of photopigments that enable cone cells to capture certain wavelengths of light.

Color Vision Basics

Cone cells are of three types, defined by the presence of any of three types of photopigments. An object appears colored because it reflects certain wavelengths of light, and each cone type captures a particular range of wavelengths with its photopigment. The brain then interprets the incoming information as a visual perception, much as an artist mixes the three primary colors to create many hues and shadings.

Each photopigment has a vitamin A–derived portion called retinal and a protein portion called an opsin. The three types of opsins correspond to short, middle, and long wavelengths of light. Mutations in opsin genes cause three different types of colorblindness. A gene on chromosome 7 encodes shortwave opsins, and mutations in it produce the rare autosomal "blue" form of colorblindness (MIM 190900). Dalton had deuteranopia (green colorblindness); his eyes lacked the middle-wavelength opsin. In the third type, protanopia (red colorblindness), long-wavelength opsin is absent. Deuteranopia (MIM 303800) and protanopia (MIM 303900) are X-linked.

A Molecular View

Jeremy Nathans of Johns Hopkins University also personally contributed to understanding color vision. First, he used a cow version of a protein called rhodopsin that provides black-and-white vision to identify the human version of the gene. Then Nathans used the human rhodopsin gene as a "probe" to search his own DNA for genes with similar sequences. He found three. One was on chromosome 7, the other two on the X chromosome.

Although Nathans can see colors, his opsin genes are not entirely normal, which provided a clue to how colorblindness arises and why it is so common. On his X chromosome, Nathans has one red opsin gene and two green genes, instead of the normal one of each. Because the red and green genes have similar sequences, Nathans reasoned, they can misalign during meiosis in the female (**figure 1**). The resulting oocytes would then have either two or none of one opsin gene type. An oocyte lacking either a red or a green opsin gene would, when fertilized by a Y-bearing sperm, give rise to a colorblind male.

People who are colorblind must function in a multicolored world. To help them, computer algorithms can convert colored video images into shades that they can distinguish.

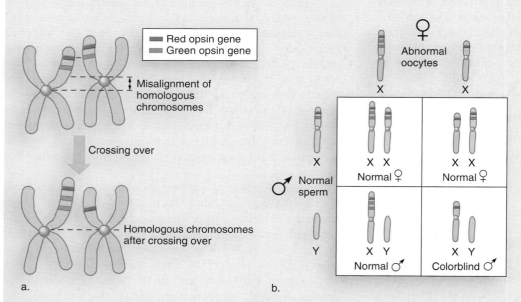

Figure 1 How colorblindness arises. (a) The sequence similarities among the opsin genes responsible for color vision may cause chromosomes to misalign during meiosis in the female. Offspring may inherit too many, or too few, opsin genes. A son inheriting an X chromosome missing an opsin gene would be colorblind. A daughter, unless her father is colorblind, would be a carrier. (b) A missing gene causes X-linked colorblindness.

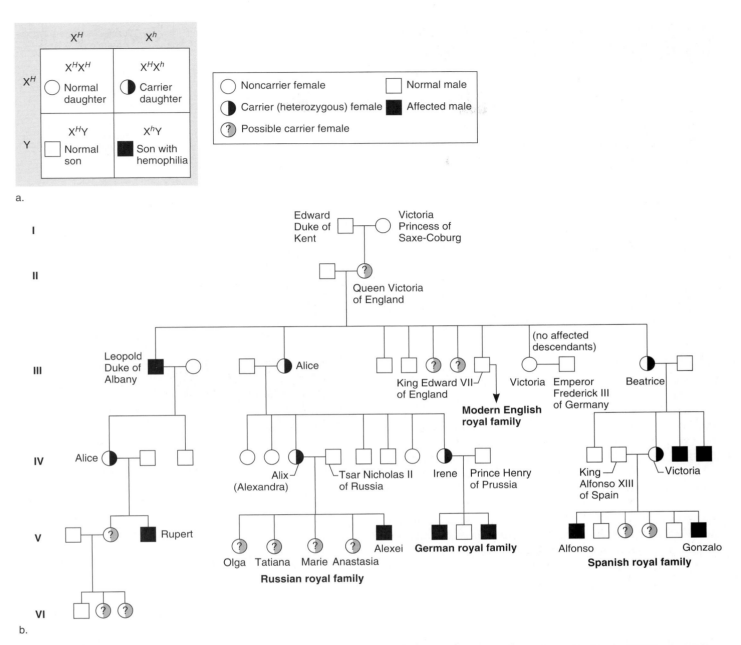

Figure 6.6 **Hemophilia B.** **(a)** This X-linked recessive disease usually passes from a heterozygous woman (designated $X^H X^h$, where h is the hemophilia-causing allele) to heterozygous daughters or hemizygous sons. The father is normal. **(b)** The disorder has appeared in the royal families of England, Germany, Spain, and Russia. The modern royal family in England does not carry hemophilia.

She passed it on through carrier daughters and one mildly affected son.

For many years, historians assumed that the royal families of Europe suffered from the more common form of hemophilia, type A (MIM 306700). However, researchers recently tested bits of bone from Tsarina Alexandra and Crown Prince Alexei for the wild type and mutant alleles for the genes that cause hemophilia A as well as B. Both genes are on the X chromosome. The result: Mother Alexandra was a carrier of the rarer hemophilia B, and son Crown Prince Alexei had the

disease, which had been known from documents describing his severe bleeding episodes.

The transmission pattern of hemophilia B is consistent with the criteria for an X-linked recessive trait listed in table 6.2. A daughter can inherit an X-linked recessive disorder or trait if her father is affected and her mother is a carrier, because the daughter inherits one affected X chromosome from each parent. Without a biochemical test, though, an unaffected woman would not know she is a carrier for an X-linked recessive trait unless she has an affected son.

Table 6.3	Criteria for an X-Linked Dominant Trait

1. Expressed in females in one copy

2. Much more severe effects in males

3. High rates of miscarriage due to early lethality in males

4. Passed from male to all daughters but to no sons

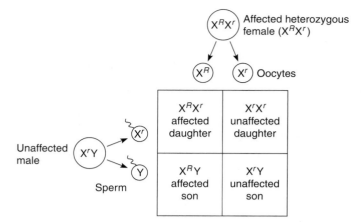

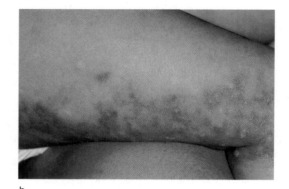

a.

b.

Figure 6.7 X-linked dominant inheritance. (a) A female who has an X-linked dominant trait has a 1 in 2 probability of passing it to her offspring, male or female. Males are generally more severely affected than females. **(b)** Note the characteristic patchy pigmentation on the leg of a girl who has incontinentia pigmenti.

A woman whose brother has hemophilia B has a 1 in 2 risk of being a carrier. Both her parents are healthy, but her mother must be a carrier because her brother is affected. Her risk is the chance that she has inherited the X chromosome bearing the hemophilia allele from her mother. The chance of the woman conceiving a son is 1 in 2, and of that son inheriting hemophilia is 1 in 2. Using the product rule, the risk that she is a carrier and will have a son with hemophilia, out of all the possible children she can conceive, is $1/2 \times 1/2 \times 1/2$, or 1/8.

X-Linked Dominant Inheritance

Dominant X-linked conditions and traits are rare. Again, gene expression differs between the sexes (**table 6.3**). A female who inherits a dominant X-linked allele has the associated trait or illness, but a male who inherits the allele is usually more severely affected because he has no other allele to mask its effect. The children of a normal man and a woman with a dominant, disease-causing allele on the X chromosome face the risks summarized in **figure 6.7**.

An example of an X-linked dominant condition is incontinentia pigmenti (IP) (MIM 308300). In affected females, swirls of skin pigment arise when melanin penetrates the deeper skin layers. A newborn girl with IP has yellow, pus-filled vesicles on her limbs that come and go over the first few weeks. Then the lesions become warty and eventually give way to brown splotches that may remain for life, although they fade with time. Males with the condition are so severely affected that they do not survive to be born. This is why women with the disorder have a miscarriage rate of about 25 percent. The gene that causes IP (called *NEMO*) activates genes that carry out the immune response and apoptosis in structures that derive from ectoderm, such as skin, hair, nails, eyes, and the brain.

Another X-linked dominant condition, congenital generalized hypertrichosis (MIM 307150), produces many extra hair follicles, and hence denser and more abundant upper body hair (**figure 6.8**). Hair growth is milder and patchier in females because of hormonal differences and the presence of a second X chromosome. Figure 6.8*b* shows part of a pedigree of a large Mexican family with 19 members who have this X-linked dominant condition. The affected man in the pedigree passed the trait to all four daughters, but to none of his nine sons. Because sons inherit the X chromosome from their mother, and only the Y from their father, they could not have inherited the hairiness from their father.

Solving a Problem of X-Linked Inheritance

Mendel's first law (segregation) applies to genes on the X chromosome. The same logic is used to solve problems as to trace traits transmitted on autosomes, with the added step of considering the X and Y chromosomes in Punnett squares. Follow these steps:

1. Look at the pattern of inheritance. Different frequencies of affected males and females in each generation may suggest X linkage. For an X-linked recessive trait:
   - An affected male has a carrier mother.
   - An unaffected female with an affected brother has a 50 percent (1 in 2) chance of being a carrier.
   - An affected female has a carrier or affected mother *and* an affected father.
   - A carrier (female) has a carrier mother or an affected father.

   For an X-linked dominant trait:
   - There may be no affected males, because they die early.
   - An affected female has an affected mother.

2. Draw the pedigree.

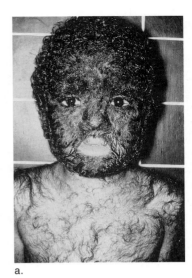

a.

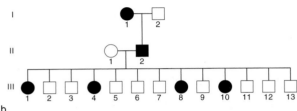

b.

Figure 6.8 **An X-linked dominant condition.** **(a)** This 6-year-old child has congenital generalized hypertrichosis. **(b)** In this partial pedigree of a large Mexican family, the affected male in the second generation passed the condition to all of his daughters and none of his sons. This is because he transmits his X chromosome only to females.

3. List all genotypes and phenotypes and their probabilities.
4. Assign genotypes and phenotypes to the parents. Consider clues in the phenotypes of relatives.
5. Determine how alleles separate into gametes for the genes of interest on the X and Y chromosomes.
6. Unite the gametes in a Punnett square.
7. Determine the phenotypic and genotypic ratios for the F_1 generation.
8. To predict further generations, use the genotypes of the F_1 and repeat steps 4 through 6.

Consider as an example Kallmann syndrome (MIM 308700), which causes very poor or absent sense of smell and small testes or ovaries. It is X-linked recessive. Tanisha does not have Kallmann syndrome, but her brother Jamal and her maternal cousin Malcolm (her mother's sister's child) have it. Tanisha's and Malcolm's parents are unaffected, as is Tanisha's husband Sam. Tanisha and Sam wish to know the risk that a son would inherit the condition. Sam has no affected relatives.

Solution

Mode of inheritance: The trait is X-linked recessive because males are affected through carrier mothers.

K = wild type k = Kallmann syndrome

Genotypes	Phenotypes
$X^K X^K$, $X^K X^k$, $X^K Y$	normal
$X^k X^k$, $X^k Y$	affected

Individual	Genotype	Phenotype	Probability
Tanisha	$X^K X^k$ or $X^K X^K$	normal (carrier)	50% each
Jamal	$X^k Y$	affected	100%
Malcolm	$X^k Y$	affected	100%
Sam	$X^K Y$	normal	100%

Tanisha's gametes

if she is a carrier: X^K X^k

Sam's gametes: X^K Y

Punnett square

	X^K	X^k
X^K	$X^K X^K$	$X^K X^k$
Y	$X^K Y$	$X^k Y$

Interpretation: If Tanisha is a carrier, the probability that their son will have Kallmann syndrome is 50 percent, or 1 in 2. (This is a conditional probability. The chance that any son will have the condition is actually 1 in 4, because Tanisha also has a 50 percent chance of being genotype $X^K X^K$ and therefore not a carrier.)

Key Concepts

1. Y-linked traits are passed on the Y chromosome, and X-linked traits on the X.

2. Because a male is hemizygous, he expresses all the genes on his X chromosome, whereas a female expresses recessive alleles on the X chromosome only if she is homozygous recessive.

3. X-linked recessive traits have a 50 percent probability of passing from carrier mothers to sons.

4. X-linked dominant conditions are expressed in both males and females but are more severe in males.

5. Mendel's first law can be used to solve problems involving X-linked genes.

6.3 Sex-Limited and Sex-Influenced Traits

An X-linked recessive trait generally is more prevalent in males than females. Other situations, however, can affect gene expression in the sexes differently.

Sex-Limited Traits

A **sex-limited trait** affects a structure or function of the body that is present in only males or only females. The gene for such a trait may be X-linked or autosomal.

Understanding sex-limited inheritance is important in animal breeding. For example, a New Zealand cow named Marge, who has a mutation that makes her milk very low in saturated fat, is founding a commercial herd. Males play their part by transmitting the mutation, even though they do not make milk. In humans, beard growth is sex-limited. A woman does not grow a beard because she does not manufacture the hormones required for facial hair growth. She can, however, pass to her sons the genes specifying heavy beard growth.

An inherited medical condition that arises during pregnancy is obviously sex-limited, but the male genome contributes to the development of supportive structures, such as the placenta. This is the case for preeclampsia, a sudden rise in blood pressure late in pregnancy. It kills 50,000 women worldwide each year. A study of 1.7 million pregnancies in Norway found that if a man's first wife had preeclampsia, his second wife had double the risk of developing the condition, too. Another study found that women whose mothers-in-law developed preeclampsia when pregnant with the women's husbands had approximately twice the rate of developing the condition themselves. Perhaps a gene from the father affects the placenta in a way that elevates the pregnant woman's blood pressure.

Sex-Influenced Traits

In a **sex-influenced trait,** an allele is dominant in one sex but recessive in the other. Such a gene may be X-linked or autosomal. The difference in expression can be caused by hormonal differences between the sexes. For example, an autosomal gene for hair growth pattern has two alleles, one that produces hair all over the head and another that causes pattern baldness. The baldness allele is dominant in males but recessive in females, which is why more men than women are bald. A heterozygous male is bald, but a heterozygous female is not. A bald woman is homozygous recessive. Even a bald woman tends to have some wisps of hair, whereas an affected male may be completely hairless on the top of his head.

Key Concepts

1. A sex-limited trait affects body parts or functions present in only one gender.
2. A sex-influenced allele is dominant in one sex but recessive in the other.

6.4 X Inactivation

Females have two alleles for every gene on the X chromosomes, whereas males have only one. In mammals, a mechanism called **X inactivation** balances this apparent inequality in the expression of genes on the X chromosome.

Equaling Out the Sexes

By the time a female mammalian embryo consists of 8 cells, about 75 percent of the genes on one X chromosome in each cell are inactivated, and the remaining 25 percent are expressed to different degrees in different women. Which X chromosome is mostly turned off in each cell—the one inherited from the mother or the one from the father—is usually random. As a result, a female mammal expresses the X chromosome genes inherited from her father in some cells and those from her mother in others. She is, therefore, a mosaic for expression of most genes on the X chromosome (**figure 6.9**).

By studying rare human females who have lost a small part of one X chromosome, researchers identified a specific region, the X inactivation center, that shuts off much of the chromosome. Genes in the pseudoautosomal regions and some other genes escape inactivation. A gene called *XIST* controls X inactivation. It encodes an RNA that binds to a specific site on the same (inactivated) X chromosome. From this point out to the chromosome tip, the X chromosome is silenced.

Once an X chromosome is inactivated in one cell, all its daughter cells have the same X chromosome shut off. Because the inactivation occurs early in development, the adult female has patches of tissue that differ in their expression of X-linked genes. With each cell in her body having only one active X chromosome, she is roughly equivalent to the male in terms of gene expression.

X inactivation can alter the phenotype (gene expression), but not the genotype. The change is reversed in germline cells destined to become oocytes, and this is why a fertilized ovum does not have an inactivated X chromosome. X inactivation is an example of an **epigenetic** change—one that is passed from one cell generation to the next but that does not alter the DNA base sequence.

We can observe X inactivation at the cellular level because the turned-off X chromosome absorbs a stain much faster than the active X. This differential staining occurs because inactivated DNA has chemical methyl (CH_3) groups bound to it that prevents it from being transcribed into RNA and also enables DNA to absorb stain.

X inactivation can be used to check the sex of an individual. The nucleus of a cell of a female, during interphase, has one dark-staining X chromosome called a Barr body. A cell from a male has no Barr body because his one X chromosome remains active. The term honors English geneticist Mary Lyon, who proposed in 1961 that the spot is the X chromosome that is turned off in early development.

Effect on the Phenotype

The consequence of X inactivation on the phenotype can be interesting. For homozygous X-linked genotypes, X inactivation has no effect. No matter which X chromosome is turned off, the same allele is left to be expressed. For heterozygotes, however, X inactivation leads to expression of one allele or the other. This doesn't affect health if enough cells express

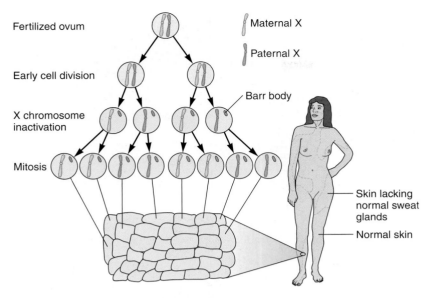

Figure 6.9 **X inactivation.** A female is a mosaic for expression of genes on the X chromosome because of the random inactivation of either the maternal or paternal X in each cell early in prenatal development. In anhidrotic ectodermal dysplasia, a woman has patches of skin that lack sweat glands and hair. (Colors distinguish cells with the inactivated X, not to depict skin color.)

the functional gene product. However, some traits reveal the X inactivation. The swirls of skin color in incontinentia pigmenti (IP) patients reflect patterns of X inactivation in skin cells (see figure 6.7*b*). Where the normal allele for melanin pigment is shut off, pale swirls develop. Where pigment is produced, brown swirls result.

A female who is heterozygous for an X-linked recessive gene can express the associated condition if the normal allele is inactivated in the tissues that the illness affects. Consider a carrier of hemophilia A. If the X chromosome carrying the normal allele for the clotting factor is turned off in the liver, then the woman's blood will clot slowly enough to cause mild hemophilia. (Luckily for her, slowed clotting time also greatly reduces her risk of cardiovascular disease caused by blood clots blocking circulation.) A carrier of an X-linked trait who expresses the phenotype is called a **manifesting heterozygote**.

Whether or not a manifesting heterozygote results from X inactivation depends upon how adept cells are at sharing. Consider two lysosomal storage disorders, which are deficiencies of specific enzymes that normally dismantle cellular debris in lysosomes. In Hunter syndrome (MIM 309900, also called mucopolysaccharidosis II), cells that make the enzyme readily send it to neighboring cells that do not, essentially correcting the defect in cells that can't make the enzyme. Carriers of Hunter syndrome do not have symptoms because cells get enough enzyme. Boys with Hunter syndrome are deaf, intellectually disabled, have dwarfism and abnormal facial features, heart damage, and enlarged liver and spleen. In contrast, in Fabry disease (MIM 301500), cells do not readily release the enzyme alpha-galactosidase A, so a female who is a heterozygote

may have cells in the affected organs that lack the enzyme. She may develop mild symptoms of this disorder that causes skin lesions, abdominal pain, and kidney failure in boys.

A familiar example of X inactivation is the coat colors of tortoiseshell and calico cats. An X-linked gene confers brownish-black (dominant) or yellowish-orange (recessive) color. A female cat heterozygous for this gene has patches of each color, forming a tortoiseshell pattern that reflects different cells expressing either of the two alleles (**figure 6.10**). The earlier the X inactivation, the larger the patches, because more cell divisions can occur after the event, producing more daughter cells. White patches may form due to epistasis by an autosomal gene that shuts off pigment synthesis. A cat with colored patches against such a white background is a calico. Tortoiseshell and calico cats are nearly always female. A male can have these coat patterns only if he inherits an extra X chromosome.

In humans, X inactivation can be used to identify carriers of some X-linked disorders. This is the case for Lesch-Nyhan syndrome (MIM 300322),

Figure 6.10 **Visualizing X inactivation.** X inactivation is obvious in a calico cat. X inactivation is rarely observable in humans because most cells do not remain together during development, as a cat's skin cells do.

in which an affected boy has cerebral palsy; bites his fingers, toes, and lips to the point of mutilation; is intellectually disabled and passes painful urinary stones. Mutation results in defective or absent HGPRT, an enzyme. A woman who carries Lesch-Nyhan syndrome can be detected when hairs from widely separated parts of her head are tested for HGPRT. (Hair is used for the test because it is accessible and produces the enzyme.) If some hairs contain HGPRT but others do not, she is a carrier. The hair cells that lack the enzyme have turned off the X chromosome that carries the normal allele; the hair cells that manufacture the normal enzyme have turned off the X chromosome that carries the disease-causing allele. The woman is healthy because her brain has enough HGPRT, but each son has a 50 percent chance of inheriting the disease. **Reading 6.2** discusses another syndrome affected by X inactivation.

Subtle Effects of X Inactivation

Theoretically, X inactivation evens out the sexes for expression of X-linked genes. In actuality, however, a female may *not* be equivalent, in gene expression, to a male because she has two cell populations, whereas a male has only one. One of a female's two cell populations has the X she inherited from her father active, and the other has the X chromosome she inherited from her mother active. For heterozygous X-linked genes, she would have some cells that manufacture the protein encoded by one allele, and some cells that produce the protein encoded by the other allele. Although most heterozygous genes have the alleles about equally represented, sometimes X inactivation can be skewed. That is, most cells express the X inherited from the same parent. This can happen if one of the X chromosomes includes an expressed allele that confers a greater rate of cell division than the different allele from the other parent, giving certain cells a survival advantage.

Another way that X inactivation makes a female different from a male is seen when the proteins encoded by different alleles interact. This can be beneficial or harmful. A beneficial example of dual expression of alleles occurs in certain types of monkeys in which an X-linked visual pigment gene has two alleles. Females who are homozygous for this gene and males have two-color vision, but lucky female monkeys who are heterozygous for this gene enjoy three-color vision.

A situation in which being a heterozygote for an X-linked gene is harmful is craniofrontonasal syndrome (MIM 304110) (**figure 6.11**). Males and homozygous females have asymmetrical facial features. However, heterozygous females have a much more severe phenotype, with very abnormal faces resulting from abnormal fusing of the skull bones. (It is highly unusual for the heterozygote to be more severely affected than the homozygous recessive individual.) An explanation is that the encoded protein is part of a signal transduction pathway that controls the bone fusion, and when two forms of that protein are made in the female heterozygote, the signal is disrupted

Figure 6.11 An unusual case in which being heterozygous is not protective. Craniofrontonasal syndrome is more severe in females because of an unusual interaction between two forms of a protein in a heterozygote.

in a way that blocks the cells that form the sutures of the skull from joining cleanly.

Key Concepts

1. In female mammals, X inactivation compensates for differences between males and females in the numbers of gene copies on the X chromosome.
2. Early in development, one X chromosome in each cell of the female is turned off.
3. The effects of X inactivation can be noticeable when heterozygous alleles are expressed in certain tissues.

6.5 Genomic Imprinting

In Mendel's pea experiments, it didn't matter whether a trait came from the male or female parent. For certain genes in mammals, however, parental origin does influence the phenotype. These genes are said to be imprinted. In **genomic imprinting,** methyl (CH_3) groups cover a gene or several linked genes and prevent them from being accessed to synthesize protein (**figure 6.12**).

For a particular imprinted gene, the copy inherited from either the father or the mother is always covered with methyls, even in different individuals. The result of this gene cloaking is that a disease may be more severe, or different, depending upon which parent transmitted the mutant allele. That is, a particular gene might function if it came from the father, but not if it came from the mother, or vice versa.

Rett Syndrome—A Curious Inheritance Pattern

Before the age of the Internet, identifying and describing a new syndrome could take years, or even decades. This was the case for Rett syndrome, a neurological condition that affects females.

In 1954, Austrian pediatrician Andreas Rett and his nurse noticed that eight young female patients moved their hands uncontrollably (**figure 1**). They'd tap objects, clap, put their hands in their mouths, and most commonly, wring their hands. The girls all had been developing normally, but then gradually lost muscle tone. Growth of their heads slowed. As time went on, seizures began, they lost the ability to speak, and they became completely disabled.

Dr. Rett filmed the girls and went around Europe looking for other cases. Meanwhile, other pediatricians were noting the symptoms in their patients. Dr. Rett published his observations in European journals, but they did not attract attention until 1983, when the syndrome was named after him.

In 1999, Ruthie Amir, at the Baylor College of Medicine in Texas, discovered the gene behind the disorder—*MECP2*, for methyl-CpG-binding protein 2 (MIM 312750), on the X chromosome. The syndrome affects several organ systems because the gene adds methyl groups to other genes, silencing them.

Rett syndrome is dominant. In 99 percent of cases, it arises anew, from a mutation in an X-bearing sperm cell. Rarely, Rett syndrome may be inherited from a woman who has a very mild case because, by chance, the X chromosomes bearing the mutation are silenced in her brain cells.

Rett syndrome has been difficult to study because nerve cells do not divide, and so researchers could not sustain them in culture. In 2010, researchers reprogrammed skin fibroblast cells from four girls with Rett syndrome, providing enough nerve cells to study the disease at the cellular level. Rett syndrome neurons are

Figure 1 Rett syndrome affects girls. One sign of Rett syndrome is holding and wringing the hands.

too small, with too few connections and abnormal signaling. Most importantly, drugs that helped mice with a form of Rett syndrome also corrected the defect in the reprogrammed cells from the four little girls. A treatment may be near!

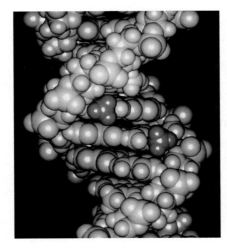

Figure 6.12 Methyl (CH_3) groups (red) "silence" certain genes.

Silencing the Contribution from One Parent

Imprinting is an epigenetic alteration. It is a layer of meaning stamped upon a gene without changing its DNA sequence. The imprinting pattern is passed from cell to cell in mitosis, but not from individual to individual through meiosis. When silenced DNA is replicated during mitosis, the pattern of blocked genes is exactly placed, or imprinted, on the new DNA, covering the same genes as in the parental DNA (**figure 6.13**). In this way, the "imprint" of inactivation is perpetuated, as if each such gene "remembers" which parent it came from. In meiosis, however, imprints are removed and reset. As oocyte and sperm form, the CH_3 groups shielding their imprinted genes are stripped away, and new patterns are set down, depending upon whether the fertilized ovum chromosomally is male (XY) or female (XX). In this way, women can have sons and men can have daughters without passing on their sex-specific parental imprints.

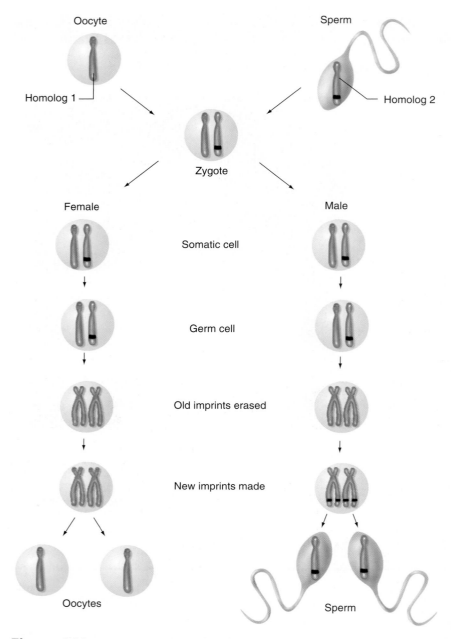

Figure 6.13 Genomic imprinting. Imprints are erased during meiosis, then reinstituted according to the sex of the new individual.

Labels within figure: Oocyte, Sperm, Homolog 1, Homolog 2, Zygote, Female, Male, Somatic cell, Germ cell, Old imprints erased, New imprints made, Oocytes, Sperm

and placenta. This apparent requirement for opposite-sex parents was discovered in the early 1980s, through experiments on early mouse embryos and examination of certain rare pregnancy problems in humans.

Researchers created fertilized mouse ova that contained two male pronuclei or two female pronuclei, instead of one from each. Results were strange. When the fertilized ovum had two male genomes, a normal placenta developed, but the embryo was tiny and quickly stopped developing. A zygote with two female pronuclei, developed into an embryo, but the placenta was grossly abnormal. Therefore, the male genome controls placenta development, and the female genome, embryo development.

The mouse results were consistent with abnormalities of human development. When two sperm fertilize an oocyte and the female pronucleus degenerates, an abnormal growth of placenta-like tissue called a hydatidiform mole forms. If a fertilized ovum contains two female genomes but no male genome, a mass of random differentiated tissue, called a teratoma, grows. A teratoma, which means "monster cancer," may consist of a strange mix of tissues, such as skin and teeth. With either a hydatidiform mole or a teratoma, no embryo results.

Genomic imprinting can explain incomplete penetrance, in which an individual is known to have inherited a genotype associated with a particular phenotype, but has no signs of the trait. This is the case for a person with normal fingers whose parent and child have polydactyly. An imprinted gene silences the dominant mutant allele.

Imprinting may be an important concern in assisted reproductive technologies that manipulate gametes to treat infertility. For example, the otherwise very rare Beckwith-Wiedemann syndrome (MIM 130650) is more prevalent among the offspring of people who used *in vitro* fertilization and intracytoplasmic sperm injection (discussed in chapter 21) to become pregnant.

The function of genomic imprinting isn't well understood. However, because many imprinted genes take part in early development, particularly of the brain, it may be a way to finely regulate the abundance of key proteins in the embryo. The fact that some genes lose their imprints after birth supports this idea of early importance. Also, imprinted genes are in clusters along a chromosome, and are controlled by other regions of DNA called imprinting centers. Perhaps one gene in a cluster is essential for early development, and the others become imprinted simply because they are nearby—a bystander effect.

Genomic imprinting has implications for understanding early human development. It suggests that for mammals, two opposite-sex parents are necessary to produce a healthy embryo

Imprinting Disorders in Humans

The number of imprinted genes in the human genome exceeds 156, and at least 60 of them affect health when abnormally expressed. The effects of genomic imprinting are revealed only when an individual has one copy of a normally imprinted allele and the other, active allele is inactivated or deleted.

Imprinting disorders can be dramatic, such as two different syndromes that arise from small deletions in the same region of chromosome 15 (**figure 6.14**). A child with Prader-Willi syndrome (MIM 176270) is small at birth and in infancy has difficulty gaining weight. Between ages 1 and 3, the child

Chromosome 15

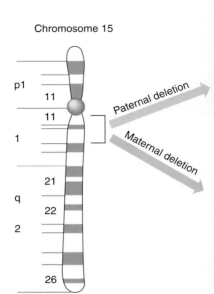

p1

11

11

1

q

21

22

2

26

a.

Paternal deletion

Maternal deletion

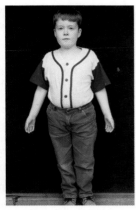

b. Prader-Willi syndrome

c. Angelman syndrome

Figure 6.14 **Prader-Willi and Angelman syndromes result when the nonimprinted copy of a gene is deleted.**
(a) Two distinct syndromes result from missing genetic material in the same region of chromosome 15. **(b)** Tyler has Prader-Willi syndrome, due to a deletion in the copy of the chromosome he inherited from his father. Note his small hands. **(c)** This child has Angelman syndrome, caused by a deletion in the chromosome 15 that he inherited from his mother. He is intellectually disabled.

develops an obsession with eating. Unless the diet can be controlled, severe obesity results because another symptom is a very slow metabolism. Parents actually lock kitchen cabinets and refrigerators to keep their children from literally eating themselves to death by bursting digestive organs. The other condition, Angelman syndrome (MIM 105830), causes autism and intellectual disability, an extended tongue, large jaw, poor muscle coordination, and convulsions that make the arms flap. In many cases of Prader-Willi syndrome, only the mother's chromosome 15 region is expressed; the father's chromosome is deleted in that region. In Angelman syndrome, the reverse occurs: The father's gene (or genes) is expressed, and the mother's chromosome has the deletion.

Symptoms of Prader-Willi syndrome arise because several paternal genes that are not normally imprinted (that is, that are normally active) are missing. In Angelman syndrome, a normally active single maternal gene is deleted. This part of chromosome 15 is especially unstable because highly repetitive DNA sequences bracket the genes that cause the symptoms.

Imprinting gone awry is associated with forms of diabetes mellitus, autism, Alzheimer disease, schizophrenia, and male homosexuality. A clue that indicates a condition is associated with genomic imprinting is differing severity, depending upon whether it is inherited from the father or mother.

Key Concepts

1. In genomic imprinting, the phenotype differs depending upon whether a gene is inherited from the mother or the father.

2. Methyl groups may bind to DNA and temporarily suppress gene expression in a pattern determined by the individual's sex.

3. Imprinting may be a normal process in mammalian embryos.

Summary

6.1 Our Sexual Selves

1. Sexual identity includes sex chromosome makeup; gonadal specialization; phenotype (reproductive structures); and gender identity.

2. The human male is the **heterogametic sex,** with an X and a Y chromosome. The female, with two X chromosomes, is the **homogametic sex**.

3. The human Y chromosome includes two pseudoautosomal regions and a large, male-specific region that does not recombine. Y-linked genes may correspond to X-linked genes, be similar to them, or be unique. Palindromic DNA sequences or inverted repeats can promote gene loss on the Y.

4. If the *SRY* gene is expressed, undifferentiated gonads develop as testes. If *SRY* is not expressed, the gonads develop as ovaries, under the direction of other genes.

5. Starting 8 weeks after fertilization, the testes secrete a hormone that prevents development of female structures and testosterone, which triggers development of male structures.

6. Testosterone converted to DHT controls development of the urethra, prostate gland, penis, and scrotum. If *SRY* is not turned on, the Müllerian ducts continue to develop into female reproductive structures.

7. Evidence points to an inherited component to homosexuality.

8. **Sex ratio** is the number of males divided by the number of females multiplied by 1,000, for people of a particular age. Interfering with pregnancy outcomes can skew sex ratios.

6.2 Traits Inherited on Sex Chromosomes

9. Y-linked traits are rare and are passed from fathers to sons only.

10. Males are **hemizygous** for genes on the X chromosome and express phenotypes associated with these genes because they do not have another allele on a homolog. An X-linked trait passes from mother to son because he inherits his X from his mother and his Y from his father.

11. An X-linked allele may be dominant or recessive. X-linked dominant traits are more devastating to males.

6.3 Sex-Limited and Sex-Influenced Traits

12. **Sex-limited traits** may be autosomal or sex-linked, but they only affect one sex because of anatomical or hormonal gender differences.

13. A **sex-influenced trait** is dominant in one sex but recessive in the other.

6.4 X Inactivation

14. **X inactivation** shuts off one X chromosome in each cell in female mammals, making them mosaics for heterozygous genes on the X chromosome. It evens out the dosages of genes on the sex chromosomes between the sexes.

15. A female who expresses the phenotype corresponding to an X-linked gene she carries is a **manifesting heterozygote**.

6.5 Genomic Imprinting

16. In **genomic imprinting,** the phenotype corresponding to a particular genotype differs depending upon whether the parent who passes the gene is female or male.

17. Imprints are erased during meiosis and reassigned based upon the sex of a new individual.

18. Methyl groups that temporarily suppress gene expression are the physical basis of genomic imprinting.

www.mhhe.com/lewisgenetics10

Answers to all end-of-chapter questions can be found at **www.mhhe.com/lewisgenetics10.** You will also find additional practice quizzes, animations, videos, and vocabulary flashcards to help you master the material in this chapter.

Review Questions

1. How is sex expressed at the chromosomal, gonadal, phenotypic, and gender identity levels?

2. How do genes in the pseudoautosomal region of the Y chromosome differ from genes in the male-specific region (MSY)?

3. Describe the phenotypes of
   a. a person with a deletion of the *SRY* gene.
   b. a normal XX individual.
   c. an XY individual with a block in testosterone synthesis.

4. List the events required for a fetus to develop as a female.

5. Cite evidence that may point to a hereditary component to homosexuality.

6. Why is it unlikely one would see a woman who is homozygous for an X-linked dominant condition?

7. What is the basis of sex ratio at birth?

8. Traits that appear more frequently in one sex than the other may be caused by genes that are inherited in an X-linked,

sex-limited, or sex-influenced fashion. How might you distinguish among these possibilities in a given individual?

9. Why are male calico cats very rare?

10. How might X inactivation cause patchy hairiness in women who have congenital generalized hypertrichosis, even though the disease-causing allele is dominant?

11. How does X inactivation even out the "doses" of X-linked genes between the sexes?

12. Cite evidence that genetic contributions from both parents are necessary for normal prenatal development.

13. Prader-Willi and Angelman syndromes are more common in children conceived with certain assisted reproductive technologies (*in vitro* fertilization and intracytoplasmic sperm injection) than among the general population. What process may these procedures disrupt?

Applied Questions

1. To answer the following questions, consider these population data on sex ratios:

Selected sex ratios at birth		Selected sex ratios after age 65	
Nation	Sex ratio	Nation	Sex ratio
Costa Rica	970	Rwanda	620
Tanzania	1,000	South Africa	630
Liechtenstein	1,010	France	700
South Africa	1,020	United States	720
United States	1,050	Qatar	990
Sweden	1,060	Montserrat	1,060
Italy	1,070	Bangladesh	1,160
China	1,130	Nigeria	990

   a. In Rwanda, South Africa, France, and the United States, males die, on average, significantly younger than females. What types of information might explain the difference?

   b. In Costa Rica, how many males at birth are there for every 100 females?

   c. In which country listed do males tend to live the longest?

2. In severe Hunter syndrome, lack of the enzyme iduronate sulfate sulfatase leads to buildup of certain carbohydrates swelling the liver, spleen, and heart. In mild cases, deafness may be the only symptom. Intellect is usually unimpaired, and life span can be normal. Hunter syndrome is X-linked recessive. A man with mild Hunter syndrome has a child with a woman who is a carrier.

   a. What is the probability that a son inherits Hunter syndrome?

   b. What is the chance that a daughter inherits Hunter syndrome?

   c. What is the chance that a daughter is a carrier?

3. Amelogenesis imperfecta (MIM 301200) is X-linked dominant. Affected males have extremely thin enamel on each tooth. Female carriers have grooved teeth from uneven deposition of enamel. Why might the phenotype differ between the sexes?

Web Activities

1. Visit the National Center for Biotechnology Information (NCBI) website. Identify an X-linked disorder, then find it in MIM and describe it.

2. At the Imprinted Gene Catalogue website, click on "search by species name" and then click on "complete list." Find two disorders that involve imprinting, one transmitted from the mother and one from the father, and use MIM to describe them.

Case Studies and Research Results

1. For each case description, identify the principle at work from the list that follows. More than one answer per case may apply.

   A. Y-linked inheritance

   B. X-linked recessive

   C. X-linked dominant inheritance

   D. Sex-limited inheritance

   E. Sex-influenced inheritance

   F. X inactivation or manifesting heterozygote

   G. Uniparental disomy

   H. Imprinting abnormality

   a. In a three-generation family, sixteen members have speech-language disorder (MIM 602081) and cannot speak. The gene that is mutant is called *FOXP2*. The speechless family members inherited both copies of the gene from their mothers and none from their fathers.

   b. Six-year-old LeQuan inherited Fabry disease (MIM 301500) from his mother, who is a heterozygote for the causative mutation. The gene, on the X chromosome, encodes a lysosomal enzyme. LeQuan would die before age 50 of heart failure, kidney failure, or a stroke, but fortunately he can be treated with twice-monthly infusions of the enzyme. His mother, Echinecea, recently began experiencing recurrent fevers, a burning pain in her hands and feet, a rash, and sensitivity to cold. She is experiencing mild Fabry disease.

   c. The Chandler family has many male members who have a form of retinitis pigmentosa (RP) in which the cells that capture light energy in the retina degenerate, causing gradual visual loss. Several female members of the family presumed to be carriers because they have affected sons are tested for RP genes on chromosomes 1, 3, 6, and the X, but do not carry these RP genes. Many years ago, Rachel married her cousin Ross, who has the family's form of RP.

They had six children. The three sons are all affected, but their daughters all have normal vision.

d. Simon's mother and her sister are breast cancer survivors, and their mother died of the disease. Simon's sister Maureen has a genetic test and learns that she, too, has inherited the *BRCA1* gene. Simon has two daughters, but doesn't want to be tested because he thinks a man cannot transmit a trait that affects a body part that is more developed in females.

e. Tribbles are extraterrestrial mammals that long ago invaded a starship on the television program *Star Trek*. A gene called *frizzled* causes kinky hair in female tribbles who inherit just one allele. However, two mutant alleles must be inherited for a male tribble to have kinky hair.

f. Prozac died at age 16 of Lowe syndrome (MIM 309000). He was slightly intellectually disabled, had visual problems (cataracts and glaucoma), seizures, poor muscle tone, and progressive kidney failure, which was ultimately fatal. His sister Lunesta is pregnant, and wonders whether she is a carrier of the disease that killed her brother. She remembers a doctor saying that her mother Yaz was a carrier. Lunesta's physician determines that she is a carrier because she has cataracts, which is a clouding of the lenses. It has not yet affected her vision. When a prenatal test reveals that Lunesta's fetus is a female, her doctor tells her not to worry about Lowe syndrome.

g. Mating among Texas field crickets depends upon females responding to a male mating call. The sounds must arrive at a particular frequency to excite the females, who do not sing back in response. However, females can pass on a trait that confers frequency of singing.

h. When Winthrop was a baby, he was diagnosed with "failure to thrive." At 14 months of age, he suddenly took an interest in food, and his parents couldn't feed him fast enough. By age 4, Winthrop was obese, with disturbing behavior. He was so hungry that after he'd eaten his meal and everyone else's leftovers, he'd hunt through the garbage for more. Finally a psychiatrist who had a background in genetics diagnosed Prader-Willi syndrome. Testing showed that the allele for the Prader-Willi gene that Winthrop had inherited from his father was abnormally methylated.

i. Certain breeds of dogs have cryptorchidism, in which the testicles do not descend into the scrotum. The trait is passed through females.

2. Reginald has mild hemophilia A that he can control by taking a clotting factor. He marries Lydia, whom he met at the hospital where he and Lydia's brother, Marvin, receive their treatment. Lydia and Marvin's mother and father, Emma and Clyde, do not have hemophilia. What is the probability that Reginald and Lydia's son will inherit hemophilia A?

3. Harold works in a fish market, but the odor does not bother him because he has anosmia (MIM 301700), an X-linked recessive lack of sense of smell. Harold's wife, Shirley, has a normal sense of smell. Harold's sister, Maude, also has a normal sense of smell, as does her husband, Phil, and daughter, Marsha, but their identical twin boys, Alvin and Simon, cannot detect odors. Harold and Maude's parents, Edgar and Florence, can smell normally. Draw a pedigree for this family, indicating people who must be carriers of the anosmia gene.

4. Caster Semenya is a South African sprinter who won in the 800-meter race at the World Championships in 2009. Because her time had dropped significantly from earlier races, the International Association of Athletics Federations asked her to take a gender test. The implication of the much-publicized request was that Semenya was really a male. For reasons of privacy, the results of her test were never released. Which gene would have been tested to determine Semenya's gender?

Should parents submit their children's DNA for genetic testing to determine which sports they should play?

Multifactorial Traits

Learning Outcomes

7.1 Genes and the Environment Mold Traits

1. Distinguish between single-gene and polygenic traits.
2. Define multifactorial traits.

7.2 Polygenic Traits Are Continuously Varying

3. Explain how continuously varying traits reflect genes and the environment.

7.3 Traditional Ways to Investigate Multifactorial Traits

4. Explain how empiric risk differs from calculating a Mendelian frequency.
5. Define *heritability*.
6. Discuss what studies on adopted individuals and twins can reveal.

7.4 Genome-Wide Association Studies

7. Explain what a genome-wide association study can reveal.
8. Discuss limitations of genome-wide association studies.

7.5 A Closer Look: Body Weight

9. Discuss tools and approaches used to study body weight.

The Big Picture: Who we are and how we feel arises from an intricate interplay among our genes and environmental influences. Understanding genetic contributions to traits and illnesses can suggest how we can alter our environments.

The Genetics of Athletics

A website offers a single-gene test that "gives parents and coaches early information on success in team or individual speed/power or endurance sports." The test is for variants of a gene that encodes a protein called actinin 3 that is expressed in skeletal muscle. One genotype is overrepresented among elite sprinters, another among world-class endurance runners. The test takes a simplistic view of a complex trait.

Athletic ability is multifactorial—determined by many genes as well as the environment. Any single gene is unlikely to have a great influence. Environmental factors include exposure to pollution and toxins, as well as opportunities to participate in sports. The ability to work well on a team cannot be reduced to a simple string of DNA letters.

The idea to market athletic genes may have come from rare mutations in other genes that bestow great physical prowess. Members of a German family with a mutation in a "double muscle" gene are amazing weight lifters, while a Scandinavian family of Olympic skiers has a mutation that increases the number of red blood cells. Genes also influence metabolic rate, bone mineral density, fat storage, glucose use, and lung function.

Genetic testing to predict athletic success is genetic determinism—the idea that our genes solely determine who we are. Using such test results to choose a child's sport can stress a child with no interest in competing, or discourage a child who loves a sport.

7.1 Genes and the Environment Mold Traits

A woman who is a prolific writer has a daughter who becomes a successful novelist. An overweight man and woman have obese children. A man whose father suffers from alcoholism has the same problem. Are these characteristics—writing talent, obesity, and alcoholism—inherited or learned? These traits, and nearly all others, are not the result of an "either/or" mechanism, but reflect the input of many genes as well as environmental influences. Even single-gene disorders are modified by environmental factors and/or other genes (called modifier genes). A child with cystic fibrosis, for example, has inherited a single-gene disorder, but her experiences reflect which variants of the gene she has, other genes that affect her immune system, the pathogens to which she is exposed, and the quality of the air she breathes. This chapter considers characteristics that represent input from many genes, and the tools used to study them.

A trait can be described as either single-gene (Mendelian or monogenic) or **polygenic**. As its name implies, a polygenic trait reflects the activities of more than one gene. Both single-gene and polygenic traits can also be **multifactorial,** which means they are influenced by the environment. Lung cancer is a multifactorial trait (**figure 7.1**). Purely polygenic traits—those not influenced by the environment at all—are very rare. Eye color, discussed in chapter 4, is close to being purely polygenic.

Polygenic multifactorial traits include common ones, such as height, skin color, body weight, many illnesses, and behavioral conditions and tendencies. Behavioral traits are not inherently different from other types of traits; they involve the functioning of the brain, rather than another organ. Chapter 8 discusses them. A more popular term for "multifactorial" is complex, but we use multifactorial here because it is more precise and is not confused with the general definition of "complex." The genes of a multifactorial trait are not more complicated than others. They follow Mendel's laws, but expression of any one gene is more difficult to predict because of the combined actions of genes and the environment.

Lung cancer caused by smoking illustrates the complexity of multifactorial traits. Variants of genes that increase the risk of becoming addicted to nicotine and of developing cancer come into play—but may not ever be expressed if a person never smokes or breathes polluted air.

A polygenic multifactorial condition reflects additive contributions of several genes. Each gene confers a degree of susceptibility, but the input of these genes is not necessarily equal. Often a rare allele may exert a large influence, but several common alleles each contribute only slightly to a trait. For example, three genes contribute significantly to the risk of developing type 2 diabetes mellitus, but other genes exert smaller effects.

Different genes may contribute different aspects of a phenotype that was once thought to be due to the actions of a single gene. Consider migraine, a condition that many a sufferer will attest is more than just a headache. A gene on chromosome 1

Figure 7.1 Genetic and environmental factors contribute to lung cancer risk. Genes raise lung cancer risk in several ways: impairing DNA repair, promoting inflammation, blocking detoxification of carcinogens, keeping telomeres long, promoting resistance to cancer-fighting drugs, and promoting addiction. These genetic risk factors interact with each other and with environmental influences, such as smoking and breathing polluted air.

contributes sensitivity to sound; a gene on chromosome 5 produces the pulsating headache and sensitivity to light; and a gene on chromosome 8 is associated with nausea and vomiting. In addition, certain environmental influences trigger migraine in some people. **Reading 7.1** takes a closer look at heart health, which reflects several multifactorial traits.

7.2 Polygenic Traits Are Continuously Varying

For a polygenic trait, the combined action of many genes often produces a "shades of grey" or "continuously varying" phenotype, also called a quantitative trait. DNA sequences that contribute to polygenic traits are called **quantitative trait loci,** or **QTLs**. A multifactorial trait is continuously varying if it is also polygenic. That is, it is the multi-gene component of the trait

Many Genes Control Heart Health

Many types of cells and processes must interact for the heart and blood vessels (the cardiovascular system) to circulate blood, and many genes maintain the system. Effects of the environment are great, too, even on single-gene cardiovascular diseases. For example, intake of vitamin K, necessary for blood to clot, influences the severity of single-gene clotting disorders. Cardiovascular disease affects one in three individuals.

Genes control the heart and blood vessels in several ways: transporting lipids; blood clotting; blood pressure; and how well white blood cells stick to blood vessel walls. Lipids can only move in the circulation when bound to proteins to form large molecules called lipoproteins. Several genes encode the protein parts of lipoproteins, which are called apolipoproteins. Some types of lipoproteins carry lipids in the blood to tissues, where they are used, and other types of lipoproteins take lipids to the liver, where they are broken down into biochemicals that the body can excrete more easily. One allele of a gene that encodes apolipoprotein E, called *E4*, increases the risk of a heart attack threefold in people who smoke.

Maintaining a healthy heart and blood vessels requires a balance between enough lipids inside cells but not an excess outside cells. Several dozen genes control lipid levels in the blood and tissues by specifying enzymes that process lipids, proteins that transport them, or receptor proteins that admit lipids into cells.

An enzyme, lipoprotein lipase, lines the walls of the smallest blood vessels, where it breaks down fat packets released from the small intestine and liver. Lipoprotein lipase is activated by high-density lipoproteins (HDLs), and it breaks down low-density lipoproteins (LDLs). This is why high HDL levels and low LDL levels are associated with a healthy cardiovascular system.

The fluidity of the blood is also critical to health. Overly active clotting factors or extra sticky white blood cells can cause clots to form that block blood flow, usually in blood vessels in the heart or in the legs. Poor clotting causes dangerous bleeding. Because clotting factors are proteins, clotting is genetically controlled.

Single-gene "inborn errors of metabolism" are responsible for a minority of cases of cardiovascular disease. One is Fabry disease (MIM 301500), an X-linked disorder in which a deficient enzyme enlarges the left ventricle, the strongest heart chamber, in adulthood. Fabry disease is treatable by replacing the enzyme. Genetic testing of people with an enlarged left ventricle can identify those with the inherited condition, who can then be treated.

Genetic test panels detect alleles of dozens of genes that each contributes risk to developing cardiovascular disease. More than 50 genes regulate blood pressure, and more than 95 contribute to inherited variation in blood cholesterol and triglyceride levels. Tests of gene expression can indicate which cholesterol-lowering drugs are most likely to be effective and tolerable for a particular individual. Computer analysis of multigene tests accounts for controllable environmental factors, such as exercising, not smoking, and maintaining a healthy weight (**table 1**). **Figure 1** shows an artery blocked by fatty plaque. Diet and medication can counter an inherited tendency to deposit cholesterol-rich material on the interior linings of arteries.

Table 1	Risk Factors for Cardiovascular Disease	
Uncontrollable	**Controllable**
Age	Fatty diet
Male sex	Hypertension
Genes	Smoking
Lipid metabolism	High serum cholesterol
Apolipoproteins	Low serum HDL
Lipoprotein lipase	High serum LDL
Blood clotting	Stress
Fibrinogen	Insufficient exercise
Clotting factors	Obesity
Inflammation	
C-reactive protein	
Homocysteine metabolism	Diabetes
Leukocyte adhesion	

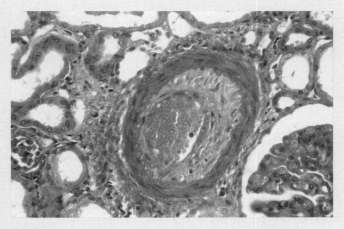

Figure 1 **Cardiovascular disease.** Genetic and dietary factors contribute to clogged arteries.

that contributes the continuing variation of the phenotype. The individual genes that confer a polygenic trait follow Mendel's laws, but together they do not produce single-gene phenotypic ratios. They all contribute to the phenotype, but without being dominant or recessive to each other. Single-gene traits are instead discrete or qualitative, often providing an "all-or-none" phenotype such as "normal" versus "affected."

A polygenic trait varies in populations, as our many nuances of hair color, body weight, and cholesterol levels demonstrate. Some genes contribute more to a polygenic trait than others. Within genes, alleles can have differing impacts depending upon exactly how they alter an encoded protein and how common they are in a population. For example, a mutation in the gene that encodes the receptor that takes LDL cholesterol into cells greatly raises blood serum cholesterol level. But because fewer than 1 percent of the individuals in most populations have this mutation, it contributes very little to the variation in cholesterol level at the population level. However, the mutation has a large impact on the person who has it.

Although the expression of a polygenic trait is continuous, we can categorize individuals into classes and calculate the frequencies of the classes. When we do this and plot the frequency for each phenotype class, a bell-shaped curve results. Even when different numbers of genes affect the trait, the curve takes the same shape, as is evident in the following examples.

Fingerprint Patterns

The skin on the fingertips is folded into patterns of raised skin called dermal ridges that align to form loops, whorls, and arches. This pattern is a fingerprint. A technique called dermatoglyphics ("skin writing") compares the number of ridges that comprise these patterns to identify and distinguish individuals (**figure 7.2**). Dermatoglyphics is part of genetics because certain disorders (such as Down syndrome) include unusual ridge patterns. Forensic fingerprint analysis is also an application of dermatoglyphics.

The number of ridges in a fingerprint is largely determined by genes, but also arises from the environment. During weeks 6 through 13 of prenatal development, the ridge pattern can be altered as the fetus touches the finger and toe pads to the wall of the amniotic sac. This early environmental effect explains why the fingerprints of identical twins, who share all genes, are in some cases not exactly alike.

We can quantify a fingerprint with a measurement called a total ridge count, which tallies the numbers of ridges in whorls, loops, or arches. The average total ridge count in a male is 145, and in a female, 126. Plotting total ridge count reveals the bell curve of a continuously varying trait.

Height

The effect of the environment on height is obvious—people who do not eat enough do not reach their genetic potential for height. Students lined up according to height, but raised in two different decades and under different circumstances, vividly reveal the effects of genes and the environment on this continuously varying trait. Part *a* of **figure 7.3** depicts students from 1920, and part *b,* students from 1997. Also note that the tallest people in the old photograph are 5′9″, whereas the tallest people in the more recent photograph are 6′5″. The difference is attributed to improved diet and better overall health.

Genome-wide association studies have identified more than 50 genes that affect height. Such a study compares genetic markers (see table 5.3) in two large groups of individuals who, ideally, differ only in the characteristic of interest. An association between the trait and the marker is then inferred, which, after much more investigation, may uncover a gene that contributes to the characteristic. For example, the Northern Finland Birth Cohort study periodically records heights of more than

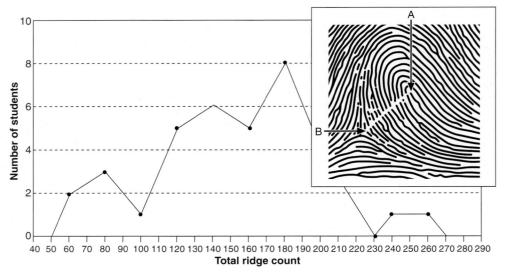

Figure 7.2 Anatomy of a fingerprint. Total ridge counts for a number of individuals, plotted on a bar graph, form an approximate bell-shaped curve. The number of ridges between landmark points A and B on this loop pattern is 12. Total ridge count includes the number of ridges on all fingers.

a.

b.

Figure 7.3 **The inheritance of height.** The photograph in **(a)** illustrates the continuously varying nature of height. In the photo, taken around 1920, 175 cadets at the Connecticut Agricultural College lined up by height. **(b)** In 1997, professor Linda Strausbaugh asked her genetics students at the school (today the University of Connecticut at Storrs) to re-create the scene.

5,000 people born in 1966. Certain patterns of SNPs in particular genes are found significantly more in individuals who experienced periods of rapid height increase. These findings led to identification of a variant of a gene that may have persisted because it confers early puberty, which enables people to have more children.

Skin Color

More than 100 genes affect pigmentation in skin, hair, and the irises. Melanin pigments color the skin to different degrees in different individuals. In the skin, as in the iris (see chapter 4),

melanocytes contain melanin in melanosomes. Melanocytes extend between the tile-like skin cells, distributing pigment granules through the skin layers. Some melanin exits the melanocytes and enters the hardened cells in the skin's upper layers. Here the melanin breaks into pieces, and as the skin cells are pushed up toward the skin's surface as stem cells beneath them divide, the melanin bits provide color. The pigment protects against DNA damage from ultraviolet radiation. Exposure to the sun increases melanin synthesis. **Figure 7.4***a* shows a three-gene model for human skin color. This is an oversimplification, but it illustrates how several genes can contribute to a very variable trait.

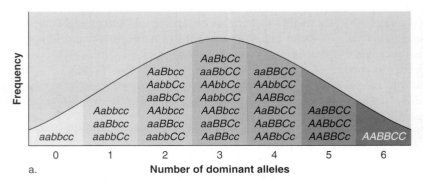

a.

b.

Figure 7.4 **Variations in skin color.** **(a)** A model of three genes, with two alleles each, can explain broad hues of human skin. In actuality, this trait likely involves many more than three genes. **(b)** Humans come in a great variety of skin colors. Skin color genes can assort in interesting ways. These beautiful young ladies, Alicia and Jasmin, are twins! Their father is German and their mother is Jamaican-English.

People come in many hues, but we all have about the same number of melanocytes per unit area of skin. How people differ is in melanosome number, size, and density of distribution. Different skin colors arise from the number and distribution of melanin pieces in the cells in the uppermost skin layers.

Skin color is one physical trait that is used to distinguish race. The definition of race based largely on skin color is more a social construct than a biological concept, because skin color is only one of thousands of traits whose frequencies vary in different populations. From a genetic perspective, when referring to nonhumans, races are groups within species that are distinguished by different allele frequencies. Humans are actually a lot less variable in appearance than other mammals, even though all chimps look alike to us. We may classify people by skin color because it is an obvious visible way to distinguish individuals, but this trait is *not* a reliable indicator of ancestry.

The concept of race based on skin color falls apart when considering many genes. That is, two people with very dark skin may be less alike than either is to another person with very light skin. For example, sub-Saharan Africans and Australian aborigines have dark skin, but are very dissimilar in other inherited characteristics. Their dark skins may reflect adaptation to life in a sunny, tropical climate rather than recent shared ancestry. Overall, 93 percent of varying inherited traits are no more common in people of one skin color than any other.

Testing DNA indicates that biologically speaking, it makes more sense to classify people by ancestry rather than by the color of their skin. In a sociology class at Pennsylvania State University, 100 students had their DNA tested for percent contribution from "European white," "black African," "Asian," and "Native American" gene variants. Many students were surprised at what their DNA revealed about their ancestry. A light-skinned black man learned that genetically he is approximately half black, half white. Another student who considered herself black was actually 58 percent white European. The U.S. census, in recognition of the complexity of classifying people into races based on skin color, began to allow "mixed race" as a category in 2000. Many of us fall into this category, including President Obama.

In a genetic sense the concept of race based on skin color has little meaning, but in a practical sense, racial groups *do* have different incidences of certain diseases. This reflects the tendency to choose partners within a group, which retains certain alleles. However, racial differences in disease prevalence may also result from social inequities, such as access to good nutrition or health care. Observations that populations of particular races have a higher incidence of certain illnesses have fueled "race-based prescribing." For example, certain hypertension and heart disease drugs are specifically marketed to African Americans, because this group has a higher incidence of these conditions than do people in other groups. Offering medical treatments based on skin color may make sense on a population level, but on the individual level it may lead to errors. A white person might be denied a drug that would work, or a black person given one that doesn't, if the treatment decision is based on a superficial trait not directly related to how the body responds to a particular drug.

Prescribing drugs is beginning to be based on personal genotypes that determine whether or not a particular drug will work or have side effects, rather than by the color of a person's skin. For example, a gene called *MDR* (for multidrug resistance) encodes a protein that pumps poisons out of certain white blood cells and intestinal lining cells. Some variants of the pump work too well, sending drugs used to treat cancer and AIDS out of the cell. Researchers have found this protein variant in 83 percent of West Africans, 61 percent of African Americans, 26 percent of Caucasians, and 34 percent of Japanese. Therefore, a person in whom these drugs do not work could come from any of these populations. MDR genotype can be used to prescribe certain drugs only for individuals whose cells would not pump the drugs out. This is a more biologically meaningful basis for prescribing a drug than skin color.

In an even more compelling study, researchers cataloged 23 markers for genes that control drug metabolism in 354 people representing eight races: black (Bantu, Ethiopian, and Afro-Caribbean), white (Norwegian, Armenian, and Ashkenazi Jews), and Asian (Chinese and New Guinean). The genetic markers fell into four very distinct groups that predict which of several blood thinners, chemotherapies, and painkillers will be effective—and these response groups did not at all match the traditional racial groups.

The premises behind race-based prescribing are far more complex than black versus white. Although some genes and their variants are not distributed along racial lines, such as the 23 markers of drug metabolism just discussed, others apparently are. This is the case for a gene that encodes an enzyme called leukotriene A4 hydrolase. The enzyme is necessary to produce leukotrienes, which inflame arteries as part of the immune response to infection. Excess leukotrienes increase the risk of heart attack. An allele in European Americans and European populations for many years increases heart attack risk only slightly, possibly because enough time has passed that variants of other genes that temper the negative effects of excess leukotrienes have accumulated in the genomes of these light-skinned groups. The overactive leukotriene A4 hydrolase allele, however, has only recently been introduced into the African American population. Without enough time for genetic protection to have arisen, the excess leukotrienes elevate risk of heart attack among African Americans five-fold compared to 16 percent elevation among whites.

Key Concepts

1. Polygenic traits are determined by more than one gene and vary continuously in expression.
2. Multifactorial traits are determined by a combination of a gene or genes and the environment.
3. A bell curve describes the distribution of phenotypic classes of a polygenic trait.

7.3 Traditional Ways to Investigate Multifactorial Traits

Predicting recurrence risks for polygenic traits is much more challenging than doing so for single-gene traits. This section reviews traditional approaches to evaluating polygenic multifactorial traits, and the next section examines a newer approach, genome-wide association studies.

Empiric Risk

Using Mendel's laws, it is possible to predict the risk that a single-gene trait will recur in a family from knowing the mode of inheritance—such as autosomal dominant or recessive. To predict the chance that a polygenic multifactorial trait will occur in a particular individual, geneticists use **empiric risk,** which is based on incidence in a specific population. **Incidence** is the rate at which a certain event occurs, such as the number of new cases of a disorder diagnosed per year in a population of known size. **Prevalence** is the proportion or number of individuals in a population who have a particular disorder at a specific time, such as during one year.

Empiric risk is not a calculation, but a population statistic based on observation. The population might be broad, such as an ethnic group or community, or genetically more well defined, such as families that have cystic fibrosis. Empiric risk increases with the severity of the disorder, the number of affected family members, and how closely related a person is to affected individuals. For example, empiric risk is used to predict the likelihood of a child being born with a neural tube defect (NTD). In the United States, the overall population risk of carrying a fetus with an NTD is about 1 in 1,000 (0.1 percent). For people of English, Irish, or Scottish ancestry, the risk is about 3 in 1,000. However, if a sibling has an NTD, for any ethnic group, the risk of recurrence increases to 3 percent, and if two siblings are affected, the risk to a third child is even greater.

If a trait has an inherited component, then it makes sense that the closer the relationship between two individuals, one of whom has the trait, the greater the probability that the second individual has the trait, too, because they share more genes. Studies of empiric risk support this logic. **Table 7.1** summarizes empiric risks for relatives of individuals with cleft lip (**figure 7.5**).

Because empiric risk is based solely on observation, we can use it to derive risks for disorders with poorly understood transmission patterns. For example, certain multifactorial disorders affect one sex more often than the other. Pyloric stenosis, an overgrowth of muscle at the juncture between the stomach and the small intestine, is five times more common among males than females. The condition must be corrected surgically shortly after birth, or the newborn will be unable to digest foods. Empiric data show that the risk of recurrence for the brother of an affected brother is 3.8 percent, but the risk for the brother of an affected sister is 9.2 percent. An empiric risk, then, is based on real-world observations. The cause of the illness need not be known.

Table 7.1	Empiric Risk of Recurrence for Cleft Lip
Relationship to Affected Person	**Empiric Risk of Recurrence**
Identical twin	40.0%
Sibling	4.1%
Child	3.5%
Niece/nephew	0.8%
First cousin	0.3%
General population risk (no affected relatives)	0.1%

Heritability

Charles Darwin noted that some of the variation of a trait is due to inborn differences in populations, and some to differences in environmental influences. A measurement called **heritability,** designated H, estimates the proportion of the phenotypic

Figure 7.5 Cleft lip. Cleft lip is more likely in a person who has a relative with the condition. This child has had corrective surgery.

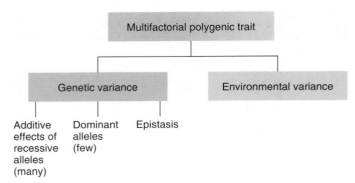

Figure 7.6 **Heritability estimates the genetic contribution to the variability of a trait.** Observed variance in a polygenic, multifactorial trait or illness reflects genetic and environmental contributions.

variation for a trait that is due to genetic differences in a certain population at a certain time. The distinction between empiric risk and heritability is that empiric risk could result from non-genetic influences, whereas heritability focuses on the genetic component of the variation in a trait. Heritability refers to the degree of *variation* in a trait due to genetics, and not to the proportion of the trait itself attributed to genes.

Figure 7.6 outlines the factors that contribute to observed variation in a trait. Heritability equals 1.0 for a trait whose variability is completely the result of gene action, such as in a population of laboratory mice whose environment is controlled. Without environmental variability, genetic differences alone determine expression of the trait in the population. Variability of most traits, however, reflects a combination of differences among genes and environmental components. **Table 7.2** lists some traits and their heritabilities.

Heritability changes as the environment changes. For example, the heritability of skin color is higher in the winter

Table 7.2	Heritabilities for Some Human Traits
Trait	**Heritability**
Clubfoot	0.8
Height	0.8
Blood pressure	0.6
Body mass index	0.5
Verbal aptitude	0.7
Mathematical aptitude	0.3
Spelling aptitude	0.5
Total fingerprint ridge count	0.9
Intelligence	0.5–0.8
Total serum cholesterol	0.6

Table 7.3	Coefficient of Relatedness for Pairs of Relatives	
Relationship	**Degree of Relationship**	**Percent Shared Genes (Coefficient of Relatedness)**
Sibling to sibling	1°	50% (1/2)
Parent to child	1°	50% (1/2)
Uncle/aunt to niece/ nephew	2°	25% (1/4)
Grandparent to grandchild	2°	25% (1/4)
First cousin to first cousin	3°	12 1/2% (1/8)

months, when sun exposure is less likely to increase melanin synthesis. The same trait may be highly heritable in two populations, but certain variants much more common in one group due to long-term environmental differences. Populations in equatorial Africa, for example, have darker skin than sun-deprived Scandinavians.

Researchers use several statistical methods to estimate heritability. One way is to compare the actual proportion of pairs of people related in a certain manner who share a particular trait, to the expected proportion of pairs that would share it if it were inherited in a Mendelian fashion. The expected proportion is derived by knowing the blood relationships of the individuals and using a measurement called the **coefficient of relatedness,** which is the proportion of genes that two people related in a certain way share (**table 7.3**).

A parent and child share 50 percent of their genes, because of the mechanism of meiosis. Siblings share on average 50 percent of their genes, because they have a 50 percent chance of inheriting each allele for a gene from each parent. Genetic counselors use the designations of primary (1°), secondary (2°), and tertiary (3°) relatives when calculating risks (table 7.3 and **figure 7.7**). For extended or complicated pedigrees, the value of 1 in 2 or 50 percent between siblings and between parent-child pairs can be used to trace and calculate the percentage of genes shared between people related in other ways.

If the heritability of a trait is very high, then of a group of 100 sibling pairs, nearly 50 would be expected to have the same phenotype, because siblings share on average 50 percent of their genes. Height is a trait for which heritability reflects the environmental influence of nutrition. Of 100 sibling pairs in a population, for example, 40 might be the same number of inches tall. Heritability for height among this group of sibling pairs is .40/.50, or 80 percent, which is the observed phenotypic variation divided by the expected phenotypic variation if environment had no influence.

Genetic variance for a polygenic trait is mostly due to the additive effects of recessive alleles of different genes. For some

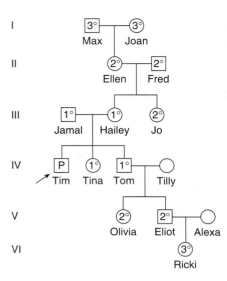

Figure 7.7 **Tracing relatives.** Tim has an inherited illness. A genetic counselor drew this pedigree to explain the approximate percentage of genes Tim shares with relatives. This information can be used to alert certain relatives to their risk.

("P" is the proband, or affected individual who initiated the study. See table 7.3 for definitions of 1°, 2°, and 3° relationships.)

traits, a few dominant alleles can greatly influence the phenotype, but because they are rare, they do not contribute greatly to heritability. This is the case for heart disease caused by a faulty LDL receptor. Heritabilities for some traits or diseases may be underestimates if only mutations are compared, and not also copy number variants (CNVs, which are differences in the numbers of copies of a particular DNA sequence).

Epistasis (interaction between alleles of different genes) can also influence heritability. To account for the fact that different genes affect a phenotype to differing degrees, geneticists calculate a "narrow" heritability that considers only additive recessive effects, and a "broad" heritability that also considers the effects of rare dominant alleles and epistasis. For LDL cholesterol level, for example, the narrow heritability is 0.36, but the broad heritability is 0.96, reflecting the fact that a rare dominant allele has a large impact.

Understanding multifactorial inheritance is important in agriculture. A breeder needs to know whether genetic or environmental influences contribute to variability in such traits as birth weight, milk yield, and egg hatchability. It is also valuable to know whether the genetic influences are additive or epistatic. The breeder can control the environment by adjusting the conditions under which animals are raised and crops grown, and control genetic effects by setting up crosses between particular individuals.

Studying multifactorial traits in humans is difficult, because information must be obtained from many families. Two special types of people, however, can help geneticists to tease apart the genetic and environmental components of the variability of multifactorial traits—adopted individuals and twins.

Adopted Individuals

An adopted person typically shares environmental influences, but not many gene variants, with the adoptive family. Conversely, adopted individuals share genes, but not the exact environment, with their biological parents. Therefore, biologists assume that similarities between adopted people and adoptive parents reflect mostly environmental influences, whereas similarities between adoptees and their biological parents reflect mostly genetic influences. Information on both sets of parents can reveal how heredity and the environment contribute to a trait.

Many early adoption studies used a database of all adopted children in Denmark and their families from 1924 to 1947. One study examined correlations between causes of death among biological and adoptive parents and adopted children. If a biological parent died of infection before age 50, the adopted child was five times more likely to die of infection at a young age than a similar person in the general population. This may be because inherited variants in immune system genes increase susceptibility to certain infections. In support of this hypothesis, the risk that an adopted individual would die young from infection did not correlate with adoptive parents' death from infection before age 50. Researchers concluded that genetics mostly determines length of life, but they did find evidence of environmental influences. For example, if adoptive parents died before age 50 of cardiovascular disease, their adopted children were three times as likely to die of heart and blood vessel disease as a person in the general population. What environmental factor might explain this correlation?

Twins

Studies that use twins to separate the genetic from the environmental contribution to a phenotype provide more meaningful information than studying adopted individuals. Twin studies have largely replaced adoption methods. However, twin studies are not perfect experiments either. The genomes of identical twins are not really identical. They differ in CNVs, which are discussed further in chapter 12.

Using twins to study genetic influence on traits dates to 1924, when German dermatologist Hermann Siemens reported that grades and teachers' comments were much more alike for identical twins than for fraternal twins. He proposed that genes contribute to intelligence based on this observation.

A trait that occurs more frequently in both members of identical (monozygotic or MZ) twin pairs than in both members of fraternal (dizygotic or DZ) twin pairs is at least partly controlled by heredity. Geneticists calculate the **concordance** of a trait as the percentage of pairs in which both twins express the trait among pairs of twins in whom at least one has the trait. Twins who differ in a trait are said to be discordant for it. In one study, 142 MZ twin pairs and 142 DZ twin pairs took a "distorted tunes test," in which 26 familiar songs were played, each with at least one note altered. A person was considered "tune deaf" if he or she failed to detect the mistakes in three or more tunes. Concordance for "tune deafness" was 67 percent

Table 7.4	Concordance Values for Some Traits in Twins	
Trait	**MZ (Identical) Twins**	**DZ (Fraternal) Twins**
Acne	14%	14%
Alzheimer disease	78%	39%
Anorexia nervosa	55%	7%
Autism	90%	4.5%
Bipolar disorder	33–80%	0–8%
Cleft lip with or without cleft palate	40%	3–6%
Hypertension	62%	48%
Schizophrenia	40–50%	10%

for MZ twins, but only 44 percent for DZ twins, indicating a considerable inherited component in the ability to accurately perceive musical pitch. **Table 7.4** compares twin types for a variety of hard-to-measure traits. (Figure 3.16 shows how DZ and MZ twins arise.)

Diseases caused by single genes that approach 100 percent penetrance, whether dominant or recessive, also approach 100 percent concordance in MZ twins. That is, if one twin has the disease, so does the other. However, among DZ twins, concordance generally is 50 percent for a dominant trait and 25 percent for a recessive trait. These are the Mendelian values that apply to any two siblings. For a polygenic trait with little environmental input, concordance values for MZ twins are significantly greater than for DZ twins. A trait molded mostly by the environment exhibits similar concordance values for both types of twins.

Comparing twin types assumes that both types of twins share similar experiences. In fact, MZ twins are often closer emotionally than DZ twins. This discrepancy between the closeness of the two types of twins can lead to misleading results. A study from the 1940s, for example, concluded that tuberculosis is inherited because concordance among MZ twins was higher than among DZ twins. Actually, the infectious disease more readily passed between MZ twins because their parents kept them closer. However, the 1940s study wasn't totally off the mark. We do inherit susceptibilities to some infectious diseases. MZ twins would share such genes, whereas DZ twins would only be as likely as any sibling pairs to do so.

For some traits that begin to manifest before birth, the type of MZ twin may be important. That is, MZ twins with the same amnion may share more environmental factors than MZ twins who have separate amnions (see figure 3.16). Schizophrenia is a condition that may begin subtly, before birth, and later become obvious when environmental factors come into play. Schizophrenia is discussed in chapter 8.

A more informative way to assess the genetic component of a multifactorial trait is to study MZ twins who were separated at birth, then raised in very different environments.

Much of the work using this "twins reared apart" approach has taken place at the University of Minnesota. Here, since 1979, hundreds of sets of twins and triplets who were separated at birth have visited the laboratories of Thomas Bouchard. For a week or more, the twins and triplets are tested for physical and behavioral traits, including 24 different blood types, handedness, direction of hair growth, fingerprint pattern, height, weight, functioning of all organ systems, intelligence, allergies, and dental patterns. Researchers videotape facial expressions and body movements in different circumstances and probe participants' fears, interests, and superstitions.

Twins and triplets separated at birth provide natural experiments for distinguishing nature from nurture. Many of their common traits can be attributed to genetics, especially if their environments have been very different. By contrast, their differences tend to reflect differences in upbringing, since their genes are identical (MZ twins and triplets) or similar (DZ twins and triplets).

MZ twins and triplets separated at birth and reunited later are remarkably similar, even when they grow up in very different adoptive families (**figure 7.8**). Idiosyncrasies are particularly striking. One pair of twins who met for the first

Separated at birth, the Mallifert twins meet accidentally.

Figure 7.8 MZ twins separated at birth and reunited as adults may have astounding similarities.

Originally published in the 4 May 1981 issue of *The New Yorker* Magazine, p. 43. © Tee and Charles Addams Foundation. Reprinted by permission.

time when they were in their thirties responded identically to questions; each paused for 30 seconds, rotated a gold necklace she was wearing three times, and then answered the question. Coincidence, or genetics?

The "twins reared apart" approach is not an ideal way to separate nature from nurture. MZ twins and other multiples share an environment in the uterus and possibly in early infancy that may affect later development. Siblings, whether adoptive or biological, do not always share identical home environments. Differences in sex, general health, school and peer experiences, temperament, and personality affect each individual's perception of such environmental influences as parental affection and discipline.

Key Concepts

1. Empiric risk applies population incidence data to predict risk of recurrence for a multifactorial trait or disorder.

2. Heritability measures the genetic contribution to the variability of a multifactorial trait; it is specific to a particular population at a particular time.

3. Coefficient of relatedness, the proportion of genes that individuals related in a certain way are expected to share, is used to calculate heritability.

4. Adopted individuals and twins are used in studies to separate environmental from inherited components to traits.

7.4 Genome-Wide Association Studies

The pedigrees in chapter 4 and linkage studies described in chapter 5 follow single-gene traits from generation to generation. Empiric risk and heritability calculations, and adoptee and twin studies, address conditions to which many genes contribute. A newer tool to analyze multifactorial traits and diseases is a **genome-wide association study (GWAS)**. The older techniques search for known gene variants, typically in only a few people. In contrast, genome-wide association studies look at signposts throughout the genome in many individuals to identify common variants behind a particular phenotype. An assumption is that more common disorders with inherited causes remain with us because they affect health later in life, after a person has had children and passed on those gene variants.

The National Institutes of Health (NIH) defines a genome-wide association study as "any study of genetic variation across the entire human genome that is designed to identify genetic associations with observable traits (such as blood pressure or weight), or the presence or absence of a disease or condition." Genetic markers are used to follow variation. These are landmarks across the genome that form patterns that researchers compare between two groups of people—one with a particular trait or disease and one without it.

Table 7.5	Types of Information Used in Genome-Wide Association Studies
Marker Type	**Definition**
SNP	A single nucleotide polymorphism is a site in the genome that is a different DNA base in >1% of a population.
CNV	A copy number variant is a tandemly repeated DNA sequence, such as CGTA CGTA CGTA
Gene expression	The pattern of genes that are overexpressed and/or overexpressed in people with a particular trait or disease.
	Epigenetic signature

The pattern of sites in the genome to which methyl groups bind.

Genome-wide association studies use several types of genetic markers (**table 7.5;** see also table 5.3). SNPs and copy number variants describe the DNA base sequence (**figure 7.9**). Gene expression patterns reflect which proteins are overproduced or underproduced in people with the trait or illness, compared to unaffected controls. Another way to compare genomes is by the sites to which methyl (CH_3) groups bind, shutting off gene expression. This is an epigenetic change because it doesn't affect the DNA base sequence.

To achieve statistical significance, a genome-wide association study must include at least 100,000 markers. It is the association of markers to a trait or disease that is informative (**figure 7.10**). That is, if a marker pattern nearly always appears in individuals who share a specific trait, but only very rarely appears in people with it, then it may do so because it lies in or near a gene that directly causes the trait. Typically,

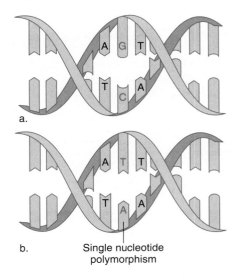

Figure 7.9 **SNPs are sites of variability in genomes.** The DNA base pair in red represents a SNP—a site that differs in more than 1 percent of a population.

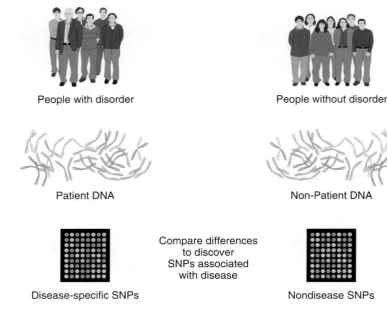

People with disorder

People without disorder

Patient DNA

Non-Patient DNA

Compare differences to discover SNPs associated with disease

Disease-specific SNPs

Nondisease SNPs

Figure 7.10 **Tracking genes in groups.** Genome-wide association studies seek DNA sequence variants that are shared with much greater frequency among individuals with the same illness or trait than among others.

genome-wide association studies use a million or more SNPs, grouped into half a million or so haplotypes. A specific "tag SNP" is sometimes used to identify a haplotype.

Designing a Genome-Wide Association Study

A GWAS is a stepwise focusing in on parts of the genome responsible to some degree for a trait (**figure 7.11**). In general, a group of people with the same condition or trait and a control group have their DNA isolated and genotyped for the 500,000 tag SNPs. Statistical algorithms identify the uniquely shared SNPs in the group with the trait or disorder. Repeating

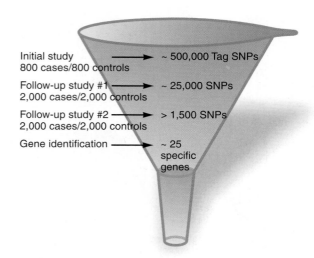

Initial study
800 cases/800 controls — ~ 500,000 Tag SNPs

Follow-up study #1
2,000 cases/2,000 controls — ~ 25,000 SNPs

Follow-up study #2
2,000 cases/2,000 controls — > 1,500 SNPs

Gene identification — ~ 25 specific genes

Figure 7.11 **A stepwise approach to gene discovery.** Genome-wide association study results must be validated in several different populations. Further research is necessary to go beyond association to demonstrate correlation and cause.

the process on additional populations narrows the SNPs and strengthens the association. It is important to validate a SNP association in different population groups, to be certain that it is the trait of interest that is being tracked, and not another part of the genome that members of one population share due to their common ancestry.

Several different study designs are used in these investigations. In a **cohort study,** researchers follow a large group of individuals over time and measure many aspects of their health. The most famous is the Framingham Heart Study, which began tracking thousands of people and their descendants in Massachusetts in 1968. Nine thousand of them are participating in a genome-wide association study.

In a **case-control study,** each individual in one group is matched to an individual in another group who shares as many characteristics as possible, such as age, sex, activity level, and environmental exposures. SNP differences are then associated with the presence or absence of the disorder. For example, if 5,000 individuals with hypertension (high blood pressure) have particular DNA bases at six sites in the genome, and 5,000 matched individuals who do not have hypertension have different bases at only these six sites, then these genome regions may include genes whose protein products control blood pressure.

The **affected sibling pair** strategy borrows from the older technique of tracking linkage in families. Researchers scan genomes for SNPs that most siblings who have the same condition share, but that siblings who do not both have the condition do not often share. Such genome regions may harbor genes that contribute to the condition. The logic is that because siblings share 50 percent of their genes, a trait or condition that many siblings share is likely to be inherited.

A variation on the affected sibling pair strategy is **homozygosity mapping,** which is performed on families that are consanguineous—that is, the parents are related. The genomes of children whose parents share ancestors have more homozygous regions than do other children, and therefore greater likelihood that they will inherit two copies of a susceptibility or disease-causing mutation. Homozygosity mapping was used to identify genes that cause autism, discussed in the next chapter.

Once a SNP association has been validated in diverse and large populations, gene identification can begin. This might entail consulting the human genome sequence for "candidate" genes near associated SNPs whose known functions make sense in explaining the condition. Or, understanding the SNP association might require exploring the phenotype. For example, a set of SNPs might track with breast cancer. Examination of the tumors at the molecular, cell, and tissue levels might reveal a subtype of the cancer that the SNP signature identifies. Then perhaps a test for the SNPs might be used to refine diagnosis.

Common characteristics, such as height, are also investigated with genome-wide association studies. The results estimate the degree to which a particular allele contributes to the variation of a trait in a population. For example, a genome-wide association study of height identified alleles of a gene called *HMGA2* that explains about 0.3 percent of population variation in height, or about 0.4 centimeters of height per allele. In addition to all of the data, common sense comes into play. Inheriting several gene variants that each add to height is meaningless if a person doesn't eat enough to grow. In another study of a common characteristic, researchers examined facial features, evaluating people by eye and nose positions and dimensions, twenty lip descriptors, and the length of the space between the nose and the upper lip. The study identified variants of several genes already known to be mutant in specific syndromes that disrupt development of facial features.

Limitations of Genome-Wide Association Studies

Genome-wide association studies are prone to error simply because they include so many data points, but, ironically, the large numbers of markers, measurements, and people are necessary for accuracy in pointing toward genes that affect health. To get an idea of the computational magnitude of such a study, consider a small version: An investigation of 20 genes, each with four SNPs. That equals 160 data points per individual (20 genes × 4 SNPs/gene × 2 copies of each gene). With thousands of possible combinations (genotypes) of those 160 points, many thousands of individuals would have to be examined to note SNP patterns that people who share symptoms have in common. Imagine the complexity of a study with a million SNPs, as is common today.

A conceptual limitation of genome-wide association studies is that they reveal associations between two types of information, and not causes. An *association* only means that one event or characteristic occurs when another occurs. A *correlation* is a directional association: If one measurement increases, so does the other, such as stress and blood pressure. In contrast, establishing a cause requires that a specific mechanism explains how one event makes another happen: *How* does stress elevate blood pressure? An association study does not provide information on a gene's function—it is more a discovery tool.

A practical limitation of genome-wide association studies is that they often identify parts of the genome that contribute only slightly to the risk of developing a disease. A genetic test that indicates a 1 percent increase in risk of developing cancer, for example, would not matter much to a smoker whose environmental risk is much higher.

The way that a patient population is selected can introduce bias into a genome-wide association study. Samples drawn from clinics, for example, would not include the very mildly affected who are unlikely to show up, or those who have died. It would also miss individuals for whom a contributing gene variant is nonpenetrant, or is associated with a late-onset disorder. Another source of error is that individuals in the control population might not actually be healthy. They might have problems other than the one being investigated.

The complicating factors discussed in chapter 5 also affect the accuracy of genome-wide association studies. Recall that a phenocopy is a trait or illness that resembles an inherited one, but has an environmental cause. Placing a person with anemia due to a drug reaction in a group with people who have an inherited anemia would be misleading. Genetic heterogeneity,

in which different genes cause the same trait or condition, could also be a source of error. Epistasis, when one gene masks the effect of another, also confounds these studies, but as we learn more, these interactions are being taken into account.

Yet another source of error arises from what a genome-wide association study misses, such as extremely rare SNPs or CNVs that cause or contribute to a disease. Or, the people who share symptoms and a SNP pattern may share something *else* that accounts for the association, such as an environmental exposure. Such a gene/environment interaction (or "G × E") can generate a false positive result. For example, mutations cause or contribute to atherosclerosis, but so do infection, smoking, lack of exercise, and a fatty diet. These environmental factors are so common that if a GWAS isn't large enough, it might not correctly identify a genetic influence.

Many of these limitations are overcome by expanding the numbers, and pooling data. Indeed, when these studies began in the early part of the century, associations would often fall apart as the numbers grew. It was not uncommon to see results reported one year retracted the next, as accumulating data destroyed the supposed association. A common complication is when heritability calculated using more traditional approaches is much higher than the genetic contribution revealed in a genome-wide association study. This may just mean that we haven't yet found all the contributing genes.

Often, the old and the new techniques for dissecting multifactorial traits work well together. This is the case for stuttering. Concordance for MZ twins ranges from 20 to 83 percent, and for DZ twins, from 4 to 9 percent, suggesting a large inherited component. The risk of a first-degree relative of a person who stutters also stuttering is 15 percent based on empiric evidence, compared to the lifetime risk of stuttering in the general population of 5 percent, although part of that increase could be due to imitating an affected relative. A genome-wide association study on 100 families who have at least two members who stutter identified candidate genes on three chromosomes that contribute to the trait.

Genome-wide association studies have examined hundreds of conditions and genes, giving drug developers many new targets to work with, and explaining how certain conditions arise. **Table 7.6** reviews terms used to study multifactorial traits. The next section probes an example of such a trait—body weight.

Key Concepts

1. Researchers compare traits in adopted individuals to those in their adoptive and biological parents to assess the genetic contribution to a trait.

2. Concordance is the percentage of twin pairs in which both express a trait. For a trait largely determined by genes, concordance is higher for MZ than DZ twins.

3. Genome-wide association studies seek correlations between SNP patterns and phenotypes in large groups of individuals.

Table 7.6	Terms Used in Evaluating Multifactorial Traits

Coefficient of relatedness The proportion of genes shared by two people related in a particular way. Used to calculate heritability.

Concordance The percentage of twin pairs in which both twins express a trait.

Empiric risk The risk of recurrence of a trait or illness based on known incidence in a particular population.

Genome-wide association study Detecting association between marker patterns and increased risk of developing a particular medical condition.

Heritability The percentage of phenotypic variation for a trait that is attributable to genetic differences. It equals the ratio of the observed phenotypic variation to the expected phenotypic variation for a population of individuals.

7.5 A Closer Look: Body Weight

Unlike rare genetic disorders, body weight is a multifactorial trait that we all have. Body weight reflects energy balance, which is the rate of food taken in versus the rate at which the body uses it for fuel. Excess food means, ultimately, excess weight. Being overweight or obese raises the risk of developing hypertension, diabetes, stroke, gallstones, sleep apnea, and some cancers.

Scientific studies of body weight use a measurement called body mass index (BMI), which is weight in proportion to height (**figure 7.12**). BMI makes sense—a person who weighs 170 pounds and is 6 feet tall is slim, whereas a person of the same weight who is 5 feet tall is obese. The tall person's BMI is 23; the short person's is 33.5.

Heritability for BMI is 0.55, which leaves room for environmental influences on our appetites and sizes. Some genes implicated in determining body weight have been known for a long time. Genome-wide association studies have identified more than 50 genome regions that harbor genes that affect how much we eat, how we use calories, and how fat is distributed in the body. The biochemical pathways and hormonal interactions that control weight may reveal points for drug intervention (**table 7.7**).

Leptin and Associated Proteins

Obesity research first embraced genetics in 1994, when Jeffrey Friedman at Rockefeller University discovered a gene that encodes the protein hormone leptin in mice and in humans. Normally, eating stimulates fat cells (adipocytes) to secrete leptin, which travels in the bloodstream to a region of the brain's hypothalamus, where it binds to receptors on nerve cells (neurons). The binding signals the neurons to release another type of hormone that binds yet other types of receptors, which ultimately function as an appetite "brake," while speeding digestion of food already eaten. When a person hasn't eaten

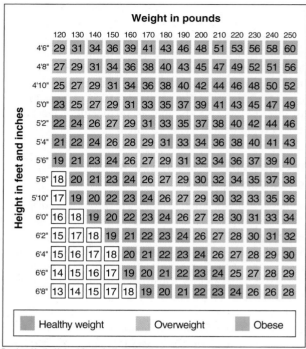

Figure 7.12 Body mass index (BMI). BMI equals weight/height^2, with weight measured in kilograms and height measured in meters. This chart provides a shortcut—the calculations have been done and converted to the English system of measurement. Squares that are not filled in indicate underweight.

in several hours, leptin levels ebb, which triggers the release of an appetite "accelerator." Table 7.7 lists the details of some proteins that affect eating behavior.

The discovery of genes and proteins that affect appetite led to great interest in targeting them with drugs to either lose or gain weight. When Friedman gave mice extra leptin, they ate less and lost weight. Headlines soon proclaimed the new magic weight loss elixir, a biotech company paid $20 million for rights to the hormone, and clinical trials ensued. The idea was to give obese people leptin, assuming that they had a deficiency, to trick them into feeling full. Only about 15 percent of the people lost weight, but the other 85 percent didn't actually lack leptin. Instead, most of them had leptin resistance, which is a diminished ability to recognize the hormone due to defective leptin receptors. Giving these people leptin had no effect on their appetites. However, the discovery helped a few severely obese children with true leptin deficiency attain normal weights after years of daily leptin injections.

The stomach is another source of obesity-related proteins. Ghrelin is a peptide (small protein) hormone produced in the stomach that responds to hunger, signaling the hypothalamus to produce more of the appetite accelerator. A different peptide hormone opposes ghrelin, signaling satiety to the brain. A drug is being developed to block the effects of ghrelin, which may one day replace weight-loss surgery with a pill. While leptin acts in the long term to maintain weight, the stomach's appetite

Table 7.7	Some Sites of Genetic Control of Body Weight		
Protein	**Function**	**MIM**	**Effect on Appetite**
Leptin	Stimulates cells in hypothalamus to decrease appetite and metabolize nutrients	164160	↓
Leptin transporter	Enables leptin to cross from bloodstream into brain	601694	↓
Leptin receptor	Binds leptin on hypothalamus cell surfaces, triggering hormone's effects	601007	↓
Neuropeptide Y	Produced in hypothalamus when leptin levels are low and the individual loses weight	162640	↑
Melanocortin-4 receptor	Activated when leptin levels are high and the individual gains weight	155541	↓
Ghrelin	Signals hunger from stomach to brain in short term, stimulating neuropeptide Y	605353	↑
PYY	Signals satiety from stomach to brain	660781	↓
Stearoyl-CoA desaturase-1	Controls whether body stores or uses fat	604031	↑

control hormones function in the short term. All of these hormonal signals are integrated to finely control appetite in a way that maintains weight.

Beyond Single Weight-Control Genes

Identifying single genes that influence weight paved the way for considering the trait to be multifactorial. Researchers are investigating combinations of known genes as well as many newly discovered genes.

One study looked at 21 genes in which mutations cause syndromes that include obesity, as well as 37 genes whose products participate in biochemical pathways related to weight. This approach identified many rare gene variants that could, in combinations, explain many people's tendency to gain weight. In another study of known genes, the effectiveness of a weight loss drug (Meridia) was associated with certain variants of the genes that encode the receptors that two appetite-controlling neurotransmitters bind.

Genome-wide association studies that compare gene expression patterns have also enhanced understanding of body weight. One study compared the sets of genes that are expressed in adipose (fat) tissue to other tissues. Samples from more than 1,600 people in Iceland revealed a set of genes whose products take part in inflammation and the immune response, but also contribute obesity-related traits. This was not known, and can suggest new drug targets.

Environmental Influences on Weight

Many studies on adopted individuals and twins suggest that obesity has a heritability of 75 percent. Because the heritability for BMI is lower than this, the discrepancy suggests that genes play a larger role in those who tend to gain weight easily. This becomes obvious when populations that have a genetic tendency to obesity experience a large and sudden plunge in the quality of the diet.

On the tiny island of Naura, in Western Samoa, the residents' lifestyles changed greatly when they found a market for the tons of bird droppings on their island as commercial fertilizer. The influx of money led to inactivity and a high-calorie, high-fat diet, replacing an agricultural lifestyle and a diet of fish and vegetables. Within just a generation, two-thirds of the population had become obese, and a third suffered from diabetes.

The Pima Indians offer another example of environmental effects on body weight. These people separated into two populations during the Middle Ages, one group settling in the Sierra Madre mountains of Mexico, the other in southern Arizona. By the 1970s, the Arizona Indians no longer farmed nor ate a low-calorie, low-fat diet, but instead consumed 40 percent of their calories from fat. With this extreme change in lifestyle, they developed the highest prevalence of obesity of any population on earth. Half of the Arizona group had diabetes by age 35, weighing, on average, 57 pounds (26 kilograms) more than their southern relatives, who still eat a low-fat diet and are very active.

The Pima Indians demonstrate that future obesity is not sealed in the genes at conception, but instead is much more likely to occur if the environment provides too many calories and too much fat. They illustrate what geneticist James Neel termed the "thrifty gene hypothesis" in 1962. He suggested that long ago, the hunter-gatherers who survived famine had genes that enabled them to efficiently conserve fat. Today, with food plentiful, the genetic tendency to retain fat is no longer healthful, but harmful. Unfortunately, for many of us, our genomes hold an energy-conserving legacy that works too well—it is much easier to gain weight than to lose it, for sound evolutionary reasons.

The thrifty gene hypothesis is also seen in people who were born after a full-term pregnancy, but were very low weight. To compensate for starvation conditions in the uterus, metabolism shifts in a way that conserves calories—and the person pays for it later with elevated risk of heart disease, stroke, obesity, osteoporosis, and type 2 diabetes. These are multifactorial conditions that, instead of arising from mutations, reflect epigenetic alterations of gene expression.

Another environmental influence on weight is the types of bacteria in our bodies. Bacterial cells in our bodies actually outnumber our own cells. The actions of certain types of bacteria affect the number of calories that we extract from particular foods. Researchers demonstrated this by controlling the diets of a group of obese individuals and monitoring the bacterial species in their feces. The investigators identified nine species of bacteria that enable a human body to extract maximal calories from food.

Perhaps nowhere are the complexities and challenges of gene-environment interactions more profound than in behavioral characteristics, nuances, quirks, and illnesses. The next chapter looks at a few of them.

Key Concepts

1. Genes that encode leptin, the leptin receptor, and proteins that transmit or counter leptin's signals affect body weight.

2. Studies on adopted individuals and twins indicate a heritability of 75 percent for obesity.

3. Populations that suddenly become sedentary and switch to a fatty diet reflect environmental influences on body weight.

Summary

7.1 Genes and the Environment Mold Traits

1. **Multifactorial traits** reflect the environment and genes. A **polygenic trait** is determined by more than one gene and varies continuously in expression.

2. Single-gene traits are rare. For most traits, many genes contribute to a small, but not equal, degree.

7.2 Polygenic Traits Are Continuously Varying

3. The frequency distribution of phenotypes for a polygenic trait forms a bell curve.

7.3 Traditional Ways to Investigate Multifactorial Traits

4. **Empiric risk** measures the likelihood that a multifactorial trait will recur based on prevalence. The risk rises with genetic closeness, severity, and number of affected relatives.

5. **Heritability** estimates the proportion of variation in a multifactorial trait due to genetics in a particular population at a particular time.

6. Characteristics shared by adopted people and their biological parents are mostly inherited, whereas similarities between adopted people and their adoptive parents reflect environmental influences.

7. **Concordance** measures the frequency of expression of a trait in both members of MZ or DZ twin pairs. The more influence genes exert over a trait, the higher the differences in concordance between MZ and DZ twins.

7.4 Genome-Wide Association Studies

8. **Genome-wide association studies** correlate SNP patterns to increased disease risk. They may use a **cohort study** to follow a large group over time, or a **case-control study** on matched pairs. The affected sibling pair strategy can identify homozygous regions that may harbor genes of interest.

7.5 A Closer Look: Body Weight

9. Many genes affect weight. Leptin and associated proteins affect appetite. Fat cells secrete leptin in response to eating, which acts in the hypothalamus to decrease appetite. Populations that switch to a fatty, high-calorie diet and a less-active lifestyle reveal effects of the environment on weight.

www.mhhe.com/lewisgenetics10

Answers to all end-of-chapter questions can be found at **www.mhhe.com/lewisgenetics10.** You will also find additional practice quizzes, animations, videos, and vocabulary flashcards to help you master the material in this chapter.

Review Questions

1. Explain how Mendel's laws apply to multifactorial traits.

2. Explain the difference in the genetic contribution to sickle cell disease or cystic fibrosis, compared to that for hypertension not due to another illness.

3. What is the difference between a Mendelian multifactorial trait and a polygenic multifactorial trait?

4. How can skin color have a different heritability at different times of the year?

5. Explain how the twins in figure 7.4 have such different skin colors.

6. In a large, diverse population, why are medium brown skin colors more common than very white or very black skin?

7. Explain how the Connecticut and Northern Finland investigations of height differ in how they were conducted and in the type of information revealed.

8. Describe the type of information in a(n)
   a. empiric risk calculation.
   b. twin study.
   c. adoption study.
   d. genome-wide association study.

9. Which has a greater heritability—eye color or height? State a reason for your answer.

10. Why does SNP mapping require extensive data?

11. How can older techniques to study multifactorial traits be combined with newer techniques?

12. How can a genome-wide association study overcome the bias of looking only at genes already known to be involved in a trait, or whose known function makes them likely candidates?

13. Name three types of proteins that affect cardiovascular functioning and three that affect body weight.

Applied Questions

1. Rebecca breeds Maine coon cats. The partial pedigree below describes how her current cats are related—the umbrella-like lines indicate littermates, which are the equivalent of fraternal (DZ) twins in humans.

   Cat lover Sam wishes to purchase a pair of Rebecca's cats to breed, but wants them to share as few genes as possible to minimize the risk that their kittens will inherit certain multifactorial disorders. Sam is quite taken with Farfel, but can't decide among Marbles, Juice, or Angie for Farfel's mate.

   Calculate the percentage of genes that Farfel shares with each of these female relatives. With which partner would the likelihood of healthy kittens be greatest?

2. Cite an example from chapter 5 of a single-gene trait or condition that is affected by an environmental influence.

3. Would you take a drug that was prescribed to you based on your race? Cite a reason for your answer.

4. The incidence of obesity in the United States has doubled over the past two decades. Is this due more to genetic or environmental factors? Cite a reason for your answer.

5. One way to calculate heritability is to double the difference between the concordance values for MZ versus DZ twins. For multiple sclerosis, concordance for MZ twins is 30 percent, and for DZ twins, 3 percent. What is the heritability? What does the heritability suggest about the relative contributions of genes and the environment in causing MS?

6. Devise a genome-wide association study to assess whether restless legs syndrome is inherited, and if it is, where susceptibility or causative genes may be located.

7. In chickens, high body weight is a multifactorial trait. Heritability accounts for several genes that contribute a small effect additively, as well as a few genes that exert a great effect. Is this an example of narrow or broad heritability?

8. A study that analyzed the genomes of more than 100,000 people from all over the world found that genes account for a fourth to a third of the variability of blood cholesterol and triglyceride levels. Explain why you think this is either good news or bad news.

9. The environmental epigenetics hypothesis states that early negative experiences, such as neglect, abuse, and extreme stress, increase the risk of developing depression, anxiety disorder, addictions, and obesity later in life, through effects on gene expression that persist. Suggest an experiment to test this hypothesis.

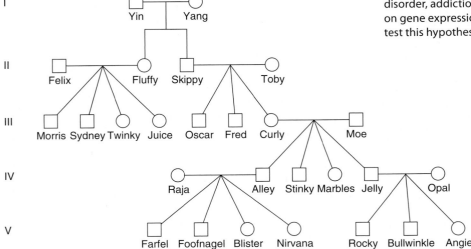

Maine coon cats

Web Activities

1. Many genes contribute to lung cancer risk, especially among people who smoke. These genes include *p53*, *IL1A* and *IL1B*, *CYP1A1*, *EPHX1*, *TERT*, and *CRR9*. Search for one of these genes on the Internet and describe how mutations in it may contribute to causing lung cancer, or polymorphisms may be associated with increased risk.

2. Locate a website that deals with breeding show animals, farm animals, or crops to produce specific traits, such as litter size, degree of meat marbling, milk yield, or fruit ripening rate. Identify three traits with heritabilities that indicate a greater contribution from genes than from the environment.

3. Visit the Centers for Disease Control and Prevention (CDC) website. From the leading causes of death, list three that have high heritabilities, and three that do not. Base your decisions on common sense or data, and explain your selections.

4. Use MIM to look up any of the following genes that encode proteins that affect cardiovascular health and explain what the proteins do: apolipoprotein E; LDL receptor; apolipoprotein A; angiotensinogen; beta-2 adrenergic receptor; toll-like receptor 4; C-reactive protein.

Case Studies and Research Results

1. Marla and Anthony enjoy hiking and mountain climbing. They have a 2-year-old son, Spencer, and want to know whether he will excel at these activities too. They find a company that offers genetic tests for athletic prowess, and send in a sample of Spencer's cheek cells, asking for a test on the angiotensin I–converting enzyme (*ACE*) gene. They read a study of 40 elite British mountaineers, many of whom had a genotype that is rare among the general, non-mountain-climbing population. Look up what the gene does, and suggest how variants of it might affect athletic ability. Do you think that the parents should make decisions about which sports Spencer tries based on the genetic test results? Explain your answer.

2. Concordance for the eating disorder anorexia nervosa for MZ twins is 55 percent, and for DZ twins, 7 percent. Ashley and Maggie are DZ twins. Maggie has anorexia nervosa. Should Ashley worry about an inherited tendency to develop the condition? Explain your answer.

3. Lydia and Reggie grew up poor in New York City in the 1960s. Both went for free to the City University of New York, then to medical school in Boston, where they met. Today, each has a thriving medical practice, and they are the parents of 18-year-old Jamal and 20-year-old Tanya.

   Jamal, taking a genetics class, wonders why he and Tanya do not resemble each other, or their parents, for some traits. The family is African American. Lydia and Reggie are short, 5'2" and 5'7" respectively, and each has medium brown eyes and skin, and dark brown hair. Tanya and Jamal are 5'8" and 6'1", respectively, and were often in the highest height percentiles since they were toddlers. Jamal has very dark skin, darker than his parents' skin, while Tanya's skin is noticeably lighter than that of either parent. Tanya's eyes are so dark that they appear nearly black.

   a. Why do Tanya's eyes appear darker than those of her parents or brother?

   b. How can Jamal's skin be darker than that of his parents, and Tanya's be lighter?

   c. Which of the traits considered is most influenced by environmental factors?

   d. What is the evidence that Jamal and Tanya's height is due to environmental and genetic factors?

   e. Which of the four traits has the highest heritability?

4. A study looked at 200,000 SNPs throughout the genome for 1,820 people with premature hair graying and 1,820 without this trait. Those with the trait shared several SNPs on chromosome 9. What type of study is this?

5. "Elite controllers" are people with HIV infection whose immune systems naturally keep levels of the virus extremely low. Researchers are conducting genome-wide SNP analyses of these people. Explain how the results of this study might be used to help people who more easily develop AIDS.

6. An affected sibling pair study identified areas of chromosomes 1, 14, and 20 that are likely to harbor genes that predispose individuals toward or cause schizophrenia. Explain how such an investigation is conducted.

7. Researchers compared the expression levels of 9,156 genes that could affect drug metabolism among 60 mother-father-child trios from Utah who were Caucasian, and 30 similar families who were black and from Nigeria. About 5 percent of the genes differed significantly in expression between the two groups. What further information would be helpful in applying this information in prescribing drugs?

8. A study is examining the expression of certain genes in people about to undergo weight loss surgery (gastric bypass) to see if these patterns predict individuals most likely to benefit by achieving long-term weight loss.

   a. Name three genes, or types of genes, that might provide valuable information for this analysis.

   b. What is a limitation of this study design?

   c. Do you think that this study has value?

9. A study in England tested 20,430 people for alleles of 12 genes known to increase risk of becoming obese. (Each person had 24 alleles assessed, 12 from each parent.) Although the number of risk alleles correlated to BMI, the more a person exercised, the lower the BMI. Do these findings support a genetic deterministic view of the trait of body weight, or not? Cite a reason for your answer.

Chronic fatigue syndrome is not "all in one's head." It is likely due to an inherited extreme response to stress.

Genetics of Behavior

Learning Outcomes

8.1   Genes and Behavior

1. Identify the physical basis of behavioral traits in the brain.

2. Discuss the difficulties in analyzing behavioral traits and disorders.

3. Explain how genetics may be used to better understand behavioral disorders.

8.2   Eating Disorders; 8.3 Sleep; 8.4 Intelligence; 8.5 Drug Addiction; 8.6 Mood Disorders; 8.7 Schizophrenia; 8.8 Autism

4. Discuss the genetic and environmental influences on eating disorders, sleep, intelligence, addiction, mood, schizophrenia, and autism.

8.9   How Are Behavioral Disorders Related?

5. Explain how autism, bipolar disorder, and schizophrenia may be related.

The Big Picture: Behavioral traits and disorders reflect effects and interactions of genes and environmental factors on the nervous system.

Chronic Fatigue Syndrome

The disabling, relentless exhaustion of chronic fatigue syndrome (CFS) usually begins with a flu-like illness, and lasts at least six months. Until recently, many people did not believe CFS was real. Genetics research suggests that it is.

Bestselling author Laura Hillenbrand has CFS. Ill since 1987, Laura has been bedbound for months on end, had to drop out of college, and lost a great deal of weight. At first, doctors blamed infection, and sent her to a psychiatrist, who ruled out mental illness. Until she finally found a physician familiar with CFS, Hillenbrand's doctors said "the symptoms are all in your head." They weren't.

Clues to the cause of CFS lay in the observation that many people report a severe physical illness or emotional trauma before the fatigue began. Also, people with CFS make too little of the stress hormone cortisol, and too much of the nervous system chemical serotonin, which induces sleep and calms mood. Could an abnormal and persistent response to a trauma—physical or emotional—cause CFS? Hormones from the brain's hypothalamus and pituitary gland, and the adrenal glands, control responses to trauma. Searching for variants of genes whose encoded proteins affect these hormones led to the discovery of three genes with variants that differ in people with CFS. One gene encodes the receptor that binds stress hormones and the other two affect the availability of serotonin. A physical explanation for CFS may lead to treatment.

8.1 Genes and Behavior

Behavior is a complex continuum of emotions, moods, intelligence, and personality that drives how we function on a daily basis. We are, to an extent, defined and judged by our many behaviors. They control how we communicate, cope with negative feelings, and react to stress. Behavioral disorders are common, with wide-ranging and sometimes overlapping symptoms. Many of our behaviors are in response to environmental factors, but *how* we respond has genetic underpinnings. Understanding the biology behind behavior can help to develop treatments for behavior-based disorders. Several are discussed in this chapter.

Behavioral genetics considers nervous system function and variation, including the hard-to-define qualities of mood and mind. The human brain weighs about 3 pounds and resembles a giant gray walnut, but with the appearance and consistency of pudding. It consists of 100 billion nerve cells, or **neurons**, and at least a trillion other cells called **neuroglia**, which support and nurture the neurons.

Neurons connect and interact in complex networks that make the brain a coordinated, functioning organ that controls all of the body. Branches from each of the 100 billion neurons

in the brain form close associations, called synapses, with 1,000 to 10,000 other neurons. Neurons communicate across these tiny spaces using chemical signals called neurotransmitters. Networked neurons oversee broad functions such as sensation and perception, memory, reasoning, and muscular movements.

Genes control the production and distribution of neurotransmitters. **Figure 8.1** indicates the points where genes control the sending and receiving of nervous system information. Enzymes oversee the synthesis of neurotransmitters and their transport from the sending (presynaptic) neuron across the synapse to receptors on the receiving (postsynaptic) neuron. Proteins called transporters ferry neurotransmitters from sending to receiving neurons, and proteins also form the subunits of receptors. Genes control the synthesis of myelin, a fatty substance that coats neuron extensions called axons. Myelin insulates the neuron, which speeds neurotransmission. Signal transduction is also a key part of the function of the nervous system (see figure 2.20). Therefore, candidate genes for the inherited components of a variety of mood disorders and mental illnesses—as well as of normal variations in temperament and personality—affect neurotransmission and signal transduction.

Researchers investigate genes known to encode proteins that take part in neurotransmission and signal transduction. They also use genome-wide association studies to identify genes not already suspected of affecting behavior.

Identifying the inherited and environmental contributors to a behavioral disorder is very challenging, partly because of the traditional way that psychiatrists diagnose such conditions. The widely used *Diagnostic and Statistical Manual* (DSM) categorizes and distinguishes mental and behavioral conditions based on symptoms. However, different syndromes share symptoms, and the same disorder can have different causes in different people.

Discoveries in genetics are challenging some DSM classifications. Many behavioral traits and disorders, like other characteristics, probably reflect a major influence from a single gene, perhaps one whose protein product takes part directly in neurotransmission, but also small inputs from common gene variants. Researchers envision from 100 to 300 genes at play. Inheriting certain subsets of variants of these genes makes an individual susceptible to developing a certain

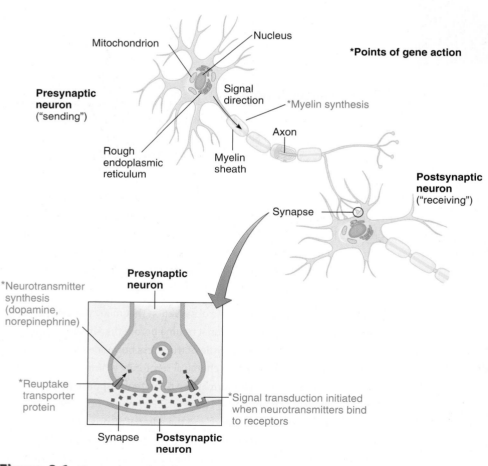

Figure 8.1 Neurotransmission. Many genes that affect behavior produce proteins that affect neurotransmission, which is sending a signal from one neuron to another across a synapse via a neurotransmitter molecule.

disorder in the presence of a particular environmental stimulus. Some genes are implicated in more than one behavioral disorder.

In contrast to the separate designations in the phenotype-based DSM, genetics studies are showing that behavioral and mental disorders lie more across a continual spectrum. The Psychiatric Genome-Wide Association Study Consortium, for example, is looking among 59,000 people for SNP patterns associated with autism, attention deficit hyperactivity disorder, bipolar disorder, depression, and schizophrenia. Another study is examining copy number variants (repeated short DNA sequences) and mutations that arise spontaneously, rather than being inherited, that may explain behavioral disorders that arise in someone without a family history.

Deciphering the genetic components of behavioral traits and disorders uses traditional empiric risk, adoptee, and twin study data and, more and more, molecular methods that determine genotypes, gene expression patterns, and sets of susceptibility gene variants. The chapter sections that follow explore these various sources of data, some beginning with case histories. Identifying gene variant combinations behind behavioral disorders may make it possible to subtype these conditions in a biologically meaningful way so that diagnoses will be more personalized and effective treatments begun sooner.

Table 8.1 lists the prevalence of some behavioral disorders.

Table 8.1	Prevalence of Behavioral Disorders in the U.S. Population
Condition	**Prevalence (%)**
Alzheimer disease	4.0
Anxiety	8.0
Phobias	2.5
Posttraumatic stress disorder	1.8
Generalized anxiety disorder	1.5
Obsessive compulsive disorder	1.2
Panic disorder	1.0
Attention deficit hyperactivity disorder	2.0
Autism spectrum disorders	0.1
Drug addiction	4.0
Eating disorders	3.0
Mood disorders	7.0
Major depressive disorder	6.0
Bipolar disorder	1.0
Schizophrenia	1.3

Key Concepts

1. Behavioral traits reflect genetic and environmental influences, and arise from connections among the brain's 100 billion neurons.

2. The genes behind behavior affect neurotransmitter production and transport across synapses; myelin synthesis; and signal transduction.

3. Genetic classification of behavioral disorders is challenging traditional psychiatric classification. Behavioral disorders may lie on a continuum, with many contributing genes.

8.2 Eating Disorders

In anorexia nervosa, a person perceives herself or himself as obese, even when obviously not, and intentionally starves, exercises obsessively, and/or takes laxatives to hasten weight loss (**figure 8.2**).

For economically advantaged females in the United States, the lifetime risk of developing anorexia nervosa is 0.5 percent. Anorexia has the highest risk of death of any psychiatric disorder—15 to 21 percent. The same population group has a lifetime risk of 2.5 percent of developing another eating disorder, bulimia. A person with bulimia eats huge amounts but exercises and vomits to maintain weight.

Five to 10 million people in the United States have eating disorders. About 10 percent of them are male. One survey of 8-year-old boys revealed that more than a third of them had

Figure 8.2 An eating disorder. A person who has anorexia nervosa perceives herself as obese, although she is the opposite.

attempted to lose weight. In an eating disorder called muscle dysmorphia, boys and young men take amino acid supplements to bulk up. Just as the person with anorexia looks in a mirror and sees herself as too large, a person with muscle dysmorphia sees himself as too small.

Because eating disorders were once associated almost exclusively with females, most available risk estimates exclude males. Twin studies reveal a considerable genetic component, with heritability ranging from 0.5 to 0.8. Studies of eating disorders that recur in siblings can be difficult to interpret. Is a young girl imitating her older sister by starving because genes predispose her to develop an eating disorder, or because she wants to be like her sister?

Genes that encode proteins that control appetite are candidate genes for developing eating disorders (see table 7.7). Japanese researchers have performed a case-control, genome-wide association study for anorexia nervosa and identified four chromosomal regions that include genes that are active in the part of the brain that regulates eating behavior. It will be interesting to learn which genes affect body image, and how they do so.

Key Concepts

1. Eating disorders are common and have high heritability.
2. Genes whose products control appetite or regulate certain neurotransmitters may elevate the risk of developing an eating disorder.

8.3 Sleep

Sleep has been called "a vital behavior of unknown function," and, indeed, without sleep, animals die. We spend a third of our lives in this mysterious state.

Genes influence sleep characteristics. When asked about sleep duration, schedule, quality, nap habits, and whether they are "night owls" or "morning people," MZ twins report significantly more in common than do DZ twins. This is true even for MZ twins separated at birth. Twin studies of brain wave patterns through four of the five stages of sleep confirm a hereditary influence. The fifth stage, REM sleep, is associated with dreaming and therefore may reflect the input of experience more than genes.

Narcolepsy

Researchers discovered the first gene related to sleep in 1999, for a condition called "narcolepsy with cataplexy" in dogs. Humans have the disorder (MIM 161400), but it is rarely inherited as a single-gene trait—it is more often polygenic requiring an environmental trigger.

A person (or dog) with narcolepsy falls asleep suddenly several times a day. Extreme daytime sleepiness greatly disrupts daily activities. People with narcolepsy have a tenfold higher rate of car accidents. Another symptom is sleep paralysis, which is the inability to move for a few minutes after awakening. The most dramatic manifestation of narcolepsy is cataplexy. During these short and sudden episodes of muscle weakness, the jaw sags, the head drops, knees buckle, and the person falls to the ground. This often occurs during a bout of laughter or excitement—which can be quite disturbing both for the affected individual and bystanders. People with narcolepsy and cataplexy cannot participate in even the most mundane of activities for fear of falling and injuring themselves. Narcolepsy with cataplexy affects only 0.02 to 0.06 percent of the general populations of North America and Europe, but the fact that it is much more common in certain families suggests a genetic component.

Studies on dogs led the way to discovery of a human narcolepsy gene. In 1999, Emmanuel Mignot and his team at Stanford University identified mutations in a gene that encodes a receptor for a neuropeptide called hypocretin (MIM 602358). In Doberman pinschers and Labrador retrievers, the receptor does not reach the cell surfaces of certain brain cells, and as a result, the cells cannot receive signals to promote a state of awakeness. Dachshunds have their own mutation—they make a misshapen, nonfunctional receptor. **Figure 8.3** shows a still frame of a film that Mignot made of narcoleptic dogs playing. Suddenly, they all collapse! A minute later, they get up and resume their antics. "You can't make dogs laugh, but you can make them so happy that they have attacks," says Mignot. To induce a narcoleptic episode in puppies, he lets them play with each other. He feeds older dogs meat, which excites them so much that they can take a while to finish a meal because they fall down in delight so often. Getting narcoleptic dogs to breed is difficult, too, for sex is even more exciting than play or food!

In 1998, Masahi Yanagisawa, at the University of Texas Southwestern Medical Center in Dallas, discovered a protein

Figure 8.3 Letting sleeping dogs lie. These Doberman pinschers have inherited narcolepsy. They suddenly fall into a short but deep sleep while playing. Research on dogs with narcolepsy led to the discovery of a gene that affects sleep in humans.

called orexin, but thought it only sent signals to eat. Yanagisawa's orexin turned out to bind Mignot's hypocretin receptor. Yanagisawa bred mice that lacked the orexin gene, and then noticed something odd while watching the animals feed at night—the rodents suddenly fell down fast asleep! Researchers are now trying to figure out how one molecule controls feeding as well as wakefulness. The hypocretin/orexin receptor gene, found on dog chromosome 12, is on human chromosome 6. The brains of humans with narcolepsy and cataplexy are remarkably deficient in hypocretin/orexin, but it isn't the orexin gene that causes the condition in humans—it is likely another gene that controls it. Researchers are now studying hypocretin/orexin and the molecules with which they interact to develop drugs for insomnia.

Familial Advanced Sleep Phase Syndrome

Daily rhythms such as the sleep-wake cycle are set by cells that form a "circadian pacemaker" in two clusters of neurons in the brain called the suprachiasmatic nuclei. In these cells, certain "clock" genes are expressed in response to light or dark in the environment.

The function of clock genes is most obvious in families that have a mutation. For example, five generations of a family in Utah have familial advanced sleep phase syndrome (MIM 604348). Affected individuals promptly fall asleep at 7:30 each night and awaken suddenly at 4:30 A.M., thanks to a mutation in a gene on chromosome 2 called *period* (**figure 8.4**). People with the condition have a single DNA base substitution that prevents the encoded protein from binding a phosphate chemical group, which it must do to signal to the brain in a way that synchronizes the sleep-wake cycle with daily sunrise and sunset. At least three other genes are associated with insomnia.

8.4 Intelligence

Intelligence is a vastly complex and variable trait that is subject to many genetic and environmental influences, and also to great subjectivity. Sir Francis Galton, a half first cousin of Charles Darwin, investigated genius, which he defined as "a man endowed with superior faculties." He identified successful and prominent people in Victorian-era English society, and then assessed success among their relatives. In his 1869 book, *Hereditary Genius,* Galton wrote that relatives of eminent people were more likely to also be successful than people in the general population. The closer the blood relationship, he concluded, the more likely the person was to succeed. This, he claimed, established a hereditary basis for intelligence.

Definitions of intelligence vary. In general, intelligence refers to the ability to reason, learn, remember, connect ideas, deduce, and create. The first intelligence tests, developed in the late nineteenth century, assessed sensory perception and reaction times to various stimuli. In 1904, Alfred Binet in France developed a test with questions based on language, numbers, and pictures to predict the success of developmentally disabled youngsters in school. The test was modified at Stanford University to assess white, middle-class Americans. An average score on this "intelligence quotient," or IQ test, is 100, with two-thirds of all people scoring between 85 and 115 in a bell curve or normal distribution (**figure 8.5**). An IQ between 50 and 70 is considered mild intellectual disability, and below 50, severe intellectual disability. In the United States, 3 in 100 individuals have intellectual disability. Causes include single-gene and chromosomal disorders, problems during pregnancy or birth, and infection, poisoning, and malnutrition.

IQ has been a fairly accurate predictor of success in school and work. However, low IQ also correlates with many societal situations, such as poverty, a high divorce rate, failure to complete high school, incarceration (males), and having a child out of wedlock (females). Opportunity has a lot to do with intellectual development.

The IQ test consists of short exams that measure verbal fluency, mathematical

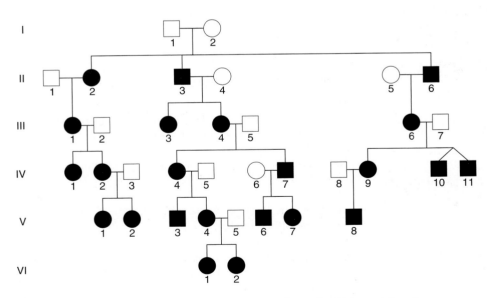

Figure 8.4 **Inheritance of a disrupted sleep-wake cycle.** This partial pedigree depicts affected members of a large family with an autosomal dominant form of familial advanced sleep phase syndrome.

reasoning, memory, and spatial visualization skills. Because people tend to earn similar scores in all these areas, psychologists use a general or global intelligence ability, called "*g*," to represent the four basic skills that IQ encompasses. In contrast, analysis of personality reveals five contributing factors. The *g* value is the part of IQ that accounts for differences between individuals based on a generalized intelligence, rather than on enhanced opportunities such as attending classes to boost test-taking skills.

Environment does not seem to play too great a role in IQ differences. With time, the IQ scores of adoptees become closer to those of their biological parents than to those of their adoptive parents. Heritability studies also reveal a declining environmental impact with age (**table 8.2**). This makes sense. As a person ages, he or she has more control over the environment, so genetic contributions to intelligence become more prominent.

Researchers have long recognized a genetic explanation for intelligence differences because nearly all syndromes that result from atypical chromosomes include some degree of intellectual disability. The search for single genes that contribute to intelligence differences focuses on proteins that control neurotransmission. For example, a certain variant of a gene encoding neural cellular adhesion molecule (N-CAM) correlates strongly with high IQ. Perhaps this gene variant eases certain neural connections that enhance learning ability. Genome-wide association studies have identified regions of several chromosomes that likely include genes that affect the hard-to-assess trait of intelligence.

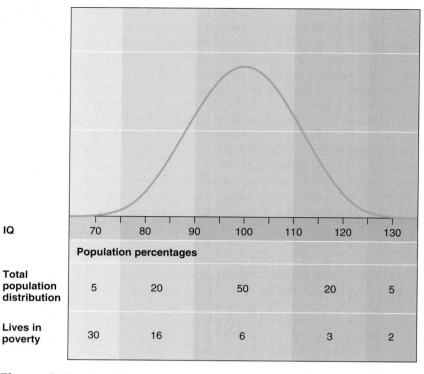

Figure 8.5 **Success and IQ.** IQ scores predict success in school and the workplace in U.S. society. The bell curve for IQ indicates that most people fall between 85 and 115. However, those living in poverty tend to have lower IQs.

IQ	70	80	90	100	110	120	130
Population percentages							
Total population distribution	5	20		50		20	5
Lives in poverty	30	16		6		3	2

Table 8.2	**Heritability of Intelligence Changes Over Time**
Age Group	**Heritability**
Preschoolers	0.4
Adolescents	0.6
Adults	0.8

Key Concepts

1. Intelligence is the use of mental skills to complete complex tasks or solve problems.
2. IQ assesses verbal fluency, mathematical reasoning, memory, and spatial visualization ability.
3. The "*g*" value measures a general intelligence factor that represents the inherited portion of IQ.
4. Environment has less of an influence on intelligence as a person ages. Individual genes affect intelligence.

8.5 Drug Addiction

One person sees what a loved one goes through in battling lung cancer and never picks up another cigarette. Another person actually has lung cancer, yet refuses to stop smoking, taking breaks from using her oxygen tank so that she can light up. Addiction is powerful. Evidence is mounting that genes play a large role in making some individuals prone to addiction, and others not.

The Definition of Addiction

Drug addiction is compulsively seeking and taking a drug despite knowing its dangers. The two identifying characteristics are tolerance and dependence. Tolerance is the need to take more of the drug to achieve the same effects as time goes on. Dependence is the onset of withdrawal symptoms upon stopping use of the drug. Both tolerance and dependence contribute to the biological and psychological components of craving the drug. The behavior associated with drug addiction can be extremely difficult to break.

Drug addiction produces long-lasting brain changes. Craving and high risk of relapse remain even after a person has abstained for years. Heritability is 0.4 to 0.6, with a two- to threefold increase in risk among adopted individuals who have one affected biological parent. Twin studies also indicate an inherited component to drug addiction.

Brain imaging techniques localize the "seat" of drug addiction in the brain by highlighting the cell surface receptors

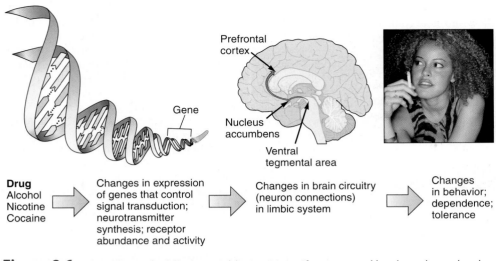

Figure 8.6 The events of addiction. Addiction is manifest at several levels: at the molecular level, in neuron-neuron interactions in the brain, and in behavioral responses.

Drug
Alcohol
Nicotine
Cocaine

Changes in expression of genes that control signal transduction; neurotransmitter synthesis; receptor abundance and activity

Changes in brain circuitry (neuron connections) in limbic system

Changes in behavior; dependence; tolerance

Gene

Prefrontal cortex

Nucleus accumbens

Ventral tegmental area

that bind neurotransmitters when a person craves the drug. The brain changes that contribute to addiction occur in parts called the nucleus accumbens, the prefrontal cortex, and the ventral tegmental area, which are part of a larger set of brain structures called the limbic system (**figure 8.6**). The effects of cocaine seem to be largely confined to the nucleus accumbens, whereas alcohol affects the prefrontal cortex.

The specific genes and proteins that are implicated in addiction to different substances may vary, but several general routes of interference in brain function are at play. Proteins involved in drug addiction are those that

- are part of the production lines for neurotransmitters, such as enzymes;
- form reuptake transporters, which remove excess neurotransmitters from the synapse;
- form receptors on the postsynaptic neuron that are activated or inactivated when specific neurotransmitters bind; and
- are part of the signal transduction pathway in the postsynaptic neuron.

Drugs of Abuse

Our ancient ancestors must have discovered that ingesting certain natural substances, particularly from plants, provided a feeling of well-being. That tendency persists today. The degree to which a particular drug is addictive has nothing to do with how a society controls access to it.

Abused drugs are often plant-derived chemicals, such as cocaine, opium, and tetrahydrocannabinol (THC), the main active ingredient in marijuana. These substances bind to receptors on human neurons, which indicates that our bodies have versions of these substances. The human equivalents of the opiates are the endorphins and enkephalins, and the equivalent of THC is anandamide. The endorphins and enkephalins relieve pain. Anandamide modulates how brain cells respond

to stimulation by binding to neurotransmitter receptors on presynaptic (sending) neurons. In contrast, neurotransmitters bind to receptors on postsynaptic neurons.

Amphetamines and LSD are drugs that produce their effects by binding to receptors on neurons that normally bind neurotransmitters called trace amines. Trace amines are found throughout the brain at low levels, compared to the more abundant neurotransmitters such as dopamine and serotonin. LSD causes effects similar to the symptoms of schizophrenia (see section 8.7), suggesting that the trace amine receptors, which are proteins, may be implicated in the illness.

Candidate gene studies as well as genome-wide association studies are revealing gene variants that people addicted to various drugs share. These inherited factors must be paired with environmental stimuli for addiction to occur. For example, people who are homozygous for the *A1* allele of the dopamine D(2) receptor gene variant are overrepresented among people with alcoholism and people with other addictions. Genome-wide association studies have found 51 chromosomal regions that may include genes that contribute to craving. Finally, studies of gene expression flesh out this picture by providing a real-time view of biochemical changes when a person craves a drug, and then takes it.

Discovering the genetic underpinnings of nicotine addiction is increasing our knowledge of addiction in general, and may have practical consequences. Two facts are clear:

1. The nicotine in tobacco products causes lung cancer.
2. Nicotine is highly addictive.

Genetics explains both.

Each year, 35 million people try to quit smoking, yet only 7 percent succeed. It is easy to see on a whole-body level how this occurs: Nicotine levels peak 10 seconds after an inhalation and the resulting pleasurable release of dopamine in brain cells fades away in minutes. To keep the feeling that researchers call the "reward," smoking must continue. On a molecular level, the tendency to be harmed by nicotine can be traced to a five-part molecular assembly called a nicotinic receptor. The receptor normally binds the neurotransmitter acetylcholine, but it also binds the similarly shaped nicotine molecule. Certain versions of the receptor bind nicotine very strongly, which triggers a nerve impulse that, in turn, stimulates the pleasurable dopamine release. That may explain the addiction. These receptors are also located on several types of lung cells, where they bind carcinogens. So the nicotine in tobacco causes addiction and susceptibility to lung cancer, and it delivers the carcinogens.

Genetics explains nicotine addiction in several ways. A different gene encodes each of the five parts of the nicotinic receptor. If two of the five parts are certain variants, then a person experiences desire to continue smoking after the first cigarette. Risk for addiction is high. Susceptibility alleles are more common in some populations than others. In one study, they were present in 37 percent of Europeans but in none of the Africans. Another gene encodes a protein, called neurexin-1, that ferries the nicotinic receptors to the neuron's surface. Genes also control dopamine synthesis and transport, and how readily the carcinogens cause cancer. One gene variant, for example, greatly increases risk of developing lung cancer, only if the person smokes.

Key Concepts

1. Drug addiction is dependency on a drug despite knowing the activity is harmful.
2. Structures in the limbic system are directly involved in drug addiction.
3. A candidate gene for addiction encodes the dopamine D(2) receptor.
4. Nicotine binds a receptor that normally binds acetylcholine, causing dopamine release and pleasure.

8.6 Mood Disorders

We all have moods, but mood disorders, which affect millions of people, impair the ability to function on a day-to-day basis. Context is important in evaluating extreme moods. For example, the same symptoms that may lead to a diagnosis of depression are normal in the context of experiencing profound grief. The two most prevalent mood disorders are **major depressive disorder** and **bipolar disorder** (also called manic-depression).

Major Depressive Disorder

Sheeva couldn't exactly remember how her depression had begun. She'd always loved her job at the bookstore after classes at the community college, but it began to annoy her, just as her classes began to bore her. She had trouble falling and staying asleep, which made her more tired and withdrawn the next day. A few times she'd started to cry while stuck in traffic, not sure why. Then she started having to drag herself out of bed in the morning, completely unmotivated to do anything. She often forgot to eat—food was no longer enjoyable. When she asked her friend Juanita to make some routine phone calls for her, Juanita became alarmed, and gently suggested that Sheeva see a doctor about depression. Sheeva was surprised, but her friend was insistent. After a thorough interview by the physician, Sheeva was indeed diagnosed with major depressive disorder. Within a few weeks of starting treatment with a drug that increases the abundance of the neurotransmitter serotonin in her brain, the fog she had been living with for months began to lift—and she realized she had been sick.

Depression affects 6 percent of the U.S. population at any given time, and affects more women than men. Lifetime risk for the general population is 5 to 10 percent. Often depression is chronic, with acute episodes provoked by stress. It is a serious illness. Fifteen percent of people hospitalized for severe, recurrent depression ultimately end their lives. About half of all people who experience a depressive episode will suffer others. Half of affected individuals do not seek medical help, and among those who do, a third do not respond to drug therapy; those who do respond may relapse if they stop taking an effective drug. Electroconvulsive (shock) therapy can help some patients for whom drugs do not work. For many people, antidepressant treatment is very helpful if paired with psychotherapy.

A likely cause of depression is deficiency of the neurotransmitter serotonin, which affects mood, emotion, appetite, and sleep. Levels of norepinephrine, another type of neurotransmitter, are important as well. In depression, these neurotransmitters become depleted in synapses. Antidepressant drugs called selective serotonin reuptake inhibitors (SSRIs) prevent presynaptic neurons from admitting serotonin from the synapse, leaving more of it available to stimulate the postsynaptic cell (**figure 8.7**). This action apparently offsets the neurotransmitter deficit. Other antidepressants target norepinephrine or

Nondepressed individual

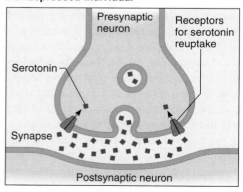

Depressed individual, untreated

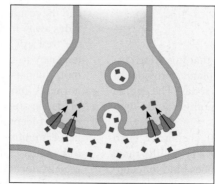

Depressed individual, treated with SSRI

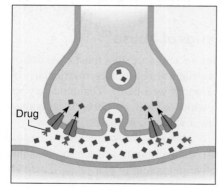

Figure 8.7 How an antidepressant works. Selective serotonin reuptake inhibitors (SSRIs) block the reuptake of serotonin, leaving more of the neurotransmitter in synapses. This corrects a neurotransmitter deficit that presumably causes the symptoms. Overactive or overabundant reuptake receptors can cause the deficit. The precise mechanism of SSRIs is not well understood, and the different drugs may work in slightly different ways.

both serotonin and norepinephrine. SSRIs may begin to produce effects after one week, often enabling a person with moderate or severe depression to return to some activities, but full response can take up to eight weeks.

In the past, physicians would try one antidepressant drug after another, based on experience with other patients. This trial-and-error approach would often take months. Tests of gene expression enable clinicians to choose an antidepressant in a personalized manner, based on which drug is likely to be most effective, with the fewest side effects, in a particular patient.

Bipolar Disorder

Phil Ochs was a political activist in the 1960s who wrote some of the most passionate and compelling antiwar songs of that tumultuous era. Yet his detailed, evocative lyrics were in sharp contrast to his inability to sustain personal relationships. Ochs suffered from bipolar disorder, which made his short life a lonesome highway of extreme ups and downs. His mood first plunged in 1968, in response to the disturbing events of the times—the Democratic National Convention, where he was arrested for protesting, and the assassinations of Martin Luther King Jr. and Robert Kennedy. By 1970, he emerged in full manic mode at Carnegie Hall, dressed in a gold Elvis costume, much to the shock of long-time fans of his folk music (the author was there.) Shortly after this odd performance, he declined again, eventually becoming unable to perform, a prisoner to his uncontrollable moods. Ochs committed suicide in 1976.

Bipolar disorder is much rarer than depression. It affects 1 percent of the population and has a general population lifetime risk of 0.5 to 1.0 percent. Weeks or months of depression alternate with periods of mania, when the person is hyperactive and restless, and may experience a rush of ideas and excitement. Ideas may be fantastic, and behavior reckless. For example, a person who is normally quiet and frugal might, when manic, suddenly make large monetary donations and spend lavishly—very out-of-character behavior. In one subtype of bipolar disorder, the "up" times are termed hypomania, and they seem more a temporary reprieve from the doldrums than the bizarre behavior of full mania. Bipolar disorder with hypomania may appear to be depression. This is an important distinction because different drugs are used to treat depression and bipolar disorder.

Many gene variants contribute to bipolar disorder. Early genetic studies looked at large Amish families, in whom the manic phase of the disorder was very obvious amid their restrained lifestyle. But studies in different families pointed to different genes behind bipolar disorder. Researchers now realize that the reason why studies disagree is that there are many gene variant combinations that cause or contribute to the phenotype of bipolar disorder, but only a few such variants are seen in any one family. A systematic scrutiny of all studies of the genes behind bipolar disorder suggests as much as 10 percent of the genome—that is, hundreds of genes—are part of the picture. One researcher compares the approach of considering many studies to a Google search: "The more links there are to a page on the Internet, the more likely it is to come up

at the top of your search list. The more experimental lines of evidence for a gene, the higher it comes up in your priority list of genes involved in the disorder." Question 2 in Web Activities at the chapter's end lists some of these genes.

Key Concepts

1. Major depressive disorder is more common than bipolar disorder, and is likely caused by deficits of serotonin, norepinephrine, or both.
2. Bipolar disorder is associated with several chromosomal sites, and its genetic roots are difficult to isolate.

8.7 Schizophrenia

Mathematician John Nash Jr. won the Nobel Prize in 1994 for game theory based on a short PhD thesis he wrote in 1950, at age 22. He is better known for the portrayal of his paranoid schizophrenia in a book and a film called A Beautiful Mind.

Symptoms began in 1959, when Nash's lectures and conversations would ramble. His thoughts tumbled out so fast, and were so disconnected, that people could not understand him. He left academia for the first of several stays in a psychiatric facility. Between hospitalizations, Nash traveled. He heard voices and imagined he was being hunted. Numbers, he thought, carried secret meanings. Nash had delusions that he was a slave, the messiah, the biblical Job, and the emperor of Antarctica.

John Nash has provided insights into schizophrenia because he got better. In his mid-fifties, he became able to ignore the voices, concentrate, and think mathematically again. Looking back, Nash attributes his professional success to the strange way of thinking that was his schizophrenia. He isn't alone. Perhaps a way to make schizophrenia easier to live with, for some people, may be to channel the heightened perceptions into art, writing, or music.

Schizophrenia is a debilitating loss of the ability to organize thoughts and perceptions, which leads to a withdrawal from reality. It is a form of psychosis, which is a disorder of thought and sense of self. In contrast, the mood disorders are emotional, and the dementias are cognitive. Various forms of schizophrenia together affect 1 percent of the world's population. Ten percent of affected individuals commit suicide.

Identifying genetic contributions to schizophrenia illustrates the difficulties in analyzing a behavioral condition. Some of the symptoms are also associated with other illnesses; many genes cause or contribute to schizophrenia; and several environmental factors mimic the condition.

Signs and Symptoms

The first signs of schizophrenia often affect thinking. In late childhood or early adolescence, a person might develop trouble paying attention in school, and learning may become difficult as memory falters and information-processing skills diminish. Symptoms of psychosis begin between ages 17 and 27 for males and 20 and 37 for females. These include delusions and

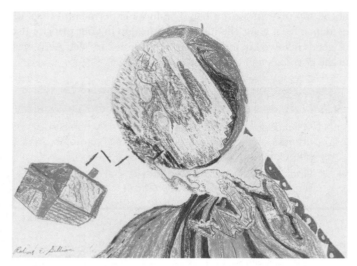

Figure 8.8 **Schizophrenia alters thinking.** People with schizophrenia communicate the disarray of their thoughts in characteristically disjointed drawings.

Table 8.3	Empiric Risks for Schizophrenia
Relationship	**Risk**
MZ twin	48%
DZ twin	17%
Child	13%
Sibling	9%
Parent	6%
Half sibling	6%
Grandchild	5%
Niece/nephew	4%
Aunt/uncle	2%
First cousin	2%
General population	1%

hallucinations, sometimes heard, sometimes seen. A person with schizophrenia may hear a voice giving instructions. What others perceive as irrational fears, such as being followed by monsters, are very real to the person with schizophrenia. Meanwhile, cognitive skills continue to decline. Speech reflects the garbled thought process, as the person skips from topic to topic with no obvious thread of logic, or responds inappropriately, such as laughing at sad news. Artwork by a person with schizophrenia can display the characteristic fragmentation of the mind (**figure 8.8**). (Schizophrenia means "split mind," but it does not cause a split or multiple personality.)

The course of schizophrenia often plateaus (evens out) or becomes episodic. It is not a continuous decline, as is the case for dementia. Schizophrenia may be misdiagnosed as depression or bipolar disorder—and, in fact, some of the same gene variants can underlie these conditions. However, schizophrenia is a very distinctive mental illness that primarily affects thinking. Depression and bipolar disorder mostly affect mood.

Genetic Associations

A heritability of 0.8 and empiric risk values indicate a strong role for genes in causing schizophrenia (**table 8.3**). However, it is possible to develop some of the symptoms, such as disordered thinking, from living with and imitating people who have schizophrenia. Although concordance is high, a person who has an identical twin with schizophrenia has a 52 percent chance of *not* developing it. Therefore, the condition has a significant environmental component, too.

Dozens of genes may interact with an environmental trigger or triggers to cause schizophrenia. Genome-wide screens of families with schizophrenia reveal at least twenty-four sites where affected siblings share alleles much more often than the 50 percent of the time that Mendel's first law predicts. These regions may therefore harbor genes that contribute to development of the pattern of behaviors that we call schizophrenia. People with schizophrenia are particularly likely to have

rare duplications or deletions of DNA that arise *de novo*—in them—rather than being inherited. Rare mutations contribute to many if not most cases of schizophrenia, along with environmental triggers, but any particular mutant gene accounts for less than 1 percent of cases.

Environmental Influences

Several environmental factors may increase the risk of developing schizophrenia. These include birth complication, fetal oxygen deprivation, herpesvirus infection at birth, and malnutrition or traumatic brain injury in the mother. Infection during pregnancy is a particularly well-studied environmental factor in elevating risk of schizophrenia. When a pregnant woman is infected, her immune system bathes the brain of the embryo or fetus with cytokines, which cause inflammation and can subtly alter brain development. It is likely the mother's immune response that increases the risk because many pathogens cannot cross the placenta and get to the fetus.

The idea that maternal infection can sow the seeds for schizophrenia came from observations that an unusually high percentage of people with schizophrenia were born in the winter, especially in the years of flu pandemics. Research supports this observation. Blood stored from 12,000 pregnant women during the 1950s and 1960s shows an association between high levels of the cytokine interleukin-8 and having a child who developed schizophrenia.

Environmental influences after birth are also important in elevating the risk of developing schizophrenia. A study in Europe of 7,500 people with schizophrenia identified four such risk factors: living in cities; migration; regular marijuana use; and victimization during childhood, such as abuse, neglect, and bullying. The opening essay to chapter 11 discusses another environmental factor that can greatly increase the risk of schizophrenia: starvation.

8.8 Autism

Stephen seemed a normal baby in every way. By a year of age, he was cruising along the furniture, babbling, smiling, and even saying a few words. By 18 months of age, his vocabulary had reached a hundred or so words—plenty to navigate among his toddling friends in daycare. Then, gradually, his language and social skills began to vanish. He stopped initiating conversations, and wouldn't play with the others, preferring to sit in a corner holding a large rubber ball and rocking back and forth, singing to himself. By age 2, he only used a few words, and referred to himself as Stephen, rather than "me" or "I." He did not make eye contact with people, nor did he appear to understand nonverbal cues, such as facial expressions, although he sometimes communicated by using his hands. The preschool teacher alerted the parents as soon as she suspected autism, and a work-up by a medical team confirmed her suspicions. Stephen was able to stay at preschool, as long as he followed a very rigid routine and was permitted to hold his ball. Special education programs in the public school greatly helped.

The autism spectrum disorders (MIM 290850) range from the classical form that Stephen has, to milder forms, including Asperger syndrome, in which language skills remain. These disorders affect one child out of every 110, striking about four times as many boys as girls. Onset is typically before age 3. About 25 percent of affected children develop seizures as they grow older. Although 70 percent of people with autism have intellectual disability, others may be very intelligent. (The term "intellectual disability" is now sometimes used instead of mental retardation.)

There are many genetic causes and contributing factors to autism spectrum disorders. Known environmental triggers include prenatal exposures to rubella (German measles) and the drug valproate. Scientific evidence does not support a link to the mercury compound once used in vaccines. Autism has actually increased since that ingredient has been removed. A better way to identify environmental risk factors for autism spectrum disorders may be to first discover the different genetic subtypes, so that studies compare individuals with the same underlying problem.

Heritability of autism is high: about 90 percent. Although siblings of affected children are at a 15 percent risk of being affected, compared to the less than 1 percent for the general population, there are MZ (identical) twins in whom one has autism and one does not. So far more than 30 susceptibility or causative genes have been identified. In about 15 percent of people with autism, the condition is part of a syndrome, including chromosomal disorders such as fragile X syndrome (see *Bioethics: Choices for the Future* in chapter 12) and Down syndrome (see Reading 12.1) or single-gene disorders, such as Rett syndrome (see Reading 6.2). A diagnostic workup for autism includes tests for these conditions, as well as a test called chromosomal microarray analysis, which detects copy number variants (tiny deletions and duplications of DNA).

In the past, chromosomal abnormalities in people with autism led researchers to specific genes. Today genome-wide association studies and homozygosity mapping in families where relatives have children are revealing genes behind autism.

Studies on two classes of cell adhesion proteins that function at the sites where two neurons meet—the synapse—may explain how autism develops. These proteins are called **neurexins** and **neuroligins**. They are embedded in the membranes of presynaptic and postsynaptic neurons, respectively, which use a neurotransmitter called glutamate that spreads excitation (**figure 8.9**). Mutations in these genes misfold the encoded proteins, which impairs communication across

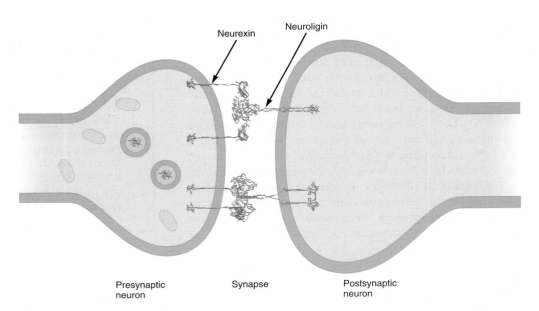

Figure 8.9 **Understanding autism.** An emerging view of a cause of autism points to changes in the nervous system that occur as a child begins to explore the environment. One such system consists of two types of proteins that strengthen synaptic connections among neurons associated with learning and memory. The proteins are called neurexins (in presynaptic neurons) and neuroligins (in postsynaptic neurons).

synapses in a way that causes autism. Perhaps these genes—and related ones that control the organization of outgrowths on dendrites—affect the synapses that naturally form in early childhood in response to experiences. This is in contrast to synapses that form during prenatal (before birth) development. The autism spectrum, therefore, may arise from interference with impaired ability, early in life, to form synapses in key brain areas that are necessary for learning and memory.

8.9 How Are Behavioral Disorders Related?

Bipolar disorder, schizophrenia, and autism are presented in their own parts in this chapter, but genetic research is increasingly indicating that these conditions are very much related to each other. They may share symptoms and therefore overlap, or seem to be opposites. Diagnosis can be confusing. Is a sudden fantastical idea an expression of mania in someone with bipolar disorder, or the disordered thinking of someone with schizophrenia? Discoveries of the genes that cause and contribute to behavioral disorders (also called neuropsychiatric disorders) are countering the decades-old attempts of psychiatric researchers to neatly contain these conditions as distinct and separate.

The new view of behavioral disorders that reflects genetics is especially interesting for autism and schizophrenia. They have some characteristics in common: Many patients have new mutations, many of which are tiny deletions or duplications of DNA sequences. These mutations tend to be very rare, affecting less than one percent of patients. The characteristics of the two disorders, however, appear to be opposite. Children with autism avoid making eye contact with people and speak of themselves in the third person, while people with paranoid schizophrenia imagine they are being followed and have deluded ideas of their own importance. Such comparisons led to the hypothesis that a certain mutation might cause autism when it is underexpressed or deleted, or schizophrenia when it is overexpressed or the gene present in extra copies. At least four genes have been implicated in both autism and schizophrenia—two encode neurexins, one encodes a protein that anchors neurexins into the receiving cell's side of the synapse, and the fourth is an enzyme. According to this "diametric" model, autism and schizophrenia lie at the extreme opposite ends of a normal scale.

Figure 8.10 illustrates the diametric model and three others to depict the relationships among bipolar disorder, autism, and schizophrenia. When autism was first recognized in the 1950s, it was described as a subset of schizophrenia, the

Figure 8.10 **The relationship between autism and schizophrenia.** Against a backdrop of schizophrenia symptoms overlapping those of bipolar disorder, autism may be subsumed by (**a**), separate from (**b**), opposite to (**c**), or overlapping (**d**) schizophrenia.

"subsumed" model. It has also been considered as completely separate from schizophrenia, and overlapping. The fourth view is the diametric one.

The new gene discoveries about behavioral disorders are confusing, because they are muddying long-held definitions. However, they all support what has been known for a long time: The traits that we call multifactorial, or complex, reflect a genetic contribution that is only realized following an environmental triggering event. Genes and the environment can combine in ways that transform a proneness to act a certain way into a persistence of that behavior, which, if it continues, becomes an impairment. Identifying and understanding these factors can suggest new ways to treat behaviors that interfere with daily living.

Key Concepts

1. Autism is a loss of language, communication, and social skills beginning in early childhood. Seizures and intellectual disability may occur.

2. Many genes contribute to autism risk. Two in particular, which encode the receptor proteins neuroligin and neurexin, may explain how the condition arises from failure of synapses to form that enable a child to integrate experiences.

3. Behavioral disorders are related to each other, sharing symptoms as well as being part of a continuum.

Summary

8.1 Genes and Behavior

1. Genes contribute to how the brain responds to environmental stimuli. **Neurons** and **neuroglia** are the two types of brain cells.

2. Candidate genes for behavioral traits and disorders affect neurotransmission and signal transduction.

3. Analyzing behaviors is difficult because symptoms of different syndromes overlap, study participants can provide biased information, and behaviors can be imitated.

4. Genetic subtypes of behavioral disorders may alter standard psychiatric diagnoses.

8.2 Eating Disorders

5. Eating disorders affect both sexes. Twin studies indicate high heritability.

6. Candidate genes for eating disorders include those whose protein products control appetite and the neurotransmitters dopamine and serotonin. Genome-wide association studies point to four genome regions.

8.3 Sleep

7. Twin studies and single-gene disorders that affect the sleep-wake cycle reveal a large inherited component to sleep behavior.

8. The *period* gene enables a person to respond to day and night environmental cues.

8.4 Intelligence

9. The general intelligence (*g*) value measures the inherited portion of IQ that may underlie population variance in IQ test performance.

10. Heritability for intelligence increases with age, suggesting that environmental factors are more important early in life.

11. Many chromosomal disorders affect intelligence, suggesting high heritability.

8.5 Drug Addiction

12. Drug addiction arises from tolerance and dependence. Addiction produces stable changes in certain brain parts.

13. Proteins involved in drug addiction affect neurotransmission and signal transduction.

14. Candidate genes for drug addiction include the dopamine D(2) receptor and variants in nicotinic receptor parts.

8.6 Mood Disorders

15. **Major depressive disorder** is relatively common and associated with deficits of serotonin and/or norepinephrine.

16. **Bipolar disorder** is depressive periods and periods of mania or hypomania. Hundreds of genes may raise the risk of developing this disorder. Different families have different combinations of these gene variants, some of which, under certain environmental conditions, can lead to the disorder.

8.7 Schizophrenia

17. Schizophrenia greatly disrupts the ability to think and perceive the world. Onset is typically in early adulthood, and the course is episodic or steady.

18. Empiric risk estimates and heritability indicate a large genetic component.

19. Many genes and environmental influences are associated with schizophrenia.

8.8 Autism

20. Autism is a loss of communication and social skills beginning in early childhood.

21. **Neuroligins** and **neurexins** are types of proteins embedded in the plasma membranes of certain brain neurons that join across synapses, permitting neural connections to form in response to environmental stimuli. These proteins are abnormal in some cases of autism.

8.9 How Are Behavioral Disorders Related?

22. Schizophrenia and bipolar disorder symptoms overlap, and schizophrenia and autism may be opposite extremes.

www.mhhe.com/lewisgenetics10

Answers to all end-of-chapter questions can be found at **www.mhhe.com/lewisgenetics10.** You will also find additional practice quizzes, animations, videos, and vocabulary flashcards to help you master the material in this chapter.

Review Questions

1. What are the two major types of cells in the brain, and what do they do?

2. Why are behavioral traits nearly always multifactorial?

3. List the pathways or mechanisms that include proteins that, when absent or atypical, cause variations in behavior.

4. What is the evidence that the Utah family with familial advanced sleep phase syndrome has a genetic condition rather than them all just becoming used to keeping weird hours?

5. Choose a behavior discussed in the chapter and identify the region of the brain and/or a specific molecule that is affected.

6. Why is identifying a candidate gene only a first step in understanding how behavior arises and varies among individuals?

7. Describe three factors that can complicate the investigation of a behavioral trait.

8. Why does the heritability of intelligence decline with age? What were some of the prejudices that were part of studying the inheritance of intelligence?

9. What are the two defining characteristics of drug addiction?

10. Select a drug mentioned in the chapter and explain what it does to the nervous system.

11. Explain how an SSRI antidepressant drug is thought to work.

12. Distinguish between the symptoms of bipolar disorder and schizophrenia.

13. What is the evidence that our bodies have their own uses for cocaine, THC, and opium?

14. What is an environmental factor that may influence the development of schizophrenia?

15. Describe how cell surface receptors are implicated in one of the conditions discussed in the chapter.

16. Describe four ways that schizophrenia and autism may be related.

17. Select a behavioral disorder mentioned in this text, or another. Compare and contrast the type of information derived from classical studies on families versus genome-wide association studies.

Applied Questions

1. Abnormal serotonin levels contribute to or cause eating disorders, major depressive disorder, and bipolar disorder.

    a. How can an abnormality in one type of neurotransmitter contribute to different disorders?

    b. What other neurotransmitter is involved in more than one behavioral disorder?

2. Many older individuals experience advanced sleep phase syndrome. Even though this condition is probably a normal part of aging, how might research on the Utah family with an inherited form of the condition help researchers develop a drug to help the elderly sleep through the night and awaken later in the morning?

3. What nongenetic factor might account for the overrepresentation of minority groups among people with low IQ scores in the United States?

4. How does the subunit structure of the nicotinic receptor provide a mechanism for epistasis? (Epistasis is one gene affecting expression of another; see chapter 5.)

5. A television and film star went into rehab for "sex addiction," much to the embarrassment of his wife. Describe how you would evaluate whether this diagnosis is valid, either in an individual or in a large study.

6. What might be the advantages and disadvantages of a SNP profile done at birth that indicates whether a person is at high risk for developing a drug addiction?

7. In some studies of depression and bipolar disorder, correlations to specific gene variants are only evident when participants are considered in subgroups based on symptoms. What might be a biological basis for this finding?

8. A study found that the risk of schizophrenia among spouses of people with schizophrenia who have no affected blood relatives is 2 percent. What might this indicate about the causes of schizophrenia?

9. In the United States, the incidence of autism has dramatically increased since 1990.

    a. Does this finding better support a genetic cause or an environmental cause for autism?

    b. What is a nongenetic factor that might explain the increased incidence of autism?

10. A "markers for addiction" gene panel scans people's DNA for variants in the following genes that are associated with addiction tendency. The eleven genes encode proteins that are ion channels, cell surface molecules, receptors, enzymes, or cell adhesion molecules. Look up one of them in *Mendelian Inheritance in Man* (or elsewhere) and explain its role in increasing the risk of nicotine addiction. The markers are: *CHRNA3, CHRNA5, CHRNB3, CLCA1, CTNNA3, GABRA4, KCNJ6, NRXN1,* and *TRPC7.*

Web Activities

1. Using the *Diagnostic and Statistical Manual of Mental Disorders,* choose a behavioral disorder, identify a candidate gene, and explain how a mutation in it might cause the phenotype.

2. Researchers working for an online dating organization analyze information on thousands of individuals for gene variants that affect behaviors, such as empathy, risk-taking, excitability, and loyalty, and categorize the people into four groups. The dating company matches people based on these profiles.

    a. What information would you like to see to determine whether this approach is valid or not for individuals?

    b. What is a benefit of this approach?

    c. What is a limitation or danger of the approach?

3. Two genome-wide association studies sought genes that contribute to intelligence. The first considered 1,000 Mexican-American participants in the San Antonio Family Heart Study, and found evidence pointing to genes on chromosomes 3, 6, and 18. The second study looked at 1,094 children with attention deficit hyperactivity disorder and their siblings, and found gene candidates for intelligence on chromosomes 7, 9, and 14. Suggest two possible reasons why the two studies had different results.

4. Researchers used a version of the placebo effect (being given a sugar pill and thinking it is a drug) on twenty-five people who have social anxiety disorder. The participants had to give a speech at the start and at the end of an 8-week period during which they took a pill that they thought was

a medication for their anxiety. Ten of the people improved dramatically, as measured by test scores that rated their anxiety, and brain scans confirmed falling activity in the amygdala, the "fear" area. Next, the participants were tested for variants of the gene that encodes tryptophan hydroxylase-2, required to synthesize the mood-related neurotransmitter serotonin. Eight of the ten responders had two copies of a gene variant previously associated with calmness. None of the people who did not respond to the placebo, remaining petrified to speak publicly, had the calm version of the gene.

a. What are the limitations of this study?

b. Describe a genome-wide approach to building on this work.

c. Suggest a practical application of this finding.

Forensics Focus

1. When 43-year-old F.F. discovered his soon-to-be-ex wife in bed with her boyfriend, he shot and killed them both. The defense ordered a pre-trial forensic psychiatric work-up that included genotyping for the enzyme monoamine oxidase A (*MAOA*, MIM 309850), which breaks down the neurotransmitters serotonin, dopamine, and noradrenalin, and for the serotonin transporter (*SLC6A4*, MIM 182138). A "high-activity" allele for *MAOA* is associated with violence in people who also suffered child abuse. Inheriting one or two "short" alleles of *SLC6A4* is associated with depression and suicidal thoughts, in people who have experienced great stress. F.F. had the high-activity MAOA genotype, two short *SLC6A4* alleles, and a lifetime of abuse and stress. However, the judge ruled that the science was not far enough along to admit the genotyping results as evidence.

a. Under what conditions or situations do you think it is valid to include genotyping results in cases like this?

b. In a Dutch family, a mutation disables *MAOA*, causing "a syndrome of borderline mental retardation and abnormal behavior," according to one report. Family members had committed arson, attempted rape, and shown exhibitionism. How can both high and low levels of an enzyme each cause behavioral problems?

c. What is a limitation to use of behavioral genotyping in a criminal trial?

Case Studies and Research Results

1. Psychiatrists consult the DSM to diagnose mental illness, treating such disorders as depression, bipolar disorder, autism, and schizophrenia as distinct and separate. Geneticists look for subtypes of these categories and for distinctive features that might allow better discrimination of the phenotype. Does the fact that the gene *DISC1* seems to be involved in causing both schizophrenia and bipolar disorder argue more for the psychiatrist's approach or the geneticist's approach?

2. Until the 1990s, bipolar disorder was thought not to affect children under age 18. Psychiatrists today maintain that fewer than 1 percent of children have bipolar disorder. However, the percentage of children being diagnosed with bipolar disorder has soared since 2000, along with the publication of many books written by parents of affected children, and appearances on TV talk shows of affected children and their parents.

a. Does this pattern of increasing disease incidence suggest a genetic or an environmental cause?

b. Suggest another explanation for the recent apparent increase in incidence of bipolar disorder in children.

c. Most children are diagnosed with bipolar disorder based on their answers to questions. What might a genetic diagnosis entail?

3. On the island of Fiji, women once valued having a full figure. Then, in 1995, television arrived, and with it, the show "Melrose Place," depicting skinny women as the ideal. Within three years, the incidence of eating disorders doubled, with a frightening percentage of the female population regularly vomiting on purpose so that they could continue to eat. Does this information argue more for a genetic or nongenetic cause of eating disorders? How could both influences contribute?

4. Variants of a gene called *CYP2A6* are associated with the speed at which a person metabolizes nicotine. A "fast-metabolizer" smokes more. Design a clinical trial using this information to test whether the *CYP2A6* genotype can be used to predict which individuals will respond to specific smoking cessation medications.

5. Psychologists set up an experiment in which volunteers who thought they were being asked to review TV commercials were offered alcoholic drinks while waiting to view the commercials. People "planted" in the room were drinking. The volunteers provided saliva samples, from which the researchers typed a gene that encodes part of the dopamine receptor DRD4. The individuals who drank the alcohol had seven repeats in the gene, and those who did not drink did not have the repeats. The researchers concluded: "Whether or not people are wired to adapt their drinking to the choice and pace of others may partly depend on their genetic susceptibility to drinking cues." How is this study flawed, or the conclusion unwarranted?

DNA is the genetic material. DNA bursts from this treated bacterial cell. The DNA in a human cell would unravel to nearly 6 feet, yet it is wound into the nucleus of a microscopic cell.

CHAPTER

9

DNA Structure and Replication

Learning Outcomes

9.1 Experiments Identify and Describe the Genetic Material

1. Describe the experiments that showed that DNA and not protein is the genetic material.

2. Explain how Watson and Crick deduced the structure of DNA.

9.2 DNA Structure

3. List the components of a DNA nucleotide building block.

4. Explain how nucleotides join in two chains to form a DNA molecule.

9.3 DNA Replication—Maintaining Genetic Information

5. Explain how researchers deduced that DNA replication is semiconservative.

6. List the steps of DNA replication.

The Big Picture: DNA is the basis of life because of three qualities: It holds information, it copies itself, and it changes.

On the Meaning of Gene

To a biologist, *gene* has a specific definition—a sequence of DNA that tells a cell how to assemble amino acids into a particular protein. To others, "gene" has different meanings:

To folksinger Arlo Guthrie, *gene* means aging without signs of the Huntington disease that claimed his father, legendary folksinger Woody Guthrie.

To rare cats in New England, *gene* means extra toes.

To Adolph Hitler and others who have dehumanized those not like themselves, the concept of *gene* was abused to justify genocide.

To a smoker, a *gene* may mean lung cancer develops.

To a redhead in a family of brunettes, *gene* means an attractive variant.

To a woman whose mother and sisters had breast cancer, *gene* means escape from their fate—and survivor guilt.

To a lucky few, *gene* means a mutation that locks HIV out of their cells.

To people with diabetes, *gene* means safer insulin.

To a forensic entomologist, *gene* means a clue in the guts of maggots devouring a corpse.

To scientists-turned-entrepreneurs, *gene* means money.

Collectively, our genes mean that we are very much more alike than different from one another.

9.1 Experiments Identify and Describe the Genetic Material

"A genetic material must carry out two jobs: duplicate itself and control the development of the rest of the cell in a specific way," wrote Francis Crick, codiscoverer with James Watson of the three-dimensional structure of DNA in 1953. Only DNA can do this.

DNA was first described in the mid-eighteenth century, when Swiss physician and biochemist Friedrich Miescher isolated nuclei from white blood cells in pus on soiled bandages. In the nuclei, he discovered an unusual acidic substance containing nitrogen and phosphorus. He and others found it in cells from a variety of sources. Because the material resided in cell nuclei, Miescher called it *nuclein* in an 1871 paper; subsequently, it was called a nucleic acid. Few people appreciated the importance of Miescher's discovery at the time, when inherited disease was widely blamed on protein.

In 1902, English physician Archibald Garrod was the first to provide evidence linking inherited disease and protein. He noted that people who had certain inborn errors of metabolism lacked certain enzymes. Other researchers added evidence of a link between heredity and enzymes from other species, such as fruit flies with unusual eye colors and bread molds with nutritional deficiencies. Both organisms had absent or malfunctioning specific enzymes. As researchers wondered what, precisely, was the connection between enzymes and heredity, they returned to Miescher's discovery of nucleic acids.

DNA *Is* the Hereditary Molecule

In 1928, English microbiologist Frederick Griffith took the first step in identifying DNA as the genetic material. He was studying pneumonia in the years after the 1918 flu pandemic. Griffith noticed that mice with a certain form of pneumonia harbored one of two types of *Streptococcus pneumoniae* bacteria. Type R bacteria were rough in texture. Type S bacteria were smooth because they were enclosed in a polysaccharide (a type of carbohydrate) capsule. Mice injected with type R bacteria did not develop pneumonia, but mice injected with type S did. The polysaccharide coat shielded the bacteria from the mouse immune system, enabling them to cause severe (virulent) infection.

When type S bacteria were heated, which killed them, they no longer could cause pneumonia in mice. However, when Griffith injected mice with a mixture of type R bacteria plus heat-killed type S bacteria—neither of which, alone, was deadly to the mice—the mice died of pneumonia (**figure 9.1**). Their bodies contained live type S bacteria, encased in polysaccharide. Griffith termed the apparent conversion of one bacterial type into another "transformation." How did it happen? What component of the dead, smooth bacteria transformed type R to type S?

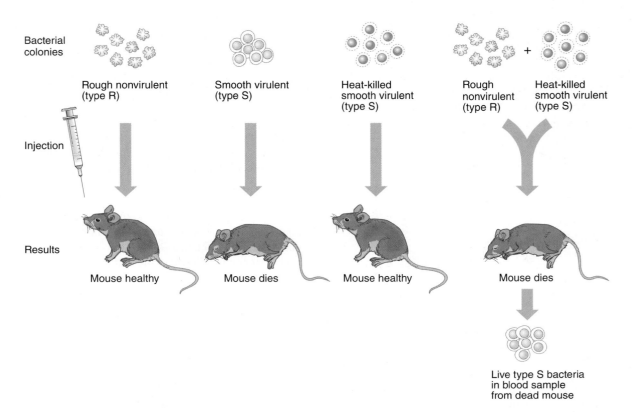

Figure 9.1 Discovery of bacterial transformation. Griffith's experiments showed that a molecule in a lethal type of bacteria can transform nonkilling (nonvirulent) bacteria into killers (virulent).

U.S. physicians Oswald Avery, Colin MacLeod, and Maclyn McCarty hypothesized that a nucleic acid might be the "transforming principle." They observed that treating broken-open type S bacteria with a protease—an enzyme that dismantles protein—did not prevent the transformation of a nonvirulent to a virulent strain, but treating it with deoxyribonuclease (or DNase), an enzyme that dismantles DNA only, did disrupt transformation. In 1944, they confirmed that DNA transformed the bacteria. They isolated DNA from heat-killed type S bacteria and injected it with type R bacteria into mice (**figure 9.2**). The mice died, and their bodies contained active type S bacteria. The conclusion: DNA passed from type S bacteria into type R, enabling the type R to manufacture the smooth coat necessary for infection. Once type R bacteria encase themselves in smooth coats, they are no longer type R.

Protein *Is Not* the Hereditary Molecule

Science seeks answers by eliminating explanations. It provides evidence in support of a hypothesis, not proof, because conclusions can change when new data become available. To identify the genetic material, researchers also had to show that protein does *not* transmit genetic information. To do this, in 1953, U.S. microbiologists Alfred Hershey and Martha Chase used *E. coli* bacteria infected with a virus that consisted of a protein "head" surrounding DNA. Viruses infect bacterial cells by injecting their DNA (or RNA) into them. Infected bacteria may then produce many more viruses. The viral protein coats remain outside the bacterial cells.

Researchers can analyze viruses by growing them on culture medium that contains a radioactive chemical that the viruses take up. The "labeled" viral nucleic acid then emits radiation, which can be detected in several ways. Hershey and Chase knew that protein contains sulfur but not phosphorus, and that nucleic acids contain phosphorus but not sulfur. Both elements also come in radioactive forms. When Hershey and Chase grew viruses in the presence of radioactive sulfur, the viral *protein coats* took up and emitted radioactivity, but when

they ran the experiment using radioactive phosphorus, the viral *DNA* emitted radioactivity. If protein is the genetic material, then the infected bacteria would have radioactive sulfur. But if DNA is the genetic material, then the bacteria would have radioactive phosphorus.

Hershey and Chase grew one batch of virus in a medium containing radioactive sulfur (designated ^{35}S) and another in a medium containing radioactive phosphorus (designated ^{32}P). The viruses grown on sulfur had their protein marked, but not their DNA, because protein incorporates sulfur but DNA does not. Conversely, the viruses grown on labeled phosphorus had their DNA marked, but not their protein, because this element is found in DNA but not protein.

After allowing several minutes for the virus particles to bind to the bacteria and inject their DNA into them, Hershey and Chase agitated each mixture in a blender, shaking free the empty virus protein coats. The contents of each blender were collected in test tubes, then centrifuged (spun at high speed). This settled the bacteria at the bottom of each tube because the lighter virus coats drift down more slowly than bacteria.

At the end of the procedure, Hershey and Chase examined fractions containing the virus coats from the top of each test tube and the infected bacteria that had settled to the bottom (**figure 9.3**). In the tube containing viruses labeled with sulfur, the virus coats were radioactive, but the virus-infected bacteria, containing viral DNA, were not. In the other tube, where the virus had incorporated radioactive phosphorus, the virus coats carried no radioactive label, but the infected bacteria were radioactive. Therefore, the part of the virus that could enter bacteria and direct them to mass produce more virus was the part that had incorporated phosphorus—the DNA. The genetic material was DNA, and not protein.

Discovering the Structure of DNA

In 1909, Russian-American biochemist Phoebus Levene identified the 5-carbon sugar **ribose** as part of some nucleic acids, and in 1929, he discovered a similar sugar—**deoxyribose**—in other nucleic acids. He had revealed a major chemical distinction between RNA and DNA: RNA has ribose, and DNA has deoxyribose.

Levene then discovered that the three parts of a nucleic acid—a sugar, a nitrogen-containing base, and a phosphorus-containing component—are present in equal proportions. He deduced that a nucleic acid building block must contain one of each component. Furthermore, although the sugar and phosphate portions were always the same, the nitrogen-containing bases were of four types. Scientists at first thought that the bases were present in equal amounts, but if this were so, DNA could not encode as much information as it could if the number of each base type varied. Imagine how much less useful a written language would be if it had to use all the letters with equal frequency.

In the early 1950s, two lines of experimental evidence converged to provide the direct clues that finally revealed DNA's structure. Austrian-American biochemist Erwin Chargaff

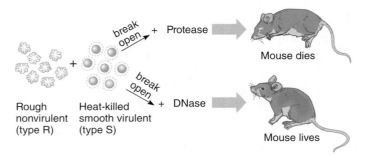

Figure 9.2 DNA is the "transforming principle." Avery, MacLeod, and McCarty identified DNA as Griffith's transforming principle. By adding enzymes that either destroy proteins (protease) or DNA (deoxyribonuclease or DNase) to bacteria that were broken apart to release their contents, they demonstrated that DNA transforms bacteria—and that protein does not.

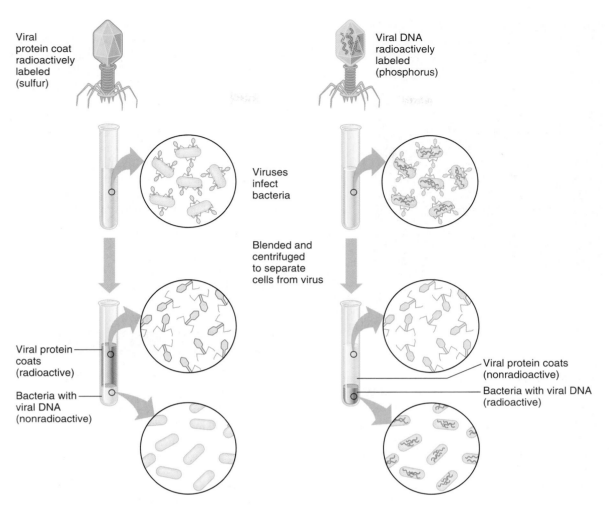

Viral protein coat radioactively labeled (sulfur)

Viral DNA radioactively labeled (phosphorus)

Viruses infect bacteria

Blended and centrifuged to separate cells from virus

Viral protein coats (radioactive)

Bacteria with viral DNA (nonradioactive)

Viral protein coats (nonradioactive)

Bacteria with viral DNA (radioactive)

Figure 9.3 **DNA is the hereditary material; protein is not.** Hershey and Chase used different radioactive molecules to distinguish the viral protein coat from the genetic material (DNA). These "blender experiments" showed that the virus transfers DNA, and not protein, to the bacterium. Therefore, DNA is the genetic material. The blender experiments used particular types of sulfur and phosphorus atoms that emit detectable radiation.

showed that DNA in several species contains equal amounts of the bases **adenine** (A) and **thymine** (T) and equal amounts of the bases **guanine** (G) and **cytosine** (C). Next, English physicist Maurice Wilkins and English chemist Rosalind Franklin bombarded DNA with X rays using a technique called X-ray diffraction, then deduced the overall structure of the molecule from the patterns in which the X rays were deflected.

Rosalind Franklin provided a pivotal clue to deducing the three-dimensional structure of DNA. She distinguished two forms of DNA—a dry, crystalline "A" form, and the wetter type seen in cells, the "B" form. It took her 100 hours to obtain "photo 51" of the B form in May 1952 (**figure 9.4**). When a graduate student showed photo 51 to Wilkins, who showed it to Watson

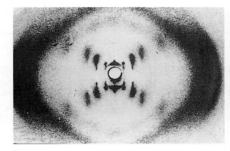

a.

b. Rosalind Franklin 1920–1958

Figure 9.4 **Deciphering DNA structure.** **(a)** Rosalind Franklin's "photo 51" of B DNA was critical to Watson and Crick's deduction of the three-dimensional structure of the molecule. The "X" in the center indicates a helix, and the darkened regions reveal symmetrically organized subunits. **(b)** Franklin died very young, of cancer.

Figure 9.5 Watson and Crick. Prints of this famed, if posed, photo fetched a high price when signed and sold at celebrations of DNA's fiftieth anniversary in 2003. Crick, told to point to the model, picked up a slide rule.

at the end of January in 1953, the men realized that the remarkable symmetry of the molecule, indicated by the positions of the phosphates in the photograph, revealed a sleek helix.

The race was on. During February, famed biochemist Linus Pauling suggested a triple helix structure for DNA. Meanwhile, Watson and Crick, certain of the sugar-phosphate backbone largely from photo 51, turned their attention to the bases. Ironically, their eureka moment occurred not with sophisticated chemistry or crystallography, but while working with cardboard cutouts of the DNA components.

On Saturday morning, February 28, Watson arrived early for a meeting with Crick. While he was waiting, he played with cardboard cutouts of the four DNA bases, pairing A with A, then A with G. When he assembled A next to T, and G next to C, he noted the similar shapes, and suddenly all the pieces

fit. He had been modeling the chemical attractions between the bases that create the helix. When Crick arrived 40 minutes later, the two quickly realized they had solved the puzzle of DNA's structure (**figure 9.5**). They published their structure of DNA in the April 25, 1953 issue of *Nature* magazine, without ever having done an experiment.

Watson, Crick, and Wilkins eventually received the Nobel Prize. In 1958, Franklin died at the age of 37 from ovarian cancer, and the Nobel can only be awarded to a living person. In recent years, she has become a heroine for her long-unappreciated role in deciphering the structure of DNA. In 2010, researchers at the Cold Spring Harbor laboratory on Long Island, where James Watson still works (Crick died in 2004), found six boxes of lost correspondence from Crick, mostly to Wilkins. The letters, photos, and postcards reveal how upset the elder Wilkins was to be in a race with the young upstarts Watson and Crick. **Table 9.1** summarizes some of the experiments that led to the discovery of the structure of the genetic material.

Key Concepts

1. DNA replicates, and contains information for protein synthesis.
2. Miescher isolated DNA in 1869.
3. Garrod linked heredity to enzymes.
4. In the 1940s, Griffith identified something that could be passed from one bacterium to another, which Avery, MacLeod, and McCarty showed was DNA.
5. Hershey and Chase confirmed that DNA, and not protein, is the genetic material.
6. Using Chargaff's discovery that the number of As equals the number of Ts, and the number of Gs equals the number of Cs, with Franklin's discovery that DNA is regular and symmetrical, Watson and Crick deciphered the structure of DNA.

Table 9.1	The Road to the Double Helix	
Investigator	**Contribution**	**Timeline**
Friedrich Miescher	Isolated nuclein in white blood cell nuclei	1869
Frederick Griffith	Transferred killing ability between types of bacteria	1928
Oswald Avery, Colin MacLeod, and Maclyn McCarty	Discovered that DNA transmits killing ability in bacteria	1940s
Alfred Hershey and Martha Chase	Determined that the part of a virus that infects and replicates is its nucleic acid and not its protein	1950
Phoebus Levene, Erwin Chargaff, Maurice Wilkins, and Rosalind Franklin	Discovered DNA components, proportions, and positions	1909–early 1950s
James Watson and Francis Crick	Elucidated DNA's three-dimensional structure	1953
James Watson	Had his genome sequenced	2008

9.2 DNA Structure

A **gene** is a section of a DNA molecule whose sequence of building blocks specifies the sequence of amino acids in a particular protein. The activity of the protein imparts the phenotype. The fact that different building blocks combine to form nucleic acids enables them to carry information, as the letters of an alphabet combine to form words. DNA also encodes RNA that does not specify a protein, but instead assists in protein synthesis or controls gene expression. These non-protein-encoding DNA and RNA sequences are discussed in chapters 10 and 11.

Inherited traits are diverse because proteins have diverse functions (see table 10.1). Malfunctioning or inactive proteins, which reflect genetic defects, can devastate health. Most of the amino acids that are assembled into proteins come from the diet or from breaking down proteins in the cell. The body synthesizes the others.

The structure of DNA is easiest to understand if we begin with the smallest components. A single building block of DNA is a **nucleotide**. It consists of one deoxyribose sugar, one phosphate group (a phosphorus atom bonded to four oxygen atoms), and one nitrogenous base. **Figure 9.6** shows the chemical structures of the four types of bases, and **figure 9.7** shows one of them as part of a nucleotide. Adenine (A) and guanine (G) are **purines,** which have a two-ring structure. Cytosine (C) and thymine (T) are **pyrimidines,** which have a single-ring structure.

The bases are the information-containing parts of DNA because they form sequences. DNA sequences are measured in

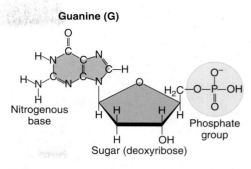

Figure 9.7 **Nucleotides.** A nucleotide of a nucleic acid consists of a 5-carbon sugar, a phosphate group, and an organic, nitrogenous base (G, A, C, or T).

numbers of base pairs. The terms kilobase (kb) and megabase (mb) are used to abbreviate a thousand and a million DNA bases, respectively. (It is a little like the terms kilobyte and megabyte used to measure capacity of a computerized document.) A particular gene, for example, may be "1,400 bases long."

Nucleotides join into long chains when chemical bonds form between the deoxyribose sugars and the phosphates. This creates a continuous **sugar-phosphate backbone** (**figure 9.8**). Two such chains of nucleotides align head-to-toe, as **figure 9.9a** depicts. M. C. Escher's drawing of hands in **figure 9.9b** resembles the spatial relationship of the two strands of the DNA double helix.

The opposing orientation of the two nucleotide chains in a DNA molecule is called **antiparallelism**. It derives from the structure of the sugar-phosphate backbone. We can follow antiparallelism by assigning numbers to the carbons of the sugars based on their relative positions in the molecule (**figure 9.10**).

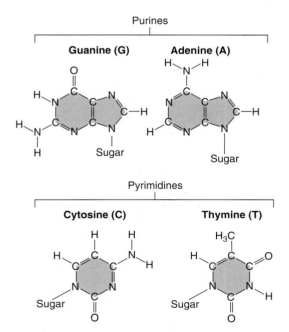

Figure 9.6 **DNA bases are the informational parts of nucleotides.** Adenine and guanine are purines, each composed of a six-membered organic ring plus a five-membered ring. Cytosine and thymine are pyrimidines, each with a single six-membered ring. (Within the molecules, C, H, N, O, and P are atoms of carbon, hydrogen, nitrogen, oxygen, and phosphorus, respectively.)

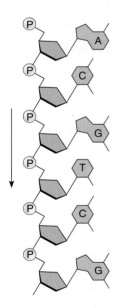

Figure 9.8 **A chain of nucleotides.** A single DNA strand consists of a chain of nucleotides that forms when the deoxyribose sugars (green) and phosphates (yellow) bond to create a sugar-phosphate backbone. The bases A, C, G, and T are blue.

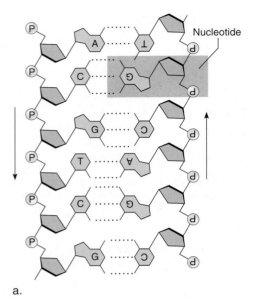

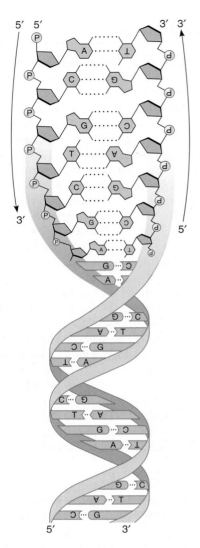

a.

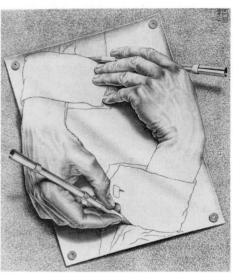

b.

Figure 9.9 **DNA consists of two chains of nucleotides.**
(a) Hydrogen bonds hold the nitrogenous bases of one strand to the nitrogenous bases of the second strand (dotted lines). Note that the sugars point in opposite directions—that is, the strands are antiparallel. **(b)** Artist M. C. Escher captured the essence of antiparallelism in his depiction of hands.

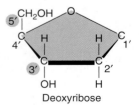

Deoxyribose

Figure 9.10 **Antiparallelism.** The antiparallel nature of the DNA double helix becomes apparent when the carbons in the sugar are numbered.

Figure 9.11 **DNA is directional.** Antiparallelism in a DNA molecule arises from the orientation of the deoxyribose sugars. One-half of the double helix runs in a 5′ to 3′ direction, and the other half runs in a 3′ to 5′ direction.

The carbons are numbered from 1 to 5, starting with the first carbon moving clockwise from the oxygen in each sugar in **figure 9.11**. One chain runs from the #5 carbon (top of the figure) to the #3 carbon, but the chain aligned with it runs from the #3 to the #5 carbon. These ends are depicted as "5′" and "3′", pronounced "5 prime" and "3 prime."

The symmetrical DNA double helix forms when nucleotides containing A pair with those containing T, and nucleotides containing G pair with those carrying C. Because purines have two rings and pyrimidines one, the consistent pairing of a purine with a pyrimidine ensures that the double helix has the same width throughout, as Watson discovered using cardboard cutouts. These specific purine-pyrimidine couples are called **complementary base pairs**. Chemical attractions called hydrogen bonds hold the base pairs together. They are weak individually, but over the many bases of a DNA molecule impart great strength. Two hydrogen bonds join A and T, and three hydrogen

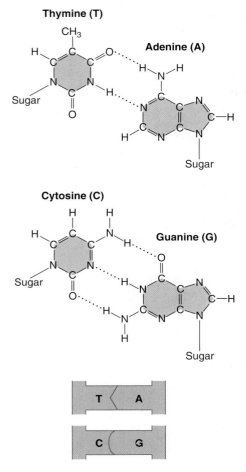

Figure 9.12 **DNA base pairs.** The key to the constant width of the DNA double helix is the pairing of purines with pyrimidines. Two hydrogen bonds join adenine and thymine; three hydrogen bonds link cytosine and guanine.

bonds join G and C, as **figure 9.12** shows. Finally, DNA forms a double helix when the antiparallel, base-paired strands twist about one another in a regular fashion. The double-stranded, helical structure of DNA gives it 50 times the strength of single-stranded DNA. A single strand of DNA would not form a helix.

DNA molecules are very long. The DNA of the smallest human chromosome, if stretched out, would be 14 millimeters long, but it is packed into a chromosome that, during cell division, is only 2 micrometers long. This means that the DNA molecule must fold so tightly that its compacted length shrinks by a factor of 7,000:

$$\left(\frac{14 \times 10^{-3} \text{ meters}}{2 \times 10^{-6} \text{ meters}} \right)$$

Various types of proteins compress the DNA without damaging or tangling it. Scaffold proteins form frameworks that guide DNA strands. Then, the DNA coils around proteins called **histones,** forming a structure that resembles a string of beads. The bead part is called a **nucleosome**. The compaction of a molecule of DNA is a little like wrapping a very long, thin piece of thread around your fingers, to keep it from unraveling

and tangling. DNA wraps at several levels, until it is compacted into a chromosome (**figure 9.13**). Specifically, a nucleosome forms around packets of eight histone proteins (a pair of each of four types). A fifth type of histone protein anchors nucleosomes to short "linker" regions of DNA, which then tighten the nucleosomes into fibers 30 nanometers (nm) in diameter. As a result, at any given time, only small sections of the DNA double helix are exposed, like holding wound up string in both hands with a section in between that is a single, outstretched strand. Chemical modification of the histones controls when particular DNA sequences unwind and become accessible. (This is discussed further in chapter 11.) DNA also unwinds locally when it replicates.

Altogether, the chromosome substance is called **chromatin,** which means "colored material." Chromatin is not just DNA; it is about 30 percent histone proteins, 30 percent DNA scaffold and other proteins that bind DNA, 30 percent DNA, and 10 percent RNA. Specific points along the chromatin attach it, in great loops, to the inner face of the nuclear membrane, placing particular chromosome parts in particular locations in the nucleus. Chromatin attachment sites are not random, and may reflect which genes a cell is using.

Key Concepts

1. The DNA double helix's backbone is alternating deoxyribose and phosphate held together by complementary pairs of A-T and G-C bases. A and G are purines; T and C are pyrimidines.

2. The DNA double helix is antiparallel, its strands running in an opposite head-to-toe manner.

3. DNA winds tightly about histone proteins, forming nucleosomes, which wind tighter, forming chromatin.

9.3 DNA Replication— Maintaining Genetic Information

As soon as Watson and Crick deciphered the structure of DNA, its mechanism for replication became obvious. They ended their report on the structure of DNA with the statement, *"It has not escaped our notice that the specific pairing we have postulated immediately suggests a possible copying mechanism for the genetic material."*

Replication Is Semiconservative

Watson and Crick envisioned the two halves of the DNA double helix unwinding and separating, exposing unpaired bases that would attract their complements. In this way, two double helices would form from one, like a line of dance partners separating and choosing new partners, forming two double lines. This route to replication is called **semiconservative,** because each new DNA double helix conserves half of the original. However, separating the long strands posed a great physical challenge. It

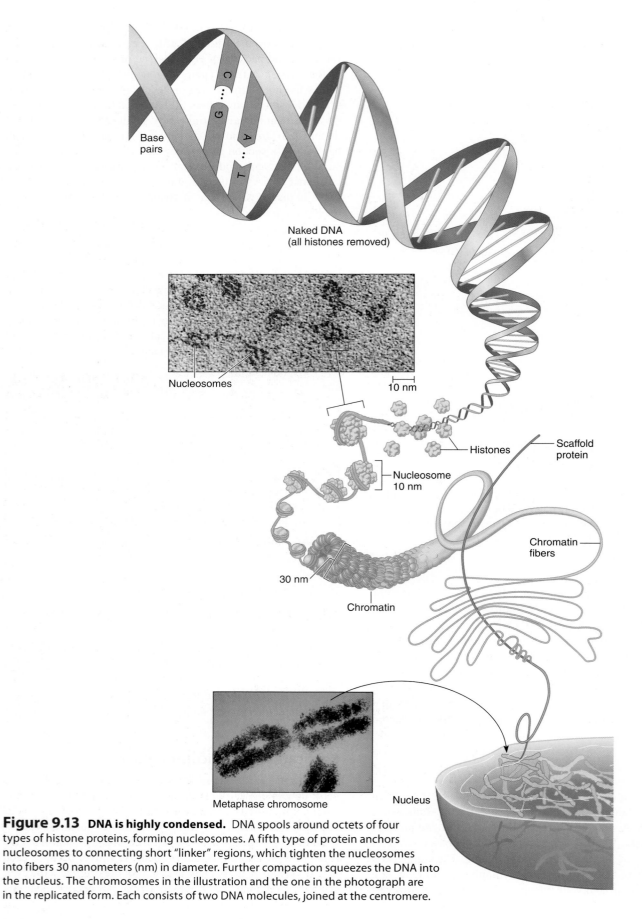

Base pairs

Naked DNA
(all histones removed)

Nucleosomes

10 nm

Histones

Nucleosome
10 nm

Scaffold
protein

Chromatin
fibers

30 nm

Chromatin

Metaphase chromosome

Nucleus

Figure 9.13 **DNA is highly condensed.** DNA spools around octets of four types of histone proteins, forming nucleosomes. A fifth type of protein anchors nucleosomes to connecting short "linker" regions, which tighten the nucleosomes into fibers 30 nanometers (nm) in diameter. Further compaction squeezes the DNA into the nucleus. The chromosomes in the illustration and the one in the photograph are in the replicated form. Each consists of two DNA molecules, joined at the centromere.

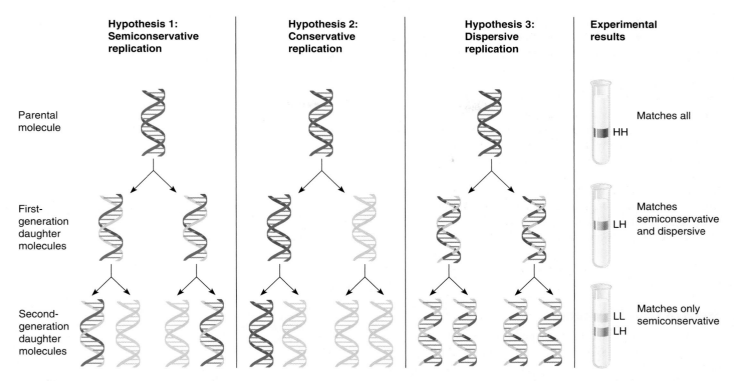

| Hypothesis 1: Semiconservative replication | Hypothesis 2: Conservative replication | Hypothesis 3: Dispersive replication | Experimental results |

Parental molecule — HH — Matches all

First-generation daughter molecules — LH — Matches semiconservative and dispersive

Second-generation daughter molecules — LL, LH — Matches only semiconservative

Figure 9.14 **Three models for DNA replication.** Density shift experiments distinguished the three hypothesized mechanisms of DNA replication. DNA molecules containing light nitrogen are designated "LL" and those with heavy nitrogen, "HH." Molecules containing both isotopes are designated "LH." These experiments established that DNA replication is semiconservative. The first three columns illustrate how parental and daughter DNA would be distributed in each of the three mechanisms of DNA replication. The fourth column depicts the density of the DNA at each stage for each of the three hypothesized replication mechanisms.

was a little like having to keep two pieces of thread the length of a football field from tangling.

At first, researchers suggested that DNA might replicate in any of three possible ways:

- semiconservative;
- conservative, with one double helix specifying creation of a second double helix;
- dispersive, with a double helix shattering into pieces that join new DNA pieces to form two molecules.

Figure 9.14 shows these three models.

Experiments showed which of the three possible ways that DNA can replicate really happens. In 1941, English geneticist J. B. S. Haldane had an idea. He wrote: *"How can one distinguish between model and copy? Perhaps you could use heavy nitrogen atoms in the food supplied to your cell, hoping that the 'copy' genes would contain it while the models did not."*

In 1957, two young researchers, Matthew Meselson and Franklin Stahl, tried Haldane's experiment using bacteria. The results of their "density shift" experiments not only supported one hypothesis (semiconservative replication), but disproved the other two (conservative replication and dispersive replication).

Meselson and Stahl labeled DNA newly made by bacteria by growing the microbes in medium containing heavy nitrogen (^{15}N). The "heavy" DNA could then be separated from older "light" DNA that had been made with the more common lighter form of nitrogen, ^{14}N, by its greater density.

DNA in which one-half of the double helix was light and one-half heavy would be of intermediate density. Meselson and Stahl grew cells with heavy nitrogen and then shifted the cells to media with light nitrogen. They traced replicating DNA through several cell divisions by growing cells, breaking them open, extracting the DNA, and spinning it in a centrifuge. The heavier DNA sank to the bottom of the centrifuge tube, the light DNA rose to the top, and the heavy-light double helices settled in the middle area of the tube. They grew *E. coli* on media containing ^{15}N for several generations, making only "heavy-heavy" molecules appear in the tube after centrifugation. They then shifted the bacteria to media containing ^{14}N, allowing enough time for the bacteria to divide only once, about 30 minutes.

When Meselson and Stahl collected the DNA after one generation and centrifuged it, the double helices were all of intermediate density. The DNA settled in the middle of the tube, indicating that the molecules contained half ^{14}N and half ^{15}N. This pattern was consistent with either semiconservative DNA replication or a dispersive mechanism. In contrast, the result of conservative replication would have been one band of material in the tube completely labeled with ^{15}N, corresponding to one double helix, and another totally "light" band containing ^{14}N only, corresponding to the other double helix. This did not happen, eliminating conservative replication.

To distinguish among the three routes to DNA replication, Meselson and Stahl extended the experiment one more generation. If the semiconservative mechanism held up, each hybrid (half ^{14}N and half ^{15}N) double helix present after the first generation following the shift to ^{14}N medium would separate and assemble a new half from bases labeled only with ^{14}N. This would produce two double helices with one ^{15}N (heavy) and one ^{14}N (light) chain, plus two double helices containing only ^{14}N. The tube would have one heavy-light band and one light-light band. This is indeed what Meselson and Stahl saw.

The conservative mechanism would have yielded two bands in the tube in the third generation, indicating three completely light double helices for every completely heavy one, as the bottom portion of the hypothesis 2 column indicates in figure 9.14. The third generation for the dispersive model would have been a single large band, somewhat higher than the second-generation band because additional ^{14}N would have been randomly incorporated into the DNA.

After experiments demonstrated the semiconservative nature of DNA replication, the next challenge was to decipher the steps of the process.

Steps of DNA Replication

DNA replication occurs during S phase of the cell cycle (see figure 2.14). When DNA replicates, it unwinds, locally separates, breaks, builds a new nucleotide chain, and then mends (**figure 9.15**). Enzymes called helicases bind to the start or origin of a DNA segment and unwind and hold apart replicating DNA. This action enables other enzymes to guide the assembly of a new DNA strand.

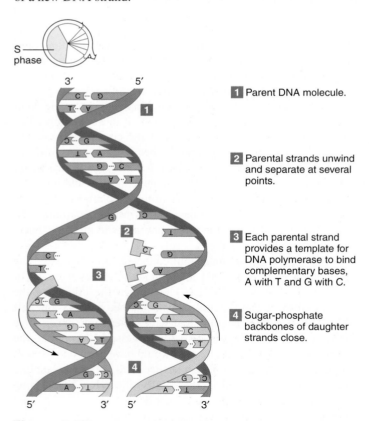

1 Parent DNA molecule.

2 Parental strands unwind and separate at several points.

3 Each parental strand provides a template for DNA polymerase to bind complementary bases, A with T and G with C.

4 Sugar-phosphate backbones of daughter strands close.

Figure 9.15 **Overview of DNA replication.** After experiments demonstrated the semiconservative nature of DNA replication, the next challenge was to decipher the steps of the process.

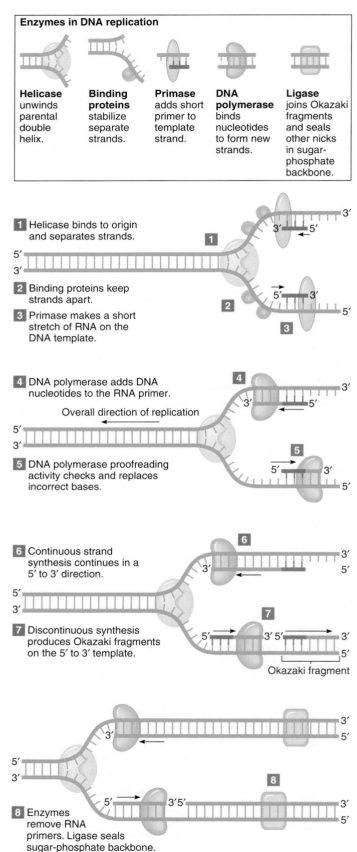

Enzymes in DNA replication

Helicase unwinds parental double helix.

Binding proteins stabilize separate strands.

Primase adds short primer to template strand.

DNA polymerase binds nucleotides to form new strands.

Ligase joins Okazaki fragments and seals other nicks in sugar-phosphate backbone.

1 Helicase binds to origin and separates strands.

2 Binding proteins keep strands apart.

3 Primase makes a short stretch of RNA on the DNA template.

4 DNA polymerase adds DNA nucleotides to the RNA primer.

Overall direction of replication

5 DNA polymerase proofreading activity checks and replaces incorrect bases.

6 Continuous strand synthesis continues in a 5′ to 3′ direction.

7 Discontinuous synthesis produces Okazaki fragments on the 5′ to 3′ template.

Okazaki fragment

8 Enzymes remove RNA primers. Ligase seals sugar-phosphate backbone.

Figure 9.16 **Activities at the replication fork.** DNA replication takes many steps.

Human DNA replicates about 50 bases per second. To get the job done, a human chromosome replicates simultaneously at hundreds of points along its length, and the pieces join. A site where DNA is locally opened, resembling a fork, is called a **replication fork**.

DNA replication begins when a helicase breaks the hydrogen bonds that connect a base pair **(figure 9.16)**. Binding proteins hold the two strands apart. Another enzyme, primase, then attracts complementary RNA nucleotides to build a short piece of RNA, called an RNA primer, at the start of each segment of DNA to be replicated. The RNA primer is required because the major replication enzyme, **DNA polymerase** (DNAP), can only add bases to an existing nucleic acid strand. (A polymerase is an enzyme that builds a polymer, which is a chain of chemical building blocks.) Next, the RNA primer attracts DNAP, which brings in DNA nucleotides complementary to the exposed bases on the parental strand; this strand serves as a mold, or template. New bases are added one at a time, starting at the RNA primer. The new DNA strand grows as hydrogen bonds form between the complementary bases. The nucleotides are abundant in cells, and are synthesized from dietary nutrients.

DNAP works directionally, adding new nucleotides to the exposed 3′ end of the sugar in the growing strand. Overall, replication proceeds in a 5′ to 3′ direction, because this is the only chemical configuration in which DNAP can add bases. How can the growing fork proceed in one direction, when both parental strands must be replicated? The answer is that on at least one strand, replication is discontinuous. It is accomplished in small pieces from the inner part of the fork outward, in a pattern similar to backstitching. Next, an enzyme called a **ligase** seals the sugar-phosphate backbones of the pieces, building the new strand. These pieces, up to 150 nucleotides long, are called Okazaki fragments, after their discoverer (see figure 9.16).

DNA polymerase also "proofreads" as it goes, excising mismatched bases and inserting correct ones. It also removes the RNA primer and replaces it with the correct DNA bases. Yet another enzyme, called an annealing helicase, rewinds any sections of the DNA molecule that remain unwound. Finally, ligases seal the entire sugar-phosphate backbone. Ligase comes from a Latin word meaning "to tie."

As a human body grows to 100 trillion or so cells, DNA replication occurs about 100 quadrillion times. Some of the DNA sequence encodes protein, but most does not. The genome is in this sense a little like a booklet describing the parts of a machine, such as a car, that comes with a much more extensive manual that explains the details of keeping it in working order. The next chapters in this part of the book explain how the human instruction manual is accessed, and explores some of the nuances of its use. The *Bioethics: Choices for the Future* box discusses a controversial use of DNA testing.

Key Concepts

1. Experiments that followed the distribution of labeled DNA showed that DNA replication is semiconservative, not conservative or dispersive.

2. Enzymes replicate DNA.

3. DNA replication occurs simultaneously at several points on each chromosome, and the pieces join.

4. At each initiation site, primase directs synthesis of a short RNA primer, which DNA eventually replaces. DNA polymerase adds complementary bases to the RNA primer. Ligase joins the sugar-phosphate backbone.

5. DNA is synthesized in a 5′ to 3′ direction, discontinuously on one strand.

Bioethics: Choices for the Future

Infidelity Testing

Afraid your significant other is cheating? Send us a DNA sample, and we'll find the proof.

Bridgette came home a day early from a business trip to find her husband Roy drinking coffee in their kitchen with Tiffany, his business associate. They were laughing so hard that it took a few moments for them to notice Bridgette standing there. When they did, Tiffany blushed and Roy knocked over her coffee mug, then they both stammered that they were discussing an acquisition. Bridgette didn't buy it.

She went upstairs to unpack. Flinging her purse on the bed, she noticed several strands of brown hair on her pillow. Bridgette's hair was brown, too, but she never left it on her pillow like that. She also noticed a crumpled tissue on the floor, partway under the bed.

Bridgette knew just what to do. She'd recently read an article about companies that test "abandoned DNA," so she went back downstairs for some plastic bags, and nonchalantly picked up Tiffany's coffee mug, carrying it all back upstairs. In the bedroom, she quickly

collected her evidence—the telltale hairs, the discarded used tissue, and on a cotton swab she rubbed along the inside rim of the mug. She e-mailed gotchaDNA.com and received a cheek swab collection kit a few days later, which she used to send in her own DNA for comparison, plus the $600 fee. Then she waited.

The technicians at gotchaDNA.com extracted the DNA from the samples. First they checked for Y chromosome markers, found on the crumpled tissue. Then they looked for several STR (short tandem repeats—see table 5.3) markers and found what Bridgette had feared—the DNA on the mug that Tiffany had used and in the hair cells matched each other, and not Bridgette's DNA. Tiffany, or at least her hair, had somehow found its way onto Bridgette's pillow.

Cells use DNA to manufacture protein. People use DNA to identify people. Chapter 1 introduced uses of DNA testing in several settings. Another is "infidelity DNA testing," which dozens of companies offer on the Internet. Although a few websites provide

(Continued)

documents for attesting that the samples are given willingly, many do not, and even list suggested sources of DNA for "adultery tracing." These sources include underwear, toothbrushes, dental floss, nail clippings, gum, cigarette butts, and razor clippings.

Questions for Discussion

1. In the United Kingdom, a law was enacted to prohibit sampling of a celebrity's DNA after someone tried to steal hair from Prince Harry to determine whether or not Prince Charles is his biological father. The United States has no such law. Do you think that one is warranted? (A half waffle whose other half was consumed by Barack Obama was auctioned on eBay, with claims that it contained the presidential DNA.)

2. Do you think that DNA data obtained without consent should be admissible in a court of law? State a reason for your answer.

3. Discuss one reason in support of infidelity testing of DNA and one reason against it.

4. Identify the individuals in the scenario whom you believe behaved unethically.

Summary

9.1 Experiments Identify and Describe the Genetic Material

1. DNA encodes information that the cell uses to synthesize protein. DNA can also replicate, passing on its information.

2. Many experimenters described DNA as the hereditary material. Miescher identified DNA in white blood cell nuclei. Garrod connected heredity to enzyme abnormalities. Griffith identified a "transforming principle" that transmitted infectiousness in pneumonia-causing bacteria; Avery, MacLeod, and McCarty discovered that the transforming principle is DNA; and Hershey and Chase confirmed that the genetic material is DNA and not protein.

3. Levene described the three components of a DNA building block and found that they appear in DNA in equal amounts. Chargaff discovered that the amount of **adenine** (A) equals the amount of **thymine** (T), and the amount of **guanine** (G) equals that of **cytosine** (C). A and G are **purines;** C and T are **pyrimidines.** Rosalind Franklin showed that the molecule is a certain type of helix. Watson and Crick deduced DNA's structure.

9.2 DNA Structure

4. A **nucleotide** is a DNA building block. It consists of a **deoxyribose,** a phosphate, and a nitrogenous base.

5. The rungs of the DNA double helix consist of hydrogen-bonded **complementary base pairs** (A with T, and C with G). The rails are chains of alternating sugars and phosphates that run **antiparallel** to each other. DNA is highly coiled, and complexed with protein to form **chromatin.**

9.3 DNA Replication—Maintaining Genetic Information

6. Meselson and Stahl demonstrated the **semiconservative** nature of DNA replication with density shift experiments.

7. During replication, the DNA unwinds locally at several sites. **Replication forks** form as hydrogen bonds break between base pairs. Primase builds short RNA primers, which DNA sequences eventually replace. Next, **DNA polymerase** fills in DNA bases, and **ligase** seals the sugar-phosphate backbone.

8. Replication proceeds in a 5′ to 3′ direction, so the process must be discontinuous in short stretches on one strand.

www.mhhe.com/lewisgenetics10

Answers to all end-of-chapter questions can be found at **www.mhhe.com/lewisgenetics10.** You will also find additional practice quizzes, animations, videos, and vocabulary flashcards to help you master the material in this chapter.

Review Questions

1. List the components of a nucleotide.

2. How does a purine differ from a pyrimidine?

3. Why must DNA be replicated?

4. Why would a DNA structure in which each base type could form hydrogen bonds with any of the other three base types not produce a molecule that is easily replicated?

5. What part of the DNA molecule encodes information?

6. Explain how DNA is a directional molecule in a chemical sense.

7. Match the experiment described in the left column to a concept it illustrates in the right column (more than one answer may be possible).

1. Density shift experiments
2. Discovery of an acidic substance that includes nitrogen and phosphorus on dirty bandages
3. "Blender experiments" that showed that the part of a virus that infects bacteria contains phosphorus, but not sulfur
4. Determination that DNA contains equal amounts of guanine and cytosine, and of adenine and thymine
5. Discovery that bacteria can transfer a "factor" that transforms a harmless strain into a lethal one

a. First experiments in identifying hereditary material.
b. Complementary base pairing is part of DNA structure and maintains a symmetrical double helix
c. Identification of nuclein
d. DNA, not protein, is the hereditary material
e. DNA replication is semiconservative, not conservative or dispersive

8. Place the following enzymes in the order in which they begin to function in DNA replication: ligase, primase, helicase and DNA polymerase.

9. How can very long DNA molecules fit into a cell's nucleus?

10. Place in increasing size order: nucleosome, histone protein, and chromatin.

11. How are very long strands of DNA replicated without twisting into a huge tangle?

12. List the steps in DNA replication.

13. Why must DNA be replicated continuously as well as discontinuously?

14. How does RNA participate in DNA replication?

15. Describe two experiments that supported one hypothesis while also disproving another.

16. Is downloading a document from the Internet analogous to replicating DNA? Cite a reason for your answer.

Applied Questions

1. In Bloom syndrome, ligase malfunctions. As a result, replication forks move too slowly. Why?

2. DNA contains the information that a cell uses to synthesize a particular protein. How do proteins assist in DNA replication?

3. A person with deficient or abnormal ligase may have an increased cancer risk and chromosome breaks that cannot heal. The person is, nevertheless, alive. Why are there no people who lack DNA polymerase?

4. Write the sequence of a strand of DNA replicated from each of the following base sequences:

   a. T C G A G A A T C T C G A T T
   b. C C G T A T A G C C G G T A C
   c. A T C G G A T C G C T A C T G

5. Which do you think was the more far-reaching accomplishment, determining the structure of DNA, or sequencing the human genome? State a reason for your answer.

6. Describe a recent news event, feature article, film, or television program that mentions a DNA sequence.

7. Cite an example of how knowing a DNA sequence could be abused, and an example of how knowing a DNA sequence could be helpful.

8. People often use the phrase "the gene for" to describe traits that do not necessarily or directly arise from a protein's actions, such as "a gene for jealousy" or "a gene for acting." How would you explain to them what a gene actually is?

Web Activities

1. The Frozen Ark project is an international consortium of zoos, laboratories, and museums that is preserving DNA samples from endangered animal species. Consult http://www.frozenark.org

   a. Follow one of the links and describe an endangered species. What do you think is the value of this project?
   b. Do you think the project should be extended to include organisms other than animals? Cite a reason for your answer.

   c. What would be the difficulties encountered in attempting to increase population sizes of endangered species using stored DNA?

2. Visit the Cystic Fibrosis Mutation Database website. Select twenty contiguous bases of the sequence for the cystic fibrosis gene and write the complementary sequence.

Case Studies and Research Results

1. Researchers at the University of Rochester studied children born with an infection called roseola. The children also have the causative herpesvirus inserted into their chromosomes. At least one parent of each child had the virus in a chromosome of a sampled hair cell.

   a. How did the children likely become infected?

   b. What does the transmission of the viral DNA from generation to generation reveal about the structure of DNA and its replication?

2. A team of astrobiologists collected bacteria from a lake that has very high levels of arsenic and nearly no phosphorus, and concluded that the bacteria had substituted arsenic for phosphorus in their DNA. If this is true (there are other interpretations), would the difference in the DNA molecule affect its information carrying capacity? State a reason for your answer.

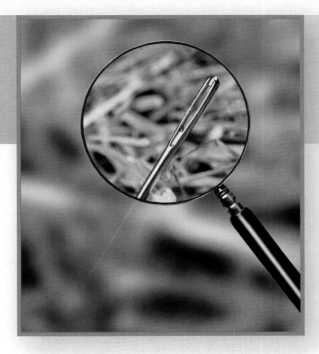

Discovering a mutation behind a rare disease is like searching for the proverbial needle-in-a-haystack. Whole exome sequencing significantly reduces the size of the metaphorical haystack by focusing on only protein-encoding parts of genes.

Gene Action: From DNA to Protein

Learning Outcomes

10.1 From DNA to Protein

1. State how much of the human genome encodes protein.

10.2 Transcription

2. List the major types of RNA molecules and their functions.

3. Explain the importance of transcription factors.

4. List the steps of transcription.

10.3 Translation of a Protein

5. Discuss how researchers deduced the genetic code.

6. List the steps of protein synthesis.

10.4 Processing a Protein

7. Define the four components of a protein's shape.

8. Explain the importance of protein folding.

The Big Picture: DNA sequences are the blueprints of life. Cells must maintain this information, yet also access it to manufacture proteins. RNA acts as the go-between, linking DNA to protein.

Whole Exome Sequencing

The phrase "needle in a haystack" refers to searching for something that is very rare among similar somethings. This is the case for identifying the genes that cause very rare single-gene disorders. A shortcut is to look only among DNA sequences that encode protein, which is a subset of the genome called the exome. Only 1.5 percent of the genome is the exome.

Whole exome sequencing surveys the 30 million or so DNA bases that account for 85 percent of mutations that affect our health. One of the first applications was for a collection of brain disorders called "malformations of cortical development," or MCD. The cortex is the outer part of the brain that consists of coils of nerve tissue. In MCD, the cortex is either too small, riddled with fluid-filled holes, or has too many or too few folds. By describing the forms of MCD anatomically, researchers had thought they were distinct disorders. Whole exome sequencing showed otherwise.

Researchers from Yale University and in Turkey studied the condition in families with cousin-cousin marriages, in which brain abnormalities were more likely to be due to inheriting shared mutations than to an environmental exposure or a birth defect. The investigation began with siblings who have microcephaly—small brains. They shared a gene called *WDR62*. Then scans of the siblings' brains revealed the other types of abnormalities too. When the researchers looked at more families with these other forms of the condition, they found mutations in the *WDR62* gene. The encoded protein is made in neural stem cells—which makes sense in terms of the phenotype.

Whole exome sequencing is cheaper and faster than whole genome sequencing. It is a powerful tool to rapidly locate the genes behind rare disorders—which number more than 6,000.

10.1 From DNA to Protein

Imagine downloading a 500-page novel on an e-reader, and getting only 8 pages of story. The rest is computer code for how the e-reader operates. The human genome is similar. Only about 1.5 percent of the DNA sequence encodes protein. This part is called the **exome**. Much of the rest of the genome controls how, where, and when proteins are made.

Our genes encode, at last count, 20,325 proteins. A protein consists of one or more long chains of amino acids called polypeptides. A short sequence of amino acids is called a peptide, and the bonds that join amino acids are called peptide bonds. Proteins have a great variety of functions, some of which are listed in **table 10.1**.

To use the genetic information in the nucleus to synthesize proteins, the process of **transcription** first makes a copy of a gene that is an RNA molecule complementary to one strand of the DNA double helix. The RNA copy is taken out of the nucleus and into the cytoplasm. There, the process of **translation** uses the information in the RNA to manufacture a protein by aligning and joining specified amino acids. Finally, the protein must fold into a specific three-dimensional form in order to function.

Accessing the genome is a huge, ongoing task. Cells replicate their DNA only during S phase of the cell cycle. In contrast, transcription and translation occur continuously, except during M phase. Transcription and translation supply the proteins essential for life, as well as those that give a cell its specialized characteristics.

10.2 Transcription

Shortly after Watson and Crick published their structure of DNA in 1953, they described the relationship between nucleic acids and proteins as a directional flow of information called the "central dogma" (**figure 10.1**). As Francis Crick explained in 1957, *"The specificity of a piece of nucleic acid is expressed solely by the sequence of its bases, and this sequence is a code for the amino acid sequence of a particular protein."* This statement inspired more than a decade of intense research to identify the participants in protein synthesis and discover how they interact. The process centers around RNA.

RNA is the bridge between gene and protein. RNA and DNA share an intimate relationship, as **figure 10.2** depicts. The bases of an RNA sequence are complementary to those of one strand of the double helix, which is called the **template strand**. An enzyme, **RNA polymerase,** assists the construction of an RNA molecule. The other strand of the DNA double helix is called the **coding strand**.

Table 10.1	Protein Diversity in the Human Body
Protein	**Function**
Actin, myosin, dystrophin	Muscle contraction
Antibodies, antigens, cytokines	Immunity
Carbohydrases, lipases, proteases, nucleases	Digestion (digestive enzymes)
Casein	Milk protein
Collagen, elastin, fibrillin	Connective tissue
Colony-stimulating factors, erythropoietin	Blood cell formation
DNA and RNA polymerase	DNA replication, gene expression
Ferritin	Iron transport in blood
Fibrin, thrombin	Blood clotting
Growth factors, kinases, cyclins	Cell division
Hemoglobin, myoglobin	Oxygen transport
Insulin, glucagon	Control of blood glucose level
Keratin	Hair structure
Tubulin, actin	Cell movements
Tumor suppressors	Cancer prevention

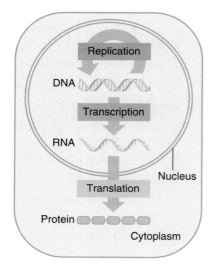

Figure 10.1 DNA to RNA to protein. Some of the information stored in DNA is copied to RNA (transcription), some of which is used to assemble amino acids into proteins (translation). DNA replication perpetuates genetic information. This figure repeats throughout the chapter, with the part under discussion highlighted.

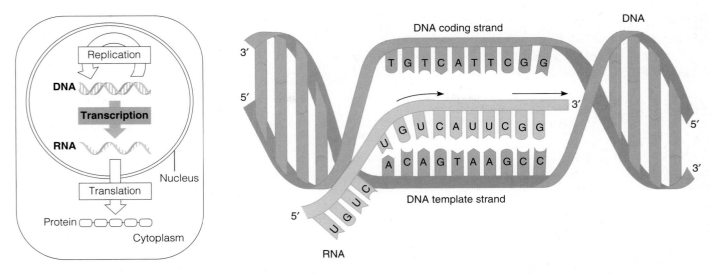

DNA coding strand

DNA

3'
5'

T G T C A T T C G G

3'

U G U C A U U C G G

A C A G T A A G C C

5'
3'

DNA template strand

5'

RNA

Figure 10.2 **The relationship among RNA, the DNA template strand, and the DNA coding strand.** The RNA sequence is complementary to the DNA template strand. This is the same sequence as the DNA coding strand, with uracil (U) in place of thymine (T).

RNA Structure and Types

RNA and DNA have similarities and differences (**figure 10.3** and **table 10.2**). Both are nucleic acids, consisting of sequences of nitrogen-containing bases joined by sugar-phosphate backbones. However, RNA is usually single-stranded, whereas DNA is double-stranded. Also, RNA has the pyrimidine base **uracil** where DNA has thymine. As their names imply, RNA nucleotides include the sugar ribose, rather than DNA's deoxyribose. Functionally, DNA stores genetic information, whereas RNA controls how that information is used. The presence of the—OH at the 5' position of ribose makes RNA much less stable than DNA, which is critical in its function as a short-lived carrier of genetic information.

As RNA is synthesized along DNA, it folds into a three-dimensional shape, or **conformation,** that is determined by complementary base pairing within the same RNA molecule. For example, a sequence of AAUUUCC might hydrogen bond to a sequence of UUAAAGG—its complement—elsewhere in the same molecule, a little like touching elbows to knees. Conformation is very important for RNA's functioning. The three

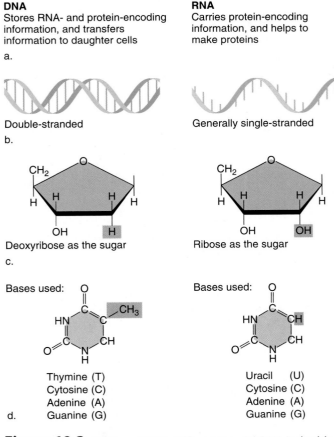

DNA
Stores RNA- and protein-encoding information, and transfers information to daughter cells

a.

RNA
Carries protein-encoding information, and helps to make proteins

Double-stranded
b.

Generally single-stranded

Deoxyribose as the sugar
c.

Ribose as the sugar

Bases used:

Thymine (T)
Cytosine (C)
Adenine (A)
d.   Guanine (G)

Bases used:

Uracil   (U)
Cytosine (C)
Adenine (A)
Guanine (G)

Figure 10.3 **DNA and RNA differences.** (a) DNA is double-stranded; RNA is usually single-stranded (b). DNA nucleotides include deoxyribose, whereas RNA nucleotides have ribose (c). Finally, DNA nucleotides include the pyrimidine thymine, whereas RNA has uracil (d).

Table 10.2	How DNA and RNA Differ
DNA	**RNA**
1. Usually double-stranded	1. Usually single-stranded
2. Thymine as a base	2. Uracil as a base
3. Deoxyribose as the sugar	3. Ribose as the sugar
4. Maintains protein-encoding information	4. Carries protein-encoding information and controls how information is used
5. Cannot function as an enzyme	5. Can function as an enzyme
6. Persists	6. Transient

Table 10.3	Major Types of RNA	
Type of RNA	**Size (number of nucleotides)**	**Function**
mRNA	500–4,500 +	Encodes amino acid sequence
rRNA	100–3,000	Associates with proteins to form ribosomes, which structurally support and catalyze protein synthesis
tRNA	75–80	Transports specific amino acids to the ribosome for protein synthesis

major types of RNA are messenger RNA, ribosomal RNA, and transfer RNA (**table 10.3**). Other classes of RNA control which genes are expressed (transcribed and translated) under specific circumstances. Table 11.2 describes them.

Messenger RNA (mRNA) carries the information that specifies a particular protein. Each three mRNA bases in a row form a genetic code word, or **codon,** that specifies a certain amino acid. Because genes vary in length, so do mature mRNA molecules. Most mRNAs are 500 to 4,500 bases long. Differentiated cells can carry out specialized functions because they express certain subsets of genes—that is, they produce certain mRNA molecules, which are also called transcripts. The information in the transcripts is then used to manufacture the encoded proteins. A muscle cell, for example, has many mRNAs that specify the contractile proteins actin and myosin, whereas a skin cell contains many mRNAs that specify scaly keratin proteins.

To use the information in an mRNA sequence, a cell requires the two other major classes of RNA. **Ribosomal RNA** (rRNA) molecules range from 100 to nearly 3,000 nucleotides long. Ribosomal RNAs associate with certain proteins to form a ribosome. Recall from chapter 2 that a ribosome is an organelle made up of many different protein and RNA subunits. Overall, a ribosome functions as a machine to attach amino acids to form proteins (**figure 10.4**).

A ribosome has two subunits that are separate in the cytoplasm but join at the site of initiation of protein synthesis. The larger ribosomal subunit has three types of rRNA molecules, and the small subunit has one. Ribosomal RNA, however, is more than a structural support. Certain rRNAs catalyze the formation of the peptide bonds between amino acids. Such an

RNA with enzymatic function is called a ribozyme. Other rRNAs help to align the ribosome and mRNA.

The third major type of RNA molecule, **transfer RNA** (tRNA), binds an mRNA codon at one end and a specific amino acid at the other. A tRNA molecule is only 75 to 80 nucleotides long. Some of its bases form weak chemical bonds with each other, folding the tRNA into loops in a characteristic cloverleaf shape (**figure 10.5**). One loop of the tRNA has three bases in a row that form the **anticodon,** which is complementary to an mRNA codon. The end of the tRNA opposite the anticodon strongly bonds to a specific amino acid. A tRNA with a particular anticodon sequence always carries the same amino acid. (Organisms have 20 types of amino acids.) For example, a tRNA with the anticodon sequence GAA always picks up the amino acid phenylalanine. Enzymes attach amino acids to tRNAs that bear the appropriate anticodons, where they form chemical bonds (**figure 10.6**).

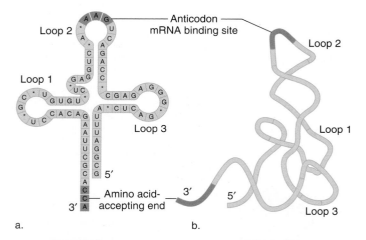

a.   b.

Figure 10.5 **Transfer RNA.** **(a)** Certain nucleotide bases within a tRNA hydrogen bond with each other to give the molecule a "cloverleaf" conformation that can be represented in two dimensions. The darker bases at the top form the anticodon, the sequence that binds a complementary mRNA codon. Each tRNA terminates with the sequence CCA, where a particular amino acid covalently bonds. A three-dimensional representation of a tRNA **(b)** depicts the loops that interact with the ribosome.

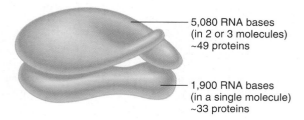

Figure 10.4 **The ribosome.** A ribosome from a eukaryotic cell has two subunits; together, they consist of 82 proteins and four rRNA molecules.

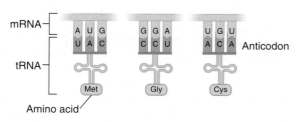

Figure 10.6 **A tRNA with a particular anticodon sequence always binds the same type of amino acid.**

Transcription Factors

If all of the genes in the human genome were being transcribed and translated at the same time, chaos would result. Even a simple bacterium must control which genes are expressed under which conditions. In 1961, French biologists François Jacob and Jacques Monod described the remarkable ability of *E. coli* bacteria to produce the enzymes to metabolize the sugar lactose—but only when lactose is in the cell's surroundings. What "tells" a simple bacterial cell to transcribe the proteins it needs, at exactly the right time?

Jacob and Monod discovered that a modified form of lactose activated the genes whose encoded proteins break it down. They named the set of genes that are coordinately controlled an operon, writing in 1961, *"The genome contains not only a series of blueprints, but a coordinated program of protein synthesis and means of controlling its execution."*

In bacteria, operons turn transcription of a few genes on or off. In more complex organisms, different cell types express different subsets of genes. To manage this, groups of proteins called **transcription factors** come together, forming an apparatus that binds DNA at certain sequences and initiates transcription at specific sites on chromosomes. The transcription factors responding to signals from outside the cell such as hormones and growth factors form a pocket for RNA polymerase to bind and begin building an RNA chain. Transcription factors include regions called binding domains that guide them to the genes they control. The DNA binding domains have very colorful names, such as "helix-turn-helix," "zinc fingers," and "leucine zippers," that reflect their distinctive shapes.

The few types of transcription factors work in combinations, providing great specificity in controlling gene expression. Overall, transcription factors link the genome to the environment. For example, lack of oxygen, such as from choking or smoking, sends signals that activate transcription factors to turn on dozens of genes that enable cells to handle the stress of low-oxygen conditions.

Mutations in transcription factor genes can have wide-ranging effects, because the factors control many genes. The varied symptoms of Rett syndrome, discussed in Reading 6.2, arise from a mutation in the gene that encodes a transcription factor called MECP2. Transcription factors are themselves controlled by each other and by other classes of molecules.

Steps of Transcription

Transcription and translation are each described in three steps: initiation, elongation, and termination.

How do transcription factors and RNA polymerase (RNAP) "know" where to bind to DNA to begin transcribing a specific gene? In transcription initiation, transcription factors and RNA polymerase are attracted to a **promoter,** which is a special sequence that signals the start of the gene, like a capital letter at the start of a sentence. Signals from outside the cell alter the chromatin structure in a way that exposes the promoter of a gene whose transcription is required under the particular conditions (see figure 11.6).

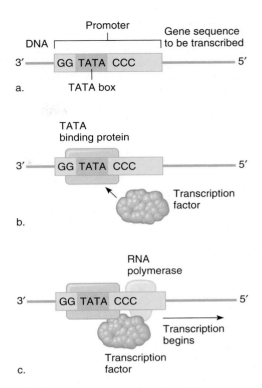

Figure 10.7 **Setting the stage for transcription to begin. (a)** Proteins that initiate transcription recognize specific sequences in the promoter region of a gene. **(b)** A binding protein recognizes the TATA region and binds to the DNA. This allows other transcription factors to bind. **(c)** The bound transcription factors form a pocket that allows RNA polymerase to bind and begin making RNA.

Figure 10.7 shows a simplified view of transcription factor binding, which sets up a site called a preinitiation complex to receive RNA polymerase. The first transcription factor to bind, called a TATA binding protein, is chemically attracted to a DNA sequence called a TATA box—the base sequence TATA surrounded by long stretches of G and C. Once the first transcription factor binds, it attracts others in groups. Finally RNA polymerase joins the complex, binding just in front of the start of the gene sequence. The assembly of these components is transcription initiation.

In the next stage, transcription elongation, enzymes unwind the DNA double helix locally, and free RNA nucleotides bond with exposed complementary bases on the DNA template strand (see figure 10.2). RNA polymerase adds the RNA nucleotides in the sequence the DNA specifies, moving along the DNA strand in a 3′ to 5′ direction, synthesizing the RNA molecule in a 5′ to 3′ direction. A terminator sequence in the DNA indicates where the gene's RNA-encoding region ends, like the period at the end of a sentence. When this spot is reached, the third stage, transcription termination, occurs **(figure 10.8).** A typical rate of transcription in humans is 20 bases per second.

RNA is typically transcribed using only a gene's template strand. However, different genes on the same chromosome may

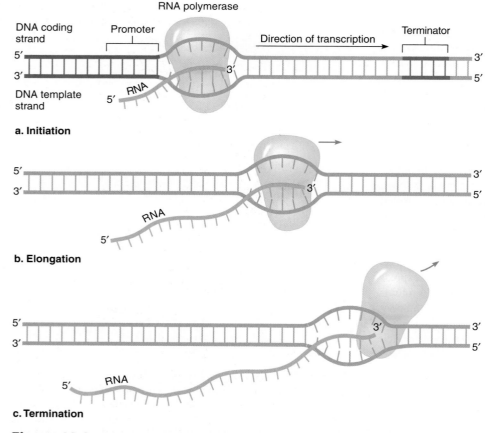

a. Initiation

b. Elongation

c. Termination

Figure 10.8 **Transcription of RNA from DNA.** Transcription occurs in three stages: initiation, elongation, and termination. Initiation is the control point that determines which genes are transcribed. RNA nucleotides are added during elongation. A terminator sequence in the gene signals the end of transcription.

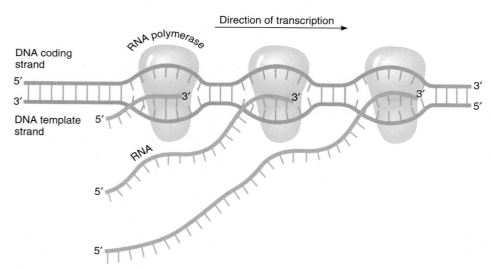

Figure 10.9 **Many identical copies of RNA are transcribed simultaneously.** Usually 100 or more DNA bases lie between RNA polymerases.

be transcribed from different strands of the double helix. The coding strand of the DNA is so-called because its sequence is identical to that of the RNA, except with thymine (T) in place of uracil (U). Several RNAs may be transcribed from the same DNA template strand simultaneously (**figure 10.9**). Since mRNA is short-lived, with about half of it degraded every 10 minutes, a cell must constantly transcribe certain genes to maintain supplies of essential proteins.

To determine the sequence of RNA bases transcribed from a gene, write the RNA bases that are complementary to the template DNA strand, using uracil opposite adenine. For example, a DNA template strand that has the sequence

C C T A G C T A C

is transcribed into RNA with the sequence

G G A U C G A U G

and the coding DNA sequence is

G G A T C G A T G.

RNA Processing

In bacteria, RNA is translated into protein as soon as it is transcribed from DNA because a nucleus does not physically separate the two processes. In eukaryotic cells, mRNA must first exit the nucleus to enter the cytoplasm, where ribosomes are located. Messenger RNA is altered in several steps before it is translated in these more complex cells.

First, after mRNA is transcribed, a short sequence of modified nucleotides, called a cap, is added to the 5′ end of the molecule. The cap consists of a backwardly inserted guanine (G), which attracts an enzyme that adds methyl groups (CH_3) to the G and one or two adjacent nucleotides. This methylated cap is a recognition site for protein synthesis. At the 3′ end, a special polymerase adds about 200 adenines, forming a "poly A tail." The poly A tail is necessary for protein synthesis to begin, and may also stabilize the mRNA so that it stays intact longer.

Further changes occur to the capped, poly A tailed mRNA before it is translated into protein. Parts of mRNAs called **introns** (short for "intervening sequences") that

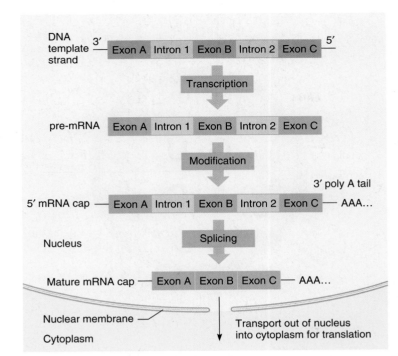

Figure 10.10 **Messenger RNA processing—the maturing of the message.** Several steps process pre-mRNA into mature mRNA. First, a large region of DNA containing the gene is transcribed. Then a modified nucleotide cap and poly A tail are added and introns are spliced out. Finally, the intact, mature mRNA is sent out of the nucleus.

were transcribed are removed. The ends of the remaining molecule are spliced together before the mRNA is translated. The parts of mRNA that remain and are translated are called **exons** **(figure 10.10)**.

Once introns are spliced out, enzymes check, or proofread, the remaining mRNA. Messenger RNAs that are too short or too long may be held in the nucleus. Proofreading also monitors tRNAs, ensuring that they assume the correct cloverleaf shape.

Prior to intron removal, the mRNA is called pre-mRNA. Introns control their own removal. They associate with certain proteins to form small nuclear ribonucleoproteins (snRNPs), or "snurps." Four snurps form a structure called a spliceosome that cuts introns out and attaches exons to form the mature mRNA that exits the nucleus. The introns cut themselves out of the RNA.

Introns range in size from 65 to 10,000 or more bases; the average intron is 3,365 bases. The average exon, in contrast, is only 145 bases long. The number, size, and organization of introns vary from gene to gene. The coding portion of the average human gene is 1,340 bases, whereas the average total size of a gene is 27,000 bases. The dystrophin gene is 2,500,000 bases, but its corresponding mRNA sequence is only 14,000 bases! The gene contains 80 introns.

The discovery of introns in the 1970s surprised geneticists, who had thought genes were like sentences in which all of the information has meaning. At first, some geneticists called introns "junk DNA"—a term that has unfortunately persisted even as researchers have discovered the functions of many introns. Some introns encode RNAs that control gene expression, whereas others are actually exons on the complementary strand of DNA. Introns may also be vestiges of ancient genes that have lost their original function, or are remnants of the DNA of viruses that once integrated into a chromosome.

The intron/exon organization of most genes provides a way to maximize genetic information. Different combinations of exons of a gene encode different versions of the protein product, termed isoforms. From 40 to 60 percent of human genes encode isoforms, and the mechanism of combining exons of a gene in different ways is called alternate splicing. This may explain how cell types use the same protein in slightly different ways in different tissues (see figure 11.9). For example, a protein that transports fats is shorter in the small intestine, where it carries dietary fats, than it is in the liver, where it carries fats made in the body.

Key Concepts

1. RNA is single-stranded, has uracil and ribose, and has different functions than DNA.

2. Messenger RNA transmits information to build proteins. Each three mRNA bases in a row form a codon that specifies a particular amino acid.

3. Ribosomal RNA and proteins form ribosomes, which physically support protein synthesis and help catalyze bonding between amino acids.

4. Transfer RNAs connect mRNA codons to amino acids.

5. Bacterial operons are simple gene control systems. In more complex organisms, cascades of transcription factors control gene expression.

6. RNA polymerase inserts complementary RNA bases opposite the DNA template strand.

7. Messenger RNA (mRNA) gains a modified nucleotide cap and a poly A tail.

8. Introns are transcribed and cut out, and exons are reattached. Introns are common and large in human genes.

9. Certain genes are processed into different-sized RNAs in different cell types.

10.3 Translation of a Protein

Transcription copies the information in DNA into the complementary language of RNA. The next step is translating mRNA into the precise sequence of amino acids that forms a protein. Particular mRNA codons correspond to particular amino acids

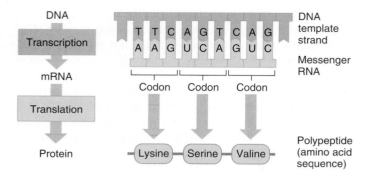

Figure 10.11 **From DNA to RNA to protein.** Messenger RNA is transcribed from a locally unwound portion of DNA. In translation, transfer RNA matches mRNA codons with amino acids.

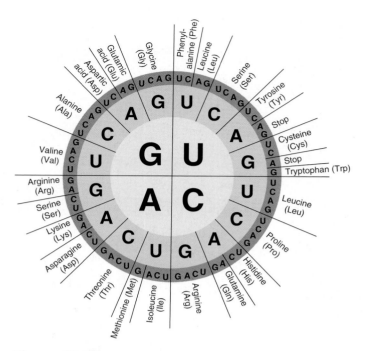

Figure 10.12 **The genetic code.** The first codon position is at the center of the circle. For example, the mRNA codon CUA encodes the amino acid leucine.

(**figure 10.11**). This correspondence between the chemical languages of mRNA and protein is the **genetic code**. Translation occurs on free ribosomes in the cytoplasm as well as on ribosomes that are embedded in the endoplasmic reticulum (ER).

Francis Crick hypothesized that an "adaptor" molecule would enable the RNA message to attract and link amino acids into proteins. He envisioned *"20 different kinds of adaptor molecule, one for each amino acid, and 20 different enzymes to join the amino acid [sic] to their adaptors."* In the 1960s, researchers deciphered the genetic code, determining which mRNA codons correspond to which amino acids. **Figure 10.12** displays the genetic code.

Deciphering the Genetic Code

The researchers who deciphered the genetic code used logic and experiments. More recently, annotation of the human genome sequence has confirmed and extended the earlier work, revealing new nuances in the genetic code. To understand how the genetic code works, it is helpful to ask the questions researchers asked in the 1960s.

Question 1—How Many RNA Bases Specify One Amino Acid?

The number of different protein building blocks (20) exceeds the number of different mRNA building blocks (4). Therefore, each codon must include more than one mRNA base. If a codon consisted of only one mRNA base, then codons could specify only four different amino acids, one corresponding to each of the four bases: A, C, G, and U. If each codon consisted of two bases, then only 16 (4^2) different amino acids could be specified, one corresponding to each of the 16 possible combinations of two RNA bases. If a codon consisted of three bases, then the genetic code could specify as many as 64 (4^3) different amino acids, sufficient to encode the 20 different amino acids that make up proteins. Therefore, the minimum number of bases in a codon is three.

Francis Crick and his coworkers experimented on a virus called T4 to confirm the triplet nature of the genetic code. They exposed the virus to chemicals that add or remove one, two, or three bases, and examined a viral gene with a well-known

sequence and protein product. Altering the DNA sequence by one or two bases produced a different amino acid sequence. This happened because the change disrupted the **reading frame,** which is the sequence of amino acids encoded from a certain starting point in a DNA sequence. However, adding or deleting three contiguous bases added or deleted only one amino acid in the protein without disrupting the reading frame. The rest of the amino acid sequence was retained. The code, the researchers deduced, is triplet (**figure 10.13**).

Further experiments confirmed the triplet nature of the genetic code. Adding a base at one point in the gene and deleting a base at another point disrupted the reading frame only between these sites. The result was a protein with a stretch of the wrong amino acids, like a sentence with a few misspelled words in the middle.

Question 2—Does the Information in a DNA Sequence Overlap?

Consider a hypothetical mRNA sequence:

AUGCCCAAG

If the genetic code is triplet and a DNA sequence is "read" in a nonoverlapping manner, then this sequence has only three codons and specifies three amino acids:

AUGCCCAAG

AUG (methionine)

CCC (proline)

AAG (lysine)

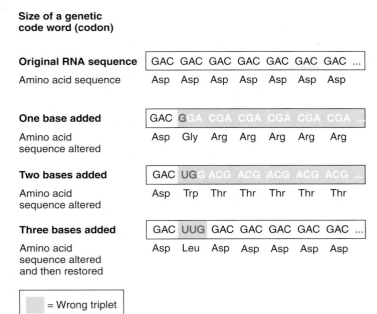

Size of a genetic code word (codon)

Original RNA sequence	GAC	GAC	GAC	GAC	GAC	GAC	GAC ...
Amino acid sequence	Asp	Asp	Asp	Asp	Asp	Asp	Asp
One base added	GAC	GGA	CGA	CGA	CGA	CGA	CGA ...
Amino acid sequence altered	Asp	Gly	Arg	Arg	Arg	Arg	Arg
Two bases added	GAC	UGG	ACG	ACG	ACG	ACG	ACG ...
Amino acid sequence altered	Asp	Trp	Thr	Thr	Thr	Thr	Thr
Three bases added	GAC	UUG	GAC	GAC	GAC	GAC	GAC ...
Amino acid sequence altered and then restored	Asp	Leu	Asp	Asp	Asp	Asp	Asp

☐ = Wrong triplet

Figure 10.13 **Three at a time.** Adding or deleting one or two nucleotides in a DNA sequence results in a frameshift that disrupts the encoded amino acid sequence. Adding or deleting three bases does not disrupt the reading frame because the code is triplet. This is a simplified representation of the Crick experiment.

If the DNA sequence is overlapping, however, the sequence specifies seven codons:

AUGCCCAAG

AUG (methionine)

UGC (cysteine)

GCC (alanine)

CCC (proline)

CCA (proline)

CAA (glutamine)

AAG (lysine)

An overlapping DNA sequence seems to pack maximal information into a limited number of bases. However, certain amino acids would always follow certain others, constraining protein structure. For example, AUG would always be followed by an amino acid whose codon begins with UG. This does not happen. Therefore, the protein-encoding DNA sequence is not overlapping.

Even though the genetic code is nonoverlapping, any DNA or RNA sequence can be read in three different reading frames, depending upon the "start" base. **Figure 10.14** depicts the three reading frames for the sequence just discussed, slightly extended. It encodes three different trios of amino acids.

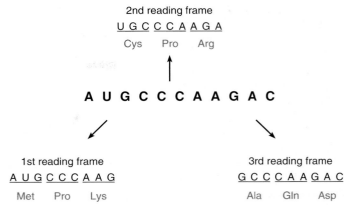

Figure 10.14 **Reading frames—where the sequence begins.** A sequence of DNA has three reading frames.

Question 3—Can mRNA Codons Specify Anything Other Than Amino Acids?

Chemical analysis eventually showed that the genetic code includes directions for starting and stopping translation. The codon AUG signals "start," and the codons UGA, UAA, and UAG signify "stop." Another form of "punctuation," a short sequence of bases at the start of each mRNA, enables the mRNA to form hydrogen bonds with rRNA in a ribosome. It is called a leader sequence.

Question 4—Do All Species Use the Same Genetic Code?

All species use the same mRNA codons to specify the same amino acids. References to the "human genetic code" usually mean a human genome sequence. The simplest explanation for the "universality" of the genetic code is that all life evolved from a common ancestor. No other mechanism as efficient at directing cellular activities has emerged and persisted.

The only known exceptions to the universality of the genetic code are a few codons in mitochondria and in certain single-celled eukaryotes (ciliated protozoa). These deviations may be tolerated because they do not affect the major repositories of DNA. The mitochondrial genome is small, and the affected ciliated protozoa have a second, smaller nucleus that houses some genes with one or two alternate codon-amino acid associations. In both cases, the major DNA sites adhere to the universal genetic code. Some types of single-celled organisms translate a stop codon into a twenty-first type of amino acid. Overall, however, the genetic code is considered to be universal.

Question 5—Which Codons Specify Which Amino Acids?

The number of bases in a codon, the nonoverlapping reading frame, and universality are general features of the genetic code. The big question, back in the 1960s, was the code itself: Which codons specify which amino acids? In 1961, Marshall

Nirenberg and Heinrich Matthaei at the National Institutes of Health used a precise and logical series of experiments to "crack the code." They synthesized very simple mRNA molecules and added them to test tubes that contained all the chemicals and structures needed for translation, extracted from *E. coli* cells. Which amino acid would each synthetic RNA specify?

The first synthetic mRNA they made had the sequence UUUUUU. . . . In the test tube, this was translated into a peptide consisting entirely of one amino acid type: phenylalanine. This was the first entry in the genetic code dictionary: The codon UUU specifies the amino acid phenylalanine. The next experiments revealed that AAA codes for the amino acid lysine and CCC for proline. (GGG was unstable, so this part of the experiment could not be done.)

To reveal other codon-amino acid pairs, researchers synthesized chains of alternating bases. Synthetic mRNA of sequence AUAUAU . . . introduced codons AUA and UAU. When translated, the mRNA yielded an amino acid sequence of alternating isoleucines and tyrosines. But was AUA the code for isoleucine and UAU for tyrosine, or vice versa? Another experiment with a more complex sequence answered the question. The mRNA UUUAUAUUUAUA, when translated from the first U of a UUU, encoded alternating phenylalanine and isoleucine. Because the first experiment had showed that UUU codes for phenylalanine, AUA must code for isoleucine. If AUA codes for isoleucine, then UAU must code for tyrosine (**table 10.4**).

By the end of the 1960s, researchers had used logic to decipher the entire genetic code. Sixty of the possible 64 codons specify particular amino acids, three indicate "stop," and one encodes both the amino acid methionine and "start." This means that some amino acids are specified by more than one codon. For example, both UUU and UUC encode phenylalanine.

Different codons that specify the same amino acid are termed **synonymous codons,** just as synonyms are words with the same meaning. The genetic code is said to be degenerate because most amino acids are not uniquely specified. Synonymous codons often differ from one another by the base in the third position. The corresponding base of a tRNA's anticodon is called the "wobble" position because it can bind to more than one type of base in synonymous codons. The degeneracy of the genetic code protects against mutation, because changes in the DNA that substitute a synonymous codon do not alter the protein's amino acid sequence. **Nonsynonymous codons** encode different amino acids.

In the 1950s and 1960s, molecular genetics was still a very young science, and so the code breakers came mostly from the ranks of chemistry, physics, and math. Some of the more exuberant personalities organized an "RNA tie club" and inducted a member whenever someone added a piece to the puzzle of the genetic code, anointing him (there was no prominent "her") with a tie and tie pin emblazoned with the structure of the specified amino acid.

The human genome project picked up where the genetic code experiments of the 1960s left off by identifying the DNA sequences that are transcribed into tRNAs. That is, 61 different tRNAs could theoretically exist, one for each codon that specifies an amino acid (the 64 triplets minus 3 stop codons). However, only 49 different genes encode tRNAs. This is because the same type of tRNA can detect synonymous codons that differ only in whether the wobble (third) position is U or C. The same type of tRNA, for example, binds to both UUU and UUC codons, which specify the amino acid phenylalanine. Synonymous codons ending in A or G use different tRNAs. Sequencing of other genomes reveals that some types of organisms preferentially use particular codons for amino acids specified by more than one type of codon. Researchers do not yet understand the significance, if any, of such "codon usage bias."

Building a Protein

Protein synthesis requires mRNA, tRNA molecules carrying amino acids, ribosomes, energy-storing molecules such as adenosine triphosphate (ATP) and guanosine triphosphate (GTP), and various protein factors. These pieces meet in a stage called translation initiation (**figure 10.15**). Chemical bonds hold the different components together.

First, the mRNA leader sequence forms hydrogen bonds with a short sequence of rRNA in a small ribosomal subunit. The first mRNA codon to specify an amino acid is always AUG, which attracts an initiator tRNA that carries the amino acid methionine (abbreviated Met). This methionine signifies the start of a polypeptide. The small ribosomal subunit, the mRNA bonded to it, and the initiator tRNA with its attached methionine form the initiation complex at the appropriate AUG codon of the mRNA.

Table 10.4	Deciphering RNA Codons and the Amino Acids They Specify	
Synthetic RNA	**Encoded Amino Acid Chain**	**Puzzle Piece**
UUUUUUUUUUUUUUUUUU	Phe-Phe-Phe-Phe-Phe-Phe	UUU = Phe
AAAAAAAAAAAAAAAAAA	Lys-Lys-Lys-Lys-Lys-Lys	AAA = Lys
CCCCCCCCCCCCCCCCCC	Pro-Pro-Pro-Pro-Pro-Pro	CCC = Pro
AUAUAUAUAUAUAUAUAU	Ile-Tyr-Ile-Tyr-Ile-Tyr	AUA = Ile or Tyr
		UAU = Ile or Tyr
UUUAUAUUUAUAUUUAUA	Phe-Ile-Phe-Ile-Phe-Ile	AUA = Ile
		UAU = Tyr

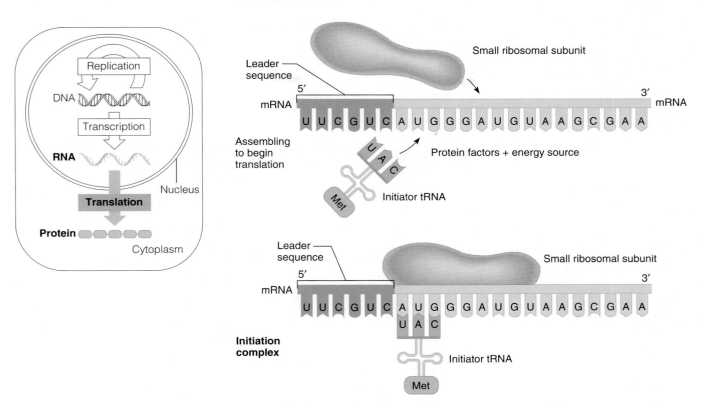

Figure 10.15 Translation begins as the initiation complex forms. Initiation of translation brings together a small ribosomal subunit, mRNA, and an initiator tRNA, and aligns them in the proper orientation to begin translation.

To start the next stage, elongation, a large ribosomal sub-unit bonds to the initiation complex. The codon adjacent to the initiation codon (AUG), which is GGA in **figure 10.16,** then bonds to its complementary anticodon, which is part of a free tRNA that carries the amino acid glycine. The two amino acids (Met and Gly in the example), still attached to their tRNAs, align.

The part of the ribosome that holds the mRNA and tRNAs together can be described as having two sites. The positions of the sites on the ribosome remain the same with respect to each other as translation proceeds, but they cover different parts of the mRNA as the ribosome moves. The P site holds the growing amino acid chain, and the A site right next to it holds the next amino acid to be added to the chain. In figure 10.16, when the forming protein consists of only the first two amino acids, Met occupies the P site and Gly the A site.

The amino acids link by forming a specific type of chemical bond called a peptide bond, with the help of rRNA that functions as a ribozyme. Then the first tRNA is released to pick up another amino acid of the same type and be used again. Special enzymes ensure that tRNAs always bind the correct amino acids, which is crucial to the accuracy of translation. The ribosome and its attached mRNA are now bound to a single tRNA, with two amino acids extending from it at the P site. This is the start of a polypeptide.

Next, the ribosome moves down the mRNA by one codon. The region of the mRNA that was at the A site is thus now at the P site. A third tRNA enters at the now-vacated new A site, corresponding to the next codon, carrying its amino acid (Cys in figure 10.16b). This third amino acid aligns with the other two and forms a peptide bond to the second amino acid in the growing chain, now extending from the P site. The tRNA attached to the second amino acid is released and recycled. The polypeptide continues to build, one amino acid at a time. Each piece is brought in by a tRNA whose anticodon corresponds to a consecutive mRNA codon as the ribosome moves down the mRNA (figure 10.16c).

Elongation halts when the A site of the ribosome has a "stop" codon (UGA, UAG, or UAA), because no tRNA molecules correspond to it. A protein release factor starts to free the polypeptide. The last tRNA leaves the ribosome, the ribosomal subunits separate and are recycled, and the new polypeptide is released (**figure 10.17**). **Table 10.5** reviews the forms of information encountered in transcription and translation.

Protein synthesis is economical. A cell can produce large amounts of a particular protein from just one or two copies of a gene. A plasma cell in the immune system, for example, manufactures 2,000 identical antibody molecules per second. To mass produce proteins at this rate, RNA, ribosomes, enzymes, and other proteins are continually recycled. In addition, transcription

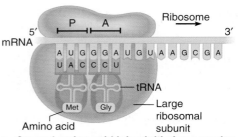

a. **Second amino acid joins initiation complex.**

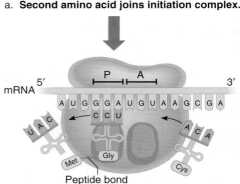

b. **First peptide bond forms as new amino acid arrives.**

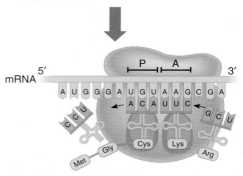

c. **Amino acid chain extends.**

Figure 10.16 **Building a polypeptide.** **(a)** A large ribosomal subunit binds to the initiation complex, and a tRNA bearing a second amino acid (glycine, in this example) forms hydrogen bonds between its anticodon and the mRNA's second codon at the A site. The first amino acid, methionine, occupies the P site. **(b)** The methionine brought in by the first tRNA forms a peptide bond with the amino acid brought in by the second tRNA, and a third tRNA arrives, in this example carrying the amino acid cysteine, at the temporarily vacated A site. **(c)** A fourth and then fifth amino acid are linked to the growing polypeptide chain. The process continues until a termination codon is reached.

Table 10.5	Information in Nucleic Acid Molecules
Type of Molecule	**Rules and Relationships**
DNA coding strand	Coding and template strands have complementary DNA bases.
	mRNA is complement of DNA template strand, with U for T.
DNA template strand	mRNA is same as DNA coding strand, with U for T.
	tRNA anticodons are complement of mRNA.
mRNA codons	tRNA anticodons are same as DNA template strand, with U for T.
	tRNA anticodons are complement of DNA coding strand, with U for T.
tRNA anticodons	tRNA translates genetic code, bringing together amino acids specified by DNA coding strand.
Amino acids (protein)	Amino acids bond to form a protein.

always produces many copies of a particular mRNA, and each mRNA may bind dozens of ribosomes, as **figure 10.18** shows. As soon as one ribosome has moved far enough along the mRNA to leave space, another ribosome attaches. In this way, many copies of the encoded protein are made from the same mRNA.

As complex as protein synthesis is, linking amino acids is only a first step. The chain must fold in a precise sequence of steps for the protein to assume its three-dimensional form, which is essential for it to function. This may occur in the cytoplasm or on the membranes of the ER, as discussed in the next section.

Some proteins undergo further alterations, called posttranslational modifications, before they can function. For example,

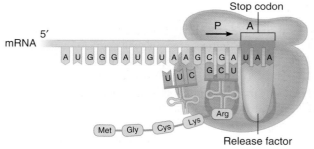

a. Ribosome reaches stop codon.

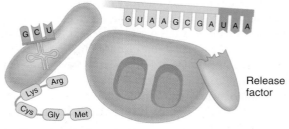

b. Components disassemble.

Figure 10.17 **Terminating a polypeptide.** **(a)** A protein release factor binds to the stop codon, releasing the completed polypeptide from the tRNA and **(b)** freeing all of the components of the translation complex.

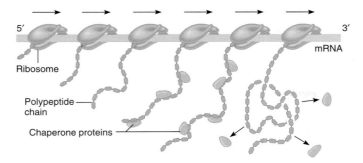

Figure 10.18 **Making many copies of a protein.** Several ribosomes can simultaneously translate a protein from a single mRNA. These ribosomes hold different-sized polypeptides—the closer to the end of a gene, the longer the polypeptide. Proteins called chaperones help fold the polypeptide.

insulin, which is 51 amino acids long, is initially translated as the polypeptide proinsulin, which is 80 amino acids long. Enzymes cut it to 51. Some proteins must have sugars attached for them to become functional, or polypeptides must aggregate.

Key Concepts

1. The genetic code is triplet, nonoverlapping, continuous, universal, and degenerate.

2. As translation begins, mRNA, tRNA with bound amino acids, ribosomes, energy molecules, and protein factors assemble. The mRNA binds to rRNA in a small ribosomal subunit. The first codon attracts a tRNA bearing methionine.

3. The large ribosomal subunit attaches and the tRNA anticodons bind to successive codons. Aligned amino acids form peptide bonds. A polypeptide forms.

4. The ribosome moves down the mRNA to the amino acid chain (the P site) and to where a new tRNA binds (the A site).

5. When the ribosome reaches a "stop" codon, protein synthesis ceases. Components are recycled.

10.4 Processing a Protein

Proteins fold into one or more three-dimensional shapes, or **conformations**. This folding is based on chemistry: attraction and repulsion between atoms of the proteins as well as inter-actions of proteins with chemicals in the immediate environment. For example, thousands of water molecules surround a growing chain of amino acids. Because some amino acids are attracted to water and some are repelled by it, the water contorts the protein's shape. Sulfur atoms also affect protein conformation by bridging the two types of amino acids that contain them.

The conformation of a protein is described at several levels (**figure 10.19**). The amino acid sequence of a polypeptide chain is its **primary (1°) structure**. Chemical attractions between amino acids that are close together in the 1° structure fold the polypeptide chain into its **secondary (2°) structure,** which may form loops, coils, barrels, helices, or sheets. Two common secondary structures are an alpha helix and a beta-pleated sheet. Secondary structures wind into larger **tertiary (3°) structures** as more widely separated amino acids attract or repel in response to water molecules. Finally, proteins consisting of more than one polypeptide form a **quaternary (4°) structure**. Hemoglobin, the blood protein that carries oxygen, has four polypeptide chains (see figure 11.1). The liver protein ferritin has 20 identical polypeptides of 200 amino acids each. In contrast, the muscle protein myoglobin is a single polypeptide chain.

Mutations or polymorphisms may alter the primary structure of a protein if the genetic change is nonsynonymous, which means that it changes the amino acid, just as a nonsynonym in the English language is a word that does not have the same meaning as another. In contrast, more than one tertiary or quaternary structure may be possible if an amino acid chain can fold in different ways.

Protein Folding

Proteins begin to fold within a minute after the amino acid chain winds away from the ribosome. A small protein might contort into its final, functional form in one quick step, taking only a few microseconds. Larger proteins may fold into a series of short-lived intermediates before assuming their final, functional forms.

Proteins start to move toward their destinations as they are being synthesized. In some proteins, part of the start of the amino acid chain forms a tag of sorts that helps direct the protein in the cell. The first few amino acids in a protein that will be secreted or lodge in a membrane form a "signal sequence" that leads it and the ribosome to which it binds into a pore in the ER membrane. Once in the ER, the protein enters the secretory network (see figure 2.5). Proteins destined for the mitochondria bear a different signal sequence. (Mitochondria manufacture their own proteins but also use many proteins that are encoded in DNA sequences in the nucleus.)

Signal sequences are not found on proteins synthesized on free ribosomes in the cytoplasm. These proteins may function right where they are made, such as the protein tubules and filaments of the cytoskeleton (see figure 2.10) or enzymes that take part in metabolism. Some proteins travel to and function in the nucleus, such as transcription factors. Proteins destined for the nucleus are synthesized on free ribosomes.

Various proteins assist in this precise folding, whatever the destination. **Chaperone proteins** stabilize partially folded regions in their correct form, and prevent a protein from getting "stuck" in an intermediate form, which would affect its function. Other proteins help new chemical bonds to form as the final shape arises, and yet others monitor the accuracy of folding.

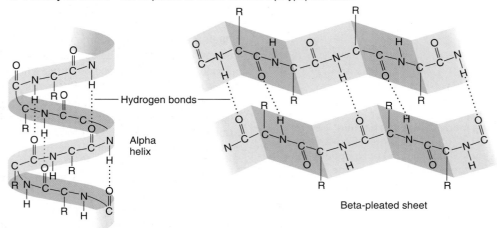

H$_2$N—Ala—Thr—Cys—Tyr—Glu—Gly—COOH

a. **Primary structure**—the sequence of amino acids in a polypeptide chain

Hydrogen bonds

Alpha helix

Beta-pleated sheet

b. **Secondary structure**—loops, coils, sheets, or other shapes formed by hydrogen bonds between neighboring carboxyl and amino groups

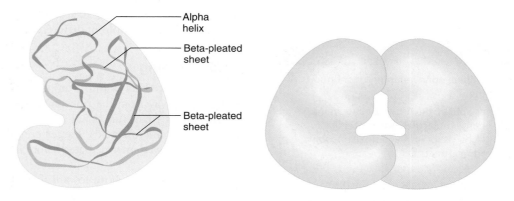

Alpha helix

Beta-pleated sheet

Beta-pleated sheet

c. **Tertiary structure**—three-dimensional forms shaped by bonds between R groups, interactions between R groups and water

d. **Quaternary structure**—protein complexes formed by bonds between separate polypeptides

Figure 10.19 Four levels of protein structure. (a) The amino acid sequence of a polypeptide forms the primary structure. Each amino acid has an amino end (NH) and a carboxyl end (COOH), and each of the 20 types of amino acids is distinguished by an R group. **(b)** Hydrogen bonds between non–R groups create secondary structures such as helices and sheets. The tertiary structure **(c)** arises when R groups interact, folding the polypeptide in three dimensions and forming a unique shape. **(d)** If different polypeptide units must interact to be functional, the protein has a quaternary structure.

Should a protein misfold, an "unfolded protein response" occurs in which protein synthesis slows or even stops, transcription of genes that encode chaperone proteins and the other folding proteins speeds up, and proper protein folding is quickly restored.

Protein Misfolding

If a misfolded protein is made despite these protections, other actions ensue. Misfolded proteins are sent out of the ER back into the cytoplasm, where they are "tagged" with yet another protein, called ubiquitin. A misfolded protein bearing just one ubiquitin tag may straighten and refold correctly, but a protein

with more than one tag is taken to another cellular machine called a **proteasome (figure 10.20)**. A proteasome is a tunnel-like multi-protein structure. As a protein moves through the opening, it is stretched out, chopped up, and its peptide pieces degraded into amino acids, a little like a wood chipper. The amino acids may be recycled to build new proteins.

Proteasomes also destroy properly folded proteins that are in excess or no longer needed, perhaps because their job is done. For example, a cell must dismantle excess transcription factors, or the genes that they control may remain activated or repressed for too long. Proteasomes also dismantle proteins from pathogens, such as viruses.

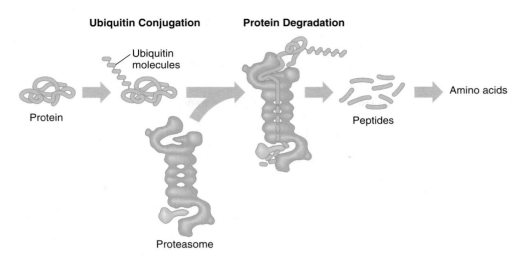

Ubiquitin Conjugation   **Protein Degradation**

Protein   Peptides   Amino acids

Proteasome

Figure 10.20 **Proteasomes provide quality control.** Ubiquitin binds to a misfolded protein and escorts it to a proteasome. The proteasome, which is composed of several proteins, encases the misfolded protein, straightening and dismantling it.

Most misfolded proteins are the result of errors in protein synthesis and processing. The primary structure (amino acid sequence) may be wild type or a functional variant, but the process goes awry as the proteins fold. Or, a mutation may alter the primary structure in a way that affects attractions and repulsions between parts of the protein. A clear example of protein misfolding in a single-gene disorder is sickle cell disease (see figure 12.2). Instead of the normal globular conformation of hemoglobin (see figure 11.1), the protein forms sheets that bend the red blood cells that contain them out of shape. Some mutations that cause cystic fibrosis prevent CFTR protein from folding and anchoring in the plasma membrane, where it normally controls the flow of chloride ions. Instead, the misfolded protein builds up in the cell. **Table 10.6** lists some disorders that, in some cases, involve protein misfolding.

In several disorders that affect the brain, misfolded proteins aggregate, forming masses that clog the proteasomes and block them from processing any malformed proteins. Different proteins are affected in different disorders. In Huntington disease, for example, extra glutamines in the protein huntingtin cause it to obstruct proteasomes. Misfolded proteins that clog proteasomes also form in the disorders listed in Table 10.6, but it isn't always clear whether the accumulated proteins cause the disease or are a response to it. Some of these disorders are discussed further in chapter 12.

Understanding how protein misfolding causes diseases can lead to development of new treatments, or recognition that existing drugs may help. This is the case for phenylketonuria (PKU).

PKU

The story of PKU began in Oslo, Norway, in 1934, when a mother of two children with extreme intellectual disability noticed that the children's soiled diapers had an odd, musty odor. She mentioned this to Ivar Fölling, a relative who was a physician and a biochemist. Intrigued, Fölling analyzed the children's urine and found excess phenylalanine, an amino acid usually scant in urine because an enzyme, phenylalanine hydroxylase (PAH), breaks it down. The children's cells lacked enzyme activity because the children had inherited PKU from their carrier parents. The buildup of phenylalanine affected their brains.

In 1963, a dietary treatment was devised that supplies protein in a synthetic food that is very low

Table 10.6	Disorders Associated with Protein Misfolding	
Disease	**Misfolded Protein**	**MIM (protein)**
Alzheimer disease	Amyloid beta precursor protein	104760
Familial amyotrophic lateral sclerosis	Superoxide dismutase	147450
Huntington disease	Huntingtin	143100
Parkinson disease	Alpha synuclein	163890
Lewy body dementia	Alpha synuclein	163890
PKU	Phenylalanine hydroxylase	261600
Prion disorders	Prion protein	176640

(All but Huntington disease are genetically heterogeneic; that is, abnormalities in different proteins cause similar syndromes.)

in phenylalanine. Begun at birth, the diet prevents symptoms. However, it is difficult to follow, and older children and teens often "cheat." When they eat protein, symptoms can surface, such as slowed reaction times, memory deficits, anxiety, depression, and irritability. In 2002, another treatment was found to help people with mild cases of PKU—the vitamin derivative tetrahydrobiopterin, or BH_4. This small molecule usually nestles into the four proteins that assemble to form PAH. In patients with mild cases of PKU, taking BH_4 increases enzyme levels, presumably restoring function in some of the enzyme. Researchers then discovered that even some patients with more serious PKU were helped with a drug version of BH_4. What was happening? Protein misfolding explains the success of the treatment.

Each of the four subunits of PAH, the enzyme thought to be missing in severe cases of PKU, may not actually be missing. Instead, enzyme molecules are so misfolded that they cannot work. When a person has a mutation, shown in the black outlines in **figure 10.21,** localized misfolding occurs. Because of the tertiary structure, the disturbance spreads to amino acids that are actually far away in the primary sequence, but close by when the protein is folded. Misfolding spreads until the entire, four-protein PAH enzyme can no longer function. The

peculiarities of the tertiary structure of this particular molecule enable it to assume any of several conformations, making it prone to misfolding. Taking BH_4 apparently stabilizes a functional form of the enzyme.

Prion Diseases

Several disorders that affect the brain reflect alternate folding of a glycoprotein called a prion, pronounced *pree-on* (*PRNP,* MIM 176640). Prion protein (PrP) can fold into any of at least eight conformations. One conformation is "infectious," causing other prions that it contacts to also assume the infectious form. Another distinguishing feature of the prion diseases is that the infectious forms can arise from the wild type primary structure.

Prion diseases are called transmissible spongiform encephalopathies (TSEs) (**figure 10.22** and **Reading 10.1**). In the 85 species in which TSEs are known, the brains become riddled with holes, resembling a sponge. Nerve cells die, neuroglia overgrow, and death occurs within 18 months of the first symptom. The location of the damage determines the specific symptoms, but most TSEs cause extreme weight loss, poor coordination, and dementia. One disease, fatal familial insomnia, causes death in months from inability to sleep. It also causes tremors, fever, sweats, muscle aches, joint pain, seizures, and a dreamlike trance.

TSEs were discovered in sheep, which develop a disease called scrapie when they eat prion-infected brains from other sheep. Ten percent of TSE cases in humans are inherited, such as Creutzfeldt-Jakob disease and Gerstmann-Straussler disease. Other cases are acquired from ingesting prions, or are sporadic, occurring without a family history and with no apparent environmental source. In acquired TSEs, exposure to prions in the infectious conformation triggers conversion of the person's own normal prions into the disease-causing conformation.

The "rules" by which DNA sequences specify protein conformations are still not well understood. Even though the central dogma remains true—genes are transcribed into mRNA, which, in turn, is translated into protein—many transcripts are not represented in the proteins encoded in the genome. What do they do? The next chapter offers some ideas.

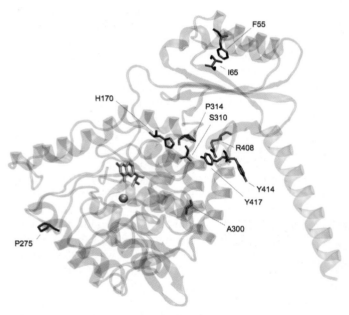

Figure 10.21 **Absence of activity of an enzyme (phenylalanine hydroxylase) causes PKU.** Intellectual disability results unless the person follows a diet that limits the amino acid phenylalanine, which the enzyme normally breaks down. In the illustration, the superimposed black regions are the parts where mutations affect the protein.

Key Concepts

1. Protein folding begins as translation proceeds, with enzymes and chaperone proteins assisting.
2. Misfolded proteins are tagged with ubiquitin and sent through a proteasome for dismantling.
3. A protein can fold in more than one way. Some conformations cause disease.

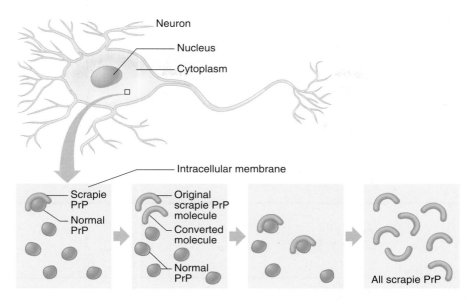

Neuron

Nucleus

Cytoplasm

Intracellular membrane

Scrapie PrP

Normal PrP

Original scrapie PrP molecule

Converted molecule

Normal PrP

All scrapie PrP

Figure 10.22 **Prions change shape.** A prion disease may begin when a single scrapie prion protein (PrP) contacts a normal PrP and changes it into the scrapie conformation. As the change spreads, disease results, usually with accumulated scrapie prion proteins clogging brain tissue.

Reading 10.1

Considering Kuru

A few rare and strange disorders are caused by *prions*—proteins that can infect. The first recognized prion disease of humans was kuru, which struck the Foré people in a remote mountainous area of Papua New Guinea **(figure 1)**. In the Foré language, *kuru* means to shake. The disease began with wobbling legs, quickly followed by trembling hands and fingers, and then body-wide shaking. An odd sign was uncontrollable laughter, leading to the nickname "laughing disease." Speech slurred and faded, thinking slowed, and after several months, the person could no longer walk or eat. Death typically came within a year.

One in 10 of the 35,000 tribe members had kuru. The fact that they were women or young children at first suggested that the disease might be inherited. Then D. Carleton Gajdusek, a physician who spent much of his lifetime studying the Foré, learned that the preparation of human brains for a cannibalism ritual probably passed on the infectious prions. People ate their relatives killed in battle to honor them. After the ritual was banned in 1959, the disease gradually disappeared. Gajdusek vividly described the Foré preparation of human brains at a time when he thought the cause was viral:

> Children participated in both the butchery and the handling of cooked meat, rubbing their soiled hands in their armpits or hair, and elsewhere on their bodies. They rarely or never washed. Infection with the kuru virus was most probably through the cuts and abrasions of the skin or from nose picking, eye rubbing, or mucosal injury.

Figure 1 **Kuru.** Kuru is a prion disease that affected the Foré people of New Guinea until they gave up a cannibalism ritual that spread an infectious form of prion protein.

(Continued)

Although kuru vanished, other prion diseases surfaced (**table 1**). In the 1970s and 1980s, several people acquired Creutzfeldt-Jakob disease (CJD). This time, the route of transmission was either through corneal transplants, in which infectious prions entered the brain through the optic nerve, or from human growth hormone taken from cadavers and used to treat short stature in children. The most familiar prion disease, "mad cow disease," caused CJD in more than 120 people in the United Kingdom since 1995. People likely acquired the infectious prions by eating infected beef.

Table 1	Prion Disorders
Disorder	**MIM #**
Creutzfeldt-Jakob disease	123400
Fatal familial insomnia	600072
Gerstmann-Straussler disease	137440

Summary

10.1 From DNA to Protein

1. A small part of the genome encodes protein. Much of the rest controls protein synthesis.

10.2 Transcription

2. Some DNA is **transcribed** into RNA, which is then **translated** into protein.

3. RNA is transcribed from the **template strand** of DNA. The other DNA strand is called the **coding strand.**

4. RNA is a single-stranded nucleic acid similar to DNA but containing uracil and ribose rather than thymine and deoxyribose.

5. **Messenger RNA** (mRNA) carries a protein-encoding gene's information. **Ribosomal RNA** (rRNA) associates with certain proteins to form ribosomes, which physically support protein synthesis. **Transfer RNA** (tRNA) is cloverleaf-shaped, with a three-base **anticodon** that is complementary to mRNA on one end and bonds to a particular amino acid on the other end.

6. Operons control gene expression in bacteria. In more complex organisms, **transcription factors** regulate which genes are transcribed in a particular cell type.

7. Transcription begins when transcription factors help **RNA polymerase** (RNAP) bind to a gene's **promoter.** RNAP then adds RNA nucleotides to a growing chain, in a sequence complementary to the DNA template strand.

8. After a gene is transcribed, the mRNA receives a "cap" of modified nucleotides at the 5' end and a poly A tail at the 3' end.

9. Many genes do not encode information in a continuous manner. After transcription, **exons** are translated into protein and **introns** are removed. Introns may outnumber and outsize exons. **Alternate splicing** increases protein diversity.

10.3 Translation of a Protein

10. Each three consecutive mRNA bases form a **codon** that specifies a particular amino acid. The **genetic code** is the correspondence between each codon and the amino acid it specifies. Of the 64 different possible codons, 60 specify amino acids, one specifies the amino acid methionine and "start," and three signal "stop."

11. Because 61 codons specify the 20 amino acids, more than one type of codon may encode a single amino acid. The genetic code is nonoverlapping, triplet, universal, and degenerate.

11. In the 1960s, researchers used logic and clever experiments using synthetic RNAs to decipher the genetic code.

12. Translation requires tRNA, ribosomes, energy-storage molecules, enzymes, and protein factors. An initiation complex forms when mRNA, a small ribosomal subunit, and a tRNA carrying methionine join. The amino acid chain elongates when a large ribosomal subunit joins the small one. Next, a second tRNA binds by its anticodon to the next mRNA codon, and its amino acid bonds with the first amino acid. Transfer RNAs add more amino acids, forming a polypeptide. The ribosome moves down the mRNA as the chain grows. The P site bears the amino acid chain, and the A site holds the newest tRNA. When the ribosome reaches a "stop" codon, it falls apart into its two subunits and is released. The new polypeptide breaks free.

13. After translation, some polypeptides are cleaved, have sugars added, or aggregate. The cell uses or secretes the protein.

10.4 Processing a Protein

14. A protein must fold into a particular **conformation** to be active and functional.

15. A protein's **primary structure** is its amino acid sequence. Its **secondary structure** forms as amino acids close in the primary structure attract one another. **Tertiary structure** appears as more widely separated amino acids attract or repel in response to water molecules. **Quaternary structure** forms when a protein consists of more than one polypeptide.

16. **Chaperone proteins** help conformation arise. Other proteins help new bonds form and oversee folding accuracy.

17. Ubiquitin attaches to misfolded proteins and escorts them to **proteasomes** for dismantling. Protein misfolding is associated with certain diseases.

18. Some proteins can fold into several conformations, some of which can cause disease.

19. At least one conformation of prion protein is infectious, causing transmissible spongiform encephalopathies.

Review Questions

1. Explain how complementary base pairing is responsible for
   a. the structure of the DNA double helix.
   b. DNA replication.
   c. transcription of RNA from DNA.
   d. the attachment of mRNA to a ribosome.
   e. codon/anticodon pairing.
   f. tRNA conformation.

2. A retrovirus has RNA as its genetic material. When it infects a cell, it uses enzymes to copy its RNA into DNA, which then integrates into the host cell's chromosome. Is this flow of genetic information consistent with the central dogma? Why or why not?

3. What are the functions of these proteins?
   a. RNA polymerase
   b. ubiquitin
   c. a chaperone protein
   d. a transcription factor

4. Explain where a hydrogen bond forms and where a peptide bond forms in the transmission of genetic information.

5. List the differences between RNA and DNA.

6. Where in a cell do DNA replication, transcription, and translation occur?

7. How does transcription control cell specialization?

8. How can the same mRNA codon be at an A site on a ribosome at one time, but at a P site at another time?

9. Describe the events of transcription initiation.

10. List the three major types of RNA and their functions.

11. Describe three ways RNA is altered after it is transcribed.

12. What are the components of a ribosome?

13. Why would an overlapping genetic code be restrictive?

14. Why would two-nucleotide codons be insufficient to encode the number of amino acids in biological proteins?

15. How are the processes of transcription and translation economical?

16. Explain how protein misfolding conditions and illnesses that result from abnormal transcription factors might each produce many different symptoms.

17. What factors determine how a protein folds into its characteristic conformation?

18. How do a protein's primary, secondary, and tertiary structures affect conformation? Which is the most important determinant of conformation?

Applied Questions

1. List the RNA sequence transcribed from the DNA template sequence TTACACTTGCTTGAGAGTC.

2. Reconstruct the corresponding DNA template sequence from the partial mRNA sequence GCUAUCUGUCAUAAAGAGGA.

3. List three different mRNA sequences that could encode the amino acid sequence

   histidine-alanine-arginine-serine-leucine-valine-cysteine.

4. Write a DNA sequence that would encode the amino acid sequence valine-tryptophan-lysine-proline-phenylalanine-threonine.

5. In the film *Jurassic Park,* which is about cloned dinosaurs, a cartoon character named Mr. DNA talks about the billions of genetic codes in DNA. Why is this statement incorrect?

6. Titin is a muscle protein named for its size—its gene has the largest known coding sequence of 80,781 DNA bases. How many amino acids long is it?

7. An extraterrestrial life form has a triplet genetic code with five different bases. How many different amino acids can this code specify, assuming no degeneracy?

8. In malignant hyperthermia, a person develops a life-threateningly high fever after taking certain types of anesthetic drugs. In one family, mutation deletes three contiguous bases in exon 44. How many amino acids are missing from the protein?

9. The protein that serves as a receptor that allows insulin to enter cells has a different number of amino acids in a fetus and in an adult. Explain how this may happen.

Web Activities

1. Go to http://www.mcb.harvard.edu/BioLinks/gencode.html. Scroll down to the lists of "noncanonical" codes in organisms other than humans. (*Noncanonical* means it differs from the universal genetic code.) Find three examples of deviations from the universal code, and list what the codon-amino acid assignment is in most organisms. (Replace the T's on the website with the U's to correspond to the genetic code chart in the textbook.)

2. Use the Web to find out how the ubiquitin-proteasome system is overtaxed or disabled in a neurodegenerative disease such as Alzheimer disease, Parkinson disease, Huntington disease, amyotrophic lateral sclerosis, or Lewy body dementia. (Find websites for these disorders and discuss how the mechanism involves proteasomes.)

Case Studies and Research Results

1. Five patients meet at a clinic for families in which several members have early-onset Parkinson disease. This condition causes rigidity, tremors, and other motor symptoms. Only 2 percent of cases of Parkinson disease are inherited. The five patients all have mutations in a gene that encodes the protein parkin, which has 12 exons. For each patient, indicate whether the mutation shortens, lengthens, or does not change the size of the protein.

   a. Manny Filipo's *parkin* gene is missing exon 3.
   b. Frank Myer's *parkin* gene has a duplication in intron 4.
   c. Theresa Ruzi's *parkin* gene lacks six contiguous nucleotides in exon 1.
   d. Elyse Fitzsimmon's *parkin* gene has an altered splice site between exon 8 and intron 8.
   e. Scott Shapiro's *parkin* gene is deleted.

2. A research project called ENCODE (ENCyclopedia Of DNA Elements) took 1 percent of the human genome and cataloged 487 genes and 2,608 mRNA transcripts. Why isn't the number the same?

3. Kabuki syndrome takes its name from the resemblance of the face to those of actors in the Japanese form of theater called Kabuki. People with this condition also have intellectual disability and birth defects. The syndrome is rare, affecting 1 in 32,000 births worldwide. Researchers identified causative genes using whole exome sequencing. What does this mean?

The Allies dropped food over the Netherlands, stopping the Dutch Hunger Winter in just two days. Starvation before birth led to schizophrenia years later by altering gene expression.

Gene Expression and Epigenetics

Learning Outcomes

11.1 Gene Expression Through Time and Tissue

1. Define *epigenetics*.

2. Explain how globin chain switching, development of organs, and the types of proteins cells make over time illustrate gene expression.

11.2 Control of Gene Expression

3. Explain how the interaction of methyl groups with histone proteins controls gene expression.

4. Explain how microRNAs control transcription.

11.3 Maximizing Genetic Information

5. Explain how division of genes into exons and introns maximizes the number of encoded proteins.

11.4 Most of the Human Genome Does *Not* Encode Protein

6. Discuss how viral DNA, noncoding RNAs, and repeated sequences account for large proportions of the human genome.

The Big Picture: Discovering the nature of the genetic material, determining the structure of DNA, cracking the genetic code, and sequencing the human genome were steps on the way toward today's challenge: deciphering how the information in the human genome is accessed and used, through tissue and time.

The Dutch Hunger Winter

"Nature versus nurture" implies that genes and the environment work separately, but this is not true. Environmental conditions can greatly affect gene expression, and this can affect health. For example, starvation before birth can alter gene expression in a way that shows up as schizophrenia years later.

From February through April 1945, the "Dutch Hunger Winter," the Nazis blocked all food supplies from entering six large cities in western Holland. As malnutrition weakened and killed people, a cruel experiment took place. The children starved before birth were much more likely to develop schizophrenia years later than their siblings born in better times. The key factor in setting the stage for future poor health was not birth weight, as had been thought, but exposure to dangerous conditions during the first weeks of pregnancy.

In the ongoing Dutch Famine Study, researchers at Columbia University and Leiden University in the Netherlands discovered the link between prenatal malnutrition and schizophrenia because they knew the exact time of the starvation, and the exact calorie intake, from food ration records. They obtained the schizophrenia diagnoses from psychiatric registries and military induction records.

Prenatal nutrition affects an adult phenotype because starvation alters the pattern in which methyl groups (CH_3) bind DNA, selectively silencing genes. Schizophrenia apparently develops in people born into famine when one particular gene has too few methyl groups and is overexpressed in the brain.

11.1 Gene Expression Through Time and Tissue

A genome is like an orchestra. Just as not all of the musical instruments play with the same intensity at every moment, not all genes are expressed continually at the same levels. Before the field of genomics began in the 1990s, the study of genetics proceeded one gene at a time, like hearing the separate contributions of a violin, a viola, and a flute. Many genetic investigations today, in contrast, track the crescendos of gene activity that parallel events in an organism's life. This new view has introduced the element of time to genetic analysis. Unlike the gene maps of old, which ordered genes linearly on chromosomes, new types of maps are more like networks that depict the timing of gene expression in unfolding programs of development and response to the environment.

The discoveries of the 1950s and 1960s on DNA structure and function answered some questions about the control of gene expression while raising many more. How does a bone cell "know" to transcribe the genes that control the synthesis of collagen and not to transcribe genes that specify muscle proteins? What causes the proportions of blood cell types to shift into leukemia? How do chemical groups "know" to shield DNA from transcription in one circumstance, yet expose it in others?

Changes to the chemical groups that associate with DNA greatly affect which parts of the genome are accessible to transcription factors and under which conditions. Such changes to the molecules that bind to DNA that are transmitted to daughter cells when the cell divides are termed **epigenetic,** which means "outside the gene."

Epigenetic changes do not alter the DNA base sequence, although they are passed from one cell generation to the next. These changes may affect the next generation if the conditions to which a fetus is exposed become dangerous. This is what happened to the survivors of the Dutch Hunger Winter described in the chapter opener. For a few sites in the genome, an epigenetic change may persist through meiosis to a third generation, but this appears to be rare. Specific classes of proteins and RNA molecules carry out epigenetic changes. Much of the genome encodes these modifiers of gene expression.

This chapter looks at how cells access the information in DNA. We begin with three examples of gene expression at the molecular, tissue, and organ levels: (1) hemoglobin switching during development; (2) the composition of blood plasma, and; (3) specialization of the two major parts of the pancreas.

Globin Chain Switching

The globin proteins transport oxygen in the blood. They vividly illustrate control of gene expression because they assemble into different hemoglobin molecules

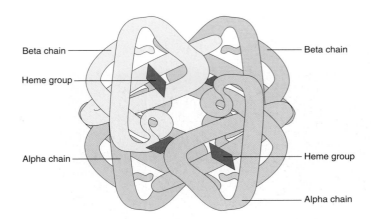

Figure 11.1 **The structure of hemoglobin.** A hemoglobin molecule is made up of two globular protein chains from the beta (β) globin group and two from the alpha (α) globin group. Each globin surrounds an iron-containing chemical group called a heme.

depending upon stage of development (**figures 11.1** and **11.2**). A hemoglobin molecule in the blood of an adult has four polypeptide chains, each wound into a globular conformation. Two of the chains are 146 amino acids long and are called "beta" (β). The other two chains are 141 amino acids long and are termed "alpha" (α).

The subunits of the hemoglobin molecule are replaced as the oxygen concentration in the body changes. This depends upon whether oxygen arrives to an embryo or fetus through the placenta or to a newborn's lungs from breathing. The chemical basis for "globin chain switching" is that different polypeptides attract oxygen molecules to different degrees. In the embryo, as the placenta forms, hemoglobin consists first

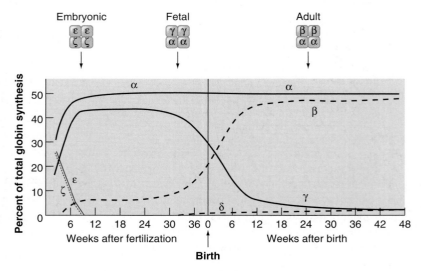

Figure 11.2 **Globin chain switching.** The subunit composition of human hemoglobin changes as the concentration of oxygen in the environment changes. With the switch from the placenta to the newborn's lungs to obtain oxygen, beta (β) globin begins to replace gamma (γ) globin.

of two epsilon (ε) chains, which are in the beta globin group, and two zeta (ζ) chains, which are in the alpha globin group. About 4 percent of the hemoglobin in the embryo includes beta chains. This percentage gradually increases.

As the embryo develops into a fetus, the epsilon and zeta chains decrease in number, as gamma (γ) and alpha chains accumulate. Hemoglobin consisting of two gamma and two alpha chains is called fetal hemoglobin. The gamma globin subunits bind very strongly to oxygen released from maternal red blood cells into the placenta, so that fetal blood carries 20 to 30 percent more oxygen than an adult's blood. As the fetus matures, beta chains gradually replace the gamma chains. At birth, however, the hemoglobin is not fully of the adult type—fetal hemoglobin (two gamma and two alpha chains) comprises from 50 to 85 percent of the blood. By 4 months of age, the proportion drops to 10 to 15 percent, and by 4 years, less than 1 percent of the child's hemoglobin is the fetal form.

Building Tissues and Organs

The globin chains affect one type of molecule, hemoglobin. Gene expression also changes on a larger scale, seen as the different types and amounts of proteins in particular tissues. For example, blood plasma, the liquid portion of blood, contains about 40,000 different types of proteins. Ten types of proteins account for 90 percent of all the plasma protein molecules, and nearly half of those are one type, albumin. Many thousands of types of proteins make up the rest, and are present in very small amounts. If conditions change, such as a person contracting an infection or having an allergic reaction, the protein profile of the plasma can change dramatically. This ability of the tissue to adapt to a changing environment is possible because of changes in gene expression—that is, how much of each protein is made.

Blood is a structurally simple tissue that is easy to obtain and study. A solid gland or organ, constructed from specialized cells and tissues, is much more complex. Its solid organization must be maintained throughout a lifetime of growth, repair, and changing external conditions.

Stem cell biology is shedding light on how genes are turned on and off during the development of an organ or gland. Researchers isolate individual stem cells and then see which combinations of growth factors, hormones, and other biochemicals must be added to steer development toward a particular cell type.

Consider the pancreas. It is a dual gland, with two types of cell clusters. The exocrine part releases digestive enzymes into ducts, whereas the endocrine part secretes polypeptide hormones that control nutrient use directly into the bloodstream. The endocrine cell clusters are called pancreatic islets.

The complexity of the pancreas unfolds in the embryo, when ducts form. Within duct walls reside rare stem cells and progenitor cells (see figure 2.22). A transcription factor is activated and controls expression of other genes in a way that stimulates some progenitor cells to divide. Certain daughter cells follow an exocrine pathway and will produce digestive enzymes **(figure 11.3)**. Other progenitor cells respond to different signals and divide to yield daughter cells that follow the endocrine pathway. The most familiar pancreatic hormone is insulin. Its absence, or the inability of cells to recognize it, causes diabetes mellitus. If pancreatic stem cells can be isolated and cultured, it might be possible to coax a person with diabetes to produce new and functional pancreatic beta cells.

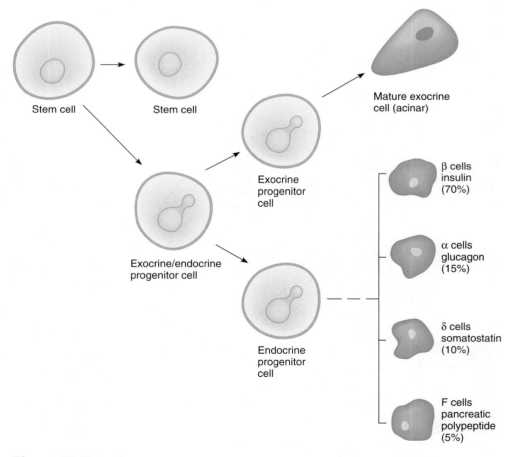

Figure 11.3 Building a pancreas. A single type of stem cell theoretically gives rise to an exocrine/endocrine progenitor cell that in turn divides to yield more restricted progenitor cells that give rise to both mature exocrine and endocrine cells. The endocrine progenitor cell in turn divides to give rise to cells that are specialized to produce particular hormones.

Proteomics

A more complete portrait of gene expression emerges through **proteomics,** which is an area of genetics that identifies and analyzes all the proteins made in a cell, tissue, gland, organ, or body. **Figure 11.4** depicts a global way of comparing the relative contributions of major categories of proteins from conception through birth and from conception through old age. The differences in proteins made at these times make sense. Transcription factors are more abundant before birth because of the extensive cell differentiation of this period, as organs form. During the prenatal period, enzymes are less abundant, perhaps because the fetus receives some enzymes through the placenta. Immunoglobulins appear after birth, when the immune system begins to function.

Another way to look at the proteome is by specific functions, which has led to the creation of various "ome" words. Genes whose encoded proteins control lipid synthesis, for example, constitute the "lipidome," and those that monitor carbohydrates form the "glycome." "Omics" designations are helpful in sorting out the thousands of proteins a human cell can manufacture. However, identifying proteins is only a first step. The next hurdle is to determine how proteins with related functions interact—forming "interactomes."

Gene expression profiles for different cell types under various conditions can provide valuable medical information and are the basis for many new tests that assess risk, diagnose disease, or monitor response to treatment. For example, 55 genes are overexpressed and 480 underexpressed in cells of a prostate cancer that has a very high likelihood of spreading, but not in a prostate cancer that will not spread. A test based on such findings assists physicians in deciding which patients can safely delay or avoid invasive and risky treatment.

Key Concepts

1. Gene expression patterns change over time and in different cell types.
2. The subunit composition of hemoglobin changes in the embryo, fetus, and after birth.
3. As a pancreas forms, progenitor cells diverge from shared stem cells and their daughter cells specialize.
4. Proteomics tracks all of the proteins in a cell, tissue, organ, or organism under specific conditions.

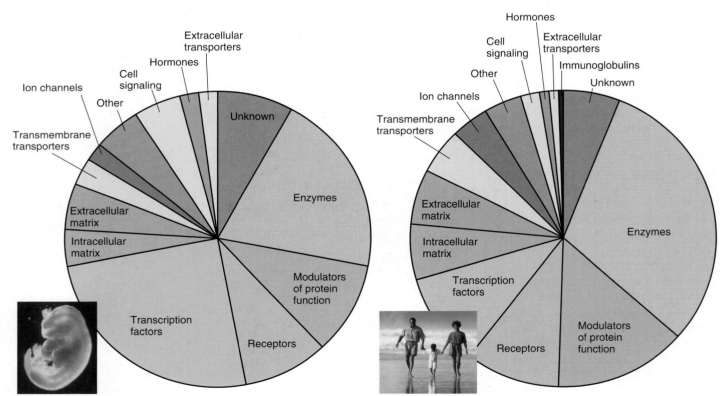

a. Distribution of health-related proteins from conception to birth

b. Distribution of health-related proteins from conception through old age

Figure 11.4 Proteomics meets medicine. We can categorize genes by their protein products, and chart the relative abundance of each class at different stages of development. The pie chart in **(a)** considers 13 categories of proteins that when abnormal or missing cause disease, and their relative abundance from conception to birth. The pie chart in **(b)** displays the same protein categories from conception to old age, plus one activated after birth, the immunoglobulins.

11.2 Control of Gene Expression

Purchasing a smartphone with many applications but no clear way to download them can be very frustrating. Fortunately, this is not true of the human genome. Embedded in the blueprints to build a human body are several types of instructions for its use.

A protein-encoding gene contains controls of its own expression. One is the promoter sequence. Recall from chapter 10 that the promoter is part of the 3′ end of a gene where RNA polymerase and transcription factors bind, marking the start point of transcription. Variations in the promoter sequence of a gene can affect how quickly the encoded protein is synthesized. For example, a form of early-onset Alzheimer disease is caused by a mutation in the promoter for the gene that encodes amyloid precursor protein (see Reading 5.1). In people who have the mutation, the sticky protein accumulates in the brain twice as fast as normal because it isn't cleared quickly enough. A second way that expression of a gene can exceed normal pace is if a person has more than one copy of it.

Control of gene expression happens in several steps. In **chromatin remodeling,** the histone proteins associated with DNA interact with other chemical groups in ways that expose some sections of DNA to transcription factors and shield other sections, blocking their expression. Later in the protein production process, small RNAs called, appropriately, **microRNAs,** bind the 3′ ends of certain mRNAs, preventing their translation into protein. Overall, these two processes determine the ebb and flow of different proteins, enabling cells to adapt to changing conditions.

Chromatin Remodeling

Recall that DNA associates with proteins and RNA to form chromatin, as figure 9.13 shows. For many years, biologists thought that the histone proteins that wind long DNA molecules into nucleosomes were little more than tiny spools. However, histones do much more. Enzymes add or delete small organic chemical groups to histones, affecting expression of the protein-encoding genes that histone proteins bind.

The three major types of small molecules that bind to histones are acetyl groups, methyl groups, and phosphate groups (**figure 11.5**). The key to the role histones play in controlling gene expression lies in acetyl groups (CH_3CO_2). They bind to very specific sites on certain histones, particularly to the amino acid lysine. The supposedly anti-aging sirtuin proteins, discussed

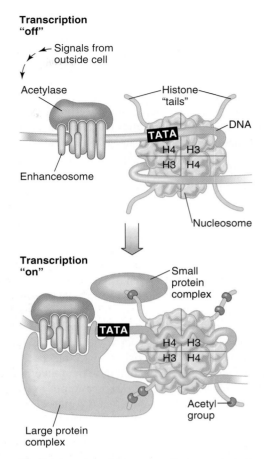

Transcription "off"

Transcription "on"

Figure 11.6 **Acetylated histones allow transcription to begin.** Once acetyl groups are added to particular amino acids in the tails of certain histones, the TATA box becomes accessible to transcription factors. H3 and H4 are histone types.

in chapter 3, act by controlling the placement of acetyl groups on histones to control gene expression.

Figure 11.6 shows how acetyl binding can shift histone interactions in a way that eases transcription. A series of proteins moves the histone complex away from the TATA box, exposing it enough for RNA polymerase to bind and transcription to begin (see figure 10.7). First, a group of proteins called an enhanceosome attracts the enzyme (acetylase) that adds acetyl groups to specific lysines on specific histones, which neutralizes their positive charge. Because DNA carries a negative charge, this change moves the histones away from the DNA, making room for transcription factors to bind and begin transcription. Enzymes called deacetylases remove acetyl groups, which shuts off gene expression.

Methyl groups (CH_3) are also added to or taken away from histones. When CH_3 binds to a specific amino acid in a specific histone type, a protein is attracted that shuts the DNA off. As CH_3 groups are added, methylation spreads from the tail of one histone to the adjacent histone, propagating the gene silencing. Methyl groups also control gene expression by binding to cytosines at about 16,000 places in the genome. The "methylome" is the collection of all the methylated sites in

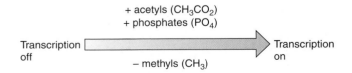

Figure 11.5 **Chromatin remodeling.** Chromatin remodeling adds or removes certain organic chemical groups to or from histones. The pattern of binding controls whether the DNA wrapped around the histones is transcribed or not.

the genome. Different cell types have distinctive subsets of the methylome.

The modified state of chromatin can be passed on when DNA replicates. These changes in gene expression are heritable from cell generation to cell generation, but they do not alter the DNA base sequence—that is, they are epigenetic. Addition and removal of acetyl groups, methyl groups, or phosphate groups are examples of epigenetic changes. Effects of methylation can sometimes be seen when MZ (identical) twins inherit the same disease-causing genotype, but only one twin is sick. The reason for the discordance may be different patterns of methylation of the gene.

Enzymes that add or delete acetyl, methyl, and phosphate groups must be in a balance that controls which genes are expressed and which are silenced. Upset this balance, and disease can result. In a blood cancer called mixed lineage leukemia (see figure 18.16), for example, a single abnormal protein binds to more than 150 genes and alters their associated chromatin. Among these genes are several that normally stimulate frequent cell division in the stem cells that give rise to blood cells. In the leukemia, these overexpressed genes send the affected white blood cells back in developmental time, to a state in which the rapid cell division causes cancer. One limitation to altering chromatin remodeling to treat inherited disease is that this action could affect the expression of many genes—not just the one implicated in the disease. **Table 11.1** lists some disorders that result from abnormal chromatin remodeling.

MicroRNAs

Chromatin remodeling determines which genes are transcribed. MicroRNAs act later in gene expression, preventing the translation of mRNA transcripts into protein. If chromatin remodeling is considered as an on/off switch to transcription, then the control of microRNAs is more like that of a dimmer switch—fine-tuning gene expression at a later stage. **Figure 11.7** schematically compares chromatin modeling and the actions of microRNAs, and **figure 11.8** places them in the overall flow of genetic information.

MicroRNAs are a type of noncoding RNA, so-named because they were not among the first three major classes of

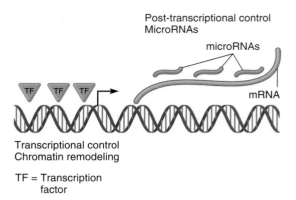

Figure 11.7 **Control of gene expression.** Chromatin opens to allow transcription factors to bind, whereas microRNAs bind to specific mRNAs, blocking their translation into protein.

RNA described (mRNA, tRNA, and rRNA). MicroRNAs are so small—just 21 or 22 bases long—that for many years, researchers accidentally threw them out when doing experiments searching for longer molecules. In the cell, microRNAs are cut from precursors.

The human genome has close to 1,000 distinct microRNAs that regulate at least one-third of the protein-encoding genes. The DNA sequences that encode microRNAs are found in protein-encoding parts of the genome as well as in the vast regions that do not encode protein. A typical human cell has from 1,000 to 200,000 microRNAs.

Each type of microRNA binds to parts of the initial control regions (corresponding to DNA promoters) of a particular set of mRNAs, by complementary base pairing. When a microRNA binds a "target" mRNA, it prevents translation. Because a single type of microRNA has many targets, it controls the expression of sets of genes. In turn, a single type of mRNA can bind several different microRNAs. Researchers use computational tools (bioinformatics) to analyze these complex interactions.

Within the patterns of microRNA function may lie clues to developing new ways to fight disease, because these controls of gene expression have stood the test of evolutionary time.

Table 11.1	Disorders of Chromatin Remodeling			
Disease	**MIM**	**Protein**	**Symptoms**	**Defect**
α-thalassemia	301040	ATRX	Anemia	Undermethylation of heterochromatin
ICF syndrome	242860	DNMT3B	Immunodeficiency, unstable centromeres, facial anomalies	Undermethylation of DNA repeats
Rett syndrome	312750	MECP2	Repetitive movements, irregular breathing, seizures, loss of motor control, neurodegeneration	Failure to remove acetyls from histones on gene *DLX5* expressed in brain
Rubinstein-Taybi syndrome	180849	CBP	Intellectual disability, short stature, facial anomalies	Adds acetyl groups to certain histones, causing inappropriate transcription of some genes

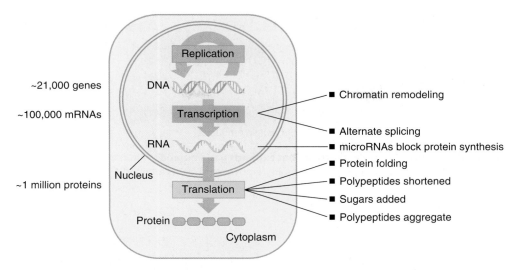

Figure 11.8 **A summary of the events of gene expression.** At the level of transcription, chromatin remodeling determines which genes are transcribed. Alternate splicing creates different forms of a protein by combining exons in different ways. MicroRNAs bind to the 3′ ends of mRNAs by complementary base pairing, blocking translation. At the level of translation, a protein must fold a certain way. Certain polypeptides must be shortened, attached to sugars, or aggregated.

That is, discovering the mRNAs that a microRNA targets may reveal genes that function together. This can suggest new uses for existing drugs.

The first practical applications of identifying specific microRNA activities are in cancer. Certain microRNAs are either more or less abundant in cancer cells than in healthy cells of the same type from which the cancer cells formed. Restoring the levels of microRNAs that normally suppress the too-rapid cell cycling of cancer, or blocking production of microRNAs that are too abundant in cancer, could help to return cells to normal. In a related technology called RNA interference (RNAi), small, synthetic, double-stranded RNA molecules are introduced into selected cell types. Here they block gene expression in the same manner as the naturally-occurring microRNAs.

Key Concepts

1. Acetyl, phosphate, and methyl groups bind to histone proteins, controlling transcription.
2. Acetyl and phosphate groups turn on transcription; methyl groups turn it off.
3. MicroRNAs bind to the control regions of specific mRNAs, blocking translation.

11.3 Maximizing Genetic Information

The human genome maximizes information in the 20,000 or so genes that encode about 100,000 mRNAs, which in turn specify more than a million proteins. Figure 11.8 depicts this increase in information from gene to RNA to protein on the left, and the mechanisms that maximize the information on the right. Maximization of genetic information occurs at the levels of transcription and translation.

Several events account for the fact that proteins outnumber genes. The "genes in pieces" pattern of exons and introns and alternate splicing make it possible for one store of information—the gene—to encode different versions of a protein, much as a few items of clothing can be assembled into many outfits by combining them in different ways (**figure 11.9**). The different proteins that result from different uses of the information in a gene are called isoforms. The driving force is circumstance. For example, when an infection begins, an immune system cell first secretes a short version on an antibody molecule that is presented on the cell's surface, where it alerts other cells. As the infection progresses, the cell transcribes an additional exon that extends the antibody in a way that enables it to be secreted into the bloodstream, where it attacks the pathogen.

Alternate splicing explains how a long sequence of DNA can specify more mRNAs than genes. On a part of chromosome 22, for example, 245 genes yield 642 mRNA transcripts. About 90 percent of all human genes are alternately spliced.

A DNA sequence that is an intron in one context may encode protein in another. Consider prostate specific antigen (PSA), which is a protein on certain cell surfaces that is overproduced in some prostate cancers. The gene for PSA has five exons and four introns. However, it is alternately spliced to encode a different protein, called PSA-linked molecule (PSA-LM), that consists of the first exon and the fourth intron. The two proteins (PSA and PSA-LM) work against one another. When the level of one is high, the other is low. Blood tests that measure levels of both proteins may more accurately assess the risk of developing prostate cancer than testing PSA alone.

Another way that introns may increase the number of proteins compared to genes is that a DNA sequence that is an

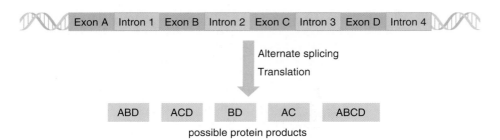

possible protein products

Figure 11.9 **Exons provide flexibility in gene structure that expands gene number.** Alternate splicing enables a cell to manufacture different versions of a protein by adding or deleting parts. Introns are removed and exons retained.

intron in one gene's template strand may encode protein on the coding strand. This is so for the gene for neurofibromin, which, when mutant, causes neurofibromatosis type 1, an autosomal dominant condition that causes benign tumors beneath the skin and spots on the skin surface. Within an intron of the gene, but on the coding strand, are instructions for three other genes. Finding such dual meaning in a gene is a little like reading a novel backwards and discovering a second story!

At the level of translation, information is maximized when a protein is modified into different forms by adding sugars or lipids to create glycoproteins and glycolipids. Another way that one gene can encode more than one protein is if the protein is cut in two. This happens in dentinogenesis imperfecta (MIM 125490), which is an autosomal dominant condition that causes discolored, misshapen teeth with peeling enamel (**figure 11.10**). The dentin, which is the bonelike substance beneath the enamel that forms the bulk of the tooth, is abnormal. Dentin is a complex mixture of extracellular matrix proteins, 90 percent of which are collagens. However, two proteins are unique to dentin: the abundant dentin phosphoprotein (DPP) and the rare dentin sialoprotein (DSP). A single gene encodes both. DPP and DSP are translated from a single mRNA molecule as the precursor protein dentin sialophosphoprotein (DSPP). DPP is more abundant because DSP is degraded much faster.

Key Concepts

1. Only a tiny proportion of the genome encodes protein, yet the number of proteins greatly outnumbers known protein-encoding genes.

2. Alternate splicing, introns that encode protein, and cutting a precursor protein maximize the number of proteins that DNA encodes.

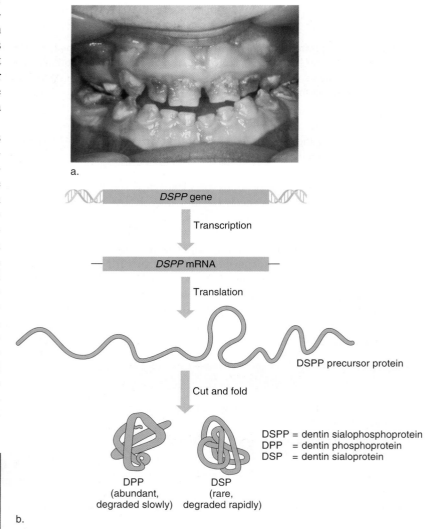

a.

DSPP precursor protein

DSPP = dentin sialophosphoprotein
DPP = dentin phosphoprotein
DSP = dentin sialoprotein

DPP (abundant, degraded slowly)   DSP (rare, degraded rapidly)

b.

Figure 11.10 **Another way to encode two genes in one.** **(a)** The misshapen, discolored, and enamel-stripped teeth of a person with dentinogenesis imperfecta were at first associated with deficiency of the protein DPP. Then researchers discovered that DSP is deficient, too, but is very scant. **(b)** Both DPP and DSP are cut from the same larger protein, but DSP is degraded faster.

11.4 Most of the Human Genome Does *Not* Encode Protein

Only 1.5 percent of human DNA encodes protein. The rest includes viral sequences, sequences that encode RNAs other than mRNA (called noncoding or ncRNAs), introns, promoters and other control sequences, and repeated sequences (**table 11.2**). In fact, most of the genome is transcribed—it isn't "junk." However, we do not yet know the functions of many transcripts.

Viral DNA

Our genomes include DNA sequences that represent viruses. Viruses are nonliving infectious particles that consist of a nucleic acid (DNA or RNA) encased in a protein coat (see Reading 17.1). A virus replicates using a cell's transcriptional and translational machinery to mass-produce itself. New viruses may exit the cell, or the viral nucleic acid remain in a host cell. A DNA virus may take over directly, inserting into a chromosome or remaining outside the nucleus in a circle called an episome. An RNA virus first uses an enzyme (reverse transcriptase) to copy its genetic material into DNA, which then inserts into a host chromosome.

About 100,000 sequences in our DNA, of varying length and comprising about 8 percent of the genome, were once a type of RNA virus called a retrovirus. The name refers to a retrovirus' direction of genetic information transfer, which is opposite DNA to RNA to protein. Retroviral sequences in our chromosomes are termed "endogenous" because they are carried from generation to generation of the host, rather than acquired as an acute infection. The retroviruses whose genetic material is in our chromosomes are called human endogenous retroviruses, or HERVs.

By comparing HERV sequences to similar viruses in other primates, researchers traced HERVs to a sequence representing a virus that infected our ancestors' genomes about 5 million years ago. Since then, HERV sequences have exchanged parts (recombined) and mutated to the extent that they no longer make us sick. Harmless HERVs silently pass from human generation to generation as parts of our chromosomes. They increase in number with time, as **figure 11.11** shows.

Noncoding RNAs

More of the human genome is transcribed than would be predicted based on the number and diversity of proteins that a human body can produce—in fact, nearly all of it. The two general classes of RNAs are the coding RNAs (the mRNAs) and the noncoding RNAs (ncRNAs), which include everything else. The best-studied noncoding RNAs are the tRNAs and rRNAs, because they were discovered first.

The rate of transcription of a cell's tRNA genes is attuned to cell specialization. The proteins of a skeletal muscle cell, for example, require different amounts of certain amino acids than the proteins of a white blood cell, and therefore different amounts of the corresponding tRNAs, too. Human tRNA genes are dispersed among the chromosomes in clusters. Altogether, our 500 or so types of tRNA genes account for 0.1 percent of the genome.

Table 11.2	Some Nonprotein-Encoding Parts of the Human Genome
Type of Sequence	**Function or Characteristic**
Viral DNA	Evidence of past infection
Noncoding RNA genes	
tRNA genes	Connect mRNA codon to amino acid
rRNA genes	Parts of ribosomes
Pseudogenes	DNA sequences very similar to known genes that are not translated
Small nucleolar RNAs (snoRNAs)	Process rRNA in nucleolus
Small nuclear RNAs (snRNAs)	Parts of spliceosomes
Telomerase RNA	Adds bases to chromosome tips
Xist RNA	Inactivates one X chromosome in cells of females
Introns	Parts of genes that are cut out of mRNA
Promoters and other control sequences	Guide enzymes that carry out DNA replication, transcription, or translation
Small interfering RNAs (siRNAs)	Control transcription
MicroRNAs (miRNAs)	Control transcription of many genes
Repeats	
Transposons	Repeats that move around the genome
Telomeres	Chromosome tips that control the cell cycle
Centromeres	Largest constriction in a chromosome, providing attachment points for spindle fibers
Duplications of 10 to 300 kilobases	Unknown
Simple short repeats	Unknown

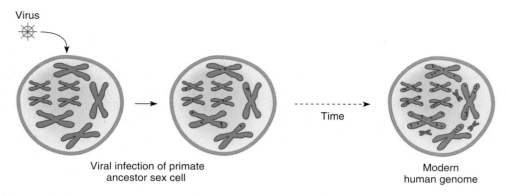

Virus

Viral infection of primate
ancestor sex cell

Time

Modern
human genome

Figure 11.11 **The human genome includes viral DNA sequences.** Most, if not all, of them do not harm us.

The 243 types of rRNA genes are grouped on six chromosomes, each cluster harboring 150 to 200 copies of a 44,000-base repeat sequence. Once transcribed from these clustered genes, the rRNAs go to the nucleolus, where another type of ncRNA called small nucleolar RNA (snoRNA) cuts them into their final forms.

Hundreds of thousands of ncRNAs are neither tRNA nor rRNA, nor snoRNAs, nor microRNAs, nor the other less abundant types described in table 11.2. Instead, they are transcribed from DNA sequences called **pseudogenes**. A pseudogene is very similar in sequence to a particular protein-encoding gene, and it may be transcribed into RNA, but it is not translated into protein. Presumably a pseudogene is altered in sequence from an ancestral gene in a way that impairs its translation—perhaps the encoded amino acids cannot fold into a functional protein. Pseudogenes may be remnants of genes past, variants that diverged from the normal sequence too greatly to encode a working protein.

Repeats

The human genome is riddled with highly repetitive sequences that may be a different type of information than a protein's amino acid sequence. Perhaps repeat size or number constitute another type of molecular language. Or, perhaps some types of repeats help to hold a chromosome together.

The most abundant type of repeat is a sequence of DNA that can move about the genome. It is called a transposable element, or **transposon** for short. Geneticist Barbara McClintock originally identified transposons in corn in the 1940s, and they were rediscovered in bacteria in the 1960s. Transposons comprise about 45 percent of the human genome sequence, and typically repeat in many copies. Some transposons include parts that encode enzymes that cut them out of one chromosomal site and integrate them into another. Unstable transposons may lie behind inherited diseases that have several symptoms,

because they insert into different genes. This is the case for Rett syndrome (see Reading 6.2).

An example of a specific type of repeat is an Alu sequence. Each Alu repeat is about 300 bases long, and a human genome may contain 300,000 to 500,000 of them. Alu repeats comprise 2 to 3 percent of the genome, and they have been increasing in number over time because they can copy themselves. We don't know exactly what these common repeats do, if anything. They may serve as attachment points for proteins that bind newly replicated DNA to parental strands before anaphase of mitosis, when replicated chromosomes pull apart.

Rarer classes of repeats comprise telomeres, centromeres, duplications of 10,000 to 300,000 bases, copies of pseudogenes, and simple repeats of one, two, or three bases. In fact, the entire human genome may have duplicated once or even twice.

Repeats may make sense in light of evolution, past and future. Pseudogenes are likely vestiges of genes that functioned in our nonhuman ancestors. Perhaps the repeats that seem to have no obvious function today will serve as raw material from which future genes may arise.

Discovery of the intricate controls of gene expression has led to a new definition of a gene, greatly expanded from the one-gene, one-protein idea of years past. A gene is a DNA sequence that contributes to a phenotype or function, plus the sequences, both in the gene and outside it, that control its expression. The next chapter looks at different types of changes in the DNA sequence—mutations—and their consequences.

Key Concepts

1. Most of the genome encodes many types of RNA as well as viral sequences, introns, promoters, and other control sequences and repeats.
2. We do not know the functions of some repeats.

Summary

11.1 Gene Expression Through Time and Tissue

1. Changes in gene expression occur over time at the molecular level (globin switching), at the tissue level (blood plasma), and at the organ/gland level (pancreas development).

2. **Proteomics** catalogs the types of proteins in particular cells, tissues, organs, or entire organisms under specified conditions.

11.2 Control of Gene Expression

3. Acetylation of certain histones enables the transcription of associated genes. Phosphorylation and methylation are also important in **chromatin remodeling**.

4. **MicroRNAs** bind to certain mRNAs, blocking translation.

11.3 Maximizing Genetic Information

5. A small part of the genome encodes protein, but these genes specify a much greater number of proteins.

6. Alternate splicing, use of introns, and cutting proteins translated from a single gene contribute to protein diversity.

11.4 Most of the Human Genome Does *Not* Encode Protein

7. The nonprotein-encoding part of the genome includes viral sequences, noncoding RNAs, **pseudogenes,** introns, promoters and other controls, and repeats.

www.mhhe.com/lewisgenetics10

Answers to all end-of-chapter questions can be found at **www.mhhe.com/lewisgenetics10.** You will also find additional practice quizzes, animations, videos, and vocabulary flashcards to help you master the material in this chapter.

Review Questions

1. Why is control of gene expression necessary?

2. Define *epigenetics*.

3. Distinguish between the type of information that epigenetics provides and the information in the DNA sequence of a protein-encoding gene.

4. Describe three types of cells and how they differ in gene expression from each other.

5. Explain how a mutation in a promoter can affect gene expression.

6. What is the environmental signal that stimulates globin switching?

7. How does development of the pancreas illustrate differential gene expression?

8. How do histones control gene expression, yet genes also control histones?

9. Name a mechanism that silences transcription of a gene and a mechanism that blocks translation of an mRNA.

10. What controls whether histones enable DNA wrapped around them to be transcribed?

11. What are two ways that microRNA functioning is complex?

12. Describe three ways that the number of proteins exceeds the number of protein-encoding genes in the human genome.

13. How can alternate splicing generate more than one type of protein from the information in a gene?

14. In the 1960s, a gene was defined as a continuous sequence of DNA, located permanently at one place on a chromosome, that specifies a sequence of amino acids from one strand. List three ways this definition has changed.

15. Give an example of a discovery mentioned in the chapter that changed the way we think about the genome.

Applied Questions

1. Several new drugs inhibit the enzymes that either put acetyl groups on histones or take them off. Would a drug that combats a cancer caused by too little expression of a gene that normally suppresses cell division add or remove acetyl groups?

2. Chromosome 7 has 863 protein-encoding genes, but many more proteins. The average gene is 69,877 bases, but the average mRNA is 2,639 bases. Explain both of these observations.

3. CHARGE syndrome (MIM 214800) causes heart defects, visual problems, facial palsy, blocked nostrils, and difficulty swallowing. A mutation in a gene called *Chd1* causes the condition. The protein product of this gene recognizes and binds methyl groups on certain histones. Explain how this mutation leads to pleiotropy (multiple symptoms).

4. How many different proteins encompassing two exons can be produced from a gene that has three exons?

5. Many people with trisomy 21 Down syndrome (an extra chromosome 21; see section 13.3) who survive into adulthood develop early-onset Alzheimer disease. The APP gene which when mutant causes this form of Alzheimer disease is on chromosome 21. Explain how this form of Alzheimer disease in trisomy 21 individuals differs from the same disorder caused by a mutation in the APP promoter in a person who has the normal two copies of chromosome 21.

6. Researchers took pieces of the aorta (the largest blood vessel) from patients undergoing a heart transplant that were to be discarded as medical waste. They grew the pieces in lab dishes and added chemicals that stimulate inflammation, inducing atherosclerosis. The researchers identified the genes that were transcribed when the cells were treated, and concluded that about 1,000 genes are expressed in atherosclerosis. Explain how expression of a thousand genes differs from inheriting thousands of mutations.

Web Activities

1. Gene expression profiling tests began to be marketed just a few years ago. Google "Oncotype DX," "MammaPrint," or simply "gene expression profiling in cancer" and describe how classifying a particular cancer based on gene expression profiling can improve diagnosis and/or treatment. (Or apply this question to a different type of disease.)

Forensics Focus

1. Establishing time of death is critical information in a murder investigation. Forensic entomologists can estimate the "postmortem interval" (PMI), or the time at which insects began to deposit eggs on the corpse, by sampling larvae of specific insect species and consulting developmental charts to determine the stage. The investigators then count the hours backwards to estimate the PMI. Blowflies are often used for this purpose, but their three larval stages look remarkably alike in shape and color, and development rate varies with environmental conditions. With luck, researchers can count back 6 hours from the developmental time for the largest larvae to estimate the time of death.

   In many cases, a window of 6 hours is not precise enough to narrow down suspects when the victim visited several places and interacted with many people in the hours before death. Suggest a way that gene expression profiling might be used to more precisely define the PMI and extrapolate a probable time of death.

Case Studies and Research Results

1. Jerrold is 38 years old. His body produces too much of the hormone estrogen, which has enlarged his breasts. He had a growth spurt and developed pubic hair by age 5, and then his growth dramatically slowed so that his adult height is well below normal. He has a very high-pitched voice and no facial hair, which reflect the excess estrogen. Jerrold's son, Timmy, is 8 years old and has the same symptoms.

   Jerrold and Timmy have an overactive gene for aromatase, an enzyme required to synthesize estrogen. Five promoters control expression of the gene in different tissues, and each promoter is activated by a different combination of hormonal signals. The five promoters lead to estrogen production in skin, fat, brain, gonads (ovaries and testes) and placenta. In premenopausal women, the ovary-specific promoter is highly active, and estrogen is abundant. In men and postmenopausal women, however, only small amounts of estrogen are normally produced, in skin and fat. The father and son have a wild type aromatase gene, but high levels of estrogen in several tissues, particularly fat, skin, and blood. They do, however, have a mutation that turns around an adjacent gene so that the aromatase gene falls under the control of a different promoter. Suggest how this phenotype arises.

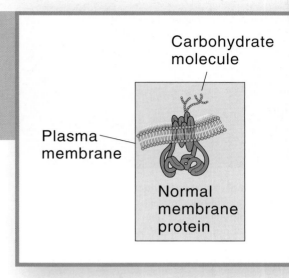

Carbohydrate
molecule

Plasma
membrane

Normal
membrane
protein

The cystic fibrosis transmembrane regulator protein—CFTR—must not only have the correct amino acid sequence, but fold properly and be able to reach the cell surface.

Gene Mutation

Learning Outcomes

12.1 The Nature of Mutations

  1. Distinguish between mutation and mutant.

  2. Distinguish between mutation and polymorphism.

12.2 Two Mutations

  3. Describe mutations in the beta globin gene and in collagen.

12.3 Causes of Mutation

  4. Explain the chemical basis of a spontaneous mutation.

  5. Describe other ways that mutations occur.

12.4 Types of Mutations

  6. Describe the two types of single-base mutations.

  7. Explain the consequences of a splice-site mutation.

  8. Discuss mutations that add, remove, and move DNA nucleotides.

12.5 The Importance of Position

  9. Give examples of how the location of a mutation in a gene affects the phenotype.

  10. Describe a conditional mutation.

12.6 DNA Repair

  11. What types of damage do DNA repair mechanisms counter?

  12. Describe the types of DNA repair.

> **The Big Picture:** Mutations provide the variation necessary for life to persist. Usually DNA repair protects against harmful mutations, but some mutations are helpful.

Cystic Fibrosis Revisited: Counteracting a Mutation

Treating the symptoms of CF has come a long way since the techniques described in Reading 4.2. Breaking up mucus and fighting infection extend life but they do not correct the underlying problem—an ion channel protein that can be abnormal in a number of ways. Most often the protein is so misfolded that it cannot make its way from inside a lining cell to the surface of the cell, where it helps control what enters and leaves. In about 5 percent of cases, the ion channel protein (CFTR) folds properly and reaches the cell's surface, but then the channel opens only fleetingly.

Pharmaceutical researchers designed a chemical that keeps the ion channel open longer. Analysis of the sweat glands and nasal secretions of people who have taken the drug in a clinical trial show that on the cellular level the ion channels are working, allowing certain charged atoms (ions) into and out of the cells. The patients can force more air out of their lungs, a sign of improved function.

For the 70 percent of patients with CF who have the delta F508 mutation, new drugs can unfold and refold the errant ion channels in a way that restores some lung function. Perhaps some day many patients can take a daily "cocktail" of these new drugs, fighting their disease at the cellular source. Because newborns are routinely tested for CF, it may soon be possible to use these new drugs to prevent symptoms.

12.1 The Nature of Mutations

A **mutation,** used as a noun, is a change in a DNA sequence that is present in less than 1 percent of individuals in a population. Used as a verb, "mutation" or "mutate" refers to the process of altering a DNA sequence. Mutations range in size from alteration of a single DNA base; to deletion or duplication of tens, hundreds, thousands, or even millions of bases; to missing or extra entire chromosomes. This chapter discusses smaller-scale mutations, and chapter 13 considers mutation at the chromosomal level. However, these are artificial distinctions. The extent of mutation is a continuum.

Mutation can affect any part of the genome. This includes sequences that encode proteins or control transcription; introns; repeats; and sites critical to intron removal and exon splicing. However, not all DNA sequences are equally likely to mutate.

The effects of mutation vary. Mutations may impair a function, have no effect, or even be beneficial. A deleterious (harmful) mutation can stop or slow production of a protein, overproduce it, or impair the protein's function—such as altering its secretion, location, or interaction with another protein. The effect of a mutation is called a "loss of function" when the gene's product is reduced or absent, or a "gain of function" when the gene's action changes in some way. Most mutations are recessive and cause a loss of function. Gain-of-function mutations tend to be dominant and are also called "toxic."

The terms *mutation* and *polymorphism* can be confusing, because both denote a genetic change. Recall from chapter 7 that a single nucleotide polymorphism, or SNP, is a single base change. So are many mutations. The distinction between mutation and polymorphism is largely artificial, reflecting frequency. A mutation is present in *less* than 1 percent of a population and a polymorphism is present in *more* than 1 percent of a population. If a genetic change greatly impairs health, individuals with it are unlikely to reproduce, and the mutant allele remains uncommon. A polymorphism that does not harm health, elevates risk of illness only slightly, or is even beneficial, will remain prevalent in a population or even increase in frequency. This chapter deals mostly with mutations that alter the phenotype.

Not all mutations are harmful, in contrast to their depiction in science fiction. For example, a mutation protects against HIV infection. About 1 percent of the general population is homozygous for a recessive allele that encodes a cell surface protein called CCR5 (see figure 17.12). To infect an immune system cell, HIV must bind CCR5 and another protein. Because the mutation prevents CCR5 from moving to the cell surface from inside the cell, HIV cannot bind. Heterozygotes for this mutation are partially protected against HIV infection.

The term *mutation* refers to genotype—that is, a change at the DNA or chromosome level. The familiar term **mutant** refers to phenotype. The nature of a mutant phenotype depends upon how the mutation affects the gene's product or activity, and usually connotes an abnormal or unusual characteristic. However, a mutant phenotype may also be an uncommon variant that is nevertheless "normal," such as red hair.

Figure 12.1 Animal models of human disease. This dog has amyotrophic lateral sclerosis (Lou Gehrig's disease), which also affects humans. Mutation in the same gene—superoxide dismutase 1—causes about 2 percent of human cases, as well as the canine cases. The dogs serve as models of the human disease. Chapter 19 discusses animal models that bear human genes.

In an evolutionary sense, mutation has been essential to life, because it produces individuals with variant phenotypes who are better able to survive specific environmental challenges, including illnesses. Our evolutionary relatedness to other species enables us to study many mutations in nonhuman species, which can provide information on our own (**figure 12.1**).

A mutation may be present in all the cells of an individual or just in some cells. In a **germline mutation,** the change occurs during the DNA replication that precedes *meiosis*. The resulting gamete and all the cells that descend from it after fertilization have the mutation—that is, every cell in the body. In contrast, a **somatic mutation** happens during DNA replication before a *mitotic* cell division. All the cells that descend from the original changed cell are altered, but they might only comprise a small part of the body. Somatic mutations are more likely to occur in cells that divide often, such as hair root cells, because there are more opportunities for replication errors. Reading 18.1 and figure 18.4 concern somatic mutations and cancer.

12.2 Two Mutations

Identifying how a mutation causes symptoms has clinical applications, and also reveals the workings of biology. Following are two examples of well-studied mutations that cause disease.

The Beta Globin Gene Revisited

The first genetic illness understood at the molecular level was sickle cell disease (**figure 12.2**). In 1904, young medical intern Ernest Irons noted "many pear-shaped and elongated forms" in a blood sample from a dental student in Chicago who had anemia. Irons sketched this first view of sickle cell disease at the cellular level, and reported his findings to his supervisor,

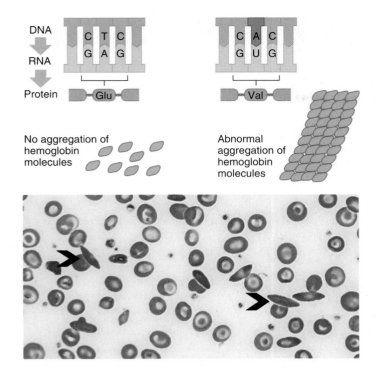

Figure 12.2 **Sickle cell disease results from a single DNA base change that substitutes one amino acid in the protein (valine replaces glutamic acid).** This changes the surfaces of the molecules, and they aggregate into long, curved rods that deform the red blood cell. Arrows in the photograph point to sickled cells.

physician James Herrick. Alas, Herrick published the work without including Irons and has been credited with the discovery ever since.

In 1949, Linus Pauling discovered that hemoglobin from healthy people and from people with the anemia, when placed in a solution in an electrically charged field moved to different positions. Hemoglobin from the parents of people with the anemia, who were carriers, moved to both positions.

The difference between the two types of hemoglobin lay in beta globin. Recall from figure 11.1 that adult hemoglobin consists of two alpha polypeptide subunits and two beta subunits. Protein chemist V. M. Ingram took a shortcut to localize the mutation in the 146-amino-acid-long beta subunit. He cut normal and sickle hemoglobin with a protein-digesting enzyme, separated the pieces, stained them, and displayed them on filter paper. The patterns of fragments—known as peptide fingerprints—were different for the two types of hemoglobin. This meant, Ingram deduced, that the two molecules differ in amino acid sequence. Then he discovered the difference. One piece of the molecule in the fingerprint, fragment four, occupied a different position for each of the two types of hemoglobin. Because this peptide was only 8 amino acids long, Ingram needed to decipher only that short sequence to find the site of the mutation. It was a little like finding which sentence on a page contains a typographical error.

Ingram identified the tiny mutation responsible for sickle cell disease: a substitution of the amino acid valine for the glutamic acid that is normally the sixth amino acid in the beta globin polypeptide chain. At the DNA level, the change was even smaller—a CTC to a CAC, corresponding to RNA codons GAG and GUG. This was learned after researchers deciphered the genetic code. The valine at this position changes the surfaces of hemoglobin molecules so that in low-oxygen conditions they attach at many more points than they would if the wild type glutamic acid were at the site. The aggregated hemoglobin molecules form ropelike cables that bend red blood cells into rigid, fragile, sickle-shaped structures. The misshapen cells lodge in narrow blood vessels, cutting off local blood supplies. Once a blockage occurs, sickling speeds up and spreads, as the oxygen level falls. The result is great pain in the blocked body parts, particularly the hands, feet, and intestines. The bones ache, and depletion of normal red blood cells causes the great fatigue of anemia.

Sickle cell disease was the first inherited illness linked to a molecular abnormality, but it wasn't the first known condition that results from a mutation in the beta globin gene. In 1925, Thomas Cooley and Pearl Lee described severe anemia in Italian children, and in the decade following, others described a milder version of "Cooley's anemia," also in Italian children. The disease was named thalassemia, from the Greek for "sea," in light of its high prevalence in the Mediterranean area. The two disorders turned out to be the same. The severe form, sometimes called thalassemia major, results from a homozygous mutation in the beta globin gene. The milder form, called thalassemia minor, affects some individuals who are heterozygous for the mutation.

Once researchers had worked out the structure of hemoglobin, and learned that different globins function in the embryo and fetus, the molecular basis of thalassemia became clear. The disorder that is common in the Mediterranean is more accurately called beta thalassemia (MIM 141900), because the symptoms result from too few beta globin chains. Without them, not enough hemoglobin molecules are assembled to effectively deliver oxygen to tissues. Fatigue and bone pain arise during the first year of life as the child depletes fetal hemoglobin, and the "adult" beta globin genes are not transcribed and translated on schedule.

As severe beta thalassemia progresses, red blood cells die because the excess of alpha globin chains prevents formation of hemoglobin molecules. Liberated iron slowly destroys the heart, liver, and endocrine glands. Periodic blood transfusions can control the anemia, but they hasten iron buildup and organ damage. Drugs called chelators that entrap the iron can extend life past early adulthood, but they are very costly and not available in developing nations.

Disorders of Orderly Collagen

Much of the human body consists of the protein collagen, which is a major component of connective tissue. Collagen accounts for more than 60 percent of the protein in bone and cartilage and provides 50 to 90 percent of the dry weight of skin, ligaments, tendons, and the dentin of teeth. Collagen is in parts of the eyes and the blood vessel linings, and it separates cell types in tissues.

Table 12.1		Some Collagen Disorders	
Disorder	**MIM**	**Genetic Defect (Genotype)**	**Signs and Symptoms (Phenotype)**
Alport syndrome	203780	Mutation in type IV collagen interferes with tissue boundaries	Deafness and inflamed kidneys
Aortic aneurysm	100070	Missense mutation substitutes Arg for Gly in α1 gene	Aorta bursts
Chondrodysplasia	302950	Deletion, insertion, or missense mutation replaces Gly with bulky amino acids	Stunted growth, deformed joints
Dystrophic epidermolysis bullosa	226600	Collagen fibrils that attach epidermis to dermis break down	Skin blisters on any touch
Ehlers-Danlos syndrome	130050	Missense mutations replace Gly with bulky amino acids; deletions or missense mutations disrupt intron/exon splicing	Stretchy, easily scarred skin, lax joints
Osteoarthritis	165720	Missense mutation substitutes Cys for Arg in α1 gene	Painful joints
Osteogenesis imperfecta type I	166200	Inactivation of α allele reduces collagen triple helices by 50%	Easily broken bones; blue eye whites; deafness
Stickler syndrome	108300	Nonsense mutation in procollagen	Joint pain, degeneration of vitreous gel and retina

Genetic control of collagen synthesis and distribution is complex; more than thirty-five collagen genes encode more than twenty types of collagen molecules. Other genes affect collagen, too. Mutations in the genes that encode collagen, not surprisingly, lead to a variety of medical problems (**table 12.1**). These disorders are particularly devastating, not only because collagen is nearly everywhere, but because collagen has an extremely precise conformation that is easily disrupted, even by slight alterations that might have little effect in proteins with other shapes (**figure 12.3**).

Collagen is sculpted from a longer precursor molecule called procollagen, which consists of many repeats of the amino acid sequence glycine-proline-modified proline. Three procollagen chains entwine. Two of the chains are identical and are encoded by one gene, and the other is encoded by a second gene and therefore has a different amino acid sequence.

The electrical charges and interactions of these amino acids with water coil the procollagen chains into a very regular triple helix, with space in the middle only for tiny glycine. The ragged ends of the polypeptides are snipped off by enzymes to form mature collagen. The collagen fibrils continue to associate with each other outside the cell, building the fibrils and networks that hold the body together. So important is the precision of collagen formation that a mutation that controls placement of a single hydroxyl chemical group ($-OH^-$) on collagen causes a severe form of osteogenesis imperfecta ("brittle bone disease").

The boy shown in **figure 12.4** has a form of Ehlers-Danlos syndrome. A mutation prevents his procollagen chains from being cut, and, as a result, collagen molecules cannot assemble. Instead, they form ribbonlike fibrils that lack the tensile strength to keep the skin from becoming too stretchy.

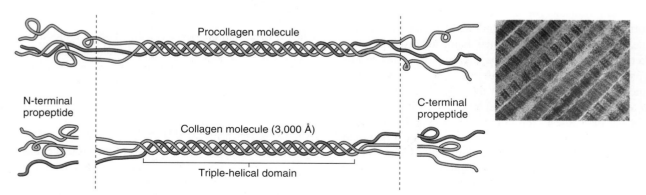

Figure 12.3 **Collagen has a very precise conformation (inset).** The α1 collagen gene encodes the two blue polypeptide chains, and the α2 procollagen gene encodes the third (red) chain. The procollagen triple helix is shortened before it becomes functional, forming the fibrils and networks that comprise much of the human body.

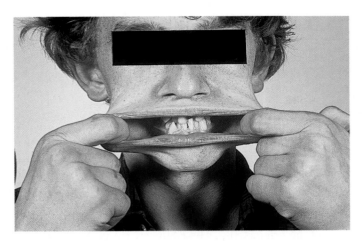

Figure 12.4 **A disorder of connective tissue produces stretchy skin.** A mutation that blocks trimming of procollagen chains to produce collagen causes the stretchy skin of Ehlers-Danlos syndrome type I.

Other collagen mutations cause missing procollagen chains, kinks in the triple helix, and defects in aggregation outside the cell.

Aortic aneurysm is a serious connective tissue disorder. It is part of Marfan syndrome. Detection of mutations that cause Marfan syndrome before symptoms arise can be lifesaving, because frequent ultrasound exams can detect aortic weakening early enough to patch the vessel before it bursts. **Table 12.2** describes how a few other mutations impair health.

Allelic Disorders

Geneticists can be inconsistent when assigning disease names to mutations. For one gene, different mutations may cause differing degrees or different subsets of symptoms of one syndrome. Yet for another gene, different mutations cause different disorders. For example, all mutations in the *CFTR* gene cause cystic fibrosis, which may include the full spectrum of impaired breathing and digestion, or just male infertility or frequent bronchitis. CF can affect different tissues in different individuals. Yet different mutations in the beta globin gene cause sickle cell disease and beta thalassemia. They have been considered two different disorders, although they both affect the same molecule in the blood.

Adding to the inconsistency in distinguishing diseases is a gene such as *lamin A*. Mutations in *lamin A* cause different disorders that affect very different tissues. They include the rapid-aging disorder Hutchinson-Gilford progeria syndrome (see figure 3.22 and table 3.3), muscular dystrophies, and a heart disorder. Lamin A proteins form a network beneath the inner nuclear membrane that interacts with chromatin. Different mutations affect *lamin A*'s interactions with chromatin in ways that cause the diverse associated disorders.

As researchers discovered more cases of different diseases arising from mutations in the same gene, it became clear that the phenomenon is not unusual, and is not merely a matter of how we name diseases. Different disease phenotypes caused by mutations in the same gene are termed **allelic disorders**.

The same gene can underlie different diseases in different ways (**table 12.3**). A pair of allelic disorders may result from

Table 12.2	How Mutations Cause Disease			
Disorder	**MIM**	**Protein**	**Genetic Defect (Genotype)**	**Signs and Symptoms (Phenotype)**
Cystic fibrosis	602421	Cystic fibrosis transmembrane regulator (CFTR)	Missing amino acid or other defect alters conformation of chloride channels in certain epithelial cell plasma membranes. Water enters cells, drying out secretions.	Frequent lung infection, pancreatic insufficiency
Duchenne muscular dystrophy	310200	Dystrophin	Deletion eliminates dystrophin, which normally binds to inner face of muscle cell plasma membranes, maintaining cellular integrity. Cells and muscles weaken.	Gradual loss of muscle function
Familial hypercholesterolemia	143890	LDL receptor	Deficient LDL receptors cause cholesterol to accumulate in blood.	High blood cholesterol, early heart disease
Hemophilia A	306700	Factor VIII	Absent or deficient clotting factor causes hard-to-control bleeding.	Slow or absent blood clotting
Huntington disease	143100	Huntingtin	Extra bases in the gene add amino acids to the protein product, which impairs certain transcription factors and proteasomes.	Uncontrollable movements, personality changes
Marfan syndrome	154700	Fibrillin or transforming growth factor β receptor	Deficient connective tissue protein in lens and aorta.	Long limbs, weakened aorta, spindly fingers, sunken chest, lens dislocation
Neurofibromatosis type 1	162200	Neurofibromin	Defect in protein that normally suppresses activity of a gene that causes cell division.	Benign tumors of nervous tissue beneath skin

Table 12.3 Allelic Diseases

Gene	Function	Associated Diseases
ATP7A	Copper transport	Menkes ("kinky hair") disease; peripheral neuropathy
DMD	Dystrophin muscle protein	Duchenne and Becker muscular dystrophy
FBN1	Encodes fibrillin-1, which forms tiny fibrils outside cells; a connective tissue protein	Marfan syndrome; stiff skin syndrome (scleroderma)
FGFR3	Fibroblast growth factor	2 types of dwarfs
GBA	Glucocerebrosidase	Gaucher disease; Parkinson disease
PSEN1	Presenilin 1 (enzyme part that trims membrane proteins)	Acne inversa; Alzheimer disease
RET	Oncogene (causes cancer)	Multiple endocrine neoplasia; Hirschsprung disease
TRPV4	Calcium channel	Peripheral neuropathy; spinal muscular atrophy

mutations in different parts of the gene; be localized (a single base change) or catastrophic (a missing gene); or alter the protein in ways that affect its interactions with other proteins.

An example of a gene that can be altered in different ways, causing different phenotypes, encodes presenilin 1. This protein forms part of an enzyme called a secretase that, when abnormal, cuts beta amyloid protein into the sticky fragments that accumulate to toxic levels in Alzheimer disease. The mutations that cause Alzheimer disease are all missense—that is, the protein is altered, not missing. However, nonsense and frameshift mutations in this gene cause an entirely different disorder, acne inversa. Unlike familiar acne that affects oil glands in the skin, acne inversa affects hair follicles and sweat glands, causing outbreaks in the groin, armpits, beneath the breasts and on the inner thighs. The presenilin mutations illustrate the "toxic gain of function" of missense mutations versus the "loss of function" mutations that shorten or remove the protein.

Key Concepts

1. Mutations add, delete, or rearrange genetic material in a germline cell or somatic cell.

2. In sickle cell disease, a mutation causes hemoglobin to crystallize in a low-oxygen environment, bending red blood cells into sickle shapes and impairing circulation. In beta thalassemia, beta globin is absent or scarce, depleting hemoglobin molecules.

3. Mutations in collagen genes can disrupt the protein's precise organization.

4. Mutations in a gene may cause either different versions of the same disease or distinct illnesses.

12.3 Causes of Mutation

A mutation can occur spontaneously or be induced by exposure to a chemical or radiation. An agent that causes mutation is called a **mutagen**.

Spontaneous Mutation

A spontaneous mutation can be a surprise. For example, two healthy people of normal height have a child with achondroplasia, an autosomal dominant form of dwarfism. How could this happen when no other family members are affected? If the mutation is dominant, why are the parents of normal height? The child has a genetic condition, but he did not inherit it. Instead, he originated it. His siblings have no higher risk of inheriting the condition than anyone in the general population, but each of his children will face a 50 percent chance of inheriting it. The boy's achondroplasia arose from a *de novo,* or new, mutation in a parent's gamete. This is a spontaneous mutation—that is, it is not caused by a mutagen. A spontaneous mutation usually originates as an error in DNA replication.

One cause of spontaneous mutation stems from the chemical tendency of free DNA bases to exist in two slightly different structures, called tautomers. For extremely short times, each base is in an unstable tautomeric form. If, by chance, such an unstable base is inserted into newly forming DNA, an error will be generated and perpetuated when that strand replicates. **Figure 12.5** shows how this can happen.

Spontaneous Mutation Rate

The spontaneous mutation rate varies for different genes. The gene that, when mutant, causes neurofibromatosis type 1 (NF1), for example, has a very high mutation rate, arising in 40 to 100 of every million gametes. NF1 affects 1 in 3,000 births, about half in families with no prior cases. The gene's large size may contribute to its high mutability—there are more ways for its sequence to change, just as there are more opportunities for a misspelling to occur in a long sentence than in a short one. In contrast, the mutation that causes the clotting disease of the royal families of Europe, hemophilia B, happens only one to ten times out of every million gametes formed.

Sequencing of human genomes has revealed that each of us has about 175 spontaneously mutated alleles. Mitochondrial genes mutate at a higher rate than genes in the nucleus because they cannot repair DNA (see section 12.6).

Estimates of the spontaneous mutation rate for a particular gene are usually derived from observations of new, dominant conditions, such as achondroplasia in the boy. This is possible because a new dominant mutation is detectable simply by observing the phenotype. In contrast, a new recessive mutation would not be obvious until two heterozygotes produced a homozygous recessive offspring with a noticeable phenotype.

The spontaneous mutation rate for autosomal genes can be estimated using the formula: number of *de novo* cases/2X, where X is the number of individuals examined. The denominator

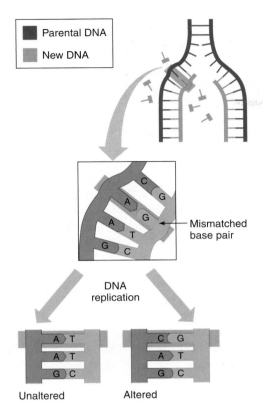

Parental DNA

New DNA

Mismatched base pair

DNA replication

Unaltered

Altered

Figure 12.5 **Spontaneous mutation.** DNA bases are very slightly chemically unstable, and for very short times they exist in alternate forms. If a replication fork encounters a base in its unstable form, a mismatched base pair can result. After another round of replication, one of the daughter cells has a different base pair than the one in the corresponding position in the original DNA. (This figure depicts two rounds of DNA replication.)

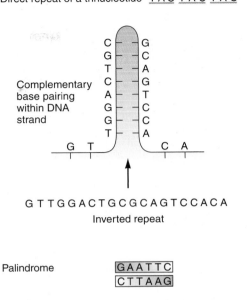

Repeat of a nucleotide   A A A A A A A A A

Direct repeat of a dinucleotide   G C G C G C G C

Direct repeat of a trinucleotide   T A C T A C T A C

Complementary base pairing within DNA strand

GTTGGACTGCGCAGTCCACA
Inverted repeat

Palindrome   GAATTC
              CTTAAG

Figure 12.6 **DNA symmetry may increase the likelihood of mutation.** These examples show repetitive and symmetrical DNA sequences that may "confuse" replication enzymes, causing errors.

has a factor of 2 to account for the nonmutated homologous chromosome.

Spontaneous mutation rates in human genes are difficult to assess because our generation time is long—usually 20 to 30 years. In bacteria, a new generation arises every half hour or so, and mutation is therefore much more frequent. The genetic material of viruses also spontaneously mutates rapidly.

Mutational Hot Spots

In some genes mutations are more likely to occur in regions called hot spots, where sequences are repetitive. It is as if the molecules that guide and carry out replication become "confused" by short repeated sequences, much as an editor scanning a manuscript might miss the spelling errors in the words "happiness" and "bananana" (**figure 12.6**).

The increased incidence of mutations in repeats has a physical basis. Within a gene, when DNA strands locally unwind to replicate in symmetrical or repeated sequences, bases located on the same strand may pair. A stretch of ATATAT might pair with TATATA elsewhere on the same strand, creating a loop that interferes with replication and repair enzymes, increasing risk of error. For example, mutations in the gene for clotting factor IX, which causes hemophilia B, occur 10 to 100

times as often at any of 11 sites in the gene that have extensive direct repeats of CG.

Small additions and deletions of DNA bases are more likely to occur near sequences called palindromes (figure 12.6). These sequences read the same, in a 5′ to 3′ direction, on complementary strands. Put another way, the sequence on one strand is the reverse of the sequence on the complementary strand. Palindromes probably increase the spontaneous mutation rate by disturbing replication.

The blood disorder alpha thalassemia (MIM 141800) illustrates the confusing effect of direct (as opposed to inverted) repeats of an entire gene. A person who does not have the disorder has four genes that specify alpha globin chains, two next to each other on each chromosome 16. Homologs with repeated genes can misalign during meiosis when the first sequence on one chromosome lies opposite the second sequence on the homolog. Crossing over can result in a sperm or oocyte that has one or three alpha globin genes instead of the normal two (**figure 12.7**). Fertilization with a normal gamete then results in a zygote with one extra or one missing alpha globin gene. At least three dozen conditions result from this unequal crossing over, including colorblindness (see Reading 6.1).

The number of alpha globin genes affects health. A person with only three alpha globin genes produces enough hemoglobin, and is a healthy carrier. Individuals with only two copies of the gene are mildly anemic and tire easily, and a person with a single alpha globin gene is severely anemic. A fetus lacking alpha globin genes does not survive.

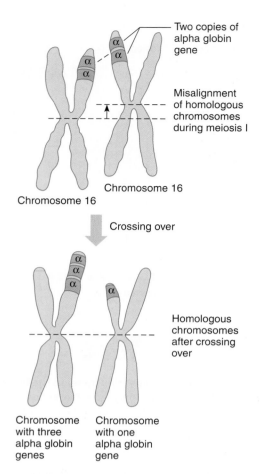

Two copies of alpha globin gene

Misalignment of homologous chromosomes during meiosis I

Chromosome 16

Chromosome 16

Crossing over

Homologous chromosomes after crossing over

Chromosome with three alpha globin genes

Chromosome with one alpha globin gene

Figure 12.7 Gene duplication and deletion. The repeated alpha globin genes are prone to mutation by mispairing during meiosis.

Induced Mutation

Researchers can infer a gene's normal function by observing what happens when mutation alters it. Because the spontaneous mutation rate is much too low to be a practical source of genetic variants for experiments, researchers make mutants, using mutagens on model organisms. These types of experiments can yield insights into human health.

Intentional Use of Mutagens

Chemicals or radiation are used to induce mutation. Alkylating agents, for example, are chemicals that remove a DNA base, which is replaced with any of the four bases—three of which are a mismatch against the complementary strand. Dyes called acridines add or remove a single DNA base. Because the DNA sequence is read three bases in a row, adding or deleting a single base can destroy a gene's information, altering the amino acid sequence of the encoded protein. Several other mutagenic chemicals alter base pairs, so that an A-T replaces a G-C, or vice versa. X rays and other forms of radiation delete a few bases or break chromosomes.

Researchers have developed several ways to test the mutagenicity of a substance. The best known, the Ames test, assesses how likely a substance is to harm the DNA of rapidly reproducing bacteria. One version of the test uses a strain of *Salmonella* that cannot grow when the amino acid histidine is absent from its medium. If exposure to a substance enables bacteria to grow on the deficient medium, then a gene has mutated that allows it to do so.

In a variation of the Ames test, researchers exposed human connective tissue cells growing in culture to liquefied cigarette smoke. The chemicals from the smoke cut chromosomes through both DNA strands. Broken chromosomes can join with each other in different ways that can activate cancer-causing genes. Hence, the experiment may have modeled one way that cigarettes cause cancer. Because many mutagens are also carcinogens (cancer-causing agents), the substances that the Ames test identifies as mutagens may also cause cancer.

A limitation of using a mutagen is that it cannot cause a specific mutation. In contrast, a technique called site-directed mutagenesis changes a gene in a desired way. A gene is mass-produced, but it includes an intentionally substituted base, just as an error in a manuscript is printed in every copy of a book. Site-directed mutagenesis is faster and more precise than waiting for nature or a mutagen to produce a useful variant. Common products that contain mutagens are hair dye, smoked meats, certain flame retardants used in children's sleepwear, and food additives.

Accidental Exposures to Mutagens

Some mutagen exposure is unintentional. This occurs from workplace contact before the danger is known; from industrial accidents; from medical treatments such as chemotherapy and radiation; from exposure to weapons that emit radiation; and from natural disasters that damage radiation-emitting equipment.

When a magnitude 9.0 earthquake struck Japan on March 11, 2011 and generated a tsunami, the upheaval damaged the cooling systems for several nuclear reactors, exposing surrounding populations to dangerous levels of radiation. To estimate the health effects that likely lie ahead for these people, experts are examining the long-term aftermath of another environmental disaster. On April 25, 1986, between 1:23 and 1:24 A.M., Reactor 4 at the Chernobyl Nuclear Power Station in Ukraine exploded, sending a great plume of radioactive isotopes into the air that spread for thousands of miles. The reactor had been undergoing a test, its safety systems temporarily disabled, when it overloaded and rapidly flared out of control. Twenty-eight people died of acute radiation exposure in the days following the explosion.

Acute radiation poisoning is not genetic. Evidence of a mutagenic effect is the increased rate of thyroid cancer among children who were living in nearby Belarus. Rates have multiplied tenfold. The thyroid glands of young people soak up iodine, which in a radioactive form bathed Belarus in the days after the explosion. In the days following the Japan earthquake of 2011, panicked people all over the world tried to take potassium iodide tablets to prevent radiation-induced thyroid cancer.

One way that researchers tracked mutation rates after the Chernobyl explosion was to compare the lengths of short DNA repeats, called minisatellite sequences, in children born in 1994 and in their parents, who lived in the Mogilev district of Belarus at the time of the accident and have remained there. Minisatellites are the same length in all cells of an individual. A minisatellite size in a child that does not match the size of either parent indicates that a mutation occurred in a parent's gamete. Such a mutation was twice as likely to occur in exposed families as in families living elsewhere. Because mutation rates of nonrepeated DNA sequences are too low to provide useful information on the effects of radiation exposure, investigators track minisatellites as a sensitive test of change.

Natural Exposure to Mutagens

Simply being alive exposes us to radiation that can cause mutation. Natural environmental sources of radiation account for 81% of our radiation exposure. They include cosmic rays, sunlight, and radioactive minerals in the earth's crust, such as radon. Contributions from medical X rays and occupational radiation hazards are comparatively minor. Job sites with increased radiation exposure include weapons facilities, research laboratories, health care facilities, nuclear power plants, and certain manufacturing plants (**figure 12.8**). Radiation exposure is measured in units called millirems; the average annual exposure in the northern hemisphere is 360 millirems.

Most of the potentially mutagenic radiation we are exposed to is ionizing, which means that it has sufficient energy to remove electrons from atoms. Unstable atoms that emit ionizing radiation exist naturally, and we make them. Ionizing radiation breaks the DNA sugar-phosphate backbone.

Ionizing radiation is of three major types. Alpha radiation is the least energetic and most short-lived, and the skin absorbs most of it. Uranium and radium emit alpha radiation. Beta radiation can penetrate the body farther, and emitters include tritium (a form of hydrogen), carbon-14, and strontium-70. Both

alpha and beta rays tend not to harm health, although they can do damage if inhaled or eaten. In contrast is the third type of ionizing radiation, gamma rays. These can penetrate the body, damaging tissues. Plutonium and cesium isotopes used in weapons emit gamma rays, and this form of radiation is used to kill cancer cells.

X rays are the major source of exposure to human-made radiation, and they are not a form of ionizing radiation. They have less energy and do not penetrate the body to the extent that gamma rays do.

The effects of radiation damage to DNA depend upon the functions of the mutated genes. Mutations in oncogenes or tumor suppressor genes, discussed in chapter 18, can cause cancer. Radiation damage can be widespread, too. Exposing cells to radiation and then culturing them causes a genome-wide destabilization, so that mutations may occur even after the cell has divided a few times. Cell culture studies have also identified a "bystander effect," when radiation harms cells not directly exposed.

Chemical mutagens are in the environment, too. Evaluating the risk that a specific chemical exposure will cause a mutation is very difficult, largely because people vary greatly in inherited susceptibilities, and are exposed to many chemicals. The risk that exposure to a certain chemical will cause a mutation is often less than the natural variability in susceptibility within a population, making it nearly impossible to track the true source and mechanism of any mutational event. Genetic tests can be used to determine specific inherited risks for specific employees who might encounter a mutagen in the workplace. However, in the United States, the Genetic Information Nondiscrimination Act (GINA) prevents employers from using such information to discriminate against employees.

Key Concepts

1. Genes have different mutation rates.
2. Spontaneous mutations result when rare bases are incorporated during replication.
3. Spontaneous mutations are more frequent in viruses and microorganisms because they reproduce often.
4. Mutations are more likely in repetitive DNA.
5. Mutagens are chemicals or radiation that increase the risk of mutation. Researchers use mutagens to more quickly obtain mutants, which reveal normal gene function. Site-directed mutagenesis creates and amplifies specific mutations.
6. Mutagens are encountered in the natural environment and may be released in industrial accidents.

Figure 12.8 Radiation and DNA. Exposure to radiation can cause mutation, whether from an unnatural source, such as this nuclear power plant, or a natural source, such as the sunlight that enables the field of rapeseed in the foreground to grow.

12.4 Types of Mutations

Mutations are classified by whether they remove, alter, or add a function, or by how they structurally alter DNA. The same single-gene disorder can result from different types of mutations, as **figure 12.9** shows for familial hypercholesterolemia.

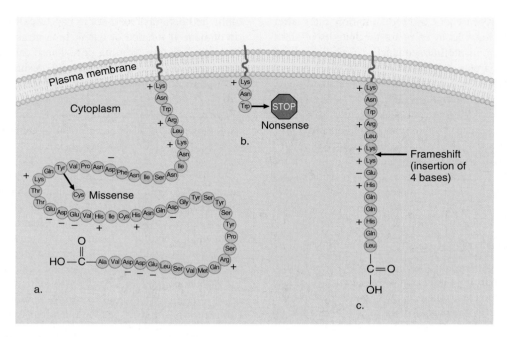

Figure 12.9 **Different mutations in a gene can cause the same disorder.** In familial hypercholesterolemia, several types of mutations alter the LDL receptor normally anchored in the plasma membrane. LDL receptor **(a)** bears a missense mutation—a cysteine substitutes for a tyrosine, bending the receptor enough to impair its function. The short LDL receptor in **(b)** results from a nonsense mutation, in which a stop codon replaces a tryptophan codon. In **(c),** a 4-base insertion alters the reading frame.

Table 12.4 summarizes types of mutations using an analogy to an English sentence.

Point Mutations

A **point mutation** is a change in a single DNA base. It is a **transition** if a purine replaces a purine (A to G or G to A) or a pyrimidine replaces a pyrimidine (C to T or T to C). It is a **transversion** if a purine replaces a pyrimidine or vice versa (A or G to T or C). A point mutation can have any of several consequences—or it may have no obvious effect at all on the phenotype, acting as a silent mutation.

Missense and Nonsense Mutations

A point mutation that changes a codon that normally specifies a particular amino acid into one that codes for a different amino acid is called a **missense mutation**. If the substituted amino acid alters the protein's conformation significantly or occurs at a site critical to its function, signs or symptoms of disease or an observable variant of a trait may result. About a third of missense mutations harm health.

The point mutation that causes sickle cell disease (see figure 12.2) is a missense mutation. The DNA sequence CTC encodes the mRNA codon GAG, which specifies glutamic acid. In sickle cell disease, the mutation changes the DNA

sequence to CAC, which encodes GUG in the mRNA, which specifies valine. This mutation changes the protein's shape, which alters its function.

A point mutation that changes a codon specifying an amino acid into a "stop" codon—UAA, UAG, or UGA in mRNA—is a **nonsense mutation**. A premature stop codon is

Table 12.4	Types of Mutations
A sentence comprised of three-letter words is analogous to the effect of mutations on a gene's DNA sequence:	
Normal	THE ONE BIG FLY HAD ONE RED EYE
Missense	THQ ONE BIG FLY HAD ONE RED EYE
Nonsense	THE ONE BIG
Frameshift	THE ONE QBI GFL YHA DON ERE DEY
Deletion	THE ONE BIG HAD ONE RED EYE
Insertion	THE ONE BIG WET FLY HAD ONE RED EYE
Duplication	THE ONE BIG FLY FLY HAD ONE RED EYE
Expanding mutation	
generation 1	THE ONE BIG FLY HAD ONE RED EYE
generation 2	THE ONE BIG FLY FLY FLY HAD ONE RED EYE
generation 3	THE ONE BIG FLY FLY FLY FLY HAD ONE RED EYE

one that occurs before the natural end of the gene. It shortens the protein product, which can greatly influence the phenotype. For example, in factor XI deficiency (MIM 264900), which is a blood clotting disorder, a GAA codon specifying glutamic acid is changed to UAA, signifying "stop." The shortened clotting factor cannot halt the profuse bleeding that occurs during surgery or from injury. Nonsense mutations are predictable by considering which codons can mutate to a "stop" codon.

In the opposite of a nonsense mutation, a normal stop codon mutates into a codon that specifies an amino acid. The resulting protein is longer than normal, because translation continues through what is normally a stop codon.

Point mutations can control transcription, affecting the quantity rather than the quality of a protein. For example, in 15 percent of people who have Becker muscular dystrophy (MIM 310200)—a milder adult form of the condition—the muscle protein dystrophin is normal, but its levels are reduced. The mutation is in the promoter for the dystrophin gene. This slows transcription, and dystrophin protein is scarce. The other 85 percent of individuals who have Becker muscular dystrophy have shortened proteins, not too few normal-length proteins.

Another way that point mutations can affect protein production is to disrupt the trimming of long precursor molecules. Such a mutation causes the type of Ehlers-Danlos syndrome that affects the boy shown in figure 12.4.

Splice-Site Mutations

A point mutation can greatly affect a gene's product if it alters a site where introns are normally removed from the mRNA. This is called a **splice-site mutation**. It can affect the phenotype if an intron is translated into amino acids, or if an exon is skipped instead of being translated, shortening the protein.

Retaining an intron is unusual because most introns have stop codons in all reading frames. However, if a stop codon is not encountered, a retained intron adds bases to the protein-coding part of an mRNA. For example, in a family with severe cystic fibrosis, a missense mutation alters an intron site so that it is not removed from the mRNA. The encoded protein is too bulky to move to its normal position in the plasma membrane.

A missense mutation can cause **exon skipping,** which removes a few amino acids. The mutation creates an intron splicing site where there should not be one, and an entire exon is "skipped" when the mRNA is translated into protein, as if it were an intron. An exon-skipping mutation is a deletion at the mRNA level, but a missense mutation at the DNA level. For example, a disorder called familial dysautonomia (MIM 223900) (FD) usually results from exon skipping in the gene encoding an enzyme necessary for the survival of certain neurons that control sensation and involuntary responses. The *In Their Own Words* box on page 222 describes life with FD.

A peculiarity of some disorders caused by exon skipping is that some cells ignore the mutation and manufacture a normal protein from the affected gene—after all, the amino acid sequence information is still there. Depending upon which cells actually make the encoded protein, the phenotype may be less severe than in individuals with the same disorder but with a different type of mutation in an exon.

Studies on cells from individuals who have or have died from FD reveal that the cells in which the exon is skipped are the cells that contribute to symptoms. That is, many cells from the brain and spinal cord skip the exon, but cells from muscle, lung, liver, white blood cells, and glands produce normal-length proteins. There may be a way to coax nervous system cells in affected children to also produce the protein. Current clinical trials are examining the ability of several natural compounds to restore normal processing of the FD gene's information. Some boys with Duchenne muscular dystrophy have splice-site mutations, and drugs that override these mutations may help them too.

Deletions and Insertions

In genes, the number 3 is very important, because triplets of DNA bases specify amino acids. Adding or deleting a number of bases that is not a multiple of three devastates a gene's function because it disrupts the gene's reading frame, which refers to the nucleotide position where the DNA begins to encode protein (see figure 10.14). Most exons are "readable" (have no stop codons) in only one of the three possible reading frames. A change that alters the reading frame is called a **frameshift mutation**.

A **deletion mutation** removes DNA. A deletion that removes three or a multiple of three bases will not cause a frameshift, but can still alter the phenotype. Deletions range from a single DNA nucleotide to thousands of bases to large parts of chromosomes. Many common inherited disorders result from deletions, including male infertility caused by tiny deletions in the Y chromosome.

An **insertion mutation** adds DNA and it, too, can offset a gene's reading frame. In one form of Gaucher disease, for example, an inserted single DNA base prevents production of an enzyme that normally breaks down glycolipids in lysosomes. The resulting buildup of glycolipid enlarges the liver and spleen and causes easily fractured bones and neurological impairment.

Another type of insertion mutation repeats part of a gene's sequence. The insertion is usually adjacent or close to the original sequence, like a typographical error repeating a word word. Two copies of a gene next to each other is a type of mutation called a **tandem duplication**. A form of Charcot-Marie-Tooth disease (MIM 118200), which causes numb hands and feet, results from a one-and-a-half-million-base-long tandem duplication.

Figure 12.9 compares the effects on protein sequence of missense, nonsense, and frameshift mutations in the gene that encodes the LDL receptor, causing familial hypercholesterolemia (see figure 5.2). These three mutations exert very different effects on the protein. A missense mutation replaces one amino acid with another, bending the protein in a way that impairs its function. A nonsense mutation greatly shortens the protein. A frameshift mutation introduces a section of amino acids that is not in the wild type protein.

Familial Dysautonomia: Rebekah's Story

Familial dysautonomia (FD) is a rare genetic disorder that affects the autonomic and peripheral nervous systems. It causes pneumonia, vomiting and retching, extremely high fevers, chills, rapid heartbeat, rashes, and seizures. FD also reduces sensation of pain, heat, and cold. Problems with balance and coordination include motor difficulties that affect feeding, swallowing, and breathing. Most people with FD have a feeding tube, and some have learning disabilities. Most individuals develop scoliosis, usually requiring corrective spine surgery. In short, FD affects every organ and system in the body.

FD is difficult to diagnose because so few physicians have ever seen it. Lynn Lieberman describes how doctors finally figured out why her daughter Rebekah was so sick. "After more than twelve local hospitalizations and a variety of tests, we traveled to a major children's teaching hospital, hoping that a fresh team of doctors would identify Rebekah's condition. One doctor knew immediately that she had FD. He recognized the pattern of 'dysautonomic crises.' Two more symptoms, which we hadn't even noticed, were diagnostic indicators. Individuals with FD do not cry tears, and they lack papillae (bumps) on the tip of the tongue. Our Eastern European, Jewish heritage was also a clue, because FD is one of a number of diseases primarily affecting this population."

Rebekah has done remarkably well, and describes her experience with FD.:

Being a girl with FD has many challenges. There are medical problems and so much more. Like most people with FD, I have had scoliosis. That's when your spine becomes all twisted. When I was in middle school, I had a very bad curve in my back and it affected my breathing and health. I have had surgery to fix my spine. I had to stay in the hospital for a while and it took a long time to recover. But, now, my spine is as straight as an arrow. Well, almost, anyway.

A lot of people with FD have problems with breathing. When I was younger, I had problems getting enough oxygen while I slept, which is called sleep apnea. I used oxygen while I was sleeping, but since I had my tonsils out, I haven't had to use the oxygen that much. Some of my friends with FD use oxygen at night, or even all the time. The first of them died when I was in the seventh grade. That is very hard.

I am not ashamed of having FD. I take my medication and do everything else I need to do so I can stay as healthy as possible. My friends with FD and I have a lot of things in common, so we understand what each other are going through. We are not treated like everyone else except by our close friends and family. Sometimes, when people don't really know us, they treat us like we are "special." We do not like that. We would just like to be treated like everyone else.

Like any other teenager, I like to go shopping, spend time with friends, and go to movies. I also like it when I go to camp so I am able to see all of my friends. I have trouble working in school, but I do my best. I am a senior this year, and I'm excited about graduating. I love working with children and someday I would love to own my own daycare. I want to go to college to study Early Childhood Education. So, I take one step at a time so I am able to make my goal."

© Rebekah Lieberman

Rebekah Lieberman loves children and writing.

Pseudogenes and Transposons Revisited

Recall from chapter 11 that a pseudogene is a DNA sequence that is very similar to the sequence of a protein-encoding gene. A pseudogene is not translated into protein, although it may be transcribed. The pseudogene may have descended from the original gene sequence, which was duplicated when DNA strands misaligned during meiosis, similar to the situation depicted in figure 12.7 for the alpha globin gene. When this happens, a gene and its copy end up right next to each other on the chromosome. The original gene or the copy then mutates to such an extent that it is no longer functional and becomes a pseudogene. Its duplicate lives on as the functional gene.

A pseudogene is not translated, but its presence can interfere with the expression of the functional gene and cause a mutation. For example, some cases of Gaucher disease result from a crossover between the working gene and its pseudogene, which has 96 percent of the same sequence and is located 16,000 bases away. The result is a fusion gene, which is a DNA sequence containing part of the functional gene and part of the pseudogene. The fusion gene does not retain enough of the normal gene sequence to enable the cell to synthesize the encoded enzyme, and Gaucher disease results. Gaucher disease is a lysosomal storage disease that causes fatigue, bruising, anemia, and weak bones. The phenotype is very variable, and for many patients, supplying the enzyme eliminates symptoms.

Chapter 11 also considered transposons, or "jumping genes." Transposons can alter gene function in several ways. They can disrupt the site they jump from, shut off transcription of the gene they jump into, or alter the reading frame of their destination if they are not a multiple of three bases. For example, a boy with X-linked hemophilia A had a transposon in his factor VIII gene—a sequence that was also in his carrier mother's genome, but on her chromosome 22. Apparently, in the oocyte, the transposon jumped into the factor VIII gene on the X chromosome, causing the boy's hemophilia.

Expanding Repeats

Myotonic dystrophy is an inherited disease that begins at an earlier age and causes more severe symptoms from one generation to the next. A grandfather might experience only mild weakness in his forearms, and cataracts. His daughter might have more noticeable arm and leg weakness, and a flat facial expression. Her affected children might experience severe muscle weakness.

For many years, the "anticipation"—the worsening of symptoms over generations—was thought to be psychological. Then, with the ability to sequence genes, researchers found that myotonic dystrophy indeed worsens with each generation because the gene expands! Myotonic dystrophy is caused by a type of mutation called an **expanding triplet repeat**. The gene, on chromosome 19, has an area rich in repeats of the DNA triplet CTG. A person who does not have myotonic dystrophy usually has from 5 to 37 copies of the repeat; a person with the disorder has from 50 to thousands of copies (**figure 12.10**).

Expanding triplet repeats have been discovered in more than fifteen human inherited disorders. Usually, a repeat number

Myotonic Dystrophy

Pedigree	Age of onset	Phenotype	Number of copies of GAC mRNA repeat
I 1 ○ 2 ■	Older adulthood	Mild forearm weakness, cataracts	50–80
II 1 ● 2 □	Mid-adulthood	Moderate limb weakness	80–700
III 1 ■ 2 ○ 3 ○	Childhood	Severe muscle impairment, respiratory distress, early death	700+

Figure 12.10 Expanding genes explain anticipation. In some disorders, symptoms that worsen from one generation to the next—termed *anticipation*—have a physical basis: The gene is expanding as the number of repeats grows.

of fewer than 40 copies is stably transmitted to the next generation and doesn't produce symptoms. Larger repeats are unstable, growing with each generation and causing symptoms that are more severe and begin sooner. **Reading 12.1** describes the triplet repeat disorder fragile X syndrome.

The mechanism behind triplet repeat disorders lies in the DNA sequence. The bases of the repeated triplets implicated in the expansion diseases, unlike others, bond to each other in ways that bend the DNA strand into shapes, such as hairpins. These shapes then interfere with replication, which causes the expansion. Once these repeats are translated, the extra-long proteins shut down cells in various ways:

- binding to parts of transcription factors that have stretches of amino acid repeats similar to or matching the expanded repeat;
- blocking proteasomes and thereby enabling misfolded proteins to persist; and
- directly triggering apoptosis.

Triplet repeat proteins may also enter the nucleus, even though their wild type versions function only in the cytoplasm, or vice versa.

The triplet repeat disorders are said to cause a "dominant toxic gain of function." This means that they cause something novel to happen, rather than removing a function, such as is often associated with recessive enzyme deficiencies. The idea of a gain of function arose from the observation that deletions of these genes do not cause symptoms. Particularly common among the triplet repeat disorders are the "polyglutamine diseases" that have repeats of the mRNA codon CAG, which encodes the amino acid glutamine.

For some triplet repeat disorders, the mutation blocks gene expression before a protein is even made. In myotonic dystrophy type 1, the expansion is in the initial untranslated region of a gene on chromosome 19, resulting in a huge mRNA. When genetic testing became available for the disorder, researchers discovered a second form of the illness in patients who had wild type alleles for the chromosome 19 gene. These people have myotonic dystrophy type 2, which is caused by a mutation in a second gene. It is an expanding *quadruple* repeat of CCTG in a gene on chromosome 3. Affected individuals have more than 100 copies of the repeat, compared to the normal maximum of 10 copies.

When researchers realized that this second repeat mutation for myotonic dystrophy was also in a non-protein-encoding part of the gene—an intron—a mechanism of disease became apparent: The mRNA is not processed normally and as a result cannot exit the nucleus. In myotonic dystrophy type 1, the excess material is added to the start of the gene; in type 2, it appears in an intron that is not excised. The bulky mRNAs bind to a protein that, in turn, alters intron splicing in several other genes. Deficiency of the proteins encoded by these final affected genes causes the symptoms.

A lesson learned from the expanding repeat disorders is that a DNA sequence is more than just one language that can be translated into another. Whether a sequence is random—CGT CGT ATG CAT CAG, for example—or

Fragile X Mutations Affect Boys and Their Grandfathers

Fragile X syndrome is the most common inherited form of intellectual disability and also accounts for 3 percent of all cases of autism. In the 1940s, geneticists thought that a gene on the X chromosome caused intellectual disability, which was then termed mental retardation, because more affected individuals were male. In 1969, a clue emerged to the genetic basis of X-linked intellectual disability. Two brothers with the condition and their mother had an unusual X chromosome. The tips at one chromosome end dangled by a thin thread (**figure 1a**). When grown in culture lacking folic acid, this part of the X chromosome was very prone to breaking—hence, the name fragile X syndrome. Worldwide, it affects 1 in 2,000 males, accounting for 4 to 8 percent of all males with intellectual disability. One in 4,000 females is affected. They usually have milder cases because their cells have normal X chromosomes too.

Youngsters with fragile X syndrome look normal, but MRI scans show brain differences as early as age 2. By young adulthood, their faces are very long and narrow, with long jaws and protruding ears. The testicles are very large. They may have learning disabilities, repetitive speech, hyperactivity, shyness, social anxiety, a short attention span, language delays, and temper outbursts.

Fragile X syndrome is inherited in an unusual pattern. The syndrome should be transmitted as any X-linked trait is, from carrier mother to affected son. However, penetrance is incomplete. One-fifth of males who inherit the chromosomal abnormality have no symptoms. Because they pass on the affected chromosome to all their daughters—half of whom have some mental impairment—they are called "transmitting males." A transmitting male's grandchildren may inherit fragile X syndrome.

A triplet repeat mutation causes fragile X syndrome. In unaffected individuals, the fragile X area contains about 30 repeats of the sequence CGG, in a gene called the fragile X mental retardation gene (*FMR1*). In people who have the fragile chromosome and show its effects, this region is expanded to 200 to 2,000 CGG repeats. Transmitting males, as well as females with mild symptoms, or who have affected sons, often have a premutation consisting of 50 to 200 repeats.

The *FMR1* gene encodes fragile X mental retardation protein (FMRP). This protein, when abnormal, binds to and disables several different mRNA molecules whose encoded proteins are crucial for brain neuron function.

Mysteries remain about fragile X syndrome. A distinct type of disorder is known in the maternal grandfathers of boys who have fragile X syndrome. Clinicians noticed that mothers of boys with fragile X syndrome very often reported the same symptoms in their fathers—tremors, balance problems, and then cognitive or psychiatric difficulties. The grandfathers were sometimes misdiagnosed with Parkinson disease due to the tremors. However, Parkinson's patients can walk a straight line, while the grandfathers cannot. The grandfathers' symptoms worsen with time and can lead to premature death (**table 1**).

Further investigation led to the description of the new condition, called fragile X-associated tremor/ataxia syndrome (FXTAS, MIM 300623). (Ataxia is poor balance and coordination.) The disorder has been studied in brains obtained after the grandfathers died and in a mouse model. Like the granddads, the mice are fine until middle age. Then they, too, develop tremors and balance problems as well as nervousness and memory impairment. Perhaps the symptoms of FXTAS arise from excess FMR1 mRNA, which attracts and disables other mRNAs.

The discovery of FXTAS has genetic counseling implications. As neurologists learn to distinguish this disorder from others, such as Parkinson disease, daughters can be counseled that they might pass on the condition to sons, and be offered testing.

Table 1	Prevalence of FXTAS in Grandfathers of Fragile X Syndrome Grandsons	
Age		**Prevalence**
50s		17%
60s		38%
80+		75%

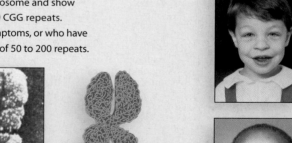

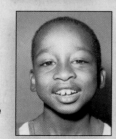
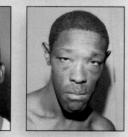

— Fragile site

a.

b.

Figure 1 **Fragile X syndrome.** A fragile site on the tip of the long arm of the X chromosome **(a)** is associated with intellectual disability and a characteristic long face that becomes pronounced with age **(b)**.

highly repetitive—such as CAG CAG CAG CAG and on and on—can affect transcription, translation, or the ways that proteins interact.

Copy Number Variants

In addition to differing slightly in our DNA sequences, we differ in the numbers of copies of particular DNA sequences. These sequences that vary in number from person to person are called **copy number variants (CNVs)**. Our genomes have hundreds to thousands of them, and they account for about a quarter of the genome. Sequencing the human genome missed CNVs because the technology used at that time detected any DNA sequence only once. It was a little like searching this book for the word "variant," and not the number of times it is used.

Copy number is a different form of information than DNA sequence differences. A language metaphor is useful to distinguish point mutations and single nucleotide polymorphisms (SNPs) from CNVs. If a wild type short sequence and a variant with two SNPs are written as:

The fat rat sat on a red cat   (wild type)

The fat rat sat in a red hat   (two SNPs)

then the sequence with two CNVs might be:

The fat fat rat sat on a red red red cat

CNVs may contribute significantly to the differences among us. A CNV can range in size from a few DNA bases to millions, and copies may lie next to each other on a chromosome ("tandem") or might be far away—even parts of other chromosomes. A duplication is a type of copy number variant.

CNVs may have no effect on the phenotype, or they can disrupt a gene's function and harm health. A CNV may have a direct effect by inserting into a protein-encoding gene and offsetting its reading frame, or have an indirect effect by destabilizing surrounding sequences. They are particularly common among people who have behavioral disorders, such as attention deficit hyperactivity disorder (ADHD), autism, and schizophrenia.

Detecting CNVs may be useful in health care even if we do not yet know exactly what they do, if specific CNVs correlate to specific phenotypes in many people. Such an approach is similar to genome-wide association studies based on SNPs—the more people examined, the more powerful the connection. **Figure 12.11** depicts, schematically, how CNVs correlated to cholesterol level might be used to guide medical advice.

12.5 The Importance of Position

The degree to which a mutation alters a phenotype depends upon where in the gene the change occurs, and how the mutation affects the conformation, activity, or abundance of an encoded protein. A mutation that replaces an amino acid with a very similar one would probably not affect the phenotype greatly, because it wouldn't substantially change the conformation of the protein. Even substituting a very different amino acid would not have much effect if the change is in part of the protein not crucial to its function.

The effects of specific mutations are well-studied in hemoglobin. They are less understood, but still fascinating, in the gene that encodes prion protein.

Globin Variants

Hundreds of globin gene mutations have been known for years. Mutations in these genes can cause anemia with or without sickling, or cause cyanosis (a blue pallor due to poor oxygen

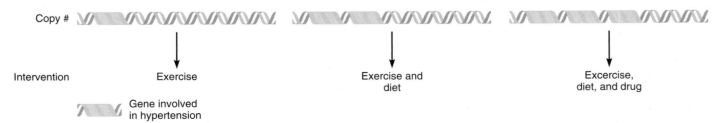

Figure 12.11 Using copy number variants in health care. If large-scale studies support a correlation of copy number of a particular DNA sequence with cholesterol level, then perhaps the number can be used to give medical advice: exercise, exercise plus dietary changes, exercise plus diet plus a cholesterol-lowering drug.

Table 12.5 Globin Mutations

Associated Phenotype	Name	Mutation
Clinically silent	Hb Wayne	Single-base deletion in alpha gene causes frameshift, changing amino acids 139–141 and adding amino acids
	Hb Grady	Nine extra bases add three amino acids between amino acids 118 and 119 of alpha chain
Oxygen binding	Hb Chesapeake	Change from arginine to leucine at amino acid 92 of beta chain
	Hb McKees Rocks	Change from tyrosine to STOP codon at amino acid 145 in beta chain
Anemia	Hb Constant Spring	Change from STOP codon to glutamine elongates alpha chain
	Hb S	Change from glutamic acid to valine at amino acid 6 in beta chain causes sickling
	Hb Leiden	Amino acid 6 deleted from beta chain
Protection against malaria	Hb C	Change from glutamic acid to lysine at amino acid 6 in beta chain causes sickling

binding). Rarely, a mutation boosts the molecule's affinity for oxygen. Some globin gene variants exert no effect at all and are considered "clinically silent" (**table 12.5**).

Different mutations at the same site in a gene can have different effects. For example, hemoglobin S and hemoglobin C result from mutations that change the sixth amino acid in the beta globin polypeptide, but in different ways. Homozygotes for hemoglobin S have sickle cell disease, yet homozygotes for hemoglobin C are healthy. Both types of homozygotes are resistant to malaria because the unusual hemoglobin alters the shapes and surfaces of red blood cells in ways that keep out the parasite that causes the illness.

An interesting consequence of certain mutations in either the alpha or beta globin chains is hemoglobin M. Normally, the iron in hemoglobin is in the ferrous form, which means that it has two positive charges. In hemoglobin M, the mutation stabilizes the ferric form, which has three positive charges and cannot bind oxygen. Fortunately, an enzyme converts the abnormal ferric iron to the normal ferrous form, so that the only symptom is usually cyanosis. The condition has been known for more than 200 years in a small town in Japan, where many people have autosomal dominant "blackmouth."

Even more noticeable than people with blackmouth are the "blue people of Troublesome Creek." Seven generations ago, in 1820, a French orphan named Martin Fugate who settled in this area of Kentucky brought in a recessive gene that causes a form of methemoglobinemia. He was missing an enzyme (cytochrome b_5 reductase) that normally catalyzes a reaction that converts a type of hemoglobin with poor oxygen affinity, called methemoglobin, back into normal hemoglobin by adding an electron. Martin's wife was a carrier for this very rare disease. After extensive inbreeding in the isolated community—their son married his aunt, for example—a large pedigree of "blue people" of both sexes arose.

In "blue person disease," excess oxygen-poor hemoglobin causes a dark blue complexion. Carriers may have bluish lips and fingernails at birth, which usually improve. Treatment is simple: A tablet of methylene blue, a commonly used dye, adds the electron back to methemoglobin, converting it to normal hemoglobin.

Susceptibility to Prion Disorders

For the prion protein gene, as with the globin genes, certain mutations exert extreme effects, while others do not. Recall from chapter 10 that a prion is a protein that assumes both stable and infectious conformations. A prion disease can be inherited, such as fatal familial insomnia, or acquired, such as developing variant Creutzfeldt-Jakob disease from eating beef from a cow that had bovine spongiform encephalopathy ("mad cow disease"). The prion protein has at least eight distinct conformations. The normal form of the protein has a central core made up of helices. In a disease-causing form, the helices open into a sheet.

Precise genetic changes control the plasticity of the prion protein. The 129th amino acid is particularly important. In people who inherit prion disorders, amino acid 129 is either valine in all copies of the protein (genotype *VV*, causing the insomnia) or methionine in all copies (genotype *MM*, causing a form of Creutzfeldt-Jakob disease). These people are homozygous for this small part of the gene. Most people, however, are heterozygous, with valine in some prion proteins and methionine in others (genotype *VM*). Perhaps having two different amino acids at this position enables the proteins to assemble and to carry out their normal functions without damaging the brain.

A mutation at a different site in the prion protein gene raises the risk of brain disease even higher. Normally prion protein folds so that amino acid 129 is near amino acid 178, which is aspartic acid. People who inherit prion diseases are homozygous for the gene at position 129, and have another mutation that changes amino acid 178 to asparagine.

Factors That Lessen the Effects of Mutation

Mutation is a natural consequence of DNA's ability to change. This flexibility is essential for evolution because it generates new variants, some of which may resist environmental change and enable a population or even a species to survive. However, many factors minimize the negative effects of mutations on phenotypes.

The genetic code protects against mutation. Recall from chapter 10 that synonymous codons specify the same amino acid. Mutation in the third codon position is "silent" because the two codons are synonymous. For example, a change from RNA codon CAA to CAG does not change the amino acid, glutamine, so a protein whose gene contains the change would not change. A change from one codon to a nonsynonymous one could affect the phenotype.

Other genetic code nuances prevent synthesis of very altered proteins. For example, mutations in the second codon position sometimes replace one amino acid with another that has a similar conformation, minimizing disruption of the protein's shape. GCC mutated to GGC, for instance, replaces alanine with equally small glycine.

A **conditional mutation** affects the phenotype only under certain circumstances. This can be protective if an individual avoids the exposures that trigger symptoms. Consider a common variant of the X-linked gene that encodes glucose 6-phosphate dehydrogenase (G6PD), an enzyme that immature red blood cells use to extract energy from glucose. One hundred million people worldwide have G6PD deficiency (MIM 305900). It can cause life-threatening hemolytic anemia, but only under rather unusual conditions—eating fava beans or taking certain antimalarial drugs (**figure 12.12**).

In the fifth century B.C., the Greek mathematician Pythagoras wouldn't allow his followers to consume fava beans, because he had discovered that it would sicken some of them. During the Second World War, several soldiers taking the antimalarial drug primaquine developed hemolytic anemia. What fava beans, antimalarial drugs, and dozens of other triggering substances have in common is that they "stress" red blood cells by exposing them to oxidants, chemicals that strip electrons from other compounds. Without the enzyme, the stress bursts the red blood cells.

Another protection against mutation occurs in stem cells. When a stem cell divides to yield another stem cell and a progenitor or differentiated cell, the oldest DNA strands segregate with the stem cell, and the most recently replicated DNA strands go to the more specialized daughter cells (see figure 2.22) This makes sense in organs where stem cells very

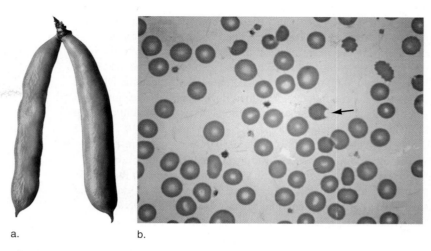

a.   b.

Figure 12.12 Sickness and circumstance. A conditional mutation causes some cases of G6PD deficiency hemolytic anemia. Exposure to two biochemicals in fava beans **(a)** unfolds hemoglobin molecules, bending red blood cells out of shape. They then burst **(b)**.

actively yield specialized daughter cells, such as the skin and small intestine. Because mutations occur when DNA replicates, this skewed distribution of chromosomes sends the DNA most likely to harbor mutations into cells that will soon be shed (from a towel rubbed on skin or in a bowel movement) while keeping mutations away from the stem cells that must continually regenerate the tissues.

Key Concepts

1. How a mutation alters the phenotype depends upon its location in the gene.
2. Mutations in globin genes have a variety of effects. Mutations in two parts of the prion protein gene predispose to developing a prion disorder.
3. Genetic code degeneracy ensures that some third-codon-position mutations do not alter the specified amino acid. Changes in the second codon position often substitute a structurally similar amino acid.
4. Conditional mutations are expressed only in certain environments.
5. Preferential segregation of the oldest DNA strands to stem cells rather than daughter cells protects against mutation.

12.6 DNA Repair

Any manufacturing facility tests a product in several ways to see whether it has been assembled correctly. Mistakes in production are rectified before the item goes on the market—most of the time. The same is true for a cell's manufacture of DNA.

Damage to DNA becomes important when the genetic material is replicated, because the error is passed on to daughter cells. In response to damage, the cell may die by apoptosis or it may repair the error. If the cell doesn't die or the error is not repaired, cancer may result. Fortunately, DNA replication is very accurate—only 1 in 100 million or so bases is incorrectly incorporated. This is quite an accomplishment, because DNA replicates approximately 10^{16} times during an average human lifetime. However, most such mutations occur in somatic cells, and do not affect the phenotype.

DNA polymerase as well as DNA damage response genes oversee the accuracy of replication. DNA repair consists of a cell's detecting damage and signaling systems in the cell that respond with death or repair. More than 50 DNA damage response genes have been identified. Mitochondrial DNA cannot repair itself, which accounts for its higher mutation rate.

Many types of organisms repair their DNA, some more efficiently than others. The master at DNA repair is a large, reddish microbe. *Deinococcus radiodurans* was discovered in a can of spoiled ground meat at the Oregon Agricultural Experiment Station in Corvallis in 1956, where it had withstood radiation used to sterilize the food. It tolerates 1,000 times the radiation level that a person can, and it can even live amidst the intense radiation of a nuclear reactor. The bacterium realigns its radiation-shattered pieces of DNA. Then enzymes bring in new nucleotides and assemble the pieces.

The discovery of DNA repair systems began with observations in the late 1940s that when fungi were exposed to ultraviolet (UV) radiation, those cultures later placed nearest a window grew best. The researchers who noted these effects were not investigating DNA repair, but were using UV light in other experiments. Therefore, DNA repair was inadvertently discovered before the structure of DNA was. The DNA-damaging effect of UV radiation, and the ability of light to correct it, was soon observed in a variety of organisms. (UV radiation has a shorter wavelength than visible light. They are both types of electromagnetic radiation.)

Types of DNA Repair

Exposure to radiation is a fact of life. The Earth, since its beginning, has been periodically bathed in UV radiation. Volcanoes, comets, meteorites, and supernovas all depleted ozone in the atmosphere, which allowed ultraviolet wavelengths of light to reach organisms. The shorter wavelengths—UVA—are not dangerous, but the longer UVB wavelengths damage DNA by forming an extra covalent bond between adjacent (same-strand) pyrimidines, particularly thymines (**figure 12.13**). The linked thymines are called thymine dimers. Their extra bonds kink the double helix enough to disrupt replication and permit insertion of a noncomplementary base. For example, an A might be inserted opposite a G or C, instead of opposite a T. Thymine dimers also disrupt transcription.

Early in the evolution of life, organisms that could survive UV damage had an advantage. Enzymes enabled them to do this, and because enzymes are gene-encoded, DNA repair came to persist.

In many modern species, three types of DNA repair peruse the genetic material for mismatched base pairs. In the

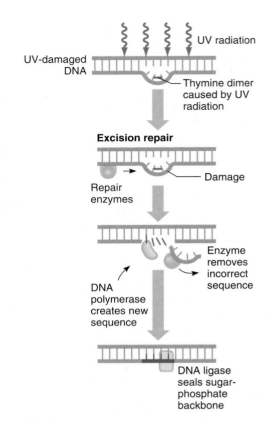

Figure 12.13 Excision repair. Human DNA damaged by UV light is repaired by excision repair, which removes and replaces the pyrimidine dimer and a few surrounding bases.

first type, enzymes called photolyases absorb energy from visible light and use it to detect and bind to pyrimidine dimers, then break the extra bonds. This type of repair, called photoreactivation, enables UV-damaged fungi to recover from exposure to sunlight. Humans do not have this type of DNA repair.

In the early 1960s, researchers discovered a second type of DNA self-mending, called **excision repair,** in mutant *E. coli* that were unable to repair UV-induced DNA damage. Enzymes cut the bond between the DNA sugar and base and snip out—or excise—the pyrimidine dimer and surrounding bases (see figure 12.13). Then, a DNA polymerase fills in the correct nucleotides, using the exposed template as a guide. DNA polymerase also detects and corrects mismatched bases in newly replicated DNA.

Humans have two types of excision repair. **Nucleotide excision repair** replaces up to 30 nucleotides and removes errors that result from several types of insults, including exposure to chemical carcinogens, UVB in sunlight, and oxidative damage. Thirty different proteins carry out nucleotide excision repair.

The second type of excision repair, **base excision repair,** replaces one to five nucleotides at a time, but specifically corrects errors that result from oxidative damage. Oxygen free radicals are highly reactive forms of oxygen that arise during chemical reactions such as those of metabolism and transcription. Free radicals damage DNA. Genes that are very actively

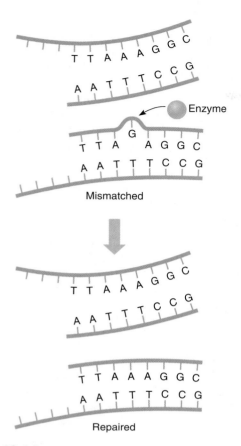

Figure 12.14 Mismatch repair. In this form of DNA repair, enzymes detect loops and bulges in newly replicated DNA that indicate mispairing. The enzymes correct the error. Highly repeated sequences are more prone to this type of error.

transcribed face greater oxidative damage from free radicals; base excision repair targets this type of damage.

A third mechanism of DNA sequence correction is **mismatch repair**. Enzymes "proofread" newly replicated DNA for small loops that emerge from the double helix. The enzymes excise the mismatched base so that it can be replaced (**figure 12.14**). These loops emerge from where the two strands do not precisely align, but instead slip and misalign. This occurs where very short DNA sequences repeat. These sequences, called microsatellites, are scattered throughout the genome. Like minisatellites, microsatellite lengths can vary from person to person, but within an individual, they are usually the same length. Excision and mismatch repair differ in the cause of the error—UV-induced pyrimidine dimers versus replication errors—and in the types of enzymes involved.

Excision repair and mismatch repair in human cells relieve the strain on thymine dimers or replace incorrectly inserted bases. Another form of repair can heal a broken sugar-phosphate backbone in both strands, which can result from exposure to ionizing radiation or oxidative damage. Such a double-stranded break is especially damaging because it breaks a chromosome, which can cause cancer. At least two types of multiprotein complexes reseal the sugar-phosphate backbone, either by rejoining the broken ends or recombining with DNA on the unaffected homolog.

In yet another type of DNA repair called damage tolerance, a "wrong" DNA base is left in place, but replication and transcription proceed. "Sloppy" DNA polymerases, with looser adherence to the base-pairing rules, read past the error, randomly inserting any other base. It is a little like retaining a misspelled word in a sentence—usually the meaning remains clear. **Figure 12.15** summarizes DNA repair mechanisms.

DNA Repair Disorders

The ability to repair DNA is crucial to health. If both copies of a repair gene are mutant, a disorder can result. Heterozygotes who have one mutant repair gene may be more sensitive to damage from environmental toxins.

A well-studied DNA repair gene encodes a protein called p53. It controls whether DNA is repaired and the cell salvaged, or the cell dies by apoptosis (see figure 18.2). Signals from outside the cell activate p53 protein to aggregate into complexes of four proteins. These quartets bind certain genes that slow the cell cycle, enabling repair to take place. If the damage is too severe, the p53 protein quartets instead increase the rate of transcription of genes that promote apoptosis, and the cell dies.

In DNA repair disorders, chromosome breakage caused by factors such as radiation cannot be repaired. Mutations in repair genes greatly increase susceptibility to certain types of cancer following exposure to ionizing radiation or chemicals that affect cell division. These conditions develop because errors in the DNA sequence accumulate and are perpetuated to a much greater extent than they are in people with functioning repair systems. We conclude this chapter with a closer look at repair disorders.

Trichothiodystrophy (MIM 601675)

At least five genes can cause trichothiodystrophy. At its worst, this condition causes dwarfism, intellectual disability, and failure to develop, in addition to brittle hair and scaly skin, both with low sulfur content. Although the child may appear to be normal for a year or two, growth soon slows dramatically, signs of premature aging begin, and life ends early. Hearing and vision may fail. Interestingly, the condition does not increase the risk of cancer. Symptoms reflect accumulating oxidative damage. Individuals have faulty nucleotide excision repair, base excision repair, or both. This repair disorder is unusual in that it does not increase cancer risk.

Inherited Colon Cancer

Hereditary nonpolyposis colon cancer (HNPCC) (MIM 120435 and others, also known as Lynch syndrome) was linked to a DNA repair defect when researchers discovered different-length microsatellites within an individual. People with this type of colon cancer have a breakdown of mismatch repair, which normally keeps a person's microsatellites all the same length. HNPCC affects 1 in 200 people, and mutations in any of at least 7 genes can cause it.

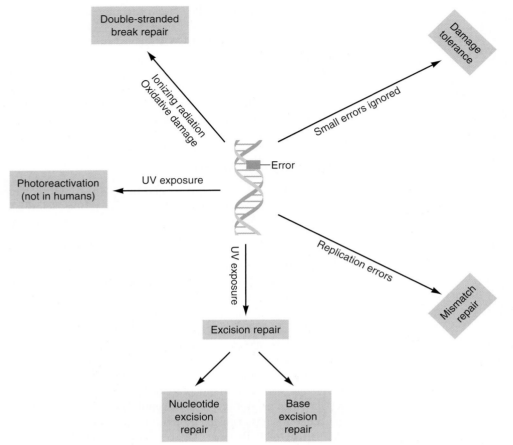

Figure 12.15 **DNA repair mechanisms.**

HNPCC accounts for 3 percent of newly diagnosed colorectal cancers. Genetic testing for this condition in all people newly diagnosed with colon cancer is advised because if they have a mutation, their relatives can be tested. If healthy relatives test positive, frequent colonoscopies can detect disease early, at a more treatable stage. Penetrance of HNPCC is about 45 percent by age 70—considered a high cancer risk.

Xeroderma Pigmentosum (XP) (MIM 278700)

A child with XP must stay indoors in artificial light, because even the briefest exposure to sunlight causes painful blisters. Failing to cover up and use sunblock can result in skin cancer (**figure 12.16**). More than half of all children with XP develop the cancer before they reach their teens. People with XP have a 1,000-fold increased risk of developing skin cancer compared to others, and a 10-fold increased risk of developing internal tumors.

XP is autosomal recessive, and results from mutations in any of seven genes. It can reflect malfunction of nucleotide excision repair or deficient "sloppy" DNA polymerase, both of which allow thymine dimers to stay and block replication.

One of the genes that causes XP, when mutant, also causes trichothiodystrophy and another disorder, Cockayne syndrome. The different symptoms arise from the different ways that mutations disrupt the encoded protein, which is a helicase that helps unwind replicating DNA.

Only about 250 people in the world are known to have XP. A family living in upstate New York runs a special summer camp for children with XP, where they turn night into day. Activities take place at night, or in special areas where the windows are covered and light comes from low-ultraviolet incandescent lightbulbs.

Ataxia Telangiectasis (AT) (MIM 208900)

This multisymptom disorder is the result of a defect in a kinase that functions as a cell cycle checkpoint (see figure 2.14). In AT, cells proceed through the cell cycle without pausing just after replication to inspect the new DNA and to repair any mispaired bases. Some cells die through apoptosis if the damage is too great to repair. Because of the malfunctioning cell cycle, individuals who have this autosomal recessive disorder have 50 times the risk of developing cancer, particularly of the blood. About 40 percent of individuals with ataxia telangiectasis have cancer by age 30. Additional symptoms include poor balance and coordination (ataxia), red marks on the face (telangiectasia), delayed sexual maturation, and high risk of infection and diabetes mellitus.

AT is rare, but heterozygotes are not. They make up from 0.5 to 1.4 percent of various populations. Carriers may have mild radiation sensitivity, which causes a two- to sixfold increase in cancer risk over that of the general population. For people who are AT carriers, dental or medical X rays may cause cancer.

DNA's changeability, so vital for evolution of a species, comes at the cost of occasional harm to individuals. Each of us harbors about 175 new mutations, many old ones, and many polymorphisms, although most are hidden in the recessive state. Individuals whose mutations cause illness or deformity can face hardships, both medical and social. Perhaps we can learn from the ancient Egyptians, who honored people who were genetically different.

The dry air of ancient Egypt and the meticulous recording of daily life in art and burial places reveal that people with short stature, particularly those with autosomal dominant achondroplasia (a form of dwarfism), were accepted, important, and even revered as gods. Ancient Egyptian "little people" were jewelers, animal keepers, entertainers, and personal attendants, often to royalty. High-ranking people with dwarfism were given

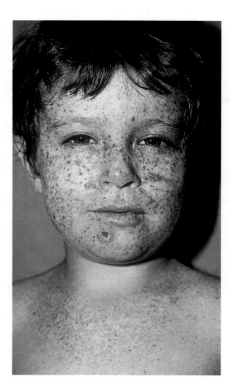

Figure 12.16   A DNA repair disorder. The marks on this child's face result from sun exposure. He is highly sensitive because he has inherited xeroderma pigmentosum (XP), an impairment of excision repair. The large lesion on his chin is a skin cancer.

special burial places near the pyramids. Wrote Chahira Kozma, a professor of pediatrics at Georgetown University who studies how the ancient Egyptians regarded unusual people, "Dwarfs were accepted in ancient Egypt; their recorded daily activities suggest assimilation into daily life, and their disorder was not shown as a physical handicap. Wisdom writings and moral teachings in ancient Egypt commanded respect for dwarfs and other individuals with disabilities."

Key Concepts

1. Many genes encode enzymes that locate and correct errors in replicating DNA.
2. A common cause of noncomplementary base insertion is a UV-induced pyrimidine dimer.
3. Photoreactivation or excision repair can unkink pyrimidine dimers.
4. Mismatch repair corrects noncomplementary base pairs that are inserted into newly replicated DNA.
5. Repair seals broken sugar-phosphate backbones.
6. DNA damage tolerance allows replication to proceed past a mismatched base.
7. Abnormal repair genes cause disorders usually associated with chromosome breaks and predisposition to cancer.

Summary

12.1 The Nature of Mutations

1. A **mutation** is a change in a gene's nucleotide base sequence that affects less than 1 percent of a population and can cause a **mutant** phenotype. A polymorphism is more common and may not alter the phenotype.
2. A **germline mutation** originates in meiosis and affects all cells of an individual. A **somatic mutation** originates in mitosis and affects a subset of cells.
3. A mutation disrupts the function or abundance of a protein or introduces a new function. Loss-of-function mutations are usually recessive, and altered or gain-of-function mutations are dominant.

12.2 Two Mutations

4. Mutations in the beta globin and collagen genes cause a variety of disorders.
5. Whether different mutations in a gene cause the same or distinct illnesses varies; nomenclature is inconsistent.

12.3 Causes of Mutation

6. A spontaneous mutation arises due to chemical phenomena or to an error in DNA replication. Spontaneous mutation rate is characteristic of a gene and is more likely in repeats.
7. **Mutagens** are chemicals or radiation that delete, substitute, or add bases. An organism may be exposed to a mutagen intentionally, accidentally, or naturally.

12.4 Types of Mutations

8. A **point mutation** alters a single DNA base. It may be a **transition** (purine to purine or pyrimidine to pyrimidine) or a **transversion** (purine to pyrimidine or vice versa). A **missense mutation** substitutes one amino acid for another, while a **nonsense mutation** substitutes a "stop" codon for a codon that specifies an amino acid, shortening the protein product. Point mutations in splice sites can add or delete amino acids.
9. Adding or deleting genetic material may upset the reading frame or otherwise alter protein function.
10. A pseudogene results when a duplicate of a gene mutates. It may disrupt chromosome pairing, causing mutation.
11. Transposons may disrupt the functions of genes they jump into.
12. Expanding triplet repeat mutations add stretches of the same amino acid to a protein. They expand because they attract each other, which affects replication.
13. **Copy number variants** are DNA sequences that are repeated a different number of times in different individuals. They may have no effect on phenotype or may directly or indirectly cause disease.

12.5 The Importance of Position

14. Several types of mutations can affect a gene.
15. Mutations in the globin genes may affect the ability of the blood to transport oxygen, or they may have no effect.

16. Susceptibility to prion disorders requires two mutations that affect different parts of the protein that interact as the amino acid chain folds.

17. Synonymous codons limit the effects of mutation. Changes in the second codon position often substitute a similarly shaped amino acid.

18. **Conditional mutations** are expressed only in response to certain environmental triggers.

19. Sending the most recently replicated DNA into cells headed for differentiation, while sending older strands into stem cells, protects against mutation.

12.6 DNA Repair

20. DNA polymerase proofreads DNA, but repair enzymes correct errors in other ways.

21. Photoreactivation repair uses light energy to split pyrimidine dimers.

22. In **excision repair,** pyrimidine dimers are removed and the area filled in correctly. **Nucleotide excision repair** replaces up to 30 nucleotides from various sources of mutation. **Base excision repair** fixes up to five bases that paired incorrectly due to oxidative damage.

23. **Mismatch repair** proofreads newly replicated DNA for loops that indicate noncomplementary base pairing.

24. DNA repair also fixes the sugar-phosphate backbone. Damage tolerance enables replication to continue beyond a mismatch.

25. Mutations in repair genes break chromosomes and increase cancer risk.

www.mhhe.com/lewisgenetics10

Answers to all end-of-chapter questions can be found at **www.mhhe.com/lewisgenetics10.** You will also find additional practice quizzes, animations, videos, and vocabulary flashcards to help you master the material in this chapter.

Review Questions

1. How do a mutation and a SNP differ?

2. Distinguish between a germline and a somatic mutation.

3. Why is the collagen molecule especially likely to be altered by mutation?

4. Describe a pair of allelic disorders.

5. How can DNA spontaneously mutate?

6. Compare the effects of alpha, beta, and gamma radiation on the human body.

7. What is the physical basis of a mutational hot spot?

8. List two types of mutations that can alter the reading frame.

9. List four ways that DNA can mutate without affecting the phenotype.

10. Cite two ways a jumping gene can disrupt gene function.

11. What is a molecular explanation for the worsening of an inherited illness over generations?

12. How can short repeats within a gene, long triplet repeats within a gene, and repeated genes cause disease?

13. How does a copy number variant differ from a missense mutation?

14. How can a mutation that retains an intron's sequence and a triplet repeat mutation have a similar effect on a gene's encoded protein?

15. Cite three ways in which the genetic code protects against the effects of mutation.

16. What is a conditional mutation?

17. How do excision and mismatch repair differ?

18. Explain how semiconservative DNA replication makes it possible for stem cells to receive the DNA least likely to bear mutations.

Applied Questions

1. Consider the following sequence of part of an mRNA molecule:

    A U G U U G U C A A A A G C A U G G C G G C C A

    Introduce the following changes to the sequence, and indicate the effect, if any, on the encoded amino acid sequence:

    a. a missense mutation
    b. a nonsense mutation
    c. a frameshift mutation
    d. a silent mutation
    e. a transversion
    f. a transition
    g. a tandem duplication
    h. a deletion

2. Retinitis pigmentosa causes night blindness and loss of peripheral vision before age 20. A form of X-linked retinitis pigmentosa (MIM 300455) is caused by a frameshift mutation that deletes 199 amino acids. How can a simple mutation have such a great effect?

3. A mutation that changes a C to a T causes a form of Ehlers-Danlos syndrome, forming a "stop" codon and shortened procollagen. Consult the genetic code and suggest one way that this can happen.

4. Part of the mRNA sequence of an exon of a gene that encodes a blood protein is:

   A U G A C U C A U C G C U G U A G U U U A C G A

   Consult the genetic code to answer the following questions:

   a. What is the sequence of amino acids that this mRNA encodes?

   b. What is the sequence if a point mutation changes the tenth base from a C to an A?

   c. What is the effect of a point mutation that changes the fifteenth base from a U to an A?

   d. How does the encoded amino acid sequence change if a C is inserted between the fourth and fifth bases?

   e. Which would be more devastating to the encoded amino acid sequence, insertion of three bases in a row, or insertion of two bases in a row?

5. Susceptibility to developing prion diseases entails a mutation from aspartic acid (Asp) to asparagine (Asn). Which nucleotide base changes make this happen?

6. Two teenage boys meet at a clinic to treat muscular dystrophy. The boy who is more severely affected has a two-base insertion at the start of his dystrophin gene. The other boy has the same two-base insertion but also has a third base inserted a few bases away. Why is the second boy's illness milder?

7. Two missense mutations in the gene that encodes an enzyme called superoxide dismutase cause amyotrophic lateral sclerosis (ALS, or Lou Gehrig's disease). This disorder causes loss of neurological function over a 5-year period. One mutation alters the amino acid asparagine (Asn) to lysine (Lys). The other changes an isoleucine (Ile) to a threonine (Thr). List the codons involved and describe how single-base mutations alter the amino acids they specify.

8. The new drug described in the chapter opener changes the amino acid glycine into aspartic acid at the 551st DNA base in the gene encoding the cystic fibrosis transmembrane conductance regulator, an ion channel.

   a. Identify a single DNA base change that could produce this amino acid change.

   b. What type of mutation does the chapter opener describe?

9. Certain antibiotic drugs suppress nonsense mutations by inserting a random amino acid into the protein corresponding to the site of the mutation in the gene. Explain how this happens, and how this finding might be applied to treat a genetic disease.

10. In one family, Tay-Sachs disease stems from a four-base insertion, which changes an amino-acid-encoding codon into a "stop" codon. What type of mutation is this?

11. Epidermolytic hyperkeratosis (MIM 607602) is an autosomal dominant condition that produces scaly skin. It can be caused by a missense mutation that substitutes a histidine (His) amino acid for an arginine (Arg). Write the mRNA codons that could account for this change.

12. A biotechnology company has encapsulated DNA repair enzymes in fatty bubbles called liposomes. Why would this be a valuable addition to a suntanning lotion?

Web Activities

1. Children with Hutchinson-Gilford progeria syndrome (figure 3.22) age extremely rapidly. In 18 of 20 children, a single base change in the *lamin A* gene alters a C to a T, but this mutation removes 50 amino acids from the encoded protein. In all 20 children, the parents do not have the mutation.

   a. Is the mutation in the 18 children *de novo* or induced? What is the evidence for this distinction?

   b. How can a change in a single base remove 50 amino acids?

   c. Using MIM, list and describe six other disorders caused by mutation in the *lamin A* gene.

2. Go to http://www.lpaonline.org/resources_faq.html and read the Little People of America's Position Statement on Genetic

Discoveries in Dwarfism. Cite an application of genetic testing for achondroplasia that could be construed as beneficial, and one that could be thought of as harmful.

3. Select an image at positiveexposure.org and describe the symptoms of the photographed disorder.

4. Look up three of the following triplet repeat disorders at www.omim.org (or elsewhere) and compare the actual repeat, the normal copy number, and the disease copy number: fragile X syndrome, Friedreich ataxia, Haw River syndrome, Jacobsen syndrome, spinal and bulbar muscular atrophy, and spinocerebellar ataxia (any type).

Forensics Focus

1. Late one night, a man broke into Colleen's apartment and raped her. He wore a mask and it was dark, so she couldn't see his face, but she did manage to yank out some of his long, greasy hair. The forensic investigator, after examining the hair,

asked Colleen if her boyfriend had been in the bed or with her earlier in the evening, because the hairs were of two genotypes for one of the repeated sequences that was analyzed. What is another explanation for the finding of two genotypes?

Case Studies and Research Results

1. Latika and Keyshauna meet at a clinic for college students who have cystic fibrosis. Latika's mutation results in exon skipping. Keyshauna's mutation is a nonsense mutation. Which young woman probably has more severe symptoms? Cite a reason for your answer.

2. Marshall and Angela have skin cancer resulting from xeroderma pigmentosum. They meet at an event for teenagers with cancer. However, their mutations affect different genes. They decide to marry but not to have children because they believe that each child would have a 25 percent chance of inheriting XP because it is autosomal recessive. Are they correct? Why or why not?

3. Two girls and a boy in a Pakistani family have a form of deafness caused by a mutation in the gene that encodes a protein called tricellulin (MIM 610153). The normal protein attaches epithelial (lining) cells in groups of three in the inner ear in a way that is crucial to hearing. Following is a pedigree for the family:

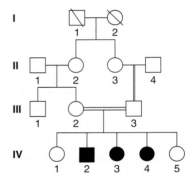

   a. What is the mode of inheritance for this form of deafness, and how do you know this?

   b. This form of deafness is rare worldwide, but more common among Pakistani families, many of whose pedigrees have double horizontal lines. What does the double line mean, and how does it account for the increased prevalence of this form of deafness in the population?

   c. The affected children have a partial sequence for the tricellulin gene of C T G C A A T G T. Unaffected family members have the sequence C T G C A G T G T. What are the amino acid differences encoded in these sequences?

4. Presenilin 1 is one of the genes that, when mutant, causes familial Alzheimer disease. It is also expressed in the heart. Certain mutations cause a condition that leads to heart failure. In the Esposito family, all of the relatives who have or had heart failure have the following partial sequence for the presenilin 1 gene: G A T G A T G G C G G G. Family members with healthy hearts have the sequence G A T G G T G G C G G G. How do the encoded amino acid sequences differ between the healthy and sick family members for this part of the gene?

5. At the Center for Applied Genomics at the Children's Hospital of Philadelphia, 100,000 young children are having their genomes scanned for copy number variants. Researchers are comparing the resulting CNV profiles with multifactorial disorders that the children have, such as asthma, obesity, cancers, and diabetes, to determine whether copy number variation is important to health. The idea is that the children are so young that environmental influences are minimized, compared, for example, to people who have asthma but have smoked or lived with pollution. Explain how this project might be helpful and how it might be harmful.

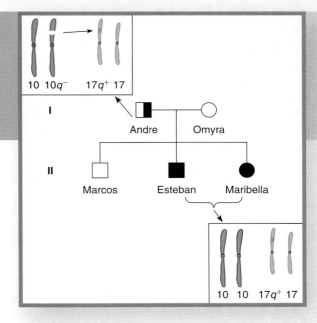

Too much genetic material. A piece of the short arm of Andre's chromosome 10 (10q⁻) has moved to one chromosome 17. He is healthy. However, Esteban and Maribella have each inherited the copy of the chromosome 17 with the extra material (17q⁺) as well as two normal chromosome 10s. The extra chromosome 10 DNA caused their symptoms.

Chromosomes

Learning Outcomes

13.1 Portrait of a Chromosome

1. List the major parts of a chromosome.
2. List the types of chromosomes based on centromere position.

13.2 Visualizing Chromosomes

3. Describe ways that chromosomes are obtained, prepared, and visualized.

13.3 Atypical Chromosome Number

4. Explain how atypical chromosome numbers arise.
5. Describe syndromes associated with incorrect chromosome number.

13.4 Atypical Chromosome Structure

6. Explain how atypical chromosome structures arise.
7. Describe syndromes associated with specific variants of chromosome structure.

13.5 Uniparental Disomy—A Double Dose from One Parent

8. Explain how a person could inherit both copies of a DNA sequence from one parent.
9. Describe how inheriting both copies of DNA from one parent can affect health.

The Big Picture: A human genome has 20,000-plus protein-encoding genes dispersed among 24 chromosome types. Abnormalities in chromosome number or structure can have sweeping effects, but mutation is a continuum. Chromosomal-level illnesses reflect disruption of individual genes.

A Late Diagnosis

Esteban was 17 years old when he learned that he had an unusual chromosome. The discovery explained a lot.

As a baby and toddler, Esteban had been much slower to walk and talk than his brother Marcos. When he started school, problems emerged: He was learning disabled and had difficulty interacting with others. Esteban had to repeat the third grade so that he could learn to read. He was also very tall and thin, causing teachers to think he was older than he really was and therefore to expect more of him. Still, Esteban was able to learn in regular classrooms with weekly visits to a resource room, and he made a few friends.

Esteban's mother, Omyra, became pregnant when Esteban was in the tenth grade. She hadn't had a fetal chromosome check when she was pregnant with Esteban, but this time she did. The test showed one chromosome 17 with a bit of another chromosome stuck into it. A geneticist at the medical center where Omyra's doctor practiced requested DNA samples from the family. She found that the fetus had inherited the unusual chromosome from the father, Andre, but the chromosome charts from father and fetus differed. Andre's cells had two unusual chromosomes but a normal amount of genetic material. A bit of chromosome 10 had inserted into chromosome 17, and the fetus received only the unusual chromosome 17 from Andre. Esteban's chromosomes had the same abnormality as his future sister's. His symptoms may have stemmed from the extra DNA. His little sister Maribella was slow, too, but is today a very happy preschooler.

13.1 Portrait of a Chromosome

Mutations range from single-base changes to entire extra sets of chromosomes. A mutation is considered a chromosomal aberration if it is large enough to see with a light microscope using stains and/or fluorescent probes to highlight missing, extra, or moved genetic material.

In general, too much genetic material has milder effects on health than too little. Still, most extensive chromosomal abnormalities present in all cells of an embryo or fetus disrupt or halt prenatal development. As a result, only 0.65 percent of all newborns have chromosomal abnormalities that produce symptoms. An additional 0.20 percent of newborns have chromosomal rearrangements in which chromosome parts have flipped or swapped, but they do not produce symptoms unless they disrupt genes that affect health.

Cytogenetics is the area of genetics that links chromosome variations to specific traits, including illnesses. This chapter explores several ways that chromosomes can be atypical (used synonymously with abnormal) and affect health. Actual cases introduce some of them.

Required Parts: Telomeres and Centromeres

A chromosome consists primarily of DNA and proteins with a small amount of RNA. It is duplicated and transmitted, via mitosis or meiosis, to the next cell generation. Chromosomes have long been described and distinguished by size and shape, using stains and dyes to contrast dark **heterochromatin** with the lighter **euchromatin (figure 13.1)**. Heterochromatin consists mostly of highly repetitive DNA sequences, whereas euchromatin has more protein-encoding sequences.

A chromosome must include structures that enable it to replicate and remain intact. Everything else is informational cargo (protein-encoding genes and their controls). The essential parts of a chromosome are:

- telomeres;
- origin of replication sites, where replication forks begin to form; and
- the centromere.

Recall from figure 2.18 that **telomeres** are chromosome tips. In humans, each telomere repeats the sequence TTAGGG. In most cell types, telomeres shorten with each mitotic cell division.

The **centromere** is the largest constriction of a chromosome. It is where spindle fibers attach when the cell divides. A chromosome without a centromere is no longer a chromosome. It vanishes from the cell as soon as division begins because there is no way to attach to the spindle.

Centromeres, like chromosomes, are made up mostly of DNA and protein. Many of the hundreds of thousands of DNA bases that form the centromere are repeats of a specific 171-base DNA sequence. The size and number of repeats are similar in many species, although the sequence differs. This suggests that these repeats have a structural role in maintaining chromosomes rather than an informational role from their sequence. Certain centromere-associated proteins are synthesized only when mitosis is imminent, forming a structure called a kinetochore that contacts the spindle fibers, enabling the cell to divide.

Centromeres replicate toward the end of S phase. A protein that may control the process is centromere protein A, or CENP-A. Molecules of CENP-A stay with centromeres as chromosomes replicate, covering about half a million DNA base pairs. When the replicated (sister) chromatids separate at anaphase, each member of the pair retains some CENP-A. The protein therefore passes to the next cell generation, but it is *not* DNA. This is another example of an epigenetic change.

Centromeres lie within vast stretches of heterochromatin. The arms of the chromosome extend outward from the centromere. Gradually, the DNA includes more protein-encoding sequences with distance from the centromere. Gene density varies greatly among chromosomes. Chromosome 21 is a gene "desert," harboring a million-base stretch with no protein-encoding genes at all. Chromosome 22, in contrast, is a gene "jungle." These two tiniest chromosomes are remarkably similar in size, but chromosome 22 contains 545 genes to chromosome 21's 225!

The chromosome parts that lie between protein-rich areas and the telomeres are termed subtelomeres (**figure 13.2**). These areas extend from 8,000 to 300,000 bases inward toward the centromere from the telomeres. Subtelomeres include some protein-encoding genes and therefore bridge the gene-rich regions and the telomere repeats. The transition is gradual. Areas of 50 to 250 bases, right next to the telomeres, consist of 6-base repeats, many of them very similar to the TTAGGG of the telomeres. Then, moving inward from the 6-base zone are many shorter repeats. Their function isn't known. Finally the sequence diversifies and protein-encoding genes appear.

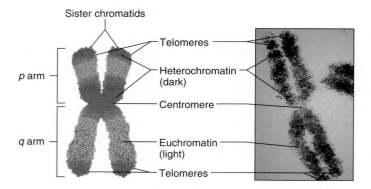

Figure 13.1 **Portrait of a chromosome.** Tightly wound, highly repetitive heterochromatin forms the centromere (the largest constriction) and the telomeres (the tips) of chromosomes. Elsewhere, lighter-staining euchromatin includes many protein-encoding genes. The centromere divides this chromosome into a short arm (*p*) and a long arm (*q*). This chromosome is in the replicated form.

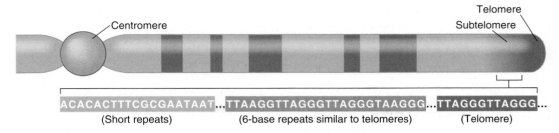

Figure 13.2 Subtelomeres. The repetitive sequence of a telomere gradually diversifies toward the centromere. The centromere is depicted as a buttonlike structure to more easily distinguish it, but it is composed of DNA like the rest of the chromosome.

At least 500 protein-encoding genes lie in the total sub-telomere regions. About half are members of multigene families (groups of genes of very similar sequence next to each other) that include pseudogenes. These multigene families may reflect recent evolution: Apes and chimps have only one or two genes for many of the large gene families in humans. Such gene organization is one explanation for why our genome sequence is so very similar to that of our primate cousins—but we are clearly different animals. Our genomes differ more in gene copy number and chromosomal organization than in DNA base sequence.

Karyotypes Chart Chromosomes

Even in this age of genomics, the standard chromosome chart, or **karyotype,** remains a major clinical tool. A karyotype displays chromosomes in pairs by size and by physical landmarks that appear during mitotic metaphase, when DNA coils tightly. **Figure 13.3** shows a karyotype with an extra chromosome.

The 24 human chromosome types are numbered from largest to smallest—1 to 22. The other two chromosomes are the X and the Y. Early attempts to size-order chromosomes resulted in generalized groupings because many of the chromosomes are of similar size. Use of dyes and stains made it easier to distinguish chromosomes because they form patterns of bands.

Centromere position is one physical feature of chromosomes. A chromosome is **metacentric** if the centromere divides it into two arms of approximately equal length. It is **submetacentric** if the centromere establishes one long arm and one short arm, and **acrocentric** if it pinches off only a small amount of material toward one end (**figure 13.4**). Some species have telocentric chromosomes that have only one arm, but humans do not. The long arm of a chromosome is designated *q*, and the short arm *p* (*p* stands for "petite").

Five human chromosomes (13, 14, 15, 21, and 22) have bloblike ends, called satellites, that extend from a thinner, stalk-like bridge from the rest of the chromosome. The stalk regions do not bind stains well. The stalks carry many copies of genes encoding ribosomal RNA and ribosomal proteins. These areas coalesce to form the nucleolus, a structure in the nucleus where ribosomal building blocks are produced and assembled (see figure 2.3).

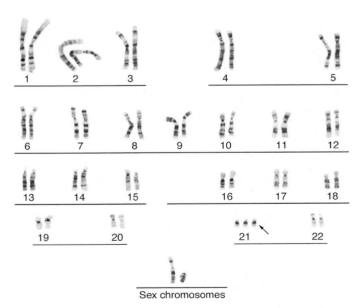

Figure 13.3 A karyotype displays chromosome pairs in size order. Note the extra chromosome 21 that causes trisomy 21 Down syndrome.

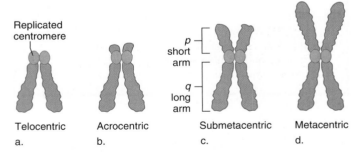

Figure 13.4 Centromere position distinguishes chromosomes. (a) A telocentric chromosome has the centromere toward one end although telomere DNA sequences are still at the tip. Humans do not have any telocentric chromosomes. **(b)** An acrocentric chromosome has the centromere near an end. **(c)** A submetacentric chromosome's centromere creates a long arm (*q*) and a short arm (*p*). **(d)** A metacentric chromosome's centromere establishes equal-sized arms.

Karyotypes are useful at several levels. When a baby is born with the distinctive facial features of Down syndrome, a karyotype confirms the clinical diagnosis. Within families, karyotypes are used to identify relatives with a particular chromosomal aberration that can affect health. In one family, several adults died from a rare form of kidney cancer. Karyotypes revealed that the affected individuals all had an exchange, called a **translocation,** between chromosomes 3 and 8. When karyotypes showed that two healthy young family members had the translocation, physicians examined and monitored their kidneys. Cancer was found very early and successfully treated.

Karyotypes of individuals from different populations can reveal the effects of environmental toxins, if abnormalities appear only in a group exposed to a particular contaminant. Because chemicals and radiation that can cause cancer and birth defects often break chromosomes into fragments or rings, detecting this genetic damage can alert physicians to the possibility that certain cancers may appear in the population.

Karyotypes compared among species can clarify evolutionary relationships. The more recent the divergence of two species from a common ancestor, the more closely related we presume they are, and the more alike their chromosome banding patterns should be. Our closest relative, according to karyotypes, is the pygmy chimpanzee (bonobo). The human karyotype is also remarkably similar to that of the domestic cat, and somewhat less similar to those of mice, pigs, and cows. Among mammals, it is least like the karyotype of the aardvark, indicating that this is a primitive placental mammal.

Key Concepts

1. A chromosome minimally includes telomeres, origins of replication, and centromeres.
2. A centromere consists of DNA repeats and associated proteins, some of which bind spindle fibers. Centromere protein A enables the centromere to replicate.
3. Subtelomeres contain telomere-like repeats and protein-encoding multigene families.
4. Chromosomes differ by size, centromere location, satellites, and staining. Karyotypes are size-order chromosome charts.

13.2 Visualizing Chromosomes

Extra or missing chromosomes are detected by counting a number other than 46. Identifying chromosome rearrangements, such as an inverted sequence or an exchange of parts between two chromosomes, requires a way to distinguish among the chromosomes. A combination of stains and DNA probes applied to chromosomes allows this. A **DNA probe** is a labeled piece of DNA that binds to its complementary base sequence on a particular chromosome.

Obtaining Cells for Chromosome Study

Any cell other than a mature red blood cell (which lacks a nucleus) can be used to examine chromosomes, but some cells are easier to obtain and culture than others. Skinlike cells collected from the inside of the cheek are the easiest to obtain for a chromosome test; white blood cells are used too. A person might require a chromosome test if he or she has a family history of a chromosomal abnormality or seeks medical help because of infertility.

Chromosome tests are commonly performed on cells from fetuses. Couples who receive a prenatal diagnosis of a chromosome abnormality can arrange for treatment of the newborn, if possible; learn more about the condition and contact support groups and plan care; or terminate the pregnancy. These choices are best made after a genetic counselor or physician provides information on the medical condition and treatment options.

Chromosomes of a fetus are checked in several ways. **Amniocentesis** and **chorionic villus sampling** have been available for many years. They sample fetal cells from the amniotic fluid and chorionic villi, respectively, and detect large-scale chromosomal abnormalities. A newer technique called **chromosome microarray analysis** can be paired with the older techniques to detect copy number variants, which include extremely small sections of missing or extra DNA. Chromosome microarray analysis probes and displays specific sequences, detecting many disorders that other techniques miss.

Amniocentesis

The first fetal karyotype was constructed in 1966 using amniocentesis. In this procedure, a doctor removes a small sample of fetal cells and fluids from the uterus with a needle passed through the woman's abdominal wall (**figure 13.5a**). The cells are cultured for a week to 10 days, and typically 20 cells are karyotyped. The sampled amniotic fluid may also be examined for deficient, excess, or abnormal biochemicals that could indicate an inborn error of metabolism. Tests for specific single-gene disorders are based on family history and may be done on cells in the amniotic fluid sample as well. Ultrasound is used to follow the needle's movement and to visualize fetal parts, such as the profile in **figure 13.6**.

Amniocentesis can detect approximately 1,000 of the more than 5,000 known chromosomal and biochemical problems. The most common chromosomal abnormality detected is one extra chromosome, called a **trisomy**. Amniocentesis is usually performed between 14 and 16 weeks gestation, when the fetus isn't yet very large but amniotic fluid is plentiful. Amniocentesis can be carried out anytime after this point.

Doctors recommend amniocentesis if the risk that the fetus has a detectable condition exceeds the risk that the procedure will cause a miscarriage. Until recently, this risk cutoff was thought to be about age 35 in the woman, when the risk to the fetus of a detectable chromosome problem about equals the risk of amniocentesis causing pregnancy loss—1 in 350. While it is still true that the risk of a chromosomal problem rises

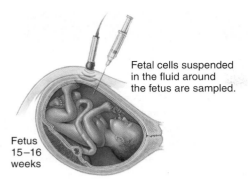

Fetal cells suspended in the fluid around the fetus are sampled.

Fetus 15–16 weeks

a. Amniocentesis

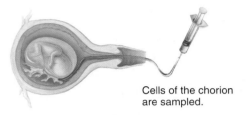

Cells of the chorion are sampled.

b. Chorionic villus sampling

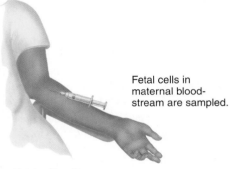

Fetal cells in maternal bloodstream are sampled.

c. Fetal cell sorting

Figure 13.5 **Three ways to check a fetus's chromosomes.** **(a)** Amniocentesis draws out amniotic fluid. Fetal cells shed into the fluid are collected and their chromosomes examined. **(b)** Chorionic villus sampling removes cells that would otherwise develop into the placenta. Since these cells descended from the fertilized ovum, they should have the same chromosomal constitution as the fetus. **(c)** Researchers can detect fetal cells, DNA, or mRNA in a sample of blood from a pregnant woman.

steeply after maternal age 35, amniocentesis has become much safer in the 30 or so years since the statistics were obtained that have been used for most risk estimates (**figure 13.7**). In 2007 a large study found the risk of amniocentesis causing miscarriage to be about 1 in 1,600, leading some physicians and organizations to offer amniocentesis to younger women too. The procedure is also warranted if a couple has had several spontaneous abortions or children with birth defects or a known chromosome abnormality, irrespective of maternal age.

Another reason to seek amniocentesis is if screening tests on a pregnant woman indicate elevated risk for a trisomy

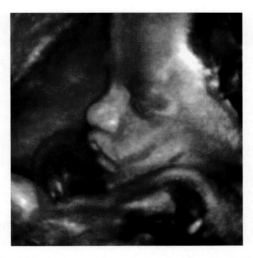

Figure 13.6 **A sonogram is an image obtained with ultrasound.** In an ultrasound exam, sound waves bounced off the embryo or fetus are converted into a three-dimensional-appearing image. "4D ultrasound" provides a video of an embryo or fetus. (The fourth dimension is time.)

(extra chromosome) of the fetus. These "multiple maternal serum marker" tests, discussed in *Bioethics: Choices for the Future* on page 240, are offered to all pregnant women. Cutoff levels for the results based on population statistics are used to identify fetuses at elevated risk, and the women are then offered the more invasive amniocentesis and chorionic villus sampling, which yield a definitive diagnosis. Screening tests consider maternal age, ultrasound findings, and levels of certain proteins in the woman's blood at certain times in the pregnancy.

Chorionic Villus Sampling

During the 10th through 12th week of pregnancy, chorionic villus sampling (CVS) obtains cells from the chorionic villi, which are fingerlike structures that develop into the placenta

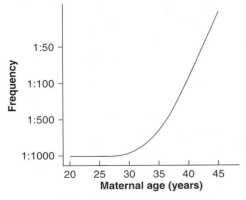

Trisomy 21 in liveborn infants

Figure 13.7 **The risk of conceiving an offspring with trisomy 21 rises dramatically with maternal age.**

Down Syndrome Ups and Downs

Prenatal tests for trisomy 21 Down syndrome are of two general types. *Screening tests* identify fetuses that are at increased risk of having trisomy 21. These tests consider an ultrasound finding (excess fluid at the back of the neck) and abnormal levels of certain proteins in the pregnant woman's blood (multiple maternal serum markers) and maternal age. (**table 1**). Screening tests that include serum markers are routinely offered to pregnant women of any age, and much more accurately predict risk than maternal age alone.

If screening tests find a fetus to be at elevated risk for trisomy 21, more invasive *diagnostic tests* are offered—chorionic villus sampling or amniocentesis. Both are highly accurate, but introduce a small risk of miscarriage. In contrast to the screening tests, the diagnostic tests actually visualize the extra chromosome.

In 2004, the government of Denmark issued new guidelines offering prenatal screening for trisomy 21 to all pregnant women. Earlier guidelines offered diagnostic tests only to women over age 35, based on assessment by age alone and using old statistics. Informed consent was required—that is, all women who took the screening tests were told the risks and benefits.

Researchers have tracked number of trisomy 21 births and use of diagnostic testing (CVS and amniocentesis) in Denmark from 2000 until 2006 to assess impact of the new guidelines. The number of infants born with trisomy 21 has been halved, the number diagnosed before birth increased by 30 percent, and the number of invasive prenatal diagnostic tests done each year has decreased by 50 percent.

In the United States, the number of people who have Down syndrome is increasing, for two reasons: the increase in number of people having children later in life, and the greater life span of people with Down syndrome, thanks to new and better treatments.

Table 1	Commonly Used Maternal Serum Markers	
Marker		**Elevated Risk**
Alpha fetoprotein (AFP)		< normal
Human chorionic gonadotropin (hCG)		> normal
Estriol		< normal
Inhibin A		> normal
Pregnancy-associated plasma protein A		< normal

Questions for Discussion

1. What is a medical benefit of the guidelines in Denmark?

2. What do you think potential patients should be told during the informed consent process in Denmark?

3. How might you feel about the program in Denmark if you had a child with Down syndrome?

4. How can the Danish government prevent women from feeling pressured to have prenatal testing for trisomy 21?

5. Explain how the number of infants born with Down syndrome in Denmark declined by more than half, but additional prenatal diagnosis increased only by about a third.

6. What should the role of the father be in deciding whether a woman should have prenatal screening or diagnosis for trisomy 21?

7. Discuss societal factors that affect the incidence (new cases) and prevalence (total number of cases) of Down syndrome in the United States.

(figure 13.5*b*). A karyotype is prepared directly from the collected cells, rather than first culturing them, as in amniocentesis. Results are ready in days.

Because chorionic villus cells descend from the fertilized ovum, their chromosomes should be identical to those of the embryo and fetus. Occasionally, a chromosomal aberration occurs only in a cell of the embryo, or only in a chorionic villus cell. This results in chromosomal mosaicism—the karyotype of a villus cell differs from that of an embryo cell. Chromosomal mosaicism has great clinical consequences. If CVS indicates an abnormality in villus cells that is not also in the fetus, then a couple may elect to terminate the pregnancy when the fetus is actually chromosomally normal. In the opposite situation, the results of the CVS may be normal, but the fetus has atypical chromosomes.

CVS is slightly less accurate than amniocentesis, and in about 1 in 1,000 to 3,000 procedures, it halts development of the feet and/or hands and may be lethal. Also, CVS does not sample amniotic fluid, so tests for inborn errors of metabolism are not possible. The advantage of CVS is earlier results, but the disadvantage is a greater risk of spontaneous abortion. However, CVS has become much safer in recent years.

Fetal Cells, DNA, and RNA

Detecting fetal cells or nucleic acids in the pregnant woman's bloodstream is safer than amniocentesis and CVS, but is still experimental in the United States (figure 13.5*c*). The technique traces its roots to 1957, when a pregnant woman died when cells from a very early embryo lodged in a major blood vessel in her lung, blocking blood flow. The fetal cells were detectable because they were from a male, and contained the telltale Y chromosome. This finding meant that fetal cells could enter a woman's circulation.

Researchers found that fetal cells enter the maternal circulation in up to 70 percent of pregnancies. Cells from female embryos, however, cannot easily be distinguished from the cells of the pregnant woman on the basis of sex chromosome analysis. Fetal cells from either sex can be distinguished from maternal cells using a device called a fluorescence-activated cell sorter. It separates fetal cells from maternal blood by identifying surface characteristics that differ from those on the woman's cells. The fetal cells are then karyotyped and specific gene tests performed on fetal DNA.

Free fetal DNA is also in a pregnant woman's bloodstream, but is difficult to detect because it is so rare. Instead, a new technique detects fetal mRNA in the woman's bloodstream. If fetal mRNA transcribed from a specific DNA sequence that differs between the parents is not present in equal amounts, then an extra copy of the sequence in the fetus is inferred. If the sequence is part of chromosome 21, for example, the technique can diagnose the extra chromosome of Down syndrome.

Preparing Cells for Chromosome Observation

Cytogeneticists have tried to describe and display human chromosomes since the late nineteenth century (**figure 13.8**). Then, the prevailing view held that humans had an XO sex determination system, with females having an extra chromosome (XX). Estimates of the human chromosome number ranged from 30 to 80. In 1923, Theophilus Painter published sketches of human chromosomes from three patients at a Texas state mental hospital. The patients had been castrated in an attempt to control their abusive behavior, and Painter was able to examine the removed tissue. He could not at first tell whether the cells had 46 or 48 chromosomes, but finally decided that he saw 48. Painter later showed that both sexes have the same chromosome number.

The difficulty in distinguishing between 46 or 48 chromosomes was physical—it is challenging to prepare a cell in which chromosomes do not overlap. To count chromosomes, scientists had to find a way to capture them when they are most condensed—during cell division—and also spread them apart. Since the 1950s, cytogeneticists have used colchicine, an extract of the crocus plant, to arrest cells during division.

Swelling, Squashing, and Untangling

How to untangle the spaghetti-like mass of chromosomes was solved by accident in 1951. A technician mistakenly washed white blood cells being prepared for chromosome analysis in a salt solution that was less concentrated than the interiors of the cells. Water rushed into the cells, swelling them and separating the chromosomes. Then cell biologists found that drawing cell-rich fluid into a pipette and dropping it onto a microscope slide prepared with stain burst the cells and freed the mass of chromosomes. Adding a glass coverslip spread the chromosomes so they could be counted. Researchers finally could see that the number of chromosomes in a diploid human cell is 46, and that the number in gametes is 23.

Karyotypes were once constructed using a microscope to locate a cell where the chromosomes were not touching, photographing the cell, developing a print, cutting out the individual chromosomes, and arranging them into a size-ordered chart. Today, a computer scans ruptured cells in a drop of stain and selects one in which the chromosomes are the most visible and well-spread. Then image analysis software recognizes the band patterns of each stained chromosome pair, sorts the structures into a size-ordered chart, and prints the karyotype. If the software recognizes an atypical band pattern, a database pulls out identical or similar karyotypes from records of other patients.

Staining

In the earliest karyotypes, dyes were used to stain chromosomes a uniform color. Chromosomes were grouped into size classes, designated A through G, in decreasing size order. In 1959,

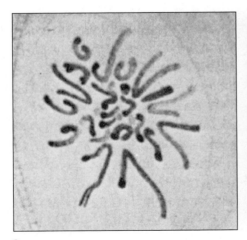
a.

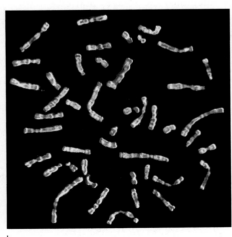
b.

Figure 13.8 **Viewing chromosomes, then and now.** **(a)** The earliest drawings of chromosomes, by German biologist Walter Flemming, date from 1882. His depiction captures the random distribution of chromosomes as they splash down on a slide. **(b)** A micrograph of actual stained human chromosomes.

scientists described the first chromosomal abnormalities— Down syndrome (an extra chromosome 21), Turner syndrome (also called XO syndrome, a female with only one X chromosome), and Klinefelter syndrome (also called XXY syndrome, a male with an extra X chromosome). These first chromosome stains could highlight large deletions and duplications, but usually researchers only vaguely understood the nature of a chromosomal syndrome. In 1967, an intellectually disabled child with material missing from chromosome 4 would have been diagnosed as having a "B-group chromosome" disorder.

Describing smaller chromosomal aberrations required better ways to distinguish chromosomes. By the 1970s, new stains created banding patterns unique to each chromosome. These stains are specific for AT-rich or GC-rich stretches of DNA, or for heterochromatin, which is dark-staining. A band represents at least 5 to 10 million DNA bases.

The ability to detect missing, extra, inverted, or misplaced bands allowed researchers to link many more syndromes with specific chromosome aberrations. Then researchers found that synchronizing the cell cycle of cultured cells revealed more bands per chromosome. Another technique, fluorescence *in situ* hybridization (FISH), introduced the ability to highlight individual genes.

FISHing

FISH is more precise and targeted than conventional chromosome staining because it uses DNA probes that are complementary to specific DNA sequences. The probes are attached to molecules that fluoresce when illuminated, producing a flash of color precisely where the probe binds to a chromosome in a patient's sample. Using a FISH probe is a little like a search engine finding the word "hippopotamus" in a book compared to pulling out all words that have the letters *h, p,* and *o.*

FISH can "paint" entire karyotypes by probing each chromosome with several different fluorescent molecules. A computer integrates the images and creates a unique false color for each chromosome. Many laboratories that perform amniocentesis or CVS use FISH probes for chromosomes 13, 18, 21, and the X and Y to quickly identify the most common problems. In **figure 13.9,** FISH reveals the extra chromosome 21 in cells from a fetus with trisomy 21 Down syndrome.

A new type of prenatal chromosome analysis amplifies certain repeated sequences on chromosomes 13, 18, 21, X, and Y. The technique distinguishes paternally derived from maternally derived repeats on each homolog for these five chromosomes. An atypical ratio of maternal to paternal repeats indicates a numerical problem, such as two copies of one parent's chromosome 21. Combined with the one chromosome 21 from the other parent, this situation would produce a fertilized ovum with three copies of chromosome 21, which causes Down syndrome.

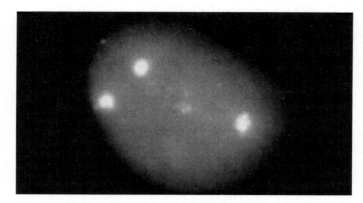

Figure 13.9 FISHing for genes and chromosomes. FISH shows three fluorescent dots that correspond to three copies of chromosome 21. Each dot represents a specific DNA sequence to which the fluorescently-labeled probe complementary base pairs.

Chromosomal Shorthand

Geneticists abbreviate the pertinent information in a karyotype by listing chromosome number, then sex chromosome constitution, then atypical autosomes. Symbols describe the type of aberration, such as a deletion or translocation; numbers correspond to specific bands. A normal male is 46,XY; a normal female is 46,XX. Geneticists use band notation to describe gene locations. For example, the gene encoding the β-globin subunit of hemoglobin is located at 11p15.5. **Table 13.1** gives some examples of chromosomal shorthand.

Chromosome information is displayed in a graphical representation called an ideogram (**figure 13.10**). The chromosome is divided into arms and numbered regions and subregions. Specific gene loci are sometimes listed on the right side. The sequencing and annotation (localization) of specific genes made possible from the human genome project is making

Table 13.1	Chromosomal Shorthand
Abbreviation	**What It Means**
46,XY	Normal male
46,XX	Normal female
45,X	Turner syndrome (female)
47,XXY	Klinefelter syndrome (male)
47,XYY	Jacobs syndrome (male)
46,XY, del (7q)	A male missing part of the long arm of chromosome 7
47,XX, + 21	A female with trisomy 21 Down syndrome
46,XY, t(7;9)(p21.1; q34.1)	A male with a translocation between the short arm of chromosome 7 at band 21.1 and the long arm of chromosome 9 at band 34.1
48, XXYY	A male with an extra X and an extra Y

22.3
22.2
22.1

21.3
21.2
21.1

15.3
15.2
15.1
14.3
14.2
14.1
13.0

12.3
12.2
12.1
11.2
11.1
11.1
11.21
11.22

11.23

p

q

Figure 13.10 Ideogram. An ideogram is a schematic chromosome map. It indicates chromosome arm (*p* or *q*) and major regions delineated by banding patterns. This figure is repeated in the context of more specific ways to depict chromosomes in figure 22.1.

chromosomal depictions, such as ideograms, less useful if not obsolete.

Key Concepts

1. Karyotypes display chromosomes in size order.

2. Chromosomes can be visualized in any cell that has a nucleus and can be cultured.

3. Fetal karyotypes are made from cells obtained by amniocentesis, CVS, or fetal cell sorting from maternal blood. Maternal serum marker patterns and ultrasound scans can reveal increased risk of an atypical chromosome number.

4. Cytogeneticists obtain cells; display, stain, and probe chromosomes with fluorescent molecules; and then arrange them in a karyotype.

5. Chromosomal shorthand summarizes the number of chromosomes, sex chromosome constitution, and type of aberration. Ideograms display features of individual chromosomes.

13.3 Atypical Chromosome Number

A human karyotype is atypical (abnormal) if the number of chromosomes in a somatic cell is not 46, or if individual chromosomes have extra, missing, or rearranged genetic material.

Atypical chromosomes account for at least 50 percent of spontaneous abortions, yet only 0.65 percent of newborns have them. Therefore, most embryos and fetuses with atypical chromosomes stop developing before birth. **Table 13.2** summarizes the types of chromosome variants in the order in which they are discussed.

Improved technology has made it possible to detect extremely small amounts of extra or missing genetic material and, as a result, more people are being diagnosed with chromosomal abnormalities. Today, many families whose members have the same chromosome abnormalities are finding each other through organizations on the Internet, communicating by e-mail or meeting in person, and sharing experiences. **Table 13.3** lists some of these organizations. Often families with the same chromosomal disorder form organizations, such as the "International 22q11.2 Deletion Syndrome Foundation," a group of families with members missing DNA on the short arm of chromosome 22.

Polyploidy

The most extreme upset in chromosome number is an entire extra set. A cell with extra sets of chromosomes is **polyploid**. An individual whose cells have three copies of each chromosome is a triploid (designated 3N, for three sets of chromosomes). Two-thirds of all triploids result from fertilization of an oocyte

Table 13.2	Chromosome Abnormalities
Type of Abnormality	**Definition**
Polyploidy	Extra chromosome sets
Aneuploidy	An extra or missing chromosome
Monosomy	One chromosome absent
Trisomy	One chromosome extra
Deletion	Part of a chromosome missing
Duplication	Part of a chromosome present twice
Translocation	Two chromosomes join long arms or exchange parts
Inversion	Segment of chromosome reversed
Isochromosome	A chromosome with identical arms
Ring chromosome	A chromosome that forms a ring due to deletions in telomeres, which cause ends to adhere

Table 13.3	Organizations for Families with Chromosome Abnormalities
Hope for Trisomy 13 + 18	www.hopefortrisomy13and18.org/
Rainbows Down Under	http://members.optushome.com.au/karens/
Support Organization for Trisomy 18, 13 and Related Disorders (SOFT)	http://www.trisomy.org/
Tracking Rare Incidence Syndromes (TRIS)	http://web.coehs.siu.edu/Grants/TTRIS/
UNIQUE—the Rare Chromosome Disorder Support Group	http://www.rarechromo.org/html/home.asp

by two sperm. The other cases arise from formation of a diploid gamete, such as when a normal haploid sperm fertilizes a diploid oocyte. Triploids account for 17 percent of spontaneous abortions (**figure 13.11**). Very rarely, an infant survives as long as a few days, with defects in nearly all organs. However, certain human cells may be polyploid. The liver, for example, has some tetraploid (4N) and even octaploid (8N) cells.

Polyploids are very common among flowering plants, including roses, cotton, barley, and wheat, and in some insects. Fish farmers raise triploid salmon, which cannot breed.

Aneuploidy

Cells missing a single chromosome or having an extra one are **aneuploid,** which means "not good set." Rarely, aneuploids can have more than one missing or extra chromosome, indicating

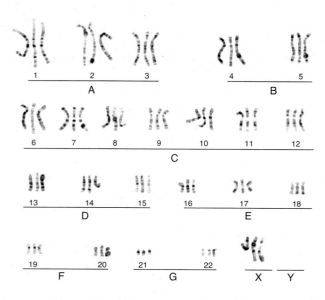

Figure 13.11 Polyploids in humans are nearly always lethal. Individuals with three copies of each chromosome (triploids) in every cell account for 17 percent of all spontaneous abortions and 3 percent of stillbirths and newborn deaths.

defective meiosis in a parent. A normal chromosome number is **euploid,** which means "good set."

Most autosomal aneuploids (with a missing or extra nonsex chromosome) are spontaneously aborted. Those that survive have specific syndromes, with symptoms depending upon which chromosomes are missing or extra. Intellectual disability is common in aneuploidy because development of the brain is so complex and of such long duration that nearly any chromosome-scale disruption affects genes whose protein products affect the brain. Sex chromosome aneuploidy usually produces milder symptoms.

Most children born with the wrong number of chromosomes have an extra chromosome (a **trisomy**) rather than a missing one (a **monosomy**). Most monosomies are so severe that an affected embryo ceases developing. Trisomies and monosomies are named for the chromosomes involved, and in the past the associated syndromes were named for the discoverers. Today, cytogenetic terminology is used because it is more precise. For example, Down syndrome can result from a trisomy or a translocation. The distinction is important in genetic counseling. Translocation Down syndrome, although accounting for only 4 percent of cases, has a much higher recurrence risk within a family than the trisomy form, a point we return to later in the chapter.

The meiotic error that causes aneuploidy is called **nondisjunction**. Recall that in normal meiosis, homologs separate and each of the resulting gametes receives only one member of each chromosome pair. In nondisjunction, a chromosome pair fails to separate at anaphase of either the first or second meiotic division. This produces a sperm or oocyte that has two copies of a particular chromosome, or none, rather than the normal one copy (**figure 13.12**). When such a gamete fuses with its partner at fertilization, the zygote has either 45 or 47 chromosomes, instead of the normal 46. Different trisomies tend to be caused by nondisjunction in the male or female, at meiosis I or II.

A cell can have a missing or extra chromosome in 49 ways—an extra or missing copy of each of the 22 autosomes, plus the five abnormal types of sex chromosome combinations—Y, X, XXX, XXY, and XYY. (Some individuals have four or even five sex chromosomes.) However, only nine types of aneuploids are recognized in newborns. Others are seen in spontaneous abortions or fertilized ova intended for *in vitro* fertilization.

Most of the 50 percent of spontaneous abortions that result from extra or missing chromosomes are 45, X individuals (missing an X chromosome), triploids, or trisomy 16. About 9 percent of spontaneous abortions are trisomy 13, 18, or 21. More than 95 percent of newborns with atypical chromosome numbers have an extra 13, 18, or 21, or an extra or missing X or Y chromosome. These conditions are all rare at birth—together they affect only 0.1 percent of all children. However, nondisjunction occurs in 5 percent of recognized pregnancies.

Types of chromosome abnormalities seem to differ between the sexes. Atypical oocytes mostly have extra or missing chromosomes, whereas atypical sperm more often have structural

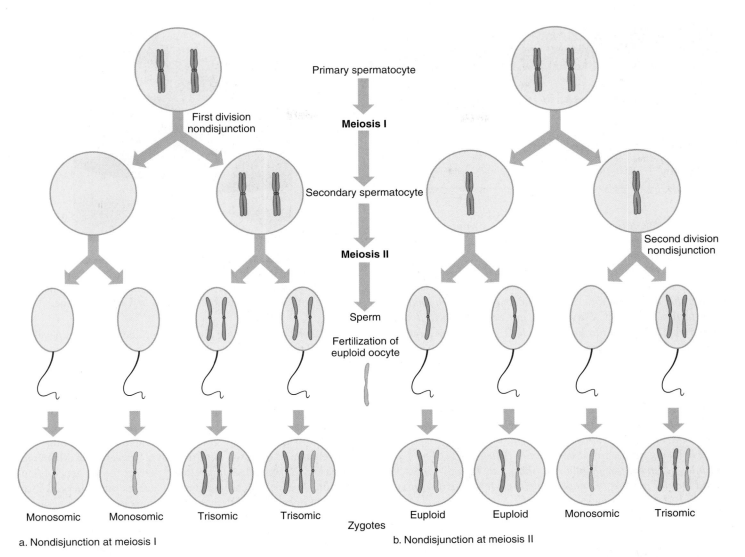

Figure 13.12 Extra and missing chromosomes—aneuploidy. Unequal division of chromosome pairs can occur at either the first or second meiotic division. **(a)** A single pair of chromosomes is unevenly partitioned into the two cells arising from meiosis I in a male. The result: Two sperm cells have two copies of the chromosome, and two sperm cells have no copies. When a sperm cell with two copies of the chromosome fertilizes a normal oocyte, the zygote is trisomic; when a sperm cell lacking the chromosome fertilizes a normal oocyte, the zygote is monosomic. **(b)** This nondisjunction occurs at meiosis II. Because the two products of the first division are unaffected, two of the mature sperm are normal and two are aneuploid. Oocytes can undergo nondisjunction as well, leading to zygotes with extra or missing chromosomes when normal sperm cells fertilize them.

In the figure labels:

Primary spermatocyte

Meiosis I

Secondary spermatocyte

Meiosis II

Sperm

Fertilization of euploid oocyte

First division nondisjunction

Second division nondisjunction

Monosomic   Monosomic   Trisomic   Trisomic   Euploid   Euploid   Monosomic   Trisomic

Zygotes

a. Nondisjunction at meiosis I

b. Nondisjunction at meiosis II

variants, such as inversions or translocations, discussed later in the chapter.

Aneuploidy and polyploidy also arise during mitosis, producing groups of somatic cells with the extra or missing chromosome. An individual with two chromosomally distinct cell populations is a mosaic. If only a few cells are altered, health may not be affected. However, a mitotic abnormality that occurs early in development, so that many cells descend from the unusual one, can affect health. A chromosomal mosaic for a trisomy may have a mild version of the associated condition. This is usually the case for the 1 to 2 percent of people with Down syndrome who are mosaic. The phenotype depends upon which cells have the extra chromosome. Unfortunately, prenatal testing cannot reveal which cells are affected.

Autosomal Aneuploids

Most autosomal aneuploids cease developing long before birth. Following are cases and descriptions of the most common autosomal aneuploids among liveborns. The most frequently seen extra autosomes in newborns are chromosomes 21, 18, and 13 because these chromosomes carry many fewer protein-encoding genes than the other autosomes, compared to their total amount of DNA. Therefore, extra copies of these chromosomes are tolerated well enough for some fetuses with them to survive to be born (**table 13.4**).

Trisomy 21—David's Story

When David G. was born in 1994, doctors told his 19-year-old mother, Toni, to put him into an institution. "They said he

Table 13.4	Comparing and Contrasting Trisomies 13, 18, and 21	
Type of Trisomy	**Incidence at Birth**	**Percent of Conceptions That Survive 1 Year After Birth**
13 (Patau)	1/12,500–1/21,700	<5%
18 (Edward)	1/6,000–1/10,000	<5%
21 (Down)	1/800–1/826	85%

wouldn't walk, talk, or do anything. Today, I want to bring him back and say look, he walks and talks and has graduated high school," says Toni.

Like other teens, David has held part-time jobs, gone to dances, and uses a computer. But he is unlike most other teens in that his cells have an extra chromosome 21, which limits his intellectual abilities. "Maybe he's not book smart, but when you look around at what he can do, he's smart," Toni says. His speech is difficult to understand, and he has facial features characteristic of Down syndrome, but he has a winning personality and close friends.

Sometimes David gets into unusual situations because he thinks literally. He once dialed 911 when he stubbed his toe, because he'd been told to do just that when he was hurt.

David has done so well that he moved into a group home and is taking courses at a community college. Still, Toni fears that he will meet few people like himself as he gets older. She read that 90 percent of couples told their fetus has Down syndrome end the pregnancy.

The most common autosomal aneuploid among liveborns is trisomy 21, because this chromosome has the fewest genes. The extra folds in the eyelids, called epicanthal folds, and flat face prompted Sir John Langdon Haydon Down to term the condition *mongoloid* when he described it in 1866. As the medical superintendent of a facility for the profoundly intellectually disabled, Down noted that about 10 percent of his patients resembled people of Mongolian heritage. The resemblance is superficial. People of all ethnic groups are affected.

A person with Down syndrome is usually short and has straight, sparse hair and a tongue protruding through thick lips. The hands have an atypical pattern of creases, the joints are loose, and poor reflexes and muscle tone give a "floppy" appearance. Developmental milestones (such as sitting, standing, and walking) come slowly, and toilet training may take several years. Intelligence varies greatly. Parents of a child with Down syndrome can help their child reach maximal potential by providing a stimulating environment (**figure 13.13**).

Many people with Down syndrome have physical problems, including heart and kidney defects and hearing and visual loss. A suppressed immune system can make influenza deadly. Digestive system blockages are common and may require surgical correction. A child with Down syndrome is 15 times more likely to develop leukemia than a child who does not have the syndrome, but this is still only a 1 percent risk. However, people

Figure 13.13 **Trisomy 21.** Many years ago, people with Down syndrome were often institutionalized. Today, thanks to tremendous strides in both medical care and special education, people with the condition can hold jobs and attend college.

with Down syndrome are somewhat protected against developing solid tumors. Many of the medical problems associated with Down syndrome are treatable, so that average life expectancy is now in the fifties. In 1910, life expectancy was only 9 years.

Some people with Down syndrome who pass age 40 develop the black fibers and tangles of amyloid beta protein in their brains characteristic of Alzheimer disease, although they usually do not become severely demented (see Reading 5.1). The chance of a person with trisomy 21 developing Alzheimer disease is 25 percent, compared to 6 percent for the general population. A gene on chromosome 21 causes one inherited form of Alzheimer disease. Perhaps the extra copy of the gene in trisomy 21 has a similar effect to a mutation in the gene that causes Alzheimer disease, such as causing amyloid beta buildup. However, Alzheimer disease doesn't affect everyone with Down syndrome. A medical journal described a 70-year-old man with Down syndrome who does not have any signs of dementia at all.

Before the human genome sequence became available, researchers studied people who have a third copy of only part of chromosome 21 to identify specific genes that could cause symptoms. This led to the discovery that genes near the tip of the long arm of chromosome 21, called the Down syndrome critical region, contribute most of the abnormalities. Two genes in particular control many aspects of Down syndrome by controlling a third gene, which encodes a transcription factor (**table 13.5**).

The likelihood of giving birth to a child with trisomy 21 Down syndrome increases dramatically with the age of the mother (see figure 13.7). However, 80 percent of children with trisomy 21 are born to women under age 35, because younger women are more likely to become pregnant and less likely to have amniocentesis. About 90 percent of trisomy 21 conceptions are due to nondisjunction during meiosis I in the female. The 10 percent of cases due to the male result from nondisjunction during meiosis I or II. The chance that trisomy 21 will recur in a family, based on empirical data (how often it actually does recur in families), is 1 percent.

Table 13.5	Some Genes Associated with Trisomy 21 Down Syndrome
Gene	**Effect in Excess**
DYRK1A	Expressed in brain (hippocampus, cortex, and cerebellum) and heart; starts amyloid formation
DSCR1	Alters synapses
NFATc	A transcription factor that affects expression of genes important in central nervous system development; regulated by levels of *DYRK1A* and *DSCR1*

The age factor in trisomy 21 Down syndrome and other trisomies may reflect the fact that the older a woman is, the longer her oocytes have been arrested on the brink of completing meiosis. This is a time period of 15 to 45 years, when oocytes may have been exposed to toxins, viruses, and radiation. A second explanation for the maternal age effect is that females have a pool of immature aneuploid oocytes resulting from spindle abnormalities that cause nondisjunction. As a woman ages, selectively releasing normal oocytes each month, the atypical ones remain, much as black jelly beans accumulate as people preferentially eat the colored ones. Yet a third possible explanation for the maternal age effect is that trisomies result from gametes in which a homolog pair do not extensively cross over during meiosis I. Such chromosomes tend to migrate to the same pole, packaging an extra chromosome into a gamete.

The association between maternal age and Down syndrome has been recognized for a long time, because affected individuals were often the youngest children in large families. Before the chromosome connection was made in 1959, the syndrome was thought to be caused by syphilis, tuberculosis, thyroid malfunction, alcoholism, emotional trauma, or "maternal reproductive exhaustion." The increased risk of Down syndrome correlates to maternal age, not to the number of children in the family.

Trisomy 18—Anthony's Story

When an ultrasound scan early in pregnancy revealed a small fetus with low-set ears, a small jaw, a pocket of fluid in the brain, and a clenched fist, the parents-to-be, Elisa and Brendan, were advised to have amniocentesis to view the fetus's chromosomes. The signs on the scan suggested an extra chromosome 18, and amniocentesis confirmed it.

Although Elisa and Brendan were upset to learn what lay ahead, they continued the pregnancy. The fetus remained small, as Elisa swelled hugely with three times the normal volume of amniotic fluid. Further ultrasound scans revealed that only one of the kidneys worked, the heart had holes between the chambers, and part of the intestine lay outside the stomach in a sac. The child would be severely developmentally delayed and intellectually disabled. Anthony was delivered at 36 weeks gestation, after his heart rate became erratic during a routine prenatal visit. He lived only 22 days.

Trisomies 18 and 13 were described in a research report in 1960 (**figure 13.14**). Trisomy 18 is also called Edward syndrome and trisomy 13 is also known as Patau syndrome. Most affected individuals do not survive to be born.

Children who have trisomy 18 have great physical and mental disabilities, with developmental skills usually stalled at the 6-month level. Major abnormalities include heart defects, a displaced liver, growth retardation, and oddly clenched fists. Milder signs include overlapping placement of fingers, a narrow and flat skull, unusually shaped and low-set ears, a small mouth and face, unusual or absent fingerprints, short, large toes with fused second and third toes, and "rocker-bottom" feet. Most cases of trisomy 18 are traced to nondisjunction in meiosis II of the oocyte.

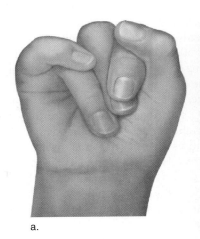

a.

b.

Figure 13.14 **Trisomies 18 and 13.** **(a)** An infant with trisomy 18 clenches the fists in an odd way, with fingers overlapping. **(b)** Very few babies with trisomy 13 are as healthy as Hazel. Most die in infancy.

Trisomy 13—Tykesia's Story

At 15 months of age, Tykesia is a "long-term survivor" of trisomy 13. About 92 percent of infants born with an extra chromosome 13 do not live to see their first birthdays.

Tykesia is small for her age, at the 5th percentile for weight, but she is happy, curious and playful. Her physical skills, however, lag. She can finally, with great effort, sit up, but cannot yet crawl. She has about 20 minor seizures a day, which look like jerks or startles, and has difficulty eating because of persistent acid reflux. She is also missing a rib. Early surgeries corrected a cleft lip and palate, removed an extra finger and toe, and corrected a hernia. Blood vessels leading from the heart to the lungs that did not close as they normally should before birth did so by the time Tykesia was 6 months old. She is intellectually disabled, but her parents hope she will live long enough to attend preschool. Despite these challenges, Tykesia's case is mild—she has her sight and hearing, unlike many others with trisomy 13.

Trisomy 13 has a different set of signs and symptoms than trisomy 18. Most striking is a fusion of the developing eyes, so that a fetus has one large eyelike structure in the center of the face. More common is a small or absent eye. Major abnormalities affect the heart, kidneys, brain, face, and limbs. The nose is often malformed, and cleft lip and/or palate is present in a small head. There may be extra fingers and toes. Ultrasound examination of an affected newborn often reveals other problems, such as an extra spleen, atypical liver, rotated intestines, and an abnormal pancreas. A few individuals have survived until adulthood, but they do not progress developmentally beyond the 6-month level.

Sex Chromosome Aneuploids: Female

People with sex chromosome aneuploidy have extra or missing sex chromosomes. **Table 13.6** indicates how these aneuploids can arise. Note that some conditions can result from nondisjunction in meiosis in the male *or* female. These conditions are generally much less serious than autosomal aneuploids, which is why this book has no photos of affected individuals—they look like anyone else.

XO Syndrome—Miranda's Story

Miranda was well into her teen years, but still looked about 12. She was short, her breasts had never developed, and she still had never menstruated. Her sister Charlotte, 2 years younger, actually looked older. When Miranda turned 16, her physician suggested that she have her chromosomes checked. While reassuring Miranda that delayed puberty could be treated, the doctor also explained what she was looking to rule out—absence of an X chromosome, called Turner or XO syndrome.

Miranda indeed lacked a second X chromosome. The diagnosis explained other problems, such as poor hearing, high blood pressure, low thyroid function, and the "beauty marks" that dotted her skin.

In 1938, at a medical conference, a U.S. endocrinologist named Henry Turner described seven young women, aged 15 to 23, who were sexually undeveloped, short, had folds of skin on

Table 13.6	How Nondisjunction Leads to Sex Chromosome Aneuploids			
Situation		**Oocyte**	**Sperm**	**Consequence**
Normal		X	Y	46,XY normal male
		X	X	46,XX normal female
Female nondisjunction		XX	Y	47,XXY Klinefelter syndrome
		XX	X	47,XXX triplo-X
			Y	45,Y nonviable
			X	45,X Turner syndrome
Male nondisjunction (meiosis I)		X		45,X Turner syndrome
		X	XY	47,XXY Klinefelter syndrome
Male nondisjunction (meiosis II)		X	XX	47,XXX triplo-X
		X	YY	47,XYY Jacobs syndrome
		X		45,X Turner syndrome
Male and female nondisjunction		XX	YY	48, XXYY syndrome

the back of their necks, and had malformed elbows. (Eight years earlier, an English physician had described the syndrome in young girls, so it is called Ullrich syndrome in the U.K.) Alerted to what would become known as Turner syndrome in the United States, other physicians soon began identifying such patients. Physicians assumed that a hormonal insufficiency caused the symptoms. They were right, but there was more to the story—a chromosomal imbalance caused the hormone deficit.

In 1954, at a London hospital, a physician discovered that cells from Turner patients lacked a Barr body, the dark spot that indicates a second X chromosome. Might lack of a sex chromosome cause the symptoms, particularly failure to mature sexually? By 1959, karyotyping confirmed the presence of only one X chromosome. Later, researchers learned that only 50 percent of affected individuals are XO. The rest are missing only part of an X chromosome or are mosaics, with only some cells missing an X.

Like the autosomal aneuploids, Turner syndrome, now called XO syndrome, is found more frequently among spontaneously aborted fetuses than among newborns—99 percent of XO fetuses are not born. The syndrome affects 1 in 2,500 female births. However, if amniocentesis or CVS was not done, a person with XO syndrome would likely not know she has a chromosome abnormality until she lags in sexual development. Two X chromosomes are necessary for normal sexual development in females.

At birth, a girl with XO syndrome looks normal, except for puffy hands and feet caused by impaired lymph flow. In childhood, signs of XO syndrome include wide-set nipples, soft nails that turn up at the tips, slight webbing at the back of the neck, short stature, coarse facial features, and a low hairline at the back of the head. About half of people with XO syndrome have impaired hearing and frequent ear infections due to a small defect in the shape of the coiled part of the inner ear. They cannot hear certain frequencies of sound.

At sexual maturity, sparse body hair develops, but the girls do not ovulate or menstruate, and their breasts do not develop. The uterus is very small, but the vagina and cervix are normal size. In the ovaries, oocytes mature too fast, depleting the supply during infancy. Intelligence is normal. XO syndrome may impair the ability to solve math problems that entail envisioning objects in three-dimensional space, and may cause memory deficits. Hormones (estrogen and progesterone) can be given to stimulate development of secondary sexual structures for individuals diagnosed before puberty, and prompt use of growth hormone can maximize height.

Individuals who are mosaics (only some cells lack the second X chromosome) may have children, but their offspring are at high risk of having too many or too few chromosomes. XO syndrome is unrelated to the age of the mother. The effects of XO syndrome continue past the reproductive years. Life span is shortened slightly. Adults with XO syndrome are more likely to develop certain disorders than the general population, including osteoporosis, types 1 and 2 diabetes, and colon cancer. The many signs and symptoms of XO syndrome result from loss of specific genes.

Triplo-X

About 1 in every 1,000 females has an extra X chromosome in each of her cells, a condition called triplo-X. The only symptoms are tall stature and menstrual irregularities. Although triplo-X females are rarely intellectually disabled, they tend to be less intelligent than their siblings. The lack of symptoms reflects the protective effect of X inactivation—all but one of the X chromosomes is inactivated.

Sex Chromosome Aneuploids: Male

Any individual with a Y chromosome is a male.

XXY Syndrome—Stefan's Story

Looking back, Stefan Schwarz's only indication of XXY syndrome was small testes. When his extra X chromosome was detected when he was 25, suddenly his personality quirks made sense. "I was very shy, reserved, and had trouble making friends. I would fly into rages for no apparent reason. My parents knew when I was very young that there was something about me that wasn't right," he recalls.

Many psychologists, psychiatrists, and therapists diagnosed "learning disabilities," and one even told Stefan he "was stupid and lazy, and would never amount to anything." But Stefan proved them wrong. He earned two bachelor's degrees, then started a successful career as a software engineer. Today he heads a support group for men with XXY syndrome.

About 1 in 500 males has the extra X chromosome that causes XXY (Klinefelter) syndrome. Severely affected men are underdeveloped sexually, with rudimentary testes and prostate glands and sparse pubic and facial hair. They have very long arms and legs, large hands and feet, and may develop breast tissue. XXY syndrome is the most common genetic or chromosomal cause of male infertility.

Testosterone injections during adolescence can limit limb lengthening and stimulate development of secondary sexual characteristics. Boys and men with XXY syndrome may be slow to learn, but they are usually not intellectually disabled unless they have more than two X chromosomes, which is rare.

Men with XXY syndrome have fathered children, with medical assistance. Doctors select sperm that contain only one sex chromosome and use the sperm to fertilize oocytes. However, sperm from men with XXY syndrome are more likely to have extra chromosomes—usually X or Y, but also autosomes—than sperm from men who do not have XXY syndrome.

XXYY Syndrome—Devon's Story

Devon's parents suspected early on that he was different. His problems were so common, however, that it was years before a chromosome check revealed an extra X and an extra Y.

Devon was late to sit, crawl, walk, and talk. In preschool, he had frequent outbursts and made inappropriate comments. He was tall and clumsy, and drooled and choked easily. Devon would run about flapping his arms, then hide under a chair. Severe ulcers formed on his legs.

By the second grade, Devon's difficulties alarmed his special education teacher, who suggested to Devon's parents, Drucilla and Neil, that they have his chromosomes checked. Theirs were normal. Devon must have been conceived from a very unusual oocyte meeting a very unusual sperm, both arising from nondisjunction. The extra sex chromosomes explained many of the boy's problems, and even a few that hadn't been recognized, such as curved pinkies, flat feet, and scoliosis. He began receiving testosterone injections so that his teen years would be more normal than his difficult childhood had been.

A male with an extra X chromosome *and* an extra Y chromosome was until recently classified as having Klinefelter syndrome. Males with the second Y chromosome have more severe behavioral problems and tend to develop foot and leg ulcers, resulting from poor venous circulation. Attention deficit disorder, obsessive compulsive disorder, and learning disabilities typically develop by adolescence. In the teen years, testosterone level is low, development of secondary sexual characteristics is delayed, and the testes are undescended. A man with XXYY syndrome is infertile.

XYY Syndrome

In 1961, a tall, healthy man, known for his boisterous behavior, had a chromosome check after fathering a child with Down syndrome. The man had an extra Y chromosome. A few other cases were detected over the next several years.

In 1965, researcher Patricia Jacobs published results of a survey among 197 inmates at Carstairs, a high-security prison in Scotland. Of twelve men with unusual chromosomes, seven had an extra Y. Might their violent or aggressive behavior be linked to their extra Y chromosome? Jacobs's findings were repeated for mental institutions, and soon after, Newsweek magazine ran a cover story on "congenital criminals."

Having an extra Y, known as Jacobs syndrome, became a legal defense for committing a violent crime.

In the early 1970s, newborn screens began in hospital nurseries in England, Canada, Denmark, and Boston. Social workers and psychologists visited XYY boys to offer "anticipatory guidance" to the parents on how to deal with their toddling future criminals. By 1974, geneticists and others halted the program, pointing out that singling out these boys on the basis of a few statistical studies was inviting self-fulfilling prophecy.

One male in 1,000 has an extra Y chromosome. Today, we know that 96 percent of XYY males are apparently normal. The only symptoms attributable to the extra chromosome may be great height, acne, and perhaps speech and reading problems. An explanation for the continued prevalence of XYY among mental-penal institution populations may be more psychological than biological. Large body size may lead teachers, employers, parents, and others to expect more of these people, and a few XYY individuals may deal with this stress aggressively.

Jacobs syndrome can arise from nondisjunction in the male, producing a sperm with two Y chromosomes that fertilizes an X-bearing oocyte. Geneticists have never observed a sex chromosome constitution of one Y and no X. Since the Y chromosome carries little genetic material, and the gene-packed X chromosome would not be present, the absence of so many genes makes development beyond a few cell divisions in a YO embryo impossible.

Key Concepts

1. Polyploids have extra sets of chromosomes and do not survive for long.

2. Aneuploids have extra or missing chromosomes. Nondisjunction during meiosis causes aneuploidy.

3. Trisomies are less severe than monosomies, and sex chromosome aneuploidy is less severe than autosomal aneuploidy.

4. Mitotic nondisjunction produces chromosomal mosaics.

5. Down syndrome (trisomy 21) is the most common autosomal aneuploid, followed by trisomies 18 and 13.

6. Sex chromosome aneuploid conditions include XO, triplo-X, XXY, XXYY, and XYY syndromes.

13.4 Atypical Chromosome Structure

A chromosome can be structurally atypical in several ways. It may have too much genetic material, too little, or a stretch of DNA that is inverted or moved and stuck onto a different type of chromosome (**figure 13.15**). Atypical chromosomes are balanced if they have the normal amount of genetic material or unbalanced if excess or deficient DNA results.

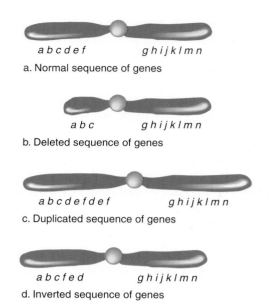

a. Normal sequence of genes

a b c d e f   g h i j k l m n

b. Deleted sequence of genes

a b c   g h i j k l m n

c. Duplicated sequence of genes

a b c d e f d e f   g h i j k l m n

d. Inverted sequence of genes

a b c f e d   g h i j k l m n

Figure 13.15 **Chromosome abnormalities.** If a hypothetical normal gene sequence appears as shown in **(a)**, then **(b)** represents a deletion, **(c)** a duplication, and **(d)** an inversion.

Deletions and Duplications

Ashley's Story

*Our daughter, Ashley Elizabeth Naylor (**figure 13.16**), was born August 12, 1988. The doctors suspected complications. Two weeks after her birth, chromosome analysis revealed cri-du-chat (cat cry) syndrome, also known as 5p⁻ syndrome because part of the short arm of one copy of chromosome 5 is missing. This is a rare disorder, we were told, and little could be offered to help our daughter. The doctors used the words "profoundly retarded."*

Ashley defied all the standard medical labels, as well as her doctors' expectations. Her spirit and determination enabled her to walk with the aid of a walker and express herself using sign language and a communication device. With early intervention and education, Ashley found the resources and additional encouragement she needed to succeed. In May of 1994, Ashley's small body could no longer support the spirit that inspired so many. She passed away after a long battle with pneumonia. Her physical presence is gone, but her message remains: hope.

Deletions and **duplications** are missing and extra DNA sequences, respectively. They are types of copy number variants (CNVs), introduced in chapter 12. The more genes involved, the more severe the associated syndrome. Ashley Naylor was missing much of her chromosome 5. **Figure 13.17** depicts a common duplication, of part of chromosome 15. Deletions and duplications often arise "*de novo*," which means that neither parent has the abnormality, and it is therefore new. In these cases, the symptoms can be attributed to the chromosomal abnormality.

A technique called comparative genomic hybridization is used to detect very small CNVs, which are also termed microdeletions and microduplications. The technique compares the

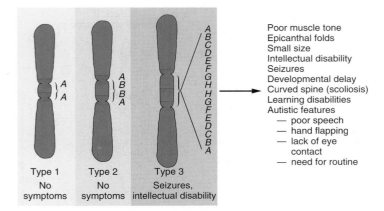

Poor muscle tone
Epicanthal folds
Small size
Intellectual disability
Seizures
Developmental delay
Curved spine (scoliosis)
Learning disabilities
Autistic features
— poor speech
— hand flapping
— lack of eye
    contact
— need for routine

Type 1
No
symptoms

Type 2
No
symptoms

Type 3
Seizures,
intellectual disability

Figure 13.17 **A duplication.** A study of duplications of parts of chromosome 15 revealed that small duplications do not affect the phenotype, but larger ones may. The letters indicate specific DNA sequences, which serve as markers to compare chromosome regions. Note that the duplication is also inverted.

Figure 13.16 **Ashley Naylor brought great joy to her family and community during her short life.** She had 5p⁻ syndrome.

Courtesy of Kathy Naylor.

abundance of copies of a particular CNV in the same amount of DNA from two sources—one with a medical condition, one healthy. Comparative genomic hybridization is being used increasingly to help narrow down diagnoses for children with autism, intellectual disability, learning disabilities, or just strange behavior. For example, the technique showed that a young boy who had difficulty concentrating and sleeping and would often scream for no reason had a small duplication in chromosome 7. A young girl plagued with head-banging behavior, digestive difficulties, severe constipation, and great sensitivity to sound had a microdeletion in chromosome 16. Other microdeletions cause male infertility.

Deletions and duplications can arise from chromosome rearrangements. These include translocations, inversions, and ring chromosomes.

Translocation Down Syndrome

Rhiannon's Story

When Rhiannon P. was born, while her parents marveled at her beauty, the obstetrician was disturbed by her facial features: the broad, tilted eyes and sunken nose looked like the face of a child with Down syndrome. The doctor might not have noticed, except that he knew that the mother, Felicia, had had two spontaneous abortions, a family history suggesting a chromosome problem. So the doctor looked for the telltale single crease in the palms of people with Down syndrome, and found it. Gently, he told Felicia and her husband Matt that he'd like to do a chromosome check.

Two days later, the new parents learned that their daughter has an unusual form of Down syndrome that she inherited from one of them, rather than the more common "extra chromosome" form. Since Matt's mother and sister had also had several miscarriages, the exchanged chromosomes likely came from his side. Karyotypes of Matt and Felicia confirmed this: Matt was a translocation carrier. One of his chromosome 14s had attached to one of his chromosome 21s, and distribution of the unusual chromosome in meiosis had led to various imbalances, depicted in figure 13.18.

Rhiannon has very mild Down syndrome. She has none of the physical problems associated with the condition, and she does well in school with the help of a special education teacher. Matt and Felicia chose to see the bright side—each conception will have a one-in-three chance of having balanced chromosomes. Some day they hope to give Rhiannon a brother or sister.

In a translocation, different (nonhomologous) chromosomes exchange or combine parts. Translocations can be inherited because they can be present in carriers, who have the normal amount of genetic material, but it is rearranged. A translocation can affect the phenotype if it breaks a gene or leads to duplications or deletions in the chromosomes of offspring.

There are two major types of translocations, and rarer types. In a **Robertsonian translocation,** the short arms of two different acrocentric chromosomes break, leaving sticky ends on the two long arms that join, forming a single, large chromosome with two long arms (chromosome 14/21 in figure 13.18). The tiny short arms are lost, but their DNA sequences are repeated elsewhere in the genome, so the loss does not cause symptoms. The person with the large, translocated chromosome, called a **translocation carrier,** has 45 chromosomes, but may not have symptoms if no crucial genes have been deleted or damaged. Even so, he or she may produce unbalanced gametes—sperm or oocytes with too many or too few genes. This can lead to spontaneous abortion or birth defects.

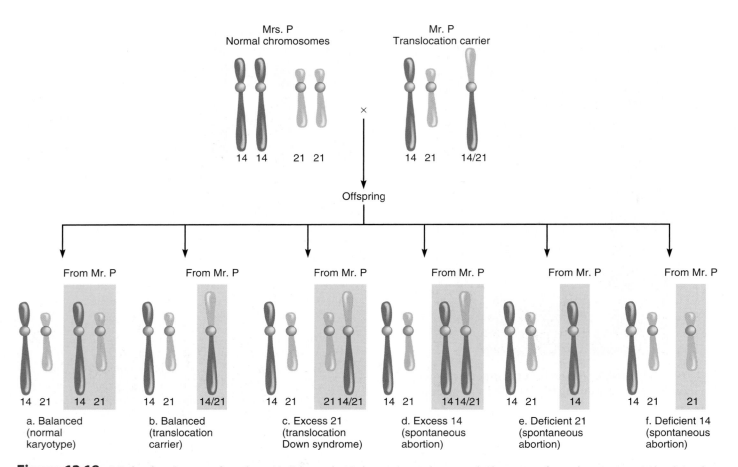

Figure 13.18 **A Robertsonian translocation.** Mr. P. has only 45 chromosomes because the long arm of one chromosome 14 has joined the long arm of one chromosome 21. He has no symptoms. Mr. P. makes six types of sperm cells, and they determine the fates of offspring. **(a)** A sperm with one normal chromosome 14 and one normal 21 yields a normal child. **(b)** A sperm carrying the translocated chromosome produces a child who is a translocation carrier, like Mr. P. **(c)** If a sperm contains Mr. P.'s normal 21 and his translocated chromosome, the child receives too much chromosome 21 material and has Down syndrome. **(d)** A sperm containing the translocated chromosome and a normal 14 leads to excess chromosome 14 material, which is lethal in the embryo or fetus. If a sperm lacks either chromosome 21 **(e)** or 14 **(f)**, it leads to monosomies, which are lethal prenatally. (Chromosome arm lengths are not precisely accurate.)

In 1 in 20 cases of Down syndrome, a parent has a Robertsonian translocation between chromosome 21 and another, usually chromosome 14. That parent produces some gametes that lack either of the involved chromosomes and some gametes that have extra material from one of the translocated chromosomes. In such a case, each fertilized ovum has a 1 in 2 chance of ending in spontaneous abortion, and a 1 in 6 chance of developing into an individual with Down syndrome. The risk of giving birth to a child with Down syndrome is theoretically 1 in 3, because the spontaneous abortions are not births. However, because some Down syndrome fetuses spontaneously abort, the actual risk of a couple in this situation having a child with Down syndrome is about 15 percent. The other two outcomes—a fetus with normal chromosomes or a translocation carrier like the parent—have normal phenotypes. Either a male or a female can be a translocation carrier, and the condition is not related to age.

In the second major type of translocation, a **reciprocal translocation,** two different chromosomes exchange parts

(**figure 13.19**). About 1 in 600 people is a carrier for a reciprocal translocation. FISH can be used to highlight the involved chromosomes. If the chromosome exchange does not break any genes, then a person who has both translocated chromosomes is healthy and a translocation carrier. He or she has the normal amount of genetic material, but it is rearranged. A reciprocal translocation carrier can have symptoms if one of the two breakpoints lies in a gene, disrupting its function. For example, a translocation between chromosomes 11 and 22 causes infertility in males and recurrent pregnancy loss in females. Sometimes, a *de novo* translocation arises in a gamete. This can lead to a new individual with a disorder, if fertilization occurs, as opposed to inheriting a translocated chromosome from a parent who is a carrier.

Reciprocal translocations do not occur at random among the chromosomes, but involve specific chromosomes that have unstable parts. It is a little like athletes who do the same sport injuring the same muscles. Vulnerable parts of chromosomes arise where the DNA is so symmetrical in sequence

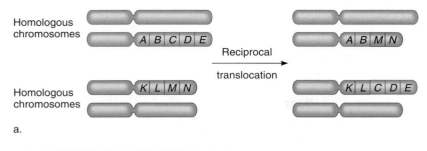

Homologous chromosomes

Homologous chromosomes

Reciprocal translocation

a.

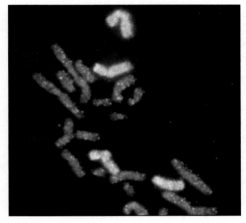

b.

Figure 13.19 A reciprocal translocation. In a reciprocal translocation, two nonhomologous chromosomes exchange parts. In **(a),** genes *C, D,* and *E* on the blue chromosome exchange positions with genes *M* and *N* on the red chromosome. Part **(b)** highlights a reciprocal translocation using FISH. The pink chromosome with the dab of blue, and the blue chromosome with a small section of pink, are the translocated chromosomes.

that complementary base pairing occurs within the same DNA strand, folding it into loops and crosses during DNA replication. These contortions can cause double-strand breaks, which enable parts of two different chromosomes to switch. A reciprocal translocation results.

A rare type of translocation is an insertional translocation. In this situation, part of one chromosome inserts into a nonhomologous chromosome. Symptoms may result if a vital gene is disrupted or if genetic material is lost or present in excess. The chapter opener describes an insertional translocation in a family.

A carrier of any type of translocation can produce some unbalanced gametes—sperm or oocytes that have deletions or duplications of some of the genes in the translocated chromosomes. The resulting phenotype depends upon the particular genes that the chromosomal rearrangement disrupts and whether they are extra or missing. A translocation and a deletion can cause the same syndrome if they affect the same part of a chromosome.

A genetic counselor suspects a translocation when a family has a history of birth defects, pregnancy loss, and/or stillbirth. Prenatal testing may also reveal a translocation in a fetus, which can then be traced back to a parent who is a translocation carrier—or sometimes it is *de novo.*

Inversions

Madison's Story

Madison and Grant were excited about getting the results of the amniocentesis because Madison had never carried a pregnancy this far before. They grew alarmed when the doctor's office called and asked them to come in for the results.

Prepared for the worst, the couple was surprised and confused to learn that the fetus had an inverted chromosome. Some of the bands that normally appear on chromosome 11 were flipped. Before the genetic counselor would give them information on genes that might be affected, she advised that the parents-to-be have their chromosomes checked. Madison had the same inversion as the fetus! Because Madison was healthy, the unusual chromosome would likely not harm their daughter. When she was older, however, she might, like her mother, experience pregnancy loss.

An inverted sequence of chromosome bands disrupts important genes and harms health in only 5 to 10 percent of cases. If neither parent has the inversion, then it arose in a gamete. Effects may depend on which genes are involved. The human genome sequence can be consulted to identify genes that might be implicated in a particular inversion.

Like a translocation carrier, an adult who is heterozygous for an inversion can be healthy, but have reproductive problems. One woman had an inversion in the long arm of chromosome 15 and had two spontaneous abortions, two stillbirths, and two children with multiple problems who died within days of birth. She did eventually give birth to a healthy child. How did the inversion cause these problems?

Inversions with such devastating effects can be traced to meiosis, when a crossover occurs between the inverted chromosome segment and the noninverted homolog. To allow the genes to align, the inverted chromosome forms a loop. When crossovers occur within the loop, some areas are duplicated and some deleted in the resulting recombinant chromosomes. In inversions, the atypical chromosomes result from the chromatids that crossed over.

Two types of inversions are distinguished by the position of the centromere relative to the inverted section. A **paracentric inversion** does not include the centromere (**figure 13.20**). A single crossover within the inverted segment gives rise to two normal and two very atypical chromosomes. The other two chromosomes are normal. One abnormal chromosome retains both centromeres and is termed dicentric. When the cell divides, the two centromeres are pulled to opposite sides of the cell, and the chromosome breaks, leaving pieces with extra or missing segments. The second type of atypical chromosome resulting from a crossover within an inversion loop is a small piece that lacks a centromere, called an acentric fragment. When the cell divides, the fragment is lost because a centromere is required for cell division.

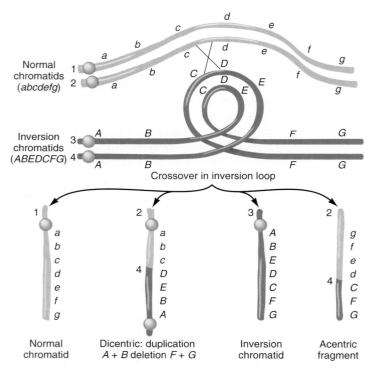

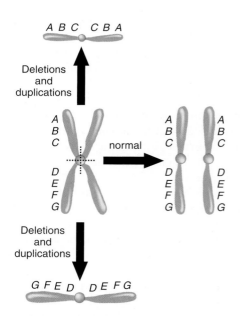

A **pericentric inversion** includes the centromere within the loop. A crossover in it produces two chromosomes that have duplications and deletions, but one centromere each (**figure 13.21**).

Isochromosomes and Ring Chromosomes

Another meiotic error that leads to unbalanced genetic material is the formation of an **isochromosome,** which is a chromosome that has identical arms. This occurs when, during division, the centromeres part in the wrong plane (**figure 13.22**). Isochromosomes are known for chromosomes 12 and 21 and for the long arms of the X and the Y. Some women with Turner syndrome are not the more common XO, but have an isochromosome with the long arm of the X chromosome duplicated but the short arm absent.

Chromosomes shaped like rings form in 1 out of 25,000 conceptions. Ring chromosomes may arise when telomeres are lost, leaving sticky ends that adhere. Exposure to radiation can form rings. They can form from any chromosome, and may be one of the 46 chromosomes or an extra.

Most ring chromosomes consist of DNA repeats and do not affect health. Some do, however. This is the case for ring chromosome 20. When 6-year-old Cara Ford lost the ability to walk, talk, or eat and developed seizures, no physician in her native United Kingdom could offer a diagnosis. Physicians at an epilepsy center in New York City detected the ring chromosome, and treated the seizures (**figure 13.23**). Cara's father, Stewart Ford, started the Ring Chromosome 20 Foundation

Figure 13.20 Paracentric inversion. A paracentric inversion in one chromosome leads to one normal chromatid, one inverted chromatid, one with two centromeres (dicentric), and one with no centromere (an acentric fragment) if a crossover occurs with the normal homolog. The letters *a* through *g* denote genes.

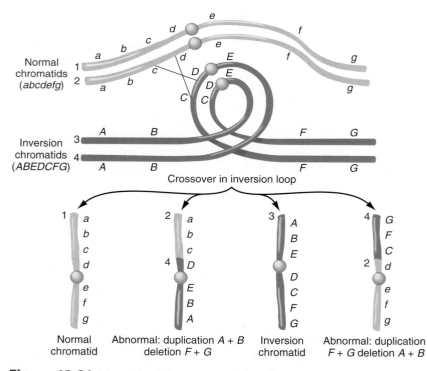

Figure 13.21 Pericentric inversion. A pericentric inversion in one chromosome leads to two chromatids with duplications and deletions, one normal chromatid, and one inverted chromatid if a crossover occurs with the normal homolog.

Figure 13.22 Isochromosomes have identical arms. They form when chromatids divide along the wrong plane (in this depiction, horizontally rather than vertically).

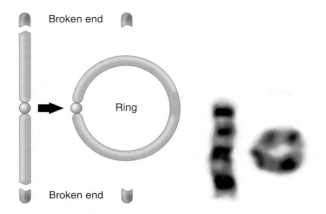

Figure 13.23 **A ring chromosome.** A ring chromosome may form if the chromosome's tips (telomeres) break, forming sticky ends. Genes can be lost or disrupted, possibly causing symptoms. Ring chromosome 20, for example, causes seizures.

to alert physicians to test patients with seizures for atypical chromosomes.

Table 13.7 summarizes causes of different types of chromosomal aberrations.

Key Concepts

1. Chromosome rearrangements can cause deletions and duplications.
2. In a Robertsonian translocation the long arms of two different acrocentric chromosomes join.
3. In a reciprocal translocation chromosomes exchange parts.
4. If a translocation leads to a deletion or duplication or disrupts a gene, symptoms may result.
5. Gene duplications and deletions can occur in isochromosomes and ring chromosomes and when inverted chromatids cross over.
6. An isochromosome has two identical arms, introducing duplications and deletions.
7. Ring chromosomes form when telomeres are missing.

13.5 Uniparental Disomy—A Double Dose from One Parent

If nondisjunction occurs in a sperm and oocyte that join, a pair of chromosomes (or their parts) can come solely from one parent, rather than one from each parent, as Mendel's law of segregation predicts. For example, if a sperm lacking a chromosome 14 fertilizes an ovum with two copies of that chromosome, an individual with the normal 46 chromosomes results, but the two chromosome 14s come only from the female.

Inheriting two chromosomes or chromosome segments from one parent is called **uniparental disomy (UPD)** ("two

Table 13.7	Causes of Chromosomal Aberrations
Abnormalities	**Causes**
Numerical Abnormalities	
Polyploidy	Error in cell division (meiosis or mitosis) in which not all chromatid pairs separate in anaphase
	Multiple fertilization
Aneuploidy	Nondisjunction (in meiosis or mitosis) leading to lost or extra chromosomes
Structural Abnormalities	
Deletions and duplications	Translocation
	Crossover between a chromosome that has a pericentric inversion and its noninverted homolog
Translocation	Exchange between nonhomologous chromosomes
Inversion	Breakage and reunion of fragment in same chromosome, but with wrong orientation
Dicentric and acentric	Crossover between a chromosome with a paracentric inversion and its noninverted homolog
Ring chromosome	A chromosome loses telomeres and the ends fuse, forming a circle

bodies from one parent"). UPD can also arise from a trisomic embryo in which some cells lose the extra chromosome, leaving two homologs from one parent. For example, an embryo may have trisomy 21, with the extra chromosome 21 coming from the father. If in some cells the chromosome 21 from the mother is lost, then both remaining copies of the chromosome are from the father.

Because UPD requires the simultaneous occurrence of two very rare events—either nondisjunction of the same chromosome in sperm and oocyte, or trisomy followed by chromosome loss—it is very rare. In addition, many cases are probably never seen, because bringing together identical homologs inherited from one parent could give the fertilized ovum a homozygous set of lethal alleles. Development would halt. Other cases of UPD may go undetected if they cause known recessive conditions and both parents are assumed to be carriers, when actually only one parent contributed to the offspring's illness. This was how UPD was discovered.

In 1988, Arthur Beaudet of the Baylor College of Medicine saw a very unusual patient with cystic fibrosis. Beaudet was comparing CF alleles of the patient to those of her parents, and he found that only the mother was a carrier—the father had two normal alleles. Beaudet constructed haplotypes for each parent's chromosome 7, which includes the CF gene, and he found that the daughter had two copies from her mother, and none from her father (**figure 13.24**). How did this happen?

Apparently, in the patient's mother, nondisjunction of chromosome 7 in meiosis II led to formation of an oocyte

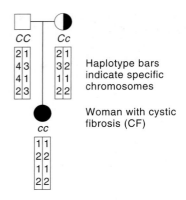

Figure 13.24 Uniparental disomy. Uniparental disomy doubles part of one parent's genetic contribution. In this family, the woman with CF inherited two copies of her mother's chromosome 7, and neither of her father's. Unfortunately, it was the chromosome with the disease-causing allele that she inherited in a double dose.

bearing two identical copies of the chromosome, instead of the usual one. A sperm that had also undergone nondisjunction and lacked a chromosome 7 then fertilized the abnormal oocyte. The mother's extra genetic material compensated for the father's deficit, but unfortunately, the child inherited a double dose of the mother's chromosome that carried the mutant CF allele. In effect, inheriting two of the same chromosome from one parent shatters the protection that combining genetic material from two individuals offers, a protection that is the defining characteristic of sexual reproduction.

UPD may also cause disease if it removes the contribution of the important parent for an imprinted gene. Recall from chapter 6 that an imprinted gene is expressed if it comes from one parent, but silenced if it comes from the other (see figure 6.13). If UPD removes the parental genetic material

that must be present for a critical gene to be expressed, a mutant phenotype results. The classic example is the 20 to 30 percent of Prader-Willi syndrome and Angelman syndrome cases caused by UPD (see figure 6.14). These disorders arise from mutations in different genes that are closely linked in a region of the long arm of chromosome 15, where imprinting occurs. They both cause intellectual disability and a variety of other symptoms, but are quite distinct.

In 1989, researchers found that some children with Prader-Willi syndrome have two parts of the long arm of chromosome 15 from their mothers. The disease results because the father's Prader-Willi gene must be expressed for the child to avoid the associated illness. For Angelman syndrome, the situation is reversed. Children have a double dose of their father's DNA in the same chromosomal region implicated in Prader-Willi syndrome, with no maternal contribution. The mother's gene must be present for health.

People usually learn their chromosomal makeup only when something goes wrong—when they have a family history of reproductive problems, exposure to a toxin, cancer, or symptoms of a known chromosomal disorder. While researchers analyze the human genome sequence, chromosome studies will continue to be part of medical care—beginning before birth.

Key Concepts

1. Uniparental disomy (UPD) results when two chromosomes or chromosome parts are inherited from the same parent.
2. It can arise from two nondisjunction events or a trisomy and subsequent chromosome loss.
3. UPD can cause disease if it creates a homozygous recessive condition, or if it disrupts imprinting.

Summary

13.1 Portrait of a Chromosome

1. **Cytogenetics** is the study of chromosome aberrations and their effects on phenotypes.
2. **Heterochromatin** stains darkly and harbors many DNA repeats. **Euchromatin** is light staining and contains many protein-encoding genes.
3. A chromosome consists of DNA and proteins. Essential parts are the **telomeres, centromeres,** and origin of replication sites.
4. Centromeres include DNA repeats and proteins that enable the cell to divide.
5. Subtelomeres have telomere-like repeats that gradually change inward toward the centromere, as protein-encoding genes predominate.
6. Chromosomes are distinguishable by size, centromere position, satellites, and staining patterns. They are displayed in **karyotypes**.

7. A **metacentric** chromosome has two fairly equal arms. A **submetacentric** chromosome has a large arm and a short arm. An **acrocentric** chromosome's centromere is near a tip, so that it has one long arm and one very short arm.

13.2 Visualizing Chromosomes

8. Chromosomes can be obtained from any cell that has a nucleus. Prenatal diagnostic techniques include **amniocentesis, chorionic villus sampling, chromosome microarray analysis,** and fetal cell and nucleic acid detection. Serum markers screen fetuses at increased risk for chromosomal abnormalities.
9. Fluorescence *in situ* hybridization provides more specific chromosome bands than dyes. Ideograms display chromosome bands.
10. Chromosomal shorthand indicates chromosome number, sex chromosome constitution, and type of abnormality.

13.3 Atypical Chromosome Number

11. A **euploid** somatic human cell has 22 pairs of autosomes and one pair of sex chromosomes.

12. **Polyploid** cells have extra chromosome sets.

13. **Aneuploids** have extra or missing chromosomes. **Trisomies** (an extra chromosome) are less harmful than **monosomies** (lack of a chromosome), and sex chromosome aneuploidy is less severe than autosomal aneuploidy. **Nondisjunction** is uneven distribution of chromosomes in meiosis. It causes aneuploidy. Most autosomal aneuploids cease developing as embryos.

13.4 Atypical Chromosome Structure

14. **Deletions** and/or **duplications** result from crossing over after pairing errors in synapsis. Crossing over in an inversion heterozygote can also generate deletions and duplications. Microdeletions and microduplications explain many disorders.

15. In a **Robertsonian translocation,** the short arms of two acrocentric chromosomes break, leaving sticky ends on the long arms that join to form an unusual, large chromosome.

16. In a **reciprocal translocation,** two nonhomologous chromosomes exchange parts.

17. An insertional translocation places a DNA sequence from one chromosome into a nonhomologous chromosome.

18. A **translocation carrier** may have an associated phenotype and produces some unbalanced gametes.

19. A heterozygote for an inversion may have reproductive problems if a crossover occurs between the inverted region and the noninverted homolog, generating deletions and duplications. A **paracentric inversion** does not include the centromere; a **pericentric inversion** does.

20. **Isochromosomes** repeat one chromosome arm but delete the other. They form when the centromere divides in the wrong plane in meiosis. Ring chromosomes form when telomeres are removed, leaving sticky ends that adhere.

13.5 Uniparental Disomy—A Double Dose from One Parent

21. In **uniparental disomy,** a chromosome or part of one doubly represents one parent. It can result from nondisjunction in both gametes, or from a trisomic cell that loses a chromosome.

22. Uniparental disomy causes symptoms if it creates a homozygous recessive state associated with an illness, or if it affects an imprinted gene.

www.mhhe.com/lewisgenetics10

Answers to all end-of-chapter questions can be found at **www.mhhe.com/lewisgenetics10.** You will also find additional practice quizzes, animations, videos, and vocabulary flashcards to help you master the material in this chapter.

Review Questions

1. What are the essential components of a chromosome? Of a centromere?

2. How does the DNA sequence change with distance from the telomere?

3. How are centromeres and telomeres alike?

4. Distinguish among a euploid, aneuploid, and polyploid.

5. What happens during meiosis to produce
   a. an aneuploid?
   b. a polyploid?
   c. increased risk of trisomy 21 in the offspring of a woman over age 40 at the time of conception?
   d. recurrent spontaneous abortions to a couple in which the man has a pericentric inversion?
   e. several children with Down syndrome in a family where one parent is a translocation carrier?

6. A human liver has patches of octaploid cells—they have eight sets of chromosomes. Explain how this might arise.

7. Describe an individual with each of the following chromosomes. List gender and possible phenotype.

   a. 47,XXX
   b. 45,X
   c. 47,XX, trisomy 21

8. Which chromosomal anomaly might you expect to find more frequently among members of the National Basketball Association than in the general public? Cite a reason.

9. Explain the difference between a fetus found to have a trisomy and one found to be triploid.

10. About 80 percent of cases of Edward syndrome are caused by trisomy 18; 10 percent are caused by mosaic trisomy 18, and 10 percent are attributed to translocation. Distinguish among these three chromosome aberrations.

11. List three examples illustrating the idea that the amount of genetic material involved in a chromosomal aberration affects the severity of the associated phenotype.

12. List three types of chromosomal aberrations that can cause duplications and/or deletions, and explain how they do so.

13. Distinguish among three types of translocations.

14. Why would having the same inversion on both members of a homologous chromosome pair *not* lead to unbalanced gametes, as having only one inverted chromosome would?

15. Define or describe the following technologies:
    a. FISH
    b. amniocentesis
    c. chorionic villus sampling
    d. fetal cell sorting

16. Why are trisomies 13 and 18 more common at birth than trisomies 5 or 16?

17. How many chromosomes would a person have who has Klinefelter syndrome and also trisomy 21?

18. Explain why a female cannot have XXY syndrome and a male cannot have XO syndrome.

19. List three causes of Turner syndrome.

20. Explain how the sequence of genes on an isochromosome differs from normal.

Applied Questions

1. Identify the structures and/or DNA sequences that must be present for a chromosome to carry information and withstand the forces of cell division.

2. Amniocentesis indicates that a fetus has the chromosomal constitution 46, XX,del(5)(p15). What does this mean? What might the child's phenotype be?

3. What type of test could determine whether a triploid infant resulted from a diploid oocyte fertilized by a haploid sperm, or from two sperm fertilizing one oocyte?

4. For an exercise in a college genetics laboratory course, a healthy student constructs a karyotype from a cell from the inside of her cheek. She finds only one chromosome 3 and one chromosome 21, plus two unusual chromosomes that do not seem to have matching partners.
   a. What type of chromosomal abnormality does she have?
   b. Why doesn't she have any symptoms?
   c. Would you expect any of her relatives to have any particular medical problems?

5. A fetus ceases developing in the uterus. Several of its cells are karyotyped. Approximately 75 percent of the cells are diploid, and 25 percent are tetraploid (four copies of each chromosome). What do you think happened? When in development did it probably occur?

6. Distinguish among Down syndrome caused by aneuploidy, mosaicism, and translocation.

7. A couple has a son diagnosed with XXY syndrome. Explain how the son's chromosome constitution could have arisen from either parent.

8. DiGeorge syndrome (MIM 188400) causes atypical parathyroid glands, a heart defect, and an underdeveloped thymus gland. About 85 percent of patients have a microdeletion of part of chromosome 22. A girl, her mother, and a maternal aunt have very mild DiGeorge syndrome. They all have a reciprocal translocation of chromosomes 22 and 2.
   a. How can a microdeletion and a translocation cause the same symptoms?
   b. Why were the people with the translocation less severely affected than the people with the microdeletion?
   c. What other problems might arise in the family with the translocation?

9. From 2 to 6 percent of people with autism have an extra chromosome that consists of two long arms of chromosome 15. It includes two copies of the chromosome 15 centromere. Two normal copies of the chromosome are also present. What type of chromosome abnormality in a gamete can lead to this karyotype, which is called isodicentric 15?

10. Consider a science fiction plot. How could Robertsonian translocations, which occur in 1 in 1,000 people, lead to formation of a new species?

Web Activities

1. Go to one of the websites listed in table 13.3, or find a similar disease organization, and learn about daily life with a particular chromosomal abnormality. Identify a challenge or problem common to several chromosomal syndromes, and describe how families cope with the problem.

2. Go to the website for the Baylor College of Medicine Medical Genetics Laboratories (http://www.bcm.edu/cma/index.htm). Select "Reference Table," and then select "Abnormalities Detected."
   a. Select an abnormality.
   b. Describe the mutation at the chromosomal level.
   c. Click on the MIM # on the left, go to MIM, and use the information to describe the disorder.

3. Go to the website for the Genetic Science Learning Center at the Eccles Institute of Human Genetics at the University of Utah. Follow the instructions to create a karyotype.

4. Visit the website for the Human Genome Landmarks poster. Select a chromosome, and use Mendelian Inheritance in Man (MIM) to describe four traits or disorders associated with it. Or, consult the website for the Human Chromosome Launchpad for information on four genes carried on a specific chromosome.

Case Studies and Research Results

1. An ultrasound of a pregnant woman detects a fetus and a similarly sized and shaped structure that has disorganized remnants of facial features at one end. Amniocentesis on both structures reveals that the fetus is 46,XX, but cells of the other structure are 47,XX,trisomy 2. No cases of trisomy 2 infants have ever been reported. However, individuals who are mosaics for trisomy 2 have a collection of defects, including a rotated and underdeveloped small intestine, a small head, a hole in the diaphragm, and seizures.

   a. How do the chromosomes of cells from the fetus and the other structure differ?

   b. What is the process that occurred during meiosis to yield the bizarre structure?

   c. Use this information to explain why children with a complete extra chromosome 2 are not seen, even though people with an extra chromosome 21 can live many years.

   d. List two factors that determine the type and severity of abnormalities in an individual who is an aneuploid mosaic.

2. Two sets of parents who have children with Down syndrome meet at a clinic. The Phelps know that their son has trisomy 21. The Watkins have two affected children, and Mrs. Watkins has had two spontaneous abortions. Why should the Watkins be more concerned about future reproductive problems than the Phelps? How are the offspring of the two families different, even though they have the same symptoms?

3. The genomes of four of 291 people with intellectual disability have a microdeletion in chromosome 17q21.3. The children have large noses, delayed speech, and mild intellectual disability. Each had a parent with an inversion in the same part of chromosome 17.

   a. Which arm of chromosome 17 is implicated in this syndrome?

   b. How can an inversion in a parent's chromosome cause a deletion in a child's chromosome?

   c. What other type of chromosome abnormality might occur in these children's siblings?

4. A 38-year-old woman, Dasheen, has amniocentesis. She learns that the fetus she is carrying has an inversion in chromosome 9 and a duplication in chromosome 18. She and her husband Franco have their chromosomes tested, and they learn that she has the duplication and Franco has the inversion. Both of the parents are healthy. Should they be concerned about the health of the fetus? Cite a reason for your answer.

5. Return to the chapter opener.

   a. What type of translocation do Esteban and Maribella have?

   b. Why doesn't their father have symptoms resulting from his atypical chromosomes?

   c. What are two possible chromosome configurations for Marcos?

A forensic scientist consults a DNA profile. The black bars represent short tandem repeats that form patterns used to exclude suspects in a crime.

Constant Allele Frequencies

Learning Outcomes

14.1 Population Genetics Underlies Evolution

1. State the unit of information of genetics at the population level.
2. Define *gene pool*.
3. List the five processes that cause microevolutionary change.
4. State the consequence of macroevolutionary change.

14.2 Constant Allele Frequencies

5. State the genotypes represented in each part of the Hardy-Weinberg equation.
6. Explain the conditions necessary for Hardy-Weinberg equilibrium.

14.3 Applying Hardy-Weinberg Equilibrium

7. Explain how the Hardy-Weinberg equilibrium uses population incidence statistics to predict the probability of a particular phenotype.

14.4 DNA Profiling Uses Hardy-Weinberg Assumptions

8. Explain how parts of the genome that are in Hardy-Weinberg equilibrium can be used to identify individuals.

The Big Picture: Human genetics at the population level considers allele frequencies. Parts of the genome that have changed over time enable us to trace our origins, migrations, and relationships. Parts of the genome that do not change provide a way to distinguish individuals.

Postconviction DNA Testing

Josiah Sutton had served 4 1/2 years of a 25-year sentence for rape when he was exonerated, thanks to the Innocence Project. The nonprofit legal clinic and public policy organization created in 1992 has so far used DNA retesting to free more than 260 wrongfully convicted prisoners, most of whom were "poor, forgotten, and have used up all legal avenues for relief," according to the website (www.innocenceproject.com). Sutton became a suspect after a woman in Houston identified him and a friend on the street 5 days after she had been raped, threatened with a gun, and left in a field. The two young men supplied saliva and blood samples, from which DNA profiles were done and compared to DNA profiles from semen found in the victim and in her car. At the trial, a crime lab employee testified that the probability that Sutton's DNA matched that of the evidence by chance was 1 in 694,000, leading to a conviction. Jurors ignored the fact that Sutton's physical description did not match the victim's description of her assailant.

The DNA evidence came from more than one individual, yielded different results when the testing was repeated, and most importantly, looked at only seven of the parts of the genome that are typically compared in a DNA profile, or fingerprint. Doing the test correctly revised the statistics dramatically: Sutton's pattern was shared not with 1 in 694,000 black men, as had originally been claimed, but with 1 in 16.

While in jail, Sutton read about DNA profiling and requested independent testing, but was refused. Then he got lucky. Journalists investigating the Houston crime laboratory became interested in his case, contacted a noted criminologist, and the Innocence Project became involved. Retesting the DNA evidence set him free. DNA profiling is a direct application of population genetics.

14.1 Population Genetics Underlies Evolution

The language of genetics at the family and individual levels is written in DNA sequences. At the population level, the language of genetics is allele (gene variant) frequencies. It is at the population level that genetics goes beyond science, embracing information from history, anthropology, human behavior, and sociology. Population genetics enables us to trace our beginnings, understand our diversity today, and imagine the future.

A **population** is any group of members of the same species in a given geographical area that can mate and produce fertile offspring (**figure 14.1**). Examples of human populations are the students in a class, a stadium full of people, and the residents of a community, state, or nation. **Population genetics** is a branch of genetics that considers all the alleles in a population, which constitute the **gene pool**. The "pool" refers to a collection of gametes; an offspring represents two gametes from the pool. Alleles can move between populations when individuals migrate and mate. This movement, termed gene flow, underlies evolution, which is explored in the next two chapters.

Thinking about genes at the population level begins by considering frequencies—that is, how often a particular gene variant occurs in a particular population. Such frequencies can be calculated for alleles, genotypes, or phenotypes. For example, an allele frequency for the cystic fibrosis (CF) gene might be the number of $\Delta F508$ alleles among the residents of San Francisco. $\Delta F508$ is the most common allele that, when homozygous, causes the disorder. The allele frequency derives from the two $\Delta F508$ alleles in each person with CF, plus alleles carried in heterozygotes, considered as a proportion of all alleles for that gene in the gene pool of San Francisco. The genotype frequencies are the proportions of heterozygotes and the two types of homozygotes in the population. Finally, a phenotypic

frequency is simply the percentage of people in the population who have CF (or who do not). With multiple alleles for a single gene, the situation becomes more complex because there are many more phenotypes and genotypes to consider.

Phenotypic frequencies are determined empirically—that is, by observing how common a condition or trait is in a population. These figures have value in genetic counseling in estimating the risk that a particular inherited disorder will occur in an individual when there is no family history of the illness. **Table 14.1** shows disease incidence for phenylketonuria (PKU), an inborn error of metabolism that causes intellectual disability unless the person follows a special, low-protein diet from birth. Note how the frequency differs in different populations.

On a broader level, shifting allele frequencies in populations reflect small steps of genetic change, called **microevolution**. These small, step-by-step changes alter genotype frequencies and underlie evolution. Genotype frequencies rarely stay constant. They can change when any of the following conditions are met:

1. Individuals of one genotype are more likely to choose to reproduce with each other than with individuals of other genotypes (*nonrandom mating*).
2. Individuals *migrate* between populations.
3. Reproductively isolated small groups form within or separate from a larger population (*genetic drift*).
4. *Mutation* introduces new alleles into a population.
5. People with a particular genotype are more likely to produce viable, fertile offspring under a specific environmental condition than individuals with other genotypes (*natural selection*).

In today's world, all of these conditions, except mutation, are quite common. Therefore, genetic equilibrium—when allele frequencies are *not* changing—is rare. Put another way, given our tendency to pick our own partners and move about, microevolution is not only possible, but also nearly unavoidable. (Chapter 15 considers these factors in depth.)

When enough microevolutionary changes accumulate to keep two fertile organisms of opposite sex from producing fertile offspring together, a new species forms. Changes that are

Figure 14.1 **A population is a group of organisms of the same species living in the same place.** Populations of sexually reproducing organisms include many genetic variants. This genetic diversity gives the group a flexibility that enhances species survival. To us, these hippos look alike, but they can undoubtedly recognize phenotypic differences in each other.

Table 14.1	Frequency of PKU in Various Populations
Population	**Frequency of PKU**
Chinese	1/16,000
Irish, Scottish, Yemenite Jews	1/5,000
Japanese	1/119,000
Swedes	1/30,000
Turks	1/2,600
U.S. Caucasians	1/10,000

great enough to result in speciation are termed **macroevolution**. This can happen through many small changes over time, and/or a few changes that greatly affect the phenotype.

Before we consider the pervasive genetic evidence for evolution, this chapter discusses the interesting, but unusual, situation in which certain allele frequencies stay constant. This is a condition called **Hardy-Weinberg equilibrium**.

> ### Key Concepts
>
> 1. Population genetics is the study of allele frequencies in groups of organisms of the same species in the same geographic area.
> 2. The genes in a population comprise its gene pool.
> 3. Microevolution reflects changes in allele frequencies in populations. It is not occurring if allele frequencies stay constant over generations (Hardy-Weinberg equilibrium).
> 4. Five factors can change genotype frequencies: nonrandom mating, migration, genetic drift, mutation, and natural selection.

14.2 Constant Allele Frequencies

Population genetics looks at phenotypes and genotypes among many individuals. Allele frequencies reveal the underlying rules. Tracking allele frequencies from one generation to the next can reveal evolution in action—or, if allele frequencies don't change, the state of Hardy-Weinberg equilibrium.

Hardy-Weinberg Equilibrium

In 1908, a Cambridge University mathematician named Godfrey Harold Hardy (1877–1947) and Wilhelm Weinberg (1862–1937), a German physician interested in genetics, independently used algebra to explain how allele frequencies can be used to predict phenotypic and genotypic frequencies in populations of diploid, sexually reproducing organisms.

Hardy unintentionally cofounded the field of population genetics with a simple letter published in the journal *Science*—he did not consider his idea to be worthy of the more prestigious British journal *Nature*. The letter began with a curious mix of modesty and condescension:

> *I am reluctant to intrude in a discussion concerning matters of which I have no expert knowledge, and I should have expected the very simple point which I wish to make to have been familiar to biologists.*

Hardy continued to explain how mathematically inept biologists had incorrectly deduced from Mendel's work that dominant traits would increase in populations while recessive traits would become rarer. At first glance, this seems logical. However, it is untrue because recessive alleles enter populations by mutation or migration and are maintained in heterozygotes.

Recessive alleles also become more common when they confer a reproductive advantage, thanks to natural selection.

Hardy and Weinberg disproved the assumption that dominant traits increase while recessive traits decrease using the language of algebra. The expression of population genetics in algebraic terms begins with the simple equation

$$p + q = 1.0$$

where p represents all dominant alleles for a gene, and q represents all recessive alleles. The expression "$p + q = 1.0$" simply means that all the dominant alleles and all the recessive alleles comprise all the alleles for that gene in a population.

Next, Hardy and Weinberg described the possible genotypes for a gene with two alleles using the binomial expansion

$$p^2 + 2pq + q^2 = 1.0$$

In this equation, p^2 represents the percentage of homozygous dominant individuals, q^2 represents the percentage of homozygous recessive individuals, and $2pq$ represents the percentage of heterozygotes (**figure 14.2**). The letter p designates the frequency of a dominant allele, and q is the frequency of a recessive allele. **Figure 14.3** shows how the binomial expansion is derived from allele frequencies. Note that the derivation is conceptually the same as tracing alleles in a monohybrid cross.

The binomial expansion used to describe genes in populations became known as the Hardy-Weinberg equation. It can

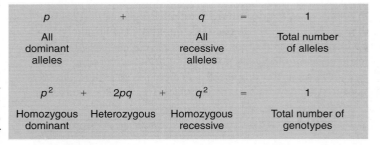

Figure 14.2 **The Hardy-Weinberg equation in English.**

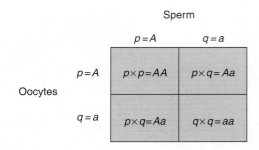

Figure 14.3 **Source of the Hardy-Weinberg equation.** A variation on a Punnett square reveals how random mating in a population in which gene *A* has two alleles—*A* and *a*—generates genotypes *aa*, *AA*, and *Aa*, in the relationship $p^2 + 2pq + q^2$.

reveal the changes in allele frequency that underlie evolution. If the proportion of genotypes remains the same from generation to generation, as the equation indicates, then that gene is not evolving (changing). This situation, called Hardy-Weinberg equilibrium, is theoretical. It happens only if the population is large, if its members mate at random, and if there is no migration, genetic drift, mutation, or natural selection.

Hardy-Weinberg equilibrium is rare for protein-encoding genes that affect the phenotype, because an organism's appearance and health affect its ability to reproduce. Harmful allele combinations are weeded out of the population while helpful ones are passed on. Hardy-Weinberg equilibrium *is* seen in DNA sequences that do not affect the phenotype.

Solving a Problem
Using the Hardy-Weinberg Equation

To understand Hardy-Weinberg equilibrium, it helps to follow the frequency of two alleles of a gene from one generation to the next. Mendel's laws underlie such calculations.

Consider an autosomal recessive trait: a middle finger shorter than the second and fourth fingers. If we know the frequencies of the dominant and recessive alleles, then we can calculate the frequencies of the genotypes and phenotypes and trace the trait through the next generation. The dominant allele D confers normal-length fingers; the recessive allele d confers a short middle finger (**figure 14.4**). We can deduce the frequencies of the dominant and recessive alleles by observing the frequency of homozygous recessives, because this phenotype— short finger—reflects only one genotype. If 9 out of 100 individuals in a population have short fingers—genotype dd—the frequency is 9/100 or 0.09. Since dd equals q^2, then q equals 0.3. Since $p + q = 1.0$, knowing that q is 0.3 tells us that p is 0.7.

Next, we can calculate the proportions of the three genotypes that arise when gametes combine at random:

Homozygous dominant = DD
= $0.7 \times 0.7 = 0.49$
= 49 percent of individuals in generation 1

Homozygous recessive = dd
= $0.3 \times 0.3 = 0.09$
= 9 percent of individuals in generation 1

Heterozygous = $Dd + dD$
= $2pq = (0.7)(0.3) + (0.3)(0.7) = 0.42$
= 42 percent of individuals in generation 1

The proportion of homozygous individuals is calculated by multiplying the allele frequency for the recessive or dominant allele by itself. The heterozygous calculation is $2pq$ because D can combine with d in two ways—a D sperm with a d egg, and a d sperm with a D egg.

In this population, 9 percent of the individuals have a short middle finger. Now jump ahead a few generations, and

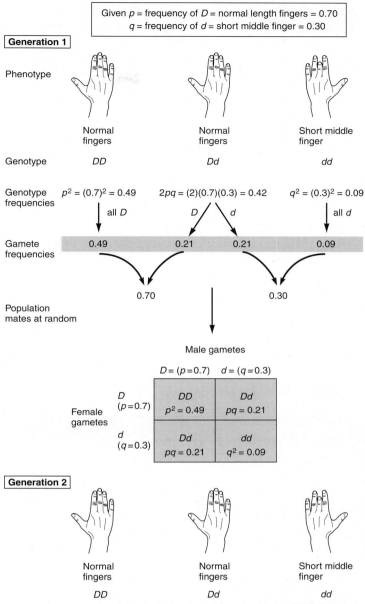

Figure 14.4 **Hardy-Weinberg equilibrium.** In Hardy-Weinberg equilibrium, allele frequencies remain constant from one generation to the next.

assume that people choose mates irrespective of finger length. This means that each genotype of a female (*DD, Dd,* or *dd*) is equally likely to mate with each of the three types of males (*DD, Dd,* or *dd*), and vice versa. **Table 14.2** multiplies the genotype frequencies for each possible mating, which leads to offspring in the familiar proportions of 49 percent *DD,* 42 percent *Dd,* and 9 percent *dd.* This gene, therefore, is in Hardy-Weinberg equilibrium—the allele and genotype frequencies do not change from one generation to the next.

Table 14.2 Hardy-Weinberg Equilibrium—When Allele Frequencies Stay Constant

POSSIBLE MATINGS			FREQUENCY OF OFFSPRING GENOTYPES		
Male	Female	Proportion in Population	DD	Dd	dd
0.49 DD	0.49 DD	0.2401 (DD × DD)	0.2401		
0.49 DD	0.42 Dd	0.2058 (DD × Dd)	0.1029	0.1029	
0.49 DD	0.09 dd	0.0441 (DD × dd)		0.0441	
0.42 Dd	0.49 DD	0.2058 (Dd × DD)	0.1029	0.1029	
0.42 Dd	0.42 Dd	0.1764 (Dd × Dd)	0.0441	0.0882	0.0441
0.42 Dd	0.09 dd	0.0378 (Dd × dd)		0.0189	0.0189
0.09 dd	0.49 DD	0.0441 (dd × DD)		0.0441	
0.09 dd	0.42 Dd	0.0378 (dd × Dd)		0.0189	0.0189
0.09 dd	0.09 dd	0.0081 (dd × dd)			0.0081
	Resulting offspring frequencies:		0.49	0.42	0.09
			DD	Dd	dd

Key Concepts

1. For any two alleles of a gene in a population, the proportion of homozygous dominants equals the square of the frequency of the dominant allele (p^2), and the proportion of homozygous recessives equals the square of the frequency of the recessive allele (q^2). The proportion of heterozygotes equals $2pq$.

2. The frequency of the recessive allele equals the proportion of homozygous recessives plus one-half that of carriers, and the frequency of the dominant allele equals the proportion of homozygous dominants plus one-half that of carriers.

3. In Hardy-Weinberg equilibrium, allele frequencies remain constant from one generation to the next.

14.3 Applying Hardy-Weinberg Equilibrium

A young woman pregnant for the first time watches a television program about cystic fibrosis. Alarmed to learn about the daily treatments and possible complications, and that CF is the most common genetic disorder in her population group (Caucasian of European descent), the woman wonders what the risk is that her child will have CF—even though there is no known history of the disorder in her or her partner's families, which are from the same ethnic group. The Hardy-Weinberg equation can help to answer that question by determining the probability that the woman and her partner are carriers. If they are, then Mendel's first law can be used to calculate the risk to offspring.

To derive carrier risks, the Hardy-Weinberg equation is applied to population statistics on genetic disease incidence. To determine allele frequencies for autosomal recessively inherited characteristics, we need to know the frequency of one genotype in the population. This is typically the homozygous recessive class, because its phenotype indicates its genotype.

The incidence (frequency) of an autosomal recessive disorder in a population is used to help calculate the risk that a particular person is a heterozygote. Returning to the example of CF, the incidence of the disease, and therefore also of carriers, may vary greatly in different populations (table 14.3).

CF affects 1 in 2,000 Caucasian newborns. Therefore, the homozygous recessive frequency—cc if c represents the disease-causing allele—is 1/2,000, or 0.0005 in the population.

Table 14.3 Carrier Frequency for Cystic Fibrosis

Population Group	Carrier Frequency
African Americans	1 in 66
Asian Americans	1 in 150
Caucasians of European descent	1 in 23
Hispanic Americans	1 in 46

Cystic Fibrosis

incidence (autosomal recessive class) = 1/2,000 = 0.0005

$\therefore\ q^2 = 0.0005$

$\therefore\ q = \sqrt{0.0005} = 0.022$

$\therefore\ p = 1 - q = 1 - 0.022 = 0.978$

$\therefore$ carrier frequency = $2pq$ = (2) (0.978) (0.022) = 0.043 = 1/23

Figure 14.5 Calculating the carrier frequency given population incidence: Autosomal recessive.

This equals q^2. The square root of q^2 is about 0.022, which equals the frequency of the *c* allele. If q equals 0.022, then p, or $1 - q$, equals 0.978. Carrier frequency is equal to $2pq$, which equals (2)(0.978)(0.022), or 0.043—about 1 in 23. **Figure 14.5** summarizes these calculations.

Since there is no CF in the woman's family, her risk of having an affected child, based on population statistics, is low. The chance of *each* potential parent being a carrier is about 4.3 percent, or 1 in 23. The chance that *both* are carriers is 1/23 multiplied by 1/23—or 1 in 529—because the probability that two independent events will occur equals the product of the probability that each event will happen alone. However, if they *are* both carriers, each of their children would face a 1 in 4 chance of inheriting the illness, based on Mendel's first law of gene segregation. Therefore, the risk that these two unrelated Caucasian individuals with no family history of CF will have an affected child is $1/4 \times 1/23 \times 1/23$, or 1 in 2,116.

For X-linked traits, different predictions of allele frequencies apply to males and females. For a female, who can be homozygous recessive, homozygous dominant, or a heterozygote, the standard Hardy-Weinberg equation of $p^2 \times 2pq \times q^2$ applies. However, in males, the allele frequency is the phenotypic frequency, because a male who inherits an X-linked recessive allele exhibits it in his phenotype.

The incidence of X-linked hemophilia A for example, is 1 in 10,000 male (X^hY) births. Therefore, q (the frequency of the *h* allele) equals 0.0001. Using the formula $p + q = 1$, the frequency of the wild type allele is 0.9999. The incidence of carriers (X^HX^h), who are all female, equals $2pq$, or (2)(0.0001)(0.9999), which equals 0.00019; this is 0.0002, or 0.02 percent, which equals about 1 in 5,000. The incidence of a female having hemophilia A (X^hX^h) is q^2, or $(0.0001)^2$, or about 1 in 100 million. **Figure 14.6** summarizes these calculations.

Neat allele frequencies such as 0.6 and 0.4, or 0.7 and 0.3, are unusual. In actuality, single-gene disorders are very rare, and so the q component of the Hardy-Weinberg equation contributes little. Because this means that the value of p approaches 1, the carrier frequency, $2pq$, is very close to $2q$. Thus, the carrier frequency is approximately twice the frequency of the rare, disease-causing allele.

Consider Tay-Sachs disease, which occurs in 1 in 3,600 Ashkenazim (Jewish people of eastern European descent). This

Hemophilia A

incidence = 1/10,000 male births = 0.0001

$\therefore\ q = 0.0001$

$\therefore\ p = 1 - q \quad 1 - 0.0001 = 0.9999$

$\therefore$ carrier frequency (females) = $2pq$ = (2) (0.9999) (0.0001) = 0.00019 = about 1/5,000

$\therefore$ affected females = q^2 = (0.0001) (0.0001) = 1/100 million

Figure 14.6 Calculating the carrier frequency given population incidence: X-linked recessive.

means that q^2 equals 1/3,600, or about 0.0003. The square root, q, equals 0.017. The frequency of the dominant allele (p) is then $1 - 0.017$, or 0.983. What is the likelihood that an Ashkenazi carries Tay-Sachs disease? It is $2pq$, or (2)(0.983)(0.017), or 0.033. This is very close to double the frequency of the mutant allele (q), 0.017. Modifications of the Hardy-Weinberg equation are used to analyze genes that have more than two alleles.

Key Concepts

1. Allele frequencies in populations can be inferred from the frequency of homozygous recessive individuals (q^2). The values of q and p can then be deduced and the Hardy-Weinberg equation applied to predict the frequency of carriers.
2. For X-linked traits, the frequency of the recessive phenotype in males is q, and in females q^2.
3. For very rare inherited disorders, p approaches 1, so the carrier frequency is approximately twice the frequency of the disease-causing allele ($2q$).

14.4 DNA Profiling Uses Hardy-Weinberg Assumptions

Hardy-Weinberg equilibrium is useful in a theoretical sense to understand the conditions necessary for evolution to occur. These calculations are also useful in a very practical sense as the foundation of **DNA profiling**. The connection is that parts of the genome that do not affect the phenotype, such as short repeated sequences that do not encode protein, are in Hardy-Weinberg equilibrium. Variability in such sequences can be used to identify individuals if the frequencies are known in particular populations, and several sites in the genome are considered at the same time.

Repeated sequences are scattered throughout the genome. Copy number variants (the number of copies of a particular repeat) can be followed, as alleles, to identify an individual.

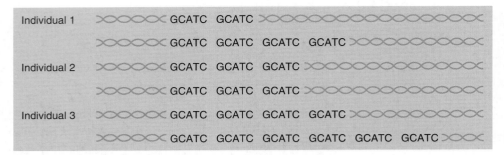

Figure 14.7 **DNA profiling detects differing numbers of repeats at specific chromosomal loci.** Individuals 1 and 3 are heterozygotes for the number of copies of a 5-base sequence at a particular chromosomal locus. Individual 2 is a homozygote, with the same number of repeats on the two copies of the chromosome. (Repeat number is considered an allele.)

The person is classified as a heterozygote or a homozygote based on the number of copies of the same repeat at the same chromosomal locus on the two homologs. A homozygote has the same number of repeats on both homologs, such as individual 2 in **figure 14.7.** A heterozygote has two different repeat sizes, such as the other two individuals in the figure. The copy numbers are distributed in the next generation according to Mendel's law of segregation. A child of individual 1 and individual 2 in figure 14.7, for example, could have any of the two possible combinations of the parental copy numbers, one from each parent: 2 repeats and 3 repeats, or 4 repeats and 3 repeats.

DNA profiling was pioneered on detecting copy number variants of very short repeats and using them to identify or distinguish individuals. In general, the technique calculates the probability that certain combinations of repeat numbers will be in two DNA sources by chance. For example, if a DNA profile of skin cells taken from under the fingernails of an assault victim matches the profile from a suspect's hair, and the likelihood is very low that those two samples would match by chance, that is strong evidence of guilt rather than a coincidental similarity. DNA evidence is more often valuable in excluding a suspect, and should be considered along with other types of evidence.

Obtaining a DNA profile is a molecular technique, but interpreting it requires statistical analysis of population data. Two types of repeats are used in forensics and in identifying victims of disasters: **variable number of tandem repeats (VNTRs),** and **short tandem repeats (STRs). Table 14.4** compares them.

DNA Profiling Began with Forensics

Sir Alec Jeffreys at Leicester University in the United Kingdom invented DNA profiling (then called DNA fingerprinting) in the 1980s. He detected differences in numbers of VNTRs among individuals by cutting DNA with restriction enzymes. These enzymes naturally protect bacteria by cutting foreign DNA, such as DNA from viruses, at specific short sequences. They are used as "molecular scissors" in biotechnology, as discussed in chapter 19. Jeffreys measured DNA fragments using a technique called agarose gel electrophoresis, described in **Reading 14.1.** The different-sized fragments that result from "digesting" DNA with these enzymes are called restriction fragment length polymorphisms (RFLPs, pronounced "riflips").

In the technique that Jeffreys used, DNA pieces migrate through a jellylike material (agarose or the more discriminating polyacrylamide) when an electrical field is applied. A positive electrode is placed at one end of the gel strip, and a negative electrode at the other. The DNA pieces, carrying negative charges because of their phosphate groups, move toward the positive pole. The pieces migrate according to size, with the shorter pieces moving faster and thus traveling farther in a given time. The pattern that forms when the different-sized fragments stop moving, with the shorter fragments closer to the positive pole and the longer ones farther away, creates a distinctive DNA pattern, or profile, that looks like a strip of black smears. An individual who is heterozygous for a repeat copy number variant will have two bands for that locus, as shown in **figure 14.8** for Individuals 1 and 3 from figure 14.7. A locus for which an individual is homozygous has only one corresponding band (Individual 2), because both DNA pieces are the same size.

Jeffreys' first cases proved that a boy was the son of a British citizen so that he could enter the country, and freed a man jailed for raping two schoolgirls. Then in 1988, Jeffreys' approach matched DNA profiles from suspect Tommie Lee Andrews' blood cells to sperm cells left on his victim in a notorious rape case. Jeffreys also used DNA profiling to demonstrate that Dolly, the Scottish sheep, was truly a clone of the 6-year-old ewe that donated her nucleus (**figure 14.9**).

DNA can be obtained from any cell with a nucleus. Common sources include cells in hair, skin, secretions, the inside of the cheek, or white blood cells. DNA sequences other than VNTRs are used when sample DNA is scarce. STRs are used when DNA is fragmented, such as from explosions and natural disasters. The smaller size of DNA pieces makes them more likely to persist in degraded DNA. STRs are mass-produced using a technique called the polymerase chain reaction (see section 19.2).

Table 14.4	Characteristics of Repeats Used in DNA Profiling			
Type	Repeat Length	Distribution	Example	Fragment Sizes
VNTRs (minisatellites)	10–80 bases	not uniform	TTCGGGTTG	50–1,500 bases
STRs (microsatellites)	2–10 bases	more uniform	ACTT	50–500 bases

DNA Profiling: Molecular Genetics Meets Population Genetics

DNA profiling is a powerful tool in forensic investigations, agriculture, paternity testing, and historical research. Until 1986, it was unheard of outside of scientific circles. A dramatic rape case changed that.

Tommie Lee Andrews watched his victims months before he attacked so that he knew when they would be home alone. On a balmy Sunday night in May 1986, Andrews awaited Nancy Hodge, a young computer operator at Disney World in Orlando, Florida. The burly man surprised her when she was in her bathroom removing her contact lenses. He covered her face, then raped and brutalized her repeatedly.

Andrews was very careful not to leave fingerprints, threads, hairs, or any other indication that he had ever been in Hodge's home. But he left DNA. Thanks to a clear-thinking crime victim and scientifically savvy lawyers, Andrews was soon at the center of a trial that would judge the technology that helped to convict him.

After the attack, Hodge went to the hospital, where she provided a vaginal secretion sample containing sperm. Two district attorneys who had read about DNA testing sent some of the sperm to a biotechnology company that extracted DNA and cut it with restriction enzymes. The sperm's DNA pieces were then mixed with labeled DNA probes that bound to complementary sequences.

The same extracting, cutting, and probing of DNA was done on white blood cells from Hodge and Andrews, who had been held as a suspect in several assaults. When the radioactive DNA pieces from each sample, which were the sequences where the probes had bound, were separated and displayed by size, the resulting pattern of bands—the DNA profile—matched exactly for the sperm sample and Andrews' blood, differing from Hodge's DNA (**figure 1**).

Andrews' allele frequencies were compared to those for a representative African American population. At the first trial the judge, fearful too much technical information would overwhelm the jury, did not allow the prosecution to cite population-based statistics. Without the appropriate allele frequencies, DNA profiling was just a comparison of smeary lines on test papers to see whether the patterns of DNA pieces in the forensic sperm sample looked like those for Andrews' white blood cells. Although population statistics indicated that the possibility that Andrews' DNA would match the evidence by chance was 1 in 10 billion, the prosecution could not mention this.

After a mistrial was declared, the prosecution cited the precedent of using population statistics to derive databases on standard blood types. When Andrews stood trial just 3 months later for raping a different woman, the judge permitted population analysis. This time, Andrews was convicted.

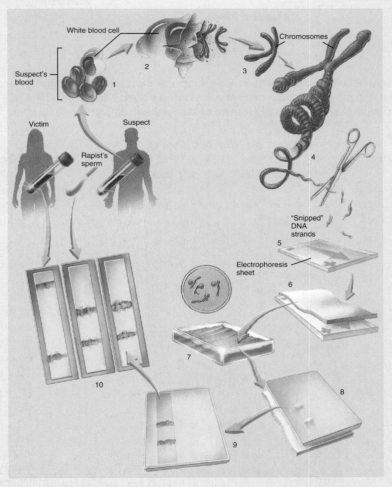

Figure 1 DNA profiling. A blood sample (1) is collected from the suspect. White blood cells are separated and burst open (2), releasing DNA (3). Restriction enzymes snip the strands into fragments (4), and electrophoresis aligns them by size in a groove on a sheet of gel (5). The resulting pattern of DNA fragments is transferred to a nylon sheet (6). It is then exposed to radioactively tagged probes (7) that bind the DNA areas used to establish identity. When the nylon sheet is placed against a piece of X-ray film (8) and processed, black bands appear where the probes bound (9). This pattern of bands is a DNA profile (10). It may be compared to the victim's DNA pattern, the rapist's DNA obtained from sperm cells, and other biological evidence. Today fluorescent labels are used.

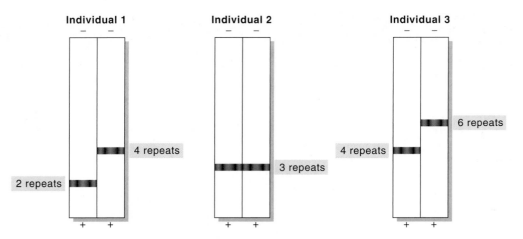

Figure 14.8 **DNA profiles.** DNA fragments that include differing numbers of copies of the same repeat migrate at different speeds and stop moving at different points on a strip of polyacrylamide gel. These gels correspond to the individuals represented in figure 14.7. Actual DNA profiles typically scan up to 25 repeats on different chromosomes.

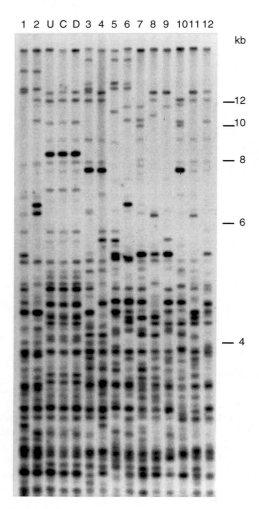

Figure 14.9 **Comparing DNA profiles.** These DNA profiles compare the DNA of Dolly the cloned sheep (lane D), fresh donor udder tissue (U), and cultured donor udder tissue (C). The other twelve lanes represent other sheep. The match between Dolly and the two versions of her nucleus donor is obvious. Dolly was born in 1996 and died in 2003—young for a ewe, perhaps due to her first cell's earlier life.

If DNA is so degraded that even STRs are destroyed, mitochondrial DNA (mtDNA) is often used instead, particularly two regions of repeats that are highly variable in populations. Because a single cell can yield hundreds or thousands of copies of the mitochondrial genome, even extremely small forensic samples can yield this DNA.

MtDNA analysis was critical in analyzing evidence from the September 11 terrorist attacks, most of which was extremely degraded. A more bizarre application was the case of the "voodoo child." The evidence was a boy's torso found floating in the River Thames in east London. The name reflects the contents of the stomach, which suggested he had been the victim of a ritualistic killing. When the DNA profile of nuclear DNA from the torso did not match that of missing English children, investigators widened the search by using a global mtDNA database. This search led to the boy's homeland, southwestern Nigeria. He had been kidnapped, enslaved, and beheaded. Several suspects were arrested, thanks to tracking the torso to Africa.

Commercially available software enables researchers to integrate different types of DNA profiling data. For forensic applications, the FBI's Combined DNA Index System (CODIS) shares DNA profiles electronically among local, state, and federal crime laboratories. More than 3 million DNA profiles are stored, and searching CODIS for DNA profiles has led to thousands of "cold hits"—identifying a suspect from DNA alone. CODIS uses the thirteen STRs shown in **figure 14.10**. The probability that any two individuals have the same thirteen STR markers by chance is 1 in 250 trillion. Therefore, identity at all thirteen sites is a virtual match, but just one mismatch disproves identity. A dog version, called Canine CODIS, helps law enforcement officials identify dogs bred to fight. Puppies sell for $5,000, and these dogs are often terribly abused.

Using Population Statistics to Interpret DNA Profiles

In forensics in general, the more clues, the better. Therefore, the power of DNA profiling is greatly expanded by tracking repeats on several chromosomes. The numbers of copies of a

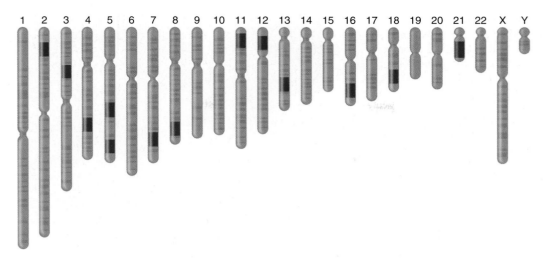

Figure 14.10 **DNA profiling.** A minimum of thirteen short tandem repeat (STR) sites in the genome are compared to rule out suspects in crimes. The green bands indicate the thirteen original CODIS sites. More are increasingly being used.

repeat are assigned probabilities (likelihood of being present) based on their observed frequencies in a particular population. Considering repeats on different chromosomes makes it possible to use the product rule to calculate the probabilities of particular combinations of repeat numbers occurring in a population, based on Mendel's law of independent assortment.

The Hardy-Weinberg equation and the product rule are used to derive the statistics that back up a DNA profile. First, the pattern of fragments indicates whether an individual is a homozygote or a heterozygote for each repeat, because a homozygote only has one band representing that locus. Genotype frequencies are then calculated using parts of the Hardy-Weinberg equation. That is, p^2 and q^2 denote each of the two homozygotes for a two-allele repeat, and $2pq$ represents the heterozygote. Then the frequencies are multiplied.

Table 14.5 shows an example of multiplying frequencies of different repeat numbers. The result is the probability

that this particular combination of repeat sizes would occur in a particular population. Logic then enters the equation. If the combination is very rare in the population the suspect comes from, and if it is found both in the suspect's DNA and in crime scene evidence, such as a rape victim's body or the stolen property in table 14.5, the suspect's guilt appears highly likely. **Figure 14.11** summarizes the procedure.

For the sequences used in DNA profiling, Hardy-Weinberg equilibrium is assumed. When it doesn't apply, problems can arise. For example, the requirement of nonrandom mating for Hardy-Weinberg equilibrium wouldn't be met in a community with a few very large families where distant relatives might inadvertently marry each other—a situation in many small towns. A particular DNA profile for one person might be shared by his or her cousins. In one case, a young man was convicted of rape based on a DNA profile—which he shared with his father, the actual rapist. Considering a larger number of repeat sites can

Table 14.5	Multiplied Frequencies of Different Repeat Numbers

The Case: A famous painting has been stolen from a gallery. The thief planned the crime carefully, but as she was removing the painting from its display, she sneezed. She averted her face, but a few tiny droplets hit the wall. Detectives obtained a DNA profile using six repeat alleles, from different chromosomes, for DNA in nose lining cells in the droplets. Then they compared the profile to those compiled for eight people in the vicinity, all women, who had been identified by hidden camera. (Assume the suspects are in the same ethnic group.) Most of the samples matched at two to four sites, but one matched at all six. She was the crook. Notice how the probability of guilt increases with the number of matches. Matching for the very rare allele #3 is particularly telling.

Allele	Repeat	Frequency	Cumulative Multiplied Frequencies
1	ACT on chromosome 4	1/60	
2	GGC on chromosome 17	1/24	1/60 × 1/24 = 1/1,440
3	AAGCTA on chromosome 14	1/1,200	1/1,440 × 1/1,200 = 1/1,728,000
4	GGTCTA on chromosome 6	1/11	1/1,728,000 × 1/11 = 1/19,008,000
5	ATACGAGG on chromosome 9	1/40	1/19,008,000 × 1/40 = 1/760,320,000
6	GTA on chromosome 5	1/310	1/760,320,000 × 1/310 = 1/235,699,200,000

Figure 14.11 **To solve a crime.** A man was found brutally murdered, with bits of skin and blood beneath a fingernail. The bits were sent to a forensics lab, where the patterns of five DNA sequences were compared to patterns in blood from the victim and blood from a suspect. The pattern for the crime scene evidence matched that for the suspect visually, but that wasn't sufficient. Allele frequencies from the man's ethnic group were used in the Hardy-Weinberg equation, yielding the probability that his DNA matched that of the skin and blood under the murdered man's fingernail by chance.

DNA collected from evidence at crime scene (blood, skin under victim's fingernail)

Cut, label, and probe selected DNA sequences (or use PCR)

Five specific DNA sequences from different chromosomes are labeled and separated by size

Blood (victim) Blood (suspect) Skin (evidence) Blood (evidence)

Visual match

Multiply genotype frequencies

DNA sequence 5	.60 } allele 1 .30 } allele 2	$2pq = (2)(.60)(.30) = 0.36$
DNA sequence 4	.50 } allele 1 .30 } allele 2	$2pq = (2)(.50)(.30) = 0.30$
DNA sequence 3	.15 } allele 1 .80 } allele 2	$2pq = (2)(.15)(.80) = 0.24$
DNA sequence 2	.20 }	$p^2 = (.2)^2 = 0.04$
DNA sequence 1	.80 } allele 1 .18 } allele 2	$2pq = (2)(.80)(.18) = 0.29$

$0.36 \times 0.30 \times 0.24 \times 0.04 \times 0.29 \approx 0.00031 \approx 1/3{,}226$

Conclusion: The probability that another person in the suspect's population group has the same pattern of these alleles is approximately 1 in 3,226.

minimize such complications. If more repeat sites had been considered in the rape case, chances are that they would have revealed a polymorphism that the son had inherited from his mother, but that the father lacked. This would have indicated that the son was not guilty, but a close male relative might be.

The accuracy and meaning of a DNA profile depend upon the population that is the source for the allele frequencies. If populations are too broadly defined, then allele frequencies are typically low, leading to very large estimates of the likelihood that a suspect matches evidence based on chance. In one oft-quoted trial, the prosecutor concluded, *The chance of the DNA fingerprint of the cells in the evidence matching blood of the defendant by chance is 1 in 738 trillion.* The numbers were accurate, but did they really reflect the gene pool compositions of actual populations?

The first DNA profiling databases relied on populations that weren't realistic enough to yield valid statistics. They neatly shoehorned many different groups into just three categories: Caucasian, black, or Hispanic. People from Poland, Greece, or Sweden were all considered white, and a dark-skinned person from Jamaica and one from Somalia would be lumped together as blacks. Perhaps the most incongruous of all were the Hispanics. Cubans and Puerto Ricans are part African, whereas people from Mexico and Guatemala have mostly Native American gene variants. Spanish and Argentinians have neither black African nor Native American genetic backgrounds. Yet these diverse peoples were considered a single population! Native Americans and Asians were left out. Analysis of these three databases revealed significantly more homozygous recessives for certain polymorphic genes than the Hardy-Weinberg equation would predict, confirming what many geneticists had suspected—allele frequencies were not in equilibrium.

Giving meaning to the allele frequencies needed to interpret DNA profiles requires more restrictive ethnic databases. A frequency of 1 in 1,000 for an allele in all whites may actually be much higher or lower in a subgroup that marry among themselves. However, narrowly defined ethnic databases may be insufficient to interpret DNA profiles from people of mixed heritage, such as someone whose mother was Scottish/French and whose father was Greek/German.

Using DNA Profiling to Identify Victims

DNA profiling was first used in criminal cases and to identify human remains from plane crashes. Then terrorist attacks and natural disasters took the scope of DNA profiling to a new level.

Identifying World Trade Center Victims

In late September 2001, a company that provides breast cancer tests received three unusual types of DNA samples:

- evidence from the World Trade Center in New York City;
- cheek brush scrapings from relatives of people missing from the site; and
- "reference samples" from the victims' toothbrushes, razors, and hairbrushes.

Technologists analyzed the DNA for copy numbers of the thirteen standard STRs and the sex chromosomes. STR analysis worked on pieces of soft tissue, but bone bits that persisted despite the ongoing fire at the site required hardier mtDNA analysis. If the DNA pattern in crime scene evidence matched DNA from a victim's toothbrush, identification was fairly certain. Forensic investigators used the DNA results to match family members to victims. DNA profiling provides much more reliable information on identity than traditional forensic identifiers such as dental patterns, scars, and fingerprints, and objects found with the evidence, such as jewelry.

Identifying Natural Disaster Victims

Different types of disasters present different challenges for DNA profiling (**table 14.6**). Whereas New York City workers searched rubble for remains, the 250,000 plus bodies strewn about by the Indian Ocean tsunami in 2004 were everywhere. Rather than hunting for tissue bits out in the open, tsunami workers had to exhume bodies that had been buried quickly to stem the spread of infectious disease. Remains that were accessible after the waves hit quickly decayed in the hot, wet climate. These conditions, combined with the lack of roads and labs, led to 75 percent of the bodies being identified by standard dental record analysis, and 10 percent from fingerprints. Fewer than half of 1 percent of the victims were identified by their DNA. Challenges to identifying remains in the aftermath of the tsunami that hit Japan in 2011 included exposure to radiation unleashed by damage from the earthquake that immediately preceded the giant waves.

Forensic scientists had learned from 9/11 the importance of matching victim DNA to that of relatives, to avoid errors when two people matched at several genome sites by chance. In New York City, many of those relatives were from nearby neighborhoods; in the 2004 tsunami, 12 countries were directly affected and victims came from 30 countries. Entire families were washed away, leaving few and many times no relatives to provide DNA, even if everyday evidence such as toothbrushes had remained.

Table 14.6	Challenges to DNA Profiling in Mass Disasters
■ Climate that hastens decay	
■ Inability to reach remains	
■ No laboratory facilities	
■ Number of casualties	
■ Lack of relatives	
■ Destruction of personal item evidence	
■ Poor DNA quality (too fragmented, scarce, degraded)	
■ Lack of availability of DNA probes and statistics for population	

To compensate for the barriers to implementing DNA profiling in mass disasters, Sir Alec Jeffreys advised assessing 15 to 20 repeat (copy number variant) sites, rather than the usual 13, and some investigators recommend using 50. Tragic as these disasters were, they have spurred forensic scientists to develop ways to better integrate many types of evidence, including that found in DNA sequences.

Reuniting Holocaust Survivors

A happier use of DNA profiling is to reunite families who were torn apart in the Holocaust of World War II. The DNA Shoah project has established a DNA database of many of the 300,000 or so survivors, including some of the 10,000-plus Holocaust orphans. (*Shoah* is Hebrew for *holocaust*.) Michael Hammer and his colleagues at the University of Arizona compare the data to DNA profiles from human remains unearthed in various building projects in parts of Europe where the mass killings occurred. The challenges in reuniting Holocaust families combine those of the 9/11 and tsunami investigations: degraded DNA and few surviving relatives and descendants. The Shoah project is linking the past to the present by matching DNA profiles. The goal: "Using science and technology to reunite families, identify the missing and educate future generations," according to the program's website.

Identifying an Individual's DNA in Mixtures

Natural disasters and crime scenes can leave investigators with DNA from many sources, and some individuals may have left more DNA than others. Researchers combined principles of population genetics with powerful algorithms to develop a way to identify an individual's DNA in a mixture. The approach uses a microarray (see figure 19.7) to probe DNA fragments for many thousands of SNPs in three sources of DNA:

- "person of interest" DNA, such as from a hair or skin cell under a clipped nail of a disaster victim;
- "mixture" DNA from a disaster or crime scene (the forensic sample); and
- "reference population" DNA from the group to which the person of interest belongs, considering ancestry, including mixed ancestry such as northern European with sub-Saharan African.

Genotypes are determined for the two copies of every SNP (from the two copies of the chromosome in a diploid cell). Researchers note places in the genome where the mixture DNA and reference population DNA SNPs differ, as well as which of those SNPs are in the person of interest's DNA (**figure 14.12**). If the person of interest's DNA has more SNPs in common with the mixture DNA than with his or her reference population DNA, then his or her DNA is likely part of the mixture. The power of the approach comes from the large number of SNPs, which provide points of comparison in the genome. Simulations and experiments found that using up to 50,000 SNPs can identify a person's DNA that contributes as little as 0.1 percent to a mixture. The technique can provide answers to families and friends of loved ones missing after a disaster.

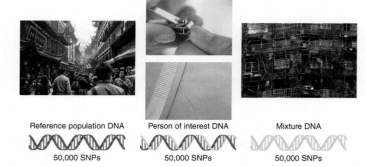

Reference population DNA   Person of interest DNA   Mixture DNA
50,000 SNPs   50,000 SNPs   50,000 SNPs

Figure 14.12 Identifying DNA in a mixture. On November 15, 2010, a 30-story apartment building in Shanghai, China, burned down. After the fire was extinguished, 58 people were missing. Forensic scientists confirmed that some of the missing "people of interest" were in the building by comparing SNP patterns from DNA taken from their homes presumed to be theirs (nail clippings and hair) to the mixture DNA collected at the disaster scene to DNA from a reference population of Chinese. The technique is based on logic: If the "person of interest" DNA is more similar to the mixture DNA than to the reference population DNA, then that person's DNA is likely in the rubble from the fire. The illustration uses dark and light colors to compare the DNA, but in actuality the different sources of DNA are compared for 50,000 single nucleotide polymorphisms (SNPs), points where the DNA sequence varies.

The ability to detect an individual's DNA in a mixture raises a bioethical concern: If this could be done for forensic samples, then people who participate in genome-wide association studies can also be identified. (See *Bioethics: Choices for the Future* in chapter 1.) When researchers at the National Institutes of Health read the study about identifying DNA in mixtures, they immediately limited Internet access to genome-wide association databases, to protect the privacy of participants.

Genetic Privacy

Before the information age, population genetics was an academic discipline that was more theoretical than practical. Today, with the combination of information technology, genome-wide association studies, genome sequencing, and shortcuts to identify people by SNP or copy number patterns, population genetics presents a powerful way to identify individuals. This new view of population genetics raises both personal and societal challenges (see *Bioethics: Choices for the Future* on page 273).

The human genome is 3.2 billion bits of information, each of which can be one of four possibilities: A, C, T, or G. That presents a huge capacity for diversity. Our genomes can vary many more ways than there are people—about 10 billion worldwide. Given these daunting numbers, one only need consider 30 to 80 genome sites to uniquely describe each person. This is why forensic tests can compare only 10 to 15 or so sites in the genome to rule out or establish identity.

The ease of assigning highly individualized genetic name tags may be helpful in forensics, but it poses privacy issues.

Population Biobanks

More than a dozen nations are recording and scrutinizing genetic, genealogical, lifestyle, and health information on their citizens to discover and archive the inherited and environmental influences on common disorders. These "biobank" projects vary in how people participate, but they raise similar concerns: Who will have access to the information? How can people benefit from providing it? How might it be abused?

Ideally, a biobank must meet several criteria. It should:

■ have data and tissue samples from at least 500,000 people;
■ draw conclusions based on a population that is representative of the nation;
■ have clinical information collected over many years;
■ include family trees that link generations; and
■ compare results to those of other populations to validate DNA-disease associations.

The first country to systematically collect genetic information in a population was Iceland, where many citizens can trace their families back more than a thousand years. In 1998, a company called deCODE Genetics received government permission to collect existing health and genealogy records and to add DNA sequence data. Participation in the database is presumed—citizens must "opt out" of the project. Despite initial concerns (mostly from outside Iceland) that the populace would feel pressured to participate, that hasn't been the case. The database includes 95 percent of everyone who has lived in the nation since 1703, when the first census was conducted. DeCODE has used the information to identify genes that contribute to more than 25 common disorders by identifying parts of the genome that people with these disorders uniquely share, then finding genes in these regions whose functions could explain the symptoms.

Some biobanks sample a population and search for connections among DNA sequence variants, health, and lifestyle characteristics as the population ages. CARTaGENE is randomly sampling 1 percent of Quebec's citizens. Researchers in the United Kingdom are recruiting 500,000 individuals between the ages of 46 and 69, when many common illnesses begin. Their information is de-identified, which means that it is studied for societal benefit, not for personal use in health care decisions. The United States is conducting a similar project that is following 500,000 citizens, including 120,000 children, to assess genetic and environmental influences on common health conditions. A smaller-scale study sampled DNA from children under age 11 at a state fair.

Some projects focus on specific medical conditions. The Estonian Genome Foundation uses registries for patients with cancer, Parkinson disease, diabetes mellitus, or osteoporosis. When patients show up for appointments, they learn about the project and are asked for details of their health histories and to donate DNA. Researchers then match variations in the DNA sequence to particular medical conditions. So far more than 50,000 people, or about 5 percent of the population, have provided samples. In the United States, a program called GAIN (Genetic Association Information Network) has amassed DNA profiles based on SNPs and copy number variants from tissue sampled from 2,000 patients who have one of six common disorders: psoriasis, schizophrenia, bipolar disorder, depression, diabetes, and attention deficit hyperactivity disorder. Comparing these profiles to DNA profiles from the same number of healthy volunteers may reveal new drug targets.

Questions for Discussion

1. What do you think the minimal age should be for participation in a biobank?

2. How should privacy of participants be maintained?

3. Do you think that participation should be mandatory, as it is in Iceland? What are the risks and benefits of requiring participation?

4. If DNA testing reveals that an individual has inherited a mutation that could affect health, should blood relatives who could also have the mutation be notified?

5. Can you foresee any dangers of a biobank that uses genetic information? How can a problem be averted?

6. If you were approached at a state fair or other public event and asked if you would allow a sample of your young child's DNA to be taken for a research study, what questions would you ask?

Consider a "DNA dragnet," which is a forensic approach that compiles DNA profiles of all residents of a town where a violent crime is unsolved. Sir Alec Jeffreys in the United Kingdom conducted some of the first DNA dragnets in the late 1980s. The largest to date occurred in 1998 in Germany, where more than 16,000 men had their DNA profiled in a search for the man who raped and murdered an 11-year-old. The dragnet indeed caught the killer.

Another highly publicized DNA dragnet happened in the small town of Truro, near the tip of Cape Cod, Massachusetts. Writer Christa Worthington was brutally murdered in January 2002. A knife went completely through her heart into the floorboards, and her toddler daughter was found at her side, trying to mop up the blood. Only 790 men lived in the seaside village in the winter. DNA from semen in her body did not match samples in any criminal databases. Three years later, on the advice of federal authorities, police began asking men at Truro's few winter gathering places to provide cheek swabs for DNA testing. There was no requirement to do so, but a record was being kept of all who refused, and everyone knew it. Several citizens filed complaints with the American Civil Liberties Union, but most of Truro's male residents complied—including the trash collector who was convicted of the crime in 2006.

A controversial application of DNA profiling is a familial DNA search, which is based on the fact that close relatives share large portions of their genomes. DNA from a crime scene is compared to DNA in databases from convicted felons. If most or all of the CODIS sites match, then a first-degree relative of the convict becomes a suspect. This might be a son, brother, or the father of a man in prison, for example.

In the case of California's "Grim Sleeper," who killed many young women over more than two decades, the suspect was finally arrested when DNA found on a victim closely matched that of a young man in prison for trafficking weapons. The young man did not have brothers, was too young to have a son old enough to be the Grim Sleeper, but did have a father who could have committed the crimes. After an intense investigation, Los Angeles police collected Lonnie David Franklin Jr.'s DNA from a discarded pizza slice. A search of his home turned up 180 photographs of potential victims, which were posted on the police department's website so that the public could help in identifying them. Yet for every criminal that familial DNA searches identify, some innocent people are accused, based on sharing CODIS markers with convicted felons.

DNA profiling is based on population genetics, but it requires logic to avoid false accusations. Investigators must consider how DNA came to be at a crime scene, which is called DNA transfer, or "touch DNA." Primary transfer occurs when the suspect's DNA is on an object, such as a glove or a weapon. Secondary transfer occurs when the person who touches the object had earlier contacted another person's DNA, such as a rapist having first shaken hands with someone. In one complicated case, a professor at an Ivy League school was accused of his wife's murder. It turned out that when he and his wife shared a towel earlier that day, his skin cells had been transferred to her face. Later, the murderer, wearing gloves, touched her face, picked up the husband's cells, and unknowingly transferred them to his weapon. Reconstructing this scenario helped to exonerate the husband.

The Hardy-Weinberg equilibrium that makes DNA profiling possible is extremely rare in the real world, for most genes. The next chapter considers the familiar circumstances that change allele frequencies.

Key Concepts

1. DNA profiles are based on copy number variants.
2. Population statistics are applied to determine the probability that the same pattern would occur by chance in two individuals.
3. A limitation of the method is that databases may not adequately represent real populations. Developing narrower ethnic databases and considering historical and social factors may make population statistics more realistic.
4. DNA profiling of nuclear and mitochondrial DNA was performed on evidence from the September 11, 2001 terrorist attacks and the 2004 tsunami, and used in reuniting relatives separated in the Holocaust.
5. Identifying DNA in a mixed sample is possible by comparing many SNPs in the forensic sample to those in the relevant reference populations, and detecting overrepresentation.
6. Each person has a unique genetic signature (except multiples).
7. DNA profiling introduces privacy issues.

Summary

14.1 Population Genetics Underlies Evolution

1. A **population** is a group of interbreeding members of the same species in a particular area. Their genes constitute the **gene pool**.
2. **Population genetics** considers allele, genotype, and phenotype frequencies to reveal microevolution. Phenotypic frequencies can be determined empirically, then used in algebraic expressions to derive other frequencies.
3. Genotype frequencies change if migration, nonrandom mating, genetic drift, mutations, or natural selection operate. In **Hardy-Weinberg equilibrium,** frequencies are not changing.

14.2 Constant Allele Frequencies

4. Hardy and Weinberg proposed an algebraic equation to explain the constancy of allele frequencies. This would show why dominant traits do not increase and recessive traits do not decrease in populations. The Hardy-Weinberg equation is a binomial expansion used to represent genotypes in a population.

5. Hardy-Weinberg equilibrium is demonstrated by following gamete frequencies as they recombine in the next generation. In equilibrium, these genotypes remain constant if evolution is not occurring. When the equation $p^2 + 2pq + q^2$ represents a gene with one dominant and one recessive allele, p^2 corresponds to the frequency of homozygous dominant individuals; $2pq$ stands for heterozygotes; and q^2 represents the frequency of the homozygous recessive class. The frequency of the dominant allele is p, and of the recessive allele, q.

14.3 Applying Hardy-Weinberg Equilibrium

6. If we know either p or q, we can calculate genotype frequencies, such as carrier risks. Often such information comes from knowing the q^2 class, which corresponds to the frequency of homozygous recessive individuals in a population.
7. For X-linked recessive traits, the mutant allele frequency for males equals the trait frequency. For very rare disorders or traits, the value of p approaches 1, so the carrier frequency ($2pq$) is approximately twice the frequency of the rare trait (q).

14.4 DNA Profiling Uses Hardy-Weinberg Assumptions

8. Repeats (**VNTRs** and **STRs**) that do not encode protein are presumably in Hardy-Weinberg equilibrium and can be compared to establish individual DNA profiles.

9. To obtain a **DNA profile,** determine repeat numbers (using RFLPs or PCR) and multiply population-based allele frequencies to derive the probability that profiles from two sources match by chance.

10. People vary genetically in more ways than there are people.

www.mhhe.com/lewisgenetics10

Answers to all end-of-chapter questions can be found at **www.mhhe.com/lewisgenetics10.** You will also find additional practice quizzes, animations, videos, and vocabulary flashcards to help you master the material in this chapter.

Review Questions

1. Define *population* and list three examples.

2. Why are Hardy-Weinberg calculations more complicated if a gene has many alleles that affect the phenotype?

3. How can evolution occur at a microscopic and macroscopic level?

4. Explain the differences among an allele frequency, a phenotypic frequency, and a genotypic frequency.

5. Why is Hardy-Weinberg equilibrium more a theoretical state than a common, real situation for genes that affect the phenotype?

6. What are the conditions under which Hardy-Weinberg equilibrium cannot be met?

7. Why is knowing the incidence of a homozygous recessive condition in a population important in deriving allele frequencies?

8. For a forensics case, why would tracking VNTR sequences provide a more reliable identification than tracking STRs?

9. Why are specific population databases needed to interpret DNA profiles?

10. How is the Hardy-Weinberg equation used to predict the recurrence of X-linked recessive traits?

11. What is the basis of assigning a probability value to a particular copy number variant?

12. Under what circumstances is analysis of repeats in mtDNA valuable?

13. Describe the following ways to identify or distinguish among individuals at the DNA level:
    a. VNTRs
    b. STRs
    c. SNPs (see chapters 7 and 12)
    d. mtDNA
    e. RFLPs

14. Explain and provide an example of how a familial DNA search can lead to a false accusation.

Applied Questions

1. "We like him, he seems to have a terrific gene pool," say the parents upon meeting their daughter's boyfriend. Why doesn't their statement make sense?

2. Two couples want to know their risk of conceiving a child with cystic fibrosis. In one couple, neither partner has a family history of the disease; in the other, one partner knows he is a carrier. How do their risks differ?

3. How does calculation of allele frequencies differ for an X-linked trait or disorder compared to one that is autosomal recessive?

4. Profiling of Y chromosome DNA implicated Thomas Jefferson in fathering a child of his slave, discussed in chapter 1. What might have been a problem with the conclusion?

5. Glutaric aciduria type I (MIM 231680) causes progressive paralysis and brain damage. It is very common in the Amish of Lancaster County, Pennsylvania—0.25 percent of newborns have the disorder. What percentage of the population are carriers for this condition?

6. Torsion dystonia (MIM 128100) is a movement disorder that affects 1 in 1,000 Jewish people of eastern European descent (Ashkenazim). What is the carrier frequency in this population?

7. Maple syrup urine disease (MSUD) (see Reading 2.1) is autosomal recessive and causes intellectual and physical disability, difficulty feeding, and a sweet odor to urine. In Costa Rica, 1 in 8,000 newborns inherits the condition. What is the carrier frequency of MSUD in this population?

8. The amyloidoses are a group of inborn errors of metabolism in which sticky protein builds up in certain organs. Amyloidosis caused by a mutation in the gene encoding a blood protein

called transthyretin (MIM 176300) affects the heart and/or nervous system. It is autosomal recessive. In a population of 177 healthy African Americans, four proved, by blood testing, to have one mutant allele of the transthyretin gene. What is the carrier frequency in this population?

9. Ability to taste phenylthiocarbamide (PTC) (MIM 607751) is mostly determined by the gene *PTC*. The letters *T* and *t* are used here to simplify analysis. *TT* individuals taste a strong, bitter taste; *Tt* people experience a slightly bitter taste; *tt* individuals taste nothing.

A fifth-grade class of 20 students tastes PTC that has been applied to small pieces of paper, rating the experience as "very yucky" (*TT*), "I can taste it" (*Tt*), and "I can't taste it" (*tt*).

For homework, the students test their parents, with these results:

Of 6 *TT* students, 4 have 2 *TT* parents; and two have one parent who is *TT* and one parent who is *Tt*.

Of 4 students who are *Tt*, 2 have 2 parents who are *Tt*, and 2 have one parent who is *TT* and one parent who is *tt*.

Of the 10 students who can't taste PTC, 4 have 2 parents who also are *tt*, but 4 students have one parent who is *Tt* and one who is *tt*. The remaining 2 students have 2 *Tt* parents.

Calculate the frequencies of the *T* and *t* alleles in the two generations. Is Hardy-Weinberg equilibrium maintained, or is this gene evolving?

Web Activities

1. Consult the website for the Innocence Project and discuss how DNA evidence exonerated someone.

2. Go to a population biobank website and describe a medical test or treatment that may be developed from its data.

Forensics Focus

1. Irene is 80 years old and lives alone with her black cat Moe. One day when she is taking out the garbage, a man jumps out from behind a garbage can and shoves her. As she struggles to get up, he runs inside, pushes open her door, enters, and grabs her purse, which is next to a slumbering Moe on the kitchen table. Moe, sensing something is wrong or perhaps just upset that his nap has been interrupted, scratches the man, who yelps and forcefully flings the cat against a wall, yanking out one of the animal's claws in the process.

   Meanwhile, outside, Irene is back on her feet, trying to get her bearings when the fleeing thief knocks her down again. This time she blacks out, and the man escapes. A few minutes later a neighbor finds her and calls the police. At the crime scene, they collect a drop of blood on the table, a curly blond hair, a few straight black hairs, several gray hairs, and a cat's claw with human skin under it. The police send Irene to the hospital, Moe to the veterinary clinic, and the claw, blood, and hairs to the state forensics lab. The police then thoroughly search the immediate area and neighborhood, but do not find any suspects. When Irene regains consciousness, she remembers nothing about her attackers.

   a. List DNA tests that could help identify the perpetrator.

   b. A familial DNA search closely matches two of the forensic samples to a 58-year-old man in prison for murder. One sample matches at 12 CODIS sites and the other at 10. The investigators find that the convict has four sons and four nephews. What should the police do with this information?

2. The captain of a commercial fishing boat in Denmark had to give his haul of Atlantic cod to the government and pay a large fine for catching fish from the North Sea, rather than from the Baltic Sea as he had claimed. Fishing in the North Sea had been banned due to overfishing. An Atlantic cod from the two seas look exactly alike, so inspectors sent the suspect fish to a geneticist, who showed different numbers of microsatellites (DNA repeats). The genetic test showed that the fisherman was lying. That work from 2003 led to

an ongoing project, FishPopTrace, that is looking at more than 1,500 SNPs in four heavily fished species: European hake, common sole, Atlantic herring, and the Atlantic cod. Researchers quickly found that looking at only 10 or 20 SNPs was sufficient to easily distinguish the four species. (The SNP tests use a DNA microarray, forming "fish 'n chips.")

   a. How did the DNA analysis of the Atlantic cod in 2003 differ from that on the four species done a few years later?

   b. Which procedure resembled CODIS testing?

   c. Explain the significance of the fact that only a few SNPs are sufficient to distinguish the four fish species.

   d. Comparing STRs or SNPs to distinguish individuals is based on the fact that these genetic markers are in Hardy-Weinberg equilibrium. Is this the case for using gene expression differences to distinguish individuals? Provide a reason for your answer.

   e. Restaurants sometimes substitute cheap fish for fancier fare, so that they can charge more. How can the "fish 'n chips" technology be used to regulate this industry? Should it be?

3. In the United Kingdom, everyone who is arrested has DNA sampled and stored in the National DNA Database. Thousands of murders, rapes, and other violent crimes have been solved using the database, and it is routinely used in familial DNA searches. Discuss the risks and benefits of using this database.

4. DNA dragnets have been so successful that some people have suggested storing DNA samples of everyone at birth, so that a DNA profile could be obtained from anyone at any time. Do you think that this is a good idea or not? Cite reasons for your answer.

5. Rufus the cat was discovered in a trash can by his owners, his body covered in cuts and bite marks and bits of gray fur clinging to his claws—gray fur that looked a lot like the coat of Killer, the huge hound next door. Fearful that Killer might attack their other felines, Rufus' distraught owners brought his body to a vet, demanding forensic analysis. The vet suggested that the hair might have come from a squirrel, but

agreed to send samples to a genetic testing lab. Identify the samples that the vet might have sent, and what information each could contribute to the case.

6. In a crime in Israel, a man knocked a woman unconscious and raped her. He didn't leave any hairs at the crime scene, but he left eyeglasses with unusual frames, and an optician helped police locate him. The man also left a half-eaten lollipop at the scene. DNA from blood taken from the suspect matched DNA from cheek-lining cells collected from the base of the telltale lollipop at four repeat loci on different chromosomes. Allele frequencies from the man's ethnic group in Israel are listed beside the profile pattern shown here:

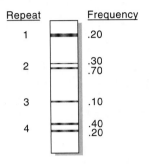

Repeat	Frequency
1	.20
2	.30 .70
3	.10
4	.40 .20

a. For which of the tested repeats is the person a homozygote? How do you know this?

b. What is the probability that the suspect's DNA matches that of the lollipop rapist by chance? (Do the calculation.)

c. The man's population group is highly inbred. How does this affect the accuracy or reliability of the DNA profile? (P.S.—He was so frightened by the DNA analysis that he confessed!)

7. Simone, a recent college graduate, was spending 2 years using her biology degree to work in a clinic in Burma, in a town called Laputta. A devastating monsoon sent 18-foot waves over many of the neighboring villages, and survivors poured into the clinic. It had been raining for 8 days straight. Suddenly, a rush of water overwhelmed the clinic, taking equipment and children in its path. Simone tried to save the children but she, too, was washed away.

Simone's family in the United States was frantic, especially when her body was never found. Her father, a geneticist, insisted that investigators attempt to identify Simone's DNA among evidence of human remains collected at the clinic. Explain how this might be done.

Case Studies and Research Results

1. An extra row of eyelashes is an autosomal recessive trait seen in 900 of the 10,000 residents of an island in the South Pacific. Greta knows that she is a heterozygote for this gene, because her eyelashes are normal, but she has an affected parent.

She wants to have children with a homozygous dominant man so that the trait will not affect her offspring. What is the probability that a person with normal eyelashes in this population is a homozygote for this gene?

The ability to digest lactose (milk sugar) became more prevalent in populations after agriculture introduced dairy foods—thanks to evolution.

Changing Allele Frequencies

Learning Outcomes

15.1 Nonrandom Mating

1. Explain how nonrandom mating changes allele frequencies in populations.

15.2 Migration

2. Explain how migration changes allele frequencies in populations.

15.3 Genetic Drift

3. Explain how genetic drift affects genetic diversity.

15.4 Mutation

4. Discuss how mutation affects populations.

15.5 Natural Selection

5. Provide examples of negative, positive, and artificial selection.

6. Explain how balanced polymorphism maintains diseases in populations.

15.6 Putting It All Together: PKU

7. Describe the forces that have affected the prevalence of PKU.

15.7 Eugenics

8. Explain how eugenics attempts to alter allele frequencies.

The Big Picture: Nonrandom mating, migration, genetic drift, mutation, and natural selection are the forces that mold populations, and drive evolution.

The Evolution of Lactose Tolerance

For millions of people who have lactose intolerance, dairy food causes cramps, bloating, gas, and diarrhea. They no longer produce lactase, an enzyme made in early childhood that breaks down the milk sugar lactose into easily digested sugars. People who can digest dairy have lactase persistence (MIM 223000). Four genes control the ability to digest milk sugar. Different populations have different proportions of lactose-intolerant versus lactase-persistent people. Clues in DNA suggest why: agriculture.

As dairy farming spread around the world, people who had gene variants enabling them to digest milk into adulthood had an advantage, and had more children. Over time, populations that consumed dairy foods had more people with lactase persistence. In contrast, in populations with few or no dairy foods, lactose intolerance was not a problem, and so those gene variants persisted.

The link between lactose intolerance and agriculture is why today, the European American population only has 10 percent lactose intolerance, whereas among Asian Americans, who eat far less dairy, 90 percent have lactose intolerance. Seventy-five percent of African Americans and Native Americans have lactose intolerance.

15.1 Nonrandom Mating

Historically, we seem to have gone out of our way to ensure that the very specific conditions necessary for Hardy-Weinberg equilibrium—unchanging allele frequencies from generation to generation—do not occur, at least for some genes. Religious restrictions and personal preferences guide our choices of mates. Wars and persecution kill certain populations. Economic and political systems enable some groups to have more children. We travel, shuttling genes in and out of populations. Natural disasters and new diseases reduce populations to a few individuals, who then rebuild their numbers, at the expense of genetic diversity. These factors, plus mutation and a reshuffling of genes at each generation, make a gene pool very fluid.

The ever-present and interacting forces of nonrandom or selective mating, migration, genetic drift, mutation, and natural selection work to differing degrees to shape populations at the allele level. Changing allele frequencies can change genotype frequencies, which in turn can change phenotype frequencies. In a series of illustrations throughout this chapter, colored shapes represent individuals who have specific genotypes. **Figure 15.1** then combines the illustrations to summarize the chapter. We begin our look at the forces that change allele frequencies in populations with nonrandom mating.

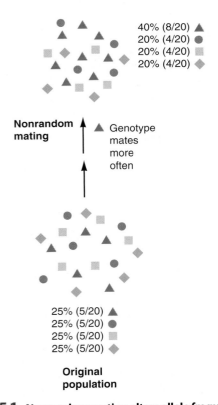

Genotype mates more often

Figure 15.1 **Nonrandom mating alters allele frequencies.** The different-colored shapes represent individuals with distinctive genotypes. If Hardy-Weinberg equilibrium exists for these genes in this population, then the percentages will remain the same through the generations. However, the blue triangle genotype is more reproductively successful, skewing the allele frequencies in the next generation.

In the theoretical state of Hardy-Weinberg equilibrium, individuals of all genotypes are presumed equally likely to successfully mate and to choose partners at random. For some traits this is true—we do not choose partners based on blood type, for example—but other traits do influence our mate choices. We choose partners based on physical appearance, ethnic background, intelligence, and shared interests. We marry people similar to ourselves about 80 percent of the time. Worldwide, about one-third of all marriages are between people who were born fewer than ten miles apart! This nonrandom mating is a major factor in changing allele frequencies in human populations.

Nonrandom mating occurs when certain individuals contribute more to the next generation than others (figure 15.1). This is common in agriculture when semen from one prize bull is used to inseminate thousands of cows, or a field of genetically identical crops is planted. Such an extreme situation can arise in a human population when a man fathers many children. A striking mutation can reveal such behavior. In the Cape population of South Africa, for example, a Chinese immigrant known as Arnold had a very rare dominant mutation that causes teeth to fall out before age 20. Arnold had seven wives. Of his 356 living descendants, 70 have the dental disorder. The frequency of this allele in the Cape population is exceptionally high, thanks to Arnold.

The high frequency of autosomal recessive albinism among Arizona's Hopi Indians also reflects nonrandom mating. Albinism is uncommon in the general U.S. population, but it affects 1 in 200 Hopi Indians. The reason for the trait's prevalence is cultural—men with albinism often stay back and help the women, rather than risk severe sunburn in the fields with the other men. They contribute more children to the population because they have more contact with the women.

The events of history reflect nonrandom mating patterns. When a group of people is subservient to another, genes tend to "flow" from one group to the other as the males of the ruling class have children with females of the underclass—often forcibly. Historical records and DNA sequences show this directional gene flow phenomenon. For example, Y chromosome analysis suggests that Genghis Khan, a Mongolian warrior who lived from 1162 to 1227, had sex with so many women that today, 1 in every 200 males living between Afghanistan and northeast China shares his Y—that's 16 million men (**figure 15.2**)! The number is so high because his many male descendants also passed on the distinctive Y.

Despite our partner preferences, many traits do mix randomly in the next generation. This may be because we are unaware of these characteristics or because we do not consider them in choosing partners. In populations where AIDS is extremely rare or nonexistent, for example, the two mutations that render a person resistant to HIV infection are in Hardy-Weinberg equilibrium. This would change, over time, if HIV arrives, because the people with these mutations would become more likely to survive to produce offspring—and some of them would perpetuate the protective mutation. Natural selection would intervene, ultimately altering allele frequencies.

Many blood types are in Hardy-Weinberg equilibrium because we do not choose partners by blood type. Yet sometimes the opposite occurs. People with mutations in the same gene

Figure 15.2 A prevalent Y. Genghis Khan left his mark on many male descendants in the form of his Y chromosome. Rape of women on a sweeping scale spread the chromosome in certain Asian populations.

meet when their families participate in programs for people with the associated disorder. For example, more than two-thirds of relatives visiting a camp for children with cystic fibrosis are likely to be carriers, compared to the 1 in 23 or fewer in large population groups.

People can avoid genetic disease with controlled mate choice and reproduction. In a program in New York City called Dor Yeshorim, for example, young people take tests for more than a dozen genetic disorders that are much more common among Jewish people of eastern European descent (Ashkenazim). Results are stored in a confidential database. Two people wishing to have children together can find out if they are carriers for the same disorder. If so, they may elect not to have children. Thousands of people have been tested, and the program is partly responsible for the near-disappearance of Tay-Sachs disease among Ashkenazi Jews. The very few cases each year are usually in non-Jews, because they have not been tested.

A population that practices consanguinity has very nonrandom mating. Recall from chapter 4 that in a consanguineous relationship, "blood" relatives have children together. On the family level, this practice increases the likelihood that harmful recessive alleles from shared ancestors will be combined and passed to offspring, causing disease. The birth defect rate in offspring is 2.5 times the normal rate of about 3 percent. On a population level, consanguinity decreases genetic diversity. The proportion of homozygotes rises as that of heterozygotes falls.

Some populations encourage marriage between cousins, which increases the incidence of certain recessive disorders. In certain parts of the middle east, Africa, and India, 20 to 50 percent of marriages are between cousins, or uncles and nieces. The tools of molecular genetics can reveal these relationships. Researchers traced DNA sequences on the Y chromosome and in mitochondria among residents of an ancient, geographically isolated "micropopulation" on the island of Sardinia, near Italy. They consulted archival records dating from the village's founding by 200 settlers around 1000 A.D. to determine familial relationships. Between 1640 and 1870, the population doubled, reaching 1,200 by 1990. Fifty percent of the present population descends from just two paternal and four maternal lines, and 86 percent of the people have the same X chromosome. Researchers are analyzing disorders that are especially prevalent in this population, which include hypertension and a kidney disorder.

Worldwide, about 960 million married couples are related, and know of their relationship. Also contributing to nonrandom mating is endogamy, which is marriage within a community. In an endogamous society, spouses may be distantly related and unaware of the connection.

Key Concepts

1. People choose mates for many reasons, and they do not contribute the same numbers of children to the next generation. These practices change allele frequencies in populations.
2. Traits lacking obvious phenotypes may be in Hardy-Weinberg equilibrium.
3. Consanguinity and endogamy increase the proportion of homozygotes in a population.

15.2 Migration

Large cities, with their pockets of ethnicity, defy Hardy-Weinberg equilibrium by their very existence. Waves of immigrants formed the population of New York City, for example. The original Dutch settlers of the 1600s had different alleles than those in today's metropolis of English, Irish, Slavics, Africans, Hispanics, Italians, Asians, and many others. **Figure 15.3** depicts the effect on allele and genotype frequencies when individuals join a migrating population. Clues to past migrations lie in historical documents as well as in differing allele frequencies in regions defined by geographical or language barriers.

The frequency of the allele that causes galactokinase deficiency (MIM 230200) in several European populations reveals how people with this autosomal recessive disorder migrated (**figure 15.4**). Galactokinase deficiency causes cataracts (clouding of the lens) in infants. It is very common among a population of 800,000 gypsies, called the Vlax Roma, who live in Bulgaria. It affects 1 in 1,600 to 2,500 people among them, and 5 percent of the people are carriers. But among all gypsies in Bulgaria as a whole, the incidence drops to 1 in 52,000. As the map in figure 15.4 shows, the disease becomes rarer to the west. This pattern may have arisen when people with the allele settled in Bulgaria, with only a few individuals or families moving westward.

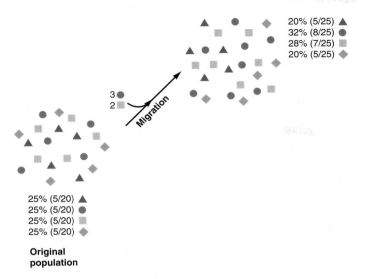

20% (5/25) ▲
32% (8/25) ●
28% (7/25) ■
20% (5/25) ◆

3 ●
2 ■

Migration

25% (5/20) ▲
25% (5/20) ●
25% (5/20) ■
25% (5/20) ◆

**Original
population**

Figure 15.3 **Migration alters allele frequencies.** If the population travels and picks up new individuals, allele (and genotype) frequencies can change.

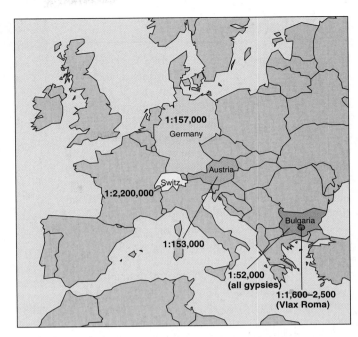

1:157,000
Germany

Austria

Switz.

1:2,200,000

Bulgaria

1:153,000

1:52,000
(all gypsies)

1:1,600–2,500
(Vlax Roma)

Figure 15.4 **Galactokinase deficiency in Europe illustrates a cline.** This autosomal recessive disorder that causes blindness varies in prevalence across Europe. It is most common among the Vlax Roma gypsies in Bulgaria. The condition becomes much rarer to the west, as indicated by the shading from dark to light green.

Allele frequencies often reflect who ruled whom. For example, the frequency of ABO blood types in certain parts of the world today mirrors past Arab rule. The distribution of ABO blood types is very similar in northern Africa, the Near East, and southern Spain. These are precisely the regions where Arabs ruled until 1492. The uneven distribution of allele frequencies can also reveal when and where nomadic peoples stopped. For example, in the eighteenth century, European Caucasians called trekboers migrated to the Cape area of South Africa. The men stayed and had children with the native women of the Nama tribe. The mixed society remained fairly isolated, leading to the distinctive allele frequencies found in the present-day people of color of the area.

Sometimes allele frequencies change from one neighboring population to another. This phenomenon is termed a **cline**. Changing allele frequencies usually reflect migration patterns, as immigrants introduced alleles and emigrants removed them. Clines may be gradual, reflecting unencumbered migration paths, but barriers often cause more abrupt changes in allele frequencies. Geographical formations such as mountains and bodies of water may block migration, maintaining population differences in allele frequencies on either side of the barrier. Language differences may also isolate alleles, if people who cannot communicate tend not to have children together.

Allele frequencies up and down the lush strip of fertile land that hugs the Nile River illustrate the concept of clines. Researchers found a gradual change in mitochondrial DNA sequences in 224 people who live on either side of the Nile, an area settled 15,000 years ago. The farther apart two individuals live along the Nile, the less alike their mtDNA. This is consistent with evidence from mummies and historical records that indicate the area was once kingdoms separated by wars and language differences. If the area had been one large interacting settlement, then the DNA sequences would have been more mixed. Instead, the Nile may have served as a "genetic corridor" between Egypt and sub-Saharan Africa.

Comparing clines is also consistent with geography and history. For example, agriculture spread faster from east to west in Eurasia because latitude does not change much. As a result, there is less evidence of gene flow in this relatively unchanging environment—even if people migrated, they tended to have the same gene variants as the people had in their new homes. In the Americas, the spread of agriculture was north to south. Differing hours of daylight and climate with longitude presented different environmental challenges, which are reflected in greater genetic change from north to south than from east to west.

Clines may also reflect human dependence on communication. Geneticists correlated twenty blood types to geographically defined regions of Italy and to areas where a single dialect is spoken. They chose Italy because it is rich in family history records and linguistic variants. Six of the blood types varied more consistently with linguistically defined subregions than with geographical regions. Perhaps differences in language prevent people from socializing, keeping alleles within groups.

Key Concepts

1. Migration alters genotype frequencies by adding and removing alleles from populations.
2. Clines are gradual changes in allele frequencies between neighboring populations.
3. Geographical barriers and language differences often create great differences in allele frequencies.

15.3 Genetic Drift

When a small group separates from a larger population, or reproduces only among themselves, allele frequencies may change as a result of chance sampling from the whole (**figure 15.5**). This change in allele frequency that occurs when a small group separates from the larger whole is termed **genetic drift**. It is like reaching into a bag of jellybeans and, by chance, grabbing only green and yellow ones. The allele frequency changes in genetic drift are random and unpredictable, just as reaching into the jellybean bag a second time might yield mostly black and orange candies.

Genetic drift occurs when the population size plummets, due either to migration, to a natural disaster or geographic barrier that isolates small pockets of people, or to the consequences of human behavior. Members of a small community might reproduce only among themselves, which keeps genetic variants within their ethnic group. For example, the skin-lightening condition vitiligo (MIM 193200) is much more common in a small community isolated in the mountains of northern Romania than elsewhere in the nation. Genetic drift is also seen within large cities. Pittsburgh, Pennsylvania and New York City are more mosaics of groups with distinct ethnic flavors than vast "melting pots" of mixed heritage.

Some groups of people become isolated in several ways—geographically, linguistically, and by choice of partners. Such populations often have a high incidence of several otherwise rare inherited conditions. The native residents of the Basque country in the western part of the Pyrenees Mountains between France and Spain, for example, still speak remnants of Euskera, a language the first European settlers brought in more than 10,000 years ago. The Basques have unusual frequencies of certain ABO and Rh blood types, rare mtDNA sequences and cell surface antigen patterns, and a high incidence of a mild form of muscular dystrophy.

Table 15.1	Founder Populations		
Population	**Number of Founders**	**Number of Generations**	**Population Size Today**
Costa Rica	4,000	12	2,500,000
Finland	500	80–100	5,000,000
Hutterites	80	14	36,000
Japan	1,000	80–100	120,000,000
Iceland	25,000	40	300,000
Newfoundland	25,000	16	500,000
Quebec	2,500	12–16	6,000,000
Sardinia	500	400	1,660,000

The Founder Effect

A common type of genetic drift in human populations is the **founder effect,** which occurs when a small group leaves home to found a new settlement and the population of the new colony has different allele frequencies than the original population (**table 15.1**). It is a sampling phenomenon. A founder effect usually appears as a community of people, known from local history to have descended from a few founders, who have inherited traits and illnesses that are rare elsewhere. This happened to the population of Norfolk Island, in the South Pacific.

On April 28, 1789, eighteen sailors aboard the British Royal Navy ship the HMS *Bounty* mutinied, leaving the ship and rowing to Tahiti, where they settled (**figure 15.6**). (The event inspired several books and films, including *Mutiny on the Bounty.*) Some of the mutineers moved to nearby Pitcairn Island. In 1856, the British government let some of the descendants of the mutineers on Pitcairn Island move to Norfolk Island, which had more space. There, twelve Tahitian women and nine Bounty mutineers had many children, and today, twelve generations after the mutiny, about half of the 1,200 people on Norfolk Island are direct descendants of the mutineers, and a quarter of them suffer from an otherwise very rare form of migraine. Apparently, among the original mutineers settling on Norfolk Island, were a few who brought in the dominant allele for the condition.

Recessive disorders are subject to founder effects too. This is striking among the Old Order Amish and Mennonites of Lancaster County, Pennsylvania. Often, worried parents would bring their ill children to medical facilities in Philadelphia. Over the years, researchers realized that these people have several conditions that are extremely rare elsewhere, including inborn errors of metabolism, a form of muscular dystrophy, and several variants of bipolar disorder (**figure 15.7**). Today, special diets, new drugs, and gene therapy can treat some of the diseases.

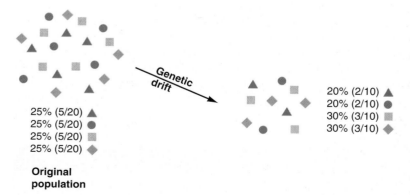

Figure 15.5 Genetic drift alters allele frequencies. If members of a population leave or do not reproduce, allele frequencies can change by chance sampling of a small population. When half of this population does not contribute to the next generation, two genotypes increase in frequency and two decrease.

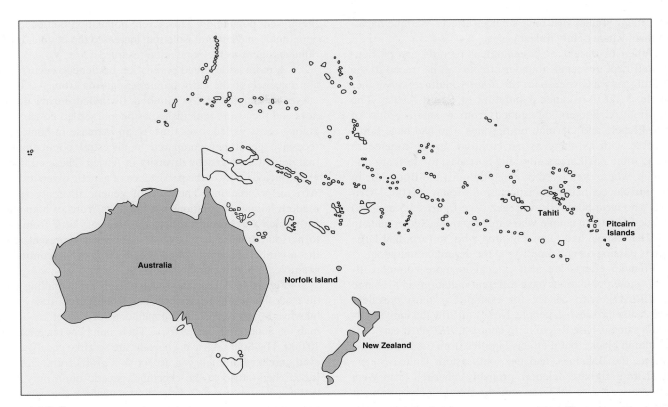

Figure 15.6 A founder effect. Migraine among mutineer descendants on Norfolk Island is very common. The affected people descend from eighteen sailors who left the HMS *Bounty* on April 28, 1789, and settled on a series of islands in the South Pacific, where they had children with native women.

Statistics readily reveal founder effects. In Lancaster County, for example, maple syrup urine disease (MIM 248600) affects 1 in 400 newborns, but in the United States' general population, only 1 in 225,000 newborns inherits the disease. Seemingly common disorders in an isolated population can actually be rare variants. A research fellow at Children's Hospital of Philadelphia discovered that cerebral palsy in several young children from Lancaster County attributed to oxygen deprivation at birth was actually an inborn error of metabolism called glutaric aciduria type 1 (MIM 231670). He went from farm to farm, tracking cases against genealogical records, and discovered that *every family* that could trace its roots back to the founders had members who had the disease! Today, 1 in 200 newborns in this population has the condition.

A mutation that is the same in all individuals with a certain illness in a population is strong evidence of a founder effect due to descent from shared ancestors. The Bulgarian gypsies who have galactokinase deficiency, for example, all have a mutation that is extremely rare elsewhere. In contrast, a population with several mutations that cause the same disorder is more likely to have picked up those variants from unrelated people joining the group, rather than from descent from shared founders.

Very often when a disease-associated allele is identical among people in the same population, so is the DNA surrounding the gene. This pattern indicates that a portion of a chromosome, rather than just the disease-causing gene, has been passed among the members of the population from its founders. For this

Figure 15.7 Ellis-van Creveld syndrome (MIM 225500). This Amish child has autosomal recessive Ellis-van Creveld syndrome, which causes dwarfism, extra fingers, heart disease, fused wrist bones, and teeth at birth. Seven percent of the people of this community have the mutant allele. Heterozygotes have the milder condition, Weyers acrofacial dysostosis (MIM 193530). The two conditions are allelic diseases (see table 12.3).

reason, many studies that trace founder effects examine haplotypes that include tightly linked genes.

When historical or genealogical records are particularly well kept or recent, founder effects can be traced to the beginning, as was the case for the *Bounty* mutineers. Another example is the Afrikaner population of South Africa. The 2.5 million Afrikaners descended from a small group of Dutch, French, and German immigrants who had huge families, often with as many as ten children. In the nineteenth century, some Afrikaners migrated northeast to the Transvaal Province, where they lived in isolation until the Boer War in 1902 introduced better transportation. Today, 30,000 Afrikaners who have porphyria variegata (see figures 5.5 and 5.6) descended from one couple who came from the Netherlands in 1688! That couple had many children who, in turn, had large families, passing on and amplifying the dominant mutation.

Founder effects are also evident in more common illnesses, where populations have different mutations in the same gene. *BRCA1* breast cancer, for example, is most prevalent among Ashkenazi Jewish people. Nearly all affected individuals have the same 3-base deletion. In contrast, *BRCA1* breast cancer is rare in blacks, but it affects families from the Ivory Coast in Africa, the Bahamas, and the southeastern United States. They share a 10-base deletion, probably inherited from West Africans ancestral to all three modern groups. Slaves brought the disease to the United States and the Bahamas between 1619 and 1808, but some of their relatives who stayed in Africa have perpetuated the mutant allele there.

Population Bottlenecks

A **population bottleneck** occurs when many members of a group die, and only a few are left to replenish the numbers. The new population has only those alleles in the small group that survived the catastrophe. An allele in the remnant population might become more common in the replenished population than it was in the original larger group. Therefore, the new population has a much more restricted gene pool than the larger ancestral population, with some variants amplified, others diminished.

Population bottlenecks can occur when people (or other animals) colonize islands. An extreme example is seen among the Pingelapese people of the eastern Caroline Islands in Micronesia. Four to 10 percent are born with "Pingelapese blindness," an autosomal recessive combination of colorblindness, nearsightedness, and cataracts also called achromatopsia (MIM 603096). Elsewhere, only 1 in 20,000 to 50,000 people inherits the condition. Nearly 30 percent of the Pingelapese are carriers. The prevalence of the blindness among the Pingelapese stems from a typhoon in 1780 that killed all but nine males and ten females who founded

the present population. This severe population bottleneck, plus geographic and cultural isolation, increased the frequency of the blindness gene as the population resurged.

A more widespread population bottleneck occurred as a consequence of the early human expansion from Africa, discussed in chapter 16. As numbers dwindled during the journeys and then were replenished as people settled down, mating among relatives led, over time, to an increase in homozygous recessive genotypes compared to ancestral populations that maintained their genetic diversity in Africa. These bottlenecks are reflected today in the persistence of genetic diversity among African populations. The lack of genetic diversity in some modern human populations is evident as "runs of homozygosity," which are chromosome regions that vary little from person to person. Runs of homozygosity generally represent regions that are inherited from shared ancestors. They are common, for example, in highly purebred dogs.

Figure 15.8 illustrates schematically the dwindling genetic diversity that results from a population bottleneck. Today's cheetahs live in just two isolated populations of a few thousand animals in South and East Africa. Their numbers once exceeded 10,000. The South African cheetahs are so alike genetically that even unrelated animals can accept skin grafts from each other. Researchers attribute the cheetahs' genetic uniformity to two bottlenecks—at the end of the most recent ice age, when habitats changed, and another following mass slaughter by humans in the nineteenth century. However, the good health of the animals today indicates that the genes that have survived enable the cheetahs to thrive in their environment.

Human-wrought disasters that kill many people can cause population bottlenecks that greatly alter gene pools because aggression is typically directed against particular groups, while a typhoon indiscriminately kills anyone in its path. For example, after the many waves of killings, called pogroms, of Jewish people, only a few thousand remained in Eastern Europe by the end of the eighteenth century. Then their numbers grew

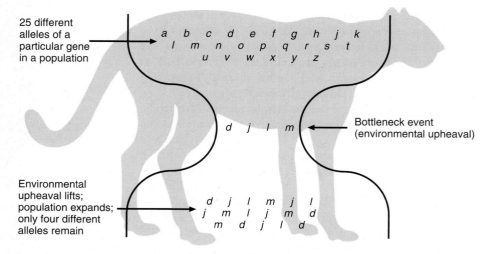

25 different alleles of a particular gene in a population

Bottleneck event (environmental upheaval)

Environmental upheaval lifts; population expands; only four different alleles remain

Figure 15.8 Population bottlenecks. A population bottleneck occurs when the size of a genetically diverse population falls, remains at this level for a time, and then expands again. The new population loses some genetic diversity if alleles are lost.

again, and from 1800 to 1939, the Jewish population in eastern Europe swelled to several million, only to be decimated again by the Holocaust.

Until recently, Jewish people tended to have children only with each other. Both of these factors—nonrandom mating and repeated population bottlenecks—changed allele frequencies and contributed to the incidence of certain inherited diseases seen among the Ashkenazi Jewish people that is ten times higher than in other populations. Several genetic testing companies offer "Jewish genetic disease" panels that are not meant to discriminate or stereotype, but are based on a genetic fact of life—some illnesses are more common in certain populations, due to human behavior. However, DNA itself does not discriminate. The "Jewish" mutations can arise anew in anyone. **Table 15.2** describes some inherited diseases that are more common among Ashkenazi (descended from eastern Europeans) Jewish populations.

Key Concepts

1. In genetic drift, a subset of a population has different allele frequencies than the larger population.

2. The founder effect occurs when a few individuals leave a community to start a new settlement. The resulting population may, by chance, either lack some alleles from the original population or have high frequencies of others.

3. In a population bottleneck, many members die, leaving only a few to contribute to the next generation.

15.4 Mutation

A major and continual source of genetic variation in populations is mutation, which changes one allele into another, and the change passes to offspring (**figure 15.9**). Genetic variability also arises from crossing over and independent assortment during meiosis, but these events recombine existing traits rather than introduce new ones.

If a DNA base change occurs in a part of a gene that encodes part of a protein necessary for its function, then an altered trait may result. Another way that genetic change can occur from generation to generation is in the numbers of repeats of copy number variants (CNVs). These function as alleles.

Natural selection, discussed in the next section, eliminates alleles that adversely affect reproduction. Yet harmful recessive alleles are maintained in heterozygotes and are reintroduced by new mutation. Therefore, all populations have some alleles that would be harmful if homozygous. The collection of such deleterious alleles in a population is called its **genetic load**.

The contribution that mutation makes to counter Hardy-Weinberg equilibrium is quite small compared to the influence of migration and nonrandom mating, because mutation is rare. Natural selection has the greatest influence. The spontaneous mutation rate is about 170 bases per haploid genome in each gamete. Each of us probably has at least five "lethal equivalents"—alleles or allele combinations that if homozygous would kill us or make us too sick to have children.

Table 15.2	Autosomal Recessive Genetic Diseases Prevalent Among Ashkenazi Jewish Populations		
Disorder	**MIM**	**Signs and Symptoms (Phenotype)**	**Carrier Frequency**
Bloom syndrome	210900	Sun sensitivity, short stature, poor immunity, impaired fertility, increased cancer risk	1/110
Breast cancer	113705, 600185	Malignant breast tumor caused by mutant *BRCA1* or *BRCA2* genes	3/100
Canavan disease	271900	Brain degeneration, seizures, developmental delay, early death	1/40
Familial dysautonomia	223900	No tears, cold hands and feet, skin blotching, drooling, difficulty swallowing, excess sweating	1/32
Gaucher disease	231000	Enlarged liver and spleen, bone degeneration, nervous system impairment	1/12
Niemann-Pick disease type A	257200	Lipid accumulation in cells, particularly in the brain; intellectual and physical disability, death by age 3	1/90
Tay-Sachs disease	272800	Brain degeneration causing intellectual disability, paralysis, blindness, death by age 4	1/26
Fanconi anemia type C	227650	Deficiencies of all blood cell types, poor growth, increased cancer risk	1/89

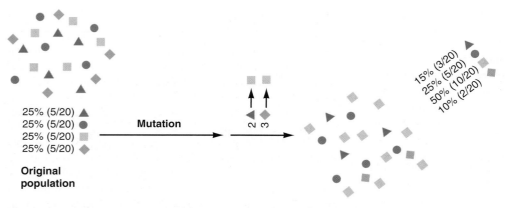

Figure 15.9 Mutation alters allele frequencies. If one allele changes into another from one generation to the next, genotype frequencies can change.

Key Concepts

1. Mutation alters genotype frequencies by introducing new alleles.
2. Heterozygotes and new mutations maintain the frequencies of deleterious alleles in populations.

15.5 Natural Selection

Environmental change can alter allele frequencies when individuals with certain phenotypes are more likely to survive and reproduce than others. This differential survival to reproduce guided by environmental change is **natural selection** (**figure 15.10**). The chapter opener chronicles natural selection acting on gene variants that enables people to digest the sugar lactose.

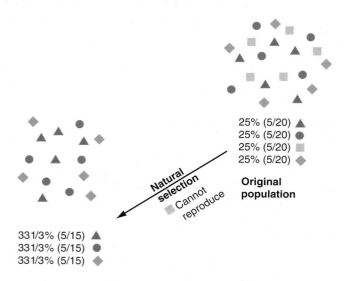

Figure 15.10 Natural selection alters allele frequencies. If health conditions impair the ability of individuals of a certain genotype to reproduce, allele frequencies can change.

Inability to digest lactose is actually the wild type condition, because it predominated before people began domesticating mammals and drinking their milk. Most of the mutations that introduced lactase persistence are point mutations. Another dietary illustration of natural selection involves copy number variants (CNVs). Members of populations that follow high-starch diets tend to have more copies of the gene that encodes salivary amylase, the digestive enzyme that begins to break down starch in the mouth. Members of populations that follow low-starch diets have fewer copies of the gene, and presumably less of the enzyme.

In natural selection, reproductive success is all-important, because this is what transmits favorable alleles and weeds out the unfavorable ones, ultimately impacting population structure and therefore driving microevolution. In the common phrase used synonymously with natural selection—"survival of the fittest"—"fit" actually refers to reproductive success, not to physical prowess or intelligence (unless those traits lead directly to reproductive success). In a Darwinian sense, an unattractive and out-of-shape parent of ten is more "fit" than a gorgeous triathlete with one child.

Negative and Positive Selection

Natural selection can retain a beneficial trait or banish one that has become dangerous in the prevailing environment. Retaining a trait is termed positive natural selection, and getting rid of a trait is termed negative natural selection, but the "positive" and "negative" are not value judgments—they merely refer to staying or leaving. In Darwin's time natural selection was thought to be primarily negative, but the ability to sequence genes has enabled us to actually measure and track the instances of positive selection that have sculpted our differences from our closest primate relatives.

Subtle nuances in DNA sequence provide a "signature" of positive selection. Specifically, a sign of positive selection is a gene in humans that has a counterpart in other primates, but in humans has at least one distinctive difference in the amino acid sequence, not just the DNA sequence. A change in the DNA sequence that does *not* substitute an amino acid does not change the protein, and therefore has no effect on the phenotype. Such a change therefore cannot be subject to natural selection, which acts on phenotypes.

Another sign of positive selection is a gene that varies little, if at all, from person to person. Positive selection is, in a sense, an evolutionary version of "if it ain't broke, don't fix it." A study of 1.2 million SNPs across the genomes of African Americans, European Americans, and Chinese revealed 101 regions that have signs of positive selection. Genes in these regions take part in the sense of smell, digestion, nervous system

Figure 15.11 **Positive selection enables the Sherpa to thrive at elevations so great that they make others very sick.** The Sherpa have migrated from their native Tibet to Nepal over the past 300 to 400 years. Many of them work as mountaineering guides, especially for climbers ascending Mt. Everest. They are short, strong, and hardy. A variant of the hypoxia inducible factor 2 (*EPAS1*) gene is responsible for their astonishing adaptation to low oxygen conditions.

Figure 15.12 **Dogs small and large.** It is hard to believe that the diminutive Chihuahua and the Great Dane are members of the same species. Thanks to artificial selection and intense inbreeding, body size in domesticated dogs varies more than it does in any other terrestrial mammal.

development, immunity, muscle contraction, and response to stress. Reading 16.1 offers several examples of traits that were positively selected in human evolution.

One of the most dramatic examples of positive selection in humans is seen in native people who live more than two miles above sea level on the Tibetan plateau (**figure 15.11**). The Tibetans thrive in the oxygen-poor air that makes others very ill and not likely to reproduce. Even the Han Chinese, who are a related population that lives in the lowlands, cannot live for long up in the mountains without developing the headache, dizziness, ringing in the ears, heart palpitations, breathlessness, fatigue, sleep disturbance, lack of appetite, and confusion of chronic mountain sickness. If they ascend to great heights, their bloodstreams become crammed with extra red blood cells that extract maximal oxygen from the thin air.

In people other than the native Tibetans exposed to low oxygen conditions, a protein called hypoxia inducible factor 2, which is a transcription factor, signals the kidneys to release the hormone EPO (erythropoietin), which stimulates the bone marrow to make more red blood cells. As a result, the hemoglobin level in the blood rises. Several genome-wide association studies showed that the native Tibetans have eight unique SNPs in the gene, called *EPAS1*, that encodes hypoxia inducible factor 2. Therefore, they have an apparently unique version of the protein that enables them to thrive in the thin air. Researchers do not yet know exactly how the Tibetans do this, but since they have resided in the area only a few thousand years, the adaptation to low oxygen levels is an example of very rapid evolution.

Artificial Selection

Natural selection acts on preexisting genetic variants. It is uncontrolled and largely unpredictable. In contrast, artificial selection is controlled breeding to perpetuate individuals with

a particular phenotype, such as a crop plant or tiny pig for a pet. Darwin's idea of natural selection grew from his observations of artificially selected pigeons.

We have created our pets by controlling their evolution. Pets arose from initial domestication from wild species, followed by artificial selection. Cats were domesticated in the Near East when agriculture began, about 10,000 years ago. They descended from one of five subspecies of wildcats. Dogs were domesticated in Southeast Asia about 40,000 years ago, from gray wolves. Most of today's dog breeds were artificially selected in the 1800s, but some, such as the Australian bulldog and the silken windhound, were bred in the 1990s. **Figure 15.12** demonstrates how different modern dog breeds can be from each other. The intense inbreeding required to fashion breeds cuts their genetic diversity, resulting in more than 300 inherited diseases, from bladder stones in Dalmatians to hip problems in St. Bernards. Yet the differences among dog breeds for some traits stem from only a handful of genes. For example, variants in only three genes account for 95 percent of the variability in the texture and length of canine coats.

Signs of our artificial selection of dog breeds are evident in their genomes. Researchers at the National Human Genome Research Institute looked at 60,000 SNPs scattered across the genomes of 80 dog breeds and 83 wolves, coyotes, jackals, and various wild dogs from museum specimens and animal shelters in Africa. When the researchers looked for associations between specific SNP patterns and specific traits, they discovered that only one, two, or three genes seem to account for the distinctions of each breed. This was surprising, since many common traits analyzed in humans, as well as agricultural traits in crops, are determined by many genes, each exerting a small effect. However, the recent history of dog domestication explains why this is so. We select oddities and quirks in our canine friends, and these are more often the consequence of a mutation. In nature, that mutant may not have survived; we intentionally breed to select it. For example, ear floppiness and dwarfism, both seen in basset hounds, are each the result of a mutation in a single gene.

Analysis of the genomes of domesticated dogs and their wild relatives confirmed what we know from history—that our pets are no longer very much like wolves. Specifically, linkage disequilibrium (blocks of SNPs next to each other on chromosomes) tends to distinguish modern breeds. The fact that breeds share few SNP blocks indicates that we have selected away many ancestral DNA sequences. In addition, runs of homozygosity—long DNA sequences identical on both chromosomes—reflect intense inbreeding. The fact that domesticated breeds differ from each other less than wild dogs differ from each other indicates population bottlenecks that accompanied dog breeding, narrowing their gene pools.

Artificial selection may be viewed as arrogance—that we think we know enough to control traits. In some situations, it may be better to trust natural selection. This is the case for native cows, pigs, and goats in Africa. Their genomes hold many gene variants that have adapted the animals, over many centuries, to living in physically challenging environments. Such genes affect milk production, ability to tolerate bacterial and viral infections and parasites, and the ability to survive temperature extremes.

The traits of the native livestock of Africa suit well the needs of the human families that care for them (**figure 15.13**). The animals provide food, labor, and transport, and are part of the cultural fabric of humanity in the region. This is not the case for livestock imported to Africa from Europe, where breeders have selected mostly for productivity traits, such as larger muscle mass. Such elite breeds raised in Africa require special food and veterinary care, and can be a financial burden compared to the hardy native species. Because of this clash between artificial and natural selection, the new field of landscape genomics has emerged. Researchers are now correlating gene variants from all over the animals' genomes to environmental situations and challenges to identify the valuable genes. The conclusion so far: We should identify, tap, and preserve what natural selection has provided, rather than importing animals with artificially selected short-term solutions.

Tuberculosis Ups and Downs—and Ups

Natural selection can be seen in the appearance or return of infectious diseases. If infection kills before reproductive age or impairs fertility, its spread will ultimately remove from the population individuals susceptible to infection. Disease incidence falls as only survivors are left. But if conditions change, the disease may resurge. This has happened with tuberculosis (TB).

When TB first appeared in the Plains Indians of the Qu'Appelle Valley Reservation in Saskatchewan, Canada in the mid-1880s, it struck swiftly, infecting many organs. Ten percent of the population died. But by 1921, TB tended to affect only the lungs, and only 7 percent of the population died annually from it. By 1950, mortality was down to 0.2 percent. Some people were symptomless carriers.

Outbreaks of TB ran similar courses in other human populations. The disease appeared in crowded settlements where the bacteria easily spread in exhaled droplets. In the 1700s, TB raged through the cities of Europe. Immigrants brought it to U.S. cities. But TB incidence and virulence fell dramatically in the cities of the industrialized world in the first half of the twentieth century—before antibiotic drugs were discovered. What tamed tuberculosis?

Natural selection, operating on both the bacterial and human populations, lessened the virulence of the infection. Some people inherited resistance and passed this beneficial trait on. At the same time, the most virulent bacteria killed their hosts so quickly that the victims had no time to spread the infection. As the deadliest bacteria were selected out of the population (negative selection), and as people who inherited resistance mutations contributed more to the next generation (positive selection), TB gradually evolved from a severe, acute, systemic infection to a rare chronic lung infection. Then in the late 1980s, conditions ideal for the infection's return arose.

At first, complacency led to the resurgence of TB. Funding for TB research was cut because the infection was considered "cured." Patients thought themselves cured when antibiotics helped in a few months, abandoning the drugs yet unknowingly continuing to spread the bacteria. With increased air travel, people began spreading different strains of the bacteria around the world. Treatment in the 1950s—isolating patients for 18 months or longer in facilities called sanitoria—was actually more effective by quarantining infectious individuals. Then AIDS happened, providing millions of vulnerable human lungs. The bacterial populations soared, mutations accumulated, and variants resistant to antibiotic drugs arose. Today, a third of all HIV-infected people also have TB. Someone with HIV faces a fifty-fold increased risk of contracting TB, and can pass it to anyone in just a sneeze or cough.

Figure 15.13 Landscape genomics. This new field mines native livestock genomes for valuable traits that natural selection has favored. Here, a Masai tribesman herds native cattle in Kenya.

Further evidence of evolution is that the bacteria that cause TB are becoming resistant to many types of antibiotic drugs and increasing in genetic diversity. At the same time, though, genetic testing for the bacteria is making it possible to diagnose TB in under 2 hours. In the past, diagnosis took up to 2 months, and during that time many patients spread the infection. **Reading 15.1** discusses antibiotic resistance in another bacterium, which causes "staph" infections.

Evolving HIV

Viral mutations accumulate rapidly because their RNA or DNA replicates often and errors are not repaired. Like bacteria, the viruses in a human body form a population, including naturally occurring genetic variants. In HIV infection, natural selection controls the diversity of viral variants in a human body as the disease progresses. The human immune system and drugs to slow the infection become the environmental factors that select (favor) resistant viral variants.

HIV infection can be divided into three stages, both from the human and the viral perspective (**figure 15.14**). A person infected with HIV may experience an initial acute phase, with symptoms of fever, night sweats, rash, and swollen glands. In a second period, lasting from 2 to 15 years, health usually returns. In a third stage, immunity collapses, the virus replicates explosively, and opportunistic infections and cancer eventually cause death.

Reading 15.1

Antibiotic Resistance: The Rise of MRSA

Many antibiotic drugs are no longer effective in treating bacterial infections. The reason is the interplay between mutation and natural selection.

Our bodies harbor populations of bacteria that have genetic variants, some of which enable the microorganisms to survive in the presence of a particular antibiotic drug. When a sick person takes the drug, symptoms abate as sensitive bacteria die. The resistant mutants reproduce, taking over the niche the antibiotic-sensitive bacteria vacated. Soon, the person has enough antibiotic-resistant bacteria to feel ill again. Usually antibiotic resistance genes already exist in the bacterial populations, and exposure to the drug selects the resistant bacteria. However, some antibiotics actually induce mutation.

Resistant bacteria circumvent antibiotic actions in several ways. Penicillin kills bacteria by tearing apart their cell walls. Resistant microbes produce enzyme variants that dismantle penicillin, or have altered cell walls that the drug cannot bind. Erythromycin, streptomycin, tetracycline, and gentamicin kill bacteria by attacking their ribosomes, which are different from ribosomes in a human. Drug-resistant bacteria have ribosomes that the drugs cannot bind.

Bacteria become resistant in two ways. Their DNA can mutate, passing the resistance from one bacterial generation to the next by cell division. Or, groups of resistance genes are passed on transposons, which move from cell to cell as part of DNA circles called plasmids. Bacteria usually pass transposons to similar bacteria, but in the unnatural environment of a hospital, genes may flit to any bacterium, and drug resistances are passed quickly. This is what has happened with infection by the bacterium *Staphylococcus aureus*.

S. aureus is normally present in low numbers in the nose and on the skin, but in high numbers it causes pimples and boils, food poisoning, toxic shock syndrome, pneumonia, and surgical wound infections (**figure 1**). *S. aureus* infection is particularly dangerous in hospitals, spreading rapidly among people unable to fight it. This common bacterium became resistant to penicillin soon after the drug was introduced in the 1940s. A related penicillin, methicillin, worked for a time, but resistant bacterial strains appeared suddenly in 2000, at such an alarming rate that the microorganism has its own acronym: MRSA, for methicillin-resistant *S. aureus*. Doctors use another antibiotic, vancomycin, to treat MRSA—for now.

DNA sequencing revealed that in one hospital, *S. aureus* picked up vancomycin resistance from another type of bacterium. A foot ulcer in a dialysis patient in Detroit harbored vancomycin-resistant *Enterococcus faecalis* as well as two types of *S. aureus,* one resistant to the antibiotic and one sensitive. By sequencing the plasmids that included the resistance gene, investigators deduced that *S. aureus* picked up an *E. faecalis* plasmid bearing a vancomycin resistance gene called *vanA*. Then the *vanA* gene jumped to an *S. aureus* plasmid. Since then, similar scenarios of drug resistance gene sharing among microorganisms have happened in many countries.

In the rise of MSRA infection, natural selection benefits the pathogen, not us. That is, bacteria that can resist the drugs that we use to fight them will survive and reproduce, ensuring that *Staphylococcus aureus* infection continues.

Figure 1 **Antibiotic resistance.** These *Staphylococcus aureus* bacteria are resistant to methicillin and several other antibiotics.

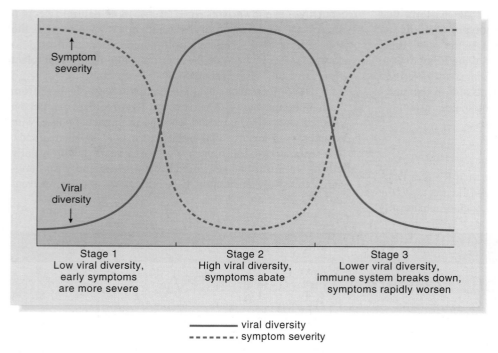

Stage 1
Low viral diversity,
early symptoms
are more severe

Stage 2
High viral diversity,
symptoms abate

Stage 3
Lower viral diversity,
immune system breaks down,
symptoms rapidly worsen

——————— viral diversity
- - - - - - - symptom severity

Figure 15.14 **Natural selection of HIV.** Natural selection controls the genetic diversity of an HIV population in a person's body. Before the immune system gathers strength, and after it breaks down, HIV diversity is low. A rapidly reproducing viral strain predominates, although new mutations continually arise. During the 2- to 15-year latency period, viral variants that can evade the immune system gradually accumulate.

The HIV population changes and expands throughout the course of infection, even when the patient seems to stay the same for a long time. New mutants continuously arise, and they alter such traits as speed of replication and the patterns of molecules on the viral surface.

In the first stage of HIV infection, as the person battles acute symptoms, viral variants that replicate swiftly predominate. In the second stage, the immune system starts to fight back and symptoms abate, as viral replication slows and many viruses are destroyed. Now natural selection acts, and certain viral variants reproduce and mutate, giving rise to a diverse viral population. Ironically, drugs used to treat AIDS may further select against the weakest HIV variants. Gradually, the HIV population overtakes the immune system cells, but years may pass before immunity begins to noticeably decline. The third stage, full-blown AIDS, occurs when the virus overwhelms the immune system. With the selective pressure off, viral diversity again diminishes, and the fastest-replicating variants predominate. HIV wins.

The entire scenario of HIV infection reflects the value of genetic diversity. It enables the survival of a population or species in the face of an environmental threat. When that threat—an immune system attack or drugs—wipes out sensitive variants, one genotype may prevail.

The fact that HIV diversifies early in the course of infection is why patients take combinations of drugs right after diagnosis. The drugs target several viral variants simultaneously in different ways, slowing the course of the infection. For many people, thanks to declining viral genetic diversity, HIV infection has become a chronic illness rather than the swift killer that it once was.

Balanced Polymorphism

If natural selection eliminates individuals with detrimental phenotypes from a population, then how do harmful mutant alleles remain in a gene pool? Harmful recessive alleles are replaced in two ways: by new mutation, and by persistence in heterozygotes.

A recessive condition can remain prevalent if the heterozygote enjoys a health advantage, such as resisting an infectious disease or surviving an environmental threat. This "heterozygous advantage" that maintains a recessive, disease-causing allele in a population is called **balanced polymorphism**. Recall that *polymorphism* means variant; the effect is *balanced* because the protective effect of the noninherited condition counters the negative effect of the deleterious allele, maintaining its frequency in the population. Balanced polymorphism is a type of balancing selection, which more generally refers to maintaining heterozygotes in a population. A few examples follow, and these and others are summarized in **table 15.3**.

Sickle Cell Disease and Malaria

Sickle cell disease is an autosomal recessive disorder that causes anemia, joint pain, a swollen spleen, and frequent, severe infections. It is the classic example of balanced polymorphism: carriers are resistant to malaria, or develop very mild cases.

Malaria is an infection by the parasite *Plasmodium falciparum* and related species that causes debilitating cycles of chills and fever. The parasite spends the first stage of its life cycle in the salivary glands of the mosquito *Anopheles gambiae*. When an infected female mosquito draws blood from a human, malaria parasites enter red blood cells, which transport the parasites to the liver. The red blood cells burst, releasing parasites throughout the body.

In sickle cell disease, many red blood cells burst too soon, expelling the parasites before they infect. The blood of a person with sickle cell disease is also thicker than normal, which may hamper the parasite's ability to infect. A sickle cell disease carrier's blood has enough atypical cells to be inhospitable to the malaria parasite, but usually not enough to cause the blocked circulation of sickle cell disease.

A clue to the protective effect of being a carrier for sickle cell disease came from striking differences in the incidence of

Table 15.3	Balanced Polymorphism				
Disease 1 (inherited, carrier)	**Protects against →**	**Disease 2**	**Because →**	**Mechanism**	**References**
Sickle cell disease		Malaria		Atypical red blood cells cannot retain parasites	Section 12.2
G6PD deficiency		Malaria		Parasite cannot reproduce in atypical red blood cells	Section 12.5
PKU		Fungal infection in fetuses		Elevated phenylalanine inactivates fungal toxin	Sections 5.2, 10.4, 14.1, 15.6
Prion protein mutation		Transmissible spongiform encephalopathy		Prion protein cannot misfold in presence of infectious prion protein	Figure 10.22, section 12.5, Reading 10.1
CF		Diarrheal disease (cholera, typhoid fever)		Fewer chloride channels in intestinal cells prevent water loss	Sections 14.1, 14.3, Readings 2.2, 4.2
Smith-Lemli-Opitz syndrome		Cardiovascular disease		Lowered serum cholesterol	MIM 270400 (multiple birth defects, intellectual disability)

the two diseases in different parts of the world (**figure 15.15**). In the United States, 8 percent of African Americans are sickle cell carriers, whereas in parts of Africa, up to 45 to 50 percent are carriers. Although Africans had known about a painful disease that shortened life, the sickled cells weren't recognized until 1910 (see section 12.2). In 1949, British geneticist Anthony Allison found that the frequency of sickle cell carriers in tropical Africa was higher in regions where malaria rages all year long. Blood tests from children hospitalized with malaria showed that nearly all were homozygous for the wild type sickle cell allele. The few sickle cell carriers among them

had the mildest cases of malaria. Was malaria enabling the sickle cell allele to persist by felling people who did not inherit it? The fact that sickle cell disease is rarer where malaria is rare supports the idea that sickle cell heterozygosity protects against the infection.

Further evidence of a sickle cell carrier's advantage in a malaria-ridden environment is the fact that the rise of sickle cell disease parallels the cultivation of crops that provide breeding grounds for *Anopheles* mosquitoes. About 1000 B.C., Malayo-Polynesian sailors from Southeast Asia traveled in canoes to East Africa, bringing new crops of bananas, yams, taros, and coconuts. When the jungle was cleared to grow these crops, the open space provided breeding grounds for the mosquitoes. The insects, in turn, offered a habitat for part of the life cycle of the malaria parasite.

The sickle cell allele may have been brought to Africa by people migrating from Southern Arabia and India, or it may have arisen directly by mutation in East Africa. However it happened, people who inherited one copy of the sickle cell allele survived or never contracted malaria—the essence of natural selection. These healthy carriers had more children and passed the protective allele to approximately half of them. Gradually, the frequency of the sickle cell allele in East Africa rose from 0.1 percent to 45 percent in 35 generations. Carriers paid the price for this genetic protection, however, whenever two of them produced a child with sickle cell disease.

A cycle set in. Settlements with large numbers of sickle cell carriers escaped malaria. They were strong enough to clear even more land to grow food, and support the disease-bearing mosquitoes.

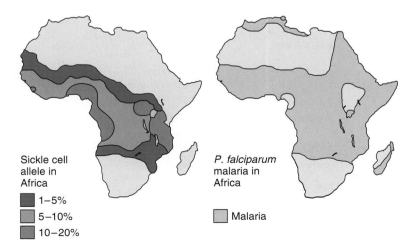

Sickle cell allele in Africa

■ 1–5%
■ 5–10%
■ 10–20%

P. falciparum malaria in Africa

■ Malaria

Figure 15.15 **Balanced polymorphism.** Comparing the distribution of people with malaria and people with sickle cell disease in Africa reveals balanced polymorphism. Carriers for sickle cell disease are resistant to malaria because changes in the blood caused by the sickle cell allele are not severe enough to impair health, but they do inhibit the malaria parasite.

Prion Disease and Cannibalism

Being a heterozygote for the prion protein gene may protect against the disorders of protein folding called transmissible spongiform encephalopathies (see Reading 10.1, and section 12.5). The best studied such illness is kuru, which caused brain degeneration among the Foré people in Papua New Guinea until the Australian government halted the practice of ritual cannibalism in the mid-1950s. A recent investigation of the prion protein gene among 30 elderly Foré women who had eaten brains revealed 23 heterozygotes. Only 15 were predicted based on Hardy-Weinberg equilibrium observed among 140 Foré who had not eaten brains. In the heterozygotes, some of the normal prion proteins have a valine at amino acid position 129, and some a methionine. The different amino acids in the same person somehow prevents infectious misfolding in the presence of abnormal prion protein—as happens in cannibalism. All of the people in the United Kingdom who developed variant CJD, the human form of "mad cow disease," had only methionine at position 129.

The overrepresentation of heterozygotes among prion disease survivors suggests that balancing selection has favored this genotype in the population, and that cannibalism may have been the driving force. That is, homozygotes who were cannibals died of a prion disorder before reproducing, leaving the resistant heterozygotes to slowly accumulate in the population.

Cystic Fibrosis and Diarrheal Disease

Balanced polymorphism may explain why CF is so common—its cellular defect protects against diarrheal illnesses such as cholera and typhus. Diarrheal disease epidemics have left their mark on many human populations, and continue to be a major killer in the developing world.

Severe diarrhea rapidly dehydrates the body and leads to shock, kidney and heart failure, and death in days. In cholera, bacteria produce a toxin that opens chloride channels in cells of the small intestine. As salt (NaCl) leaves the intestinal cells, water rushes out, producing diarrhea. The CFTR protein does just the opposite, closing chloride channels and trapping salt and water in cells, which dries out mucus and other secretions. A person with CF is very unlikely to contract cholera, because the toxin cannot open the chloride channels in the small intestine cells.

CF carriers enjoy the mixed blessing of balanced polymorphism. They do not have enough abnormal chloride channels to cause the labored breathing and clogged pancreas of CF, but they have enough of a defect to block the cholera toxin. During cholera epidemics throughout history, individuals carrying mutant CF alleles had a selective advantage, and they disproportionately transmitted those alleles to future generations.

Because CF arose in western Europe and cholera originated in Africa, an initial increase in CF heterozygosity may have been a response to a different diarrheal infection—typhoid fever. The causative bacterium, *Salmonella typhi*, rather than producing a toxin, enters cells lining the small intestine—but only if CFTR channels are present. The cells of people with severe CF manufacture CFTR proteins that never reach the cell surface, and therefore no bacteria get in. Cells of CF carriers admit some bacteria. Protection against infections that produce diarrhea may therefore have kept CF in populations.

15.6 Putting It All Together: PKU

Monitoring allele frequencies in populations can reveal the effects of nonrandom mating, migration, genetic drift, mutation, and natural selection on evolution. **Figure 15.16** summarizes the forces acting alone, and **table 15.4** lists the examples

Table 15.4	Forces that Change Allele Frequencies
Mechanism of Allele Frequency Change	**Examples**
Nonrandom mating	Agriculture
	Cape population and Arnold
	Hopi Indians with albinism
	Genghis Khan's Y chromosome
Migration	Consanguinity
	Galactokinase deficiency in Europe
	ABO blood type distribution
	Clines along the Nile and in Italy
Genetic drift	
Founder effect	Norfolk Island mutineer descendants and migraine
	Disorders among
	Old Order Amish and Mennonites
	Afrikaners and porphyria variegata
Population bottleneck	Pingelapese blindness
	Cheetahs
	Pogroms against Ashkenazi Jews
Mutation	Chapters 12 and 13
Natural selection	Lactose intolerance
	Tibetan adaptation to high altitude
	TB incidence and virulence
	HIV infection
	Antibiotic resistance in bacteria
	Sickle cell disease and malaria
	Prion disease and cannibalism
	CF and diarrheal disease
	PKU and protection against fungal infection

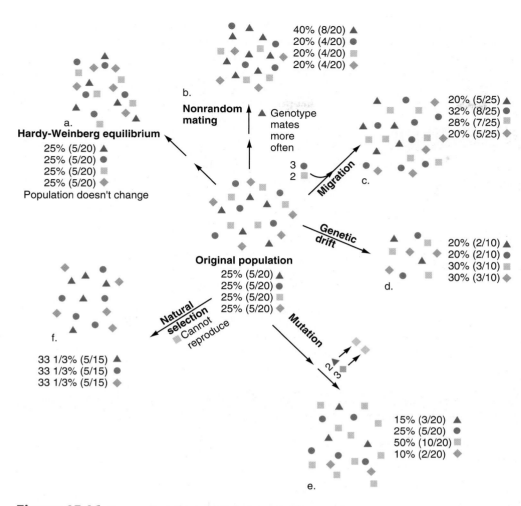

Figure 15.16 Forces that change allele frequencies.

In some isolated populations, migration and the founder effect have maintained certain PKU alleles. In most populations, point mutations in the phenylalanine hydroxylase (PAH) gene cause PKU. However, all Yemeni Jews in Israel who have PKU have a large deletion. This mutation spread from northern Africa to Israel. To track it, researchers tested for the telltale deletion in the grandparents of the 22 modern Yemeni Jewish families with PKU in Israel. The earliest court and religious records identify two families with PKU in San'a, the capital of Yemen. By 1809, religious persecution and hard economic times led nine families carrying the mutation to migrate north and settle in three towns, then spread farther (**figure 15.17**).

PKU may have persisted because being a heterozygote protects against certain fungal infections spread in tainted grains, particularly in Ireland and Scotland, where the inherited disease is the most prevalent. PKU carriers have elevated phenylalanine levels in their blood that are not enough to cause PKU symptoms, but high enough to inactivate a

in the chapter with the mechanisms that they illustrate. Often, historical, archeological, and linguistic evidence can help us to understand the complex interactions of these forces. Consider, again, PKU, which causes intellectual disability unless a specific diet is followed from birth.

The diversity of PKU mutations suggests that the disease has arisen more than once. Mutations common to many groups of people probably represent more ancient mutations that occurred before groups spread and separated. In contrast, mutations found only in a small geographical region are more likely to be of recent origin, perhaps kept apart by genetic drift. These mutations have had less time to spread. For example, Turks, Norwegians, French Canadians, and Yemeni Jews have their own PKU alleles. Analysis of the frequencies of PKU mutations in different populations, plus logic, can reveal the roles that genetic drift, mutation, and balanced polymorphism have played in maintaining the mutation.

A high mutation rate cannot be the sole reason for the continued prevalence of PKU because some countries continue to have only one or two mutations. If the gene mutated frequently, all populations would have several different types of PKU mutations. This is not so.

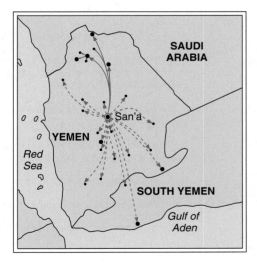

Figure 15.17 **The origin of PKU.** The deletion in Israeli Yemeni Jews probably arose in San'a, Yemen, in the mid–eighteenth century. The allele spread northward as families moved from San'a in 1809 (solid arrows) and subsequently spread to other regions (broken arrows), and eventually Israel.

Source: Data from Smadar Avigad, et al., A single origin of phenylketonuria in Yemenite Jews, Nature 344:170, March 8, 1990.

fungal poison, ochratoxin A, that harms fetuses. During the famines that have periodically plagued this part of the world, starving people ate moldy grain. If PKU carriers eating the grain were more likely to have healthy children, the mutant allele would have increased in the population, thanks to balanced polymorphism. The Vikings spread PKU eastward.

Key Concepts

1. PKU originated more than once.
2. Genetic drift, balanced polymorphism, and perhaps mutation, have affected the prevalence of PKU.

15.7 Eugenics

We usually think of artificial selection in the context of Darwin's pet pigeons or purebred cats, dogs, and horses. We also practice artificial selection through control of our own reproduction, through individual choices as well as at the societal level.

Some people attempt to control the genes in their offspring. They do this by seeking mates with certain characteristics, by choosing egg or sperm donors with particular traits, or by ending pregnancies after a test reveals a devastating disorder. On a societal level, **eugenics** refers to programs that control human reproduction with the intent of changing the genetic structure of the population. Eugenics works in two directions. Positive eugenics creates incentives for reproduction among those considered superior; negative eugenics interferes with reproduction of those judged inferior. Obviously, eugenic measures are highly subjective. **Table 15.5** lists some famous examples of eugenics, and *Bioethics: Choices for the Future* on page 295 considers a highly personal viewpoint.

The word "eugenics" was coined in 1883 by Sir Francis Galton to mean "good in birth." He defined eugenics as "the science of improvement of the human race germplasm through better breeding." One vocal supporter of the eugenics movement was Sir Ronald Aylmer Fisher. In 1930, he tried to apply the principles of population genetics to human society, writing that those at the top of a society tend to be "genetically infertile," producing fewer children than the less-affluent classes. This, he claimed, was the reason why civilizations ultimately

Table 15.5	A Chronology of Eugenics-Related Events
1883	Sir Francis Galton coins the term *eugenics*.
1889	Sir Francis Galton's writings are published in the book *Natural Inheritance*.
1896	Connecticut enacts law forbidding sex with a person who has epilepsy or is "feebleminded" or an "imbecile."
1904	Galton establishes the Eugenics Record Office at the University of London to keep family records.
1907	First eugenic law in the United States orders sterilization of institutionalized intellectually disabled males and criminal males when experts recommend it.
1910	Eugenics Record Office founded in Cold Spring Harbor, New York, to collect family and institutional data.
1924	Immigration Act limits entry into the United States of "idiots, imbeciles, feebleminded, epileptics, insane persons," and restricts immigration to 7 percent of the U.S. population from a particular country according to the 1890 census—keeping out those from southern and eastern Europe.
1927	Supreme Court (*Buck vs. Bell*) upholds compulsory sterilization of the intellectually disabled by a vote of 8 to 1, leading to many state laws.
1934	Eugenic sterilization law of Nazi Germany orders sterilization of individuals with conditions thought to be inherited, including epilepsy, schizophrenia, and blindness, depending upon rulings in Genetic Health Courts.
1939	Nazis begin killing 5,000 children with birth defects or intellectual disability, then 70,000 "unfit" adults.
1956	U.S. state eugenic sterilization laws are repealed, but 58,000 people have already been sterilized.
1965	U.S. immigration laws reformed, lifting many restrictions.
1980s	California's Center for Germinal Choice established, where Nobel Prize winners can deposit sperm to inseminate selected women.
1990s	In the U.S., state laws passed to prevent health insurance or employment discrimination based on genotype.
2003	Many governments recommend certain genetic tests, and enact legislation to prevent genetic discrimination.
2004	Genocide of black Africans in Sudan.
2009	U.S. Genetic Information Nondiscrimination Act enacted, but is limited in scope.

Two Views of Neural Tube Defects

Genetic tests enable people to make reproductive choices that can alter allele frequencies in populations. Identifying carriers of a recessive illness, who then may decide not to have children together, is one way to remove disease-causing alleles from a population. Screening pregnant women for fetal anomalies, then terminating affected pregnancies, also alters disease prevalence and, if the disorder has a genetic component, allele frequencies. This is the case for neural tube defects (NTDs), which are multifactorial.

An NTD forms at the end of the first month, when the embryo's neural tube does not completely close. An opening in the head (anencephaly) usually ends in miscarriage, stillbirth, or a newborn who dies within days. An opening in the spinal cord (spina bifida) causes paralysis but the person can live into adulthood and have normal intelligence. Surgery can help.

In 1992, the Centers for Disease Control and Prevention concluded that taking the vitamin folic acid in pregnancy lowers the risk of NTD recurrence from 3 to 4 percent to 1.5 to 2 percent. Women who had had an affected child began taking large doses before conception. But when epidemiologists tried to monitor how well it was working, they faced a problem—prevalence of NTDs was greatly underestimated. This happened because the statistics on NTD prevalence—vital to discovering whether folic acid was preventing the defect—included only newborns, stillborns, and older fetuses. Most reports did not account for pregnancies terminated following prenatal diagnosis. These pregnancies caused the underreporting of anencephaly by 60 to 70 percent, and of spina bifida, by 20 to 30 percent in some states.

A Personal View

Blaine Deatherage-Newsom has a different view of population screening for NTDs because he has one (see photo). Blaine was born in 1979 with spina bifida. Paralyzed from the armpits down, he has endured much physical pain, but he has also achieved a great deal. While in high school, he put the question, "If we had the technology to eliminate disabilities from the population, would that be good public policy?" on the Internet—initiating a global discussion. He wrote:

I was born with spina bifida and hydrocephalus. I hear that when parents have a test and find out that their unborn child has spina bifida, in more than 95 percent of the cases they choose to have an abortion. I also went to an exhibit at the Oregon Museum of Science and Industry several years ago where the exhibit described a child born with spina bifida and hydrocephalus, and . . . asked people to vote on whether the child should live or die. I voted that the child should live, but when I voted, the child was losing by quite a few votes.

When these things happen, I get worried. I wonder if people are saying that they think the world would be a better place without me. I wonder if people just think the lives of people with disabilities are so full of misery and suffering that they think we would be better off dead. It's true that my life has suffering (especially when I'm having one of my 11 surgeries so far), but most of the time I am very happy and I like my life very much. My mom says she can't imagine the world without me, and she is convinced that everyone who has a chance to know me thinks that the world is a far better place because I'm in it.

Today Blaine works for a not-for-profit organization that refurbishes computer equipment for community service organizations.

Questions for Discussion

1. Is the decision to end a pregnancy that would otherwise lead to the birth of a child with a neural tube defect a eugenic measure or not? State a reason for your answer.

2. People with certain medical conditions or limitations, such as those with hearing loss, object to genetic tests that would ultimately decrease their numbers in the population. How would you feel if you had such a condition?

3. Do you think that eugenics will resurge as personal genome sequencing and genetic testing become more widespread?

Excerpt by Blaine Deatherage-Newsom, "If we could eliminate disabilities from the population, should we? Results of a survey on the Internet." Reprinted by permission.

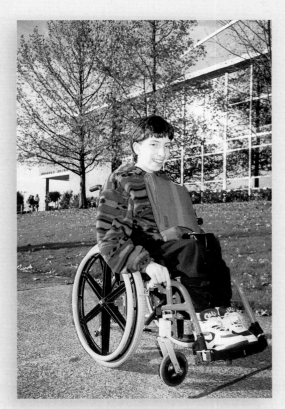

Blaine Deatherage-Newsom as a teen.

topple. He offered several practical suggestions to remedy this, including state monetary gifts to high-income families for each child born to them.

Early in the twentieth century, eugenics focused on maintaining purity. One prominent geneticist, Luther Burbank, realized the value of genetic diversity at the beginning of a eugenic effort. Known for selecting interesting plants and crossing them to breed plants with useful characteristics, Burbank in 1906 applied his ideas to people. In a book called *The Training of the Human Plant,* he encouraged immigration to the United States so that advantageous combinations of traits would appear as the new Americans interbred. Burbank's plan ran into problems, however, at the selection stage, which allowed only those with "desirable" traits to reproduce.

On the East Coast of the United States, Charles Davenport led the eugenics movement. In 1910, he established the Eugenics Record Office at Cold Spring Harbor, New York. There he headed a massive effort to compile data from institutions, prisons, circuses, and general society. He attributed nearly every trait to a single gene. "Feeblemindedness," for example, was an autosomal recessive trait. It was a catch-all phrase for a person with low intelligence (as measured on an IQ test) and such characteristics as "criminality," "promiscuity," and "social dependency." In one famous case, a young woman named Carrie Buck was ordered to be sterilized when she, her mother, and her illegitimate infant daughter Vivian were declared feebleminded. Carrie had been raped by a relative of her foster parents, and was actually an average student. **Figure 15.18** shows the pedigree for Carrie Buck and her "inherited trait" of feeblemindedness.

Other nations practiced eugenics. From 1934 until 1976, the Swedish government forced certain individuals to be sterilized as part of a "scientific and modern way of changing society for the better," according to one historian. At first, only mentally ill people were sterilized, but poor, single mothers were later included. The women's movement in the 1970s pushed for an end to forced sterilizations.

In 1994, China passed the Maternal and Infant Health Care Law, which proposes "ensuring the quality of the newborn population" and forbids procreation between two people if physical exams show "genetic disease of a serious nature . . . that may totally or partially deprive the victim of the ability to live independently, that [is] highly possible to recur in generations to come, and that [is] medically considered inappropriate for reproduction." Such "genetic diseases" include intellectual disability, mental illness, and seizures, conditions that are ill-defined in the law and are not necessarily inherited.

Another guise of eugenics is war, if the fighting groups differ genetically. Throughout history, war and conflict have altered gene pools, sometimes dramatically. These effects are eugenic when they take the form of rape of women of one group by men from another, with the intent of "diluting" the genes of the rape victims. During the Rwandan genocide of 1994, for example, Hutu policy was to rape Tutsi women—250,000

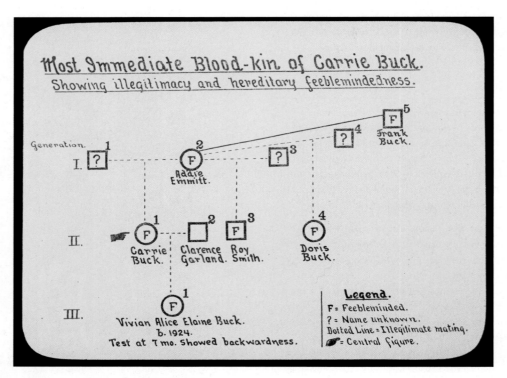

Figure 15.18 Eugenics sought to abolish "feeblemindedness." In 1927, 17-year-old Carrie Buck, of Charlottesville, stood trial for the crime of having a mother who lived in an asylum for the feebleminded, and having a daughter out-of-wedlock (following rape) also deemed feebleminded, as was Carrie herself, though she was a B student in school. Ruled Sir Oliver Wendell Holmes, Jr., "three generations of imbeciles are enough." Carrie Buck made history as the first person to be sterilized to prevent having another "socially inadequate offspring."

of them. Genocide by rape has been ongoing in the eastern Democratic Republic of the Congo since 1996. Since 2003 in Darfur, part of western Sudan in Africa, Arab militia have systematically attacked black Africans, killing men and children and repeatedly raping women. In Rwanda, Congo, and Darfur, the conquerors have claimed that their intent is to diminish the genetic contributions of their victims and spread their own genes. That's eugenics.

Modern genetics is sometimes compared to eugenics because genetic technologies may affect reproductive choices and can influence which alleles are passed to the next generation. However, medical genetics and eugenics differ in their goals. Eugenics aims to allow only people with certain "valuable" genotypes to reproduce, for the supposed benefit of the population as a whole. The goal of medical genetics, in contrast, is to prevent and alleviate suffering.

Humanity arose in Africa some 200,000 years ago. As pockets of peoples spread across the globe, our behaviors and

intelligence introduced culture. The next chapter explores some of our journeys, through clues to the past in the sequences of our DNA.

Key Concepts

1. Eugenics is the control of individual human reproduction for societal goals, maximizing the genetic contribution of those deemed acceptable (positive eugenics) and minimizing the contribution from those considered unacceptable (negative eugenics).
2. Some people consider modern genetic screening practices eugenic, but genetic testing usually aims to prevent or alleviate human suffering.
3. Wars may have eugenic consequences.

Summary

15.1 Nonrandom Mating

1. Hardy-Weinberg equilibrium assumes all individuals mate with the same frequency and choose mates without regard to phenotype. This rarely happens. We choose mates based on certain characteristics, and some people have many more children than others.
2. DNA sequences that do not cause a phenotype important in mate selection or reproduction may be in Hardy-Weinberg equilibrium.
3. Consanguinity increases the proportion of homozygotes in a population, which may lead to increased incidence of recessive illnesses or traits.

15.2 Migration

4. **Clines** are changes in allele frequencies from one area to another.
5. Clines may reflect geographical barriers or linguistic differences and may be either abrupt or gradual.
6. Human migration patterns through history explain many cline boundaries. Forces behind migration include escape from persecution and a nomadic lifestyle.

15.3 Genetic Drift

7. **Genetic drift** occurs when a small population separates from a larger one, or its members breed only among themselves, perpetuating allele frequencies not characteristic of the larger population due to chance sampling.
8. A **founder effect** occurs when a few individuals found a settlement and their alleles form a new gene pool, amplifying their alleles and eliminating others.
9. A **population bottleneck** is a narrowing of genetic diversity that occurs after many members of a population die and the few survivors rebuild the gene pool.

15.4 Mutation

10. Mutation continually introduces new alleles into populations. It occurs as a consequence of DNA replication errors.
11. Mutation does not have as great an influence on disrupting Hardy-Weinberg equilibrium as the other factors.
12. The **genetic load** is the collection of deleterious alleles in a population.

15.5 Natural Selection

13. Environmental conditions influence allele frequencies via **natural selection**. Alleles that do not enable an individual to reproduce in a particular environment are selected against and diminish in the population, unless conditions change. Beneficial alleles are retained.
14. In **balanced polymorphism,** the frequencies of some deleterious alleles are maintained when heterozygotes have a reproductive advantage under certain conditions.

15.6 Putting It All Together: PKU

15. Frequencies of different mutations in different populations provide information on the natural history of alleles and on the relative importance of nonrandom mating, migration, genetic drift, mutation, and natural selection in deviations from Hardy-Weinberg equilibrium.

15.7 Eugenics

16. **Eugenics** is the control of individual reproduction to serve a societal goal.
17. Positive eugenics encourages those deemed superior to reproduce. Negative eugenics restricts reproduction of those considered inferior.
18. Some aspects of genetic technology affect reproductive choices and allele frequencies, but the goal is to alleviate or prevent suffering, not to change society.

Answers to all end-of-chapter questions can be found at
www.mhhe.com/lewisgenetics10. You will also find
additional practice quizzes, animations, videos, and vocabulary
flashcards to help you master the material in this chapter.

Review Questions

1. Give examples of how each of the following can alter allele frequencies from Hardy-Weinberg equilibrium:
   a. nonrandom mating
   b. migration
   c. a population bottleneck
   d. mutation

2. Explain the influence of natural selection on
   a. the virulence of tuberculosis.
   b. bacterial resistance to antibiotics.
   c. the changing degree of genetic diversity in an HIV population during infection.

3. Why can increasing homozygosity in a population be detrimental?

4. How might a mutant allele that causes an inherited illness in homozygotes persist in a population?

5. Give an example of an inherited disease allele that protects against an infectious illness.

6. Explain how negative and positive selection can operate on a population at the same time.

7. How does a founder effect differ from a population bottleneck?

8. Describe two scenarios in human populations, one of which accounts for a gradual cline, and one for an abrupt cline.

9. How do genetic drift, nonrandom mating, and natural selection interact?

10. Define:
    a. founder effect
    b. balanced polymorphism
    c. genetic load

11. How does a knowledge of history, sociology, and anthropology help geneticists to interpret allele frequency data?

12. Cite three examples of eugenic actions or policies.

13. Distinguish between positive and negative selection, and between positive and negative eugenics. How do selection and eugenics differ?

Applied Questions

1. Begin with the original population represented at the center of figure 15.16, and deduce the overall, final effect of the following changes:

   ■ Two yellow square individuals join the population when they stop by on a trip and stay awhile.

   ■ Four red circle individuals are asked to leave as punishment for criminal behavior.

   ■ A blue triangle man has sex with many females, adding five blue triangles to the next generation.

   ■ A green diamond female produces an oocyte with a mutation that adds a yellow square to the next generation.

   ■ A new infectious disease affects only blue triangles and yellow squares, removing two of each from the next generation.

2. Before 1500 A.D., medieval Gaelic society in Ireland isolated itself from the rest of Europe, physically and culturally. Men in the group are called "descendants of Niall," and they have Y chromosomes inherited from a single shared ancestor. In the society, men took several partners, and sons born out of wedlock were fully accepted. Today, in a corner of northwest Ireland, one in five men has the "descendant of Niall" Y chromosome. In all of Ireland, the percentage of

Y chromosomes with the Niall signature is 8.2 percent. In western Scotland, where the Celtic language is similar to Gaelic, 7.3 percent of the males have the telltale Niall Y. In the United States, among those of European descent, it is 2 percent. Worldwide, the Niall Y chromosome makes up only 0.13 percent of the total. What concept from the chapter do the data illustrate?

3. Fred Schnee, who teaches human genetics at Loras College in Iowa, offers a good example of genetic drift: seven castaways are shipwrecked on an island. The first mate has blue eyes, the others brown. A coconut falls on the first mate, killing him. The coconut accident is a chance event affecting a small population. Explain how this event would affect allele frequency, and offer another example of genetic drift.

4. The Old Order Amish of Lancaster, Pennsylvania have more cases of polydactyly (extra fingers and toes) than the rest of the world combined. All of the affected individuals descend from the same person, in whom the dominant mutation originated. Does this illustrate a population bottleneck, a founder effect, or natural selection? Give a reason for your answer.

5. Predict how natural selection might affect the frequency of alleles that protect against HIV infection in Africa a century from now, based on what you know about TB.

6. The ability to taste bitter substances is advantageous in avoiding poisons, but might keep people from eating bitter vegetables that contain chemicals that protect against cancer. Devise an experiment, perhaps based on population data, to test either hypothesis—that the ability to taste bitter substances is either protective or harmful.

7. Define microevolution. Give three examples, either from the chapter or from the news, that show microevolution going on right now.

8. A mutation that removes the receptor for HIV on human cells also blocks infection by the bacterium that causes plague. Seven centuries ago, in Europe, the "Black Death" plague epidemic increased the protective allele in the population. Today it makes 3 million people in the United States and the United Kingdom resistant to HIV infection. Is the increase in incidence of this allele due to nonrandom mating or natural selection?

9. Use the information in chapters 14 and 15 to explain why
   a. porphyria variegata is more prevalent among Afrikaners than other South African populations.
   b. many people among the Cape population in South Africa lose their teeth before age 20.
   c. cystic fibrosis and sickle cell disease remain common.
   d. the Sherpa from Tibet can tolerate thin air.
   e. the Amish in Lancaster County have a high incidence of genetic diseases that are very rare elsewhere.
   f. the frequency of the allele that causes galactokinase deficiency varies across Europe.
   g. mitochondrial DNA sequences vary gradually in populations along the Nile River valley.
   h. disease-causing *BRCA1* alleles are different in Jewish people of eastern European descent and African Americans.

10. Which principles discussed in this chapter do the following classic science fiction plots illustrate?
    a. In the novel *The Passage*, evil researchers infect a dozen criminals bound for death row with experimental viruses that turn them into vampires. The infected criminals, called "virals," escape and ravage the Earth, killing 90 percent of the population and turning most of the remaining 10 percent into vampires. A century later, the numbers of infected humans have greatly increased, but not their genetic diversity. Each person has one of the twelve original virally infected human genomes.
    b. In *When Worlds Collide,* the Earth is about to be destroyed. One hundred people are chosen to colonize a new planet.
    c. In *The Time Machine,* set in the distant future on Earth, one group of people is forced to live on the planet's surface while another group is forced to live in caves. Over many years, they come to look and behave differently. The Morlocks that live below ground have dark skin, dark hair, and are very aggressive, whereas the Eloi that live above ground are blond, fair-skinned, and meek.
    d. In *Children of the Damned,* all of the women in a small town are suddenly impregnated by genetically identical beings from another planet.

e. In *The War of the Worlds,* Martians cannot survive on Earth because they die from infection by terrestrial microbes.
f. In Dean Koontz's novel *The Taking,* giant mutant fungi kill nearly everyone on Earth, sparing only young children and the few adults who protect them. The human race must re-establish itself from the survivors.

11. Ashkenazim, French Canadians, and people who live in southwestern Louisiana have a higher incidence of Tay-Sachs disease than other populations. Each of these groups has a different mutation. How is this possible?

12. Syndrome X consists of obesity, type 2 diabetes, hypertension, and heart disease. Researchers sampled blood from nearly all of the 2,188 residents of the Pacific Island of Kosrae, and found that 1,709 of them are part of the same pedigree. The incidence of syndrome X is much higher in this population than for other populations. Suggest a reason for this finding, and indicate why it would be difficult to study these particular traits, even in an isolated population.

13. By which mechanisms discussed in this chapter do the following situations alter Hardy-Weinberg equilibrium?
    a. In ovalocytosis (MIM 166910), a protein that anchors the red blood cell plasma membrane to the cytoplasm is abnormal, making the membrane so rigid that parasites that cause malaria cannot enter.
    b. In the mid-1700s, a multitoed male cat from England crossed the sea and settled in Boston, where he left behind many kittens, about half of whom also had extra digits. People loved the odd felines and bred them. Today, in Boston and nearby regions, multitoed cats are much more common than in other parts of the United States.
    c. Many slaves in the United States arrived in groups from Nigeria, which is an area in Africa with many ethnic subgroups. They landed at a few sites and settled on widely dispersed plantations. Once emancipated, former slaves in the South were free to travel and disperse.
    d. About 300,000 people in the United States have Alzheimer disease caused by a mutation in the presenilin-2 gene. They all belong to five families that came from two small villages in Germany that migrated to Russia in the 1760s and then to the U.S. from 1870 through 1920.

14. A challenging environment can either kill individuals whose genetic susceptibilities and characteristics make it difficult to survive or reproduce, or encourage such individuals to migrate to more comfortable surroundings. Describe the different effects of these alternatives on the genetic structure of the population.

15. African Americans develop a form of end-stage kidney disease associated with elevated blood pressure that Europeans do not. Two variants in a gene on chromosome 22, called *ApoL1*, cause the condition. The encoded protein is secreted into the blood, but only the forms in the African Americans who have the kidney disease also kill the parasites that cause African sleeping sickness. The mutations persist because they protect against African sleeping sickness. What phenomenon described in the chapter does this situation illustrate?

16. Describe an event in history that likely led to a population bottleneck.

17. Explain how dogs illustrate genetic drift and artificial selection.

Web Activities

1. Go to the Centers for Disease Control and Prevention website and access the journal *Emerging Infectious Diseases*. Using this resource, describe an infectious disease that is evolving, and cite the evidence for this.

2. Do a Google search for a pair of disorders listed in table 15.3 (balanced polymorphism) and discuss how the carrier status of the inherited disease protects against the second condition.

3. Go to the Image Archive on the American Eugenics Movement website. Look at several images, and either find one that presents a genetic disorder and describe it, or find an image that presents biologically incorrect information, and explain the error.

Forensics Focus

1. In the 1870s, prison inspector and self-described sociologist Richard Dugdale noticed that many inmates at his facility in Ulster County, New York, were related. He began studying them, calling the family the "Jukes," although he kept records of their real names. Dugdale traced the family back seven generations to a son of Dutch settlers, named Max, who was a pioneer and lived off the land. Margaret, "the mother of criminals," as Dugdale wrote in his 1877 book *The Jukes: A Study in Crime, Pauperism, Disease and Heredity,* married one of Max's sons, and the couple presumably ultimately gave rise to 540 of the 709 criminals on Dugdale's watch. Dugdale attributed the Jukes' less desirable traits to heredity.

    The Jukes study influenced social scientists to probe other families seemingly riddled with misfits—they were all Caucasian, descended from colonial settlers, and poor. Poverty was not seen as an economic problem, but as a reflection of inborn degeneracy that if left unchecked would cost society greatly.

    Dugdale's book fed the fledgling eugenics movement. In 1911, researchers at the Eugenics Record Office in Cold Spring Harbor described the Jukes' phenotype as "feeblemindedness, indolence, licentiousness, and dishonesty." The Jukes story and others were used to support compulsory sterilization of those deemed unfit. But the original research on the Jukes family was flawed, and its accuracy never questioned. Less notorious Jukes family members served in respected professions, some even holding public office. The Jukes were vindicated in 2003, when archives at the State University of New York at Albany revealed the original names of the people in Dugdale's account; most were not even related. The Jukes family curse was more legend than fact.

    a. What would have had to happen to the original jailed Jukes family members or their descendants to be considered eugenic?

    b. How could studies on one family harm others?

    c. Cite an example of an idea based on eugenics today or in the recent past.

    d. If you were a contemporary of Dugdale's, what type of evidence would you have sought to counter his ideas?

Case Studies and Research Results

1. Lana S. seemed to be a healthy newborn. In fact, she seemed to be hitting developmental milestones ahead of schedule, trying to lift her head up at only 3 weeks. But then she rapidly lost skills. Her head flopped, she stopped trying to turn over, and her arms and legs became spastic. When she no longer made eye contact, her anxious parents took her to the pediatrician. She referred the family to a pediatric neurologist, who was puzzled. "She has all the symptoms of Canavan disease, but that can't be. You're not Jewish." Explain how the neurologist was incorrect.

2. The human population of India is divided into many castes, and the people follow strict rules governing who can marry whom. Researchers compared several genes among 265 Indians of different castes and 750 people from Africa, Europe, and Asia. The study found that the genes of higher Indian castes most closely resembled those of Europeans, and that the genes of the lowest castes most closely resembled those of Asians. In addition, maternally inherited genes (mitochondrial DNA) more closely resembled Asian versions of those genes, but paternally inherited genes (on the Y chromosome) more closely resembled European DNA sequences. Construct an historical scenario to account for these observations.

3. A magazine article featured parents who filed a "wrongful birth" lawsuit against their doctor for failing to offer prenatal testing for spina bifida, which their daughter was born with in 2003—even though they love the child dearly. They will not say whether they would have ended the pregnancy had they known about the birth defect. If they had ended it, would that have been a eugenic act? Explain your answer.

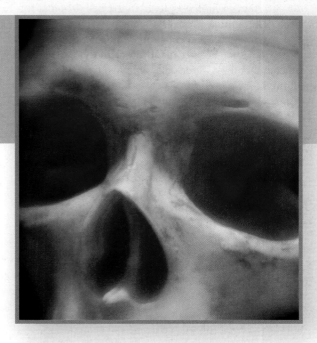

Comparing skulls among modern humans, our modern primate cousins, and fossilized hominins can reveal much about our ancestors and our evolution.

Human Ancestry

Learning Outcomes

16.1 Human Origins

1. Distinguish between hominoids and hominins.

2. Explain why more than one species of *Australopithecus* coexisted.

3. Distinguish between *Australopithecus* and *Homo.*

4. Explain what genome sequencing has revealed about the ancestry of Neanderthals and us.

16.2 Molecular Evolution

5. Explain how DNA information can be used to shed light on evolution.

6. List genes that were important in our evolution.

7. Explain how chromosome banding patterns and protein sequences reveal evolutionary trends.

8. Explain what mitochondrial DNA and Y chromosome sequences reveal about human ancestry.

16.3 The Peopling of the Planet

9. Explain what mitochondrial Eve represents.

10. Describe how people expanded out of Africa and then Eurasia, populating the world.

The Big Picture: Our genes and genomes are informational molecules, and their sequences hold clues to our deep past as well as our present diversity.

The Hobbits

It's odd to be the only ones of our kind, which may be why a dual humanity theme persists in science fiction. *The Time Machine* looked at two battling breeds of people. In *Darwin's Children,* a virus scrambles the genomes of a group of newborns, starting a new species. In other stories, a Neanderthal lives in modern-day Tajikistan and a caveman in Kenya.

Fossils indicate that from 2 to 6 million years ago, humans and prehumans overlapped, in time if not place. The discovery of preserved bones of several ancient humans on the island of Flores in Indonesia in 2004 suggested a recent coexistence of two types of people. A female skeleton found 17 feet beneath a cave floor with pieces of others nearby was named *Homo floresiensis,* popularly called the Hobbit. She was about half as tall as a modern human, with a brain about a third of the size. She lived about 18,000 years ago.

The Hobbits exhibited "island dwarfism," an effect of natural selection on small, isolated, island populations. With limited resources, those who need less food are more likely to reproduce. Who were the Hobbits? At first, researchers thought that Hobbits were direct descendants of *Homo erectus,* who lived before us. Then, analysis of limb bones revealed feet and proportions like those of an ape, despite a more humanlike skull. Therefore, the Hobbits may have been direct descendants of a primate older than *Homo erectus, w*ho evolved in a different direction on their isolated island.

16.1 Human Origins

We have sparse evidence of our beginnings—pieces of a puzzle in time, some out of sequence, many missing. Traditionally, paleontologists (scientists who study evidence of ancient life) have consulted the record in the earth's rocks—fossils—to glimpse the ancestors of *Homo sapiens,* our own species. Researchers assign approximate ages to fossils by observing which rock layers fossils are in, and by extrapolating the passage of time from the ratios of certain radioactive chemicals in surrounding rock.

Fossils aren't the only way to peek into species' origins and relationships. Modern organisms also provide intriguing clues to the past in their DNA. Sequences of DNA change over time due to mutation. Frequencies of gene variants (alleles) change over time on a population level by the forces of nonrandom mating, migration, genetic drift, and, most powerfully, natural selection.

The premise of DNA sequence comparisons is that closeness of relationship is reflected in greater similarity of sequence. The logic is that similar sequences are more likely to have arisen from individuals or species sharing ancestors than from the exact same set of spontaneous mutations occurring by chance. By analogy, it is more likely that two women wearing the same combination of clothes and accessories purchased them at the same store than that each happened to assemble the same collection of items from different sources. On rare occasions, DNA is available from ancient specimens to add to what we know from DNA sequences of modern organisms.

Treelike diagrams are used to depict evolutionary relationships, based on fossil evidence and/or inferred from DNA sequence similarities. Branchpoints on the diagrams represent divergence from shared ancestors. Overall, evolution is shown as a series of branches as species diverged, driven by allele frequencies changing in response to the forces discussed in chapter 15: nonrandom mating, genetic drift, migration, mutation, and natural selection. Evolution is *not* a linear morphing of one type of organism into another—a common misunderstanding. Humans and chimps diverged from a shared ancestor; humans didn't form directly from chimps. Similarly, two second cousins share great-grandparents, but one cousin did not descend from the other.

This chapter explores human origins and considers how genetic and genomic evidence adds to our view of our evolution. It concludes with a look at more recent events in our ancestry.

Hominoids and Hominins

A species includes organisms that can successfully produce healthy offspring only among themselves. *Homo sapiens* ("the wise human") probably first appeared during the Pleistocene epoch, about 200,000 years ago. Our ancestry reaches farther back, to about 60 million years ago when rodentlike insect eaters flourished. These first primates diverged to give rise to many new species. Their ability to grasp and to perceive depth provided the flexibility and coordination necessary to dominate the treetops.

About 30 to 40 million years ago, a monkeylike animal the size of a cat, *Aegyptopithecus,* lived in the lush tropical forests of Africa. The animal probably spent most of its time in the trees, but fossilized remains of limb bones indicate it could run on the ground, too. Fossils of different individuals found together indicate that they were social animals. *Aegyptopithecus* had fangs it might have used for defense. The large canine teeth seen only in males suggest that males may have hunted to feed their mates. *Propliopithecus* was a monkeylike contemporary of *Aegyptopithecus.* Both animals are possible ancestors of gibbons, apes, and humans.

From 22 to 32 million years ago, Africa was home to the first **hominoids,** animals ancestral to apes and humans only. One such resident of southwestern and central Europe was *Dryopithecus,* meaning "oak ape," because its fossilized bones were found with oak leaves (**figure 16.1a**). The way the bones fit together suggests that this animal lived in the trees but could swing and walk farther than *Aegyptopithecus.* More abundant fossils represent the middle-Miocene apes of 11 to 16 million years ago. These apes were about the size of a human 7-year-old and had small brains and pointy snouts. (*Miocene* refers to the geologic time period.)

Apelike animals similar to *Dryopithecus* and the middle-Miocene apes flourished in Europe, Asia, and the Middle East during the same period. Because of the large primate population in the forest, selective pressure to venture onto the grasslands in search of food and habitat space must have been intense. Many primate species probably vanished as the protective forests shrank. One type of middle-Miocene ape survived to give rise to humans and African apes. Eventually, animals ancestral to humans only, called **hominins,** arose and eventually thrived. (An older term is *hominid.*)

Hominoid and hominin fossils from 4 to 19 million years ago are scarce, and are often just fragments of tooth and jaw. About 6 million years ago, the hominin lineage split from the apes. There are at least three candidates for this first primate one step closer to humanity from the chimp: *Ardipithecus kadabba* from Ethiopia, *Sahelanthropus tchadensis* from Chad, and *Orrorin tugenensis* from Kenya. They are near the base of the evolutionary tree diagram in **figure 16.2,** which depicts probable relationships among some of our relatives, past and present. This evolutionary tree is based on fossil evidence and DNA sequence comparisons for the modern species.

Fossil evidence is more complete for our ancestors who lived 2 to 4 million years ago, who walked upright and conquered vast new habitats on the plains. Several species of a hominin called *Australopithecus* lived at this time, probably following a hunter-gatherer lifestyle long before agriculture. More than one species could coexist because they lived in small, widely separated groups that probably never came into contact. The australopithecines were gradually replaced with members of our own genus, *Homo.* The following sections introduce a few of these ancestors known from rare fossil remains and what our computer modeling and imaginations can fill in.

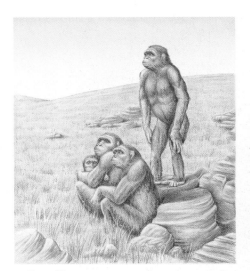

a. *Dryopithecus*

b. *Australopithecus*

c. *Homo erectus*

Figure 16.1 Human forerunners. **(a)** The "oak ape" *Dryopithecus,* who lived from 22 to 32 million years ago, was more dextrous than his predecessors. **(b)** Several species of *Australopithecus* lived from 2 to more than 4 million years ago, and walked upright. **(c)** *Homo erectus* made tools out of bone and stone, used fire, and dwelled communally in caves from 35,000 to 1.6 million years ago.

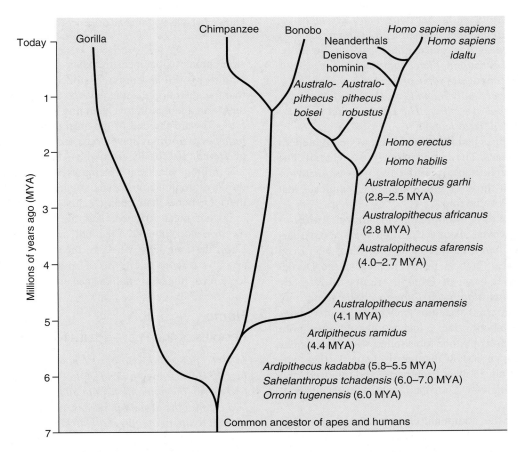

Figure 16.2 Charting the past. This evolutionary tree diagram indicates the relationships among primates, past and present. (Not all known hominin species are indicated.) Refer back to this figure as the chapter discusses the various human ancestors.

Australopithecus

Peoples of the Past—The Dikika Infant, 3.3 Million Years Ago

The 3-year-old had probably wandered away from her family when she had been swept up in the sudden flood 3.3 million years ago in Ethiopia. She liked to sit up in the trees ringing the large lake to watch the antelopes and giraffes, elephants and wildebeests. She wandered the grasslands to the small rivulets that merged into the Awash river delta, avoiding crocodiles while she waded among the reeds, picking up snails and trying to catch fish. One day, perhaps after seasonal rains, the girl was overcome by rushing waters and was quickly buried in thick sediments. Much of her skeleton remained intact.

Paleontologists discovered the top of her skull protruding from sandstone. After 5 years of meticulously removing the encasing stone with dental instruments, clues to the Dikika infant's identity began to emerge. The tiny, delicate bones of the Australopithecus afarensis *child reveal a mix of human and chimp characteristics. Her legs were shorter but her arms were longer than ours. Her upper half looked like that of an ape, with a large, protruding lower face and neck muscles to support her head. She was hairy, with arms long enough to have easily grabbed a branch, her hands and fingers curved enough to grasp and hold on. Yet her bottom half was much more humanlike. The great toes aligned with the other toes and the hip joints and leg bones were able to support walking, but she was hunched because her knee joints could not lock. She had a voice box, and while some of her teeth resembled those of apes, some were more humanlike. Her cranium was slightly larger than that of a chimp. The Dikika infant looked like an ape, but walked like a human.*

Australopithecines had flat skull bases, as do all modern primates except humans. They stood about 4 to 5 feet tall. The angle of preserved pelvic bones, plus the discovery of *Australopithecus* fossils with those of grazing animals, indicate that this ape-human had left the forest.

The oldest species of australopithecine known, *Australopithecus anamensis*, lived about 4.1 million years ago. A partial skeleton, named Lucy, represents an individual who lived about 3.6 million years ago in the same area as the Dikika infant—the Afar river basin of Ethiopia (**figure 16.3**). Lucy was a member of *Australopithecus afarensis*. She stood only about 4 feet tall, but partial skeletons from slightly older males of her species stood about 5 feet 3 inches tall, and were more humanlike than chimplike. Stone tools found with *A. afarensis* from 3.4 million years ago show distinctive cut marks that indicate these hominins sliced meat from bones and removed, and presumably ate, the marrow. Lucy died, with arthritis, at about age 20.

Other fossils offer additional clues to australopithecine life. Two parallel paths of humanlike footprints, preserved in volcanic ash in the Laetoli area of Tanzania, are contemporary with Lucy. A family may have left the prints, which are from a large and small individual walking close together, with a third following in the steps of the larger animal in front. Computer

Figure 16.3 *Australopithecus afarensis.* About 3.6 million years ago, Lucy walked upright in the grasses along a lake in the Afar region of Ethiopia, about six miles from where the Dikika infant would live 300,000 years later. She skimmed the shores for crabs, turtles, and crocodile eggs to eat. The Afar region is the only place known to have evidence of our ancestors that spans 6 million years.

simulations using measurements from modern monkeys and chimps as well as from a fossilized ancestor of Lucy revealed that the australopithecine jaw was strong enough to crack hard nuts—an adaptation to a changing environment.

Toward the end of the australopithecine reign, *Australopithecus garhi* may have coexisted with the earliest members of *Homo*. Its fossils from the Afar region date from about 2.5 million years ago. Remains of an antelope found near the australopithecine fossils suggest butchering. The ends of the long bones had been cleanly cut with tools, the marrow removed, meat stripped, and the tongue cleanly sliced off. *A. garhi* stood about 4.5 feet tall, and like the Dikika infant and Lucy, the long legs were like those of a human, but the long arms were more like those of an ape. The small cranium and large teeth hinted at apelike ancestors.

Homo

Peoples of the Past—Idaltu Man, 156,000 Years Ago

In 1997, paleoanthropologist Tim White was driving by the village of Herto, along a bend of the Awash River. Seasonal rains had driven the nomadic people and their livestock away, and had cleared the ground in places. In one such bald spot, White spotted a hippopotamus skull sticking up. Near it were tools made of obsidian, a glasslike rock. A few days later he sent two students to explore further, and they found a humanlike skull lying on its side. Soon, two other skulls were discovered. One was from another adult, and the other was a child's skull shattered and scattered into more than 200 pieces, including baby teeth. The researchers named the hominin Homo sapiens idaltu, *which in the local language means "elder."*

The most intact skull was slightly longer, and the brain slightly larger, than those of modern humans. Fine, parallel lines had been etched along the base of the skull. The dome of the skull had not been damaged, as it would have been had cannibalism been practiced. The skull was very smooth, and there were no other bones nearby. Might the skulls have been gently separated from the bodies, saved, and touched, as modern cultures do to honor the dead?

Other fossils filled in the story. Evidence of catfish and hippos indicate that the Awash River had flooded, forming a freshwater lake. The hippo and buffalo bones bore marks made with tools that had probably sliced off meat. Some bones were broken in ways that suggested that the people ate the marrow. The tools were of a sophisticated design compared to the flaked tools from a million years ago. Overall, the scene evoked an image of ancestors who not only understood the concept of death, but who practiced mortuary rituals.

Our knowledge of how *Homo* replaced *Australopithecus* is sparse. Some australopithecines were "dead ends" that died off. Clues suggest that by 2.3 million years ago, *Australopithecus* coexisted with *Homo habilis,* who was a more humanlike cave dweller that cared intensively for its young. *Habilis* means handy, and this primate was the first to use tools for tasks more challenging than stripping meat from bones. *H. habilis* may have descended from hominins who ate a more varied diet than other ape-humans, allowing them to live in a wider range of habitats.

H. habilis coexisted with and was followed by *Homo erectus* during the Paleolithic Age (**table 16.1**). One of the first *H. erectus* individuals described, Nariokotome Boy, influenced descriptions of the species as tall and thin for two decades, until discovery of more specimens revealed that *H. erectus* matured faster than we do, but their heights and builds varied. Another famed *H. erectus* fossil, named "Daka" for the place where he was found in the Afar region, is from an individual who lived about a million years ago. He had a shallow forehead, massive brow ridges, a brain about a third smaller than ours, and strong, thick legs. Daka lived on a grassland, with elephants, wildebeests, hippos, antelopes, many types of pigs, and giant hyenas. **Figure 16.4** depicts what he might have looked like.

H. erectus left fossil evidence of cooperation, social organization, tools, and use of fire. Fossilized teeth and jaws suggest that they ate meat. The distribution of fossils indicates

Figure 16.4 *Homo erectus.* This artist's rendition is based on clues from many fossils.

that they lived in families of male-female pairs (most primates have harems). The male hunted, and the female nurtured the young. They were the first to have an angled skull base that enabled them to produce a greater range of sounds, making speech possible. *H. erectus* fossils have been found in China, Java, Africa, Europe, and Southeast Asia, indicating that these animals could migrate farther than earlier primates.

The fossils of *H. sapiens idaltu* from Ethiopia reveal that by 156,000 years ago, our ancestors did not look very different from ourselves (**figure 16.5**). *H. sapiens idaltu* probably resembled an Australian aborigine, with a large and powerful build and dark skin.

Figure 16.5 *Homo sapiens idaltu.* Discovery of three skulls made possible this artist's depiction of what this early member of our species might have looked like—not very much different from us.

Table 16.1	Cultural Ages	
Age	**Time (years ago)**	**Defining Skills**
Paleolithic	750,000 to 15,000	Earliest chipped tools
Mesolithic	15,000 to 10,000	Cutting tools, bows and arrows
Neolithic	10,000 to present	Complex tools, agriculture

By 70,000 years ago, humans, still mostly confined to Africa, used more intricately carved tools made of bones, and red rock that bore highly symmetrical hatchmarks, which may indicate early counting. Groups of hominins may have been very isolated on the vast continent. The first *Homo* may have left Africa around 100,000 years ago, as a founder group of about a thousand individuals, most of them male, who died out by 70,000 years ago. Because of the isolation, it's possible that even as *H. sapiens idaltu* and perhaps others yet to be discovered were far along the road to modern humanity, pockets of *H. erectus* may have persisted, perhaps until as recently as 35,000 years ago.

A hominin that has especially fascinated us is *Homo neanderthalensis*, better known as the Neanderthals (or Neandertals). Early fossil finds, beginning when quarry workers blasted out the first bones in Neander Valley, Germany, in 1856, suggested that these hominins had skeletons wider than ours with shorter arms and legs, and distinctive faces with prominent brow ridges, sloping foreheads, and jutting faces. However, discovery of additional skeletons revealed some individuals that had traits of both Neanderthals and modern humans, such as large teeth and jaws (Neanderthal) but small chins (modern human). In addition, we both used similar tools, and Neanderthal bones buried with flowers suggests that they were capable of spirituality. Starch granules discovered within dental plaque on Neanderthal teeth indicate that these early people not only ate grains, but cooked them.

Fossils and genetic clues have allowed us to reconstruct the wanderings of the Neanderthals. In Africa, they probably descended from two species, *Homo heidelbergensis* and *Homo rhodesiensis*. About 400,000 years ago, the forebears of the Neanderthals began to leave Africa, heading toward Europe and west Asia but not establishing themselves there until about 100,000 years ago. The Neanderthals left Africa in separate migrations from our forerunners, which is why modern Africans do not have Neanderthal DNA sequence variants in their genomes, but Europeans and Asians do. We have DNA evidence of Neanderthals because they lived in cold areas, which preserved the genetic material. In contrast, the DNA of earlier *Homo* and *Australopithecus* was not only much older, but would have degraded rapidly in the heat of Africa.

From 50,000 to 80,000 years ago, our ancestors left Africa and encountered Neanderthals in the Middle East en route to Eurasia. The two types of hominins interbred—a discovery deduced from shared DNA sequence variants. When researchers first began to sequence the DNA from Neanderthal bones, they concluded that interbreeding was highly unlikely, but sequencing the entire genome changed the long-held story of our separateness. Our early views of the Neanderthals, based on a few mitochondrial DNA sequences, was like trying to guess the end of a novel by reading only a small part of the story.

The researchers who sequenced the Neanderthal genome used dental drills to delicately remove pulverized bone "dust" from three bones from females found in a cave in Croatia, which kept the bones intact for future studies. DNA sequencing in a "clean room" overcame the problem of the investigators'

DNA contaminating the samples, and other techniques avoided sequencing bacterial or fungal DNA, another source of error in ancient DNA studies. Then researchers compared the Neanderthal genome sequence to those of modern humans from Asia, Europe, and Africa. By comparing SNPs—sites where the DNA base differs—it became clear that the genomes of modern Europeans and Asians have 1 to 4 percent of their DNA in common with Neanderthals! However, the evidence cannot reveal whether the interbreeding was intense over a short period of time, or once-in-a-while mating that happened over many years. Discoveries of individual genes and their variants have added to our knowledge of these people. Variants of the *FOXP2* gene might have enabled them to vocalize and possibly speak, and some specimens had a mutation in the *MC1R* gene that gave them pale skin and red hair (**figure 16.6**).

More interesting than what our genomes share with those of the Neanderthals is how our genomes differ, because this reveals what makes us human. The genome sequencing identified 73 proteins that are different in modern humans, and they take part in such varied functions as sperm motility, wound healing, immunity, hair shaft structure, bone shapes, transcription control, and cognitive development. In addition, certain regions of the modern human genome show the signs of positive selection—linked genes that are uniquely human and not seen in the genomes of the Neanderthals or chimps, our closest relatives. These 212 regions of the human genome are said to have undergone "selective sweeps."

Neanderthals may have lived as recently as 30,000 years ago, in warm caves in Gibraltar when northern Europe was

Figure 16.6 A Neanderthal phenotype. Mutations found in Neanderthal DNA suggest that some of them may have had pale skin and red hair. Others had darker hair and skin. From 1 to 4 percent of the genomes of modern people of European or Asian ancestry arose from long-ago breeding with Neanderthals.

under ice. They might have inhabited these caves since 100,000 years ago. Then, the fossil record indicates, the Neanderthals vanished. According to their DNA, however, they may have been assimilated as different groups interbred. Without a time machine, we may never know exactly what happened.

Because DNA analysis and fossil studies depend upon finding remains, we probably do not know all of the types of hominins that shared the planet as some of our ancestors left Africa and settled in Eurasia and then the rest of the world. At least one other type of hominin coexisted with our ancestors and the Neanderthals, called the Denisova hominin. The discovery was based on analysis of DNA from a tiny finger bone thought to have come from a Neanderthal. It had a mitochondrial DNA sequence unlike those of Neanderthals and modern humans. The finger bone came from Denisova Cave in the Altai Mountains of southern Siberia, and dates to about 40,000 years ago. The Denisova hominin likely left Africa a little more than one million years ago, diverging from the ancestors of Neanderthals and modern humans, as figure 16.2 indicates. The remains of the Denisova hominin are too scant and scattered, so far, to have revealed anything about them other than who they were not. Perhaps the most important lesson from their discovery is that our view of our ancestors is, like the fossil record,

fragmented. Evolutionary trees are frequently redrawn when researchers find and analyze new bones (**figure 16.7**).

Modern Humans

Cave art from about 14,000 years ago indicates that by that time, our ancestors had developed fine-hand coordination and could use symbols. These were milestones in cultural evolution. By 10,000 years ago, people had expanded from the Middle East across Europe, bringing agricultural practices.

DNA evidence has glimpsed one scene from the spread of agriculture. Researchers sequenced mitochondrial DNA haplotypes from 21 bodies found in a graveyard in Germany, about 100 miles south of Berlin, and compared them to DNA from 36 modern Eurasian populations. The bodies were from about 7,100 years ago. The comparisons yielded a clear cline, with genetic similarities indicating a long-ago migration of early farmers from Turkey, Syria, Iraq, and other Near Eastern cultures westward from the Balkans north along the Danube into central Europe—not just the spread of their agricultural techniques by word of mouth. The migration took centuries, and, according to the DNA, once the farmers arrived in Europe and encountered hunter-gatherers descended from the original

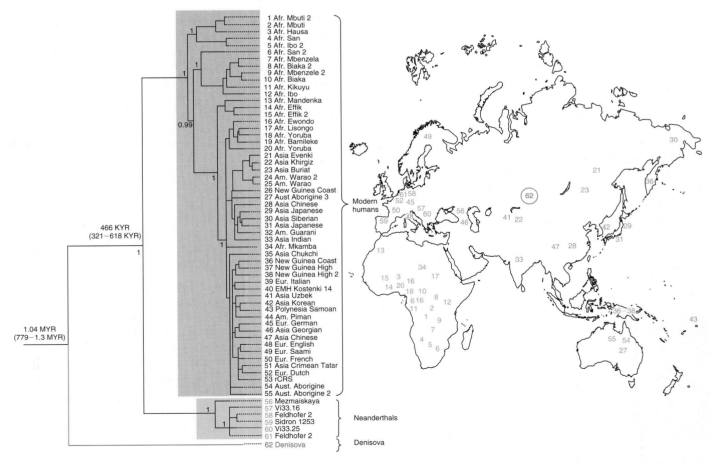

Figure 16.7 New evidence of ancestors. The Denisova hominin represents a recently discovered branch of the human family tree, coexisting with Neanderthals and our direct ancestors, in time if not also in place. Note the recent genetic diversification of humanity.

Figure 16.8 Farmers from 7,100 years ago probably looked like modern people.

Karol Schauer, State Museum of Prehistory in Halle (Saale), Saxony-Anhalt, Germany.

population from 40,000 BC, the two groups of people interbred. **Figure 16.8** shows what the ancient farmers from 7,100 years ago might have looked like.

Peoples of the Past: Ötzi

*In 1991, hikers in the Ötztaler Alps of northern Italy discovered an ancient man frozen in the ice (**figure 16.9**). Named Ötzi, the Ice Man was on a mountain more than 10,000 feet high 5,200 years ago when he perished. He wore furry leggings, leather suspenders, a loincloth, fanny pack, bearskin cap and cape, and sandal-like snowshoes. He had stained his skin to fashion tattoos, and indentations in his ears suggest that he might have worn earrings. He carried mushrooms that had antibiotic properties. Berries found with him place the season as late summer or early fall. His last meal was ibex and venison.*

Ötzi died following a fight. He had a knife in one hand, cuts and bruises, and an arrowhead embedded in his left shoulder that nicked a vital artery. The wound bore blood from two other individuals, and his cape had the blood of a third person. Mosses found on his body may have been wound

Figure 16.9 A 5,200-year-old man.
Hikers discovered Ötzi, the Ice Man, in the Otztaler Alps of northern Italy in 1991.

dressings. He likely bled to death and fell into a ditch, where snow covered him. After this safe burial, which preserved his body intact, a glacier sealed the natural tomb. DNA profiling suggests that he belonged to the same gene pool as modern people living in the area, which is near the Italian-Austrian border.

Another way that anthropologists try to envision what humans were like a few thousand years ago is by studying vanishing indigenous peoples living today, such as the Khoisan (bushmen) and Pygmies of Africa, the Etas of Japan, the Hill People of New Guinea, and a Brazilian tribe, the Arawete, who number only 130 individuals. Studying DNA sequences in these populations provides information on their origins because the people have stayed in the same geographical region and not mixed their gene pools with others. Comparisons of their genomes with those of other modern groups can reveal the adaptations to an agricultural or industrial way of life that show evolution in action. However, researchers wishing to sample DNA from these groups have run into problems when science clashes with culture. In some indigenous societies, ancestry is extremely important, and provides the basis for identity and rights. Some people fear that genetic research might be used to dispute ancestry claims. *Bioethics: Choices for the Future* on page 309 takes a closer look at the consequences of DNA testing on an indigenous people.

A powerful look into the past comes from the genomes of the Khoisan, the modern people whose roots go back the farthest. These hunter-gatherers live today, as they have for millennia, in the Kalahari Desert in southern Africa (**figure 16.10**). They are also known as "San" or "Bushmen," and they speak with a language that uses several "click" sounds.

Researchers compared the complete genome sequences of a Khoisan man named !Gubi to that of Archbishop Desmond Tutu, a well-known South African civil rights activist and a member of the majority Bantu group, as well as partial genome sequences of three other Khoisan who live near each other. The results indicate that the great genetic diversity from which humanity sprung in Africa persists in the Khoisan today, whose genomes are as different from each other as a modern European genome is from that of a modern Asian. The four Khoisan genomes and Desmond Tutu's differ from each other at more than a million places. Comparing the Khoisan genomes to those of other groups may reveal adaptations to an agricultural way of life.

The Khoisan have gene variants that reflect their lifestyle in the desert: a variant of the actinin-3 muscle gene that promotes sprinting over distance running; a gene variant that encodes a chloride channel that conserves water; and the "bitter taste" gene, something that would enable a hunter-gatherer to avoid poison and perhaps locate medicine. They lack a gene variant that in other populations protects against malaria. Selection would have ignored it in the dry climate where malaria-bearing mosquitoes cannot live.

It is a great irony that humanity began in Africa and remains the most genetically diverse there, but most drugs are

a.

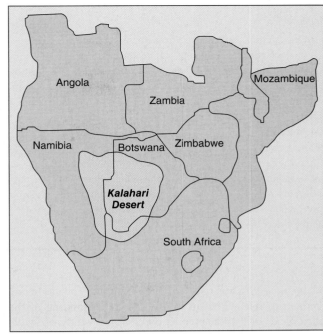

b.

Figure 16.10 **The most ancient modern humans.** The Khoisan are the modern people whose genetic roots go back the farthest. Their genomes are very diverse, with gene variants that reflect their long adaptation to a hunter-gatherer lifestyle **(a)**. They live in the Kalahari Desert in Botswana and Namibia **(b)**.

Bioethics: Choices for the Future

The Havasupai Indians: New Uses for Old DNA Samples

An indigenous group of people is one that can trace its ancestry back farther in a particular geographic region than any other group, and has retained its uniqueness in cultural practices, social organization, and/or language. The group has remained physically or culturally isolated among colonists, and has therefore kept its gene pool separate, too. In those gene pools lie clues to adaptations to past ways of life, and by comparison to other modern genomes, clues to how we are continuing to evolve.

Today less than 6 percent of the world population is indigenous, accounting for about 380 million individuals in 5,000 groups, living in 72 nations. They range from just a few dozen people to sizeable portions of a country's population. Some live in distinct tribes, such as the Maori of New Zealand, who nevertheless go to school and work and dress just like anyone else. Yet some indigenous tribes are not very different in lifestyle from their hunter-gatherer ancestors, such as the Khoisan in southern Africa.

The DNA of indigenous groups holds valuable information, but obtaining and studying that DNA raises bioethical concerns. Consider a group of Native Americans called the Havasupai Indians, who live in a village called Supai at the bottom of the Grand Canyon. They number about 500 today, but their population has been shrinking since diabetes began to plague them, and some have had to leave the community to be treated. The people are indigenous according to the strict definition, but they are hardly hunter-gatherers. They operate a tourist business and have a website.

In 1990, a researcher from Arizona State University asked the Havasupai to provide blood samples so that she could study their diabetes, thinking that they might have the same mutations as the well-studied Pima Indians. Two hundred tribe members eagerly gave the samples, although the consent forms that they signed were vague and did not promise treatments. Over the years, the researcher, as well as others, probed the DNA for several medical conditions, including behavioral ones, and for ancestry studies. They wrote dissertations and published papers.

When a member of the tribe attending Arizona State University learned of the continuing research on the Havasupai blood samples, she set into motion a protest that led to a lawsuit and a settlement that included destruction of the blood samples, which happened in 2010. Not only did the research on the Havasupai DNA not address the diabetes problem that started the study, but

(Continued)

disturbing cultural issues arose. For example, part of the genetic analysis indicated consanguinity, and the Havasupai fear the consequences of relatives marrying relatives. The DNA also revealed their origins in Siberia, as is true for all Native Americans. Tribal elders told children that they all came from the Grand Canyon. The people also objected to testing their DNA for schizophrenia genes, fearing stigmatization. They would not have given permission for such testing, had they been asked.

The case of the Havasupai extends beyond their small and isolated community. It applies to anyone who provides a cell's worth of DNA to a researcher or to a testing company. That cell includes an entire genome's worth of information that can be stored long after a sample is destroyed.

Questions for Discussion

1. Why is sampling DNA different from studying another chemical in a body fluid or other tissue sample, such as cholesterol or iron?

2. What can we learn from sampling DNA from an indigenous people?

3. Do you think that informed consent documents for studies using DNA should be phrased to account for future uses of the DNA or its information, or should researchers seek additional consent later?

4. What should geneticists know about a group of people before approaching them about donating DNA samples? Do you think that scientists should consider cultural matters when designing experiments?

---

tested on non-Africans. Sequencing genomes from as many groups as possible will lead to a better understanding of the genetics of all of us. Chapter 22 picks up the study of whole genomes.

Key Concepts

1. *Aegyptopithecus* lived 30 to 40 million years ago and was ancestral to gibbons, apes, and humans. The first hominoid, *Dryopithecus,* lived 22 to 32 million years ago.

2. Hominins appeared about 19 million years ago.

3. About 4 million years ago, bipedalism opened up new habitats for *Australopithecus. A. garhi* may have coexisted with the first *Homo.*

4. By 2 million years ago, *Australopithecus* coexisted with *Homo habilis.* Later, *H. habilis* coexisted with *H. erectus,* who used tools in more complex societies. *H. erectus* then coexisted with *H. sapiens. H. sapiens idaltu* lived 156,000 years ago.

5. The Neanderthals, our ancestors, and at least one other type of hominin (the Denisova hominin) coexisted in Eurasia. Comparison of the Neanderthal genome and ours indicates some interbreeding.

6. A preserved man from 5,200 years ago is genetically like us.

16.2 Molecular Evolution

Fossils paint an incomplete picture of the past because they are scarce and only certain parts of certain organisms were preserved. We can also glimpse the past through the informational molecules of life in organisms living today. These molecules, DNA and proteins, change in sequence over time as mutations occur and are perpetuated. The more alike a gene or protein sequence is in two species, the more closely related the two are presumed to be—that is, the more recently they shared an ancestor. The assumption is that it is highly unlikely that two unrelated species would evolve precisely the same sequence of DNA nucleotides by chance.

Comparing genome, DNA or protein sequences, and chromosome banding patterns constitute the field of **molecular evolution**. Knowing the mutation rates for specific genes provides a way to measure the passage of time using a sequence-based molecular clock, of sorts.

Comparing Genes and Genomes

We can assess similarities in DNA sequences between two species for a piece of DNA, a single gene, a chromosome segment, a chromosome, mitochondrial DNA, or an entire genome.

For some genes, similarities among species can be startling. People with Waardenburg syndrome (MIM 148820), for example, have a white forelock of hair; wide-spaced, light-colored eyes; and hearing impairment (**figure 16.11**). The mutant gene is very similar in sequence in cats, horses, mice, and minks, who have light coats and eyes and are deaf. In these species, the phenotype stems from abnormal movements of pigment cells in the embryo's outermost layer.

In general, DNA sequences that encode protein are often very similar among closely related species. The related species presumably inherited the gene from a shared ancestor, and a change in that gene would not persist in a population unless it provided a selective advantage. At the same time, natural selection weeded out proteins that did not promote survival to reproduce.

Similar DNA or amino acid sequences in different species are said to be "highly conserved." Sequences that are similar in closely related species but that do not encode protein often control transcription or translation, and so are also vital and therefore subject to natural selection. In contrast, some genome regions that vary widely among species do not affect the phenotype, and are therefore not subject to natural selection. Within a protein-encoding gene, the exons tend to be highly conserved, but the introns, which are removed from the corresponding RNA, are not.

a.

b.

c.

Figure 16.11 **The same mutation can cause similar effects in different species.** A mutation in mice **(a)**, cats **(b)**, humans **(c)**, and other types of mammals causes light eye color, hearing or other neurological impairment, and a fair forelock.

Our closest living relatives are chimpanzees and bonobos. Just how similar we are at the genome level, however, depends upon how we assess similarity—by DNA sequence, numbers of copies of sequences, or sequences missing from the human genome. We share about 98.7 percent sequence similarity, but we also differ in the number of copies of certain DNA sequences. These include insertions and deletions, which are collectively called "indels." Considering indels, our degree of genome similarity to chimps and bonobos is only about 96.6 percent. The degree of similarity may even be as low as 94 percent if sequences not in the human genome are considered. That is, what *isn't* present defines us as well as what *is* present.

Genes That Help to Define Us

Uniquely human traits include spoken language, abstract reasoning ability, highly opposable thumbs, and larger frontal lobes of the brain. One stark difference between chimp and human that could stem from a single gene is hairiness. Chimpanzees and gorillas express a keratin gene whose counterpart in humans is not expressed. When our ancestors left the forests, natural selection might have favored loss of body hair to provide more efficient cooling or as a way to shed skin parasites such as lice. Speech may also be due to a single gene difference between humans and chimps. A family in London whose members have unintelligible speech led to the discovery of a single gene (*FOXP2*, MIM 605317) that controls speaking ability. The gene is present, but different, in chimps. **Reading 16.1** takes a closer look at the traits that make us human.

Another single gene that accounts for great differences among primates controls the switch from embryonic to fetal hemoglobin (see figure 11.2). More primitive primates lack or have very little fetal hemoglobin. In more recently evolved and more complex primates, fetal hemoglobin correlates to lengthened fetal period, which extended the time for brain growth. With larger brains came greater skills. Single genes can also explain the longer childhood and adolescence in humans compared to chimpanzees.

Single genes that distinguish humans from chimps appear to be few, but they tend to be implicated in Mendelian disorders. Perhaps this reflects the fact that the genes that distinguish us have recently taken on their new functions, and the genome has not yet had time for protective redundancies to have evolved.

The major reason why humans and chimps are genetically so similar, but look and behave so differently, is differences in gene expression, not genome sequence. For example, a study contrasted gene expression in the liver and brain in the two species. The differences in the brain were far greater than in the liver. It makes sense that our livers are more alike than our brains (or at least we'd like to think so!).

Comparisons of the human genome sequence to those of other species are interesting, too (see figure 22.9). Our close relationship to the other vertebrates is revealed by comparing the human genome sequence to that of the pufferfish *Tetraodon nigroviridis*. Its genome is like ours, minus many of the repeats and introns. It is odd to think that the protein-encoding portion of our genome is nearly the same as that of a fish.

Considering Genomes

Overall, the human genome has a more complex organization of the same basic parts as the fruit fly and roundworm genomes. For example, the human genome harbors thirty copies of the gene that encodes fibroblast growth factor, compared to two copies in the fly and worm genomes. This growth factor is important for the development of highly complex organs.

Genome studies indicate that over deep evolutionary time, genes and gene pieces provided vertebrates, including humans, with certain defining characteristics:

- complex neural networks;
- blood clotting pathways;
- acquired immunity;
- refined apoptosis;
- greater control of transcription;

What Makes Us Human?

Comparison of the chimpanzee and human genomes has revealed "human accelerated regions." These are highly conserved sequences that show signs of positive selection in humans, such as an amino acid change seen in all human groups but not in the chimp or orangutan versions of the same gene. These genes may represent characteristics that distinguish us from our closest relatives. Signs of positive selection in the human genome flesh out views of our ancestry from fossils.

Tool Use

A paleontologist views the origin of humanity differently than a geneticist. University of California, Berkeley, paleontologist Tim White heads a team that explores the Afar region of Ethiopia. Here, scattered and at different levels, lie remains of our ancestors stretching back some 6 million years, teasing at the time when our forebears split from an ancestor shared with the chimpanzee. Dr. White led the teams that discovered *Ardipithecus, Australopithecus garhi,* "Daka," and *H. sapiens idaltu.* He sums up what distinguishes our species in one word: culture. His imagination takes him back in time:

> About 2.5 million years ago, guys started banging rocks together. That's what allowed the niche to expand in the beginning, the start of culture. Tool making, utilizing stone, probably began in *Australopithecus,* such as Lucy. They were very adaptable and very widespread, all over Africa. These bipeds were small-brained, and they weren't busy becoming human, but being australopithecines. A population of that highly intelligent, bipedal *Australopithecus* began to exhibit behaviors that we see in the chimp. Chimps hunt monkeys, but chimps lack tools. At some point, an early hominid didn't lack those tools anymore, and formed the beginning of the lineage that would ultimately diverge from other australopithecines that kept on being australopithecines. That lineage would go on to become early *Homo.*

Walking

Diseases of modern humanity can reveal traits of evolutionary import. Consider Joubert syndrome (MIM 608629). In this disorder, nerve cell fibers cannot cross from their origin on one side of the brain to the other, so a person cannot move just one arm or leg. In response to a command to move one limb, both move. The part of the brain that controls posture, balance, and coordination is compromised. The gene that causes Joubert syndrome, called *AHI1,* is identical in all modern human groups examined, but has different alleles in chimps, gorillas, and orangutans. Perhaps in the lineage leading to humans, the gene came to control walking by making it possible to place one foot in front of the other.

Running

Homo erectus distinguished itself in another key way: It could run for long distances, thanks to specific anatomical adaptations. The nuchal ligament that connects the skull to the neck became more highly developed in *H. erectus,* enabling the head to stay in place with the force of running. The leg muscles were also more highly developed than those of chimps or australopithecines, acting as springs. *H. erectus* originated a large buttocks, whose muscles contract during running. All three of these structures are not merely the result of being able to walk, but enabled early *Homo,* and us, to run. This skill would have helped our ancestors to escape predators, find food, and locate new homes. Today, wearing running shoes actually impedes the adaptations that have stood the test of time to enable us to run.

A Big Brain

The difference between a big-brained human and a small-brained chimp may be a few single genes. About 2.4 million years ago, a gene called *MYH16* underwent a nonsense mutation, which prevented production of a type of muscle protein called a myosin. The mutation is seen in all modern human populations, but not in other primates. Without this particular type of myosin, jaw development is not as great. With a diminished jaw, the bony plates of the skull could expand, allowing greater brain growth. Researchers nicknamed the mutation RFT, for "room for thought." Fossil evidence indicates that the switch from "big jaw, small brain" to "small jaw, big brain" happened when *Homo* gradually replaced *Australopithecus,* about 2 million years ago. The genetic analysis may be new, but the idea isn't. Charles Darwin wrote in 1871 that different-sized jaw muscles were at the root of the distinction between apes and humans.

Cognition

At the genetic level, humans and chimps may differ more in the numbers of copies of particular genes than in the nature of the genes. Researchers identified 134 genes with an increased copy number in the human genome compared to the genomes of the great apes. Many of these genes are involved in brain structure or function. Some of the genes promote the signal transduction that underlies long-term memory; others, when mutant, cause intellectual disability or impair language skills. Single genes implicated in fueling human brain growth control the migration of nerve cells in the front of the fetal brain.

Sense of Smell

Our senses of taste and smell have diminished as our reliance on them for survival has waned. The sense of smell derives from a 1-inch-square patch of tissue high in the nose that consists of 12 million cells that bear odorant receptor (OR) proteins. (In contrast, a bloodhound has 4 *billion* such cells!) Molecules given off by something smelly bind to combinations of these receptors, which then signal the brain in a way that creates the perception of an associated odor.

(Continued)

(Continued)

Our odorant receptor genes number 906, comprise about 1 percent of the genome, and occur in clusters. About 60 percent of them are pseudogenes—their sequences are similar to those of functional "smell" genes, but are riddled with mutations that prevent translation of complete proteins. Perhaps they are remnants of a distant past, when we depended more upon our chemical senses for survival, and natural selection eliminated them. Natural selection also has acted positively to retain OR genes that continue to function. While the pseudogenes harbor many diverse SNPs, the functional OR genes are remarkably alike in sequence. In addition, the nucleotide differences that persist among the retained genes actually alter the encoded amino acid, suggesting that natural selection favored these sequences.

Bare Skin

Mice, dogs, giraffes, wildebeests, and nearly all mammals are covered with thick, abundant hairs. Fur coats offer insulation, protection, and a means of social displays and species recognition. Hairless animals don't really need it. Hippos would be too hot with a coat, and it would drag them down in the water. Naked mole rats do not need coat markings or hair standing on end to exchange social cues in the darkness of their tunnels. In the seas, hair slows aquatic mammals such as dolphins, which is why swimming humans sometimes shave their legs.

Our naked skins make sense in terms of evolution—they enable us to sweat. When our forebears moved onto the plains as the forest habitat shrunk, about 2.5 to 3 million years ago, individuals with less hair and more abundant sweat glands could travel farther in search of food, or to avoid becoming food. Today, our skin is peppered with eccrine sweat glands that pump out quarts of sweat a day, compared to our furry fellow mammals, who instead have the types of sweat glands that are sparser and associated with hair follicles. Human skin is thin yet strong, thanks to a unique recipe of keratin proteins that fill the flattened bricklike skin cells cemented with a fatty mortar. **Figure 1** illustrates several of the characteristics that help to define us.

Figure 1 Which traits make us human? The runner stopping to drink illustrates several characteristics that make us human: running, sweating, tool use, hair on the head, thinking, and naked skin.

- complex development; and
- more intricate signaling within and among cells.

Comparing the human genome to itself provides clues to evolution, too. The many duplicated genes and chromosome segments in the human genome suggest that it doubled, at least once, since diverging from that of a vertebrate ancestor about 500 million years ago. Either the human genome doubled twice, followed by loss of some genes, or, more likely, one doubling was followed by additional duplication of certain DNA sequences.

The extensive duplication within the human genome distinguishes us from other primates. Some of the doublings are vast. Half of chromosome 20 repeats, rearranged, on chromosome 18. Much of chromosome 2's short arm reappears as almost three-quarters of chromosome 14, and a block on its long arm is echoed on chromosome 12. The gene-packed yet tiny chromosome 22 includes eight huge duplications and several gene families. However, the many repeated DNA sequences in the human genome may provide raw material for future evolution. A copy of a DNA sequence can mutate, allowing a cell to "try out" a new function while the old one carries on, a little like trying a new car before selling the old one. More often, the twin gene mutates into a silenced pseudogene, leaving a ghost of the gene behind as a similar but untranslated DNA sequence.

A duplication can be located near the original DNA sequence it was copied from, or away from it. A sequence repeated right next to itself is called a tandem duplication, and it usually results from mispairing during DNA replication. A copy of a gene on a different chromosome may arise when messenger RNA is copied (reverse transcribed) into DNA, which then inserts elsewhere among the chromosomes.

Duplication of an entire genome results in polyploidy, discussed in chapter 13. It is common in plants and some insects, but not vertebrates. If a polyploid event was followed by loss of some genes and duplication of others, the result would look much like the modern human genome (**figure 16.12**). The remnants of such an ancient whole-genome duplication would have become further muddled with time, as inversions and translocations altered the ancestral DNA sequence.

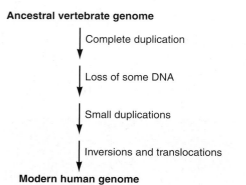

Ancestral vertebrate genome

↓ Complete duplication

↓ Loss of some DNA

↓ Small duplications

↓ Inversions and translocations

Modern human genome

Figure 16.12 Evolution of the human genome. The many duplicated DNA sequences in the human genome suggest a complete duplication followed by other chromosome-level events.

Comparing Chromosomes

Before gene and genome sequencing, researchers recognized that similarities in chromosome banding patterns reflect evolutionary relatedness. Human chromosome banding patterns most closely match those of chimpanzees, then gorillas, and then orangutans (**table 16.2**). The karyotypes of humans, chimpanzees, and apes differ from each other mostly by inversions, which occur within chromosomes.

Chromosome banding patterns are like puzzle pieces. If both copies of human chromosome 2 were broken in half, we would have 48 chromosomes, as the three species of apes do, instead of 46. The banding pattern of chromosome 1 in humans, chimps, gorillas, and orangutans matches that of two small chromosomes in the African green monkey, suggesting that this monkey was ancestral to the other primates. Karyotype differences between these three primates and more primitive primates are mostly translocations.

We can also compare chromosome patterns between species that are not as closely related as we are to other primates. All mammals, for example, have identically banded X chromosomes. One section of human chromosome 1 that is alike in humans, apes, and monkeys is remarkably similar to parts of chromosomes in cats and mice. A human even shares chromosomal segments with a horse, but our karyotype is much less like that of the aardvark, the most primitive placental mammal.

Chromosome band pattern similarities of stained material are imprecise, because a band can contain many genes that differ from those within a band at a corresponding locus in another species' genome. In contrast, DNA probes used in a FISH analysis highlight specific genes (see figure 13.9). FISH can indicate direct correspondence of gene order, or **synteny,** between species, which is solid evidence of close evolutionary relationships. For example, 11 genes are closely linked on the long arm of human chromosome 21, mouse chromosome 16, and on a chromosome called U10 in cows. However, several genes on human chromosome 3 are found near the human chromosome 21 counterpart in mice and cows. Perhaps a mammal ancestral to these three species had all of these genes together, and the genes dispersed to an additional chromosome in humans.

Comparing Proteins

Many different types of organisms use the same proteins, with only slight variations in amino acid sequence. The similarity of protein sequences is compelling evidence for descent from shared ancestors—that is, evolution. Many proteins in humans and chimps are alike in 99 percent of their amino acids, and several are identical. When analyzing a gene's function, researchers routinely consult databases of known genes in many other organisms. Two of the most highly conserved proteins are cytochrome c and homeobox proteins. Another interesting conserved protein causes "double muscles."

Cytochrome c is one of the most ancient and well-studied proteins. It helps to extract energy from nutrients in the mitochondria. Twenty of 104 amino acids occupy identical positions in the cytochrome c of all eukaryotes. The more closely related two species are, the more alike their cytochrome c amino acid sequence is (**table 16.3**). Human cytochrome c, for example, differs from horse cytochrome c by 12 amino acids, and from kangaroo cytochrome c by 8 amino acids. The human protein is identical to chimpanzee cytochrome c.

A class of genes that has changed little across evolutionary time is a **homeobox,** or *HOX* gene. These genes encode transcription factors that control the order in which an embryo

Table 16.2	Percent of Common Chromosome Bands Between Humans and Other Species
Chimpanzees	99⁺%
Gorillas	99⁺%
Orangutans	99⁺%
African green monkeys	95%
Domestic cats	35%
Mice	7%

Table 16.3	Cytochrome *c* Evolution
Organism	**Number of amino acid differences from humans**
Chimpanzee	0
Rhesus monkey	1
Rabbit	9
Cow	10
Pigeon	12
Bullfrog	20
Fruit fly	24
Wheat germ	37
Yeast	42

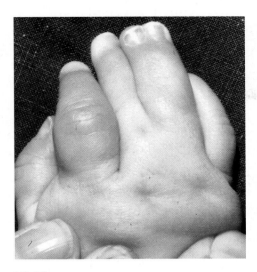

Figure 16.13 **A human *HOX* gene mutation causes synpolydactyly (MIM 186000).** Mutation in the *HOXD13* gene disrupts development of fingers and toes, causing a very distinctive phenotype. The third and fourth fingers are partially fused with an extra digit within the webbed material.

Figure 16.14 **Double muscling in cattle.** The mutation that causes "double muscling" is highly conserved.

turns on genes that ensure that anatomical parts—whether a leg, petal, or segment of a larva—develop in the right places. The highly conserved portion of a homeobox protein is a 60-amino-acid sequence. Humans and most other vertebrates have 39 *HOX* genes in four clusters. The individual genes are expressed in a sequence, in developmental time or anatomical position, that mirrors their order on the chromosome.

Mutations in the homeobox were first studied in the fruit fly *Drosophila melanogaster,* mixing up body parts. *Antennapedia,* for example, has legs in place of its antennae. In humans, mutations in homeobox genes cause various illnesses, such as DiGeorge syndrome (MIM 188400). Although affected individuals hardly sprout legs from their heads, as do *Antennapedia* flies, the missing thymus and parathyroid glands and abnormal ears, nose, mouth, and throat correspond to the sites of abnormalities in the flies. **Figure 16.13** shows another human disorder caused by a mutation in a *HOX* gene.

"Double muscle" mutations vividly demonstrate how similar genes can have similar effects on different types of organisms. The double muscle phenotype is caused by lack of the protein myostatin, which normally prevents stem cells from producing muscle cells. Several members of a German family have myostatin deficiency (MIM 601788), including a boy who, at 5 years of age, could lift weights that a normal adult couldn't, as well as several construction workers who excel in lifting boulders. Double-muscled cattle (**figure 16.14**) and chickens are prized for their extra meat, and "mighty mice" with myostatin mutations are used to study muscle overgrowth.

Molecular Clocks

A clock measures the passage of time as its hands move through a certain degree of a circle in a specific and constant interval of time—a second, a minute, or an hour. In the same way, an informational molecule can be used as a "molecular clock" if its building blocks are replaced at a known and constant rate.

The similarity of nuclear DNA sequences in different species can be used to estimate the time when the organisms diverged from a common ancestor, if the rate of base substitution mutation is known. For example, many nuclear genes studied in humans and chimpanzees differ in 5 percent of their bases, and substitutions occur at a rate of 1 percent per 1 million years. Therefore, 5 million years have presumably passed since the two species diverged. Mitochondrial DNA (mtDNA) sequences may also be tracked in molecular clock studies, as we will soon see.

Timescales based on fossil evidence and molecular clocks can be superimposed on evolutionary tree diagrams constructed from DNA or protein sequence data. However, evolutionary trees can become complex when data can be arranged into different tree configurations. A tree for seventeen mammalian species, for example, can be constructed in 10,395 different ways! The sequence in which the data are entered into tree-building computer programs influences the tree's shape, which is vital to interpreting species relationships. With new sequence information, the tree possibilities change.

Parsimony analysis is a statistical method used to identify an evolutionary tree likely to represent what really happened. An algorithm connects all evolutionary tree sequence data using the fewest possible number of mutational events to account for observed DNA base sequence differences. For the 5-base sequence in **figure 16.15,** for example, the data can be arranged into two possible tree diagrams. Because mutations are rare events, the tree that requires the fewest mutations is more likely to reflect reality.

To track ancient human migration patterns, researchers use the types of genetic markers that are used to track traits in modern families, an approach called genetic genealogy or genetic ancestry. These markers include single nucleotide polymorphisms (SNPs), short tandem repeats (STRs, or microsatellites) and other copy number variants (CNVs) (see table 5.3). Markers of mitochondrial DNA (mtDNA) are used to trace the female lineage, and markers of Y chromosome sequences to

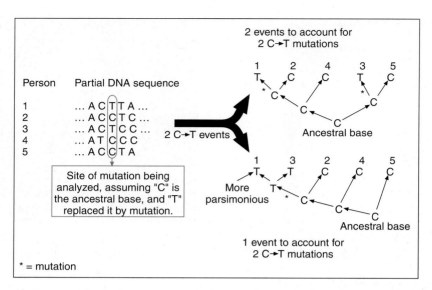

Figure 16.15 Parsimony analysis. Even a computer has trouble arranging DNA differences into an evolutionary tree showing species, population, or individual relationships. A parsimonious tree accounts for all data with the fewest number of mutations. Here, the two individuals who have a T in place of the ancestral C could have arisen in two mutational events or one, assuming that these individuals had a common ancestor. Since mutations are rare events, the more realistic scenario is one mutation.

trace the male lineage. Markers also follow DNA sequences that are part of autosomes, called "ancestry informative markers." Sequencing mtDNA and Y chromosome DNA sequences provides information on only some of a person's ancestors, as the pedigree in **figure 16.16** illustrates. It is easy to see that the contribution of a particular ancestral DNA sequence decreases as the number of generations increases.

MtDNA is ideal for monitoring recent events because it mutates faster than DNA in the nucleus. Its sequences change by 2 to 3 percent per million years. Mutations accumulate faster because mtDNA has no DNA repair. Another advantage of typing mtDNA is that it is more abundant than nuclear DNA because mitochondria have several copies of it, and a cell has many mitochondria. When researchers are lucky enough to find fossils or ancient humans, mtDNA is the most likely DNA to be recovered.

Most of the Y chromosome DNA sequence offers the advantage of not recombining. Crossing over, which it could

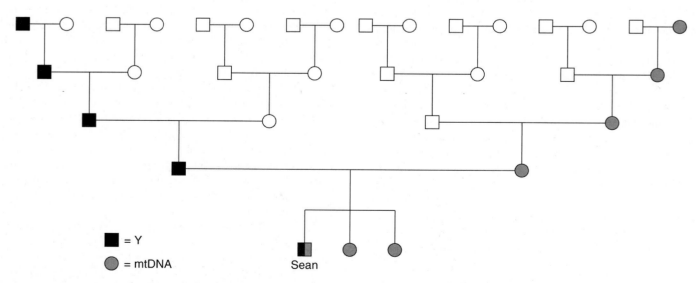

Figure 16.16 Genetic genealogy. Y chromosome and mtDNA sequences represent only some of a person's ancestors. Considering autosomal sequences can capture the contributions of other relatives, represented as the unfilled symbols in the interior of this pedigree. Sean's symbol has two shades because his Y and mtDNA were tested.

only do with an X chromosome because there is no second Y, would break the linkage from the past generation and therefore make tracing relationships impossible.

Sets of SNPs along mitochondrial and Y chromosome DNA define long DNA stretches termed **haplogroups**. (The haplotypes used to describe linkage in chapter 5 refer to shorter DNA sequences.) Y haplogroups are classified from "A" through "T," with several subgroups, called subclades, indicated by alternating letters and numbers. Haplogroups and their subgroups also describe mtDNA. Populations can be classified into both mtDNA and Y chromosome groups, indicating the sources of female and male lineages, respectively. Sub-Saharan Africans, for example, have Y haplogroups E1, E2, and E3a and mtDNA haplogroups L3. Europeans, however, have Y haplogroups R, I, E3b, and J, and their mtDNA haplogroup is R, which includes three subgroups. **Reading 16.2** discusses ancestry testing, available on the Web, that reveals individuals' mtDNA and Y chromosome haplogroups.

Key Concepts

1. The more recently two species shared an ancestor, the more alike their DNA and protein sequences and chromosome banding patterns.

2. Chimps and humans share about 98.7 percent of protein-encoding DNA. Our genomes also differ in insertions, deletions, introns, repeats, and gene expression patterns. The human genome likely duplicated.

3. Molecular clocks apply mutation rates to timescales to estimate when two individuals or types of organisms most recently shared ancestors.

4. Different genes evolve at different rates. Parsimony analysis selects likely evolutionary trees from DNA data.

5. Mitochondrial DNA clocks trace maternal lineages, and Y chromosome sequences trace paternal lineages.

16.3 The Peopling of the Planet

Fossil evidence and extrapolating and inferring relationships from DNA sequence data provide only peeks at the major movements that peopled the planet (see figure 1 in Reading 16.2). The evidence so far is like reading chapters from different parts of a novel. Three such chapters in the story of modern human origins stand out: our beginnings some 200,000 years ago; our expansion from Africa; and the populating of the New World. **Table 16.4** tells other "stories" from DNA sequence comparisons.

Mitochondrial Eve

Theoretically, if a particular sequence of mtDNA could have mutated to yield the mtDNA sequences in modern humans, then that ancestral sequence may represent a very early human or humanlike female—a mitochondrial "Eve," or metaphorical first woman. **Figure 16.17** shows how one maternal line may have persisted.

When might this theoretical "first" woman, the most recent female ancestor common to us all, have lived? In the mid 1980s, researchers compared mtDNA sequences for protein-encoding as well as noncoding DNA regions in a variety of people, including Africans, African Americans, Europeans, New Guineans, and Australians. They deduced that the hypothesized ancestral woman lived about 200,000 years ago, in Africa. More recent analysis of mtDNA from 600 living East Africans estimated 170,000 years ago for the beginning of the modern human line, which is remarkably close to the date of the *H. sapiens idaltu* fossils. One way to reach this time estimate is by comparing how much the mtDNA sequence differs among modern humans to how much it differs between humans and chimps. The differences in mtDNA sequences among contemporary humans are 1/25 the difference between humans and chimps. The two species diverged about 5 million years ago, according to extrapolation from fossil and molecular

Table 16.4		Tales of Ancestry in Genomes	
Population	**DNA**	**History**	**New Findings**
Tuscans (Italy)	mt	Tuscans descend from Etruria, an advanced culture predating the Roman Empire.	Unexpected Near East haplogroups found.
Ashkenazi Jews	mt	Originated in 7th and 8th centuries when families from north Italy migrated to the Rhine Valley.	40% of the 8 million modern Ashkenazim descend from four women.
Lebanon	Y	Modern humans arrived ~47,000 years ago, Muslims from Arabian Peninsula in 7th century CE, Christians from Crusades, 11th–13th centuries.	Current groups reflect the two religions.
Island Southeast Asia (Indonesia, East Malaysia, Philippines)	mt	Modern humans arrived 50,000 years ago; rice farmers from Taiwan came 4,000 years ago and dominate linguistic and archeological evidence.	From 5,000 to 15,000 years ago, indigenous peoples expanded; Taiwanese farmer contribution was minor.

Should You Take a Genetic Ancestry Test?

More than a million people have searched for information on their "deep ancestry" by mailing a DNA sample to a company offering genetic genealogy services. Most companies offer two basic types of tests: mitochondrial DNA to trace maternal lineages and Y chromosome tests to trace paternal lineages. Several markers are checked for each DNA source, and the results compared to growing databases to look for matches. "Deep ancestry" refers to assigning a match to a major part of the world—sub-Saharan Africa, for example—or a major population group, such as Native American or African American (**figure 1**).

DNA ancestry testing can reveal whether any two individuals living today share an ancestor, assigning an approximate generation to the "most recent common ancestor" (MRCA). The more markers tested, the more meaningful the results. If two people share all thirty-seven of thirty-seven tested markers, there is a 50 percent chance that their MRCA was no more than two generations ago. They are so alike that other DNA variants haven't intruded, because not enough generations have passed. Sharing twenty-five of twenty-five markers gives a 50 percent chance that the MRCA was not more than three generations ago. If they share twelve markers, there is a 50% chance that the MRCA was no longer ago than seven generations.

As is the case with direct-to-consumer genetic testing, discussed in chapter 1, ancestry testing websites offer a great deal of information, but people may have unrealistic expectations. The results are not nearly as specific as what President Barack Obama found when he visited Kenya to meet his half-siblings. But as genetic genealogy databases grow, so will the information available. Most companies inform clients when newcomers' DNA indicates that they are related, introducing distant cousins. Companies are also beginning to offer autosomal markers, tracing parental lines in greater detail. As a result, places of origin are becoming more specific. At first, companies tested for only four parental populations—African, European, East Asian, and Native American. Now companies report on subregions of these areas.

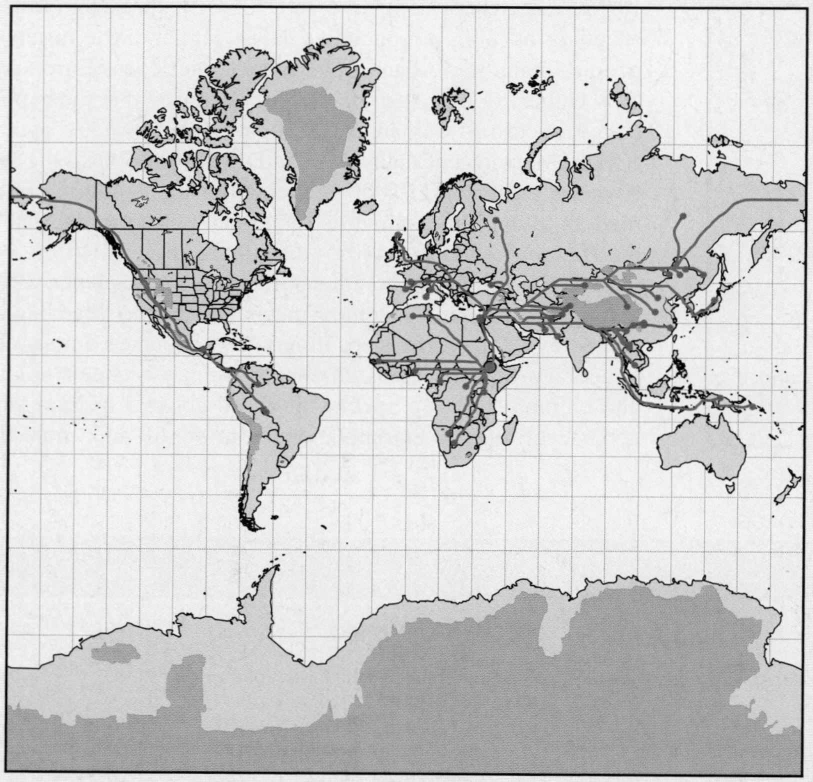

Ancestry testing comes with some caveats:

- MtDNA and Y chromosome DNA testing sample much less than 1 percent of the genome.

- MtDNA and Y chromosome DNA trace only some lineages.

- Not all of the human haplogroups in the world have been discovered. A person's reported geographical place of origin can change with new findings.

- A haplogroup need not come from only one geographic region, due to gene flow. People move, and so do their genes.

- Geographic regions that a haplogroup may point to may not reflect a particular ethnic or racial group, which are social, not genetic.

For people not seeking personal information but wishing to partake in a worldwide effort to map ancient migratory paths, the Genographic Project offers mtDNA or Y chromosome DNA testing for $99. The project began with studying indigenous peoples, but then expanded to everyone. The Genographic Project is supported by National Geographic, IBM, private foundations, and academic institutions.

Figure 1 **The peopling of the planet.** This illustration depicts major migratory paths based on DNA haplogroup information, mostly from mtDNA and the Y chromosome. Fossil and DNA evidence suggest that humanity arose in East Africa, but a southern African origin is possible too.

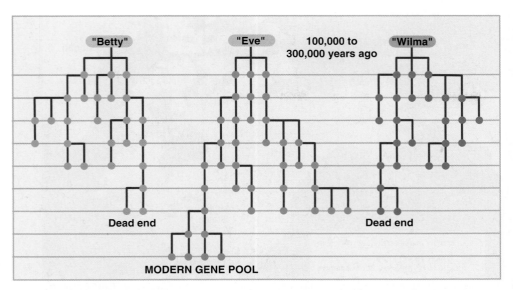

Figure 16.17 Mitochondrial Eve. According to the mitochondrial Eve hypothesis, modern mtDNA retains some sequences from a figurative first woman, "Eve," who lived in Africa 300,000 to 100,000 years ago. In this schematic illustration, the lines represent generations, and the circles, females. Lineages cease whenever a woman does not have a daughter to pass on the mtDNA.

evidence. Multiplying 1/25 by 5 million gives a value of 200,000 years ago, assuming that the mtDNA mutation rate is constant over time.

Where did Eve live? The locations of fossil evidence, such as *H. sapiens idaltu* skulls, support an African origin, and Charles Darwin suggested it, too. In addition, studies comparing mitochondrial and nuclear DNA sequences among modern populations consistently find that Africans have the most numerous and diverse mutations. For this to be so, Africans must have existed longer than other modern peoples, because it takes time for mutations to accumulate. In many evolutionary trees constructed by computer parsimony analysis, the individuals whose DNA sequences form the bases are from Africa. That is, other modern human populations all have at least part of an ancestral African genome, plus mutations that occurred after their ancestors left Africa.

Expansion Out of Africa

The idea of mitochondrial Eve is part of the "out of Africa" view, or replacement hypothesis, of human origins. It states that about 200,000 years ago, *H. sapiens* evolved from an *H. erectus* or other *Homo* population in Africa. This may have occurred quickly, in small, isolated pockets, or gradually across a broader swath of the continent. However it happened, eventually descendants of some of these early *H. sapiens* expanded out of Africa about 56,000 years ago. Some remained. An alternate view, the multiregional hypothesis, has been largely disproved. It maintained that modern humans arose in several places, gradually emerging on a global scale in which people from all over mixed—not the isolated pockets of peoples that the replacement hypothesis envisions.

Data from mtDNA, Y chromosome DNA, and markers on the autosomes indicate that the peopling of the world was a series of founder effects as groups left Africa, perhaps when the Sahara desert periodically grew wetter. These movements yielded "chains of colonies" that may have overlapped and merged when neighbors met, and genes flowed from one region to another. Geographic and climatic barriers periodically shrank human populations, while natural selection and genetic drift narrowed the African gene pool. At the same time, new mutations established the haplogroups that are consulted to trace ancestry.

Many questions remain. When did groups leave Africa? Did they meet and mate? How and when did population bottlenecks carve out modern gene pools? How many types of *Homo* made the trek to Europe and Asia? Probing genes and genomes for clues to the past is like a photo slowly coming into focus, capturing a portrait of our origins.

Populating the New World

People spread across Eurasia by 40,000 years ago, as well as elsewhere, and lastly through Siberia (**figure 16.18**). From here they could cross the Bering Land Bridge, which emerged between Siberia and Alaska during times when the glaciers had retreated. The land bridge stretched for about 1,000 miles from north to south, appearing as winds from the southwest blew snow away. The areas for several hundred miles on either side of the bridge, and the bridge itself, are called Beringia.

Sometime between 23,000 and 19,000 years ago, a severe population bottleneck affected the people in Beringia. Only about 1,000 of them survived the journey over the bridge from Siberia, and some of them continued southward along the Pacific coastline. As the ice age ended about 18,000 years ago, the tiny founding population in the Americas began a period of rapid expansion that lasted 3,000 years. This amplified alleles that had survived, as alleles unique to people who had perished vanished from the population. The people spread through the Americas as the first Native Americans. The mitochondrial DNA evidence that paints this picture is particularly valuable because the coastal migratory path is now underwater, hiding archeological clues.

Today, Native Americans carry a very distinctive genetic nametag that reflects the long-ago trek across the land bridge: five mtDNA haplogroups (A, B, C, D and X) and two Y chromosome haplogroups (C and Q). These markers are seen in all indigenous populations in southern Siberia too, indicating a single gene pool traveling in a single migration. DNA sequence information extrapolated back from present-day Native American populations is consistent with molecular clock data from ancient DNA. In addition, Native American populations have

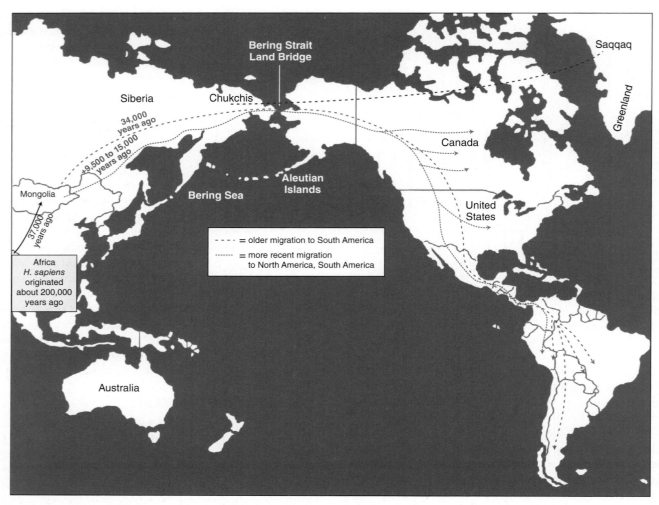

Figure 16.18 Tracing human origins. Analyses of mitochondrial DNA and Y chromosome DNA sequences reveal that the ancestors of Native Americans came from Mongolia and Siberia.

an STR marker and some mtDNA haplogroups that are not seen in eastern Siberian peoples, indicating mutations that happened after the crossing of the Bering Strait.

A comparison of 678 autosomal STR markers from 29 Native American populations and 49 other indigenous groups worldwide found that Native Americans are very different from other populations, yet are very much like each other—as might be expected from a multigenerational journey southward by a small but hardy group, along the coastline of the Americas.

By 14,000 years ago, Native Americans had arrived inland. Evidence comes from their DNA in coprolites—fossilized excrement—discovered in Oregon, Wisconsin, and even Florida. Genetic evidence also suggests that some Native American populations died out. Perhaps they were the descendants of the people who chose to stay in the harsh north of the new land.

It seems that with every new DNA discovery, we have to rewrite prehistory. This was the case for a tuft of hair from a Paleo-Eskimo discovered in Greenland from the first group of people to settle there, called the Saqqaq. The hair is about

4,000 years old. Analysis of its DNA revealed characteristics of the man whose head it graced, as well as clues to how the Saqqaq got to Greenland (**figure 16.19**).

The owner of the hair was male with coarse, dark hair and skin, brown eyes, shovel-shaped teeth, dry earwax, and, if he had lived longer, would have become bald. He had type A$^+$ blood, was at increased risk for high blood pressure and diabetes, had a high tolerance for alcohol, and would likely have become addicted to nicotine had it been possible to cultivate it in the frozen wasteland that he called home. The most revealing information, however, concerned the Saqqaq Eskimo's origins. His mitochondrial DNA was distinct from that of either Native Americans or modern Eskimos, yet was very similar to that of the Chukchis people of Siberia. Apparently, the ancestors of the Greenland Eskimos crossed the land bridge about 5,400 years ago, separate from the crossings that founded the Native Americans and modern Eskimos. The conclusion: The earliest Eskimos in Greenland came from a different migration than the one that was ancestral to Native Americans and modern

Figure 16.19 **A 4,000-year-old hair holds clues to a man's phenotype in the DNA.** This is an artist's rendition of the Paleo-Eskimo from Greenland, by way of Siberia.

Eskimos, and left no present-day descendants. The discovery of the 4,000-year-old non–Native American/non-Eskimo reveals a major limitation of genetic anthropology: We cannot fit in the puzzle pieces that we cannot find, or that no longer echo in modern gene pools.

Key Concepts

1. Molecular clocks have been used to examine the origin and migrations of modern humans.
2. Mitochondrial Eve, the most recent female ancestor common to us all, lived about 170,000 years ago in Africa.
3. *Homo sapiens* first left Africa about 56,000 years ago. People populated the planet in a series of founder effects.

Summary

16.1 Human Origins

1. The first primates were rodentlike insectivores that lived about 60 million years ago. By 30 to 40 million years ago, monkeylike *Aegyptopithecus* lived. **Hominoids,** ancestral to apes and humans, lived 22 to 32 million years ago.
2. **Hominins,** ancestral to humans only, appeared about 19 million years ago. They were more upright, dwelled on the plains, and had smaller brains than their ancestors.
3. At least three types of hominins lived about 6 million years ago, shortly after the split from the chimp lineage.
4. The australopithecines preceded and then coexisted with *Homo habilis,* who lived in caves, had strong family units, and used tools extensively. *Homo erectus* was a contemporary who outsurvived *H. habilis,* lived in societies, and used fire. *Homo sapiens idaltu* lived about 156,000 years ago, and looked like us.
5. Neanderthals left Africa before our direct ancestors did, but the human genome sequence indicates that our ancestors bred with Neanderthals, in Eurasia. The Denisova hominin was a contemporary.

16.2 Molecular Evolution

6. Molecular evolution considers differences at the genome, chromosome, protein, and DNA sequence levels with mutation rates to estimate species relatedness.
7. Humans and chimps share 98.7 percent of their protein-encoding gene sequences. Indels, introns, and repeats create genome differences between humans and chimps, which also differ in gene expression.

8. The human genome shows many signs of past duplication.
9. Genes in the same order on chromosomes in different species show **synteny**.
10. For a highly conserved gene or protein, DNA sequence is similar or identical in different species, indicating importance and shared ancestry.
11. Evolutionary tree diagrams represent gene sequence information from several species, using molecular clocks based on mutation rates.
12. Parsimony analysis selects the evolutionary trees requiring the fewest mutations, which are therefore the most likely.
13. Molecular clocks based on mtDNA date recent events through the maternal line because this DNA mutates faster than nuclear DNA. Y chromosome genes trace paternal lineage. Markers (SNPs, STRs, and CNVs) in mtDNA, Y chromosome DNA, and autosomal DNA are used to study human origins and expansions. Groups of linked markers inherited together form **haplogroups**.

16.3 The Peopling of the Planet

14. The rate of mtDNA mutation and current mtDNA diversity can be extrapolated to hypothesize that a theoretical first woman lived, in Africa, about 200,000 years ago. *Homo sapiens* began to leave Africa about 56,000 years ago.
15. A series of migrations and founder effects peopled the planet, with genetic diversity decreasing from that of the ancestral African population, but new mutations occurring.
16. After the last ice age, people crossed the Bering Strait from Siberia, occupying the Americas.

www.mhhe.com/lewisgenetics10

Answers to all end-of-chapter questions can be found at
www.mhhe.com/lewisgenetics10. You will also find
additional practice quizzes, animations, videos, and vocabulary
flashcards to help you master the material in this chapter.

Review Questions

1. Arrange the following primates in the order in which they lived, indicating any that may have overlapped in time.

    a. *Homo erectus*
    b. *Australopithecus anamensis*
    c. *Dryopithecus*
    d. Neanderthals
    e. *Ardipithecus*
    f. *Homo habilis*
    g. *Australopithecus garhi*
    h. *Homo sapiens idaltu*
    i. the Denisova hominin

2. What is the difference between a hominoid and a hominin?

3. Some anthropologists classify chimpanzees with humans in genus *Homo*. How does this conflict with fossil evidence of the *Australopithecus* species?

4. Give an example of how a single gene difference can have a profound effect on the phenotypes of two species.

5. Explain how describing a hominin from a few fossils or a few genes can lead to misleading conclusions.

6. Give an example of an investigation of a gene or genome that revealed how a group of people moved about the world.

7. Give an example of a trait in any of the hominins discussed in the chapter that illustrates positive selection.

8. Explain what comparing genomes from indigenous peoples to other modern peoples can reveal about evolution.

9. Give an example of molecular evidence that is consistent with fossil or other evidence.

10. Describe the type of information that Y chromosome and mitochondrial DNA sequences provide.

11. List three aspects of development, anatomy, or physiology that were important in human evolution.

12. Explain why a DNA sequence that is highly conserved among humans and chimps, gorillas, and orangutans is unlikely to vary greatly among modern human populations.

13. What types of information are needed to construct an evolutionary tree diagram? What assumptions are necessary? What are the limitations of these diagrams?

14. Researchers compare a number of types of information in the human and chimp genomes, including SNPs, CNVs, STRs, indels, and linkage patterns, discussed in this and other chapters. Define each of these types of information.

Applied Questions

1. Select an example from this chapter and explain how it illustrates one of the forces of evolutionary change discussed in chapter 15 (natural selection, nonrandom mating, migration, genetic drift, or mutation.)

2. Create a narrative of how a group of australopithecines might have branched off into *Homo*.

3. Hypothesize why many species of *Australopithecus* lived at one time, but only one species of *Homo* lives today.

4. A geneticist aboard a federation starship must deduce how closely related humans, Klingons, Romulans, and Betazoids are. Each organism walks on two legs, lives in complex societies, uses tools and technologies, looks similar, and reproduces in the same manner. Each can interbreed with any of the others. The geneticist finds the following data:

    ■ Klingons and Romulans each have 44 chromosomes. Humans and Betazoids have 46 chromosomes. Human chromosomes 15 and 17 resemble part of the same large chromosome in Klingons and Romulans.

    ■ Humans and Klingons have 97 percent of their chromosome bands in common. Humans and Romulans have 98 percent of their chromosome bands in common, and humans and Betazoids show 100 percent correspondence. Humans and Betazoids differ only by an extra segment on chromosome 11, which appears to be a duplication.

    ■ The cytochrome *c* amino acid sequence is identical in humans and Betazoids, differs by one amino acid between Humans and Romulans, and differs by two amino acids between humans and Klingons.

    ■ The gene for collagen contains 50 introns in humans, 50 introns in Betazoids, 62 introns in Romulans, and 74 introns in Klingons.

    ■ Mitochondrial DNA analysis reveals many more individual differences between Klingons and Romulans than between humans and Betazoids.

    a. Suggest a series of chromosomal abnormalities or variants that might explain the karyotypic differences among these four types of organisms.

b. Which are our closest relatives among the Klingons, Romulans, and Betazoids? What is the evidence for this?

c. Are Klingons, Romulans, humans, and Betazoids distinct species? What information reveals this?

d. Which of the evolutionary tree diagrams is consistent with the data?

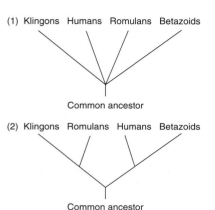

(1) Klingons Humans Romulans Betazoids

Common ancestor

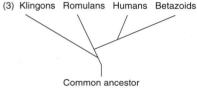

(2) Klingons Romulans Humans Betazoids

Common ancestor

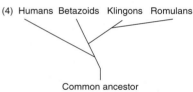

(3) Klingons Romulans Humans Betazoids

Common ancestor

(4) Humans Betazoids Klingons Romulans

Common ancestor

5. Why might it be important to identify DNA sequences that chimps have that humans do not, as well as identify sequences unique to humans?

6. Several women have offered to be inseminated with sperm from the Ice Man, who died 5,300 years ago in the Alps. If sperm could have been recovered, and a woman inseminated, what do you think the child would be like?

7. A man with white skin checks the box on forms about personal information for "African American," claiming that we are all, if we go back far enough in our family trees, from Africa. Is he correct?

8. Can ancestry testing help to dispel race-based problems, or worsen them? Cite a reason for your answer.

9. Hatshepsut was a pharaoh who ruled Egypt more than 3,000 years ago. Although she led many successful projects and the economy soared during her two-decade reign, she was erased from written records. In 2008, her mummy was found and tentatively identified from a distinctive tooth. What information was needed for DNA evidence to confirm that the mummy was, indeed, Hatshepsut, part of the royal line?

10. In Central Africa, the Mbuti Pygmies are hunter-gatherers who live amid agricultural communities of peoples called the Alur, Hema, and Nande. Researchers compared autosomal STRs, mitochondrial DNA, and Y chromosome DNA haplogroups among these four types of people. The pygmies had the most diverse Y chromosomes, including about a third of the sequence that was the same as those among the agricultural groups, who had greater mtDNA diversity than the pygmies. None of the agricultural males had pygmy Y DNA sequences. Create a narrative of gene flow to explain these findings.

Web Activities

1. Go to http://www.peoplesoftheworld.org/ or a similar website. Select an indigenous people, do further research, and describe their habitat and culture, and any distinctive health strengths or problems. To what extent do you think genetics is responsible for the state of their health? Explain your reasoning.

2. Consult the website for the Genographic Project, and read the material under "Atlas of Human Journey."

a. What does "deep ancestry" mean?

b. Explain why analyzing your mtDNA or Y chromosome DNA cannot provide a complete picture of your ancestry.

c. Explain how a female can trace her paternal lineage if she doesn't have a Y chromosome.

d. Would you want your ancestry information and identity posted on the Genographic database so that cousins can contact you?

Forensics Focus

1. How can the technology used to describe the 4,000-year-old Greenland Eskimo be applied to forensics techniques used on crime scene evidence?

2. In many African cultures, "family" is not dictated by genetics, but by who cares for whom. Any adult can be "mother" or

"father" to any child. Since the early 1990s, many parts of East Africa have been under civil war. Thousands of Africans have asked to be admitted to the United States to join relatives. The "family reunification resettlement program" enables parents, siblings, and children of U.S. citizens to come to the U.S. In

2008, addressing rumors that many people were lying that they were related to people in the United States, the State Department began to ask refugees from Kenya to voluntarily provide a DNA sample, to be compared to that of the U.S. citizen claimed to be a relative. When the DNA testing turned up many cases of people claiming to be family who were not blood relatives, the resettlement program was stopped.

a. Do you think that DNA testing should have been imposed on people seeking asylum in the United States?

b. Should the people have been compelled to have their DNA tested?

c. How should cultural definitions, such as that of "family," be handled?

d. How should the situation be resolved?

Case Studies and Research Results

1. The ancestors of woolly mammoths, African elephants, and Indian elephants originated in Africa 7 million years ago. The ancestors of modern African elephants stayed there, but between 4 and 5 million years ago, some of them left Africa. The ones that migrated toward Southeast Asia became modern Indian elephants, which are smaller than their African counterparts. From 1.2 to 2 million years ago, some of these animals made their way north, thanks to adaptations that included heavy coats and tiny tails and ears. These were the woolly mammoths. They thrived until about 10,000 years ago, although a few survived until about 3,800 years ago.

   Researchers sequenced the genomes from hairs from two young mammoths flash-frozen in Siberia 60,000 and 20,000 years ago. The genome is very similar to that of the two modern elephant species. When researchers sequenced the beta globin gene from another frozen mammoth, they found that mammoth hemoglobin releases bound oxygen more readily than does elephant hemoglobin. This likely enabled the animal to survive at high altitudes where the concentration of oxygen in the air is lower.

   a. Pose a question that comparing the mammoth genome to that of either modern elephant species might answer.

   b. Suggest a way to reconstruct a mammoth.

   c. How does mammoth hemoglobin illustrate positive selection?

   d. A father gives his young daughter a stuffed mammoth. "Annika, this is a woolly mammoth. He was an ancestor of Babar, your elephant." Is he correct? Why or why not?

   e. What does the evolution of the two elephants and the mammoth have in common with the evolution of humans?

2. "Neanderthals are not totally extinct; they live on in some of us," said Svante Paabo, the leader of the Neanderthal Genome Project. What does he mean?

3. The 4,000-year-old Saqqaq Paleo Eskimo from Greenland is known only from a tuft of hair. How did researchers learn about other characteristics?

4. For more than 20 million years, lice have lived on the skins of primates. Researchers compared a 1,525-base-pair sequence of mtDNA among modern varieties of lice, and, applying the mutation rate, derived an evolutionary tree. It depicts a split in the louse lineage, with one group of head and body lice living throughout the world, and another group of only head lice living in the Americas.

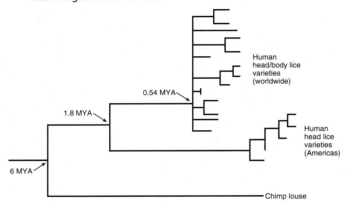

a. What events in human evolution roughly correspond to the branch points in the louse evolutionary tree?

b. What might be the significance of the similarity between the evolutionary trees for lice and humans?

c. What is the evidence that lice moved from archaic humans to modern humans?

5. A Y chromosome haplotype has mutations for the *SRY* gene and genes called *M96* and *P29*. Modern Africans have three variants of this haplotype. Two are only in Africans, but the third variant, E3, is also seen in western Asia and parts of Europe. Researchers examined specific subhaplotypes (variations of the variations) and found that one type, E-M81, accounts for 80 percent of the Y chromosomes sampled in northwest Africa, falling sharply in incidence to the east, and not present in sub-Saharan Africa. That same haplotype is found in a small percentage of the Y chromosomes in Spain and Portugal. Consult a map, and propose a scenario for this gene flow. What further information would be useful in reconstructing migration patterns?

6. Roland has always considered himself African-American, but ancestry testing of his Y chromosome indicates a Chinese background. He is very upset, concluding that he is not African-American after all. Explain how he has misinterpreted the test result.

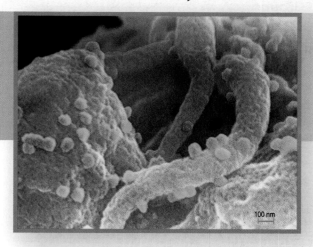

100 nm

These HIV particles (green) attach to a white blood cell (pink). In people with a certain mutation, cells lack receptors that HIV requires to enter.

Genetics of Immunity

Learning Outcomes

17.1 The Importance of Cell Surfaces

1. List the components of the immune system.
2. Describe the basis of blood groups.
3. Explain what human leukocyte antigens are and what they indicate about health.

17.2 The Human Immune System

4. Distinguish among physical barriers, innate immunity, and adaptive immunity.
5. Distinguish between the humoral and cellular immune responses.

17.3 Abnormal Immunity

6. Discuss conditions that result when the immune system is underactive, overactive, and misdirected.

17.4 Altering Immunity

7. Describe how medical technologies boost or suppress immunity to prevent or treat disease.
8. Explain the requirements for the body to accept an organ from another person.

17.5 A Genomic View of Immunity—The Pathogen's Perspective

9. Discuss what we can learn from studying the genomes of pathogens.

The Big Picture: The immune system enables us to share the planet with other organisms. Genes control the immune response. We can alter immunity to enhance health.

Changing the Genotype to Vanquish HIV

In 2008, a 40-year-old man received a stem cell transplant to treat leukemia at a hospital in Berlin. He had been HIV positive for at least a decade, and had taken anti-HIV drugs for 4 years. Leukemia was his first HIV-related illness. Known at first as "the Berlin patient," Timothy Brown became the star of a groundbreaking experiment.

Stem cell transplants had been tried before to treat HIV, but hadn't worked. This time, researchers chose a very special donor: a person who was both a tissue match for Brown, and genetically resistant to HIV infection. As a "CCR5 delta 32" homozygote, his cells were missing receptors that HIV must bind to enter. Could a transplant of the man's stem cells into Brown cure his leukemia and his HIV infection?

The answer was yes! The transplanted stem cells eventually replaced Brown's blood and bone marrow with HIV-resistant cells, and all signs of the infection vanished. As Brown made headlines, researchers were already thinking about how to recreate the success. They did, on other patients.

Although these cases shows that people with HIV can be cured with stem cells from the blood of CCR5 delta 32 homozygotes, such donors are rare—they account for less than 1 percent of most populations. A solution may be to team two technologies: stem cell therapy and gene therapy. The new approach will give patients their own blood stem cells that have been removed and their CCR5 genes replaced with mutant ones. This is one example of a helpful mutation.

17.1 The Importance of Cell Surfaces

We share the planet with plants, microbes, fungi, and other animals, but can become ill when they, or their parts, enter our bodies. The human immune system protects us against this happening. It is a mobile army of about 2 trillion cells, the biochemicals they release, and the organs where they are produced and stored.

Protection against infection is based on the ability of the immune system to recognize "foreign" or "nonself" cell surfaces that are not part of the body. These include surfaces of microorganisms such as bacteria and yeast; nonliving "infectious agents" such as viruses and prions; and tumor cells and transplanted cells. Then, the immune system launches a highly coordinated attack that includes both general and highly specific responses. Organisms or infectious agents that cause disease are called pathogens. **Reading 17.1** highlights one common type of pathogen—viruses. **Figure 17.1** shows another—bacteria.

Understanding how genes control immunity makes it possible to enhance or redirect the system's ability to fight disease. Mutations can impair immune function, causing immune deficiencies, autoimmune disorders, allergies, and cancer. Genes affect immunity by conferring susceptibilities or resistances to certain infectious diseases. Like other inherited characteristics, degree of immune protection varies from person to person. This may be why one person suffers frequent respiratory infections, whereas another is rarely ill. A study of the immune systems of people who survived the 1918 flu pandemic revealed that their antibodies, many decades later, can still rapidly destroy a flu virus. Yet 50 million people died of that flu, many in just days.

A few types of genes encode proteins that powerfully affect immunity. **Antibodies** and **cytokines** are proteins that directly attack foreign **antigens,** which are any molecules that can elicit an immune response. Most antigens are proteins or carbohydrates. Genes also specify the cell surface antigens that mark the body's cells as "self."

Blood Groups

Some of the antigens that dot our cell surfaces determine blood types. Figures 5.3 and 5.4 describe the familiar ABO blood types. We actually have twenty-nine major blood types based on protein and carbohydrate antigens on the surfaces of red blood cells. Each of these types includes many subtypes, generating hundreds of ways that the topographies of our red blood cells differ from individual to individual. **Table 17.1** lists a few blood groups.

For blood transfusions, blood is typed and matched from donor to recipient. For more than a century, an approach called serology typed blood according to red blood cell antigens. A newer way to type blood is to identify the *instructions* for the cell-surface antigens—that is, the genes that encode these proteins. This approach, termed genotyping, uses a tiny device that detects 100 distinct DNA "signatures for blood types." Genotyping is especially useful for people who have a chronic disorder that requires multiple transfusions, such as leukemia or sickle cell disease. They produce so many antibodies against so many types of donor blood that it is often difficult to determine their blood types by serology.

The Major Histocompatibility Complex

Many proteins on our cell surfaces are encoded by genes that are part of a 6-million-base-long cluster on the short arm of chromosome 6 called the **major histocompatibility complex** (**MHC**). This region includes about 70 genes, and about 50 percent of the genetic influence on immunity stems from them. MHC genes are classified into three groups based on function.

MHC class III genes encode proteins that are in blood plasma (the liquid portion of blood, discussed in section 11.1) and that provide nonspecific immune functions. Class I and II

Figure 17.1 A bacterial pathogen. *Escherichia coli* is a normal resident of the human small intestine, but under certain conditions can produce a toxin that causes severe diarrhea ("food poisoning") and can damage the kidneys.

Table 17.1	Blood Groups
Blood Group (MIM)	**Description**
MN (111300)	Codominant alleles *M, N,* and *S* determine six genotypes and phenotypes. The antigens bind two glycoproteins.
Lewis (111100)	Allele *Le* encodes fucosyltransferase (FUT3) that adds an antigen to the sugar fucose, which the product of the *H* gene places on red blood cells. *H* gene expression is necessary for the ABO phenotype (see section 5.1). People with *LeLe* or *Lele* have the Lewis antigen on red blood cells and in saliva. People of genotype *lele* do not.
Secretor (182100)	People with the *Se* allele secrete A, B, and H antigen in body fluids.

Viruses

A viral infection can make us feel terrible, but many of the aches and pains are in fact actions of the immune system.

A virus is a single or double strand of RNA or DNA wrapped in a protein coat, and in some types, an outer envelope, too. A virus can reproduce only if it enters and uses a host cell's energy resources, protein synthetic machinery, and secretion pathway. It is a stunningly streamlined structure. A virus may have only a few protein-encoding genes, but many copies of the same protein can assemble to form an intricate covering, like the panes of glass in a greenhouse. Ebola virus, for example, has only seven types of proteins, but they assemble into a structure capable of reducing a human body to little more than a bag of blood and decomposed tissue (**figure 1**). In contrast, the smallpox virus has more than 100 different types of proteins, and HIV is also complex.

Viruses are with us all the time—not only when we are ill. Part of the DNA sequence of some human chromosomes includes viral DNA sequences that are vestiges of past infections, perhaps passed, silently, from distant ancestors. Many DNA viruses reproduce by inserting their DNA into the host cell's DNA. In contrast, an RNA virus must first copy its RNA into DNA before it can insert into a human chromosome. A viral enzyme called reverse transcriptase does this. Certain RNA viruses are called retroviruses because they transmit genetic information opposite the usual direction—instead of DNA to RNA to protein, viral RNA is copied into DNA, which may then be copied back into RNA to guide the synthesis of viral proteins. HIV is a retrovirus.

Once viral DNA integrates into the host cell's DNA, it can either remain and replicate along with the host's DNA without causing harm, or it can take over and kill the cell. Activated viral genes direct the host cell to replicate viral DNA and then use it to manufacture viral proteins. The cell bursts, releasing many new virus copies into the body.

Diverse viruses infect all types of organisms. Their genetic material cannot repair itself, so the mutation rate may be high. This is one reason why we cannot develop an effective vaccine against HIV or the common cold, and why new influenza vaccines must be developed each year.

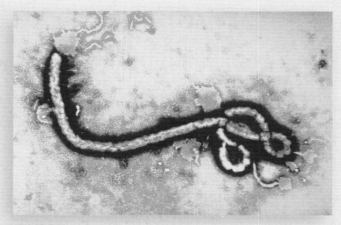

Figure 1 Ebola virus is a single strand of RNA and seven proteins. People become infected when they touch body fluids of an infected individual.

genes of the MHC encode the **human leukocyte antigens** (**HLAs**), which were first studied in leukocytes (white blood cells). The HLA proteins link to sugars, forming branchlike glycoproteins that extend from cell surfaces.

Class I and II HLA proteins differ in the types of immune system cells they alert. Some HLA glycoproteins bind bacterial and viral proteins, displaying them like badges to alert other immune system cells. This action, called antigen processing, is often the first step in an immune response. The cell that displays the foreign antigen is called an **antigen-presenting cell**. **Figure 17.2** shows how a large cell called a macrophage displays bacterial antigens. Certain white blood cells called T cells (or T lymphocytes) are also antigen-presenting cells. Dendritic cells are antigen-presenting cells found in places where the body contacts the environment, such as in the skin and in the linings of the respiratory and digestive tracts. Dendritic cells signal T cells, starting an immune response.

HLAs identify all cells of a person as "self," or belonging to the same individual. In addition to these common HLA markers are more specific markers that distinguish particular tissue types. Class I includes three genes that vary greatly and are found on all cell types, and three other genes that are more restricted in their distribution. Class II includes three major genes whose encoded proteins are found mostly on antigen-presenting cells.

Individuals have an overall HLA "type" based on the six major HLA genes. Only 2 in every 20,000 unrelated people match for the six major HLA genes by chance. When transplant physicians attempt to match donor tissue to a potential recipient, they determine how alike the two individuals are for these six genes. Usually at least four of the genes must match for a transplant to have a reasonable chance of success. Before DNA profiling, HLA typing was the predominant type of blood test used in forensic and paternity cases to rule out involvement of certain individuals. However, HLA genotyping has become very complex because hundreds of alleles are now known.

A few disorders are very strongly associated with inheriting particular HLA types. This is the case for ankylosing spondylitis, which inflames and deforms vertebrae. A person with either of two particular subtypes of an HLA called B27 is 100 times as likely to develop the condition as someone who lacks either form of the antigen. HLA-associated risks are

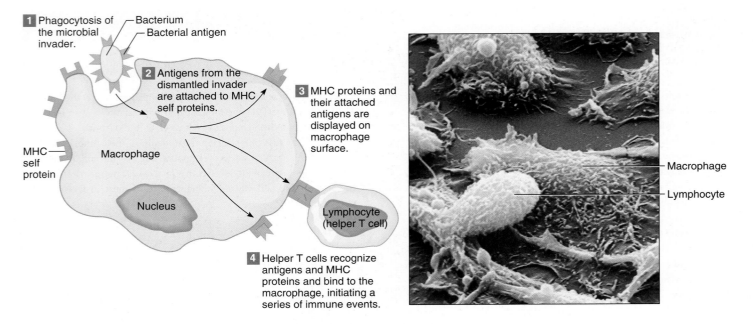

Figure 17.2 **Macrophages are antigen-presenting cells.** A macrophage engulfs a bacterium, then displays foreign antigens on its surface, which are held in place by major histocompatibility complex (MHC) self proteins. This event sets into motion many immune reactions.

not absolute. More than 90 percent of people who suffer from ankylosing spondylitis have the B27 antigen, which occurs in only 5 percent of the general population. However, 10 percent of people who have ankylosing spondylitis do *not* have the B27 antigen, and some people who have the antigen never develop the disease.

Key Concepts

1. The immune system consists of cells and biochemicals that distinguish self from nonself antigens.

2. Pathogens include microorganisms and infectious agents.

3. Blood types result from self antigen patterns on red blood cells. HLA cell surface proteins establish self and display foreign antigens.

17.2 The Human Immune System

The immune system is a network of vessels called lymphatics that transport lymph fluid to bean-shaped structures throughout the body called lymph nodes. The spleen and thymus gland are also part of the immune system (**figure 17.3**).

Lymph fluid carries white blood cells called lymphocytes and the wandering, scavenging macrophages that capture and degrade bacteria, viruses, and cellular debris. **B cells** and **T cells** are the two major types of lymphocytes.

The genetic connection to immunity is the proteins required to carry out an immune response. The immune response attacks

pathogens, cancer cells, and transplanted cells with two lines of defense—an immediate generalized **innate immunity,** and a more specific, slower **adaptive immunity**. These defenses act after various physical barriers block pathogens. **Figure 17.4** summarizes the basic components of the immune system, discussed in the following sections.

Physical Barriers and Innate Immunity

Several familiar structures and fluids keep pathogens from entering the body in the innate immune response: unbroken skin, mucous membranes such as the lining inside the mouth, earwax, and waving cilia that push debris and pathogens up and out of the respiratory tract. Most microbes that reach the stomach perish in a vat of churning acid or are flushed out in diarrhea. These physical barriers are nonspecific. That is, they keep out anything foreign, not just particular pathogens.

If a pathogen breaches these physical barriers, innate immunity provides a rapid, broad defense. The term *innate* refers to the fact that these general defenses are in the body, ready to function should infection threaten. A central part of the innate immune response is **inflammation,** a process that creates a hostile environment for certain types of pathogens at an injury site. Inflammation sends in cells that engulf and destroy pathogens. Such cells are called phagocytes, and their engulfing action is phagocytosis (**figure 17.5**). Certain types of white blood cells and the large, wandering macrophages are phagocytes. Also at the infection site, plasma accumulates, which dilutes toxins and brings in antimicrobial chemicals. Increased blood flow with inflammation warms the area, turning it swollen and red.

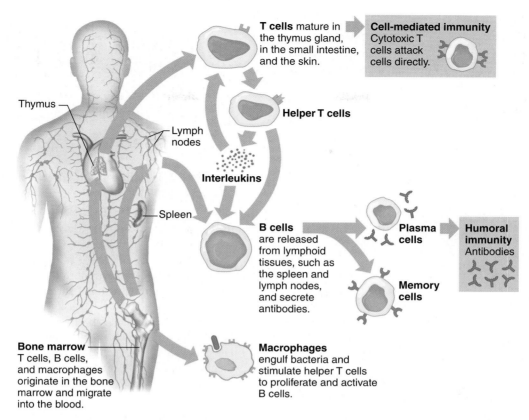

Figure 17.3 The immune system produces diverse cells. T cells, B cells, and macrophages build an overall immune response. All three types of cells originate in the bone marrow and circulate in the blood.

In addition to inflammation, three classes of proteins participate in innate immunity. These are the complement system, collectins, and cytokines. Mutations in the genes that encode these proteins increase susceptibility to infection.

The **complement system** consists of plasma proteins that assist, or complement, several other defenses. Some complement proteins puncture bacterial plasma membranes, bursting the cells. Others dismantle viruses or trigger release of histamine from mast cells, another type of immune system cell that is involved in allergies. Histamine dilates blood vessels, sending fluid to the infected or injured area. Still other complement proteins attract phagocytes to an injury site.

Collectins broadly protect against bacteria, yeasts, and some viruses by detecting slight differences in their surfaces from human cells. Groups of human collectins correspond to the surfaces of different pathogens, such as the distinctive sugars on yeast, the linked sugars and lipids of certain bacteria, and the surface features of some viruses.

Cytokines play roles in both innate and adaptive immunity. As part of the innate immune response, cytokines called **interferons** alert other components of the immune system to the presence of cells infected with viruses. These cells are then destroyed, which limits the spread of infection.

Interleukins are cytokines that cause fever, temporarily triggering a higher body temperature that directly kills some infecting bacteria and viruses. Fever also counters microbial growth indirectly, because higher body temperature reduces the iron level in the blood. Bacteria and fungi require more iron as the body temperature rises; a fever-ridden body stops their growth. Phagocytes also attack more vigorously when the temperature rises. Tumor necrosis factor is another type of cytokine that activates other protective biochemicals, destroys certain bacterial toxins, and attacks cancer cells. Many of the aches and pains we experience from an infection are actually due to the immune response, not directly to the actions of the pathogens.

Adaptive Immunity

Adaptive immunity must be stimulated into action. It may take days to respond, compared to minutes for innate immunity. Adaptive immunity is highly specific and directed.

B cells and T cells carry out adaptive immunity. In the **humoral immune response,** B cells produce antibodies in response to activation by T cells. ("Humor" means fluid; antibodies are carried in fluids.) In the **cellular immune response,** T cells produce cytokines and activate other cells. B and T cells differentiate in the bone marrow and migrate to the lymph nodes, spleen, and thymus gland, as well as circulate in the blood and tissue fluid.

The adaptive arm of the immune system has three basic characteristics. It is *diverse,* vanquishing many types of pathogens. It is *specific,* distinguishing the cells and molecules that

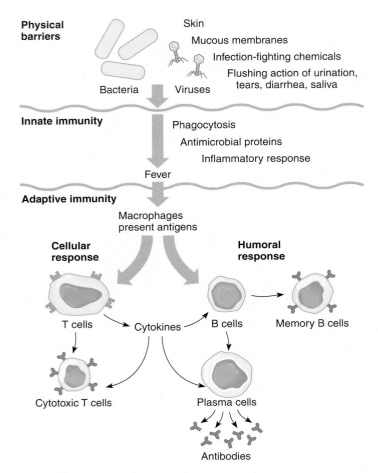

Physical barriers

Skin
Mucous membranes
Infection-fighting chemicals
Flushing action of urination, tears, diarrhea, saliva

Bacteria   Viruses

Innate immunity

Phagocytosis
Antimicrobial proteins
Inflammatory response
Fever

Adaptive immunity

Macrophages present antigens

Cellular response   **Humoral response**

T cells → Cytokines → B cells → Memory B cells

Cytotoxic T cells   Plasma cells

Antibodies

Figure 17.4 Levels of immune protection. Disease-causing organisms and viruses (pathogens) first must breach physical barriers, then nonspecific cells and molecules attack in the innate immune response. If this is ineffective, the adaptive immune response begins: Antigen-presenting cells stimulate T cells to produce cytokines, which activate B cells to divide and differentiate into plasma cells, which secrete antibodies. Once activated, these specific cells "remember" the pathogen, allowing faster responses to subsequent encounters.

cause disease from those that are harmless. The immune system also *remembers,* responding faster to a subsequent encounter with a foreign antigen than it did the first time. The first assault initiates a **primary immune response**. The second assault, based on the system's "memory," is a **secondary immune response**. This is why we get some infections, such as chickenpox, only once. However, upper respiratory infections and influenza recur because the causative viruses mutate, presenting a different face to our immune systems each season.

The Humoral Immune Response—B Cells and Antibodies

An antibody response begins when an antigen-presenting macrophage activates a T cell. This cell in turn contacts a B cell that has surface receptors that can bind the type of foreign antigen the macrophage presents. The immune system has so many B cells, each with different combinations of surface antigens,

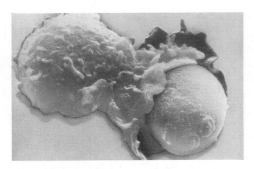

Figure 17.5 Nature's garbage collectors. A human phagocyte engulfs a yeast cell.

that there is almost always one or more available that corresponds to a particular foreign antigen. Turnover of these cells is high. Each day, millions of B cells perish in the lymph nodes and spleen, while millions more form in the bone marrow, each with a unique combination of surface molecules.

Once the activated T cell finds a B cell match, it releases cytokines that stimulate the B cell to divide. Soon the B cell gives rise to two types of cells (**figure 17.6**). The first, **plasma cells,** are antibody factories, each secreting 1,000 to 2,000 identical antibodies per second into the bloodstream. They live

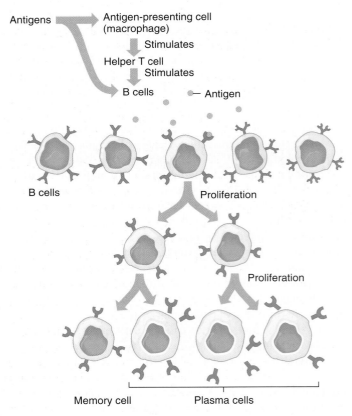

Antigens → Antigen-presenting cell (macrophage)
Stimulates
Helper T cell
Stimulates
B cells   — Antigen

B cells   Proliferation

Proliferation

Memory cell   Plasma cells

Figure 17.6 Production of antibodies. In the humoral immune response, B cells proliferate and mature into antibody-secreting plasma cells. Note that only the B cell that binds the antigen proliferates; its descendants may develop into memory cells or plasma cells. Plasma cells greatly outnumber memory cells.

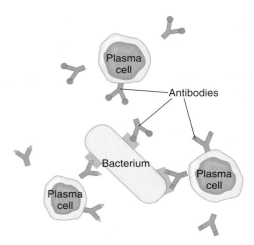

Figure 17.7 An immune response recognizes many targets. A humoral immune response is polyclonal, which means that different plasma cells produce antibody proteins that recognize and bind to different features of a foreign cell's surface.

only days. These cells provide the primary immune response. Plasma cells derived from different B cells secrete different antibodies. Each type of antibody corresponds to a specific part of the pathogen, like hitting a person in different parts of the body. This multi-pronged attack is called a polyclonal antibody response (**figure 17.7**). The second type of B cell descendant, **memory cells,** are far fewer and usually dormant. They respond to the foreign antigen faster and with more force should it appear again. This is a secondary immune response. Memory B cells are what enabled survivors of the 1918 flu pandemic to resist infection.

An antibody molecule is built of several polypeptides and is therefore encoded by several genes. The simplest type of antibody molecule is four polypeptide chains connected by disulfide (sulfur-sulfur) bonds, forming a shape like the letter Y (**figure 17.8**). A large antibody molecule might consist of three, four, or five such Ys joined.

In a Y-shaped antibody subunit, the two longer polypeptides are called **heavy chains,** and the other two **light chains**. The lower portion of each chain is an amino acid sequence that is very similar in all antibody molecules, even in different species. These areas are called constant regions, and they provide the activity of the antibody. The amino acid sequences of the upper portions of each polypeptide chain, the variable regions, can differ greatly among antibodies. These parts provide the specificities of particular antibodies to particular antigens.

Antibodies can bind certain antigens because of the three-dimensional shapes of the tips of the variable regions. These specialized ends are **antigen binding sites,** and the parts that actually contact the antigen are called **idiotypes**. The parts of the antigens that idiotypes bind are **epitopes**. An antibody contorts to form a pocket around the antigen.

Antibodies have several functions. Antibody-antigen binding may inactivate a pathogen or neutralize the toxin it produces. Antibodies can clump pathogens, making them more visible to macrophages, which then destroy them. Antibodies also activate complement, extending the innate immune response. In some situations, the antibody response can be harmful.

Antibodies are of five major types, distinguished by where they act and what they do (**table 17.2**). (Antibodies are also called immunoglobulins, abbreviated *Ig.*) Different antibody types predominate in different stages of an infection.

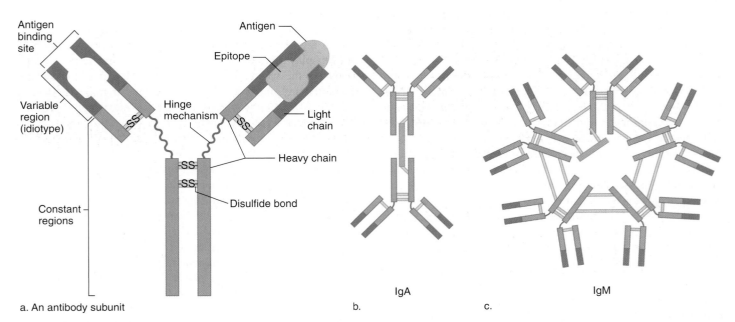

Figure 17.8 Antibody structure. The simplest antibody molecule **(a)** consists of four polypeptide chains, two heavy and two light, joined by pairs of sulfur atoms that form disulfide bonds. Part of each polypeptide chain has a constant sequence of amino acids, and the remainder varies. The tops of the Y-shaped molecules form antigen binding sites. **(b)** IgA consists of two Y-shaped subunits, and IgM **(c)** consists of five subunits.

Table 17.2	Types of Antibodies	
Type*	Location	Functions
IgA	Milk, saliva, urine, and tears; respiratory and digestive secretions	Protects against pathogens at points of entry into body
IgD	On B cells in blood	Stimulates B cells to make other types of antibodies, particularly in infants
IgE	In secretions with IgA and in mast cells in tissues	Acts as receptor for antigens that cause mast cells to secrete allergy mediators
IgG	Blood plasma and tissue fluid; passes to fetus	Protects against bacteria, viruses, and toxins, especially in secondary immune response
IgM	Blood plasma	Fights bacteria in primary immune response; includes anti-A and anti-B antibodies of ABO blood groups

*The letters A, D, E, G, and M refer to the specific conformation of heavy chains characteristic of each class of antibody.

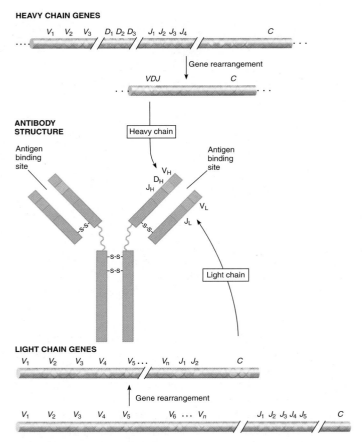

Figure 17.9 Antibody diversity. The human immune system can produce antibodies to millions of possible antigens because each polypeptide is encoded by more than one gene. That is, the many components of antibodies can combine in many ways.

The human body can manufacture seemingly limitless varieties of antibodies, though the genome has a limited number of antibody genes. This great diversity is possible because parts of different antibody genes combine. During the early development of B cells, sections of their antibody genes move to other chromosomal locations, creating new genetic instructions for antibodies.

The assembly of antibody molecules is like putting together many different outfits from the contents of a closet containing 200 pairs of pants, a drawer containing fifteen different shirts, and four belts. Specifically, each variable region of a heavy chain and a light chain consists of three sections, called V (for variable), D (for diversity), and J (for joining). The *V, D,* and *J* genes—several of each—for the heavy chains are on chromosome 14, and the corresponding genes for the light chains are on chromosomes 2 and 22. *C* (constant) genes encode the constant regions of each heavy and light chain. A promoter sequence precedes the *V* genes and an enhancer sequence precedes the *C* genes. These control sequences oversee the mixing and matching of the *V, D,* and J genes. **Figure 17.9** shows how the genetic instructions for the antibody parts are combined in different ways to encode the heavy and light polypeptide chains.

Enzymes cut and paste the pieces of antibody gene parts. The number of combinations is so great that virtually any antigen that a person with a healthy immune system might encounter will elicit an immune response.

The Cellular Immune Response—T Cells and Cytokines

T cells provide the cellular immune response. It is called "cellular" because the T cells themselves travel to where they act, unlike B cells, which secrete antibodies into the bloodstream. T cells descend from stem cells in the bone marrow, then travel to the thymus gland ("T" refers to thymus). As the immature T cells, called thymocytes, migrate toward the interior of the thymus, they display diverse cell surface receptors. Then selection happens. As the wandering thymocytes touch lining cells in the gland that are studded with "self" antigens, thymocytes that do not attack the lining cells begin maturing into T cells, whereas those that harm the lining cells die by apoptosis—in great numbers. Gradually, T cells-to-be that recognize self persist, while those that harm body cells are destroyed.

Several types of T cells are distinguished by the types and patterns of receptors on their surfaces, and by their functions. Helper T cells have many functions: They recognize foreign antigens on macrophages, stimulate B cells to produce antibodies, secrete cytokines, and activate another type of T cell called a cytotoxic T cell (also called a killer T cell). Certain T cells may help to suppress an immune response when it is no longer required. The cytokines that helper T cells secrete include interleukins, interferons, tumor necrosis factor, and colony-stimulating factors, which stimulate white blood cells in bone marrow to mature (**table 17.3**). Cytokines interact with and signal each other, sometimes in complex cascades.

Distinctive surfaces distinguish subsets of helper T cells. Certain antigens called cluster-of-differentiation antigens, or

Table 17.3	Types of Cytokines
Cytokine	**Function**
Colony-stimulating factors	Stimulate bone marrow to produce lymphocytes
Interferons	Block viral replication, stimulate macrophages to engulf viruses, stimulate B cells to produce antibodies, attack cancer cells
Interleukins	Control lymphocyte differentiation and growth, cause fever that accompanies bacterial infection
Tumor necrosis factor	Stops tumor growth, releases growth factors, stimulates lymphocyte differentiation, dismantles bacterial toxins

CD antigens, enable T cells to recognize foreign antigens displayed on macrophages. One such cell type, called a CD4 helper T cell, is an early target of HIV. Considering the critical role helper T cells play in coordinating immunity, it is little wonder that HIV infection ultimately topples the entire system, a point we will return to soon.

Cytotoxic T cells lack CD4 receptors but have CD8 receptors. These cells attack virally infected and cancerous cells by attaching to them and releasing chemicals. They do this by linking two surface peptides to form structures called T cell receptors that bind foreign antigens. When a cytotoxic T cell encounters a nonself cell—a cancer cell, for example—the T cell receptors draw the two cells into physical contact. The T cell then releases a protein called perforin, which pierces the cancer cell's plasma membrane, killing it (**figure 17.10**). Cytotoxic T cell receptors also attract cells that are covered with certain viruses, destroying the cells before the viruses on them can enter, replicate, and spread the infection. Cytotoxic T cells continually monitor body cells, recognizing and eliminating virally infected cells and tumor cells.

Table 17.4 summarizes types of immune system cells.

> ## Key Concepts
>
> 1. The immune system consists of physical barriers; an innate immune response of inflammation, phagocytosis, complement, collectins, and cytokines; and an adaptive immune response that is diverse, specific, and remembers.
>
> 2. In the humoral immune response, stimulated B cells divide and differentiate into plasma cells and memory cells. A plasma cell secretes abundant antibodies of a single type. Antibodies are Y-shaped polypeptides, each with two light and two heavy chains, each with a constant and a variable region. The tips of the Y form an antigen binding site with a specific idiotype. Antibodies make foreign antigens more visible to macrophages and stimulate complement. Shuffling gene pieces generates antibody diversity.
>
> 3. In the cellular immune response, helper T cells stimulate B cells to manufacture antibodies and cytotoxic T cells to secrete cytokines. Using T cell receptors, cytotoxic T cells bind to nonself cells and virus-covered cells and burst them.

17.3 Abnormal Immunity

The immune system continually adapts to environmental change. Because the immune response is so diverse, its breakdown affects health in many ways. Immune system malfunction may be inherited or acquired, and immunity may be too weak, too strong, or misdirected. Abnormal immune responses may be multifactorial, with several genes contributing to susceptibility to infection, or caused by mutation in a single gene. Or, susceptibility to an immune disorder may reflect abnormal gene expression.

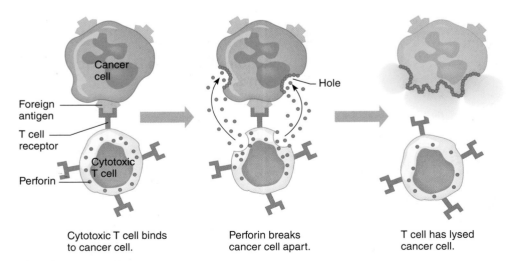

Cytotoxic T cell binds to cancer cell.
Perforin breaks cancer cell apart.
T cell has lysed cancer cell.

Figure 17.10 **Death of a cancer cell.** A cytotoxic T cell binds to a cancer cell and injects perforin, a protein that pierces (lyses) the cancer cell's plasma membrane. The cancer cell dies, leaving debris that macrophages clear away.

Inherited Immune Deficiencies

The more than twenty types of inherited immune deficiencies affect innate and/or adaptive immunity (**table 17.5**). These conditions can arise in several ways.

In chronic granulomatous disease, neutrophils can engulf bacteria, but, due to deficiency of an enzyme called an oxidase, they cannot produce the activated oxygen compounds that kill bacteria. Because this enzyme is made of four polypeptide chains, four genes encode it, and there are four ways to inherit the disease, all X-linked. A very rare autosomal recessive form is caused by a defect in the part of the host

Table 17.4 — Types of Immune System Cells

Cell Type	Function
Macrophage	Presents antigens
	Performs phagocytosis
Dendritic cell	Presents antigens
Mast cell	Releases histamine in inflammation
	Releases allergy mediators
B cell	Matures into antibody-producing plasma cell or into memory cell
T cells	
Helper	Recognizes nonself antigens presented on macrophages
	Stimulates B cells to produce antibodies
	Secretes cytokines
	Activates cytotoxic T cells
Cytotoxic	Attacks cancer cells and cells infected with viruses upon recognizing antigens
Natural killer	Attacks cancer cells and cells infected with viruses without recognizing antigens; activates other white blood cells
Suppressor	Inhibits antibody production

Table 17.5 — Inherited Immune Deficiencies

Disease	MIM	Inheritance*	Defect
Chronic granulomatous disease	306400	ar, AD, XIr	Abnormal phagocytes can't kill engulfed bacteria
Immune defect due to absence of thymus	242700	ar	No thymus, no T cells
Neutrophil immuno-deficiency syndrome	608203	ar	Deficiencies of T cells, B cells, and neutrophils
SCID			
Adenosine deaminase deficiency	102700	ar	No T or B cells
Adenosine deaminase deficiency with sensitivity to ionizing radiation	602450	ar	No T, B, or natural killer cells
IL-2 receptor mutation	300400	XIr	No T, B, or natural killer cells
X-linked lymphoproliferative disease	308240	XIr	Absence of protein that enables T cells to bind B cells
X1	300400	XIr	Abnormal interleukin-2

*ar = autosomal recessive   SCID = severe combined immune deficiency
AD = autosomal dominant
XIr = X-linked recessive

cell that encloses bacteria. Antibiotics and gamma interferon are used to prevent bacterial infections in these patients, and the disease can be cured with a bone marrow or an umbilical cord stem cell transplant.

Mutations in genes that encode cytokines or T cell receptors impair cellular immunity, which primarily targets viruses and cancer cells. Because T cells activate the B cells that manufacture antibodies, abnormal cellular immunity (T cell function) disrupts humoral immunity (B cell function). Mutations in the genes that encode antibody segments, that control how the segments join, or that direct maturation of B cells mostly impair immunity against bacterial infection. Inherited immune deficiency can also result from defective B cells, which usually increases vulnerability to certain bacterial infections.

Severe combined immune deficiencies (SCIDs) affect both humoral and cellular immunity. About half of SCID cases are X-linked. In a less severe form, the individual lacks B cells but has some T cells. Before antibiotic drugs became available, children with this form of SCID died before age 10 of overwhelming bacterial infection. In a more severe form of X-linked SCID, lack of B and T cells causes death by 18 months of age, usually of severe and diverse infections.

A young man named David Vetter taught the world about the difficulty of life without immunity years before AIDS arrived. David had an X-linked recessive form of SCID, called SCID-X1, that caused him to be born without a thymus gland. His T cells could not mature and activate B cells, leaving him defenseless in a germ-filled world. Born in Texas in 1971, David spent his short life in a vinyl bubble, awaiting a treatment that never came (**figure 17.11**). As he reached adolescence, David wanted to leave his bubble. He did, and received a bone marrow transplant from his sister. Sadly, her bone marrow contained Epstein-Barr virus. Her healthy immune system could handle it and she had no symptoms, but the virus caused lymphoma, a cancer of the immune system, in David. He died in just a few weeks. Today experimental gene therapy can treat SCID-X1, discussed in chapter 20.

Acquired Immune Deficiency Syndrome

AIDS is not inherited, but acquired by infection with HIV, a virus that gradually shuts down the immune system. The effect of HIV on a human body is especially astounding because the virus is so simple. Its genome is a millionth the size of ours, and its nine genes, consisting of about 9,000 RNA bases, encode only fifteen proteins! But HIV commandeers more than 200 human proteins as it invades the immune system.

HIV infection begins as the virus enters macrophages, impairing this first line of defense. In these cells, and later in helper T cells, the virus adheres with its surface protein, called gp120,

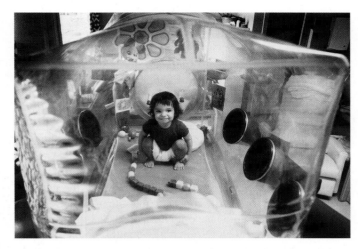

Figure 17.11 **David Vetter, the original "bubble boy," had severe combined immune deficiency X1.** Because his T cells could not mature, he was virtually defenseless against infection. Gene therapy can now treat this disease.

to two coreceptors on the host cell surface, CD4 and CCR5 (**figure 17.12**). Another glycoprotein, gp41, anchors gp120 molecules into the viral envelope. When the virus binds both coreceptors, virus and cell surface contort in a way that enables viruses to enter the cell. Once in the cell, reverse transcriptase copies the viral RNA into DNA, which replicates to form a DNA double helix. This enters the nucleus and inserts into a chromosome. The viral DNA sequences are transcribed and translated, and the cell fills with viral pieces, which are assembled into complete new viral particles that eventually bud from the cell (**figure 17.13**).

Once helper T cells start to die at a high rate, bacterial infections begin, because B cells aren't activated to produce antibodies. Much later in infection, HIV variants arise that can bind to a receptor called CXCR4 on cytotoxic T cells, killing them. Loss of these cells renders the body very vulnerable to viral infections and cancer.

HIV replicates quickly, changes quickly, and can hide. The virus mutates easily because it cannot repair replication errors and errors happen frequently—1 per every 5,000 or so bases—because of the "sloppiness" of reverse transcriptase in copying viral RNA into DNA. The immune system cannot keep up; antibodies against one viral variant are useless against the next. For several years, the bone marrow produces 2 billion new T and B cells a day. A million to a billion new HIV particles bud daily from infected cells.

So genetically diverse is the population of HIV in a human host that, within days of the initial infection, variants arise that resist the drugs used to treat AIDS (see figure 15.14). HIV's changeable nature is why combining drugs with different actions is the most effective way to slow the disease into a chronic, lifelong, but treatable illness, instead of a killer (**table 17.6**). Several classes of drugs have cut the death rate from AIDS dramatically. They work at different points of infection: blocking binding or entry of the virus into T cells, replicating viral genetic material, and processing viral proteins.

Clues to developing new drugs to treat HIV infection come from people at high risk who resist infection. Researchers identified variants of four receptors or the molecules that bind to them that block HIV from entering cells by looking at the DNA of people who had unprotected sex with many partners but who never became infected. Some of them were homozygous recessive for a 32-base deletion in the *CCR5* gene. This is the mutation that cured the HIV infection of Timothy Brown described in the chapter opener. The CCR5 coreceptors are too stunted to reach the cell's surface, so HIV has nowhere to dock. Heterozygotes, with one copy of the deletion, can become infected, but they remain healthy for several years longer than people who do not have the deletion. Curiously,

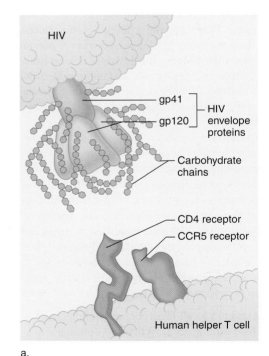

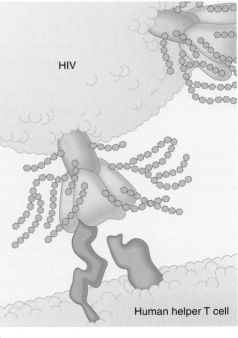

a.　　　　　　　　　　　　　　　　b.

Figure 17.12 **HIV binds to a helper T cell.** **(a)** The part of HIV that binds to helper T cells is called gp120 (gp stands for glycoprotein). **(b)** The carbohydrate chains that shield the protein part of gp120 move aside as they approach the cell surface, and the viral molecule can now bind to a CD4 receptor. Binding to the CCR5 receptor is also necessary. Then the viral envelope fuses with the plasma membrane and the virus enters. (The size of HIV is greatly exaggerated.) The man described in the chapter opener received stem cells from a donor whose cells lack CCR5, and was thus cured of both his leukemia and HIV infection.

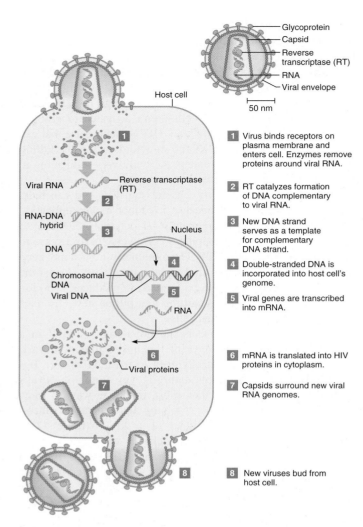

Figure 17.13 **How HIV infects.** HIV integrates into the host chromosome, then commandeers transcription and translation, ultimately producing more virus particles.

1	Virus binds receptors on plasma membrane and enters cell. Enzymes remove proteins around viral RNA.
2	RT catalyzes formation of DNA complementary to viral RNA.
3	New DNA strand serves as a template for complementary DNA strand.
4	Double-stranded DNA is incorporated into host cell's genome.
5	Viral genes are transcribed into mRNA.
6	mRNA is translated into HIV proteins in cytoplasm.
7	Capsids surround new viral RNA genomes.
8	New viruses bud from host cell.

the same mutation may have enabled people to survive various plagues in Europe during the Middle Ages. Apparently, more than one pathogen uses this entryway into human cells.

Autoimmunity

In **autoimmunity,** the immune system produces antibodies that attack the body's own tissues. These antibodies are called **autoantibodies**. About 5 percent of the population has an

Table 17.6	Anti-HIV Drugs
Drug Type	**Mechanism**
Reverse transcriptase inhibitor	Blocks copying of viral RNA into DNA
Protease inhibitor	Blocks shortening of certain viral proteins
Fusion inhibitor	Blocks ability of HIV to bind a cell
Entry inhibitor	Blocks ability of HIV to enter a cell

autoimmune disorder. The signs and symptoms resulting from autoimmune disorders reflect the cell types under attack (**table 17.7**).

Most autoimmune disorders are not inherited as single-gene diseases. However, the fact that different autoimmune disorders affect members of the same family, and may respond to the same drugs, suggests that these conditions stem from shared susceptibilities. For example, in autoimmune polyendocrinopathy syndrome type I (MIM 240300), caused by a mutation in a single gene on chromosome 21, autoantibodies attack endocrine glands in a sequence, so that different members of a family may have very different symptoms. However, the genetics of autoimmunity is usually more complex than this. Genome-wide association studies have identified dozens of genes that are each associated with more than one autoimmune disorder. On the other hand, an autoimmune disorder may result from the actions of variants of several genes that each contributes to susceptibility, perhaps in the presence of a specific environmental trigger such as diet. This is the case for the digestive disorder Crohn's disease, which has been associated with thirty-two genome regions.

Some of the more common autoimmune disorders may actually arise in several ways when parts of the immune response are overactive. This is the case for systemic lupus erythematosus, better known as "lupus." The name "lupus" comes from the characteristic butterfly-shaped rash on the cheeks, but the "systemic" is actually a more important description because the disease produces autoantibodies that affect the connective tissue of many organs. These are the kidneys, joints, lungs, brain, spinal cord, and the heart and blood vessels. A person may need dialysis when the kidneys are involved, blood pressure medication to counter increasing pressure in the lungs, and drugs to minimize buildup of fatty deposits on artery walls. Lupus can also cause strokes, memory loss, fever, seizures, headache, and psychosis.

Lupus involves several aspects of the immune response, including cell surface characteristics, secretion of interferons, production of autoantibodies, activation of B and T cells, antigen presentation, adhesion of immune system cells to blood vessel linings, inflammation, removal of complexes of immune cells and foreign antigens, and cytokine production. Therefore, it isn't surprising that variants of at least ten different genes can predispose a person to developing this condition. Perhaps inheriting susceptibility in three or four of the genes causes lupus.

How does the immune system turn against itself? Autoimmunity may arise in several ways:

- A virus replicating in a cell incorporates proteins from the cell's surface onto its own. When the immune system "learns" the surface of the virus to destroy it, it also learns to attack human cells that normally bear the protein.
- Some cells that should have died in the thymus somehow escape the massive die-off, persisting to attack "self" tissue later on.
- A nonself antigen coincidentally resembles a self antigen, and the immune system attacks both. In rheumatic fever, for example, antigens on heart valve cells resemble those

Table 17.7 — Autoimmune Disorders

Disorder	Symptoms	Autoantibodies Against
Diabetes mellitus (type 1)	Thirst, hunger, weakness, weight loss	Pancreatic beta cells
Graves disease	Restlessness, weight loss, irritability, increased heart rate and blood pressure	Thyroid gland cells
Hemolytic anemia	Fatigue, weakness	Red blood cells
Multiple sclerosis	Weakness, poor coordination, failing vision, disturbed speech	Myelin in the white matter of the central nervous system
Myasthenia gravis	Muscle weakness	Neurotransmitter receptors on skeletal muscle cells
Rheumatic fever	Weakness, shortness of breath	Heart valve cells
Rheumatoid arthritis	Joint pain and deformity	Cells lining joints
Systemic lupus erythematosus	Red facial rash, fever, weakness, joint pain	Connective tissue
Ulcerative colitis	Lower abdominal pain	Colon cells

on *Streptococcus* bacteria; antibodies produced to fight a strep throat also attack the heart valve cells.

- If X inactivation is skewed, a female may have a few cells that express the X chromosome genes of one parent. The immune system may respond to these cells as foreign if they have surface antigens that are not also on the majority of cells. Skewed X inactivation may explain why some autoimmune disorders are much more common in females.

Reading 17.2 highlights a special situation in which two immune systems must coexist—pregnancy.

Allergies

An allergy is an immune system response to a substance, called an allergen, that does not actually present a threat. Many allergens are particles small enough to be carried in the air and enter a person's respiratory tract. The size of the allergen may determine the type of allergy. For example, grass pollen is large and remains in the upper respiratory tract, where it causes hayfever. But allergens from house dust mites, cat dander, and cockroaches are small enough to infiltrate the lungs, triggering asthma. Asthma is a chronic disease in which contractions of the airways, inflammation, and accumulation of mucus block air flow.

Both humoral and cellular immunity take part in an allergic response (**figure 17.14**). Antibodies of class IgE bind to mast cells, sending signals that open the mast cells, which releases allergy mediators such as histamine and heparin. Allergy mediators cause inflammation, with symptoms that may include

runny eyes from hay fever, narrowed airways from asthma, rashes, or the overwhelming body-wide allergic reaction called anaphylactic shock. Allergens also activate a class of helper T cells that produce cytokines, whose genes are clustered on chromosome 5q. Regions of chromosomes 12q and 17q have genes that control IgE production.

The misdirected immune response of an allergy may actually be due to a mutation. Too little of a skin protein called filaggrin can indirectly cause the allergic conditions eczema, asthma, and hay fever (**figure 17.15**).

Filaggrin is a gigantic protein that binds to the keratin proteins that form most of the outermost skin layer, the epidermis. Filaggrin normally breaks down, releasing amino acids

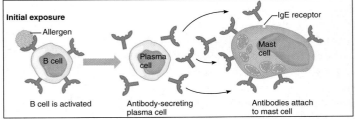

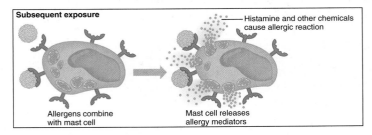

Figure 17.14 Allergy. In an allergic reaction, an allergen such as pollen activates B cells, which divide and give rise to antibody-secreting plasma cells. The antibodies attach to mast cells. When the person encounters allergens again, the allergens combine with the antibodies on the mast cells, which then burst, releasing the chemicals that cause itchy eyes and a runny nose.

A Special Immunological Relationship: Mother-to-Be and Fetus

The immune system recognizes "self" cell surfaces and protects the body from foreign, "nonself" cells and molecules. This is very helpful when the nonself triggers are parts of infecting bacteria, but what tempers the immune system of a pregnant woman to accept cells from her fetus? Half of a fetal genome comes from the father, and so fetal cell surfaces likely include some antigens from him that would be "foreign" to the mother-to-be. Similarly, some of her antigens might be foreign to the fetus. Yet pregnant woman and fetus routinely swap cells. We do not understand how the immune system evolved tolerance between pregnant woman and fetus, but following are three examples of the immunological "crosstalk" between the two.

"T Regs"

Samples of lymph nodes from fetuses indicate that up to 1 percent of the cells are maternal. The woman's cells stimulate the fetal immune system to produce "regulatory T cells," called "T Regs," which dampen the fetal immune response. The maternal immune system similarly produces T Regs that inhibit response to fetal cells. In one experiment, fetal lymph node samples did not react against cells from the mother unless the regulatory T cells were removed. Children retain these cells for several years. It may be possible to stimulate production of T Regs later in life to help a recipient's body accept an organ transplant.

Scleroderma

People who have scleroderma describe the condition as "the body turning to stone." The skin hardens into an armorlike texture (**figure 1**). Scleroderma usually begins in middle age, and affects mostly women. It was long thought to be autoimmune, but discovery of Y chromosomes in skin cells from scleroderma patients who are mothers of sons revealed a very different source of the illness—lingering cells from a fetus. Cells from female fetuses can presumably have the same effect but cannot be distinguished from the mother on the basis of a sex chromosome check.

The degree of genetic difference between a mother and a son may play a role in development of scleroderma. Mothers who have the condition tend to have cell surfaces that are more similar to those of their sons than mothers who do not have scleroderma. Perhaps the similarity of cell surfaces enabled the fetal cells to escape destruction by the mother's immune system.

Rh Incompatibility

"Rh," the rhesus factor discovered in rhesus monkeys, is a blood group (MIM 111700). A person is Rh$^+$ if red blood cells have a surface molecule called the RhD antigen. Rh type is important when an Rh$^+$ man and

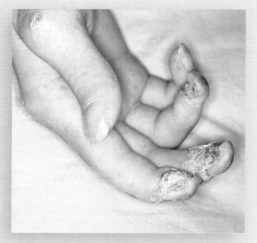

Figure 1 An autoimmune disorder—maybe. Scleroderma hardens the skin. Some cases appear to be caused by a long-delayed immune response to cells retained from a fetus decades earlier.

an Rh$^-$ woman conceive a child who is Rh$^+$ (**figure 2**). The woman's immune system manufactures antibodies against the few fetal cells that enter her bloodstream. Not enough antibodies form to harm the fetus that sets off the reaction, but the number of antibodies continues to increase. If she carries another Rh$^+$ fetus, the antibodies can attack the fetal blood supply, causing potentially fatal hemolytic disease of the fetus and newborn. It can be treated at birth with a transfusion of Rh$^-$ blood.

Fortunately, natural and medical protections make this complication rare today. If a woman's ABO blood type is O and the fetus is A or B, her anti-A or anti-B antibodies attack the fetal cells in her circulation before her immune system produces anti-Rh antibodies. Also, if a pregnant woman alerts her physician to a potential incompatibility, she can be given RhoGAM, which is antibody against the Rh antigen. It shields fetal cells so her system does not manufacture the harmful antibodies. When she becomes pregnant again, fetal DNA in her circulation can be tested to see if it is Rh$^-$ or Rh$^+$. If the second fetus is Rh$^-$, she does not need RhoGAM.

A first Rh$^+$ fetus developing in an Rh$^-$ mother can be affected if her blood has been exposed to Rh$^+$ cells in amniocentesis, a blood transfusion, an ectopic (tubal) pregnancy, a miscarriage, or an abortion.

(Continued)

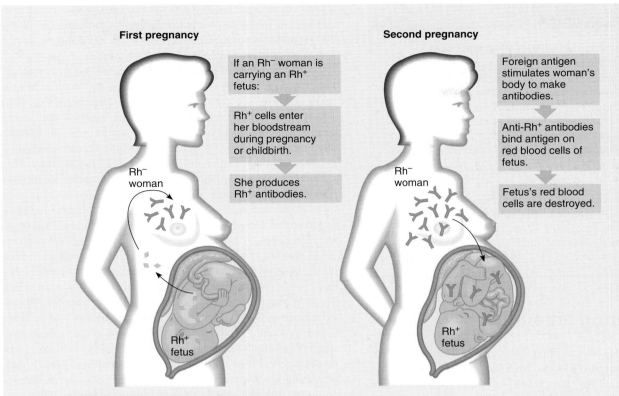

First pregnancy

If an Rh⁻ woman is carrying an Rh⁺ fetus:

Rh⁺ cells enter her bloodstream during pregnancy or childbirth.

She produces Rh⁺ antibodies.

Rh⁻ woman

Rh⁺ fetus

Second pregnancy

Foreign antigen stimulates woman's body to make antibodies.

Anti-Rh⁺ antibodies bind antigen on red blood cells of fetus.

Fetus's red blood cells are destroyed.

Rh⁻ woman

Rh⁺ fetus

Figure 2 Rh incompatibility. Fetal cells entering the pregnant woman's bloodstream can stimulate her immune system to make anti-Rh antibodies, if the fetus is Rh⁺ and she is Rh⁻. A drug called RhoGAM prevents attacks on subsequent fetuses.

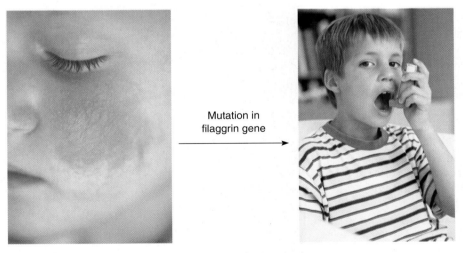

Mutation in filaggrin gene

Figure 17.15 One cause of asthma. Asthma due to skin barrier deficiency can result from the dry, cracked skin of eczema in infancy, admitting allergens. A mutation in the gene encoding the skin protein filaggrin causes the eczema.

People with the rare inherited disease ichthyosis vulgaris (MIM 146700) have two mutations in the filaggrin gene, and experience severe skin flaking. Many more individuals—possibly one in ten of us—has a mutation that causes ichthyosis that is so mild that we just treat it with skin lotion. When researchers realized that people with either form of ichthyosis very often also have the itchy red inflamed skin of eczema, they realized there could be a connection: When the epidermis is cracked, allergens enter and reach deeper skin layers, where they encounter dendritic cells, the sentries of the immune system. Once activated, these cells signal inflammation. Eczema results. The dendritic cells also activate immune memory, so that when years later the person inhales the same allergens that once crossed the broken skin, an immune response ensues in the airways, causing a form of asthma due to a skin barrier deficiency. Perhaps treating breaks in the skin early in life can prevent eczema, asthma, and other allergies.

that rise to the top of the skin and keep it moisturized. The epidermis forms a barrier that keeps out irritants, pathogens, and allergens.

17.4 Altering Immunity

Medical technologies can alter or augment immune system functions. Vaccines trick the immune system into acting early. Antibiotic drugs, which are substances derived from organisms such as fungi and soil bacteria, have been used for decades to assist an immune response. Cytokines and altered antibodies are used as drugs. Transplants require suppression of the immune system so that the body will accept a nonself replacement part.

Vaccines

A **vaccine** is an inactive or partial form of a pathogen that stimulates the immune system to alert B cells to produce antibodies. When the person then encounters the natural pathogen, a secondary immune response ensues, even before symptoms arise. Vaccines consisting of entire viruses or bacteria can, rarely, cause illness if they mutate to a pathogenic form. This was a risk of the smallpox vaccine. A safer vaccine uses only the part of the pathogen's surface that elicits an immune response. Vaccines against different illnesses can be combined into one injection, or the genes encoding antigens from several pathogens can be inserted into a harmless virus and delivered as a "super vaccine."

Vaccine technology dates back to the eleventh century in China. Because people saw that those who recovered from smallpox never got it again, they crushed scabs from pox into a powder that they inhaled or rubbed into pricked skin. In 1796, the wife of a British ambassador to Turkey witnessed the Chinese method of vaccination and mentioned it to English country physician Edward Jenner. Intrigued, Jenner was vaccinated the Chinese way, and then thought of a different approach.

It was widely known that people who milked cows contracted a mild illness called cowpox, but did not get smallpox. The cows became ill from infected horses. Since the virus seemed to jump from one species to another, Jenner wondered whether exposing a healthy person to cowpox lesions might protect against smallpox. A slightly different virus causes cowpox, but Jenner's approach worked, leading to development of the first vaccine (the word comes from the Latin *vacca,* for "cow").

Jenner tried his first vaccine on a volunteer, 8-year-old James Phipps. Jenner dipped a needle in pus oozing from a small cowpox sore on a milkmaid named Sarah Nelmes, then scratched the boy's arm with it. He then exposed the boy to people with smallpox. Young James never became ill. Eventually, improved versions of Jenner's smallpox vaccine eradicated a disease that once killed millions (**figure 17.16**). By the 1970s, vaccination became unnecessary. However, several nations have resumed smallpox vaccination, as section 17.5 discusses.

Most vaccines are injections. New delivery methods include nasal sprays (flu vaccine) and genetically modified fruits and vegetables. A banana as a vaccine makes sense in theory, but in practice it is difficult to obtain a uniform product. Edible plants are grown from cells that are given genes from pathogens that encode the antigens that evoke an immune response. When the plant vaccine is eaten, the foreign antigens stimulate phagocytes beneath the small intestinal lining to "present" the antigens to nearby T cells. From here, the antigens go to the bloodstream, where they stimulate B cells to divide to yield plasma cells that produce IgA. These antibodies coat the small intestinal lining, protecting against pathogens in food. Current research focuses on converting plant-based vaccines into powders so that doses can be regulated, but this counters the original goal of feeding bananas to easily immunize babies.

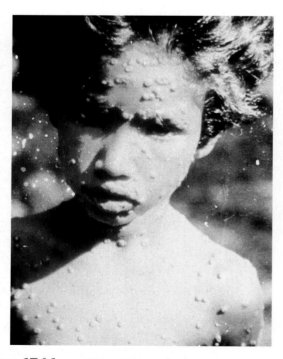

Figure 17.16 Smallpox: Gone? This boy is one of the last victims of smallpox, which has not naturally infected a human since 1977. Because many doctors are unfamiliar with smallpox, and people are no longer vaccinated, an outbreak would be a major health disaster.

Whatever the form of vaccine, it is important that a substantial proportion of a population be vaccinated to control an infectious disease. This establishes "herd immunity"—that is, if unvaccinated people are rare, then if the pathogen appears, it does not spread, because so many people are protected. If the population includes unvaccinated individuals who come into contact, the disease can spread.

Immunotherapy

Immunotherapy amplifies or redirects the immune response. It originated in the nineteenth century to treat disease. Today, a few immunotherapies are in use, with more in clinical trials.

Monoclonal Antibodies Boost Humoral Immunity

When a B cell recognizes a single foreign antigen, it manufactures a single, or monoclonal, type of antibody. A large amount of a single antibody type could target a particular pathogen or cancer cell because of the antibody's great specificity.

In 1975, British researchers Cesar Milstein and George Köhler devised monoclonal antibody (MAb) technology, which mass-produces a single B cell, preserving its specificity and amplifying its antibody type. First, they injected a mouse with a sheep's red blood cells (**figure 17.17**). They then isolated a single B cell from the mouse's spleen and fused it with a cancerous white blood cell from a mouse. The fused cell, called a hybridoma, had a valuable pair of talents. Like the B cell, it produced large numbers of a single antibody type. Like the cancer cell, it divided continuously.

Today MAbs are made to more closely resemble natural human antibodies because the original mouse preparations caused allergic reactions. MAbs are used in basic research, veterinary and human health care, agriculture, forestry, and forensics. They can diagnose everything from strep throat to turf grass disease. In a home pregnancy test, a woman places drops of her urine onto a paper strip containing a MAb that binds hCG, the "pregnancy" hormone. The color changes if the MAb binds its target. In cancer diagnosis, if a MAb attached to a fluorescent dye and injected into a patient or applied to a sample of tissue or body fluid binds its target—an antigen found mostly or only on cancer cells—fluorescence indicates disease. MAbs linked to radioactive isotopes or to drugs deliver treatment to cancer cells. The MAb drug trastuzumab (Herceptin) blocks receptors on certain breast cancer cell surfaces, preventing them from receiving signals to divide.

Cytokines Boost Cellular Immunity

As coordinators of immunity, cytokines are used to treat a variety of conditions. However, it has been difficult to develop these body chemicals into drugs because they act only for short periods. They must be delivered precisely where they are needed, or overdose or side effects can occur.

Interferons (IFs) were the first cytokines tested on a large scale, and today are used to treat cancer, genital warts, multiple sclerosis, and some other conditions. Interleukin-2 (IL-2) is a

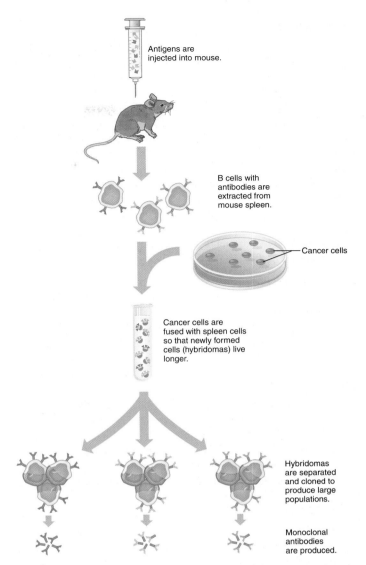

Figure 17.17 Monoclonal antibody technology. Monoclonal antibodies are pure preparations of a single antibody type that recognize a single antigen type. They are useful in diagnosing and treating disease because of their specificity.

cytokine that is administered intravenously to treat kidney cancer recurrence. Colony-stimulating factors, which cause immature white blood cells to mature and differentiate, are used to boost white blood cell levels in people with suppressed immune systems, such as individuals with AIDS or those receiving cancer chemotherapy. Treatment with these factors enables a patient to withstand higher doses of a conventional drug.

Because excess of another cytokine, tumor necrosis factor (TNF), underlies some disorders, blocking its activity treats some conditions. The drug etanercept (Enbrel), for example, consists of part of a receptor for TNF. Taking it prevents TNF from binding to cells that line joints, relieving arthritis. Excess TNF in rheumatoid arthritis prevents the joint lining cells from secreting lubricating fluid.

Transplants

When a car breaks down, replacing the damaged part often fixes the trouble. The same is sometimes true for the human body. Hearts, kidneys, livers, lungs, corneas, pancreases, skin, and bone marrow are routinely transplanted, sometimes several organs at a time. Although transplant medicine had a shaky start, many problems have been solved. Today, thousands of transplants are performed annually and recipients gain years of life. The challenge to successful transplantation lies in genetics because individual inherited differences in cell surfaces determine whether the body will accept tissue from a particular donor.

Transplant Types

Transplants are classified by the relationship of donor to recipient (**figure 17.18**):

1. An autograft transfers tissue from one part of a person's body to another. A skin graft taken from the thigh to replace burned skin on the chest, or a leg vein that replaces a coronary artery, are autografts. The immune system does not reject the graft because the tissue is self. (Technically, an autograft is not a transplant because it involves only one person.)
2. An isograft is tissue from a monozygotic twin. Because the twins are genetically identical, the recipient's immune system does not reject the transplant. Ovary isografts have been performed.
3. An allograft comes from an individual who is not genetically identical to the recipient, but is a member of the same species. A kidney transplant from an unrelated donor is an allograft.
4. A xenograft transplants tissue from one species to another. (See *Bioethics: Choices for the Future* on page 343.)

Rejection Reactions—Or Acceptance

The immune system recognizes most donor tissue as nonself, and launches a tissue rejection reaction in which T cells, antibodies, and activated complement destroy the foreign tissue.

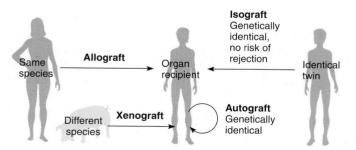

Figure 17.18 **Transplant types.** An autograft is within an individual. An isograft is between identical twins. An allograft is between members of the same species, and a xenograft is between members of different species.

The greater the difference between recipient and donor cell surfaces, the more rapid and severe the rejection reaction. An extreme example is the hyperacute rejection reaction against tissue transplanted from another species. Donor tissue from another type of animal is usually destroyed in minutes as blood vessels blacken and cut off the blood supply.

Physicians use several approaches to limit rejection to help a transplant recipient survive. These include closely matching the HLA types of donor and recipient and stripping donor tissue of antigens. Gene expression microarrays can be used to better match donors to recipients.

Immunosuppressant drugs inhibit production of the antibodies and T cells that attack transplanted tissue. If recipients get bone marrow stem cells from the donors, they need immunosuppressant drugs for only a short time, because along with the organ, the bone marrow stem cells help the recipient's body accept the transplanted tissue. Gene expression profiling can identify transplant recipients unlikely to reject their new organs. Still experimental, this approach can spare some people taking immunosuppressants, which have side effects.

Rejection is not the only problem that can arise from an organ transplant. Graft-versus-host disease develops sometimes when bone marrow transplants are used to correct certain blood deficiencies and cancers. The transplanted bone marrow, which is actually part of the donor's immune system, attacks the recipient—its new body—as foreign. Symptoms include rash, abdominal pain, nausea, vomiting, hair loss, and jaundice.

Sometimes a problem arises if a bone marrow transplant to treat cancer is too closely matched to the recipient. If the cancer returns with the same cell surfaces as it had earlier, the patient's new bone marrow is so similar to the old marrow that it is equally unable to fight the cancer. The best tissue for transplant may be a compromise: different enough to control the cancer, but not so different that rejection occurs.

Key Concepts

1. Vaccines are disabled pathogens or their parts that elicit an immune response against infection by the active pathogen.
2. Immunotherapy uses immune system components to fight disease. B cells fused with cancer cells produce MAbs that target specific antigens. Cytokines boost immune function and destroy cancer cells.
3. Autografts transfer tissue from one part of a person's body to another; isografts are between identical twins; allografts are between members of the same species; and xenografts are cross-species transplants.
4. Allografts can cause tissue rejection reactions, and xenografts can set off hyperacute rejection. In graft-versus-host disease, transplanted bone marrow rejects the recipient's tissues.

Pig Parts

In 1902, a German medical journal reported an astonishing experiment. A physician, Emmerich Ullman, had attached the blood vessels of a patient dying of kidney failure to a pig's kidney set up by her bedside. The patient's immune system rejected the attachment almost immediately.

Nearly a century later, in 1997, a similar experiment took place. Robert Pennington, a 19-year-old suffering from acute liver failure and desperately needing a transplant, survived for six and a half hours with his blood circulating outside of his body through a living liver removed from a 15-week-old, 118-pound pig named Sweetie Pie. The pig liver served as a bridge until a human liver became available. But Sweetie Pie was no ordinary pig. She had been genetically modified and bred so that her cells displayed a human protein that controlled rejection of tissue transplanted from an animal of another species. Because of this bit of added humanity, plus immunosuppressant drugs, Pennington's body tolerated the pig liver's help for the few crucial hours. Baboons have also been organ donors (**figure 1**).

Successful xenotransplants would help alleviate the organ shortage. However, some people object to raising animals to use their organs as transplants because it requires killing the donors, although many of us eat animals. A possible danger of xenotransplants is that people may acquire viruses from the donor organs. Viruses can "jump" species, and the outcome in the new host is unpredictable. For example, a virus called PERV—for "porcine endogenous retrovirus"—can infect human cells in culture. However, several dozen patients who received implants of pig tissue did not show evidence of PERV years later. That study, though, looked only at blood. We still do not know what effect pig viruses can have on a human body. Because many viral infections take years to cause symptoms, introducing a new infectious disease in the future could be the trade-off for using xenotransplants to solve the current organ shortage.

Questions for Discussion

1. Pig parts as transplants may become necessary due to the shortage of human organs. Discuss the pros and cons of the following systems for rationing human organs:
   a. first come, first served;
   b. closeness of match of cell surface antigens;
   c. ability to pay;
   d. the importance of the recipient;
   e. the youngest;
   f. the most severely ill, who will soon die without the transplant;
   g. the least severely ill, who are strong enough to survive a transplant; and
   h. those who are not responsible for their condition, such as a nonsmoker with hereditary emphysema versus a person who has emphysema caused by smoking.

2. In the novel and film *Never Let Me Go,* Kazuo Ishiguro tells of a society in which certain people are designated as organ donors. They know that at a certain age, their organs will be removed, one by one, until they die, to provide transplants for wealthy recipients. The film *The Island* has a similar plot, except that the donors do not know their fate. In Robin Cook's novel *Chromosome Six,* a geneticist places *HLA* genes into fertilized ova from bonobos (pygmy chimps), and the animals are raised to provide organs for wealthy humans.

   Choose a book or film with a transplant plot and discuss the source of the transplants from the points of view of the donor, the recipient, the families of both, and the government.

3. Discuss the issues that people might find disturbing about creating animals such as pigs or primates that have certain human molecules that make their organs more likely to be accepted as transplants.

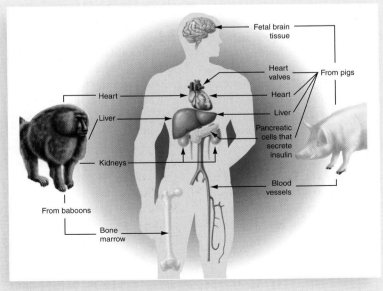

Figure 1   Baboons and pigs can provide tissues and organs for transplant.

17.5 A Genomic View of Immunity—The Pathogen's Perspective

Immunity against infectious disease involves interactions of two genomes—ours and the pathogen's. Human genome information is revealing how the immune system halts infectious disease. Information from pathogen genomes reveals how they make us sick.

Knowing the DNA sequence of a pathogen's genome, or the sequences of key genes, can reveal exactly how that organism causes illness in humans. This information can suggest new treatment strategies. The DNA sequence for *Streptococcus pneumoniae,* for example, revealed genetic instructions for a huge protein that enables the bacterium to adhere to human cells. Potential drugs could dismantle this adhesion protein.

Pathogen genome information is also used to protect against infection in an approach called reverse vaccinology. Instead of culturing hard-to-grow pathogens in the laboratory, researchers identify and use parts of genomes that encode antigens that provoke the human immune system. Researchers rapidly developed vaccines against severe acute respiratory syndrome (SARS) in 2003, H5N1 avian flu in 2005, and H1N1 swine flu in 2009 using this strategy.

Crowd Diseases

History provides clues to the complex and ever-changing relationships between humans and our pathogens. Because adaptive immunity responds to an environmental stimulus, epidemics often followed the introduction of a pathogen into a population that had not encountered it before.

When Europeans first explored the New World, they inadvertently brought bacteria and viruses to which their immune systems had adapted. The immune systems of Native Americans, however, had never encountered these pathogens. Many people died. Smallpox decimated the Aztec population in Mexico from 20 million in 1519, when conquistador Hernán Cortés arrived from Spain, to 10 million by 1521, when Cortés returned. By 1618, the Aztec nation population had fallen to 1.6 million. The Incas in Peru and northern populations were also dying of smallpox. When explorers visited what is now the southeast United States, they found abandoned towns where natives had died from smallpox, measles, pertussis, typhus, and influenza.

The diseases that so easily killed Native Americans are known as "crowd" diseases, because they arose with the spread of agriculture and urbanization and affect many people. Crowd diseases swept Europe and Asia as expanding trade routes spread bacteria and viruses along with silk and spices. Today, air travel spreads crowd diseases.

Crowd diseases tend to pass from conquerors who live in large, intercommunicating societies to smaller, more isolated and susceptible populations, not vice versa. When Columbus arrived in the New World, the large populations of Europe and Asia had existed far longer than American settlements. In Europe and Asia, infectious diseases had time to become established and for human populations to adapt to them. In contrast, an unfamiliar infectious disease can quickly wipe out an isolated tribe, leaving no one to give the illness to new invaders.

Most crowd diseases vanish quickly, for several reasons: Vaccines or treatments may stop transmission; people may alter their behaviors to avoid contracting the infection; or the disease may kill before individuals can pass it on. Sometimes, we don't know why a disease vanishes or becomes milder.

We may be able to treat and control newly evolving infectious diseases one at a time, with new drugs and vaccines. But the mutation process that continually spawns new genetic variants in microbe populations—resulting in evolution—means that new infectious diseases will continue to arise, and old ones to return or ravage new populations.

Bioweapons

Biological weapons intentionally infect people. They have existed since medieval warriors catapulted plague-ridden corpses over city walls to kill the inhabitants. During the French and Indian War, the British gave Native Americans blankets intentionally contaminated with secretions from smallpox victims. Although international law banned "germ warfare" in 1925, from 1932 until 1942 Japan field-tested bacterial bioweapons in rural China, killing thousands.

In 1973, in the Soviet Union, thousands of workers prepared anthrax bombs and other bioweapons, under the guise of manufacturing legitimate drugs, that were even more lethal than their natural counterparts. Plague bacteria, for example, were genetically modified to resist sixteen antibiotics and to strip nerve cells of their fatty coats, adding paralysis to the list of natural symptoms. But the work backfired. In 1979, a miscommunication among shift workers in charge of changing safety air filters released a cloud of dried anthrax spores over the city of Sverdlorsk. Within weeks, more than 100 people had died of inhalation anthrax. They were mostly young, healthy men who were outside on that Friday night and breathed in enough anthrax spores to cause respiratory collapse. The government blamed the deaths on eating infected meat. Workers sprayed water everywhere, stirring up the spores and spreading infection.

Genetics underlies inhalation anthrax because the toxin the bacteria produce consists of three proteins. One protein forms a barrel-like structure that binds to macrophages and admits the other two proteins. One of these components overloads signal transduction and impairs the cell's ability to engulf pathogens. The other toxin component breaks open macrophages, which release tumor necrosis factor and interleukins.

In the United States, a small-scale bioweapons effort began in 1942. A facility at Fort Detrick in Frederick, Maryland, stored 5,000 bombs loaded with anthrax spores, which were tested in Indiana, Mississippi, and Utah. President Richard Nixon halted the program in 1969 because he thought that conventional and nuclear weapons were a sufficient deterrent and defense. In 1972, political leaders in London, Moscow, and Washington signed the Biological Weapons Convention, an effort to prevent bioterrorism.

Bioweapons are still a threat, but the development of technology to detect them has had medical benefits. Consider tuberculosis. It still kills many millions of people each year, and spreads rapidly because the standard diagnostic test, on sputum, is very unreliable. A test called GeneXpert, invented to detect pathogen DNA sequences in mail following the anthrax letters in the United States in 2001, is now in use to very rapidly diagnose tuberculosis. Fast diagnosis leads to quicker treatment, stopping the spread of infection.

Key Concepts

1. Knowing the genome sequence of a pathogen can reveal how it evades the human immune system.

2. Crowd diseases happen when infectious agents are introduced into a population that hasn't encountered them before.

3. Bioterrorism is the use of pathogens—either in their natural state or genetically manipulated—to kill people.

Summary

17.1 The Importance of Cell Surfaces

1. The cells and biochemicals of the immune system distinguish self from nonself, protecting the body against infections and cancer.

2. Genes encode immune system proteins, and may confer susceptibilities to certain infectious diseases.

3. An **antigen** is a molecule that elicits an immune response. Patterns of cell surface protein and glycoprotein antigens determine blood types. **Human leukocyte antigens** genes encode cell surface antigens that present foreign antigens to the immune system.

17.2 The Human Immune System

4. If a pathogen breaches physical barriers, the **innate immune response** produces the redness and swelling of inflammation, plus **complement, collectins,** and **cytokines.** The response is broad and general.

5. The **adaptive immune response** is slower, specific, and has memory.

6. The **humoral immune response** begins when macrophages display foreign antigens near HLAs. This activates **T cells,** which activate **B cells,** which give rise to plasma cells and secrete specific **antibodies.** Some B cells give rise to **memory cells.**

7. An antibody is Y-shaped and has four polypeptide chains, two heavy and two light. Each antibody molecule has regions of constant amino acid sequence and regions of variable sequence.

8. The tips of the Y of each subunit form antigen binding sites, which include the more specific idiotypes that bind foreign antigens at their epitopes.

9. Antibodies bind antigens to form immune complexes large enough for other immune system components to detect and destroy. Antibody genes are rearranged during early B cell development, providing instructions to produce a great variety of antibodies.

10. T cells carry out the **cellular immune response**. Their precursors are selected in the thymus to recognize self. Helper T cells secrete cytokines that activate other T cells and B cells. A helper T cell's CD4 antigen binds macrophages that present foreign antigens. Cytotoxic T cells release biochemicals that kill bacteria and destroy cells covered with viruses.

17.3 Abnormal Immunity

11. Mutations in antibody or cytokine genes, or in genes encoding T cell receptors, cause inherited immune deficiencies. Severe combined immune deficiencies affect both branches of the immune system.

12. HIV binds to the coreceptors CD4 and CCR5 on macrophages and helper T cells, and, later in infection, triggers apoptosis of cytotoxic T cells. As HIV replicates, it mutates, evading immune attack. Falling CD4 helper T cell numbers allow opportunistic infections and cancers to flourish. People who cannot produce a complete CCR5 protein resist HIV infection.

13. In an **autoimmune disease,** the body manufactures **autoantibodies** against its own cells.

14. In susceptible individuals, allergens stimulate IgE antibodies to bind to mast cells, which causes the cells to release allergy mediators. Certain helper T cells release selected cytokines. Allergies may be a holdover of past immune function.

17.4 Altering Immunity

15. A **vaccine** presents a disabled pathogen, or part of one, to elicit a primary immune response.

16. Immunotherapy enhances or redirects immune function. Monoclonal antibodies are useful in diagnosing and treating some diseases because of their abundance and specificity. Cytokines are used to treat various conditions.

17. Transplant types include autografts (within oneself), isografts (between identical twins), allografts (within a species), and xenografts (between species). A tissue rejection reaction occurs if donor tissue is too unlike recipient tissue.

17.5 A Genomic View of Immunity—The Pathogen's Perspective

18. Pathogen genome sequences can reveal how they infect, which provides clues to developing new treatments.

19. Crowd diseases spread rapidly through a population that has had no prior exposure, passed from members of a population that have had time to adapt to the pathogen.

20. Throughout history, people have used bacteria and viruses as weapons.

www.mhhe.com/lewisgenetics10

Answers to all end-of-chapter questions can be found at **www.mhhe.com/lewisgenetics10.** You will also find additional practice quizzes, animations, videos, and vocabulary flashcards to help you master the material in this chapter.

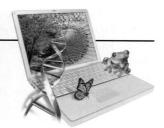

Review Questions

1. Match the cell type to the type of biochemical it produces.

   1. mast cell
   2. T cell
   3. B cell
   4. macrophage
   5. all cells with nuclei
   6. antigen-presenting cell

   a. antibodies
   b. HLA class II genes
   c. interleukin
   d. histamine
   e. interferon
   f. heparin
   g. tumor necrosis factor
   h. HLA class I genes

2. What does "nonself" mean? Give an example of a nonself cell in your own body.

3. Distinguish between viruses and bacteria.

4. What is the physical basis of a blood type?

5. Distinguish between using serology or genotyping to type blood.

6. Explain why an HLA-disease association is not a diagnosis.

7. Explain how mucus, tears, cilia, and ear wax are part of the immune response.

8. Distinguish between
   a. a T cell and a B cell.
   b. innate and adaptive immunity.
   c. a primary and secondary immune response.
   d. a cellular and humoral immune response.
   e. an autoimmune condition and an allergy.
   f. an inherited and acquired immune deficiency.

9. Which components of the human immune response explain why we experience the same symptoms of an upper respiratory infection (a "cold") when many different types of viruses can cause these conditions?

10. State the function of each of the following immune system biochemicals:
    a. complement proteins
    b. collectins
    c. antibodies
    d. cytokines

11. What does HIV do to the human immune system?

12. Cite three reasons why developing a vaccine against HIV infection has been challenging.

13. What would be the consequences of lacking
    a. helper T cells?
    b. cytotoxic T cells?
    c. B cells?
    d. macrophages?

14. Explain how the immune system can respond to millions of different nonself antigens, if there are only a few hundred antibody genes.

15. How are SCID and AIDS similar and different?

16. What part do antibodies play in allergic reactions and in autoimmune disorders?

17. What do a plasma cell and a memory cell descended from the same B cell have in common? How do they differ?

18. Why is a deficiency of T cells more dangerous than a deficiency of B cells?

19. Cite two explanations for why autoimmune disorders are more common in females.

20. How do each of the following illnesses disturb immunity?
    a. graft-versus-host disease
    b. SCID
    c. scleroderma
    d. AIDS
    e. hayfever

21. Why is a polyclonal antibody response valuable in the body, but a monoclonal antibody valuable as a diagnostic tool?

22. State how each of the following alters immune system functions:
    a. a vaccine
    b. an antibiotic drug
    c. a cytokine-based drug
    d. an antihistamine drug
    e. a transplant

Applied Questions

1. "Winter vomiting disease," a form of gastroenteritis sometimes called "stomach flu," is caused by a virus called norovirus. It makes a person miserable for 1 to 2 days. Why do some people get the illness every year?

2. Rasmussen's encephalitis causes 100 or more seizures a day. Affected children have antibodies that attack brain cell receptors that normally bind neurotransmitters. Is this condition most likely an inherited immune deficiency, an adaptive immune deficiency, an autoimmune disorder, or an allergy? State a reason for your answer.

3. In the TV program *House*, a talented physician and his staff confront difficult-to-diagnose medical cases. They often have to hypothesize whether symptoms are due to an infection, allergy, poison, autoimmunity, or genetic disease. Discuss how these alternatives might be distinguished.

4. In people with a certain HLA genotype, a protein in their joints resembles an antigen on the bacterium that causes Lyme disease. This infection is transmitted in a tick bite and causes flulike symptoms and joint pain (arthritis). When these individuals become infected, their immune systems attack the bacteria and their joints. Explain why antibiotics treat the early phase of the disease, but not the arthritis.

5. A person exposed for the first time to Coxsackie virus develops a painful sore throat. How is the immune system alerted to the exposure to the virus? When the person encounters the virus again, why doesn't she develop symptoms?

6. A young woman who has aplastic anemia will soon die as her lymphocyte levels drop sharply. What type of cytokine might help her?

7. Tawanda is a 16-year-old with cystic fibrosis. She receives a lung transplant from a woman who has just died in a car accident. Tawanda and the donor share four of the six HLA markers commonly used to match donor to recipient. What action can the transplant team take to minimize the risk that Tawanda's immune system rejects the transplanted lung?

Web Activities

1. Many websites describe products (food supplements) that supposedly "boost" immune system function. Locate such a website and identify claims that are unclear, deceptive, vague, or incorrect. Alternatively, identify a claim that *is* consistent with the description of immune system function in this chapter.

Case Studies and Research Results

1. State whether each of the following situations describes an autograft, an isograft, an allograft, or a xenograft.
   a. A man donates part of his liver to his daughter, who has a liver damaged by cystic fibrosis.
   b. A woman with infertility receives an ovary transplant from her identical twin.
   c. A man receives a heart valve from a pig.
   d. A woman who has had a breast removed has a new breast built using her fatty thigh tissue.

2. Mark and Louise are planning to have their first child, but they are concerned because they think that they have an Rh incompatibility. He is Rh$^-$ and she is Rh$^+$. Will there be a problem? Why or why not?

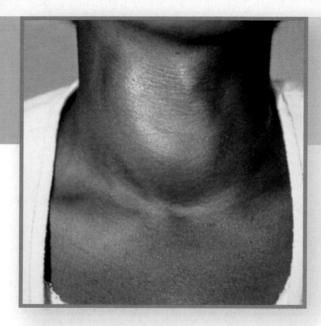

Thyroid cancer may be discovered on autopsy. This type of cancer is slow-growing and may not produce symptoms. In this person the tumor is obvious as a swelling in the neck. The author's tumor was slightly smaller than this, and she had no symptoms.

Genetics of Cancer

Learning Outcomes

18.1 Cancer Is Genetic, But Usually Not Inherited

1. Explain how loss of cell cycle control causes cancer.
2. Explain how most cancers are not inherited, but are genetic.

18.2 Characteristics of Cancer Cells

3. Describe cancer cells.

18.3 Origins of Cancer Cells

4. List ways that cancer cells arise.

18.4 Cancer Genes and MicroRNAs

5. Explain the genetic causes of cancer.

18.5 Many Genes Contribute to Cancer

6. Discuss how mutations in several genes contribute to cancer.

18.6 Environmental Causes of Cancer

7. Discuss environmental factors in cancer.

18.7 The Personalization of Cancer Diagnosis and Treatment

8. Discuss how cancer diagnosis and treatment have become personalized.

The Big Picture: A few cells probably escape the controls of the cell cycle in each of us, but are usually squelched by the immune system. In one in three of us, though, such errant cells continue to divide and invade healthy tissue, causing cancer. The many forms of cancer reflect mutations in particular cell types.

Microarrays Illuminate Thyroid Cancer

I never thought I would care much about the cells of my thyroid gland. That changed on August 4, 1993, when my physician, looking at me from across a room, said, "What's that lump in your neck?"

Soon after, a specialist stuck six thin needles into the lump to sample thyroid cells for testing, telling me that 99 percent of thyroid "nodules" are not cancerous. However, when he approached with a seventh needle for a sample for a study on something called *p53,* I began to worry, because *p53* is a gene associated with cancer. When the specialist called early on a Monday morning, I knew I was among the unlucky 1 percent. I had papillary thyroid cancer, which accounts for 80 percent of cases and is easily treated with surgery and radioactive iodine. But when I was on the operating table, the surgeon did not think my nodule looked like a papillary tumor. Off it went to the pathology lab, while I waited on the table. The results: I had two tumors, one papillary, one follicular. Treatment was successful.

Had I developed thyroid cancer today, I might not have had to wait on an operating table while a pathologist examined my cells for the telltale distinctions between tumor types. DNA microarrays can now highlight key genes that are expressed differently in papillary and follicular thyroid cancers. My physicians would have known, before surgery, the genetic nature of my tumors. This approach is very valuable for cancer in which treatment differs, depending upon the genetic profiles of the cells, such as breast or prostate cancers.

18.1 Cancer Is Genetic, But Usually Not Inherited

Cancer has been part of human existence for eons. Egyptian mummies from 3000 B.C. show evidence of cancerous tumors, and by 1600 B.C., the Egyptians were attempting to treat cancer. Papyruses illustrate them cutting or burning off growths, and using more inventive treatments for less obvious tumors. A remedy for uterine cancer, for example, introduced fresh ground dates mixed with pig's brain into the vagina!

By 300 B.C., Hippocrates had described several types of tumors, and coined the term "cancer" to describe the crablike shape of a tumor invading normal tissue. He attributed cancer to a buildup of black bile; others blamed it on fermenting lymph, injury, irritation, or simply "melancholia." Today we know that the collection of diseases called cancer reflects a profound derangement of the cell cycle that can be set into motion by environmental factors. Sequences of mutations in somatic cells and gene expression changes underlie the progression of cancer as it spreads.

Cancer has or will affect one in three of us, some more than once. Diagnosis and treatment are becoming increasingly individualized, thanks largely to genetic and genomic approaches to describing cancer cells.

Cancer is a complication of being a many-celled organism. Our specialized cells must follow a schedule of mitosis—the cell cycle—so that organs and other body parts either grow appropriately during childhood, stay a particular size and shape throughout adulthood, or repair damage by replacing tissue. If a cell in solid tissue escapes normal controls on its division rate, it forms a growth called a tumor (**figure 18.1**). In the blood, such a cell divides more frequently than others, taking over the population of blood cells.

A tumor is benign if it grows in place but does not spread into, or "invade," surrounding tissue. A tumor is cancerous, or malignant, if it infiltrates nearby tissue. A malignant tumor also sends parts of itself into the bloodstream or lymphatic vessels, either of which transports it to other areas, where the cancer cells "seed" the formation of new tumors. The process of spreading is termed **metastasis,** which means "not standing still."

Cancer is a group of disorders that arise from alterations in genes. Only about 10 percent of cases are inherited as single-gene disorders, in which the faulty instructions are in every cell. More often, mutations in cancer-causing genes occur in a few somatic cells over a lifetime. Cancer is usually a genetic disease at the cellular level, but not at the whole-body level. This is vividly seen in experiments that sequence the genome of a cancer cell and also of a normal cell right next to it. The genome of a cancer cell typically has dozens of mutations compared to the normal cell near it from which it presumably descended. Such studies look at "tumor-normal pairs" of cells.

Combinations of particular gene variants sum to increase the risk of cancer by making cells more sensitive to environmental factors that affect the cell cycle. As a result, cancer may "run in families" yet not follow a single-gene pattern of

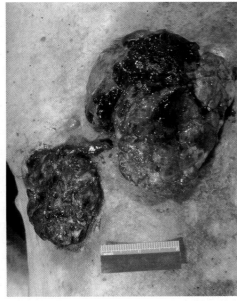
a.

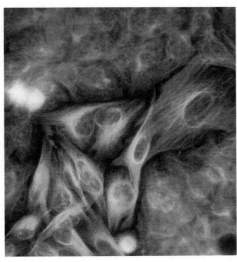
b.

Figure 18.1 **Cancer cells stand out.** **(a)** A melanoma is a cancer of the pigment-producing cells (melanocytes) in the skin. It may have any or all of four characteristics, abbreviated abcde: it is <u>a</u>symmetric, has <u>b</u>orders that are irregular, <u>c</u>olor variations, a <u>d</u>iameter of more than 5 millimeters, and <u>e</u>levation. **(b)** These melanoma cells stain orange. The different staining characteristics of cancer cells reflect differences in gene expression patterns between the normal and cancerous states.

inheritance. Cancer often takes years to develop, as a sequence of genes mutate in the affected tissue. Then, the cells whose mutations enable them to divide more often than others gradually take over the tissue. Even though a cancer may not spread for years, certain mutations or changes in gene expression can indicate that it will.

An early hint at the genetic nature of cancer was the observation that most substances known to be carcinogens (causing cancer) are also mutagens (damaging DNA). Researchers first discovered genes that could cause cancer in 1976 in viruses that cause cancer in rodents. In the 1980s and 1990s, searches for cancer-causing genes began with rare families that had many young members who had the same type of cancer and specific unusual chromosomes. Then the search focused on genes in the identified chromosome region whose protein products could affect cell cycle control, leading to the discovery of more than 100 **oncogenes**. An oncogene is a gene that causes cancer when inappropriately activated.

Family studies also identified more than 30 **tumor suppressor genes,** which cause cancer when they are deleted or inactivated. The normal function of a tumor suppressor gene is to keep the cell cycle running at the appropriate rate for a particular cell type under particular conditions.

In addition to activated oncogenes and inactivated tumor suppressor genes, changes in gene expression accompany cancer. Such changes may be epigenetic, with methyl groups turning off certain genes, or their removal turning on others. DNA microarrays highlight mutations and patterns of gene expression that paint "molecular portraits" of the disease. These views are making it possible to recognize subtypes of cancers affecting the same cell types that cannot be visualized by examining cancer cells under a microscope.

Loss of Cell Cycle Control

Cancer is a consequence of cell cycle disruption. **Figure 18.2** repeats the cell cycle diagram from chapter 2. Cancer begins when a cell divides more frequently, or more times, than the noncancerous cell it descended from (**figure 18.3**). Mitosis in a cancer cell is like a runaway train, racing along without signals and control points.

The timing, rate, and number of mitoses a cell undergoes depend on protein growth factors and signaling molecules from outside the cell, and on transcription factors from within. Because these biochemicals are under genetic control, so is the cell cycle. Cancer cells probably arise often, because mitoses are so frequent that an occasional cell escapes control. However, the immune system destroys most cancer cells after recognizing tumor-specific antigens on their surfaces.

The discovery of the checkpoints that control the cell cycle revealed how cancer can begin. A mutation in a gene that normally halts or slows the cell cycle can lift the constraint, leading to inappropriate mitosis. Failure to pause long enough to repair DNA can allow a mutation in an oncogene or tumor suppressor gene to persist.

Loss of control over telomere length may also contribute to cancer by affecting the cell cycle. Recall that telomeres, or chromosome tips, protect chromosomes from breaking (see figure 2.18). Human telomeres consist of the DNA sequence TTAGGG repeated thousands of times. The repeats are normally lost from the telomere ends as a cell matures, from 15 to 40 nucleotides per cell division. The more specialized a cell, the shorter its telomeres. The chromosomes in skin, nerve, and muscle cells, for example, have short telomeres. Chromosomes in a sperm cell or oocyte, however, have long telomeres. This makes sense—as the precursors of a new organism, gametes must retain the capacity to divide many times.

Gametes keep their telomeres long thanks to an enzyme, telomerase, that is built of RNA and protein. Part of the RNA—AAUCCC—is a template for the 6-DNA-base repeat TTAGGG that builds telomeres. Telomerase moves down the DNA like a zipper, adding six "teeth" (bases) at a time.

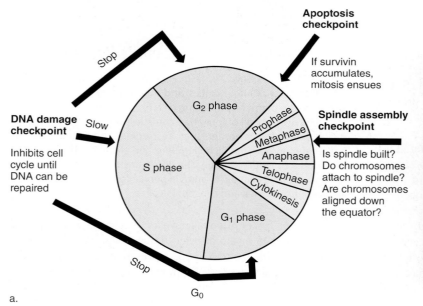

a.

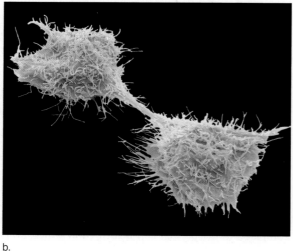

b.

Figure 18.2 Cell cycle checkpoints. **(a)** Checkpoints ensure that mitotic events occur in the correct sequence. Many types of cancer result from faulty checkpoints. **(b)** These fibrosarcoma cancer cells descend from connective tissue cells (fibroblasts) in bone. The photo captures them in telophase of mitosis.

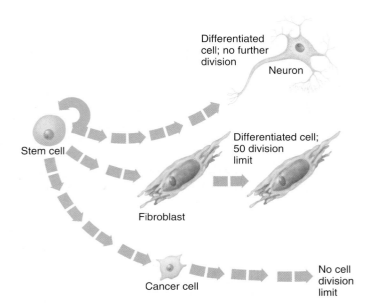

Figure 18.3 **Cancer sends a cell down a pathway of unrestricted cell division.** Cells may be terminally differentiated and no longer divide, such as a neuron, or differentiated yet still capable of limited cell division, such as a fibroblast (connective tissue cell). Cancer cells either lose specializations or never specialize; they divide unceasingly. (Arrows represent some cell divisions; not all daughter cells are shown.)

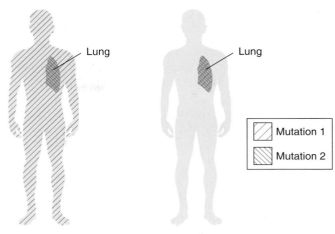

a. Germline (inherited) cancer   b. Sporadic cancer

Figure 18.4 **Germline versus sporadic cancer.** **(a)** In germline cancer, every cell has one gene variant that increases cancer susceptibility, and a second mutation in a cell of the affected tissue. This type of predisposition to cancer is inherited as a single-gene trait. **(b)** A sporadic cancer forms when a dominant mutation occurs in a somatic cell or two recessive mutations occur in the same gene in the same somatic cell. An environmental factor can cause the somatic mutations of cancer. Note that each lung has undergone both mutations 1 and 2.

In normal, specialized cells, telomerase is turned off, and telomeres shrink, signaling a halt to cell division when they reach a certain size. In cancer cells, telomerase is turned back on. Telomeres extend, and this releases the normal brake on rapid cell division. As daughter cells of the original abnormal cell continue to divide uncontrollably, a tumor forms, grows, and may spread. Usually the longer the telomeres in cancer cells, the more advanced the disease. However, turning on telomerase production in a cell is not sufficient in itself to cause cancer. Many other things must go wrong for cancer to begin.

Inherited Versus Sporadic Cancer

Most cancers are sporadic, caused by mutations found only in cells of the affected tissue. These are **somatic mutations** because they are in nonsex cells. A sporadic cancer may result from a single dominant mutation or from two recessive mutations in the same gene. The cell harboring the mutation loses control of its cell cycle, divides continuously, and a tumor forms from its daughter cells.

Susceptibility to developing a sporadic cancer is *not* directly passed on to future generations because the mutant allele or alleles are not in gametes. In contrast are **germline mutations,** in which cancer susceptibility *is* directly passed to future generations because the mutations are in every cell, including gametes. Cancer develops when a second mutation occurs in the other allele in a somatic cell in the affected body part (**figure 18.4**).

Germline mutations may explain why some heavy smokers develop lung cancer, but many do not; the unlucky ones may have inherited a susceptibility allele in every cell. Years of exposure to the carcinogens in smoke eventually cause a mutation in a tumor suppressor gene or oncogene of a lung cell, giving it a proliferative advantage. Without the susceptibility gene, two such somatic mutations are necessary to trigger the cancer. This, too, can be the result of an environmental insult, but it takes longer for two events to occur than one. Germline cancers are rare, but they have high penetrance and tend to strike earlier in life than sporadic cancers.

Key Concepts

1. Cancer is genetic, but not usually inherited.
2. Single genes (oncogenes and tumor suppressors), when mutant, can cause cancer. Cancer cells have different gene expression profiles compared to the cells from which they descend.
3. Cancer is caused by a loss of cell division control. Implicated genes encode growth factors, transcription factors, or telomerase.
4. Most cancer mutations occur in somatic cells.
5. Cancer may develop when an environmental trigger mutates a somatic cell or when a somatic mutation compounds an inherited susceptibility.

18.2 Characteristics of Cancer Cells

Cancer begins at the genetic and cellular levels. If not halted, it spreads through tissues to take over organs and organ systems. **Figure 18.5** summarizes the steps in the origin and spread of a cancer.

Cell division is rigorously controlled. Whether a cell divides or stops dividing and whether it differentiates depends upon signals from surrounding cells. A cancer cell simply stops "listening" to those signals.

Cancer cells can divide continuously if given sufficient nutrients and space. Cervical cancer cells of a woman named Henrietta Lacks, who died in 1951, vividly illustrate the hardiness of these cells. Her cells persist today as standard cultures in many research laboratories. These "HeLa" cells divide so vigorously that when they contaminate cultures of other cells they soon take over.

Cells vary greatly in their capacity to divide. Cancer cells divide more frequently or more times than the cells from which they arise. Yet even the fastest-dividing cancer cells, which complete mitosis every 18 to 24 hours, do not divide as often as some cells in a normal human embryo do. Still, some cancers

grow alarmingly fast. The smallest detectable fast-growing tumor is half a centimeter in diameter and can contain a billion cells. These cells divide, producing a million or so new cells in an hour. If 99 percent of the tumor's cells are destroyed, 10 million are left to proliferate. Other cancers develop over years. A tumor grows more slowly at first because fewer cells divide. By the time the tumor is the size of a pea—when it is usually detectable—billions of cells are actively dividing. A cancerous tumor eventually grows faster than surrounding tissue because a greater proportion of its cells is dividing.

A cancer cell looks different from a normal cell. Some cancer cells are rounder than the cells they descend from because they do not adhere to surrounding normal cells as strongly as other cells do. Because the plasma membrane is more fluid, different substances cross it. A cancer cell's surface may sport different antigens than are on other cells or different numbers of antigens that are also on normal cells. The "prostate specific antigen" (PSA) blood test that indicates increased risk of prostate cancer, for example, detects elevated levels of this protein that may come from cancer cell surfaces. Researchers found elevated PSA in a tumor-riddled male skeleton discovered in a cave in Russia dating to about 700 B.C.!

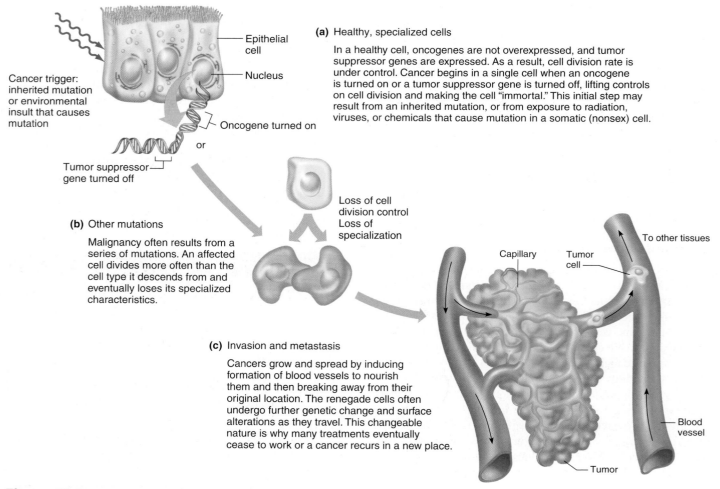

(a) Healthy, specialized cells

In a healthy cell, oncogenes are not overexpressed, and tumor suppressor genes are expressed. As a result, cell division rate is under control. Cancer begins in a single cell when an oncogene is turned on or a tumor suppressor gene is turned off, lifting controls on cell division and making the cell "immortal." This initial step may result from an inherited mutation, or from exposure to radiation, viruses, or chemicals that cause mutation in a somatic (nonsex) cell.

(b) Other mutations

Malignancy often results from a series of mutations. An affected cell divides more often than the cell type it descends from and eventually loses its specialized characteristics.

(c) Invasion and metastasis

Cancers grow and spread by inducing formation of blood vessels to nourish them and then breaking away from their original location. The renegade cells often undergo further genetic change and surface alterations as they travel. This changeable nature is why many treatments eventually cease to work or a cancer recurs in a new place.

Figure 18.5 **Steps in the development of cancer.**

When a cancer cell divides, both daughter cells are cancerous, since they inherit the altered cell cycle control. Therefore, cancer is said to be heritable because it is passed from parent cell to daughter cell. A cancer is also transplantable. If a cancer cell is injected into a healthy animal of the same species, it will proliferate there.

A cancer cell is **dedifferentiated,** which means that it is less specialized than the normal cell types near it that it might have descended from. A skin cancer cell, for example, is rounder and softer than the flattened, scaly, healthy skin cells above it in the epidermis, and is more like a stem cell in both appearance and division rate. Cancer cell growth is unusual. Normal cells in a container divide to form a single layer; cancer cells pile up on one another. In an organism, this pileup would produce a tumor. Cancer cells that grow all over one another are said to lack contact inhibition—they do not stop dividing when they crowd other cells.

Cancer cells have surface structures that enable them to squeeze into any space, a property called **invasiveness.** They anchor themselves to tissue boundaries, called basement membranes, where they secrete enzymes that cut paths through healthy tissue. Unlike a benign tumor, an invasive malignant tumor grows irregularly, sending tentacles in all directions. The cell can move. Mutations affect the cytoskeleton (see figure 2.10), breaking down actin microfilaments and releasing actin molecules that migrate to the cell surface, moving the cell from where it is anchored in surrounding tissue.

A cancer may take years or even decades to spread, as **figure 18.6** shows for lung cancer. It may do so without causing symptoms. Pancreatic cancer, for example, begins 10 to 15 years before it causes abdominal pain. At this point, when diagnosis often occurs, the cancer has usually spread, and most cases are lethal within 2 years.

Cancer cells eventually reach the bloodstream or lymphatic vessels, which take them to other parts of the body—unless treatment stops the disease process. The traveling cancer cells settle into new sites—this is **metastasis.**

Once a tumor has grown to the size of a pinhead, interior cancer cells respond to the oxygen-poor environment by secreting a protein, called vascular endothelial growth factor (VEGF). It stimulates nearby capillaries (the tiniest blood vessels) to sprout new branches that extend toward the tumor, bringing in oxygen and nutrients and removing wastes. This growth of new capillary extensions is called **angiogenesis,** and it is critical to a cancer's growth and spread. Capillaries may snake into and out of the tumor (**figure 18.7**). Cancer cells wrap around the blood vessels and creep out upon this scaffolding, invading nearby tissue. In addition to attracting their own blood supply, cancer cells may also secrete hormones that encourage their own growth. This is a new ability because the cells they descend from do not produce these hormones.

When cancer cells move to a new body part, the DNA of secondary tumor cells often mutates, and chromosomes may break or rearrange. Many cancer cells are aneuploid (with missing or extra chromosomes). The metastasized cancer thus becomes a new genetic entity that may resist treatments that were effective against most cells of the original tumor. Because

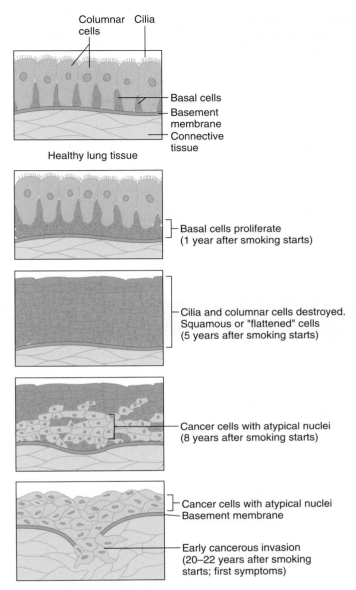

Healthy lung tissue — Columnar cells, Cilia, Basal cells, Basement membrane, Connective tissue

Basal cells proliferate (1 year after smoking starts)

Cilia and columnar cells destroyed. Squamous or "flattened" cells (5 years after smoking starts)

Cancer cells with atypical nuclei (8 years after smoking starts)

Cancer cells with atypical nuclei, Basement membrane

Early cancerous invasion (20–22 years after smoking starts; first symptoms)

Figure 18.6 **Some cancers take years to spread.** Lung cancer due to smoking begins with irritation of the lining tissue in respiratory tubes. Ciliated cells die and basal cells divide. If the irritation continues, cancerous changes occur.

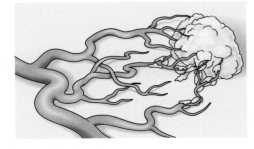

Figure 18.7 **Angiogenesis nurtures a tumor.** Cells starved for oxygen deep within a tumor secrete vascular endothelial growth factor (VEGF), which stimulates nearby capillaries to extend branches toward a tumor. A class of drugs treats cancer by blocking VEGF.

Table 18.1	Characteristics of Cancer Cells
Oilier, less adherent	
Loss of cell cycle control	
Heritable	
Transplantable	
Dedifferentiated	
Lack contact inhibition	
Induce local blood vessel formation (angiogenesis)	
Invasive	
Increased mutation rate	
Can spread (metastasize)	

Table 18.2	Processes and Pathways Affected in Cancer

	Discussed in Chapter
Angiogenesis	18
Apoptosis	7
Cell adhesion	2
Cell cycle control	2
DNA repair	12
Metabolism of carcinogens	18
Signal transduction	2

gene expression patterns associated with metastasis are detectable early, new cancer treatments may be designed to prevent metastasis.

Table 18.1 summarizes the characteristics of cancer cells. **Table 18.2** lists the processes and pathways that can be abnormal in cancer. The many mutations that cause or contribute to cancer converge in ways that affect these few pathways. Once a pathway is implicated in a particular type of cancer, researchers can look at existing drugs that affect that pathway, or develop new ones that target specific molecules.

Key Concepts

1. Cancer occurs when cells divide faster or more times than normal.
2. Cancer cells are heritable, transplantable, and dedifferentiated. They lack contact inhibition, cutting through basement membranes.
3. A cancerous growth is invasive and can metastasize and stimulate angiogenesis, spreading farther.

18.3 Origins of Cancer Cells

Mutations that turn a cell cancerous are only a first step in the disease process. Factors that influence whether or not cancer develops include how specialized the initial cell is and the location of that cell in the tissue.

Cancer can begin at a cellular level in at least four ways:

- activation of stem cells that produce cancer cells;
- dedifferentiation;
- increase in the proportion of a tissue that consists of stem or progenitor cells; and
- faulty tissue repair.

Dedifferentiation is not an all-or-none phenomenon. Most cancer cells are more specialized than stem cells, but considerably less specialized than the differentiated cells near them in a tissue. From which does the cancer cell arise? A cancer cell may descend from a stem cell that yields slightly differentiated daughter cells that retain the capacity to self-renew, or a cancer cell may arise from a specialized cell that loses some of its features and can divide. Certain stem cells, called **cancer stem cells,** veer from normal development and produce both cancer cells and abnormal specialized cells. Cancer stem cells are found in cancers of the brain, blood, and epithelium (particularly in the breast, colon, and prostate).

Figure 18.8 illustrates how cancer stem cells may cause brain tumors. In (*a*), as cancer stem cells give rise to progenitors and then differentiated cells (neurons, astrocytes, and oligodendrocytes), a cell surface molecule called CD133 is normally lost (designated CD133$^-$) at the late progenitor stage. In contrast, in (*b*), cancer cells retain the molecule (designated CD133$^+$). Some progenitor cells that descend from a cancer stem cell can relentlessly divide, and they ultimately accumulate, forming a brain tumor.

Another route to cancer may be cells that lose some of their distinguishing characteristics as mutations occur when they divide. Or, cells on the road to cancer may begin to express "stemness" genes that override signals to remain specialized (**figure 18.9**).

Another possible origin of cancer may be a loss of balance at the tissue level in favor of cells that can divide continually or frequently—like a population growing faster if more of its members are of reproductive age. Consider a tissue that is 5 percent stem cells, 10 percent progenitors, and 85 percent differentiated cells. If a mutation, over time, shifts the balance in a way that creates more stem and progenitor cells, the extra cells pile up, and a tumor forms (**figure 18.10**).

Uncontrolled tissue repair may cause cancer (**figure 18.11**). If too many cells divide to fill in the space left by injured tissue, and those cells keep dividing, an abnormal growth may result.

With so many millions of cells undergoing so many error-prone DNA replications, and so many ways that cancer can arise, it isn't surprising that cancer is so common. Yet most of the time, the immune system destroys a cancer before it progresses very far.

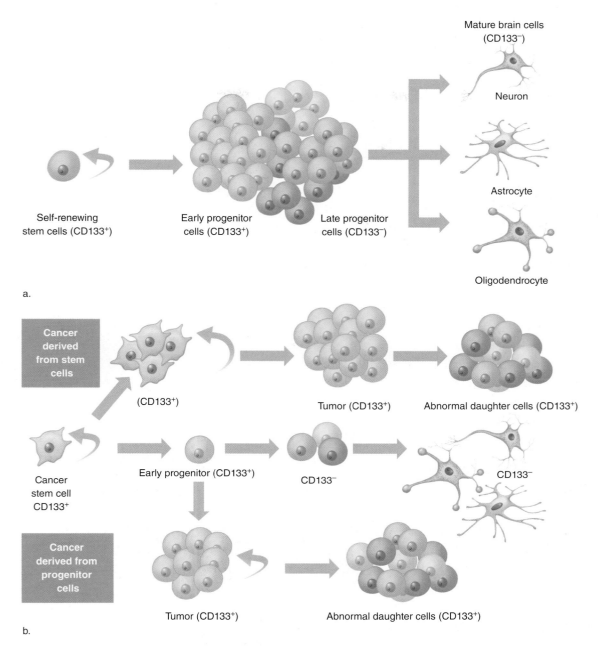

a.

b.

Figure 18.8 **Cancer stem cells.** **(a)** In the developing brain, stem cells self-renew and give rise to early progenitor cells, which divide to yield late progenitor cells. These late progenitor cells lose the CD133 cell surface marker, and they divide to give rise to daughter cells that specialize as neurons or two types of supportive cells, astrocytes or oligodendrocytes. **(b)** A cancer stem cell can divide to self-renew and give rise to a cancer cell, which in turn can also spawn abnormal daughter cells. Some early progenitors give rise to normal differentiated cells. Sometimes the cancer-causing mutations occur in the cancer stem cell–derived early progenitor cell. In this case, the early progenitors form the tumor, which may spawn some abnormal daughter cells. Note that stem cells, cancer stem cells, early progenitor cells, and abnormal daughter cells all have the CD133+ marker, but the differentiated cells do not.

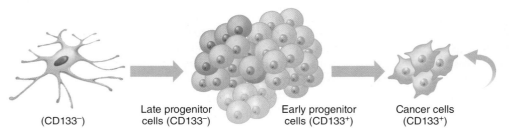

(CD133⁻) Late progenitor cells (CD133⁻) Early progenitor cells (CD133+) Cancer cells (CD133+)

Figure 18.9 **Dedifferentiation reverses specialization.** Mutations in a differentiated cell could reactivate latent "stemness" genes, giving the cell greater capacity to divide while causing it to lose some of its specializations. These are two of the defining characteristics of cancer cells.

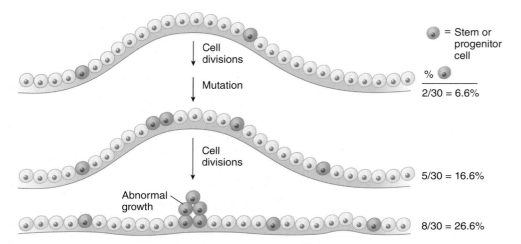

Figure 18.10 Shifting the balance in a tissue toward cells that divide. If a mutation renders a differentiated cell able to divide to yield other cells that frequently divide, then over time these cells may take over, forming an abnormal growth.

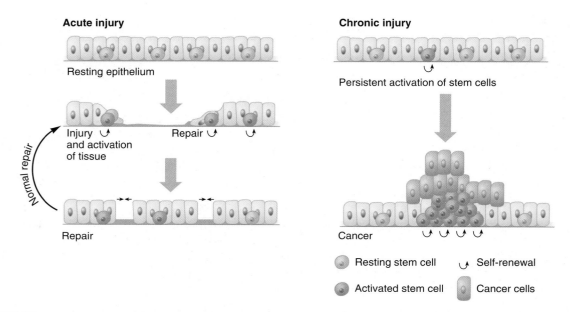

Figure 18.11 Too much repair may trigger tumor formation. If epithelium is occasionally damaged, resting stem cells can become activated and divide to fill in the tissue. If injury is chronic, the persistent activation of stem cells to renew the tissue can veer out of control, fueling an abnormal growth.

Key Concepts

1. Cancer stem cells produce cancer cells and abnormal specialized cells.

2. Dedifferentiation might occur through mutation or overexpression of "stemness" genes.

3. Upsetting the balance of stem and progenitor to differentiated cells can cause cancer as excess, fast-dividing cells accumulate.

18.4 Cancer Genes and MicroRNAs

Most mutations that cause cancer are in oncogenes or tumor suppressor genes (see figure 18.5a). A third category includes mismatch mutations in DNA repair genes (see section 12.6) that allow other mutations to persist. When such mutations activate oncogenes or inactivate tumor suppressor genes, cancer results. DNA repair disorders are often inherited in a single-gene fashion, and are quite rare. They typically cause diverse and widespread tumors.

Oncogenes

Genes that normally trigger cell division when it is appropriate are called **proto-oncogenes**. They are active where and when high rates of cell division are necessary, such as in a wound or in an embryo. When proto-oncogenes are turned on at the wrong time or place, they function as oncogenes ("onco" means cancer). Usually oncogene activation is associated with a point mutation, chromosomal translocation, or inversion, and a gain of function. In contrast, a tumor suppressor gene mutation that causes cancer is usually a deletion that removes a function.

Inappropriate activation of a proto-oncogene into an oncogene may be the result of either a mutation or a change in expression of the wild type gene. A single base change in a proto-oncogene causes bladder cancer, for example. Alternatively, a proto-oncogene may be moved near a gene that is highly expressed where it, too, is frequently transcribed. A human proto-oncogene is normally activated in cells at the site of a wound, where it stimulates production of growth factors that cause mitosis to fill in the damaged area with new cells. When that proto-oncogene is activated at a site other than a wound—as an oncogene—it still hikes growth factor production and stimulates mitosis. The new cells form a tumor.

Some proto-oncogenes encode transcription factors that, as oncogenes, are too highly expressed. (Recall from chapter 10 that transcription factors bind to specific genes and activate transcription.) The products of these activated genes contribute the cancer cell's characteristics. Oncogenes may also block apoptosis. As a result, damaged cells do not die, but divide.

Increased Expression in a New Location

A proto-oncogene can become an **oncogene** when it is placed next to a gene that boosts its expression. A virus infecting a cell, for example, may insert DNA next to a proto-oncogene. When the viral DNA is rapidly transcribed, the adjacent proto-oncogene (now an oncogene) is also rapidly transcribed. Increased production of the oncogene's encoded protein then switches on genes that promote mitosis, triggering the cascade of changes that leads to cancer. Viruses cause cervical cancer, Kaposi sarcoma, and acute T cell leukemia.

A proto-oncogene can also be activated when it is moved next to a gene that is normally very actively transcribed. This can happen when a chromosome is inverted or translocated, placing a gene in a new chromosomal environment. For example, a cancer of the parathyroid glands in the neck is associated with an inversion on chromosome 11, which places a proto-oncogene next to a DNA sequence that controls transcription of the parathyroid hormone gene. When the gland synthesizes the hormone, the oncogene is expressed, too. Cells in the gland divide, forming a tumor.

Ironically, the immune system contributes to cancer when a translocation or inversion places a proto-oncogene next to an antibody gene. Recall from chapter 17 that antibody genes normally move into novel combinations when a B cell is stimulated and they are very actively transcribed. Cancers associated with viral infections, such as cervical cancer and anal cancer following HPV infection, may begin when proto-oncogenes are mistakenly activated with antibody genes. Similarly, in Burkitt lymphoma, a cancer common in Africa, a large tumor develops from lymph glands near the jaw. People with Burkitt lymphoma are infected with the Epstein-Barr virus, which stimulates specific chromosome movements in maturing B cells to assemble antibodies against the virus. A translocation places a proto-oncogene on chromosome 8 next to an antibody gene on chromosome 14. The oncogene is overexpressed, and the cell division rate increases. Tumor cells of Burkitt lymphoma patients have the translocation (**figure 18.12**).

We can use the information of changes in gene expression that promote cancer to diagnose, treat, or track response to treatment, even without knowing what the expression patterns mean. For example, ocular melanoma affects pigment cells in the eye—it is much more deadly than common skin melanoma. Many cases of ocular melanoma spread, 95 percent to the liver. Researchers extracted mRNAs from affected eyes

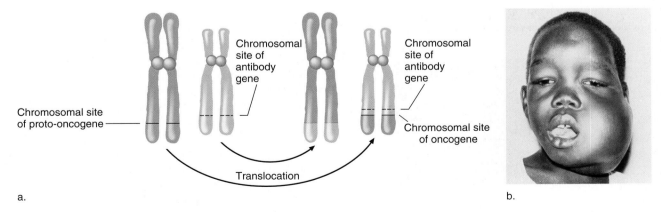

a.

b.

Figure 18.12 A translocation that causes cancer. (a) The cause of Burkitt lymphoma is translocation of a proto-oncogene on chromosome 8 to chromosome 14, next to a highly expressed antibody gene. Overexpression of the translocated proto-oncogene, now an oncogene, triggers the molecular and cellular changes of cancer. **(b)** Burkitt lymphoma often affects the jaw.

and measured the expression of ten genes. They derived two "molecular signatures"—patterns of mRNAs that are more or less abundant than normal—one predicting a low risk of spread to the liver, the other a high risk. This information is used to guide treatment choices.

Fusion Proteins with New Functions

An oncogene may be activated when a proto-oncogene moves next to another gene, and the gene pair is transcribed and translated together, as if they are one gene. The double gene product, called a **fusion protein,** activates or lifts control of cell division.

A fusion oncoprotein causes acute promyelocytic leukemia. (Leukemias differ by the type of white blood cell affected.) A translocation between chromosomes 15 and 17 brings together a gene coding for the retinoic acid cell surface receptor and an oncogene called *myl*. The fusion protein functions as a transcription factor, which, when overexpressed, causes cancer. The nature of this fusion protein explains why some patients who receive retinoid (vitamin A–based) drugs recover. Their immature, dedifferentiated cancer cells, apparently stuck in an early stage of development where they divide frequently, suddenly differentiate, mature, and die. Perhaps the cancer-causing fusion protein prevents affected white blood cells from getting enough retinoids to specialize, locking them in an embryonic-like, rapidly dividing state. Supplying extra retinoids allows the cells to resume their normal developmental pathway. **Reading 18.1** tells the story of a young woman who recovered from a different type of leukemia, thanks to a drug developed from understanding how a fusion protein causes cancer.

Receiving a Too-Strong Division Signal

An above-normal response to a growth factor lies behind some cancers. In about 25 percent of women with breast cancer, affected cells have 1 to 2 million copies of a cell surface protein called HER2 that is the product of an oncogene. The normal number of these proteins is only 20,000 to 100,000.

The HER2 proteins are receptors for epidermal growth factor. The receptors traverse the plasma membrane, extending outside the cell into the extracellular matrix and also

---

Reading 18.1

Erin's Story: How Gleevec Treats Leukemia

When 23-year-old *Glamour* magazine editor Erin Zammett Ruddy went for a routine physical in November 2001, she expected reassurance that her healthy lifestyle had indeed been keeping her healthy (**figure 1**). What she got, a few days later, was a shock. Instead of having 4,000 to 10,000 white blood cells per milliliter of blood, she had more than 10 times that number—and many of the cells were cancerous.

"I had just returned from a nice, long lunch to find a message from my doctor. Could I call back? Something had come up in my blood work," recalled Erin. "I was diagnosed with chronic myelogenous leukemia. CML is cancer, and until very recently, it proved fatal in the vast majority of cases."

Erin's diagnosis came just a few months after a landmark report of a new drug—and, ironically, an article in *Glamour* about three CML

Figure 1 "My third bone marrow biopsy—you never get used to the pain," said Erin. Gleevec has controlled her leukemia.

survivors. A successful cancer drug typically helps about 20 percent of the patients who take it, often just extending life a few months. But cancer in the blood had vanished in 53 of 54 initial patients, usually quickly. So Erin contacted the lead researcher, Brian Druker, and joined the group. Her cancer was reversed—with just a pill a day, and no side effects.

The drug, imatinib (Gleevec), is now the standard treatment for CML and a few other cancers. The story of its development illustrates how understanding the genetic events that start and propel a cancer can guide development of an effective weapon.

The tale of Gleevec began on August 13, 1958, when two men entered hospitals in Philadelphia and reported weeks of fatigue. Each had very high white blood cell counts and were diagnosed with CML. Too many immature white blood cells were crowding the healthy cells. The men's blood samples eventually fell into the hands of pathologist Peter Nowell and cytogeneticist David Hungerford. They had developed ways to stimulate white blood cells to divide in culture, and they probed the chromosomes of both leukemic and normal-appearing white blood cells in the two tired men and five others with CML.

Nowell and Hungerford discovered a small, unusual chromosome that was only in the leukemic cells. This was the first chromosome abnormality to be linked to cancer. Later, it would be dubbed "the Philadelphia chromosome" (Ph[1]). The link between the cancer and the chromosome anomaly held up in other patients.

(Continued)

With refinements in chromosome banding, important details emerged. By 1972, new stains that distinguished AT-rich from GC-rich chromosome regions revealed that Ph[1] is the result of a translocation. By 1984, researchers had homed in on the two genes juxtaposed in the translocation between chromosomes 9 and 22. Therein lay the clues that would lead to Erin's treatment.

One gene from chromosome 9 is called the Abelson oncogene (abl), and the other gene, from chromosome 22, is called the breakpoint cluster region (bcr). Two different fusion genes form. The bcr-abl fusion gene is part of the Philadelphia chromosome, and it causes CML. The encoded fusion protein, called the BCR-ABL oncoprotein, is a form of the enzyme tyrosine kinase, which is the normal product of the abl gene. The cancer-causing form of tyrosine kinase is active for too long, which sends signals into the cell, stimulating it to divide too many times. (The other fusion gene does not affect health.)

The discovery that a fusion oncoprotein started the cellular changes that cause CML gave drug researchers a target. Through the 1980s, they tested more than 400 small molecules in search of one that would block the activity of the errant tyrosine kinase, without derailing other important enzymes. When they found a candidate in 1992, Druker joined the effort and led the way in developing it into Gleevec. **Figure 2** shows how the drug works—it nestles into the pocket on the tyrosine kinase that must bind ATP to stimulate cell division. With ATP binding blocked, cancer cells do not receive the message to divide, and they cease doing so. After passing safety tests, the drug worked so dramatically that it set a new speed record for drug approval—10 weeks.

Erin and the other patients tracked their progress in several ways:

- Hematological remission: The percentage of leukemia cells in the blood fell.
- Cytogenetic remission: The percentage of cells with the Ph[1] chromosome fell.
- Molecular remission: The level of mRNA representing the fusion gene fell.

Molecular remission is the goal of CML treatment, but in actuality, fusion gene mRNA rarely reaches undetectable levels. As a result, patients can become resistant to Gleevec—relapse occurs in 3 to 16 percent of patients, depending on how sick they were when diagnosed. Resistance is a result of natural selection. Those few cancer cells able to divide in the presence of the drug eventually take over. Again, genetic research came to the rescue. By discovering how resistant cells evade the drug, researchers tweaked Gleevec, making it bind more strongly, and developed new drugs that fit the slightly altered active site in resistant cancer cells.

As for Erin, she went off the drug twice, to have her children, and today is healthy. But she knows that a few resistant cancer cells may lurk in her body, unaffected by Gleevec. Fortunately, newer drugs are tailored to cancer cells that evade Gleevec.

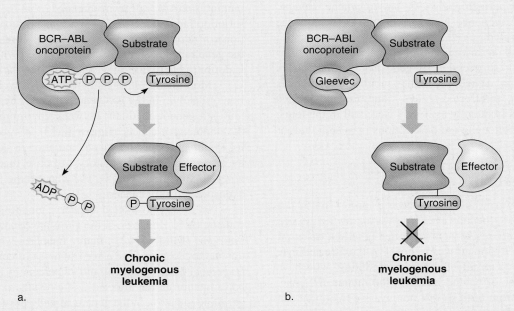

a.

b.

Figure 2 **How Gleevec treats chronic myelogenous leukemia.** In CML, a translocation forms the fusion oncoprotein BCR-ABL, which functions as a tyrosine kinase. A tyrosine (an amino acid) of a substrate molecule picks up a phosphate from the ATP nestled in the oncoprotein, making the substrate able to bind to another protein, called an effector, that triggers runaway cell division **(a).** Gleevec replaces the ATP **(b).** Without phosphorylation of the tyrosine on the substrate, division stops. As cancer progresses, mutations in the DNA of some cells make the shape of their pockets unable to bind the drug. Newer drugs can replace Gleevec once the cancer becomes resistant.

Source: Adapted from "Drug therapy: Imantinib mesylate—A new oral targeted therapy" by Savage & Antman: *New England Journal of Medicine* 346: 683–693. Copyright © 2002 Massachusetts Medical Society. All rights reserved. Reprinted by permission.

dipping into the cytoplasm. They function as a tyrosine kinase, as is the case for the leukemia described in Reading 18.1. When the growth factor binds to the tyrosine of the receptor, the tyrosine picks up a phosphate group, which signals the cell to activate transcription of genes that stimulate cell division. In HER2 breast cancer, too many tyrosine kinase receptors send too many signals to divide.

HER2 breast cancer usually strikes early in adulthood and spreads quickly. However, a monoclonal antibody-based drug called trastuzumab (Herceptin) binds to the receptors, blocking the signal to divide (see figure 17.17). Interestingly, Herceptin works when the extra receptors arise from multiple copies of the gene, rather than from extra transcription of a single *HER2* gene.

Tumor Suppressors

Some cancers result from loss or silencing of a gene that normally suppresses tumor formation by blocking the activities of other genes. Such a **tumor suppressor** gene normally inhibits expression of genes involved in all of the activities that turn a cell cancerous, listed in table 18.2. Cancer can result when a tumor suppressor's control is lifted. This can happen if the gene has a deletion, or if the promoter region binds too many methyl (CH_3) groups, which blocks transcription. Binding of CH_3 groups to "CpG islands"—regions in the starts of genes where the sequence "CG" repeats many times—turns off transcription. Such hypermethylation is an epigenetic change, because the mRNA sequence, and therefore the DNA sequence it reflects, is unaffected.

Wilms' tumor is an example of a cancer that develops from loss of tumor suppression. A gene that normally halts mitosis in the rapidly developing kidney tubules in the fetus is absent. As a result, an affected child's kidney retains pockets of cells dividing as frequently as if they were still in the fetus, forming a tumor. Following are descriptions of specific tumor suppressor genes.

Retinoblastoma (RB)

RB (MIM 180200) is a rare childhood eye tumor. In 1597, a Dutch anatomist described the eye cancer as a growth "the size of two fists." In 1886, researchers identified inherited cases. At that time, the only treatment was removal of the affected eye. Today, children with an affected parent or sibling, who have a 50 percent chance of having inherited the mutant *RB* gene, can be monitored from birth so that noninvasive treatment can begin early. Full recovery is common. Often the first abnormal sign is an unusual gray area that appears in an eye in a photograph because the tumor reflects light differently than unaffected parts of the eye.

About half of the 1 in 20,000 infants who develop RB inherit susceptibility to the disorder. They have one germline mutant allele for the *RB* gene in each of their cells, and then cancer develops in a somatic cell where the second copy of the *RB* gene mutates. Therefore, inherited retinoblastoma requires two point mutations or deletions, one germline and one somatic. In sporadic (noninherited) cases, two somatic mutations occur in the *RB* gene. Either way, RB usually starts in a cone cell of the retina, which provides color vision. Study of RB was the origin of the "two-hit" hypothesis of cancer causation—that two mutations (germline and somatic or two somatic) are required to cause a cancer related to tumor suppressor deletion or malfunction.

Many children with RB have deletions in the same region of the long arm of chromosome 13, which led researchers to the cancer-causing gene. In 1987, they found the *RB* gene and identified its protein product, which linked the cancer to control of the cell cycle. The RB protein normally binds transcription factors so that they cannot activate genes that carry out mitosis. It normally halts the cell cycle at G_1. When the *RB* gene is mutant or missing, the hold on the transcription factor is released, and cell division ensues.

Mutations in the *RB* gene cause other cancers. Children successfully treated for retinoblastoma often develop bone cancer as teens or bladder cancer as adults. Mutant *RB* genes have been found in the cells of patients with breast, lung, or prostate cancers, or acute myeloid leukemia, who never had the eye tumors. These other cancers may be caused by expression of the same genetic defect in different tissues.

p53 Normally Prevents Many Cancers

Another single gene that causes a variety of cancers when mutant is *p53*. Recall from chapter 12 that the p53 protein transcription factor "decides" whether a cell repairs DNA replication errors or dies by apoptosis. If a cell loses a *p53* gene, or if the gene mutates and malfunctions, a cell with damaged DNA is permitted to divide, and cancer may be the result.

More than half of human cancers involve a point mutation or deletion in the *p53* gene. This may be because p53 protein is a genetic mediator between environmental insults and development of cancer (**figure 18.13**). A type of skin cancer, for example, is caused by a *p53* mutation in skin cells damaged by an excessive inflammatory response that can result from repeated sunburns. That is, *p53* may be the link between sun exposure and skin cancer.

In most *p53*-related cancers, mutations occur only in somatic cells. However, in the germline condition Li-Fraumeni syndrome (MIM 151623), family members who inherit a mutation in the *p53* gene have a very high risk of developing cancer—50 percent do so by age 30, and 90 percent by age 70. A somatic *p53* mutation in the affected tissue results in cancer because a germline mutation in the gene is already present.

Stomach Cancer Lifts Cellular Adhesion

E-cadherin normally acts as a cellular adhesion protein found in tissue linings, but it is a tumor suppressor because when it is deleted, cancer results. This was the case for the Bradfield family. Golda Bradfield died of stomach cancer in 1960. By the time some of her grown children developed the cancer too, the grandchildren began to realize that their family had a terrible

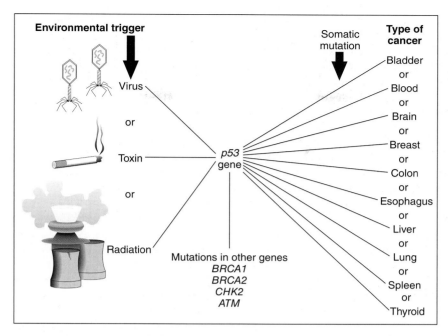

Figure 18.13 *p53* **cancers reflect environmental insults.** The environment triggers mutations or changes in gene expression that lead to cancer. The *p53* gene is a mediator—"the guardian of the genome." The protein products of many genes interact with *p53*.

legacy. Genetic testing revealed familial diffuse gastric cancer (MIM 192090), caused by an "exon skipping" mutation in the E-cadherin gene.

Golda's grandchildren had genetic tests. Eleven of them had inherited the mutant gene, but scans of their stomachs did not show any tumors. Still, they all had their stomachs removed. It was good that they did, because most of them already had hundreds of tumors, too tiny to have been seen on medical scans. The cousins without stomachs are doing well. Like people who have their stomachs surgically shrunk to lose weight, the cousins avoid hard-to-digest foods and eat a little at a time, throughout the day. The inconvenience, they say, is a fair trade for eliminating the fear of their grandmother's, parents', and aunts' and uncles' fates—stomach cancer.

BRCA1/BRCA2 Mutations Disrupt Repair

Breast cancer that runs in families may be due to inheriting a germline mutation and then having a somatic mutation occur in a breast cell (a familial form), or two somatic mutations affecting the same breast cell (a sporadic form), as figure 18.4 depicts. However, breast cancer is so common that a family with many affected members may actually have multiple sporadic cases, rather than an inherited form of the disease.

Familial breast cancer exhibits some of the complications of Mendel's laws described in chapter 5. Mutations in many genes can cause breast cancer, and many are incompletely penetrant, which means that a genotype does not always lead to the associated phenotype. That is, cancer may never develop, even if a person has inherited a breast cancer mutation. For this reason, mutations in breast cancer genes are said to confer susceptibility, rather than cause cancer directly. The genetic heterogeneity

of breast cancer also means that absence of one cancer-causing mutation is not a guarantee that breast cancer will never develop.

Only about 5 percent of breast cancers are familial, caused by mutations in any of more than a dozen different genes. Most of the genes associated with familial breast cancer encode proteins that interact in ways that enable DNA to be repaired. If DNA cannot be repaired, mutations that directly cause cancer can accumulate and persist.

The two major breast cancer susceptibility genes are *BRCA1* and *BRCA2*. Together they account for 15 to 20 percent of familial cases. *BRCA1,* which stands for "breast cancer predisposition gene 1," greatly increases the lifetime risk of inheriting breast and ovarian cancer. This risk, however, varies in different population groups because of the modifying effects of other genes. In the most common *BRCA1* mutation, deletion of two adjacent DNA bases alters the reading frame, shortening the protein. The mutation is inherited as an autosomal dominant trait, with incomplete penetrance.

BRCA1 and *BRCA2* encode proteins that join two others to form a complex that moves to areas of the genome where both DNA strands are broken at the same site. These double-stranded breaks are particularly dangerous because they cut the chromosome, making rearrangements such as deletions and translocation possible. Normally the protein complex that includes BRCA1 and BRCA2 allows *p53* and other cell cycle checkpoint genes to repair the damage or, if it is too extensive, trigger cell death.

BRCA1 mutations have different incidences in different populations. Only 1 in 833 people in the general U.S. population has a mutant *BRCA1* allele. That figure is more than 1 in 50 among Ashkenazi Jewish people, due to population bottlenecks and nonrandom mating. The *BRCA1* gene was initially discovered in Ashkenazi families in which several members developed the cancer at very young ages. In this population, a woman who inherits a BRCA1 mutation has up to an 87 percent risk of developing breast cancer over her lifetime, and a 50 percent risk of developing ovarian cancer.

Some women who learn that they have inherited an allele predisposing to breast or ovarian cancer have their breasts and/or ovaries removed. This action makes more sense for an Ashkenazi woman facing an 87 percent lifetime risk of breast or ovarian cancer than it does for a woman in a population group with a much lower risk. The general population risk of actually developing a BRCA1 cancer if one inherits a mutation is only about 10 percent, based on empirical (observational) evidence. Women with such mutations born after 1940 have a higher risk than those born earlier, suggesting that the environment also plays a role in whether inheriting a mutation causes cancer.

BRCA2 breast cancer is also more common among the Ashkenazim. Ashkenazi women who inherit a mutation in BRCA2 face a 60 to 85 percent lifetime risk of developing breast cancer and a 10 to 20 percent risk of developing ovarian cancer. Men who inherit a BRCA2 mutation have a 6 percent lifetime risk of developing breast cancer, which is 100 times the risk for men in the general population. Inheriting a BRCA2 mutation also increases the risk of developing cancers of the colon, kidney, prostate, pancreas, gallbladder, skin, or stomach.

The fact that p53, BRCA1, and BRCA2 proteins all bind to each other in the nucleus suggests that they interact to enable a cell to repair double-stranded DNA breaks. BRCA2 also pulls apart daughter cells as mitosis completes. Mutations in the *BRCA2* gene may explain the aneuploidy (extra or missing chromosomes) that is common in cancer cells.

Genes whose protein products affect those of *BRCA1*, *BRCA2*, and *p53* can cause breast cancer. For example, the product of a gene called *ATM* adds a phosphate to the product of a gene called *CHK2*, which then adds a phosphate to the BRCA1 protein. Mutations in *ATM* and *CHK2* also cause breast cancer because they affect BRCA1 protein. Another form of breast cancer results from mutations in any of five genes known to cause Fanconi anemia, a fatal blood disorder. Five of the Fanconi anemia proteins form a cluster that activates a sixth protein, which in turn binds to and inactivates the BRCA2 protein. Overall, the mutations that increase risk of breast cancer seem to have in common interference with DNA repair.

MicroRNAs Revisited

MicroRNAs are small RNA molecules that act as "dimmer switches," blocking translation of certain genes into proteins by binding to their mRNA transcripts (see section 11.2). Because microRNAs normally control the expression of certain proto-oncogenes and tumor suppressor genes, when microRNAs themselves have mutations or their expression is too high or too low, cancer can result. For example, in B cell lymphoma, a blood cancer, a B cell has thirty times the normal number of copies of a particular microRNA. Normally, this microRNA activates B cells in the lymph nodes. The extra microRNAs stimulate too much cell division among the B cells, and cancer results.

MicroRNAs, oncogenes, and tumor suppressor genes interact because they target the same pathways (see table 18.2). Their relationships are complex. For example, the tumor suppressor p53 controls a family of three microRNAs that regulate the cell cycle and cause lung cancer if disrupted.

Changes in microRNA expression also accompany cancer's progression. When cells at the center of a solid tumor become starved for oxygen before angiogenesis brings in a blood supply, a set of microRNAs different from the ones that may have started the cancer appear. These microRNAs alter the expression of yet other genes so that the cell can use metabolic pathways that enable it to survive without oxygen. Yet other combinations of microRNAs alter gene expression as a cancer invades nearby tissue and spreads.

Researchers are using the changes in microRNA expression that accompany cancer to develop new, more sensitive ways to diagnose, treat, and follow response to treatment. For example, different microRNA expression patterns indicate whether pancreatic pain is due to inflammation only, or to cancer—an important distinction. For breast cancer, a microRNA expression "signature" correlates to the likelihood of spread, helping physicians to decide which patients would benefit from further treatments.

Understanding microRNAs may solve a mystery in cancer genetics: families in which several members have or had cancer, but do not have any of the recognized, single-gene family cancer syndromes. The fact that a single sequence of microRNA can have many targets—many genes whose translation it blocks—may explain why different family members develop different cancers.

Key Concepts

1. Proto-oncogenes normally control the cell cycle. They can become oncogenes when they mutate, move next to a gene that is highly expressed, or are transcribed and translated with another gene, forming a fusion protein.

2. Mutations in tumor suppressor genes usually are deletions that cause a cell to ignore extracellular constraints on cell division, or affect DNA repair.

3. MicroRNAs alter the expression of oncogenes and tumor suppressor genes. Patterns of microRNA expression change as a cancer progresses.

18.5 Many Genes Contribute to Cancer

The field of cancer genetics has progressed much like genetics in general, from a single-gene approach to considering the sequential and combined actions of many genes.

Genes that guide a cell toward the cancerous state when mutant are considered in two broad categories, based on their effects. "Gatekeeper" genes control mitosis and apoptosis, which must be in balance to maintain the number of cells forming the affected tissue. Their effect is direct. "Caretaker" genes, in contrast, control the mutation rates of gatekeepers, and may have the overall effect, when mutant, of destabilizing the genomes of cells that will become or are cancerous.

Most, if not all, cancers are the culmination of a series of changes in several genes, including gatekeepers and caretakers. To identify the steps and reconstruct the progression of the disease, researchers examine DNA in tumor cells from people in different stages of the same type of cancer. The older the tumor, the more genetic changes have accumulated. A mutation present in all stages acts early in the disease process, whereas a mutation seen only in the tumor cells of sicker people functions late in the process. Each step provides a potential point of intervention.

Another way to identify the mutations that fuel a cancer's growth and spread is to study the cancer cells of patients who have responded to a particular drug, then relapsed. How does the cancer escape destruction by a drug that had initially worked? For example, an experimental drug effectively treated several patients with metastatic melanoma (see figure 18.1), but the cancers returned after 7 months. Examination of the cancer cells from some of these patients showed mutations in genes other than the one that had initially caused the cancer. Specifically, mutations in three genes changed the cancer cells' surfaces or metabolisms in ways that enabled them to ignore the drug and keep dividing and spreading. This information is important to drug developers because it means that a "cocktail" of several drugs, each acting on a different pathway (see table 18.2), is the best approach to treat this and possibly many cancers, just as it is for treating HIV infection.

Colon Cancer

The series of genetic changes underlying cancer is particularly well-studied for cancer of the colon (large intestine). Colon cancer does not usually occur in families with the frequency or pattern of a single-gene disorder. However, when family members with noncancerous growths (polyps) in the colon are considered with those who have colon cancer, a Mendelian pattern emerges (**figure 18.14***a*). Five percent of colon cancer cases are inherited. One in 5,000 people in the United States has precancerous colon polyps, a condition called familial adenomatous polyposis (FAP; MIM 175100).

Healthy colon lining cells typically live 3 days. In FAP, they fail to die on schedule and instead build up, forming polyps. FAP begins in early childhood with hundreds of tiny polyps that progress over many years to colon cancer. Both oncogenes and silenced tumor suppressors take part as FAP becomes colon cancer.

Our current understanding of the sequence of genetic changes that underlies FAP colon cancer comes from genealogical sleuthing as well as scientific research on a cluster of families living in Salt Lake City, Utah. Geneticists at the University of Utah compared the vast genealogical records of the Utah Population Database, maintained by the Mormon community, to state registries of cancer cases, births, and deaths. They discovered a founder effect (see section 15.3). The affected families in Utah and another in upstate New York that share the same FAP mutation also share ancestors—specifically, Mr. and Mrs. George Fry. The couple, born in Somerset, England, in the 1590s, sailed to Plymouth, Massachusetts, with their children around 1630. A son passed the colon cancer mutation to the branch of the family that settled in upstate New York, and a daughter passed it to the pioneers who left for Utah with other Mormons in the 1850s.

In the fall of 1947 at the University of Utah, young professor Eldon Gardner told his class that colon cancer might be inherited. A student, Eugene Robertson, excitedly told the class that he knew of a family in which a grandmother, her three children, and three grandchildren had colon cancer. Intrigued, Gardner delved into the family's records and began interviewing relatives. He eventually found fifty-one family members

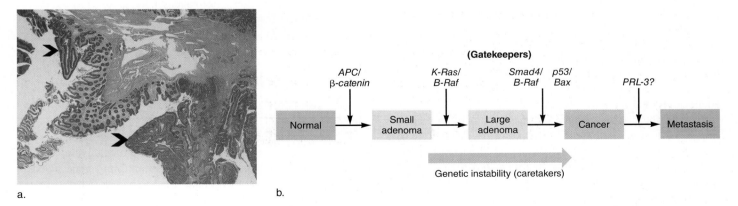

a.   b.

Figure 18.14 **Several mutations contribute to FAP colon cancer.** **(a)** In familial adenomatous polyposis (FAP), hundreds of polyps become cancerous. The arrows indicate two polyps extending from the intestinal lining **(b)**. Cells lining the colon divide more frequently when the *APC* gene on chromosome 5q undergoes a deletion or point mutation, causing small benign tumors (adenomas) to form. Activation of certain oncogenes, such as *K-Ras* and *B-Raf,* fuel growth of the adenomas. Mutations in *p53* and other genes push the adenoma cells to become cancerous. Finally, mutations in a gene called *PRL-3* trigger metastasis. Caretaker genes cause genetic instability that contributes to the disease process.

and arranged for each to be examined with a colonoscope to view the wall of the colon. The colons of six of the fifty-one people were riddled with polyps, although none of the six had symptoms. Removal of the affected tissue probably saved their lives.

In the years that followed, researchers identified other families with more than one case of colon polyps. Individuals with only polyps were diagnosed with FAP. If a person with colon polyps had cancer elsewhere, extra teeth, and pigment patches in the eye, the condition was called Gardner syndrome, named for the professor.

Researchers identified the chromosomal defect that causes Gardner syndrome in 1985 with the help of a 42-year-old man at the Roswell Park Cancer Institute in Buffalo, New York. He had several problems—no gallbladder, an incomplete liver, an abnormal kidney, intellectual disability, and Gardner syndrome. To a geneticist, a seemingly unrelated combination of symptoms suggests a chromosomal abnormality affecting several genes. Sure enough, the man's chromosome 5 had a small deletion. This was the first piece to the puzzle of colon cancer. The deletion removed a gene called *APC*. It is the main "gatekeeper" for this type of colon cancer, and is the first step depicted in figure 18.14*b*.

Normally APC protein binds to another protein, β-catenin, adding a phosphate to it that blocks its action. When the *APC* gene is deleted, β-catenin isn't silenced, and instead it enters the nucleus and activates genes that promote mitosis. The cell becomes unable to stop dividing. A tumor forms, but it is not yet malignant. Other pathways, such as those controlled by the genes *TGF* and *p53,* push the abnormal cells to become cancerous. *TGF* normally inhibits mitosis, and *p53* normally sends cells to a fate of apoptosis. *PRL-3* is a gene that acts late in the process, enabling the cancer to spread. Several caretaker genes affect the expression of the gatekeepers, so the overall picture is quite complex.

The Cancer Genome

A cancer may undergo several dozen genetic changes as well as changes in gene expression as it grows and spreads. Several large-scale projects are simultaneously analyzing many aspects of the genomes of cancer cells and constructing "atlases" that include the types of information in **table 18.3**.

A cancer genome study on a form of lung cancer called adenocarcinoma illustrates the type of insights that emerge by cataloging the genetic changes in a cancer cell. Researchers sequenced 623 genes that affect the pathways involved in cancer in tumor samples from 188 patients, and narrowed the list of mutations detected down to 26 found in a majority of the samples. Most of the 26 implicated genes had not been associated with lung cancer, but had been seen in other types of cancer. These included some already mentioned—*APC* (colon), *ATM* (breast), and *RB* (eye). This study added to the growing evidence that mutations in the same genes can cause or contribute to cancers in different parts of the body. Such genome-level looks at cancer genes enable researchers to investigate

Table 18.3	Types of Information in a Cancer "Atlas"
Mutation	
	■ Somatic or germinal
	■ Oncogene or tumor suppressor gene
	■ Chromosome level
Variation	
	■ Single nucleotide polymorphisms (SNPs)
	■ Copy number variants (CNVs)
Gene Expression	
	■ mRNA profile (transcriptomics)
	■ Protein profile (proteomics)
	■ Epigenetic profile (methylation pattern)
	■ MicroRNA pattern
Environmental Exposure (carcinogens)	

environmental effects, too. For example, in the adenocarcinoma study, cancer cells from smokers had many more mutations than cancer cells from nonsmokers, and different genes were mutant.

Key Concepts

1. Gatekeeper genes affect mitosis and apoptosis. Caretaker genes affect genome stability.
2. Determining which mutations are present in particular stages of a cancer can reveal the sequence of gene actions.
3. Large-scale studies of genetic changes in cancer are revealing shared abnormalities in the same pathways in different cancer types.

18.6 Environmental Causes of Cancer

Environmental factors contribute to cancer by mutating or altering the expression of genes that control the cell cycle, apoptosis, and DNA repair. Inheriting a susceptibility gene places a person farther along a particular road to cancer, but cancer can happen in somatic cells in anyone. It is more practical, for now, to identify environmental cancer triggers and develop ways to control them or limit our exposure to them, than to alter genes.

Looking at cancer at a population level reveals the interactions of genes and the environment. For example, researchers examined samples of non-Hodgkin's lymphoma tumors from

172 farmers, 65 of whom had a specific chromosomal translocation. The 65 farmers were much more likely to have been exposed for long times to toxic insecticides, herbicides, fungicides, and fumigants, compared to the farmers with lymphoma who did not have the translocation.

Determining precisely how an environmental factor such as diet affects cancer risk can be complicated. Consider the cruciferous vegetables, such as broccoli and brussels sprouts, which are associated with decreased risk of developing colon cancer. These vegetables release compounds called glucosinolates, which in turn activate "xenobiotic metabolizing enzymes" that detoxify carcinogenic products of cooked meat called heterocyclic aromatic amines. With a vegetable-poor, meaty diet, these amines accumulate. They cross the lining of the digestive tract and circulate to the liver, where enzymes metabolize them into compounds that cause the mutations associated with colon cancer (**figure 18.15**).

Exposure to carcinogens—in the workplace, home, or outdoors—can raise cancer risk. Chemical carcinogens were recognized as long ago as 1775, when British physician Sir Percival Potts suggested that the high rate of skin cancer in the scrotums of chimney sweeps in London was due to their exposure to a chemical in soot. Since then, epidemiological studies have identified many chemicals as possibly causing cancer in certain populations. However, most studies reveal correlations rather than cause-and-effect relationships. In the strongest cases, genetic or biochemical evidence explains the observed environmental connection.

Epidemiologists use different statistical tools and compare people in different ways to link environmental exposures and cancer. A **population study** compares the incidence of a type of cancer among very different groups of people. If the incidence differs, then some distinction among the populations may be responsible. For example, an oft-mentioned study from 1922 found that primitive societies have much lower rates of many cancers than more developed societies. The study attributed the lack of cancer to the high level of physical activity among the primitive peoples—but diet might also have explained the difference.

Population studies often have too many variables to clearly establish cause and effect. Consider the very high incidence of breast cancer on Long Island, New York. One

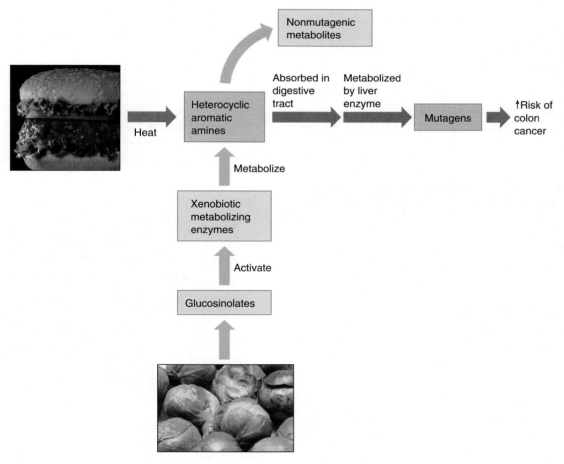

Figure 18.15 **One way that cruciferous vegetables lower cancer risk.** Compounds called heterocyclic aromatic amines form in cooking meat, are absorbed into the digestive tract, and are metabolized by a liver enzyme into mutagens, which may cause colon cancer. Broccoli and brussels sprouts produce glucosinolates, which activate xenobiotic metabolizing enzymes that block part of the pathway that leads to production of the mutagens.

hypothesis attributes the mini-epidemic to pesticide exposure from when the area was farmland. However, this population also has a high frequency of *BRCA1* mutations among its Ashkenazi citizens. Sociological factors come into play, too. In this population, women have frequent mammograms starting at a young age. As a result, the percentage of the population with recognized early stages of the disease may be higher than in other populations where women are less likely to have regular mammograms. All of these factors may contribute to the high breast cancer incidence in this area.

More informative than a population study is a **case-control study,** in which people with a type of cancer are matched with healthy individuals for age, sex, and other characteristics. Then researchers look for differences between the pairs. If, for example, the cancer patients had extensive dental X rays at a young age but the control group didn't, X-ray exposure may be a causal factor. Limitations of this type of study are that much of the information is based on recall, people make mistakes, and not all relevant factors are identified and taken into account.

The most informative type of epidemiological investigation is a **prospective study**. Two or more groups of people follow a specified activity plan, such as a dietary regimen, and are checked periodically for cancer. By looking ahead, the investigator has more control over the activities and can verify information. However, a limitation of this type of study is that cancer usually takes many years to appear and progress.

Once epidemiological studies indicate a correlation, a biological explanation is necessary to draw conclusions or suggest further studies. For example, eating certain vegetables that contain antioxidant compounds, which deactivate the free radicals that can damage DNA, may explain the fact that lifetime cancer risk for vegetarians is 29 in 100, compared to 33 in 100 for the general population.

18.7 The Personalization of Cancer Diagnosis and Treatment

Estimating the risk that a certain type of cancer will affect a particular individual is possible for a few disorders inherited in a single-gene fashion through known mutations. More often, discovery of cancer follows a screening test such as mammography or high levels of prostate specific antigen in the bloodstream, feeling a lump, or noting a sore that does not heal.

Cancers have traditionally been categorized by the body part or tissue type affected. However, when this approach lumps together cancers that actually have different causes, a patient may be prescribed a treatment that will not work. This situation lead to the discovery of mixed-lineage leukemia.

Acute lymphoblastic leukemia (ALL) causes fatigue, fever, and bruising, like other forms of this blood cancer. The cancerous white blood cells divide too frequently, crowding out red blood cells and platelets. Chemotherapy helps 90 percent of children with ALL. To discover why the others die, researchers examined what the cancer cells do, rather than what they look like under a microscope. They used DNA microarrays (chips) to compare the expression of 12,000 genes in cancer cells from both types of leukemia patients. Compared to patients who responded, the cancerous cells of the other patients made too little of 1,000 of the encoded proteins and too many of 200 others. Even though the children all had the same symptoms and their cancer cells looked alike, at the level of gene expression—the proteins manufactured in the cancer cells—the diseases were distinct (**figure 18.16**). The 10 percent of children who do not respond to ALL treatment actually have MLL (mixed-lineage leukemia). Different chemotherapy drugs help them.

The oldest cancer treatment is surgery, which prevents invasiveness by removing the tumor. Two other common treatment approaches are radiation and chemotherapy, which kill all cells that divide rapidly. These treatments also affect healthy cells in the digestive tract, hair follicles, and bone marrow, causing side effects of nausea, hair loss, great fatigue, and susceptibility to infection. Other drugs can help patients tolerate the side effects, including colony-stimulating factors to replenish bone marrow. These other drugs enable patients to withstand higher doses of chemotherapy, which may be more effective.

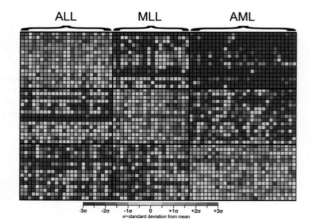

Figure 18.16 **Cancer cells that look alike may be genetically distinct.** These leukemias—ALL, MLL, and AML—differ in gene expression patterns. The columns of squares represent DNA from tumor samples, and the rows compare the activities of particular genes. Red tones indicate higher-than-normal expression and blue tones show lower-than-normal expression. The different patterns indicate distinct cancers, although the cells look alike under a microscope.

Several newer types of cancer drugs affect cancer cell characteristics or activities other than accelerated division rate. Some treatments:

- stimulate cells to regain specialized characteristics, such as drugs based on retinoic acid;
- inhibit telomerase, which prevents cancer cells from elongating their telomeres and continually dividing;
- induce apoptosis, which halts cell division; and
- inhibit angiogenesis, which robs a cancer of its blood supply, makes the tumor soak up more chemotherapy, kills tumor cells directly, and stimulates the immune response.

Table 18.4	Evolution of Treatments for Breast Cancer
Strategy	**Examples**
Remove or destroy cancerous tissue	Surgery, radiation, chemotherapy
Use phenotype to select drug	Estrogen receptor–positive women take a selective estrogen receptor modulator or an aromatase inhibitor or both
Use genotype to select drug	Women with *HER2*-positive cancers take trastuzumab (Herceptin, a monoclonal antibody)
Genomic level	Gene expression profile on DNA microarray used to guide drug choice; 70-gene signature predicts metastasis

Diagnostic tests and treatments for cancer have become more targeted to cancer cells while sparing healthy ones, thanks to genetic discoveries. For example, colonoscopy can reveal colon tumors that protrude from the intestinal wall, but not small or flat growths. Tests on feces can detect colon cancer mutations as well as gene methylation changes associated with colon cancer. Screening for colon cancer by probing feces can limit the more invasive colonoscopy to individuals with genetic evidence of colon cancer in what passes from their intestines.

Treating breast cancer illustrates how genetic information is refining management of these diseases (**table 18.4**). The first targeting of breast cancer treatment came with recognition that breast cancer cells have receptors for one or two hormones, estrogen and/or progesterone—or neither receptor type. Women with estrogen receptor–positive tumors begin a several-year course of either a drug that blocks these receptors from receiving signals to divide, or a drug that inhibits an enzyme called aromatase, which is required to produce estrogen.

Determining the estrogen receptor status of a breast tumor is subtyping the disease by phenotype. With the discovery of single genes that cause cancer, diagnosis began to include genotyping. For example, a woman might have *BRCA1* and/or *HER2* breast cancer. The monoclonal antibody–based drug trastuzumab (Herceptin) is highly effective in treating *HER2* breast cancer by blocking the receptors for a certain growth factor.

Increasingly, cancer diagnosis utilizes DNA microarrays that scan the genome for cancer-associated mutations as well as gene expression patterns. This approach of genetically characterizing tumor cells enables physicians to match a particular patient to the treatments most likely to work right from the start with the fewest side effects, and then to monitor response. DNA tests can also predict metastasis. For example, a test that evaluates the expression of seventy genes is used to identify early-stage breast cancers that are most likely to recur after treatment; those tumors are treated more aggressively. The test saves some patients from painful treatment that might not be necessary or effective.

Breast cancer is also described by *lack* of specific receptors. The cancer cells of women with "triple negative" breast cancer do not have excess estrogen, progesterone, or *HER2* receptors. Although these cancers can respond to conventional, nontargeted chemotherapy, the newer drugs that target these receptors are ineffective. Triple negative breast cancer is aggressive, strikes early, and tends to recur.

The limitation of any cancer treatment, old or new, is defined by the strength of the enemy. Cancer cells are incredibly abundant and ever-changing. Surgery followed by a barrage of drugs and radiation can slow the course of the disease, but all it takes is a few escaped cancer cells—called micrometastases—to sow the seeds of a future tumor. The DNA of cancer cells mutates in ways that enable the cells to pump out any drug sent into them. In addition, cancer cells have redundancies, so that if a drug shuts down angiogenesis or invasiveness, the cell completes the task another way. Although cancer treatments can cure, it is more likely that they kill enough cancer cells, and sufficiently slow the spread, so that it takes the remainder of a lifetime for the tumors to grow back. In this way, cancer becomes a chronic, manageable condition.

Even as targeted cancer treatments are becoming available, continuing analysis of the human genome is revealing that our view of cancer as a derangement of the cell cycle may be a great oversimplification. Scans of different types of tumors reveal hundreds of genes that mutate as the disease progresses, including well-known cancer genes such as those discussed in this chapter, but also many genes never implicated in cancer before, such as genes that control cell adhesion. In addition, the same type of cancer in different individuals often has different mutations. The overall conclusion: We still have a lot to learn about cancer.

Key Concepts

1. Treatments for cancer target the characteristics of cancer cells. Surgery removes tumors. Chemotherapy and radiation nonselectively destroy rapidly dividing cells.
2. Newer treatments target receptors on cancer cells, block telomerase, stimulate differentiation, or attack a tumor's blood supply.
3. Diagnosis and treatment of cancer will increasingly consider genetic and genomic information that enables physicians to better match patient to treatment.

Summary

18.1 Cancer Is Genetic, But Usually Not Inherited

1. Cancer is a genetically dictated loss of cell cycle control, creating a population of highly proliferative cells that invades surrounding tissue.

2. Mutations in genes that encode or control transcription factors, cell cycle checkpoint proteins, growth factors, repair proteins, or telomerase may disrupt the cell cycle, causing cancer.

3. Changing gene expression patterns also contribute to cancer, and can be used to distinguish cancer types.

4. Sporadic cancers result from two **somatic mutations** in the two copies of a gene. They are more common than cancers that are caused by **germline mutations** plus somatic mutations in affected tissue.

18.2 Characteristics of Cancer Cells

5. A tumor cell divides more frequently or more times than cells surrounding it, has altered surface properties, loses the specializations of the cell type it arose from, and produces daughter cells like itself.

6. A malignant tumor infiltrates tissues and can **metastasize** by attaching to basement membranes and secreting enzymes that penetrate tissues and open a route to the bloodstream. From there, a cancer cell can travel, establishing secondary tumors.

18.3 Origins of Cancer Cells

7. Cell specialization and position within a tissue are important determinants of whether cancer begins and persists.

8. **Cancer stem cells** can divide to yield cancer cells and abnormally differentiated cells.

9. A cell that dedifferentiates and/or turns on expression of "stemness" genes can begin a cancer.

10. A mutation that enables a cell to divide continually can alter the percentages of cells in a tissue that can divide, resulting in an abnormal growth.

11. Chronic repair of tissue damage can provoke stem cells into producing an abnormal growth.

18.4 Cancer Genes and MicroRNAs

12. Cancer is often the result of activation of **proto-oncogenes** to **oncogenes,** and inactivation of **tumor suppressor genes**. Mutations in DNA repair genes cause cancer by increasing the mutation rate.

13. Proto-oncogenes normally promote controlled cell growth, but are overexpressed because of a point mutation, placement next to a highly expressed gene, or transcription and translation with another gene, producing a **fusion protein**. Oncogenes may also be overexpressed growth factor receptors.

14. A tumor suppressor is a gene that normally enables a cell to respond to factors that limit its division.

15. MicroRNAs control expression of oncogenes and tumor suppressor genes. MicroRNA gene expression changes as a cancer progresses.

18.5 Many Genes Contribute to Cancer

16. To decipher the gene action sequences that result in cancer, researchers examine the mutations in cells from patients at various stages of the same type of cancer. Those mutations present at all stages of the cancer are the first to occur.

17. FAP colon cancer requires several mutations to develop.

18. Cancer atlases catalog all changes in gene structure and function that underlie a particular cancer.

18.6 Environmental Causes of Cancer

19. **Population, case-control,** and **prospective studies** can correlate environmental exposures to development of certain cancers. Biochemical and/or genetic evidence can sometimes explain epidemiological observations.

18.7 The Personalization of Cancer Diagnosis and Treatment

20. Traditional cancer treatments are surgery, radiation, and chemotherapy. Newer approaches block hormone receptors, stimulate cell specialization, block telomerase, and inhibit angiogenesis. Identifying mutations and gene expression patterns are used to subtype cancers and better target treatments.

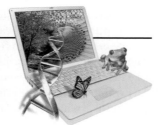

www.mhhe.com/lewisgenetics10

Answers to all end-of-chapter questions can be found at **www.mhhe.com/lewisgenetics10.** You will also find additional practice quizzes, animations, videos, and vocabulary flashcards to help you master the material in this chapter.

Review Questions

1. Explain the connection between cancer and control of the cell cycle.

2. Explain why not all cells whose chromosomes have long telomeres are cancer cells.

3. Explain how the cell cycle is controlled from both inside and outside the cell.

4. Explain why cancer is usually a genetic disease at the molecular and cellular levels, but not at the whole-body level.

5. Explain why it is important to know whether a cancer is sporadic or inherited.

6. List four characteristics of cancer cells.

7. Define *dedifferentiation*.

8. List the processes and pathways that are abnormal in cancer.

9. Describe what happens as a cancer grows beyond the original tumor.

10. Explain what is inaccurate about the statement "cancer cells are the fastest dividing cells in the body."

11. Describe four ways that cancer can originate at the cell or tissue level.

12. Define *cancer stem cell*.

13. Distinguish between a proto-oncogene and an oncogene.

14. Describe two ways that oncogenes are activated.

15. Explain how an oncogene is associated with a gain of function and a mutation in a tumor suppressor gene is associated with a loss of function.

16. Explain how retinoblastoma, a *p53*-related cancer, inherited stomach cancer, and *BRCA1* breast cancer have similar causes.

17. Explain how microRNAs can affect oncogenes and tumor suppressor genes.

18. Distinguish between gatekeeper and caretaker genes in cancer.

19. Describe how a form of inherited colon cancer illustrates that several mutations can contribute to causing the disease.

20. Explain how comparing mutations in cells from the same cancer type at different stages can reveal the sequence of genetic changes behind the cancer.

21. Name a type of cancer caused by mutation of a single gene and a cancer caused by mutations in more than one gene.

22. Distinguish among mutations, altered gene expression, and epigenetic changes in cancer.

23. Describe a way that exploring the genome of cancer cells reveals information not apparent from focusing on one type of cancer or one gene associated with cancer.

24. Distinguish among population, case-control, and prospective studies to identify environmental factors in cancer.

25. Explain how a cancer treatment that targets cell surface receptors works.

26. Explain why not all cancers affecting the same cell type respond the same way to a particular drug.

Applied Questions

1. An individual can develop breast cancer by inheriting a germline mutation, then undergoing a second mutation in a breast cell; or by undergoing two mutations in a breast cell, one in each copy of a tumor suppressor gene. Cite another type of cancer, discussed in the chapter, that can arise in these two ways.

2. How do the mechanisms of the drugs Gleevec and Herceptin differ?

3. A young black woman thinks that she cannot get a *BRCA* form of breast cancer because she isn't Jewish. Is she correct?

4. von Hippel-Lindau syndrome (MIM 193300) is an inherited cancer syndrome. The responsible mutation lifts control over the transcription of certain genes, which, when overexpressed, cause tumors to form in the kidneys, adrenal glands, and blood vessels. Is the von Hippel-Lindau gene an oncogene or a tumor suppressor? Cite a reason for your answer.

5. The *BRCA2* gene causes some cases of Wilms' tumor and some cases of breast cancer. Explain how the same tumor suppressor mutation can cause different cancers.

6. Ads for the cervical cancer vaccine present the fact that a virus can cause cancer as startling news, when in fact this has been known for decades. Explain how a virus might cause cancer.

7. A tumor is removed from a mouse and broken up into cells. Each cell is injected into a different mouse. Although all the mice used in the experiment are genetically identical and raised in the same environment, the animals develop cancers with different rates of metastasis. Some mice die quickly, some linger, and others recover. What do these results indicate about the characteristics of the original tumor cells?

8. Colon, breast, ovarian, and stomach cancers can be prevented by removing the affected organ. Why is this approach not possible for chronic myeloid leukemia?

9. A vegetarian develops pancreatic cancer and wants to sue the nutritionist who suggested she follow a vegetarian diet. Is her complaint justified? Why or why not?

10. MammaPrint is a DNA microarray-based test of the expression of seventy genes implicated in breast cancer. Certain patterns are significantly more common in cancers that spread, creating a "signature" that doctors can use to guide treatment decisions. Cite an advantage and a shortcoming of this test.

11. The discovery of cancer stem cells suggests a new type of treatment—develop a drug that stops self-renewal. Explain how such a drug might work, and what an adverse effect might be.

12. Colorectal cancer is diagnosed in half a million people worldwide each year. In 4 percent of diagnosed individuals, the cancer is part of a familial cancer syndrome, such as Lynch syndrome (MIM 114400). Genetic testing for Lynch syndrome targets mismatch repair genes, and costs about $3,000. What information would be valuable to decide if it is practical to test for Lynch syndrome for all cases of newly diagnosed colon cancer?

13. A mutation in a gene called *FLT3*, which encodes a tyrosine kinase receptor, causes acute myeloid leukemia, which has a 5-year survival rate of 20 percent. A new drug blocks the receptor on white blood cells. Explain how it works.

14. Rose and Angela are 4-year-old identical twins. Rose develops leukemia, but Angela does not. How is this possible?

15. News reports very often feature celebrities who, after a few months of cancer treatment, are "cancer-free." How likely is this to be true?

16. The media widely reported a technology that detects one cancer cell out of a billion cells in a blood sample as a breakthrough, claiming that it could be used as a test on which to base treatment choices. What might be a limitation of this technology?

Web Activities

1. Go to the Cancer Quest website (www.cancerquest.org). Select "Cancer Biology" from the drop-down menu and click on Cancer Genes. Select an oncogene or tumor suppressor gene and describe how, when mutant, the gene causes cancer.

2. Go to http://cancergenome.nih.gov and describe a recent discovery. Identify the type of cancer and explain how

genomic information either confirms what was already known about particular genes that cause the cancer, or adds to or changes what was known.

3. Consult the websites for the pharmaceutical companies that market Herceptin, Gleevec, bevacizumab (Avastin), or any other cancer drug and explain how the drug works.

Case Studies and Research Results

1. "In dedifferentiation, cancer cells forget who they are," wrote a researcher. Explain what she meant.

2. Epigenetic changes are associated with some cancers, such as certain changes in chromatin remodeling. Specifically, in some cancers several histone proteins have extra methyls and too few acetyls (see figure 11.5). Researchers are developing drug combinations that restore the correct chromatin configuration. Why might this approach be unfeasible?

3. What further experiments could be done to demonstrate how pesticide exposure caused lymphoma in farmers (section 18.6)?

4. DeShawn takes the drug Gleevec to treat his leukemia, and it has worked so well that he thinks he is cured. He stops taking the drug, and 4 months later his leukemia returns. This time, the cancer cells do not have the BCR-ABL mRNA characteristic of the disease. Explain what has happened.

5. The genomes of three patients with acute myeloid leukemia are sequenced and mutations in the following genes noted:

   patient 1   *IDH1* and *NPM1*
   patient 2   just *IDH1*
   patient 3   *IDH1*, *NPM1*, and *IDH2*
   patient 4   *IDH1*, *NPM1*, *IDH2*, and *FLT3*

Explain how these patients can have the same diagnosis, yet mutations in different genes.

6. When a medical journal published a meta-analysis (a review of many studies) indicating that evidence of cancer in mummies and other preserved ancient humans is extremely rare, the news media reported that the analysis indicated that cancer is a result of our modern way of life. Suggest two alternative explanations.

7. Elsie finds a small lump in her breast and goes to her physician, who takes a medical and family history. She mentions that her father died of brain cancer, a cousin had leukemia, and her older sister was just diagnosed with a tumor of connective tissue. The doctor assures her that the family cancer history doesn't raise the risk that her breast lump is cancerous, because the other cancers were not in the breast. Is the doctor correct?

8. Lung cancer is classified as "small cell" or "non–small cell" based on the appearance of cancer cells under a microscope. However, non–small cell lung cancers fall into three subgroups, based on gene expression patterns. Suggest two ways that this information might be used.

Transgenic pigs given a bacterial digestive enzyme excrete genetically modified, less-polluting manure. The author poses amidst a pile of the nonmodified material at the University of Georgia.

Genetic Technologies: Amplifying, Modifying, and Monitoring DNA

Learning Outcomes

19.1 Patenting DNA

1. State the criteria for a patentable invention.
2. Discuss the history of patenting organisms and DNA.
3. Identify current problems and controversies concerning patenting DNA sequences.

19.2 Amplifying DNA

4. Explain how the polymerase chain reaction makes many copies of a DNA sequence.
5. List uses of DNA amplification.

19.3 Modifying DNA

6. Distinguish between recombinant DNA and transgenic organisms.
7. Describe applications of recombinant DNA technology.

19.4 Monitoring Gene Function

8. Explain how a DNA microarray is used to monitor gene expression.

19.5 Silencing DNA

9. Describe ways to decrease gene expression.
10. List uses of gene silencing.

The Big Picture: Ancient biotechnologies gave us bakeries and breweries. Modern biotechnologies manipulate DNA to give us new ways to study, monitor, and treat disease, and alter the environment.

Improving Pig Manure

Pig manure presents a serious environmental problem. The animals do not have an enzyme that would enable them to extract the mineral nutrient phosphorus from a compound called phytate in grain, so they are given dietary phosphorus supplements. As a result, their manure is full of phosphorus. The element washes into natural waters, contributing to fish kills, oxygen depletion in aquatic ecosystems, algal blooms, and even the greenhouse effect. But biotechnology may have solved the "pig poop" problem.

In the past, pig raisers have tried various approaches to keep their animals healthy and the environment clean. Efforts included feeding animal by-products from which the pigs can extract more phosphorus, and giving supplements of the enzyme phytase, which liberates phosphorus from phytate. But consuming animal by-products can introduce prion diseases, and giving phytase before each meal is costly. A "phytase transgenic pig," however, is genetically modified to secrete bacterial phytase in its saliva, which enables it to excrete low-phosphorus manure.

A transgenic organism has a genetic change in each of its cells. The transgenic pig has a phytase gene from the bacterium *E. coli*. Its manure has 75 percent less phosphorus than normal pig excrement.

19.1 Patenting DNA

DNA is the language of life, the instruction manual for keeping an organism alive. Yet we also use DNA. Manipulating DNA is part of **biotechnology,** which is the use or alteration of cells or biological molecules for specific applications, including products and processes. Biotechnology is an ancient art as well as a modern science, and is familiar as well as futuristic. Using yeast to ferment fruit into wine is a biotechnology, as is extracting and using biochemicals from organisms.

The terminology for biotechnology can be confusing. The popular terms "genetic engineering" and "genetic modification" refer broadly to any biotechnology that manipulates DNA. It includes altering the DNA of an organism to suppress or enhance the activities of its own genes, and combining the genetic material of different species. Organisms that harbor DNA from other species are termed **transgenic** and their DNA is called **recombinant DNA.**

Creating transgenic organisms is possible because all life uses the same genetic code—that is, the same DNA triplets encode the same amino acids (**figure 19.1**). Mixing DNA from different species may seem unnatural, but in fact DNA moves and mixes between species in nature—bacteria do it, and it is why we have viral DNA sequences in our chromosomes. But human-directed genetic modification usually gives organisms traits they would not have naturally, such as fish that can tolerate very cold water, tomatoes that grow in salt water, and bacteria that synthesize human insulin.

What Is Patentable?

Creating transgenic organisms raises legal questions, because the design of novel combinations of DNA may be considered intellectual property, and therefore patentable. To qualify for patent protection, a transgenic organism, as any other invention, must be new, useful, and not obvious to an expert in the field. A corn plant that manufactures a protein naturally found in green beans but not in corn, thereby making the corn more nutritious, is an example of a patentable transgenic organism. A patent for a DNA sequence from the human genome might be used to diagnose a specific disease, but could inhibit research unless exceptions to use are made for researchers. DNA is also patentable as a research tool, as are algorithms used to extract information from DNA sequences and databases built of DNA sequences. The Technology Timeline highlights some of the events and controversies surrounding patenting of genetic material.

Patent law has had to evolve to keep up with modern biotechnology. In the 1980s, when sequencing a gene was

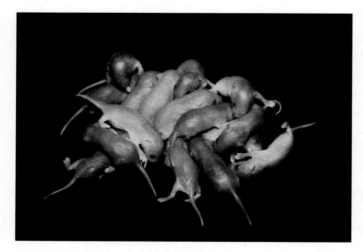

Figure 19.1 **The universality of the genetic code makes biotechnology possible.** The greenish mice contain the gene encoding a jellyfish's green fluorescent protein (GFP). Researchers use GFP to mark genes of interest. The GFP mice glow less greenly as they mature and more hair covers the skin. The non-green mice are not genetically modified.

Technology Timeline

PATENTING LIFE AND GENES

1790	U.S. patent act enacted. A patented invention must be new, useful, and not obvious.
1873	Louis Pasteur is awarded first patent on a life form, for yeast used in industrial processes.
1930	New plant variants can be patented.
1980	First patent awarded on a genetically modified organism, a bacterium given four DNA rings that enable it to metabolize components of crude oil.
1988	First patent awarded for a transgenic organism, a mouse that manufactures human protein in its milk. Harvard University granted patent for "OncoMouse" transgenic for human cancer.
1992	Biotechnology company awarded patent for all forms of transgenic cotton. Groups concerned that this will limit the rights of subsistence farmers contest the patent several times.
1996–1999	Companies patent partial gene sequences and certain disease-causing genes for developing specific medical tests.
2000	With gene and genome discoveries pouring into the Patent and Trademark Office, requirements tightened for showing utility of a DNA sequence.
2003	Attempts to enforce patents on non-protein-encoding parts of the human genome anger researchers who support open access to the information.
2007	Patent requirements must embrace new, more complex definition of a gene.
2009	Patents on breast cancer genes challenged.
2010	Direct-to-consumer genetic testing companies struggle to license DNA patents for multigene and SNP association tests. Patents on breast cancer genes invalidated.
2011	U.S. government considers changes to gene patent laws.

painstakingly slow, only a few genes were patented. In the mid-1990s, with faster sequencing technology and short-cuts to finding the protein-encoding parts of the genome, the U.S. National Institutes of Health and biotech companies began seeking patent protection for thousands of short DNA sequences, even if their functions weren't known. Because of the flood of applications, the U.S. Patent and Trademark Office tightened requirements for usefulness. Today, with entire genomes being sequenced much faster than it once took to decipher a single gene, a DNA sequence alone does not warrant patent protection. It must be useful as a tool for research or as a novel or improved product, such as a diagnostic test or a drug. In the United States, more than one in five human genes is patented in some way, yet only a few gene patents have been challenged.

DNA patenting became very controversial in 2009, when several groups, including the American Civil Liberties Union and the Public Patent Foundation, challenged patents on two breast cancer genes (*BRCA1* and *BRCA2*) held by biotechnology company Myriad Genetics and the University of Utah. The company did not license the technology to other companies, so patients were forced to take Myriad's test for increased familial breast cancer risk, costing more than $3,000. The patents discouraged research and prevented patients from getting second opinions. In 2010, a federal judge in the United States ruled seven patents on the genes "improperly granted" because they are based on a "law of nature." In 2011, the court invalidated the patents on the two genes, but a federal appeals court overruled that action, claiming that an isolated gene is not the same as a gene in a cell, which is part of a chromosome.

The Patent Thicket

Analysis of human genome information and the rise of personalized medicine based on it continues to complicate DNA patenting. One problem is redundancy. For the same gene, it is possible to patent the entire sequence (termed genomic DNA), just the protein-encoding exons, or a gene variant, such as a sequence containing a SNP or mutation. The effect of multiple patents on a gene is termed "the patent thicket." It means that a company or researcher developing a tool or test based on a particular gene or its encoded protein might infringe upon several patents that are based on essentially the same information. The patents are a barrier, just as a thicket of bushes impedes moving forward. Now, as genetics shifts from a gene-by-gene focus to analyzing expression patterns of suites of interacting genes that are highly specific to individuals, patent law will have to once again adjust to keep up with scientific developments.

A new challenge to patenting DNA stems from the shift in focus of the entire field from a single-gene to a genome-wide approach. A company seeking to use part of a gene sequence in a test must license the use of that sequence from the patent holder. The test panels that many companies market, such as for several heart-related disorders or for conditions that are more prevalent among Ashkenazi Jews, require multiple licenses. Direct-to-consumer genetic testing companies that scan clients' DNA for many thousands of SNPs face a predicament. If each SNP is patented, and requires payment of 1 to 5 percent of the profit, such tests cannot be developed unless a company owns the patents.

The direct-to-consumer companies are in an identity crisis. When these companies began to spring up a few years ago, they circumvented regulations on genetic tests for disease by claiming that they provided only information as an educational service. If the law disallows patents for use of DNA sequences in diagnostic tests, these companies would not be included because of how they identify themselves. But if the companies claim to offer tests for diseases so that they have access to patented sequences, they will be under scrutiny of the federal agencies that regulate genetic testing and products for health-related purposes. Ancestry testing is another application of using patented DNA sequences.

Because DNA testing companies are forming faster than the Patent and Trademark Office can evaluate patent applications, the U.S. government is exploring ways around the patent thicket:

- Ban the patenting of associations between DNA sequence variants and disease.
- Exempt the use of DNA in a diagnostic or risk assessment test from patent infringement.
- Exempt physicians and researchers from litigation if they use patented DNA sequences.

A broader action is the Genomic Research and Accessibility Act, which would ban patenting any DNA or its encoded proteins. The U.S. government may also follow in the footsteps of European nations that compel patent holders to license their DNA-based inventions for public health reasons. If these measures are widely enacted, in the future DNA patents may be seen more as temporary permits than long-term exclusive-use mandates.

While the laws are being worked out, companies can navigate the patent thicket by moving the parts of their operations that use the patented DNA sequences to countries where the restrictions on use do not apply. They can also tweak the recipes for a patented procedure, such as substituting a different type of cell in culture that produces a particular protein, or altering chemical protocols.

Key Concepts

1. Biotechnology is the use or modification of cells or biological molecules for a specific application.
2. DNA patenting is evolving to embrace genome-wide applications.

19.2 Amplifying DNA

Some forensic and medical tests require many copies of a specific DNA sequence from a small sample. Mass-producing a DNA sequence, called nucleic acid amplification, was invented in the 1970s and 1980s. The first and best-known technique is the **polymerase chain reaction** (PCR), which works on DNA molecules outside cells. Another approach, recombinant DNA technology, amplifies DNA from one type of organism placed in the cell of another. Recombinant DNA technology is addressed in section 19.3.

PCR is based on the natural process of DNA replication. Recall from chapter 9 that every time a cell divides, it replicates all of its DNA. In contrast, PCR uses DNA polymerase to rapidly replicate a specific DNA sequence millions of times.

Applications of PCR are eclectic (**table 19.1**). In forensics, PCR is used routinely to amplify DNA sequences that are profiled to establish blood relationships, to identify remains, and to help convict criminals or exonerate the falsely accused. In agriculture, veterinary medicine, environmental science, and human health care, PCR is used to amplify the DNA or RNA of pathogens to detectable levels. In genetics, PCR is both a crucial laboratory tool to identify genes and a component of many diagnostic tests.

PCR was born in the mind of Kary Mullis on a moonlit night in northern California in 1983. As he drove the hills, Mullis was thinking about the precision of DNA replication, and a way to tap into it popped into his mind. He excitedly explained his idea to his girlfriend and then went home to think it through. "It was difficult for me to sleep with deoxyribonuclear bombs exploding in my brain," he wrote much later.

The idea behind PCR was so simple that Mullis had trouble convincing his superiors at Cetus Corporation that he was onto something. Over the next year, he used the technique to amplify a well-studied gene. Mullis published a landmark 1985 paper and filed patent applications, launching the field of nucleic acid amplification. He received a $10,000 bonus for his invention, which the company sold to another for $300 million. Mullis did, however, win a Nobel Prize.

PCR rapidly replicates a selected sequence of DNA in a test tube (**figure 19.2**). The requirements include:

1. knowing parts of a target DNA sequence.
2. two types of lab-made, single-stranded, short pieces of DNA called primers. These are complementary in sequence to opposite ends of the target sequence.
3. a large supply of the four types of DNA nucleotide building blocks.
4. Taq1, a DNA polymerase produced by a microbe that inhabits hot springs. This enzyme is adapted to its host's hot surroundings and makes PCR easy because it does not fall apart when DNA is heated, as most proteins do.

In the first step of PCR, heat is used to separate the two strands of the target DNA. Next, the two short DNA primers and Taq1 DNA polymerase are added. The temperature is lowered. Primers bind by complementary base pairing to the separated target strands. In the third step, the Taq1 DNA polymerase adds bases to the primers and builds a sequence complementary to the target sequence. The newly synthesized strands then act as templates in the next round of replication, which is initiated

Table 19.1	**Uses of PCR**

PCR has been used to amplify DNA from:

- a cremated man, from skin cells left in his electric shaver, to diagnose an inherited disease in his children.
- a preserved quagga (a relative of the zebra) and a marsupial wolf, both extinct.
- microorganisms that cannot be cultured for study.
- the brain of a 7,000-year-old human mummy.
- the digestive tracts of carnivores, to reveal food web interactions.
- roadkills and carcasses washed ashore, to identify locally threatened species.
- products illegally made from endangered species.
- genetically altered bacteria that are released in field tests, to follow their dispersion.
- one cell of an 8-celled human embryo to detect a disease-related genotype.
- poached moose meat in hamburger.
- remains in Jesse James's grave, to make a positive identification.
- the guts of genital crab lice on a rape victim, which matched the DNA of the suspect.
- fur from Snowball, a cat that linked a murder suspect to a crime.

immediately by raising the temperature. It is a little like a square dance, with a line of couples but also many individuals standing by themselves. At intervals, the line parts and each dancer takes a new partner, until all the singles—the free nucleotides—are partnered.

PCR is done in an automated device called a thermal cycler, or in a device that uses microscopic layers of heated and cooled silicon, to control the key temperature changes. The heat-resistant DNA polymerase is crucial to the process.

The pieces of identical DNA accumulate exponentially. The number of amplified pieces of DNA equals 2^n, where n equals the number of temperature cycles. After just 20 cycles, 1 million copies of the original sequence have accumulated in the test tube.

PCR's greatest strength is that it works on crude samples of rare, old, and minute sequences. PCR's greatest weakness, ironically, is its exquisite sensitivity. A blood sample submitted for diagnosis of an infection, if contaminated by leftover DNA from a previous test, or a stray eyelash from the person running the reaction, can yield a false result.

Using layered silicon instead of a thermal cycler to amplify DNA greatly speeds PCR. Thirty cycles using the thermal cycler take ninety minutes; with the silicon layers, it takes a little over 4 minutes. The speed is valuable in situations where

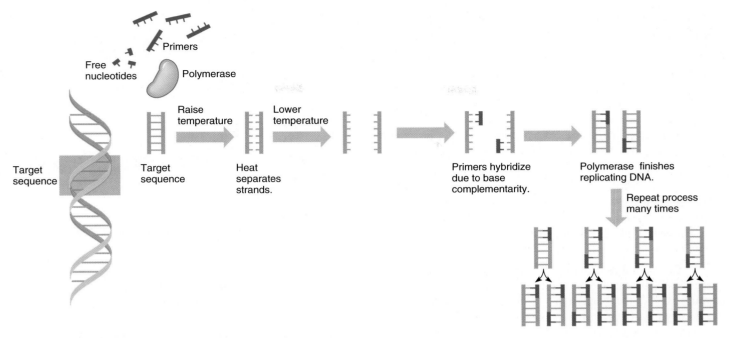

Figure 19.2 Amplifying a specific DNA sequence. In the polymerase chain reaction, specific primers, a thermostable DNA polymerase, and free nucleotides replicate a DNA sequence of interest. The reaction rapidly builds up millions of copies of the target sequence.

rapid diagnosis is important, such as the case of a person with a life-threatening infection who requires a certain antibiotic, or on a battlefield to detect biological weapons.

The invention of PCR inspired other nucleic acid amplification techniques. One is transcription-mediated amplification, which copies target DNA into RNA and then uses RNA polymerase to amplify the RNA. This procedure doesn't require temperature shifts, and it generates 100 to 1,000 copies per cycle, compared to PCR's doubling, and can yield 10 billion copies of a selected sequence in a half hour.

Key Concepts

1. PCR rapidly replicates a short DNA sequence.
2. PCR is based on DNA replication and has many uses.
3. Other nucleic acid amplification technologies followed PCR.

19.3 Modifying DNA

Recombinant DNA technology adds genes from one type of organism to the genome of another. It was the first gene modification biotechnology, and was initially done in bacteria to produce peptides and proteins useful as drugs. When bacteria bearing recombinant DNA divide, they yield many copies of the "foreign" DNA, and under proper conditions they produce many copies of the protein that the foreign DNA specifies.

Recombinant DNA technology is also known as gene cloning. "Cloning" in this context refers to making many copies of a specific DNA sequence.

Recombinant DNA

In February 1975, molecular biologists convened at Asilomar, on California's Monterey Peninsula, to discuss the safety and implications of a new type of experiment: combining genes of two species. Would experiments that deliver a cancer-causing virus be safe? The researchers discussed restricting the types of organisms and viruses used in recombinant DNA research and brainstormed ways to prevent escape of a resulting organism from the laboratory. The guidelines drawn up at Asilomar outlined measures of "physical containment," such as using specialized hoods and airflow systems, and "biological containment," such as weakening organisms so that they could not survive outside the laboratory.

Recombinant DNA technology turned out to be safer than expected, and it spread to industry faster and in more diverse ways than anyone had imagined. However, recombinant DNA-based products have been slow to reach the marketplace because of the high cost of the research and the long time it takes to develop a new drug. Today, several dozen such drugs are available, and more are in the pipeline. Recombinant DNA research initially focused on providing direct gene products such as peptides and proteins. These included insulin, growth hormone, and clotting factors. However, the technology can target carbohydrates and lipids by affecting the genes that encode enzymes required to synthesize them.

Constructing Recombinant DNA Molecules—An Overview

Manufacturing recombinant DNA molecules requires **restriction enzymes** that cut donor and recipient DNA at the same sequence; DNA to carry the donor DNA (called **cloning vectors**); and recipient cells (bacteria or other cultured single cells).

After inserting donor DNA into vectors, the procedure requires several steps to get the desired modified cell type:

- selecting cells where the genetic material includes any foreign DNA;
- selecting cells that received the gene of interest;
- stimulating transcription of the foreign gene and translation of its protein product; and
- collecting and purifying the desired protein.

The natural function of restriction enzymes is to protect bacteria by cutting DNA of infecting viruses. Methyl (CH_3) groups shield the bacterium's own DNA from its restriction enzymes. Bacteria have hundreds of types of restriction enzymes. Some of them cut DNA at particular sequences of four, five, or six bases that are symmetrical in a specific way—the recognized sequence reads the same, from the 5′ to 3′ direction, on both strands of the DNA. For example, the restriction enzyme EcoR1, shown in **figure 19.3,** cuts at the sequence GAATTC. The complementary sequence on the other strand is CTTAAG, which, read backwards, is GAATTC. (You can try this with other sequences to see that it rarely works this way.) In the English language, this type of symmetry is called a palindrome, referring to a sequence of letters that reads the same in both directions, such as "Madam, I'm Adam." Unlike the language comparison, however, palindromic sequences in DNA are on complementary strands.

The cutting action of some restriction enzymes on double-stranded DNA creates single-stranded extensions. They are called "sticky ends" because they are complementary to each other, forming hydrogen bonds as their bases pair. Restriction enzymes work as molecular scissors in creating recombinant DNA molecules because they cut at the same sequence in any DNA source. That is, the same sticky ends result from the same restriction enzyme, whether the DNA is from a mockingbird or a maple.

Another natural "tool" used in recombinant DNA technology is a cloning vector. This structure carries DNA from the cells of one species into the cells of another. A vector can be any piece of DNA into which other DNA can insert. A commonly used type of vector is a **plasmid,** which is a small circle of double-stranded DNA that exists naturally in some bacteria, yeasts, plant cells, and other types of organisms. Viruses that infect bacteria, called bacteriophages, are another type of vector, manipulated to transport DNA but not cause disease. Disabled retroviruses are used as vectors too, as are DNA sequences from bacteria and yeast called artificial chromosomes.

When choosing a cloning vector, size matters. The desired gene must be short enough to insert into the vector. Gene size is typically measured in kilobases (kb), which are thousands

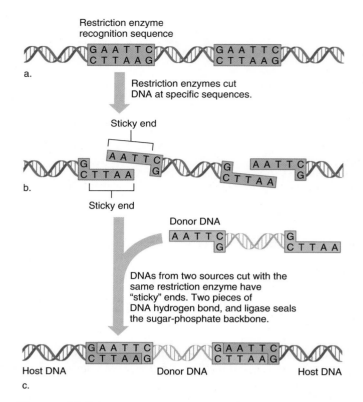

Figure 19.3 Recombining DNA. A restriction enzyme makes "sticky ends" in DNA by cutting it at specific sequences. **(a)** The enzyme EcoR1 cuts the sequence GAATTC between G and A. **(b)** This staggered cutting pattern produces "sticky ends" of sequence AATT. The ends attract through complementary base pairing. **(c)** DNA from two sources is cut with the same restriction enzyme. Pieces join, forming recombinant DNA molecules.

of bases. Various types of cloning vectors can hold up to about 2 million DNA bases.

To create a recombinant DNA molecule, a restriction enzyme cuts DNA from a donor cell at sequences known to bracket the gene of interest (**figure 19.4**). The enzyme leaves single-stranded ends on the cut DNA, each bearing a characteristic base sequence. Next, a plasmid is isolated and cut with the same restriction enzyme used to cut the donor DNA. Because the same restriction enzyme cuts both the donor DNA and the plasmid DNA, the same complementary single-stranded base sequences extend from the cut ends of each. When the cut plasmid and the donor DNA are mixed, the single-stranded sticky ends of some plasmids base pair with the sticky ends of the donor DNA. The result is a recombinant DNA molecule, such as a plasmid carrying the human insulin gene. The plasmid and its human gene can now be transferred into a cell, such as a bacterium or a white blood cell.

Isolating the Gene of Interest

Constructing recombinant DNA molecules usually begins by cutting all of the DNA in the donor cell. This genomic DNA includes protein-encoding as well as non-protein-encoding

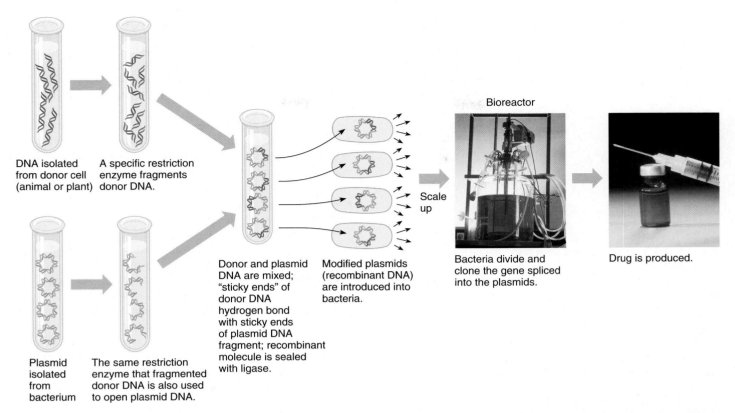

Figure 19.4 Recombinant DNA. DNA isolated from a donor cell and a plasmid are cut with the same restriction enzyme and mixed. Sticky ends from the donor DNA hydrogen bond with sticky ends of the plasmid DNA, forming recombinant DNA molecules. When such a modified plasmid is introduced into a bacterium, it is mass produced as the bacterium divides.

Labels in figure:
- DNA isolated from donor cell (animal or plant)
- A specific restriction enzyme fragments donor DNA.
- Plasmid isolated from bacterium
- The same restriction enzyme that fragmented donor DNA is also used to open plasmid DNA.
- Donor and plasmid DNA are mixed; "sticky ends" of donor DNA hydrogen bond with sticky ends of plasmid DNA fragment; recombinant molecule is sealed with ligase.
- Modified plasmids (recombinant DNA) are introduced into bacteria.
- Scale up
- Bioreactor
- Bacteria divide and clone the gene spliced into the plasmids.
- Drug is produced.

sequences. Researchers assemble collections of recombinant bacteria (or other single cells) that harbor pieces of a genome. By using several copies of a genome, the pieces overlap where sequences align. Such a collection is called a **genomic library**. For each application, such as using a human protein as a drug, a particular piece of DNA must be identified and isolated from a genomic library. There are several ways to do this "needle-in-a-haystack" type of search.

A piece of DNA that is complementary to part of the template strand of the gene in question can be linked to a label, such as a radioactive or fluorescent molecule. This labeled gene fragment is called a **DNA probe**. It emits a signal when it binds to its complement in a cell that contains a recombinant plasmid. DNA probes can also be made using genes of similar sequence from other species—they will bind the human version of the gene. Using such a probe is a little like mistakenly typing "hipropotamus" to google "hippopotamus." Google would probably still come up with a hippo.

A genomic library contains too much information for a researcher seeking a particular protein-encoding gene—it may also contain introns, the genes that encode rRNAs and tRNAs, and many repeated sequences. A shortcut is to use another type of library, called a complementary DNA, or **cDNA library,** that represents only protein-encoding genes. A cDNA library is made from the mRNAs in a differentiated cell, which represent the proteins manufactured there. For example, a muscle cell has abundant mRNAs that encode contractile proteins, whereas a fibroblast has many mRNAs that represent connective tissue proteins.

To make a cDNA library, researchers first extract the mRNAs from cells. Then, these RNAs are used to construct complementary or "c" DNA strands using reverse transcriptase, DNA nucleotide triphosphates, and DNA polymerase (**figure 19.5**). Reverse transcriptase is an enzyme that makes a DNA molecule that is complementary to a specific mRNA. DNA polymerase and the nucleotides then can synthesize the complementary strand to the single-stranded cDNA to form a double-stranded DNA. Different cell types yield different cDNA collections, or libraries, that reflect which genes are expressed. They do not, however, reveal protein abundance because in a cell mRNA molecules are transcribed and degraded at different rates.

A specific cDNA can be taken from a cDNA library and used as a probe to isolate the original gene of interest from the genomic library. If the goal is to harness the gene and eventually collect its protein product, then the genomic version is useful, because it includes control regions such as promoters. Once a gene of interest is transferred to a cell where it can be transcribed into mRNA and that RNA can be translated, the protein is collected. Such cells are typically grown in containers called bioreactors, with nutrients sent in and wastes removed. The desired product is collected from the medium in which the cells are growing.

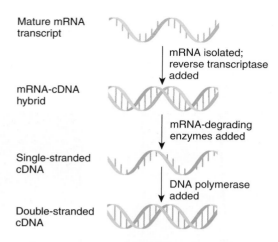

Mature mRNA
transcript

mRNA isolated;
reverse transcriptase
added

mRNA-cDNA
hybrid

mRNA-degrading
enzymes added

Single-stranded
cDNA

DNA polymerase
added

Double-stranded
cDNA

Figure 19.5 Copying DNA from RNA. Researchers make cDNA from mRNA using reverse transcriptase, an enzyme from a retrovirus. A cDNA includes codons for a mature mRNA, but not sequences for promoters and introns. Control sequences from bacteria may be added so that the eukaryotic gene can be transcribed and translated in a prokaryote (the bacterium).

Selecting Recombinant DNA Molecules

Much of the effort in recombinant DNA technology is in identifying and separating cells that contain the gene of interest, once the foreign DNA is inserted into the vector. Three types of recipient cells can result:

1. cells that lack plasmids;
2. cells that contain plasmids that do not contain a foreign gene; and
3. cells that contain plasmids that have picked up a foreign gene (the goal).

The procedure is set up to select bacteria that have taken up recombinant plasmids that carry the foreign DNA sequence of interest. One type of strategy has two steps: using an antibiotic resistance gene and a color change reaction to highlight the useful plasmids. First, human and plasmid DNA are cut with the same restriction enzymes and mixed. The plasmids are closed with ligase (the enzyme that glues the sugar-phosphate backbone when DNA replicates), and transferred to bacterial cells. When the antibiotic is applied, only cells harboring plasmids survive. The plasmids also include a gene that encodes an enzyme that catalyzes a reaction that produces a blue color. If a human gene inserts and interrupts the gene for the enzyme, the bacterial colony that grows is not blue, and is therefore easily distinguished from the blue bacterial cells that have not taken up the human gene.

When cells containing the recombinant plasmid divide, so does the plasmid. Within hours, the original cell gives rise to many cells harboring the recombinant plasmid. The enzymes, ribosomes, energy molecules, and factors necessary for protein synthesis transcribe and translate the plasmid DNA and its foreign gene, producing the desired protein.

Products from Recombinant DNA Technology

In basic research, recombinant DNA technology provides a way to isolate individual genes from complex organisms and observe their functions on the molecular level. Recombinant DNA has many practical uses, too. The first was to mass-produce protein-based drugs.

Drugs manufactured using recombinant DNA technology are pure, and are the human version of the protein. Before recombinant DNA technology was invented, human growth hormone came from cadavers, follicle-stimulating hormone came from the urine of postmenopausal women, and clotting factors were pooled from hundreds or thousands of donors. These sources introduced great risk of infection, especially after HIV and hepatitis C became more widespread.

The first drug manufactured using recombinant DNA technology was insulin. Before 1982, people with type 1 diabetes mellitus obtained the insulin that they injected daily from pancreases removed from cattle in slaughterhouses. Cattle insulin is so similar to the human peptide, different in only two of its fifty-one amino acids, that most people with diabetes could use it. However, about one in twenty patients is allergic to cow insulin because of the slight chemical difference. Until recombinant DNA technology was developed, the allergic patients had to use expensive combinations of insulin from other animals or human cadavers. **Table 19.2** lists some drugs produced using recombinant DNA technology.

Insulin is a simple peptide and is therefore straightforward to mass-produce in bacteria. Some drugs, however, require that sugars be attached, or must fold in specific, intricate ways to function. These molecules must be produced in eukaryotic cells that readily carry out these modifications.

Drugs developed using recombinant DNA technology must compete with conventional products. Deciding whether a recombinant drug is preferable to an existing similar drug is often a matter of economics. For example, interferon β-1b treats a type of multiple sclerosis, but this recombinant drug costs more than $20,000 per year. British researchers calculated that more people would be served if funds were spent on improved supportive care for many rather than on this costly treatment for a few.

Tissue plasminogen activator (tPA), a recombinant clot-busting drug, also has cheaper alternatives. If injected within 4 hours of a heart attack, tPA dramatically limits damage to the heart muscle by restoring blood flow. It costs $2,200 a shot. An older drug, streptokinase, is extracted from unaltered bacteria and is nearly as effective, at $300 per injection. Patients who have already received streptokinase and could have an allergic reaction if they were to use it again can benefit from tPA. *Bioethics: Choices for the Future* on page 380 considers another drug derived from recombinant DNA technology: erythropoietin (EPO).

An application of recombinant DNA technology in the textile industry is a novel source of indigo—the dye used to make blue jeans blue. The dye originally came from mollusks and fermented leaves of the European woad plant or Asian

Table 19.2	Drugs Produced Using Recombinant DNA Technology
Drug	**Use**
Atrial natriuretic peptide	Dilates blood vessels, promotes urination
Colony-stimulating factors	Help restore bone marrow after marrow transplant; restore blood cells following cancer chemotherapy
Deoxyribonuclease (DNase)	Thins secretions in lungs of people with cystic fibrosis
Epidermal growth factor	Accelerates healing of wounds and burns; treats gastric ulcers
Erythropoietin (EPO)	Stimulates production of red blood cells in cancer patients
Factor VIII	Promotes blood clotting in treatment of hemophilia
Glucocerebrosidase	Corrects enzyme deficiency in Gaucher disease
Human growth hormone	Promotes growth of muscle and bone in people with very short stature due to hormone deficiency
Insulin	Allows cells to take up glucose in treatment of type 1 diabetes
Interferons	Treat genital warts, hairy cell leukemia, hepatitis C and B, Kaposi sarcoma, multiple sclerosis
Interleukin-2	Treats kidney cancer recurrence
Lung surfactant protein	Helps lung alveoli to inflate in infants with respiratory distress syndrome
Renin inhibitor	Lowers blood pressure
Somatostatin	Decreases growth in muscle and bone in pituitary giants
Superoxide dismutase	Prevents further damage to heart muscle after heart attack
Tissue plasminogen activator	Dissolves blood clots in treatment of heart attack, stroke, and pulmonary embolism

indigo plant. The 1883 discovery of indigo's chemical structure led to the invention of a synthetic process to produce the dye using coal-tar. That method has dominated the industry, but it releases toxic by-products. A recombinant DNA-based solution came from the bacterium *E. coli*. It converts glucose to the amino acid tryptophan, which then forms indole, a precursor to indigo. Another type of bacterium takes the indole to indigo. Researchers altered *E. coli* to suppress alternative pathways for metabolizing glucose, coaxing the cells to synthesize excess tryptophan. They then added genes from the other bacterial species, extending the biochemical pathway all the way to produce indigo. The result: common bacteria that manufacture the blue dye of denim jeans from glucose, a simple sugar.

Transgenic Animals

Eukaryotic cells growing in culture are generally better at producing human proteins than are prokaryotic cells such as bacteria. An even more efficient way to express some recombinant genes is in a body fluid of a transgenic animal, such as milk. The fact that the cells secreting the human protein are part of an animal more closely mimics the environment in the human body.

Transgenic sheep, cows, and goats have all expressed human genes in their milk, including those that encode clotting factors, clot busters, and collagen. Production of human antibodies in rabbit and cow milk illustrates the potential value of transgenic animals. Recall from figure 17.9 that antibodies are assembled from the products of several genes. Researchers attach the appropriate human antibody genes to promoters for milk proteins. (Promoters are the short sequences at the starts of genes that control transcription rates.) These promoters normally oversee production of abundant milk proteins. The mammary gland cells of transgenic animals can assemble antibody parts to secrete the final molecules—just as if they were being produced in a plasma cell in the human immune system. Such antibodies are used to treat cancer.

Several techniques are used to insert DNA into animal cells to create transgenic animals. They include:

- chemicals that open transient holes in plasma membranes;
- liposomes (fatty bubbles) that carry DNA into cells as plasma membranes envelop them;
- brief jolts of electricity (electroporation) that open transient holes in plasma membranes;
- microscopic needles that inject DNA into cells (microinjection); and
- metal particles coated with foreign DNA shot into cells (particle bombardment).

EPO: Built-in Blood Cell Booster or Performance-Enhancing Drug?

"Cycle, run, and swim longer and faster than anyone else!" proclaims a website selling a product that supposedly boosts levels of EPO (erythropoietin), a glycoprotein hormone that the kidneys produce in response to low levels of oxygen in the blood. EPO travels to the bone marrow and binds receptors on cells that give rise to red blood cell progenitors. Soon, more red blood cells enter the circulation, carrying more oxygen to the tissues (**figure 1**). Enhanced stamina results.

The value of EPO as a drug became evident after the invention of hemodialysis to treat kidney failure in 1961. Dialysis removes EPO from the blood, causing severe anemia. But boosting EPO levels proved difficult because levels in human plasma are too low to pool from donors. Instead, in the 1970s, the U.S. government obtained EPO from South American farmers with hookworm infections and Japanese aplastic anemia patients, who secrete abundant EPO into urine. But when the AIDS epidemic came, biochemicals from human body fluids were no longer safe to use. Recombinant DNA technology solved the EPO problem. It is sold under various names to treat anemia in dialysis and AIDS patients and is given with cancer chemotherapy to avoid the need for transfusions.

EPO's ability to increase the oxygen-carrying capacity of blood under low oxygen conditions is why athletes train at high altitudes to increase endurance. Since the early 1990s, athletes have abused EPO to reproduce this effect, at great risk. EPO thickens the blood, raising the risk of a blockage that can cause a heart attack or stroke, especially when intense, grueling exercise removes water from the bloodstream. Excess EPO caused sudden death during sleep for at least eighteen cyclists. Olympic athletes now take urine tests that detect recombinant EPO, which has a slightly different configuration of sugars than the form an athlete's kidneys naturally produce. Famed cyclist Floyd Landis admitted to EPO abuse.

People with familial erythrocytosis get the effects of extra EPO naturally. Type 1 (MIM 133100) is autosomal dominant, and is caused by mutation in the EPO receptor. Affected individuals have large and abundant red blood cells, but low blood serum levels of EPO. A member of a family from Finland with this condition won several Olympic medals for skiing thanks to his inborn ability. An autosomal recessive form of the condition, type 2 (MIM 263400), increases the level of EPO in the bloodstream. Both forms of erythrocytosis usually have no symptoms, but increase the risk of circulation blocked by the sluggish, oxygen-laden blood.

Figure 1 At least two genes control EPO secretion. Certain variants of these genes increase the number of red blood cells, increasing endurance but also raising risk of heart attack and stroke.

Questions for Discussion

1. Was it ethical in the 1970s to obtain EPO from sick, poor people in South America and Japan to treat people in the United States?

2. Should using a substance made naturally in the body be considered performance enhancement?

3. Should tests be developed to identify athletes whose genes, anatomy, or physiology give them a competitive advantage? Why or why not?

4. When developing drugs that use recombinant DNA technology, should researchers consider how the product could be abused?

Getting foreign DNA into appropriate cells is only a first step in creating a transgenic organism. It is quite a technical challenge! The recombinant DNA must then enter the nucleus, replicate with the cell's own DNA, and be transmitted when the cell divides. Finally, an organism must be regenerated from the altered cell. If the trait is dominant, the transgenic organism must express it in the appropriate tissues at the right time in development to be useful. If the trait is recessive, crosses between heterozygotes may be necessary to yield homozygotes that express the trait. Then the organisms must pass the characteristic on to the next generation.

Animal Models

Herds of transgenic farm animals supplying drugs in their milk have not become important sources of pharmaceuticals—they are too difficult to maintain. Transgenic animals are more useful as models of human disease (**figure 19.6**). Inserting the mutant human beta globin gene that causes sickle cell disease into mice, for example, results in a mouse model of the disorder. Drug candidates can be tested on these animal models and abandoned if they cause significant side effects before testing in humans.

Transgenic animal models, however, have limitations. Researchers cannot control where a transgene inserts in a genome, and how many copies do so. The level of gene expression necessary for a phenotype to emerge may also differ in the model and humans. This was the case for a mouse model of familial Alzheimer disease (MIM 104760). The transgene has the exact same DNA sequence that disrupts amyloid precursor protein in a Swedish family with the condition, but apparently did nothing to the mice—until researchers increased transcription rate tenfold. Only then did the telltale plaques and tangles, and neuron cell death, appear in the mouse brains.

Figure 19.6 **Animal models mimic human disease.**
Transgenic mice that have the human mutation that causes
Huntington disease are tested for coordination on a rotating
drum with a grooved surface called a Rotarod.

Animal models might not mimic the human condition
exactly because of differences in their rates of development. For
example, for some inherited diseases that do not cause symptoms
until adulthood in humans, mice simply do not live long enough
to evaluate the phenotype. A transgenic monkey, for example, is a
more accurate model of Huntington disease than a mouse, because
as a primate the monkey is much more similar to a human in life
span, metabolism, reproduction, behavior, and cognition.

Bioremediation

Recombinant DNA technology and transgenic organisms provide
processes as well as products. In **bioremediation,** bacteria or
plants with the ability to detoxify certain pollutants are released
or grown in a particular area. Natural selection has sculpted such
organisms, perhaps as adaptations that render them unpalatable
to predators. Bioremediation uses genes that enable an organism
to metabolize a substance that, to another species, is a toxin. The
technology uses unaltered organisms, and also transfers "detox"
genes to other species so that the protein products can more easily penetrate a polluted area.

Nature offers many organisms with interesting tastes. A
type of tree that grows in a tropical rainforest on an island near
Australia, for example, accumulates so much nickel from soil
that slashing its bark releases a bright green latex ooze. This
tree can be used to clean up nickel-contaminated soil.

Bioremediation can tap the metabolisms of transgenic
microorganisms, sending them into plants whose roots then distribute the detox proteins in the soil. For example, transgenic yellow poplar trees can thrive in mercury-tainted soil if they have a
bacterial gene that encodes an enzyme, mercuric reductase, that
converts a highly toxic form of mercury in soil to a less toxic gas.
The tree's leaves then release the gas.

Bioremediation cleans up munitions dumps from wars.
One application uses bacteria that normally break down
trinitrotoluene, or TNT, the major ingredient in dynamite and
land mines. The enzyme that provides this capability is linked
to the *GFP* gene (see figure 19.1). When the bacteria are spread
in a contaminated area, they glow near land mines, revealing
the locations more clearly than a metal detector could. Once
the land mines are removed, the bacteria die as their food vanishes.

Genetically modifying an organism for use in bioremediation takes time and planning. Most disasters, however, happen
without warning. This was the case for the oil spill along the Gulf
Coast of the United States in the spring of 2010 (**figure 19.7**).

Figure 19.7 **Bioremediation.** A Coast Guard worker rescues
an oil-drenched pelican from Gulf Coast waters as clean birds
dot the background. A biotech company is using "experimental
evolution" to select bacteria that can metabolize the pollutants.

Instead of manipulating DNA in bacteria to create "oil-eaters,"
researchers at a biotech company turned to "experimental evolution." They set up large containers of the tainted water, added
diverse but unaltered bacteria, then watched as natural selection
did its work. At first nearly all the bacteria died. A few, however,
survived and reproduced. Within 2 weeks, the vats were filled
with bacteria that naturally thrived in the presence of the oil. Natural selection had acted on preexisting variants with a taste for the
organic molecules of oil. If this scenario of a population plummet-
ing and resurging sounds familiar, it should. The bacteria in the
vats at the biotech company underwent a population bottleneck in
the face of environmental stress—just like the hunted cheetahs in
Africa (see figure 15.8); the Jewish people experiencing persecution at various times and places, resulting in the "Jewish genetic
diseases" of today; and HIV in a human body (see figure 15.14).

19.4 Monitoring Gene Function

We usually cannot do very much about the gene variants that we inherit. Gene expression, in contrast, is where we can make a difference by controlling our environment. Monitoring gene expression requires detecting the mRNAs in particular cells under particular conditions. Gene expression DNA microarrays, also known as gene chips, are devices that detect and display the mRNAs in a cell. The creativity of the technique lies in choosing the types of cells to interrogate.

Evaluating a spinal cord injury illustrates the basic steps in creating a DNA microarray to assess gene expression. Researchers knew that in the hours after such a devastating injury, immune system cells and inflammatory biochemicals flood the affected area, but it took **gene expression profiling** to reveal just how fast healing begins.

A **microarray** is a piece of glass or plastic that is about 1.5 centimeters square—smaller than a postage stamp. Many small pieces of DNA (oligonucleotides) of known sequence are attached to one surface, in a grid pattern. The researcher records the position of each DNA piece in the grid. In many applications, a sample from an abnormal situation (such as disease, injury, or environmental exposure) is compared to a normal control. **Figure 19.8** compares cerebrospinal fluid (CSF; the liquid that bathes the spinal cord) from an injured person (sample A) to fluid from a healthy person (sample B). Messenger RNAs are extracted from the samples and cDNAs made (see figure 19.5). The cDNAs from the injury sample are labeled with a red fluorescent dye, and the cDNAs from the control sample are labeled with a green fluorescent dye. These labeled DNAs are then applied to the microarray, which displays thousands of genes likely to be involved in a spinal cord injury, or the entire human genome. Considering so many DNA sequences allows for surprises, avoiding the assumption that we know what to look for.

DNA that binds to complementary sequences on the grid fluoresce in place. A laser scanner then detects and converts the results to a colored image. Each position on the microarray can bind DNA pieces from both samples, either, or neither. The scanner also detects fluorescence intensities, which provides information on how strongly the gene is expressed. Then a computer algorithm interprets the pattern of gene expression. For the spinal cord example, the visual data mean the following:

- Red indicates a gene expressed in CSF only when the spinal cord is injured (and presumably leaking inflammatory molecules).
- Green indicates a gene expressed in CSF only when the spinal cord is intact.
- Yellow indicates positions where both red- and green-bound dyes fluoresce, representing genes that are expressed whether or not the spinal cord has been injured.
- Black, or a lack of fluorescence, corresponds to DNA sequences that are not expressed in CSF, since they do not show up in either sample.

The color and intensity pattern of the microarray provides a glimpse of gene expression following spinal cord injury.

Table 19.3	Gene Expression Profiling Chronicles Repair After Spinal Cord Injury
Time After Injury (rats)	**Type of Increased Gene Expression**
Day 1	Protective genes to preserve remaining tissue
Day 3	Growth, repair, cell division
Day 10	Repair of connective tissues
	Angiogenesis
Days 30–90	Blood vessels mature
	New type of connective tissue associated with healing

The technique is even more powerful when repeated at different times after injury. When researchers did exactly that on injured rats, they discovered genes expressed just after the injury whose participation they never suspected. Their microarrays, summarized in **table 19.3,** revealed waves of expression of genes involved in healing. Analysis on the first day indicated activation of the same suite of genes whose protein products heal injury to the deep layer of skin—a total surprise that suggests new points for drugs to intervene.

Key Concepts

1. DNA microarrays enable researchers to track gene expression.
2. In a DNA microarray experiment, DNA pieces of known sequence are attached to a glass or plastic chip, and differentially labeled sample cDNAs are applied.
3. The patterns and color intensities of spots indicate which genes are expressed. A laser scanner detects and computer algorithm interprets the results.

19.5 Silencing DNA

Since the 1960s, when the flow of genetic information from DNA to RNA to protein was discovered and described, researchers have been trying to manipulate the process to diminish ("knock down") or stop ("silence") the expression of specific genes. These efforts use the base pairing rules to block transcription or translation of a particular DNA sequence. Different approaches to gene silencing have had varying degrees of success. A technique called **RNA interference (RNAi)** is based on the fact that RNA molecules can fold into short, double-stranded regions where the base sequence is complementary, as **figure 19.9** shows. (Such localized base pairing enables a tRNA molecule to assume its characteristic cloverleaf shape.)

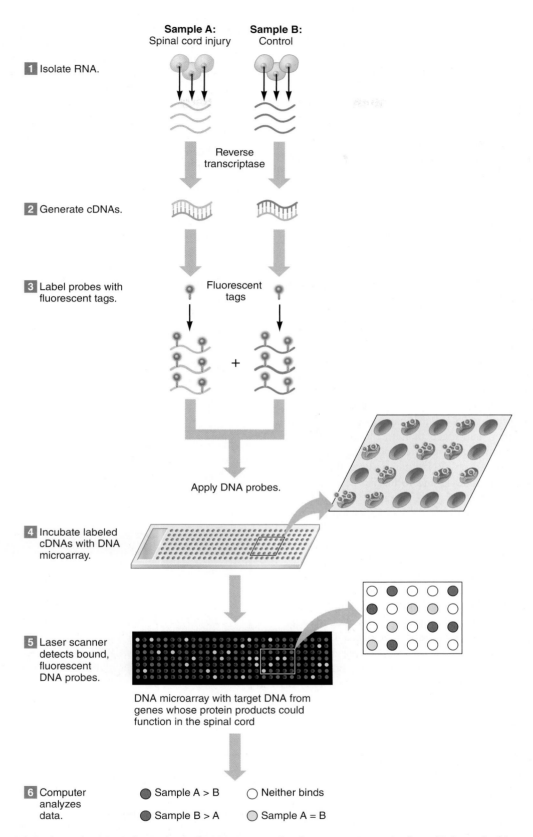

Sample A:
Spinal cord injury

Sample B:
Control

1 Isolate RNA.

Reverse
transcriptase

2 Generate cDNAs.

3 Label probes with
fluorescent tags.

Fluorescent
tags

+

Apply DNA probes.

4 Incubate labeled
cDNAs with DNA
microarray.

5 Laser scanner
detects bound,
fluorescent
DNA probes.

DNA microarray with target DNA from
genes whose protein products could
function in the spinal cord

6 Computer
analyzes
data.

● Sample A > B   ○ Neither binds

● Sample B > A   ○ Sample A = B

Figure 19.8 **A DNA microarray experiment reveals gene expression in response to spinal cord injury.** In this example, the red label represents DNA from a patient with a spinal cord injury, and the green label represents control DNA from a healthy person. DNA targets on the microarray that bind red but not green can reveal new drug targets.

GACAUUCGCAGCUAUAGCUGCGAAUAGU →

Figure 19.9 Hairpins. RNA hydrogen bonds with itself, forming hairpin loops. RNAi uses similar molecules to "knock down" expression of specific genes by binding their mRNAs.

RNA interference introduces short, double-stranded RNAs into cells, where, like microRNAs (see section 11.2), they bind to and halt the activities of mRNAs bearing complementary sequences.

Shortly after RNA interference was discovered in 1998, biotechnology and pharmaceutical companies began developing the approach to silence genes. Potential applications ranged from creating vaccines by knocking down expression of key genes in viruses that cause disease, such as AIDS, polio, and hepatitis C; to treating cancer by silencing oncogenes; to creating a better-tasting decaf coffee by silencing an enzyme required for caffeine synthesis in coffee plants. However, problems arose in clinical trials of RNAi-based drugs. In the human body, the synthetic "small interfering RNAs" that carry out RNAi tend to inflame the liver, rather than reach their intended targets. For now, they seem better suited as a research tool to see what happens when a gene is turned off. RNAi can be used this way in animal models of human disease and in human cells growing in culture.

Another approach to inhibiting gene expression is to use a type of synthetic molecule called a morpholino. It consists of a selected sequence of twenty-five DNA bases that are attached to organic groups that are not the same as the sugar-phosphate backbone of DNA. Morpholinos may be useful in blocking splice-site mutations that would otherwise delete entire exons (see section 12.4). In a clinical trial under way, morpholinos targeted to the dystrophin gene are injected into the leg muscles of boys with exon-skipping mutations that cause Duchenne muscular dystrophy (see figure 2.1). Results so far are promising. The use of morpholinos is gene silencing, but it actually restores a gene's function by silencing a certain type of mutation.

Completely silencing a gene uses a natural process called homologous recombination to replace a DNA sequence with a similar or identical sequence that cannot be transcribed or translated, like a dummy key that fits a lock but won't turn it. The technique, called gene targeting, was developed in the late 1980s by introducing an inactivated version of a gene into a mouse embryonic stem cell. It does not work on a mammal's fertilized ovum. For this reason, it has not been done on humans, but homologous recombination is used to breed "knockout" mice that are valuable in several ways:

- They are more accurate models than transgenic mice because a transgene can insert anywhere in a genome. A knockout swaps in a gene at a specific site.
- Populations are easily tested. Two groups of identical knockout mice, for example, can be exposed to different environmental factors to assess their impact.
- Knockouts for several genes can be created to observe polygenic traits and diseases.
- Mouse embryos and fetuses with diseases that humans also get can be observed.

Ironically, the greatest value of knockout mouse models became clear with what looked at first like failed experiments—the knockouts for supposedly vital genes were healthy! For example, knockout mice lacking a collagen gene thought to be essential for normal growth and development of long bones, in both mice and humans, had normal skeletons. These types of results reveal genetic heterogeneity—when different genes have redundant or overlapping effects on the phenotype. Therefore, knocking out one gene's function exposes that of a different gene.

The next chapter explores how genetic technologies are applied to people.

Key Concepts

1. RNA interference occurs when short, single-stranded RNAs are introduced into a cell and bind to their complements in mRNAs, preventing translation.
2. Morpholinos block expression of splice-site mutations, restoring translation.
3. Gene targeting uses homologous recombination to knock out a specific gene, revealing the gene's function by removing it.

Summary

19.1 Patenting DNA

1. **Biotechnology** alters cells or biochemicals to provide a product. It includes extracting natural products, altering an organism's DNA, and combining DNA from different species.
2. A **transgenic organism** has DNA from a different species. **Recombinant DNA** comes from more than one type of organism. Both are possible because of the universality of the genetic code.

3. Patented DNA must be useful, novel, and non-obvious. Patent law is evolving to ease use of multigene tests.

19.2 Amplifying DNA

4. Nucleic acid amplification, such as **PCR,** uses the power and specificity of DNA replication enzymes to selectively mass-produce DNA sequences.
5. In PCR, primers corresponding to a DNA sequence of interest direct polymerization of supplied nucleotides to make many copies.

19.3 Modifying DNA

6. Recombinant DNA technology mass-produces proteins in bacteria or other single cells. Begun hesitantly in 1975, the technology has matured into a valuable method to produce proteins.

7. To construct a recombinant DNA molecule, **restriction enzymes** cut the gene of interest and a **cloning vector** at a short palindromic sequence, creating complementary "sticky ends." The DNAs are mixed and vectors that pick up foreign DNA selected.

8. **Genomic libraries** consist of recombinant cells containing fragments of a foreign genome. **DNA probes** are used to select genes of interest from genomic libraries. DNA probes may be synthetic, taken from another species, or a **cDNA libraries,** which is reverse transcribed from mRNA.

9. Genes conferring antibiotic resistance and color changes in growth media are used to select cells harboring recombinant DNA. Useful proteins are isolated and purified.

10. A multicellular transgenic organism has an introduced gene in every cell. Heterozygotes for a transgene are bred to yield homozygotes.

19.4 Monitoring Gene Function

11. **DNA microarrays** are devices that hold DNA pieces to which fluorescently labeled DNA probes from samples are applied. They are used in **gene expression profiling**.

19.5 Silencing DNA

12. In **RNA interference,** small interfering RNAs, introduced into a cell, bind their complementary sequences on mRNA, suppressing protein production.

13. Morpholinos are short synthetic nucleic acids that block expression of splice-site mutations, restoring synthesis of a particular protein.

14. Gene targeting uses homologous recombination to knock out a specific gene in a fertilized ovum, revealing gene function.

---

www.mhhe.com/lewisgenetics10

Answers to all end-of-chapter questions can be found at **www.mhhe.com/lewisgenetics10.** You will also find additional practice quizzes, animations, videos, and vocabulary flashcards to help you master the material in this chapter.

Review Questions

1. Cite three examples of a DNA sequence that meets requirements for patentability.

2. Explain how a thicket serves as a metaphor for the current state of patents on DNA sequences.

3. Describe the roles of each of the following tools in a biotechnology:
   a. restriction enzymes
   b. cloning vectors
   c. DNA microarrays
   d. short nucleic acid sequences

4. How are cells containing recombinant DNA selected?

5. List the components of an experiment to produce recombinant human insulin in *E. coli* cells.

6. Why would recombinant DNA technology be restricted if the genetic code were not universal?

7. What is an advantage of a drug produced using recombinant DNA technology compared to one extracted from natural sources?

8. Describe three ways to insert DNA into cells.

9. Describe a biotechnology that uses an organism or its cells without modifying its DNA.

10. Explain the advantages of using a DNA microarray that covers all of the protein-encoding genes in the human genome (the "exome"), rather than selected genes whose protein products are known to take part in the disease process being investigated.

11. Explain how natural selection can be used in bioremediation.

12. Describe how a technology to knock down or silence gene expression can be used to treat a particular disease.

Applied Questions

1. Phosphorus in pig excrement pollutes aquatic ecosystems, causing fish kills and algal blooms, and contributes to the greenhouse effect. *E. coli* produces an enzyme that breaks down phosphorus. Describe the steps to create a transgenic pig that secretes the bacterial enzyme, and therefore excretes less polluting feces.

2. To diagnose a rare form of encephalitis (brain inflammation), a researcher needs a million copies of a viral gene. She decides to use PCR on a sample of the patient's cerebrospinal fluid. If one cycle takes 2 minutes, how long will it take to obtain a millionfold amplification?

3. HIV infection was once diagnosed by detecting antibodies in a person's blood or documenting a decline in the number of helper T cells. Why is PCR detection of HIV RNA more sensitive?

4. Genetic modification endows organisms with novel abilities. From the following three lists (choose one item from each list), devise an experiment to produce a particular protein, and suggest its use.

Organism	Biological Fluid	Protein Product
pig	milk	human beta globin chains
cow	semen	human collagen
goat	milk	human EPO
chicken	egg white	human tPA
aspen tree	sap	human interferon
silkworm	blood plasma	jellyfish GFP
rabbit	honey	human clotting factor
mouse	saliva	human alpha-1-antitrypsin

5. Collagen is a connective tissue protein that is used in skincare products, shampoo, desserts, and in artificial skin. For many years, it was obtained from the hooves and hides of cows collected from slaughterhouses. Human collagen can be manufactured in transgenic mice. Describe the advantages of the mouse source of collagen.

6. People did not object to the production of human insulin in bacterial cells used to treat diabetes, yet some people object to mixing DNA from different animal and plant species in agricultural biotechnology. Why do you think that the same general technique is perceived as beneficial in one situation, yet a threat in another?

7. A human oncogene called *ras* is inserted into mice, creating transgenic animals that develop a variety of tumors. Why are mouse cells able to transcribe and translate human genes?

8. In a DNA microarray experiment, researchers attach certain DNA pieces to the grid. For example, to study an injury, genes known to be involved in the inflammatory response might be attached. How might this approach be limited?

9. Devise an experiment using DNA microarrays to determine whether men and women have different hormonal responses to watching an emotional film. (A hormone is a type of messenger molecule that is carried in the blood).

Web Activities

1. Nobel Prizes have been awarded for several of the biotechnologies mentioned in this chapter. Go to Nobelprize.org, nobelprizes.com, or another website and describe the work that led to one of the following:

   a. 1993 for the polymerase chain reaction

   b. 2006 for RNA interference

   c. 2007 for gene targeting by homologous recombination

   d. 2008 for green fluorescent protein from jellyfish to mark gene expression

2. Use the Web to identify three drugs made using recombinant DNA technology, and list the illnesses they are used to treat.

3. Look at the websites for direct-to-consumer genetic testing companies, and discuss the patenting challenges that these companies face in providing tests for multiple genes or SNPs.

4. Go to the DNA Patent Database (http://dnapatents. georgetown.edu) and click on "About the DPO." Consider the number of DNA patents filed and granted each year since 2000. Explain why the number of granted patents peaked in 2003 (see chapter 22).

5. Recombinant DNA technology is used to manufacture human growth hormone (hGH), which is used to treat some forms of dwarfism. However, "anti-aging" clinics and websites sell what they claim is hGH to healthy consumers, although studies indicate that the only benefit is a slight increase in muscle mass. Possible side effects are serious, and include diabetes, breast development in men, joint pain, fluid retention, and shortened life. Legislation is pending to classify hGH as a controlled substance, limiting its distribution.

   a. Consult a website selling hGH and list the claims and warnings. Which do you think are accurate?

   b. Do you think laws should restrict access to hGH by people who do not have the medical conditions for which it is prescribed? Cite a reason for your answer.

Case Studies and Research Results

1. Nancy is a transgenic sheep who produces human alpha-1-antitrypsin (AAT) in her milk. This protein, normally found in blood serum, enables the microscopic air sacs in the lungs to inflate. Without it, inherited emphysema results. Donated blood cannot yield enough AAT to help the thousands of people who need it. Describe the steps taken to enable Nancy to secrete human AAT in her milk.

2. To investigate causes of acne, researchers used DNA microarrays that cover the entire human genome. Samples came from facial skin of people with flawless complexions and from people with severe acne. In the simplified portion of a DNA microarray shown on the next page, one sample is labeled green and comes from healthy skin; a second sample is labeled red and represents skin with acne. Sites on the

microarray where both probes bind fluoresce yellow. The genes are indicated by letter and number.

a.

b.

c.

d.

e.

f.

g.

a. Which genes are expressed in skin whether or not a person has acne?
b. Which genes are expressed only when acne develops?
c. List three DNA pieces that correspond to genes that are not expressed in skin.
d. How would you use microarrays to trace changes in gene expression as acne begins and worsens?
e. Design a microarray experiment to explore gene expression in response to sunburn.

Max Randell is one of the pioneers of gene therapy, which he received directly into his brain to treat Canavan disease. On his 13th birthday, he listened to the author read cards made by students who learned about him in past editions of this book. Max's mom Ilyce is in blue, and the woman kneeling is Paola Leone, the neuroscientist who has guided children through gene therapy. Max's grandmother Peggy pats his head.

CHAPTER

20

Genetic Testing and Treatment

Learning Outcomes

20.1 Genetic Counseling

1. Describe the services that a genetic counselor provides.

20.2 Genetic Testing

2. Explain the uses of preconception comprehensive carrier testing.

3. Explain how a fetal genome can be sequenced.

4. Describe the basis of most newborn screening tests.

5. Discuss the benefits and limitations of direct-to-consumer genetic testing.

6. Define *pharmacogenetics* and *pharmacogenomics*.

20.3 Treating Genetic Disease

7. Describe three approaches to correcting inborn errors of metabolism.

8. Explain how an existing drug can be "repurposed" to treat a genetic disease.

9. Explain what gene therapy does.

10. Discuss the ups and downs of the history of gene therapy.

The Big Picture: DNA-based tests have moved from the realm of a health care setting to wide availability, thanks to the Internet. Such tests are not simple and can have effects beyond the individual. At the same time, gene therapy has recovered from setbacks with recent successes.

Fighting Canavan Disease

Max Randell was not expected to survive his first 2 years. Today he is a teen, thanks to gene therapy.

In early 1998, Ilyce and Mike Randell, of Buffalo Grove, Illinois, were worried about their 5-month-old son. He could not hold up his head, roll over, or reach for objects, and he was not very responsive. When doctors diagnosed Canavan disease (MIM 271900), which robs brain neurons of their fatty sheaths so that the brain slowly degenerates, they suggested that the best place for the red-haired boy would be a nursing home. Instead, Max became the youngest person to receive gene therapy for a degenerative brain disease.

Max underwent his first gene therapy, to test safety, in 1998 at Yale University. Soon after, he became able to use a walker, and his vision greatly improved. He had a second gene therapy in 2001. Each time, billions of viral particles, each carrying a corrected copy of the gene that was mutant in each of Max's cells, were delivered through six holes drilled into his skull. Perhaps because he was the youngest of the fourteen patients treated, he has done the best. Although today his movements are very limited, he is alert and aware. He answers questions slowly, with a blink for "yes" and widening eyes for "no." He attends school, and his favorite activity is listening to his little brother Alex read to him.

20.1 Genetic Counseling

Over the past few decades, the field of human genetics has evolved from an academic life science, to a medical specialty, to a source of personal information that ordinary people can access. This chapter presents the types of genetic services that a health care consumer might encounter: genetic counseling, genetic testing, and protein and gene-based therapies.

Our genomes are windows into who we are, filled with clues to health, ancestry, how we differ from each other and how we are also very similar. Genetic tests provide views of our genomes at several levels. They may detect variants in a single gene, variability across chromosome arms or wide swaths of the genome, the entire protein-encoding part of the genome (the exome), or even the whole genome.

Genetic tests are administered at all stages of human existence: from 8-celled embryos, to fetuses, to newborns and older children, to couples contemplating parenthood and others of all ages who want to understand puzzling sets of symptoms. Tests are available in medical settings, as they have been for decades, but also on the Internet, "direct-to-consumer." Unlike a cholesterol check or a chest X ray, the results of a genetic test can have repercussions beyond the individual, to family members who share genotypes that affect health. Because more than 1,600 genetic tests are now available, the U.S. National Institutes of Health in 2011 introduced the Genetic Testing Registry.

This clearinghouse provides information on tests to health care consumers, physicians, and researchers, and is integrated with related databases.

A **genetic counselor** is a health care professional who helps patients and their families navigate the confusing path of genetic testing (**figure 20.1**). Such tests identify genotypes that cause, contribute to, or raise the risk of developing a specific disease. Genetic counseling addresses medical, psychological, sociological, and ethical issues, and a genetic counselor has medical, scientific, and communication skills. A counselor can interpret a DNA test, explain uncertainties, and suggest ways to cope with anxiety, fear, or guilt associated with taking genetic tests—or not taking them.

In 1947, geneticist Sheldon Reed coined the term "genetic counseling" for the advice he gave to physician colleagues on how to explain heredity to patients with single-gene diseases. In 1971, the first class of specially trained genetic counselors graduated from Sarah Lawrence College, in Bronxville, New York. Today, thirty-two programs in the United States offer a master's degree in genetic counseling, and many other countries have programs.

Genetic counseling began in pediatrics and prenatal care. Today it embraces diseases of adults too, branching into such specialties as cancer, cardiovascular disease, neurology, hematology, and ophthalmology. At the same time, genetic counselors are increasingly meeting patients who come in with results

Reasons to seek genetic counseling:

Family history of abnormal chromosomes

Elevated risk of single-gene disorder

Family history of multifactorial disorder

Family history of cancer

Genetic counseling sessions:

Family history

Pedigree construction

Information provided on specific disorders, modes of inheritance, tests to identify at-risk family members

Testing arranged, discussion of results

Links to support groups, appropriate services

Follow-up contact

Figure 20.1 **The genetic counseling process.**

of direct-to-consumer tests for dozens of conditions. The field has even infiltrated public policy, as genetic testing has become widespread. Genetic counselors in New York, for example, hold "DNA days" to educate state legislators.

Most genetic counselors work directly with patients as parts of health care teams, typically at medical centers. A consultation may entail a single visit to explore a test result, such as finding that a pregnant woman is a carrier for cystic fibrosis, or a several-month-long relationship as the counselor guides a decision to take (or not take) a test for an adult-onset disorder, such as Huntington disease.

The knowledge that a genetic counselor imparts is similar to what you have read in this book, but personalized and applied to a specific disorder. A counselor might explain Mendel's laws, but substitute a family's condition for pea color. Or, a counselor might explain how an inherited susceptibility can combine with a controllable environmental factor, such as smoking or poor diet, to affect health.

A genetic counseling session begins with a discussion of the family's health history. Using a computer program or pencil and paper, the counselor constructs a pedigree, then deduces and explains the risks of disease for particular family members (figure 20.1). She may initially present possibilities and defer discussion of specific risks and options until test results are available. The counselor also explains which second-degree relatives—aunts, uncles, nieces, nephews, and cousins—might benefit from being informed about a test result. She provides detailed information on the condition and refers the family to support groups. If a couple wants to have a biological child who does not have the illness, a discussion of assisted reproductive technologies (see chapter 21) might be in order.

A large part of the genetic counselor's job is to determine when specific biochemical, gene, or chromosome tests are appropriate, and to arrange for people to take the tests. The counselor then interprets test results and helps the patient or family choose among medical options, in consultation with the appropriate medical specialists. Until the recent broadening of genetic counseling thanks to direct-to-consumer tests, people most often sought genetic counseling for either prenatal diagnosis or a disease in the family.

Prenatal genetic counseling typically presents population (empiric) and family-based risks, explains tests, and discusses whether the benefits of testing outweigh the risks. The couple, or woman, decides whether amniocentesis, chorionic villus sampling, maternal serum screening, ultrasound, or no testing is best for them. Part of a prenatal genetic counseling session is to explain that tests that rule out some conditions do not guarantee a healthy baby. For example, amniocentesis checks only large-scale chromosome aberrations and specifically requested single-gene tests. If a test reveals that a fetus has a serious medical condition, the counselor discusses possible outcomes, treatment plans, and the option of ending the pregnancy.

Genetic counseling when an inherited disease is in a family is another matter. For recessive disorders, the affected individual is usually a child. Illness in the first affected child is often a surprise, and especially if this is a first child, recognition of a problem may take months and a diagnosis, years.

Communicating the risk subsequent children face may be difficult. Many people think that if one child has an autosomal recessive condition, then the next three will be healthy. Actually, each child has a 1 in 4 chance of inheriting the illness. Counseling for subsequent pregnancies requires great sensitivity. Some people will not terminate a pregnancy when the fetus has a condition that already affects their living child, yet some will see that as the kindest option. Genetic counselors must respect these feelings, and tailor the discussion accordingly, while still presenting all the facts.

Genetic counseling for adult-onset disorders does not have the problem of potential parents making important decisions for existing or future children, but presents the conflicting feelings of people choosing whether or not to find out if a disease is likely in their future. Often, they have seen loved ones suffer with the illness. This is the case for Huntington disease (see *Bioethics: Choices for the Future* on page 76). Predictive tests are introducing a new type of patient, the "genetically unwell" or those in a "premanifest" state—people with mutant genes but no symptoms (yet). Such a disease-associated genotype indicates elevated risk, but is not a medical diagnosis, which is based on symptoms.

When genetic counseling began in the 1970s, it was "nondirective," meaning that the practitioner presented options but did not offer an opinion or suggest a course of action. That approach is changing as the field moves from analyzing hard-to-treat, rare single-gene disorders to considering inherited susceptibilities to more common illnesses that are more treatable, and for which lifestyle changes might realistically alter the outcome. A more recent description of a genetic counseling session is "shared deliberation and decision making between the counselor and the client."

Genetic counselors regularly communicate with physicians and other health care professionals. They are important parts of teams at molecular diagnostic testing laboratories, where they guide physicians in ordering and interpreting tests. Before a test is ordered, the counselor helps to assess the patient's pedigree, discusses the pros and cons of the appropriate test, and raises ethical issues that might arise when other family members are considered. While the test is under way, the genetic counselor ensures that time constraints are respected, such as an advancing pregnancy, and updates the physician. Once test results are in, the counselor may request a repeat if they are inconsistent with the patient's symptoms; interpret the results; suggest additional tests; and alert the physician if the patient might be a candidate for participating in a research project under way or planned at the lab.

The United States has about 3,000 genetic counselors. Because there are so few, and most of them practice in urban areas, access to their services is limited. Finding a genetic counselor with a specific expertise is especially difficult. For example, only 400 genetic counselors in the United States are specially trained in cancer genetic counseling. Due to the shortage of counselors and demand for their services, sometimes other types of health care professionals, such as physicians, nurses, social workers, and PhD geneticists, provide counseling. One survey found that dietitians, physical therapists, psychologists, and speech-language pathologists also regularly discuss genetics with their patients.

Some genetic testing companies, both real and Web-based, use "virtual" genetic counseling, in which an interactive computer program calculates risks and provides basic information. Direct-to-consumer genetic testing companies vary widely in their provision of genetic counseling services. Some offer many Web pages of information, but lack a genetic counselor to interpret the information that a particular test provides in the context of a specific family. A few companies offer phone access to genetic counselors. As genetic testing becomes more commonplace, the need for genetic counselors and other genetics-savvy professionals to help individuals and families best use the new information is very likely to increase.

Key Concepts

1. A genetic counselor provides information to individuals, couples expecting children, and families about modes of inheritance, recurrence risks, genetic tests, and treatments.

2. The counselor helps people make decisions while being sensitive to individual choices.

3. The shortage of genetic counselors has led other health care professionals to provide the service. Genetic counselors help other health care professionals.

20.2 Genetic Testing

Genetic testing is used to diagnose and predict the course of a disease, select treatment, and predict and monitor response to treatment (**table 20.1**). Soon it will be routine to rapidly and inexpensively sequence one's own genome, or to focus on gene variants and markers of clinical importance. Chapter 22 considers the clinical utility of personal genome sequencing.

Using genetic tests wisely must balance our tendency to blame all ills on our genes—genetic determinism—with identifying ways to change our lifestyles to compensate for factors that we have inherited and therefore cannot control. The following sections consider types of genetic tests not already discussed.

Preconception Comprehensive Carrier Testing

In the past, genetic tests for carriers of recessive disorders have focused on specific population groups in which the disease is more common, mostly for economic reasons. In the 1970s, carrier testing for sickle cell disease targeted African Americans, while testing for Tay-Sachs disease focused on Ashkenazi Jews. Test panels screen for the dozen or so genetic diseases that are more common in the Ashkenazim (see table 15.2). However, thanks to much faster DNA sequencing, the cost of carrier tests for many diseases has plummeted so much that restricting tests to certain groups no longer makes much sense. In addition, most of us have at least one recessive disease-causing mutation.

The 1,139 single-gene (Mendelian) diseases that are recessive and have a known mechanism are individually rare,

---

Table 20.1	Types of Genetic Tests	
Type of Test	**Information Provided**	**Example**
Population carrier screen	Identifies heterozygotes—people with one copy of a mutant gene	College students are offered testing for sickle cell disease carrier status.
Preconception testing	Identifies recessive disease-causing alleles in potential parents	A couple learn that they are carriers for the same disease.
Prenatal test	Detects mutant allele in a fetus for a condition present in a family	A couple who know they are carriers of Tay-Sachs disease has a fetus tested.
Prenatal screen	Tests embryos or fetuses from a population for increased risk of a condition, not based on family history	A pregnant woman's blood is tested for elevated level of a protein indicating increased risk for a neural tube defect.
Newborn screen	Population-wide testing for several treatable inborn errors of metabolism	A child with identified sickle cell disease genes at birth can ease or delay symptoms with antibiotics.
Diagnostic test	Confirms diagnosis based on symptoms	A child with "failure to thrive" and frequent lung infections is tested for mutant alleles for CF.
Predisposition test	Detects allele(s) associated with an illness, but not absolutely diagnostic of it	A young Jewish woman with a strong family history of breast cancer has a mutant *BRCA1* allele, giving her an 86 percent lifetime risk of developing the condition.
Predictive test	Detects highly penetrant mutation with adult onset in an individual at high risk based on family history	A healthy person is tested for the Huntington disease mutation because a parent has the condition.

but collectively account for about 20 percent of infant deaths and about 10 percent of childhood hospitalizations. Researchers have developed "preconception comprehensive carrier screening," a single test that detects 448 recessive diseases that affect children. Included diseases meet certain criteria:

- the phenotype must be severe;
- several mutations must be known; and
- penetrance is high (a disease genotype causes symptoms).

The goal of comprehensive carrier screening is to prevent suffering, but the actual application of the information raises ethical concerns. The people most likely to take the test are young adults contemplating having children. If partners discover that they carry the same disease-causing mutation, they can prevent the birth of an affected child in several ways. They might adopt children, use prenatal diagnosis to select fetuses that have not inherited the disease (see figure 13.5), or use an assisted reproductive technology to replace the gamete of one parent (discussed in chapter 21). For example, a sperm donor would replace the sperm of a man who carries the same recessive disease as his partner. Another approach, called pre-implantation genetic diagnosis, samples a cell from an early embryo and allows the rest of the embryo to continue development only if it has not inherited the family's recessive disease (see figure 21.4).

Comprehensive carrier testing is likely to become widespread quickly. The average number of disease-causing recessive mutations in an individual is 2.8, and ranges from zero to 7. Indeed, people who have had their entire genomes sequenced have found out that they carry several recessive disorders.

Prenatal Testing

Checking fetal chromosomes has been routine since the 1970s, using chorionic villus sampling or amniocentesis (see figure 13.5). These techniques are invasive and therefore expose the fetus to risk. Maternal serum markers are also routinely used to assess health of a fetus (see the table in *Bioethics: Choices for the Future* in chapter 13). Such biomarker tests, however, present elevated risks, not diagnoses. A new technology that avoids the risks of invasive testing and the limitations of maternal serum marker testing determines the sequence of a fetus's genome from DNA pieces in maternal serum.

For several years, DNA from rare fetal cells in a pregnant woman's circulation has been tested for sex chromosome constitution and aneuploidy (extra or missing chromosomes), as chapter 13 mentions. Fetal DNA that is free in the serum, rather than in cells, is another significant source of material for testing (**figure 20.2**). Fetal DNA fragments make up only about 10 percent of the free DNA in maternal serum, but altogether they cover the entire genome many times over. If they can be isolated, sequenced, and overlapped to deduce the entire sequence, that would offer a wealth of health information. A key recent discovery makes this possible: Most fetal DNA fragments are smaller than most maternal fragments. Researchers can separate the pieces by size.

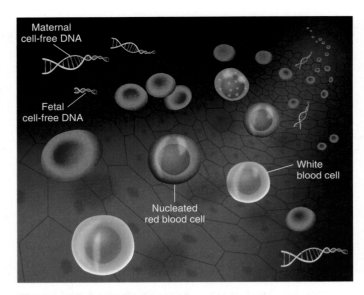

Figure 20.2 **Testing fetal DNA.** The fact that pieces of cell-free fetal DNA in the maternal circulation are much shorter than pieces of maternal DNA provides a way to collect the fetal material and sequence the genome. Comparison of a fetal genome sequence to that of the parents can reveal inherited diseases.

Genome sequencing of cell-free fetal DNA is a technology still in its infancy, but it has the potential to be extremely useful. Comparing a fetus's genome to that of the parents would reveal any inherited diseases. DNA sequences that a fetus does not share with the parents would reveal new mutations. Using this technology, any disease-causing mutations missed in preconception carrier screening could be caught at the fetal stage.

Newborn Screening

Hunter Kelly was born in 1997, the son of NFL quarterback Jim Kelly and his wife Jill. When he was diagnosed at 9 months of age with Krabbe disease (MIM 245200), also known as galactosylceramide lipidosis after the enzyme that is missing, his nervous system was already damaged. The enzyme deficiency causes a buildup of certain metabolites, yet too little myelin, the fatty substance that coats neurons. Early symptoms include crying, difficulty feeding, and stiffness. As time goes on, the back arches, the limbs jerk, tube feeding becomes necessary, and motor and mental development slow and stop. The child loses hearing and vision, and usually dies by age 2.

Hunter Kelly lived for 8 years. Had he been born today, Hunter would have been tested for Krabbe disease and dozens of other inborn errors of metabolism, with just a few drops of blood taken from his heel shortly after birth. A cord blood stem cell transplant from a compatible donor could have prevented his symptoms. **Figure 20.3** shows a family in which one child suffered, as did Hunter, yet a younger sibling was helped because testing for their disease, glutaric acidemia (MIM 231670), became available before she was born.

Most newborn screening tests for inborn errors of metabolism are not genetic tests, because they do not analyze DNA. Instead they use a technique from analytical chemistry called

Figure 20.3 Newborn screening expands. Siblings Emma and Matthew Chorey inherited the same inborn error of metabolism—glutaric acidemia—but Emma was luckier. Thanks to newborn screening, her condition was diagnosed 9 days after birth, and she began a special diet that has prevented buildup of protein in the part of the brain that controls movement. She is well. Matthew was diagnosed at 9 months, because newborn testing was not available. He cannot walk or talk but is still an important member of the Chorey family.

tandem mass spectrometry to identify unusual metabolites or chemical imbalances that indicate a particular inherited disease. "Mass spec" can detect telltale biochemicals from many disorders at a time, in blood drops from a newborn's heel. In the United States, the individual states determine which diseases to test for. The American College of Medical Genetics recommends newborn screening for a minimum of twenty-nine disorders, but the actual number of tests ranges from twenty-one to fifty-four. This spotty regulation means that a child in one state may be diagnosed in time for treatment, but not a child with the same disease living in a different state. Newborn screening for disorders that currently lack treatment is nevertheless useful, because it can help narrow down a diagnosis and alert other family members.

The field of newborn screening began in 1961, with phenylketonuria, discussed in section 5.2. The Guthrie test sampled blood from a newborn's heel and tested for the amino acid buildup that indicates PKU. In 1963, a specialized diet (legally termed a "medical food" so that insurance will cover the high cost) became available, with dramatic positive results. The diet sharply reduces the amount of the amino acid phenylalanine, which builds up in the disorder. The diet is very difficult to stick to, and must be followed for many years, but it does prevent intellectual disability. After the success of newborn screening for PKU, state testing expanded to include eight genetic conditions and a nongenetic form of hearing loss. Gradually, the offerings have grown.

Newborn screening is important because the earliest signs and symptoms of many of the disorders tested for are common in babies. An inexperienced parent may be told by a physician unfamiliar with rare diseases, "Oh, it's just colic," and sent home, only to witness declines in motor and mental skills over

the ensuing weeks. For certain diseases, treatment must begin soon after birth to help. Some states perform DNA tests on newborns as well as biochemical tests. The National Newborn Screening and Genetics Resource Center helps states to tailor programs to specific populations. For example, a test for cystic fibrosis might include 100 mutations out of the 1,600-plus that have been identified, but the 100 most common mutations in New Mexico differ from those in New Jersey. The test panels for CF in these states differ.

When the American College of Medical Genetics widened the scope of newborn testing in 2005, the goal was to save parents from months or years of searching for an accurate diagnosis for their child's condition. However, the more extensive testing has also led to false positives or "patients-in-waiting," children whose newborn screening reveals disease markers, but who do not develop the predicted symptoms. Sociologists documented this situation in 75 families, noting the constant anxiety and restricted lifestyles for parents waiting for their children to become ill. Considering the millions of newborns screened, however, "patients-in-waiting" are rare.

DTC Testing

Companies market direct-to-consumer (DTC) DNA-based tests for traits, susceptibilities, and genetic diseases to the general public (**table 20.2**). The tests range from the obvious (eye color) to the dubious (athletic ability) to the serious (cancers).

In the United States, the Clinical Laboratory Improvement Amendments, or CLIA, control genetic testing of body materials for the prevention, diagnosis, or monitoring response to treatment of a disease or health impairment. The CLIA regulations, instituted in 1988, added "specialty areas" in 1992 to cover very complex tests, such as those involving immunology or toxicology. These did not include genetic testing, which at

Table 20.2	**Direct-to-Consumer Genetic Tests**
DTC website/companies that offer information may not be held to the same standards as laboratories that provide tests used to make clinical diagnoses. Here are some offerings from two DTC companies.	

Company A: Traits	**Company B: Disorders**
Alcohol flush reaction	Alzheimer disease
Athletic performance	Brain aneurysm
Bitter taste	Cancers
Earwax type	Diabetes
Eye color	Glaucoma
Lactose intolerance	Multiple sclerosis
Muscle performance	Obesity
Skin quality	Osteoarthritis, restless legs, rheumatoid arthritis

the time was very limited. Since then, the genetics community has repeatedly asked that genetic tests be included as a specialty area, to no avail. State regulations can override CLIA, but only if they are equally or more stringent. Therefore, the regulation of genetic tests remains somewhat unclear.

Tests that offer genetic information, but are not intended to be used to diagnose a disease, do not come under the CLIA regulations. The distinction between information and diagnosis is often based on careful wording. Consider Company B in table 20.2, which "... scans your DNA for genetic risk markers associated with both common and uncommon health conditions." "Association" means a relationship between one piece of information and another—it is not a correlation or a cause, nor a diagnosis. Some companies have CLIA certification for some of their tests but not others, but even that regulation is not always clear. For example, CLIA certifies that laboratories are safe and that a particular test measures what it claims to measure. However, the regulations do not require that knowing the measurement, such as the concentration of a molecule in blood, leads to actions that improve health.

Even DTC tests for very well-studied mutations can lead to complications and confusion, especially if the consumer does not know much about genetics. This is the case for hereditary hemochromatosis (HH, MIM 235200). In this autosomal recessive "iron overload" disease, cells in the small intestine absorb too much iron from food. Early signs and symptoms of HH include chronic fatigue, increased susceptibility to infection, hair loss, infertility, muscle pain, and feeling cold. Over many years, the excess iron is deposited throughout the body, damaging vital organs. However, HH is incompletely penetrant—that is, many people who have the disease-associated genotype do not have symptoms, especially women who lose the extra iron in the monthly menstrual flow.

Diagnosis of symptomatic HH is important, because lowering the body's iron levels is easy—have blood removed periodically. However, diagnosis is based on an increase in the level of the iron-carrying protein ferritin in the blood and confirmed with a liver biopsy, rather than on a genetic test because of the incomplete penetrance. Yet a website offering genetic testing claims: "Most patients do not know that they can test themselves for hereditary hemochromatosis. Here is an easy way to get direct testing without a prescription and results sent directly to the patient only." At-home testing for HH could be confusing to consumers unfamiliar with the uncertainty of genetics.

An advantage of DTC genetic testing companies is that they are amassing tremendous data stores, which can be mined to make discoveries much faster than is possible with traditional research methods. Most company websites also provide a great deal of excellent information about genetics. However, some companies prey on consumer unfamiliarity with genetics. This was the case for certain "nutrigenetics" companies.

Nutrigenetics Testing

"Nutrigenetics" DTC websites offer genetic tests along with general questionnaires about diet, exercise, and lifestyle habits. The companies return supposedly personalized profiles with dietary suggestions, often with an offer of a pricey package of exactly the supplements that an individual purportedly needs to prevent realization of his or her genetic fate.

After the media spread the word of these services, the U.S. Government Accountability Office tested the tests. An investigator took two DNA samples—one from a 9-month-old girl and the other from a 48-year-old man—and created fourteen lifestyle/dietary profiles for these "fictitious consumers"—twelve for the female, two for the male. The samples were sent to four nutrigenetics companies. Here is an example of the information sent to the companies:

- The DNA from the man was submitted as being from a 32-year-old male, 150 pounds, 5′9″, who smokes, rarely exercises, drinks coffee, and takes vitamin supplements.
- The DNA from the baby girl was submitted as being from a 33-year-old woman, 185 pounds, 5′5″, who smokes, drinks a lot of coffee, doesn't exercise, and eats a lot of dairy, grains, and fats.
- The same baby girl DNA was also submitted as that of a 59-year-old man, 140 pounds, 5′7″, who exercises, never smoked, takes vitamins, hates coffee, and eats a lot of protein and fried foods.

The elevated risks found for the three individuals were exactly the same: osteoporosis, hypertension, type 2 diabetes, and heart disease. One company offered the appropriate multivitamin supplements for $1,200, which the investigation found to be worth about $35. Recommendations stated the obvious, such as advising a smoker to quit. The advice tracked with the fictional lifestyle/diet information, and *not* genetics. Concluded the study: "Although these recommendations may be beneficial to consumers in that they constitute common sense health and dietary guidance, DNA analysis is not needed to generate this advice." Some of the suggestions could even be dangerous, such as vitamin excesses in people with certain medical conditions. (A health history was not required.)

Matching Patient to Drug

People react differently to the same dose of the same drug because we differ in the rates at which our bodies react to drugs and break them down. Genetic tests can highlight these differences. Specifically, a **pharmacogenetic** test detects a variant of a single gene that affects drug metabolism, and a **pharmacogenomic** test detects variants of multiple genes or gene expression patterns that affect drug metabolism. Pharmaceutical and biotechnology companies now routinely use these tests in developing drugs, and physicians are increasingly using these tests in prescribing drugs. Pharmacogenetics and pharmacogenomics are often considered together under the umbrella term "personalized medicine."

Genetic testing to guide drug selection offers several advantages:

- identifying patients likely to suffer an adverse reaction to a drug;
- selecting the drug most likely to be effective;

- monitoring response to drug treatment; and
- predicting the course of the illness (prognosis).

An early use of pharmacogenetics was in breast cancer (see table 18.4), where women with the HER2 subtype respond to the drug trastuzumab (Herceptin). A pharmacogenomic example is the use of a DNA microarray depicting the expression of eighteen genes to predict whether a person is likely to respond to certain drugs used to treat hepatitis C, which have severe side effects. A pharmocogenetic/genomic approach might have averted disaster in 2004, when widespread use of a type of arthritis drug called a COX-2 inhibitor caused heart damage in some patients. Several drugs were discontinued or their use restricted, robbing many people with arthritis of their benefits. Tests that detect specific variants of genes that encode proteins called cytochromes (P450, 2D6, and 2C19) are now used to predict who will develop adverse effects from these drugs, so that prescribing can be appropriately restricted.

One of the first widespread uses of pharmacogenetics is in prescribing the blood thinner warfarin (also known as Coumadin). This drug has a very small range of concentration in which it keeps blood at a healthy consistency, but people can vary up to tenfold in the dose required. Too little allows dangerous clotting; too much causes dangerous bleeding. In the past, physicians would give an initial standard dose, based on a patient's age, gender, health status, weight, and ethnicity, then monitor the patient for a few weeks to check for too much clotting or bleeding, tweaking the dose until it was about right. But this general approach led to hospitalization for abnormal bleeding in 43,000 of the 2 million people prescribed the drug each year.

A "pharmacogenetic algorithm" is now used to prescribe warfarin. It considers two genes: two variants of *CYP2C9* and one variant of *VKORC1* are associated with increased sensitivity to the drug. People with these gene variants require lower doses of warfarin. The new, genetic way of testing for warfarin response is especially helpful for the 50 percent of patients who fall at the extremes of the range of drug concentration that is effective.

Key Concepts

1. Preconception comprehensive carrier testing checks couples for many recessive diseases before they have children.
2. Cell-free fetal DNA can be analyzed to detect mutations.
3. Newborn screening uses tandem mass spectrometry to detect dozens of telltale metabolites in a single blood sample. Newborns undergo DNA testing too.
4. Government regulations for genetic testing apply only to tests used to diagnose disease. Unawareness of incomplete penetrance is one complication of DTC genetic testing.
5. Nutrigenetics testing can be very misleading.
6. Pharmacogenetics and pharmacogenomics help physicians to select the best drugs for individual patients.

20.3 Treating Genetic Disease

The number of tests for genetic diseases is much greater than the number of treatments. A great challenge with developing such treatments is ensuring that they correct the abnormality in the appropriate cells and tissues to prevent or minimize symptoms, while not harming other parts of the body. Treatments for single-gene diseases have evolved through several stages, in parallel to development of new technologies:

- removing an affected body part;
- replacing an affected body part or biochemical with material from a donor;
- delivering pure, human proteins derived from recombinant DNA technology to compensate for the effects of a mutation; and
- gene therapy, to replace mutant alleles.

The first three approaches affect the phenotype. Only gene therapy attempts to alter the genotype.

Drugs

Preventing a disease-associated phenotype may be as straightforward as adding digestive enzymes to applesauce for a child with cystic fibrosis, or giving a clotting factor to a boy with hemophilia. Inborn errors of metabolism are particularly treatable when the biochemical pathways are well understood and enzymes can be replaced.

Lysosomal storage diseases are a subclass of inborn errors of metabolism whose effects on cells are well known. A deficient or abnormal enzyme leads to buildup of the substrate (the molecule that the enzyme acts on) as well as a deficit of the breakdown product of the substrate. Recall from figure 2.6 that a lysosome is an organelle that functions as a garbage can of sorts in a cell. It houses at least forty types of enzymes, and each breaks down a specific molecule.

Treatment of type 1 Gaucher disease (MIM 230800), a lysosomal storage disease, illustrates three general approaches to counteracting an inborn error of metabolism that affects an enzyme, summarized in **table 20.3** and **figure 20.4**. In type 1 Gaucher disease, the enzyme glucocerebrosidase is deficient or absent. As the substrate builds up because there is little or

Table 20.3	Treatments for Lysosomal Storage Diseases
Treatment	**Mechanism**
Enzyme replacement therapy	Recombinant human enzyme infused to compensate for deficient or absent enzyme
Substrate reduction therapy	Oral drug that reduces level of substrate so enzyme can function more effectively
Pharmacological chaperone therapy	Oral drug that binds to patient's misfolded protein, restoring function

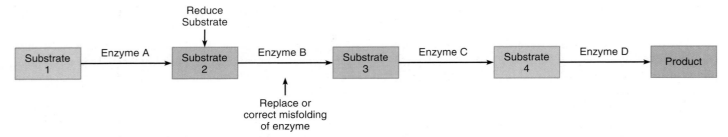

Figure 20.4 Counteracting a metabolic abnormality. Treatments of lysosomal storage diseases are based on understanding metabolic pathways, in which a series of enzyme-catalyzed reactions leads to formation of a product. If any enzyme is deficient or its activity blocked, the substrate builds up and the product is deficient. Enzyme replacement therapy delivers an absent or deficient enzyme. Substrate reduction therapy decreases the amount of substrate and pharmacological chaperone therapy corrects misfolded proteins.

no enzyme to break it down, lysosomes swell, ultimately bursting cells. Symptoms include an enlarged liver and spleen, bone pain, and deficiencies of blood cells. Too few red blood cells cause the fatigue of anemia; too few platelets cause easy bruising and bleeding; and too few white blood cells increase the risk of infection. The disease is very variable in age of onset, severity of symptoms, and rate of progression.

Early treatments for Gaucher disease corrected affected body parts: removing the spleen, replacing joints, transfusing blood, or transplanting bone marrow. In 1991, *enzyme replacement therapy* became available, which supplies recombinant glucocerebrosidase. This treatment is effective but costs about $550,000 a year, and takes several hours to infuse each week.

In 2003 came a different approach: *substrate reduction therapy.* This is a drug taken by mouth that decreases the amounts of the substrate, the molecule on which the deficient enzyme acts. A third approach is called *pharmacological chaperone therapy,* in which an oral drug binds a patient's misfolded enzyme, stabilizing it sufficiently to allow some function. Two existing drugs used to treat heart disease act in this way, restoring protein folding in cells from Gaucher patients. This repurposing of a drug is less costly and time-consuming than development of a new drug. Sometimes parents notice such an effect. For example, parents of children with Canavan disease found that while their children were taking the antibiotic amoxicillin to treat ear infections, their muscle control improved. Researchers are now investigating how this happens, and in the meantime children with Canavan disease can benefit from taking this old antibiotic.

A class of antibiotics called aminoglycosides may also be useful in treating genetic disease. These drugs treat bacterial infections by interfering with the part of a ribosome that monitors the interaction between an mRNA's codon and a tRNA's anticodon (see figure 10.6). The drugs shut down bacterial protein synthesis. However, they also affect nonsense mutations in human DNA. Recall that a nonsense mutation replaces a codon that specifies an amino acid with a "stop" codon that halts protein synthesis. These antibiotics distort the ribosome in a way that enables protein synthesis to continue past the stop codon by substituting an amino acid, like ignoring a period accidentally placed in the middle of a sentence. This phenomenon is

called nonsense suppression. Because about 12 percent of all mutations are nonsense, these drugs could prove very valuable. So far they correct the phenotype for about a dozen diseases in mice and in human cells growing in culture. (A modified short "antisense" nucleic acid is similarly being used to ignore a nonsense mutation that affects some boys with Duchenne muscular dystrophy—so far, it is helping them.) Other existing drugs are being tested to treat Hutchinson-Gilford progeria syndrome (see figure 3.22) and Marfan syndrome (see figure 5.7).

Gene Therapy

More than a thousand clinical trials of gene therapies have been conducted since 1990, yet it is still experimental. As the new millennium dawned, researchers had expected that the sequencing of the human genome would accelerate the pace of gene therapy development. Instead, new information about the complexity of how genes interact, and a few cases in which gene therapies harmed patients, led to a reevaluation of the idea that we can augment or replace a gene with predictable effects. Since 2008, the field has been reborn with a few successes.

Altering genes to treat an inherited disorder theoretically can provide a longer-lasting effect than treating symptoms. The first gene therapy efforts focused on well-studied inherited disorders, even though they are very rare. Gene therapy is now also targeting more common illnesses, such as heart disease and cancers. **Tables 20.4** and **20.5** list some general concerns and requirements related to gene therapy.

Types and Targets of Gene Therapy

Gene therapy approaches vary in the way that healing genes are delivered and where they are sent. **Germline gene therapy** alters the DNA of a gamete or fertilized ovum. As a result, all cells of the individual have the change. The transgenic organisms discussed in chapter 19 had germline gene therapy. The correction is heritable, passing to offspring. Germline gene therapy is not being done in humans.

Somatic gene therapy corrects only the cells that an illness affects. It is nonheritable: A recipient does *not* pass the genetic correction to offspring. Clearing lungs congested from cystic fibrosis with a nasal spray containing functional *CFTR* genes is an example of somatic gene therapy.

Table 20.4	Gene Therapy Concerns	
Scientific		**Bioethical**
1. Which cells should be treated, and how?		1. Does the participant in a gene therapy trial truly understand the risks?
2. What proportion of the targeted cell population must be corrected to alleviate or halt progression of symptoms?		2. If a gene therapy is effective, how will recipients be selected, assuming it is expensive at first?
3. Is overexpression of the therapeutic gene dangerous?		3. Should rare or more common disorders be the focus of gene therapy research and clinical trials?
4. Is it dangerous if the altered gene enters cells other than the intended ones?		4. What effect should deaths among volunteers have on research efforts?
5. How long will the affected cells function?		5. Should clinical trials be halted if the delivered gene enters the germline?
6. Will the immune system attack the introduced cells?		
7. Is the targeted DNA sequence in more than one gene?		

Table 20.5	Requirements for Approval of Clinical Trials for Gene Therapy
1. Knowledge of defect and how it causes symptoms	
2. An animal model	
3. Success in human cells growing *in vitro*	
4. No alternate therapies, or patients for whom existing therapies are not possible or have not worked	
5. Safe experiments	

Gene therapy approaches vary in invasiveness (**figure 20.5**). Cells can be altered outside the body and then infused into the bloodstream. This is called *ex vivo* ("outside the body") **gene therapy**. In the more invasive *in vivo* ("in the living body") **gene therapy,** the gene and its vector are introduced directly into the body. A catheter might be used to deliver the gene to the liver or the brain, for example. There, the vector must enter the appropriate cells and the human DNA be transcribed into mRNA and translated into protein. Then, the protein must do its job.

Researchers obtain therapeutic genes using the recombinant DNA and polymerase chain reaction technologies described in chapter 19. In the future, researchers—and, someday, clinicians—may deliver synthetic genes. The origin should not make a difference—DNA is DNA.

The physical, chemical, and biological methods discussed in section 19.3 are used to send DNA into cells. Physical methods include electroporation, microinjection, and particle bombardment. Chemical methods include liposomes and other types of lipids that carry DNA across the plasma membrane. The lipid carrier can penetrate the plasma membrane that DNA alone cannot cross, but it may not deliver a sufficient payload, and gene expression is only temporary.

Biological approaches to gene transfer use a vector, such as a virus. Researchers remove the viral genes that cause symptoms or alert the immune system and add the corrective gene. Different viral vectors are useful for different treatments. A certain virus may transfer its cargo with great efficiency to a specific cell type, but carry only a short DNA sequence. Another virus might carry a large piece of DNA but enter many cell types, causing side effects. Even if a viral vector goes where it is intended, it must enter enough cells to alleviate symptoms. Finally, a viral vector must not integrate into a gene that harms the patient, such as an oncogene or tumor suppressor gene, which could cause cancer.

Some gene therapies use viruses that normally infect the targeted cells. For example, a herpes simplex virus delivers the gene encoding enkephalin, a

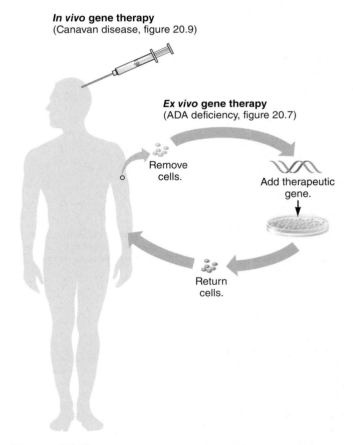

In vivo gene therapy
(Canavan disease, figure 20.9)

Ex vivo gene therapy
(ADA deficiency, figure 20.7)

Remove cells.

Add therapeutic gene.

Return cells.

Figure 20.5 Gene therapy invasiveness. Therapeutic genes are delivered to cells removed from the body that are then returned (*ex vivo* gene therapy) or delivered directly to an interior body part, such as through the skull for Canavan disease or to an artery leading to the liver (*in vivo* gene therapy).

pain-relieving peptide, to nerve endings in skin. Researchers can combine parts of viruses to target a certain cell type. Adeno-associated virus (AAV), for example, infects many cell types, but adding a promoter from a parvovirus gene restricts it to red blood cell progenitors in bone marrow. Add a human gene that encodes a protein normally found in red blood cells, and the vector can treat an inherited blood disorder, such as sickle cell disease. **Figure 20.6** shows four targets of somatic gene therapies.

Initial Success

Any new medical treatment begins with courageous volunteers who know that they may risk their health. Gene therapy, however, is unlike conventional drug therapy in that it alters an individual's genotype in part of the body. Because the potentially

Endothelium. The tile-like endothelium that forms capillaries, the tiniest blood vessels, can be genetically altered to secrete a variety of needed proteins into the circulation.

Muscle. Immature muscle cells (myoblasts) given healthy dystrophin genes may treat muscular dystrophy.

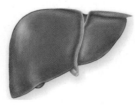

Liver. To treat certain inborn errors of metabolism, only 5 percent of the liver's 10 trillion cells would need to be genetically altered.

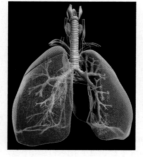

Lungs. Gene therapy can easily reach damaged lungs through an aerosol spray. Like the muscle, enough cells would have to be reached to help people with hereditary emphysema (alpha-1-antitrypsin deficiency) or cystic fibrosis.

Figure 20.6 **Some sites of gene therapy.**

therapeutic gene is usually delivered with other DNA, and it may enter cell types other than those affected in the disease, reactions are unpredictable. Following is a look at some of the pioneers of gene therapy.

In the late 1980s, the DeSilvas did not think their little girl, Ashanthi ("Ashi"), would survive. She had near-continual coughs and colds, and was so fatigued that she could walk only a few steps before becoming winded. Doctors diagnosed the obvious: asthma, an allergy, bronchitis. Then Ashi's uncle, an immunologist, suggested blood tests for inherited immune deficiencies. Ashi had severe combined immune deficiency due to adenosine deaminase (ADA) deficiency (**figure 20.7**). At age 2, she began enzyme replacement therapy, but as she neared her fourth birthday, it stopped working. She would likely die of infection.

Then her physician heard about a clinical trial of gene therapy that would patch her own white blood cells with functional ADA genes. Ashi was chosen, and on September 14, 1990, at 12:52 P.M., she sat up in bed at the National Institutes of Health in Bethesda, Maryland, and began receiving her own corrected white blood cells intravenously. To be sure she stayed healthy, the researchers gave her enzyme replacement therapy too. Ashi has done very well, but researchers feared the enzyme replacement might have masked the effect of the gene therapy. By 2005, however, thirty youngsters had been treated, successfully, with a new version of the gene therapy.

Serious Setbacks

In September 1999, 18-year-old Jesse Gelsinger died 4 days after receiving gene therapy, from an overwhelming immune response to the virus used to introduce the therapeutic gene. He had ornithine transcarbamylase deficiency (OTC) (MIM 311250). In this X-linked recessive disorder, one of five enzymes made in the liver and required to break down amino acids liberated from dietary proteins is absent (**figure 20.8**). Instead of being excreted in urine, the nitrogen released from the amino acids combines with hydrogen to form ammonia (NH_3), which rapidly accumulates in the bloodstream and travels to the brain, with devastating effects. It usually causes irreversible coma within 72 hours of birth. Half of affected babies die within a month, and another quarter by age 5. The survivors can control their symptoms by following a very restrictive low-protein diet and taking drugs that bind ammonia.

Jesse wasn't diagnosed until he was 2, when a coma landed him in the hospital. He had a mild case because he was a mosaic—some of his cells could produce the enzyme. At age 10, Jesse's diet lapsed and he was hospitalized again. When he went into a coma in December 1998 after missing a few days of his medications, he considered volunteering for a gene therapy trial the doctor had mentioned. When Jesse turned 18 the next June, he underwent tests at the University of Pennsylvania and was admitted into the clinical trial. He was jubilant. He knew he might not directly benefit, but he had wanted to help babies who die of the condition.

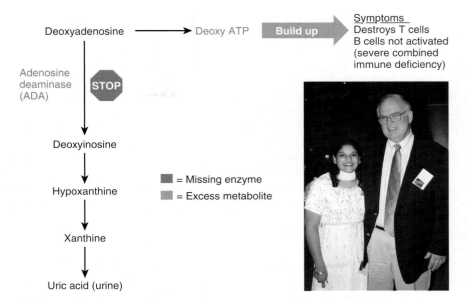

Figure 20.7 ADA deficiency. Absence of the enzyme adenosine deaminase (ADA) causes deoxy ATP to build up, which destroys T cells, which then cannot stimulate B cells to secrete antibodies. The result is severe combined immune deficiency (SCID). In 1990, Ashi DeSilva became the first person to have gene therapy. Today she is healthy. This photo, with Dr. Michael Blaese, was taken when she was 17.

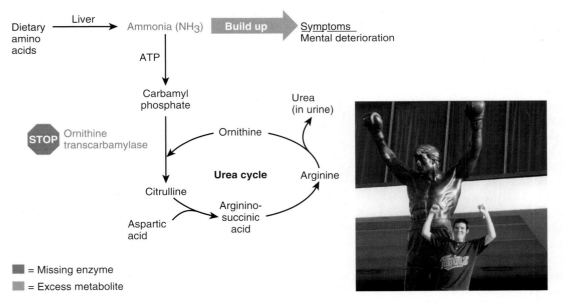

Figure 20.8 OTC deficiency. Lack of the enzyme ornithine transcarbamylase prevents nitrogen stripped from dietary proteins from leaving the body in urine. Instead, the nitrogen combines with hydrogen to form ammonia, which damages the brain, causing coma. Jesse Gelsinger had a mild case because he was a mosaic, with the mutation in only some of his cells. He had gene therapy and died 4 days later of an overwhelming immune response. The gene-bearing viruses introduced into his liver targeted the intended liver cells, but also entered immune system cells.

The gene therapy was an adenovirus carrying a functional human *OTC* gene. This virus had already been used in many gene therapy experiments and did not have the genes to replicate and cause respiratory symptoms. Seventeen patients had been treated, without serious side effects, when Jesse entered the hospital to receive a few billion altered viruses in an artery leading into his liver. It was Monday, September 13. That night, Jesse developed a high fever. By Tuesday morning, the whites of his eyes were yellow, indicating that his liver was struggling to dismantle the hemoglobin released from burst red blood cells. A flood of hemoglobin meant a flood of protein, elevating the ammonia level in his liver to ten times normal levels by mid-afternoon. Jesse became disoriented, then comatose. His lungs and then other vital organs began to fail, and by Friday he was brain dead. His dedicated and devastated medical team stood by as his father turned off life support.

The autopsy and analysis of tests done before, during, and after the procedure showed that the adenovirus had entered not only the hepatocytes as expected, but also the macrophages that function as sentries for the immune system. In response, interleukins flooded Jesse's body, and inflammation raged. Although afterward parents of children with OTC deficiency implored government officials to continue to fund the research, the death of Jesse Gelsinger led to suspension of many gene therapy trials. The death drew particular attention to safety because, unlike most other volunteers, Jesse had not been very ill. However, some gene therapy clinical trials continued in other countries, and that led to the second setback.

Since the early 1990s, researchers in France had been working on gene therapy for X-linked severe combined immune deficiency (SCID-X1). This is the disease that claimed the life of David Vetter, the "bubble boy" (see figure 17.11). In SCID-X1, T cells lack certain cytokine receptors, which prevents the immune system from recognizing infection. *Ex vivo* gene therapy would remove a boy's T cell progenitor cells from bone marrow, give them wild type alleles, and infuse the corrected cells back into his body. Instead of using adenovirus, now known to be dangerous, the French researchers used a retrovirus, which only enters dividing cells. If the healing viruses could infect the T cell progenitor cells, the cells would differentiate into mature T cells capable of doing their job of alerting the immune system.

In 1999, the French researchers gave the gene therapy for SCID-X1 to two babies. Both boys initially did well, their skin infections and persistent diarrhea clearing up. Even their tiny thymus glands hummed to life and grew, and their T cells showed the correction. Encouraged, the researchers treated more boys, and published a report in 2002 in a prominent medical journal. Soon after the article appeared, blood work on one of the first two boys showed a sharp increase in the numbers of a subtype of T cell. In a short time, this abnormality progressed to leukemia. Later, researchers discovered that some of the retroviruses had inserted into chromosome 11, activating a proto-oncogene. The little boy was cured of his inherited immune deficiency, but developed leukemia. He began chemotherapy.

The researchers hoped the leukemia was a fluke, but by the end of the year, another boy had it. By then, ten boys in France and ten in England had received the gene therapy for SCID-X1. Eventually, 17 of them regained immunity, but 5 developed leukemia and one boy died of it. Gene therapy would have to reinvent itself again using different viruses.

A Trio of Successes

Max Randell, featured in the chapter opener, became the youngest person to receive gene therapy, for Canavan disease in 1998, with fourteen other children. Canavan disease affects the brain. Lack of the enzyme aspartoacylase (ASPA) disrupts the interaction between neurons and neighboring cells called oligodendrocytes, which produce the fatty myelin that coats neurons. Myelin enables neurons to transmit impulses fast enough for the brain to function (**figure 20.9**). Specifically, the enzyme normally breaks down another chemical, N-acetylaspartate (NAA), which oligodendrocytes produce. Without the enzyme, NAA builds up and destroys the oligodendrocytes, stripping the

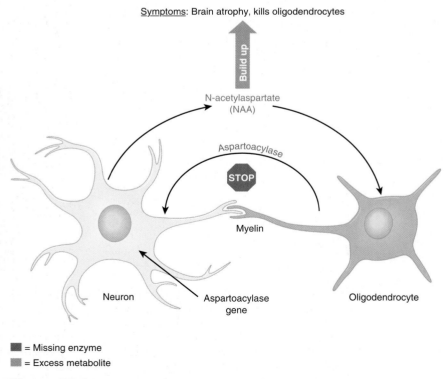

Symptoms: Brain atrophy, kills oligodendrocytes

Build up

N-acetylaspartate (NAA)

Aspartoacylase

STOP

Myelin

Neuron

Aspartoacylase gene

Oligodendrocyte

■ = Missing enzyme
■ = Excess metabolite

Figure 20.9 Canavan disease. Stripping of the lipid layer on brain neurons occurs because oligodendrocytes lack an enzyme to break down NAA, which neurons produce. Gene therapy enables brain neurons to secrete the enzyme, restoring the fatty covering that makes nerve transmission possible.

neurons of myelin. The result is "spongy degeneration of the brain," as the disease was first described in 1931.

Parents typically notice the first symptoms of Canavan disease by the child's second month. The eyes flick back and forth, the child doesn't smile, and the head is large and floppy. Excess NAA in the urine and signs of myelin loss on brain scans strongly suggest Canavan disease, and a DNA test for the *ASPA* gene establishes the diagnosis.

Before gene therapy, nearly all children with Canavan disease died before age 10. Today, several who received gene therapy are in their teens. In the months following gene therapy, levels of NAA in the urine fell and brain scans showed new myelin ensheathing neurons. The children could make new sounds, see better, and had better muscle control. However, their progress leveled off when further gene therapy was delayed after Jesse Gelsinger's death, and again as they got older. Still, today they use computers to communicate, love their iPods, and respond as best they can. Their parents attribute the fact that their children are alive and alert to the gene therapy. (A question in chapter 15 describes another of the Canavan disease gene therapy participants.) *Bioethics: Choices for the Future* on page 402 relates the battle over access to genetic tests for Canavan disease.

A striking gene therapy success is the case of Corey Haas, who inherited a form of blindness called Leber congenital amaurosis type 2 (MIM 204100). At the backs of Corey's eyes, cells in a layer called the retinal pigment epithelium (RPE) cannot make an enzyme necessary to convert vitamin A to a form that the rods and cones, the cells that transmit the energy in light to the brain, can use (**figure 20.10**). Corey was genetically destined to lose his sight by early adulthood because he had inherited two mutant copies of the *RPE65* gene.

a.

b.

Figure 20.11 **Sheepdogs and people get LCA2** **(a)** Kristina Narfström, the veterinary researcher who discovered the disease in dogs, is shown with her favorite pooch, Pluto. He had gene therapy when he was four years old. **(b)** Corey Haas lived in an ever-narrowing visual world until gene therapy restored his ability to see. He is age 10 in this photo, taken when he was between gene therapies.

Corey's parents first noticed that their baby son didn't make eye contact, yet stared at brightly lit bulbs. Once he began toddling, he was very clumsy. A series of doctors thought the problem was only extreme nearsightedness, but then an educator visiting the Haas home mentioned another child with similar visual problems who was seeing a doctor in Boston. That physician ultimately diagnosed Corey. Luckily, she had been on a team that had conducted successful gene therapy on a breed of sheepdog that had the very same disease. The doctor also knew that her colleagues in Philadelphia were looking for a youngster about Corey's age to be part of a clinical trial of gene therapy that had already restored vision to sheepdogs and several young adults (**figure 20.11**).

Three days after his eighth birthday, in September 2008, Corey had the gene therapy on his worse eye, his left; his second would be done later, if the treatment worked. The surgeon cut a tiny pocket in the eye and introduced several billion copies of adeno-associated virus (AAV), each one carrying a wild type version of the *RPE65* gene. By this time AAV had replaced the more dangerous viral vectors. Just 4 days later, on a brilliant afternoon, the Haas family visited the Philadelphia zoo. When Corey looked up at the giant balloon hovering over the entranceway, he screamed! It was the first time he had been able to see bright sunlight. It was a very different outcome than the tragedy that befell Jesse Gelsinger 4 days after his gene therapy at the same hospital. Corey's second eye was treated in fall 2011.

The researchers who restored Corey's vision have stockpiled enough of the gene therapy to treat all of the few thousand people in the world with his disease. Even more exciting is the fact that the same researchers are using the procedure to treat people with a much more common form of visual loss, age-related macular degeneration. In the "wet" form of this condition, the RPE degenerates and blood vessels grow into the photoreceptor layer, dislodging the rods and cones, destroying central vision. This is the same process of angiogenesis that brings a blood supply to a tumor (see figure 18.7). The gene therapy blocks part of the receptors on the interiors of capillaries

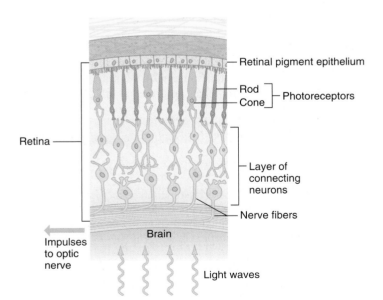

Figure 20.10 **LCA2.** In Leber congenital amaurosis type 2, a missing enzyme prevents cells of the retinal pigment epithelium from activating vitamin A so that it can nourish the rods and cones, which send visual signals to the brain.

Labels in figure 20.10:
- Retinal pigment epithelium
- Rod
- Cone
- Photoreceptors
- Retina
- Layer of connecting neurons
- Nerve fibers
- Brain
- Impulses to optic nerve
- Light waves

Canavan Disease: Patients Versus Patents

When Debbie Greenberg gave birth to Jonathan in 1981, she and her husband Dan had no idea that they would one day lead an effort to challenge how a researcher and a hospital patented a gene. Jonathan lived 11 years with Canavan disease; his brain never developed past infancy. The couple had an affected daughter, Amy, a few years after Jonathan was born, and three healthy children.

Shortly after Jonathan's diagnosis, the Greenbergs started the Chicago chapter of the National Tay-Sachs and Allied Diseases Association, through which they met Dr. Reuben Matalon. They asked him to develop a prenatal test for Canavan disease, which would entail discovering the gene. In 1987, the Greenbergs donated Jonathan and Amy's brains to Dr. Matalon, and convinced 160 families to also donate tissue, including skin, blood, and urine. The donated tissue was critical to identifying the gene and the causative mutation in 1993, which was done in Dr. Matalon's lab at Miami Children's Hospital.

Detecting the mutation is the basis of prenatal, carrier, and diagnostic tests. By 1996, the Canavan Foundation was offering free carrier testing. But unknown to the parents who had donated their children's tissues for the gene search, Dr. Matalon and Miami Children's Hospital had filed for a patent on their discovery. It was issued in 1997. A year later, Miami Children's Hospital sent letters to doctors and diagnostic laboratories stating, "We intend to enforce vigorously our intellectual property rights relating to carrier, pregnancy, and patient DNA tests." The hospital also restricted the number of tests. Suddenly, families whose donations—both monetary and biological—had made the discovery of the gene possible had to pay for tests. Many labs simply stopped testing. The families were outraged.

Students and three professors at the Chicago-Kent College of Law filed a lawsuit from parents and three nonprofit organizations on October 30, 2000. They did not challenge the gene patent, but the use of the donated tissue for profit. Although a judge supported the parents, they ran out of money and the case was settled. The outcome: Patients still have to pay for tests, but researchers can use the gene. It is a compromise.

Questions for Discussion

1. It is common today for families with genetic diseases to start organizations that help researchers better understand the disease. Go to the Genetic Alliance or National Organization of Rare Diseases or other websites and compare and contrast how different organizations have learned to protect themselves from having to pay for tests that they helped to develop.

2. Which organizations or individuals should provide oversight to the use of patients' cells and tissues in developing medical tests?

3. What is a justification for a pharmaceutical company charging families for tests that their biological materials helped to develop?

---

that bind vascular endothelial growth factor (VEGF), which is the signal that tells them to extend. Compared to drugs that must be injected several times to accomplish the same goal, gene therapy will be a one-time procedure.

A year after Corey's exciting day at the zoo, researchers reported success in gene therapy to treat adrenoleukodystrophy (ALD). This is the disease of peroxisomes that affected Lorenzo Odone, whose parents devised a dietary oil to help him, described in chapter 2. Peroxisomes are tiny sacs inside cells that house enzymes. They have porthole-like openings formed by a protein called ABCD1. A mutation in the *ABCD1* gene prevents the portholes from admitting an enzyme needed to process fats called very long chain fatty acids, which are used to make myelin. ALD, then, is a demyelinating disease like Canavan disease, but affected boys (it is X-linked) are typically healthy for their first few years. Unlike Canavan disease, ALD is easier to treat with gene therapy. This is because the affected brain cells, called microglia, descend from progenitor cells in the bone marrow.

The gene therapy for ALD fixes bone marrow cells outside the body, which are then injected into the bloodstream and find their way to the brain, where they give rise to corrected microglia. The viral vector is HIV, stripped of the genes that cause AIDS. The researchers used HIV to deliver the healthy genes because it does not insert into proto-oncogenes, is very efficient, and enters many types of cells. For several treated boys, blood levels of the crippling fats fell so greatly and brain neurons gained enough myelin so that they could attend school—with only about 15 percent of their microglia corrected! Teaming the gene therapy for ALD with newborn screening may completely prevent symptoms of this otherwise devastating disease. Conquering inherited disease is the goal of all of the tests and technologies described in this chapter.

Key Concepts

1. Protein-based therapies replace proteins, reduce substrates, and correct misfolding.

2. Gene therapies replace malfunctioning or absent genes. Germline gene therapy targets gametes or fertilized ova, is heritable, and is not done on humans. Somatic gene therapy targets somatic cells and is not heritable. Gene-carrying viral vectors may infect cells isolated from the body (*ex vivo*) or be introduced into the body (*in vivo*).

3. Physical, chemical, and biological methods are used to deliver DNA in gene therapy.

4. Gene therapy has had a rocky history. Initial success in the early 1990s was followed by several deaths. Using safer vectors, gene therapy has resurged in recent years.

Summary

20.1 Genetic Counseling

1. **Genetic counselors** provide information on inheritance patterns, disease risks and symptoms, and available tests and treatments.

2. Prenatal counseling and counseling a family coping with a particular disease pose different challenges.

3. Genetic counselors interpret direct-to-consumer genetic tests and assist other health care professionals in incorporating genetic information into their practices.

20.2 Genetic Testing

4. Preconception comprehensive carrier testing can identify couples at risk for passing on any of nearly 500 recessive diseases.

5. Fetal genomes can be sequenced and mutations identified from cell-free fetal DNA in the maternal circulation.

6. Newborns are routinely screened for several inborn errors of metabolism, some of which are treatable. The numbers of tests vary by state.

7. The Clinical Laboratory Improvement Amendments regulate some genetic tests. Direct-to-consumer tests presented as information and not diagnoses may evade regulation.

8. Nutrigenetics tests may provide inaccurate information.

9. **Pharmacogenetic** and **pharmacogenomic** tests provide information on how individuals metabolize certain drugs.

20.3 Treating Genetic Disease

10. Enzyme replacement therapy, substrate reduction therapy, and pharmacological chaperone therapy treat biochemical imbalances.

11. Drugs may be repurposed to treat genetic diseases.

12. Gene therapy delivers genes and encourages production of a needed substance at appropriate times and in appropriate tissues, in therapeutic (not toxic) amounts.

13. **Germline gene therapy** affects gametes or fertilized ova, affects all cells of an individual, and is transmitted to future generations. It is not performed in humans. **Somatic gene therapy** affects somatic cells and is not passed to offspring.

14. ***Ex vivo* gene therapy** is applied to cells outside the body that are then reimplanted or reinfused into the patient. ***In vivo* gene therapy** delivers gene-carrying vectors directly into the body.

15. Several types of vectors are used to deliver therapeutic genes. Viruses are most commonly used. Some gene therapies target stem or progenitor cells, because they can divide and move.

16. After initial success in 1990, many gene therapy clinical trials halted after a death in 1999. Safer vectors and better understanding of disease have made recent gene therapies more successful.

Review Questions

1. What is unique about the services that a genetic counselor provides compared to those of a nurse or physician?

2. What are the advantages and disadvantages of "virtual" genetic counseling that uses an interactive computer program rather than a person?

3. Compare and contrast the types of information from preconception comprehensive carrier screening, prenatal diagnosis, and newborn screening in terms of the types of actions that can be taken in response to receiving the results.

4. Why is newborn screening economically feasible?

5. Using information from this or other chapters, or the Internet, cite DNA-based tests given to a fetus, a newborn, a young adult, and a middle-aged person.

6. Explain how a pharmacogenetic test can improve quality of life for a person with cancer.

7. List an advantage and a limitation of a direct-to-consumer genetic test.

8. Explain why the removal of blood in people with hereditary hemochromatosis is not gene therapy.

9. Distinguish among enzyme replacement therapy, substrate reduction therapy, and pharmacological chaperone therapy.

10. Explain how an antibiotic drug might be used to treat a genetic disease.

11. Explain whether and how somatic gene therapy or germline gene therapy can affect evolution.

12. What factors would a researcher consider in selecting a viral vector for gene therapy?

13. Identify two complications that slowed the development of gene therapy.

14. Explain how one of the gene therapies described in the chapter works.

www.mhhe.com/lewisgenetics10

Answers to all end-of-chapter questions can be found at **www.mhhe.com/lewisgenetics10.** You will also find additional practice quizzes, animations, videos, and vocabulary flashcards to help you master the material in this chapter.

Applied Questions

1. Discuss the challenges that a genetic counselor faces in explaining to parents-to-be a prenatal diagnosis of trisomy 21 in a fetus with no family history, compared to a prenatal diagnosis of translocation Down syndrome in a family with a reproductive history of pregnancy loss and birth defects. (Chapter 13 discusses Down syndrome.)

2. Suggest how newborn screening might be changed to minimize the numbers of "patients-in-waiting."

3. A company tests for variants of the gene *ACTN3* (alpha-actinin 3), which encodes a protein that binds actin, a cytoskeletal protein. One genotype is more common among elite sprint athletes, and another among endurance athletes. Some parents are testing their young children for these gene variants and using the results to decide whether the child should pursue a sport that entails sprinting or endurance. Opponents of the test point out that simply having kids run a race would provide more meaningful information than would data on this single gene. Would you have your child tested? Cite a reason for your answer.

4. Hannah Sames, the little girl introduced in Reading 2.2, will have gene therapy for her disease, giant axonal neuropathy. Reread the essay and suggest how a gene therapy to treat her condition might work.

5. Select one of the rare genetic diseases described in this chapter and suggest how learning how to treat it can be applied to treating a more common condition.

6. The Newborn Screening Translational Research Network Coordinating Center analyzes stored bloodspots for information on treatments and outcomes. Do you think that the parents of the newborns whose blood is being analyzed in new ways should be consulted, or is their consent presumed because state law mandates newborn screening? (Many people are not aware that newborn blood is routinely sampled and tested.)

7. Describe a combination of two or three genetic tests that members of a family might take, and how they might respond to particular results.

8. Cite a reason why the American College of Medical Genetics has asked that the government mandate testing for inborn errors of metabolism that do not have treatments.

9. Explain why high penetrance is an important criterion for including a disease in newborn screening.

10. Describe how a treatment for a lysosomal storage disease works.

Web Activities

1. Go to clinicaltrials.gov and search under "gene therapy." Describe one. Include the mode of inheritance, age of onset, symptom severity, variability, existing treatments (if any), and how the gene therapy works.

2. Consult any of these websites to determine the number of newborn screening tests in your state:

    http://www.huntershope.org/UNBS/index.html

    marchofdimes.com/peristats

    genes-r-us.uthscsa.edu

Select one of the diseases, then consult the websites for the Genetic Alliance or the National Organization for Rare Diseases, and describe the symptoms and treatments.

3. Go to the Pharmacogenomics Knowledge Base (http://www.pharmgkb.org/) and click on "Important PGx gene." Select a gene and discuss its importance in health care.

4. Go to www.genedx.com. Describe a genetic test that this company offers.

Case Studies and Research Results

1. How would you, as a genetic counselor, handle the following situations? What would you tell the patients, and what tests would you suggest? (See other chapters for specific information.)

    a. A couple in their early forties is expecting their first child. Amniocentesis indicates that the fetus is XXX, which might never have been noticed without the test. When they learn of the abnormality, the couple asks to terminate the pregnancy.

    b. Two people of normal height have a child with achondroplastic dwarfism, an autosomal dominant trait.

    They are concerned that subsequent children will also have the condition.

    c. A couple has results from tests they have taken from a direct-to-consumer company. Tests based on genome-wide association studies indicate that they each have inherited susceptibility to asthma as well as gene variants that in some populations are associated with autism. Both also have several gene variants that are found in lung cancers. Each has a few recessive mutant alleles, but not in the same genes. On the basis of these results, they do not think that they are "genetically healthy" enough to have children.

2. Three-year-old Tawny Fitzgerald has been to the emergency department repeatedly for broken bones. At the last visit, a nurse questioned Tawny's parents, Donald and Rebecca, about possible child abuse. No charges were filed—the child just appeared to be clumsy. Then Tawny's brother Winston was born. When he was 6 months old, Donald found him screaming in pain one morning. A trip to the hospital revealed a broken arm. This time, a social worker visited the Fitzgerald home, interviewed Donald and Rebecca, and advised that they hire a lawyer. A relative in medical school suggested that they have the children examined for osteogenesis imperfecta, also known as "brittle bone disease."

Consult MIM and list the facts about a form of this condition that could affect both sexes, with carrier parents. If you were the genetic counselor hired to help this couple, what would you ask them, and tell them, to help them deal with the legal and social services authorities who might need a biology lesson?

3. Jill and Scott S. had thought 6-month-old Dana was developing just fine until Scott's sister, a pediatrician, noticed that the baby's abdomen was swollen and hard. Knowing that the underlying enlarged liver and spleen could indicate an inborn error of metabolism, Scott's sister suggested the child undergo several tests.

Dana had inherited sphingomyelin lipidosis, also known as Niemann-Pick disease type A (MIM 257200). Both parents were carriers, but Jill had tested negative when she took a Jewish genetic disease panel during her pregnancy because her mutation was very rare and not part of the test panel. Dana was successfully treated with a transplant of umbilical cord blood cells from a donor. She caught up developmentally and became more alert. Monocytes, a type of white blood cell, from the cord blood traveled to her brain and manufactured the deficient enzyme. Dietary therapy does not work for this condition because the enzyme cannot cross from the blood to the brain. Monocytes, however, can enter the brain.

a. Did Dana's treatment alter her phenotype, genotype, or both?

b. Why did the transplant have to come from donated cord blood, and not from Dana's own, which had been stored?

c. If you were the genetic counselor, what advice would you give this couple if they conceive again?

d. Describe how a gene therapy might help a child with this disease. Which disease mentioned in the chapter suggests a way to do it?

4. The Food and Drug Administration has approved tests for variants of two genes that encode cytochrome 450 enzymes that affect the rate at which a person metabolizes selective serotonin reuptake inhibitors, a drug class that includes widely used antidepressants. However, a study by the Centers for Disease Control and Prevention found "no evidence . . . showing that the results of CYP450 testing influenced SSRI choice or dose and improved patient outcomes." More than a dozen direct-to-consumer websites offer the tests, several of which claim that the test results can be used to choose the correct anti-depressant and dose.

a. Why might the genotypes be associated with a drug metabolism phenotype, yet not improve outcome?

b. Do you think that selling these tests directly to consumers is misleading or unethical?

c. Suggest a way to ensure that genetic tests provide useful information.

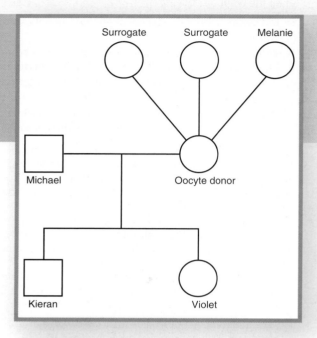

Assisted reproductive technologies provide new ways to create families. It is hard to know how to bend the pedigree rules to describe the five people who contributed to the conception, gestation, births, and raising of "twiblings" Violet and Kieran. The photograph of this extended family in The New York Times Magazine *included mother Melanie (who wrote the article), her husband Michael, and the two surrogate mothers. This pedigree includes the oocyte donor, too.*

Reproductive Technologies

Learning Outcomes

21.1 Savior Siblings and More

1. Explain how a child can be conceived to provide tissue for an older sibling.
2. Define *assisted reproductive technology.*

21.2 Infertility and Subfertility

3. Distinguish infertility from subfertility.
4. Describe causes of infertility in the male.
5. Describe causes of infertility in the female.
6. List infertility tests.

21.3 Assisted Reproductive Technologies

7. Describe assisted reproductive technologies that donate sperm, uterus, or oocyte.
8. List the steps of *in vitro* fertilization.
9. Explain how preimplantation genetic diagnosis avoids the birth of a child with a particular genetic disease.

21.4 Extra Embryos

10. Discuss uses for extra embryos resulting from assisted reproductive technologies.

The Big Picture: Assisted reproductive technologies provide intriguing and sometimes complex variations on the process of conceiving a child and carrying it to term.

The Twiblings

Violet and Kieran graced the cover of *The New York Times Magazine* when they were a year old, above the headline "Meet the Twiblings." Their births required:

- one sperm donor;
- one oocyte donor;
- two uterus donors (surrogate mothers); and
- adoptive parents.

The twiblings' parents, Melanie and Michael, had attempted *in vitro* fertilization six times. They turned to a more complicated combination of assisted reproductive technologies, asking others to help them have two children. Since a twin pregnancy is riskier than a singleton pregnancy, Melanie and Michael found two surrogate mothers, happily married women who simply enjoyed being pregnant. Because Melanie's "advanced maternal age" of 41 might explain why IVF hadn't worked, the couple also used an oocyte donor.

Michael's sperm was mixed with the donated oocytes in a lab dish, very early embryos implanted in the two surrogates, and 9 months later, Violet and Kieran were born. Genetically they are siblings. They are twins in that they were conceived at the same time and in the same place, yet not, because they were born from different women at different times. Despite the children's confusing beginnings, today Michael and Melanie are their parents.

21.1 Savior Siblings and More

A couple in search of an oocyte donor advertises in a college newspaper seeking an attractive young woman from an athletic family. A cancer patient stores her oocytes before undergoing treatment. Two years later, she has several of them fertilized in a laboratory dish with her partner's sperm, and has a cleavage embryo implanted in her uterus. She becomes a mother. A man paralyzed from the waist down has sperm removed and injected into his partner's oocyte. He, too, becomes a parent when he thought he never would.

Lisa and Jack Nash sought to have a child for a different reason. Their daughter Molly, born on July 4, 1994, had Fanconi anemia (MIM 227650). This autosomal recessive condition would destroy her bone marrow and her immunity. An umbilical cord stem cell transplant from a sibling could likely cure her, but Molly had no siblings. Nor did her parents wish to have another child with a one in four chance of inheriting the disorder, as Mendel's first law dictates. Technology offered another solution.

In late 1999, researchers at the Reproductive Genetics Institute at Illinois Medical Center mixed Jack's sperm with Lisa's oocytes in a laboratory dish. After allowing fifteen of the fertilized ova to develop to the 8-cell stage, researchers separated and applied DNA probes to one cell from each embryo. A cell that had wild type Fanconi anemia alleles and that matched Molly's human leukocyte antigen (HLA) type was identified and its seven-celled remainder implanted into Lisa's uterus. Adam was born in late summer. A month later, physicians infused his umbilical cord stem cells into Molly, saving her life (**figure 21.1**). The Nashes were initially sharply criticized for intentionally conceiving a "savior sibling." As others followed their example, conceiving and selecting a child to provide cells for a sibling became more accepted.

Increased knowledge of how the genomes of two individuals come together and interact has spawned several novel ways to have children. **Assisted reproductive technologies (ARTs)** replace the source of a male or female gamete, aid fertilization, or provide a uterus. These procedures were developed to treat infertility, but are increasingly including genetic testing too. In the United States, the government does not regulate ARTs, but the American Society for Reproductive Medicine provides voluntary guidelines. The United Kingdom has pioneered ARTs and its Human Fertilisation and Embryology Authority has served as a model for government regulation. A great advantage of the British regulation of reproductive health services and technologies is that databases include success rates of the different procedures. Another advantage is that access to reproductive technology is not limited to those who can afford it.

The landscape of assisted reproductive technologies is constantly changing, fed by imagination as well as by new discoveries. In early 2007, for example, bioethicists in the Netherlands published a controversial proposal: Select two savior sibling embryos. Permit one to continue developing, and use the other to derive and store embryonic stem cells that could one day provide healing cells to the sick older child. Later that same year, the invention of induced pluripotent stem cells (reprogramming; see figure 2.24) provided a similar source of cells from the patient, not requiring use of an embryo.

Key Concepts

1. Assisted reproductive technologies provide innovative ways to conceive offspring.
2. ARTs are used to avoid conception of a child with a particular genetic condition, or to overcome infertility.

21.2 Infertility and Subfertility

Infertility is the inability to conceive a child after a year of frequent intercourse without the use of contraceptives. Some specialists use the term *subfertility* to distinguish those individuals and couples who can conceive unaided, but for whom this may take longer than usual. On a more personal level, infertility is a seemingly endless monthly cycle of raised hopes and crushing despair. In addition to declining fertility, as a woman ages, the incidence of pregnancy-related problems rises, including chromosomal anomalies, fetal deaths, premature births, and low-birth-weight babies. For most conditions, the man's advanced age does not raise the risk of pregnancy complications, although sperm motility declines with age.

Physicians who specialize in infertility treatment can identify a physical cause in 90 percent of cases. Of these, 30 percent of the time the problem is primarily in the male, and 60 percent of the time it is primarily in the female. When a physical problem is not obvious, the cause is usually a mutation or chromosomal aberration that impairs fertility in the male. The statistics are somewhat unclear, because in 20 percent of the 90 percent, both partners have a medical condition that could contribute to infertility or subfertility. A common combination is a woman with an irregular menstrual cycle and a man with a low sperm count. One in six couples has difficulty in conceiving or giving birth to children.

Figure 21.1 **Savior siblings.** Adam Nash was conceived and selected to save his sister Molly's life. He is also a much-loved sibling and son. Several other families have since conceived a child to help another.

Male Infertility

Infertility in the male is easier to detect but sometimes harder to treat than female infertility. One in twenty-five men is infertile. Some men have difficulty fathering a child because they produce fewer than the average 20 to 200 million sperm cells per milliliter of ejaculate. This condition is called oligospermia, and it has several causes. If a low sperm count is due to a hormonal imbalance, administering the appropriate hormones may boost sperm output. Sometimes a man's immune system produces IgA antibodies that cover the sperm and prevent them from binding to oocytes. Male infertility can also be due to a varicose vein in the scrotum. This enlarged vein produces too much heat near developing sperm, which keeps them from maturing. Surgery can remove a scrotal varicose vein.

Most cases of male infertility are genetic. About a third of infertile men have small deletions of the Y chromosome that remove the only copies of key genes whose products control spermatogenesis. Other genetic causes of male infertility include mutations in genes that encode androgen receptors or protein fertility hormones, or that regulate sperm development or motility. **Reading 21.1** describes the recently identified first type of autosomal recessive male infertility that is not part of a syndrome.

For many men with low sperm counts, if they have at least 60 million sperm cells per ejaculate, fertilization is likely eventually. To speed conception, a man with a low sperm count can donate several semen samples over a period of weeks at a fertility clinic. The samples are kept in cold storage, then pooled. Some of the seminal fluid is withdrawn to leave a sperm cell concentrate, which is then placed in the woman's body. It isn't very romantic, but it is highly effective at achieving pregnancy. Men who actually want a very low sperm count—those who have just had a vasectomy for birth control—can use an at-home test kit to monitor their sperm counts. Fewer than 250,000 sperm cells per milliliter of seminal fluid makes pregnancy highly unlikely.

Sperm quality is more important than quantity. Sperm cells that are unable to move or are shaped abnormally cannot reach an oocyte. Inability to move may be due to a hormone imbalance, and abnormal shapes may reflect impaired apoptosis (programmed cell death) that normally removes such sperm. The genetic package of an immobile or abnormally shaped sperm cell can be injected into an oocyte and sometimes this leads to fertilization. However, even sperm that look and move normally may be unable to fertilize an oocyte.

Female Infertility

Abnormalities in any part of the female reproductive system can cause infertility (**figure 21.2**). Many women with subfertility or infertility have irregular menstrual cycles, making it difficult to pinpoint when conception is most likely. In an average menstrual cycle of 28 days, ovulation usually occurs

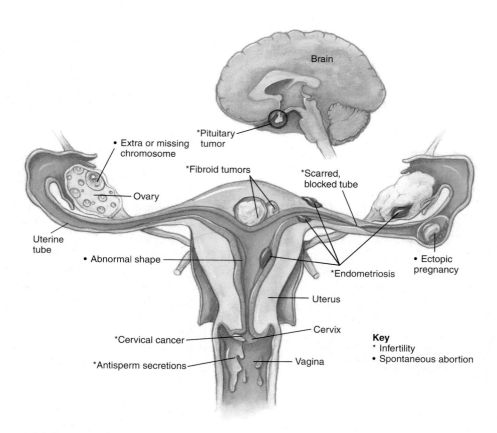

Figure 21.2 Sites of reproductive problems in the female.

The Case of the Round-Headed Sperm—and a Review of This Book

In fewer than a tenth of a percent of men who are infertile, sperm cells lack the tip, called the acrosome, where the enzymes that break through the layers surrounding an oocyte collect. This condition is called "globozoospermia" (**figure 1**). An Ashkenazi Jewish family led researchers to a gene that, when mutant, causes an autosomal recessive form of male infertility due to round-headed sperm.

The family went to a center for reproductive medicine in Brussels, the Netherlands. Of the six sons, three were infertile (**figure 2**). Three daughters were fertile. The affected sons' sperm were obviously misshapen, and the mode of inheritance obviously recessive, since the parents were fertile.

Researchers suspected that consanguinity was involved, because a shared ancestor increases risk of inheriting a very rare autosomal recessive condition if the mutation is in the family. But the family denied knowing a relative who had married a relative. Reasoning that perhaps DNA could reveal what the family did not know, researchers scanned the genomes of all six sons for regions of homozygosity. Recall from chapter 7 that these areas in a genome reflect inheritance from shared ancestors, indicating consanguinity. A region of homozygosity in this case was defined as 25 consecutive SNPs that were homozygous. The genomes of all six sons were riddled with these regions, suggesting that at some point, cousin married cousin or an aunt/uncle wed a nephew/niece. One region of homozygosity was seen in all three infertile brothers, but was heterozygous in two of the three fertile brothers. The remaining brother was homozygous wild type for the region.

Next, the researchers delved into the part of the long arm of chromosome 3 where the telltale region of homozygosity lay. It houses fifty genes, only one of which is expressed specifically in the testes. This gene is called "spermatogenesis-associated protein 16," or *SPATA16* (MIM 102530). It has eleven exons, and the mutation in the Ashkenazi family is a single base change, from G to A, at the 848th position in the gene, near the end of exon 4. The mutation likely upsets the splicing out of introns.

The wild type protein product of the *SPATA16* gene is transported from the Golgi apparatus into vesicles that take it to the acrosome as it telescopes out of the front end of a sperm cell. By attaching the gene for the jellyfish's green fluorescent protein (see figure 19.1) to the wild type *SPATA16* gene in cells growing in culture, researchers visualized the protein being transported to the forming acrosome in immature sperm.

Figure 1 A misshapen sperm cannot fertilize an oocyte.

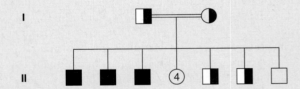

Figure 2 In a family with autosomal recessive globozoospermia, three of six sons are infertile.

around the 14th day after menstruation begins. This is when a woman is most likely to conceive.

For a woman with regular menstrual cycles who is under 30 years old and not using birth control, pregnancy typically happens within 3 or 4 months. A woman with irregular menstrual periods can tell when she is most fertile by using an ovulation predictor test, which detects a peak in the level of luteinizing hormone that precedes ovulation by a few hours. Another way to detect the onset of ovulation is to record body temperature each morning using a digital thermometer with subdivisions of hundredths of a degree Fahrenheit, which can indicate the 0.4 to 0.6 rise in temperature when ovulation starts. Sperm can survive in a woman's body for up to 5 days, but the oocyte is only viable for 24 to 48 hours after ovulation.

The hormonal imbalance that usually underlies irregular ovulation has various causes. These include a tumor in the ovary or in the pituitary gland in the brain that controls the reproductive system, an underactive thyroid gland, or use of steroid-based drugs such as cortisone. Sometimes a woman produces too much prolactin, the hormone that promotes milk production and suppresses ovulation in new mothers. If prolactin is abundant in a nonpregnant woman, she will not ovulate.

Fertility drugs can stimulate ovulation, but they can also cause women to "superovulate," producing more than one oocyte each month. A commonly used drug, clomiphene, raises the chance of having twins from 1 to 2 percent to 4 to 6 percent. If a woman's ovaries are completely inactive or absent (due to a birth defect or surgery), she can become pregnant only if she uses a donor oocyte. Some cases of female infertility are due to "reduced ovarian reserve"—too few oocytes. This is typically discovered when the ovaries do not respond to fertility drugs. Signs of reduced ovarian reserve are an ovary with too few follicles (observed on an ultrasound scan) or elevated levels of follicle-stimulating hormone on the third day of the menstrual cycle.

The uterine tubes are a common site of female infertility because fertilization usually occurs in open tubes. Blockage can prevent sperm from reaching the oocyte, or entrap a fertilized ovum, keeping it from descending into the uterus. If an embryo begins developing in a blocked tube and is not removed and continues to enlarge, the tube can burst and the woman can die. Such a "tubal pregnancy" is called an ectopic pregnancy.

Uterine tubes can also be blocked due to a birth defect or, more likely, from an infection such as pelvic inflammatory disease. A woman may not know she has blocked uterine tubes until she has difficulty conceiving and medical tests uncover the problem. Surgery can sometimes open blocked uterine tubes.

Excess tissue growing in the uterine lining may make it inhospitable to an embryo. This tissue can include benign tumors called fibroids or areas of thickened lining from a condition called endometriosis. The tissue can grow outside of the uterus too, in the abdominal cavity. In response to the hormonal cues to menstruate, the excess lining bleeds, causing cramps. Endometriosis can hamper conception, but curiously, if a woman with endometriosis conceives, the cramps and bleeding usually disappear after the birth.

Secretions in the vagina and cervix may be hostile to sperm. Cervical mucus that is thick or sticky due to infection can entrap sperm, keeping them from moving far enough to encounter an oocyte. Vaginal secretions may be so acidic or alkaline that they weaken or kill sperm. Douching daily with an acidic solution such as acetic acid (vinegar) or an alkaline solution, such as bicarbonate, can alter the pH of the vagina so that in some cases it is more receptive to sperm cells. Too little mucus can prevent conception too; this is treated with low daily doses of oral estrogen. Sometimes mucus in a woman's body harbors antibodies that attack sperm. Infertility may also result if the oocyte fails to release sperm-attracting biochemicals.

One reason the incidence of female infertility increases with age is that older women are more likely to produce oocytes with an abnormal chromosome number, which often causes spontaneous abortion because defects are too severe for development to proceed for long. The cause is usually misaligned spindle fibers when meiosis resumes, causing aneuploidy (extra or missing chromosomes). Perhaps the longer exposure of older oocytes to harmful chemicals, viruses, and radiation contributes to the risk of meiotic errors. Losing very early embryos may appear to be infertility because the bleeding accompanying the aborted embryo resembles a heavy menstrual flow.

Infertility Tests

A number of medical tests can identify causes of infertility. The man is checked first, because it is easier, less costly, and less painful to obtain sperm than oocytes.

Sperm are checked for number (sperm count), motility, and morphology (shape). An ejaculate containing up to 40 percent unusual forms is still considered normal, but many more than this can impair fertility. A urologist performs sperm tests.

A genetic counselor can evaluate Y chromosome deletions associated with lack of sperm. If a male cause of infertility is not apparent, a gynecologist checks the woman to see that reproductive organs are present and functioning.

Some cases of subfertility or infertility have no clear explanation. Psychological factors may be at play, or it may be that inability to conceive results from consistently poor timing. Sometimes a subfertile couple adopts a child, only to conceive one of their own shortly thereafter; many times, infertility remains a lifelong mystery.

Key Concepts

1. Male infertility is due to a low sperm count or sperm that cannot swim or are abnormal in structure.

2. Female infertility can be due to an irregular menstrual cycle or blocked uterine tubes. Fibroid tumors, endometriosis, or a misshapen uterus may prevent implantation of a fertilized ovum, and secretions in the vagina and cervix may inactivate or immobilize sperm. Oocytes may fail to release a sperm-attracting biochemical.

3. Early pregnancy loss due to abnormal chromosome number may be mistaken for infertility; this is more common among older women.

4. A variety of medical tests can pinpoint some causes of infertility.

21.3 Assisted Reproductive Technologies

Many people with fertility problems who do not choose to adopt children use alternative ways to conceive. Several of the ARTs were developed in nonhuman animals (see the Technology Timeline on page 411). In the United States, about 1 percent of the approximately 4 million births a year are from ARTs, and worldwide ART accounts for about 250,000 births a year.

This section describes types of ARTs. The different procedures can be performed on material from the parents-to-be ("nondonor") or from donors, and may be "fresh" (collected just prior to the procedure) or "frozen" (preserved in liquid nitrogen). Except for intrauterine insemination, the ARTs cost thousands of dollars and are not typically covered by health insurance in the United States.

Donated Sperm—Intrauterine Insemination

The oldest assisted reproductive technology is **intrauterine insemination (IUI),** in which a doctor places donated sperm into a woman's cervix or uterus. (It used to be called artificial insemination.) The success rate is 5 to 15% per attempt. The sperm are first washed free of seminal fluid, which can inflame female tissues. A woman might seek IUI if her partner is infertile or has a mutation that the couple wishes to avoid passing to

LANDMARKS IN REPRODUCTIVE TECHNOLOGY

	In Nonhuman Animals	In Humans
1782	Intrauterine insemination in dogs	
1790		Pregnancy reported from intrauterine insemination (IUI)
1890s	Birth from embryo transplantation in rabbits	IUI by donor
1949	Cryoprotectant successfully freezes animal sperm	
1951	First calf born after embryo transplantation	
1952	Live calf born after insemination with frozen sperm	
1953		First reported pregnancy after insemination with frozen sperm
1959	Live rabbit offspring produced from *in vitro* ("test tube") fertilization (IVF)	
1972	Live offspring from frozen mouse embryos	
1976		First reported commercial surrogate motherhood arrangement in the United States
1978	Transplantation of ovaries from one cow to another	Baby born after *in vitro* fertilization (IVF) in United Kingdom
1980		Baby born after IVF in Australia
1981	Calf born after IVF	Baby born after IVF in United States
1982	Sexing of embryos in rabbits Cattle embryos split to produce genetically identical twins	
1983		Embryo transfer after uterine lavage
1984		Baby born in Australia from frozen and thawed embryo
1985		Baby born after gamete intrafallopian transfer (GIFT) First reported gestational surrogacy arrangement in the United States
1986		Baby born in the United States from frozen and thawed embryo
1989		First preimplantation genetic diagnosis (PGD)
1992		First pregnancies from intracytoplasmic sperm injection (ICSI)
1994	Intracytoplasmic sperm injection (ICSI) in mouse and rabbit	62-year-old woman gives birth from donated oocyte
1995	Sheep cloned from embryo cell nuclei	Babies born following ICSI
1996	Sheep cloned from adult cell nucleus	
1998	Mice cloned from adult cell nuclei	Baby born 7 years after his twin
1999	Cattle cloned from adult cell nuclei	
2000	Pigs cloned from adult cell nuclei	
2001		Sibling born following PGD to treat sister for genetic disease Human preimplantation embryo cloned, survives to 6 cells
2003		3,000-plus preimplantation genetic diagnoses performed to date
2004	Woman pays $50,000 to have her cat cloned	First birth from a woman who had ovarian tissue preserved and implanted on an ovary after cancer treatment
2005	Dog cloned	
2009		A woman who had already given birth to six children through IVF has octuplets

their child. Women also undergo IUI to be a single parent without having sex, or a lesbian couple may use it to have a child.

The first documented IUI in humans was done in 1790. For many years, physicians donated sperm, and this became a way for male medical students to earn a few extra dollars. By 1953, sperm could be frozen and stored and IUI became much more commonplace. Today, donated sperm are frozen and stored in sperm banks, which provide the cells to obstetricians who perform the procedure. IUI costs about $125 to $615, with higher charges from some facilities for sperm from donors who have professional degrees because those men are paid more for their donations. Additional fees are charged for a more complete medical history of the donor, for photos of the man at different ages, and for participation in a "consent program" in which the donor's identity is revealed when his offspring turns 18 years old. If ovulation is induced to increase the chances of success of IUI, additional costs may exceed $3,000.

A couple who chooses IUI can select sperm from a catalog that lists the personal characteristics of donors, such as blood type, hair and eye color, skin color, build, educational level, and interests. One donor profile listed that the man enjoys spear-fishing and wrestling, listens to singer Tori Amos, and loves the film *The Princess Bride*, as if these are inherited traits. Many women selected him because he was a handsome doctor! If a couple desires a child of one sex—such as a daughter to avoid passing on an X-linked disorder—sperm can be separated into fractions enriched for X-bearing or Y-bearing sperm.

Problems can arise in IUI if a donor learns that he has an inherited disease. For example, a man developed cerebellar ataxia (MIM 608029), a movement disorder, years after he donated sperm. Eighteen children conceived using his sperm face a 1 in 2 risk of having inherited the mutant gene. Over-enthusiastic sperm donors can lead to problems. One man, listed in the Fairfax Cryobank as "Donor 401," earned $40,000 donating sperm while in law school. He was quite attractive and popular, and forty-five children were conceived with his sperm. When a few of the families he started appeared on a talk show, several other families tuning in were shaken to see so many children who resembled their own. The website http://www.donorsiblingregistry.com has enabled thousands of half-siblings who share sperm donor fathers to meet.

A male's role in reproductive technologies is simpler than a woman's. A man can be a genetic parent, contributing half of his genetic self in his sperm, but a woman can be both a genetic parent (donating an oocyte) and a gestational parent (donating the uterus). Not all "third-party pregnancies" have as joyous an outcome as those that led to the births of the babies described in the chapter opener.

A Donated Uterus—Surrogate Motherhood

If a man produces healthy sperm but his partner's uterus cannot maintain a pregnancy, a surrogate mother may help by being inseminated with the man's sperm. When the child is born, the surrogate mother gives the baby to the couple. In this variation of the technology, the surrogate is both the genetic and the gestational mother. Attorneys usually arrange surrogate relationships. The surrogate mother signs a statement signifying her intent to give up the baby. In some U.S. states, and in some nations, she is paid for her 9-month job, but in the United Kingdom compensation is illegal. This is to prevent wealthy couples from taking advantage of women who become surrogates for the money.

A problem with surrogate motherhood is that a woman may not be able to predict her responses to pregnancy and childbirth in a lawyer's office months earlier. When a surrogate mother changes her mind about giving up the baby, the results are wrenching for all. A prominent early case involved Mary Beth Whitehead, who carried the child of a married man for a fee and then changed her mind about giving up the baby. The courts eventually awarded custody to the father and his wife. The woman who raises the baby may feel badly too, especially when people say she is not the "real" mother. For this reason, surrogate mothers are also called "gestational carriers." It is difficult in this situation not to hurt someone's feelings!

Another type of surrogate mother lends only her uterus, receiving a fertilized ovum conceived from a man and a woman who has healthy ovaries but lacks a functional uterus. This variation is an "embryo transfer to a host uterus," and the pregnant woman is a "gestational-only surrogate mother." She turns the child over to the biological parents.

In Vitro Fertilization

In *in vitro* **fertilization** (**IVF**), which means "fertilization in glass," sperm and oocyte join in a laboratory dish. Soon after, the embryo that forms is placed in a uterus. If all goes well, it implants into the uterine lining and continues development until a baby is born.

Louise Joy Brown, the first "test-tube baby," was born in 1978, amid great attention and sharp criticism. A prominent bioethicist said that IVF challenged "the idea of humanness and of our human life and the meaning of our embodiment and our relation to ancestors and descendants." Yet Louise is, despite her unusual beginnings, an ordinary young woman. Today IVF accounts for one in eighty births in the United States. More than 5 million children have been born following IVF.

A woman might undergo IVF if her ovaries and uterus work but her uterine tubes are blocked. Using a laparoscope, which is a lit surgical instrument inserted into the body through a small incision, a physician removes several of the largest oocytes from an ovary and transfers them to a culture dish. If left in the body, only one oocyte would exit the ovary, but in culture, many can mature sufficiently to be fertilized *in vitro*. Chemicals, sperm, and other cell types similar to those in the female reproductive tract are added to the culture. An acidic solution may be applied to the zona pellucida, which is the layer around the egg, to thin it to ease the sperm's penetration.

Sperm that cannot readily enter the oocyte may be sucked up into a tiny syringe and microinjected into the female cell. This technique, called **intracytoplasmic sperm injection** (**ICSI**), is more effective than IVF alone and has become standard at some facilities (**figure 21.3**). ICSI is very helpful for men who have low sperm counts or many abnormal sperm, and

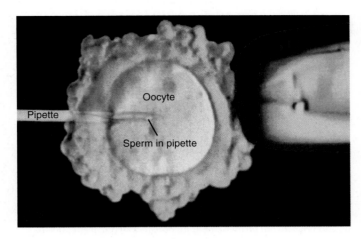

Figure 21.3 **ICSI.** Intracytoplasmic sperm injection (ICSI) enables some infertile men, men with spinal cord injuries, or men with certain illnesses to become fathers. A single sperm cell is injected into the cytoplasm of an oocyte. This photo is falsely colored.

makes fatherhood possible for men who cannot ejaculate, such as those who have suffered spinal cord injuries. ICSI has been performed on thousands of men with about a 30 percent success rate.

Five days after sperm wash over the oocytes in the dish, or are injected into them, one or two blastocysts are transferred to the uterus. If the hormone human chorionic gonadotropin appears in the woman's blood a few days later, and its level rises, she is pregnant.

IVF costs from $6,500 to $15,000 per attempt. Children born following IVF have a slight increase in the rate of birth defects (about 8 percent) compared to children conceived naturally. This may reflect medical problems of parents seeking IVF, the tendency of IVF to interfere with the parting of chromosome pairs during meiosis, closer scrutiny of IVF pregnancies, and/or effects on imprinting from the time in culture. Children born after IVF on average have higher birth weights.

In the past, several embryos were implanted to increase the success rate of IVF, but this led to many multiple births, which are riskier than single births. In many cases, physicians had to remove embryos to make room for others to survive. To avoid the multiples problem, and because IVF has become more successful as techniques have improved, guidelines now suggest transferring only one embryo.

Embryos resulting from IVF that are not soon implanted in the woman can be frozen in liquid nitrogen ("cryopreserved") for later use. Cryoprotectant chemicals are used to prevent salts from building up or ice crystals from damaging delicate cell parts. Freezing takes a few hours; thawing about a half hour. The longest an embryo has been frozen, stored, and then successfully revived is 13 years; the "oldest" pregnancy using a frozen embryo occurred 9 years after the freezing!

So many people have had IVF since Louise Joy Brown was born that researchers have developed algorithms to predict the chances that the procedure will be successful and lead to a birth for a particular couple. Overall the chances of a live birth following IVF are about 25 percent, but this prediction varies greatly, depending on certain risk factors that lower the likelihood of success. These include:

- maternal age—success is about 33 percent for women (oocyte donors) under age 30, but only 5 to 10 percent for women over 40;
- increased time being infertile;
- number of previous failed IVF attempts;
- number of previous IVF attempts;
- use of a woman's own oocytes rather than a donor's; and
- infertility with a known cause.

A website (http://www.ivfpredict.com) assesses these risks. For example, a couple who has been infertile for 11 years—attempted IVF four times that included two failures and two spontaneous abortions, and used the woman's eggs and ICSI because too many sperm were abnormal—has a chance of success of only about 8 percent. If they use a donor oocyte, the chance of success doubles.

Gamete and Zygote Intrafallopian Transfer

IVF may fail because of the artificial environment for fertilization. A procedure called **GIFT,** which stands for **gamete intrafallopian transfer,** improves the setting. (Uterine tubes are also called fallopian tubes.) Fertilization is assisted in GIFT, but it occurs in the woman's body rather than in glassware.

In GIFT, several of a woman's largest oocytes are removed. The man submits a sperm sample, and the most active cells are separated from it. The collected oocytes and sperm are deposited together in the woman's uterine tube, at a site past any obstruction that might otherwise block fertilization. GIFT is about 22 percent successful, and usually costs less than IVF.

A variation of GIFT is **ZIFT,** which stands for **zygote intrafallopian transfer**. In this procedure, an IVF ovum is introduced into the woman's uterine tube. Allowing the fertilized ovum to make its own way to the uterus increases the chance that it will implant. ZIFT is also 22 percent successful.

GIFT and ZIFT are done less frequently than IVF. They often will not work for women who have scarred uterine tubes.

The *Bioethics: Choices for the Future* on page 414 considers the unusual situation of collecting gametes from a person shortly after the person has died.

Oocyte Banking and Donation

Oocytes can be stored, as sperm are, but the procedure may create problems. Because an oocyte is the largest type of cell, it contains a large volume of water. Freezing can form ice crystals that damage cell parts. Candidates for preserving oocytes include women who wish to have children later in life and women who will contact toxins or teratogens in the workplace or in chemotherapy.

Oocytes are frozen in liquid nitrogen at −30°C to −40°C, when they are at metaphase of the second meiotic division. At this time, the chromosomes are aligned along the spindle, which

Removing and Using Gametes After Death

A gamete is a packet containing a person's genome. If a person dies, his or her gametes can be collected from the body and combined with an opposite gamete type, potentially making the deceased a new parent. This "postmortem gamete retrieval" may sound like science fiction, but it has happened for years, nearly always for men.

One of the first cases of postmortem sperm removal affected Bruce and Gaby V. In their early thirties, they had delayed becoming parents, confident that their good health would make pregnancy possible later. Then Bruce suddenly died of an allergic reaction to a drug. Gaby knew how much Bruce had wanted to be a father, so she asked the medical examiner to collect Bruce's sperm. The sample was sent to the California Cryobank, where it lay deeply frozen for more than a year. In the summer of 1978, the sperm were defrosted and used to fertilize one of Gaby's oocytes. On March 17, Bruce and Gaby's daughter was born. It was the first case of postmortem sperm retrieval in which the father did not actively participate in the decision. Since 1990, U.S. servicemen who feared infertility from exposure to chemical or biological weapons have taken advantage of sperm bank discounts to the military, preserving their sperm before deploying.

Postmortem sperm retrieval raises legal and ethical issues based on timing. In one case, a woman conceived twins using her husband's preserved sperm 16 months after he died of leukemia; he had stated his wishes for her to do so. The Social Security Administration refused to provide survivor benefits to their daughters, claiming that the husband was not a father, but a sperm donor. The Massachusetts Superior Court reversed this decision. In New Jersey, a mother claimed Social Security benefits for twins conceived after her husband's death. An appeals court upheld the denial of benefits, claiming that the children must have been dependents at the time of their father's death.

The first case of postmortem oocyte retrieval was reported in 2010. A 36-year-old woman stood up on a plane following many hours of sleeping in one position, and her heart stopped. By the time a doctor on board restarted it, the woman's brain had been robbed of oxygen for several precious minutes. The plane made an emergency landing and she was taken to a hospital, where she was placed on a respirator. Scans showed blood clots in her lungs that had caused the collapse. By the fourth day, the woman's brain was dangerously swelling, and by the ninth day, her brain activity was nearly nil,

although she could still open her eyes and move spontaneously. Her husband, parents, and in-laws asked that the tubes keeping her alive be withdrawn. Then, several hours after this was done, they changed their minds, and asked that the breathing tube be reinserted so that the woman's oocytes could be retrieved. They had no idea how difficult this would be.

The physicians, knowing the complexities of the medical situation, wanted to know one other person's opinion—the patient's. So they consulted with the young woman's gynecologist, who had no record or recollection of the patient stating she wished to have children. The young woman was not completely brain dead, so the decisions would not be the same as for donating an organ after death. If her oocytes were to be used to give her husband a child, assisted reproductive technologies would obviously be necessary—*in vitro* fertilization and a surrogate mother. Even before that could happen, though, the woman would have to undergo 2 weeks of hormone treatments to ovulate, during which time she would have to lie flat, which could cause her death. For these practical reasons, and the fact that the woman had never stated that she wished to be a parent, the family elected to turn off life support, and she quickly died.

Like other assisted reproductive technologies, postmortem gamete retrieval is not regulated at the federal level in the United States. Bioethicists have identified situations to avoid:

- someone other than a spouse wishing to use the gamete;
- a too-hasty decision based on grief; and
- use of the gamete for monetary gain.

Questions for Discussion

1. How does the case of the 36-year-old woman whose oocytes were retrieved following her brain death differ from that of a pregnant woman in a coma who is kept alive for several weeks so that her baby can be born?

2. The people described in this essay did not have other children. How might the situation differ for a couple who already have children?

3. Do you think that Social Security or another benefit system should cover fetuses, embryos, or gametes?

4. How might postmortem gamete retrieval be abused?

is sensitive to temperature extremes. If the spindle comes apart as the cell freezes, the oocyte may lose a chromosome, which would devastate development. Another problem with freezing oocytes is retention of a polar body, leading to a diploid oocyte. Only 100 babies have been born using frozen oocytes despite two decades of attempts. The probability of achieving pregnancy using a frozen oocyte with current technology is only about 3 percent, and the technique is still investigational. However, websites offering "egg freezing" claim high rates of success, which can mean anything from fertilization to a birth.

To avoid the difficulty of freezing oocytes, strips of ovarian tissue can be frozen, stored, thawed, and reimplanted at various sites, such as under the skin of the forearm or abdomen

or in the pelvic cavity near the ovaries. The tissue ovulates and the oocytes are collected and fertilized *in vitro*. The first child resulting from fertilization of an oocyte from reimplanted ovarian tissue was born in 2004. The mother, age 25, had been diagnosed with advanced Hodgkin's lymphoma. The harsh chemotherapy and radiation cured her cancer, but destroyed her ovaries. Beforehand, five strips of tissue from her left ovary were frozen. Later, several pieces of ovarian tissue were thawed and implanted in a pocket that surgeons crafted on one of her shriveled ovaries, near the entrance to a uterine tube. Menstrual cycles resumed, and shortly thereafter, the woman became pregnant with her daughter, who is healthy. Freezing ovarian tissue may become routine for cancer patients of childbearing age.

Women who have no oocytes or wish to avoid passing on a mutation can obtain oocytes from donors, who are typically younger women. Some women become oocyte donors when they undergo IVF and have "extra" (see the opening essay for chapter 3 on oocyte donation). The potential father's sperm and donor's oocytes are placed in the recipient's uterus or uterine tube, or fertilization occurs in the laboratory and a blastocyst is transferred to the woman's uterus. A program in the United Kingdom funds IVF for women who cannot afford the procedure if they donate their "extras." The success of using oocytes from younger women confirms that it is the oocyte that age affects, and not the uterine lining. The live birth rate per cycle for older women using their own oocytes in IVF is 28 percent. For older women using oocytes donated by younger women, the rate is about 50 percent.

Embryo adoption is a variation on oocyte donation. A woman with malfunctioning ovaries but a healthy uterus carries an embryo that results when her partner's sperm is used in intrauterine insemination of a woman who produces healthy oocytes. If the woman conceives, the embryo is gently flushed out of her uterus a week later and inserted through the cervix and into the uterus of the woman with malfunctioning ovaries. The child is genetically that of the man and the woman who carries it for the first week, but is born from the woman who cannot produce healthy oocytes. "Embryo adoption" also describes use of IVF "leftovers."

In another technology, cytoplasmic donation, older women have their oocytes injected with cytoplasm from the oocytes of younger women to "rejuvenate" the cells. Although resulting children conceived through IVF appear to be healthy, they are being monitored for a potential problem—heteroplasmy, or two sources of mitochondria in one cell (see figure 5.11). Researchers do not yet know the health consequences of having

mitochondria from the donor cytoplasm plus mitochondria from the recipient's oocyte. These conceptions also have an elevated incidence of XO syndrome, which often causes spontaneous abortion.

Because oocytes are harder to obtain than sperm, oocyte donation technology has lagged behind that of sperm banks, but is catching up. One IVF facility that has run a donor oocyte program since 1988 has a patient brochure that describes 120 oocyte donors of various ethnic backgrounds, like a catalog of sperm donors. The oocyte donors are young and have undergone extensive medical and genetic tests. Recipients may be up to 55 years of age.

Preimplantation Genetic Diagnosis

Prenatal diagnostic tests such as amniocentesis and chorionic villus sampling can be used in pregnancies achieved with assisted reproductive technologies. A test called **preimplantation genetic diagnosis (PGD)** detects genetic and chromosomal abnormalities *before* pregnancy starts. The couple selects a very early "preimplantation" embryo that tests show has not inherited a specific detectable genetic condition. "Preimplantation" refers to the fact that the embryo is tested at a stage prior to when it would implant in the uterus. PGD was used to select Adam Nash, whose umbilical cord stem cells cured his sister's Fanconi anemia (see figure 21.1). PGD has about a 29 percent success rate.

PGD is possible because one cell, or blastomere, can be removed for testing from an 8-celled embryo, and the remaining seven cells can complete development normally in a uterus. Before the embryo is implanted into the woman the single cell is karyotyped, or its DNA amplified and probed for genes that the parents carry. Embryos that pass these tests are selected to complete development or are stored. At first, researchers implanted the remaining seven cells, but letting the selected embryo continue developing in the dish until day 5, when it is 80 to 120 cells, is more successful. Obtaining the cell to be tested is called "blastomere biopsy" (**figure 21.4**). Accuracy in detecting a mutation or abnormal chromosome is about

Preimplantation embryo

One cell removed for genetic analysis

Seven cells complete development

DNA probes

If genetically healthy, preimplantation embryo is implanted into woman and develops into a baby.

If genetic disease is inherited, preimplantation embryo is *not* implanted into woman.

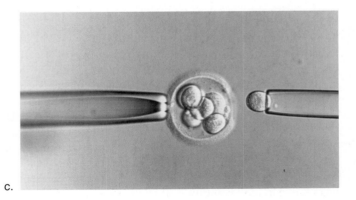

Figure 21.4 **Preimplantation genetic diagnosis (PGD) probes disease-causing genes or chromosome aberrations in an 8-celled preimplantation embryo.** **(a)** A single cell is separated and tested to see if it contains a disease-causing genotype or atypical chromosome. **(b)** If it doesn't, the remaining seven cells divide a few more times and are transferred to the oocyte donor to complete development. **(c)** This preimplantation embryo is held still by suction applied on the left. On the right, a pipette draws up a single blastomere. *In vitro* fertilization took place 45 hours previously.

97 percent. Errors generally happen when a somatic mutation affects the sampled blastomere but not the rest of the embryo. Amplification of the selected blastomere DNA may cause such a somatic mutation.

PGD is not a new technique. The first children who had PGD were born in 1989. In these first cases, probes for Y chromosome–specific DNA sequences were used to select females, who could not inherit the X-linked conditions their mothers carried. The alternative to PGD would have been to face the 25 percent chance of conceiving an affected male.

In March 1992, the first child was born who underwent PGD to avoid a specific inherited disease. Chloe O'Brien was checked as an 8-celled preimplantation embryo to see if she had escaped the cystic fibrosis that affected her brother. Since then, PGD has helped to select thousands of children free of several dozen types of inherited illnesses. It has been used for the better-known single-gene disorders as well as for rare ones.

Like most ARTs, use of PGD has expanded as it has become more accurate and more familiar. Today it has taken on a quality-control role in addition to being a tool to detect and prevent rare diseases. PGD is increasingly being used to screen early embryos derived from IVF for normal chromosome number before implanting them into women. This should increase the chances of successful live births, but in the first large trial, PGD actually lowered the birth rate—perhaps the intervention harms the embryos. In the Netherlands, researchers examined *all* of the cells of several preimplantation human embryos and found that some cells can have normal chromosomes and others not. Therefore, the assumption underlying PGD—that a sampled cell represents them all—may need to be reexamined.

PGD can introduce a bioethical "slippery slope" when it is used other than to ensure that a child is free of a certain disease, such as for gender selection. A couple with five sons might, for example, use PGD to select a daughter. But this use of technology might just be a high-tech version of age-old human nature, according to one physician who performs PGD. "From the dawn of time, people have tried to control the sex of offspring, whether that means making love with one partner wearing army boots, or using a fluorescence-activated cell sorter to separate X- and Y-bearing sperm. PGD represents a quantum leap in that ability—all you have to do is read the X and Y chromosome paints," he says.

While PGD used solely for family planning is certainly more civilized than placing baby girls outside the gates of ancient cities to perish, the American Society for Reproductive

Table 21.1	Some Assisted Reproductive Technologies
Technology	**Procedures**
GIFT	Deposits collected oocytes and sperm in uterine tube.
IVF	Mixes sperm and oocytes in a dish. Chemicals simulate intrauterine environment to encourage fertilization.
IUI	Places or injects washed sperm into the cervix or uterus.
ICSI	Injects immature or rare sperm into oocyte, before IVF.
Oocyte freezing	Oocytes retrieved and frozen in liquid nitrogen.
Ovulation induction	Drugs control timing of ovulation in order to perform a particular procedure.
PGD	Searches for specific mutant allele in sampled cell of 8-celled embryo. Its absence indicates remaining 7-celled embryo can be nurtured and implanted in woman, and child will be free of genetic condition.
Surrogate mother	Woman carries a pregnancy for another.
ZIFT	Places IVF ovum in uterine tube.

Medicine endorses the use of PGD for sex selection only to avoid passing on an X-linked disease, which was the first application of the technology. Yet even PGD to avoid disease can be controversial. In the United Kingdom, where the government regulates reproductive technology, inherited cancer susceptibility is an approved indication for PGD. These cancers do not begin until adulthood, the susceptibility is incompletely penetrant (not everyone who inherits the disease-associated genotype will actually develop cancer), and the cancer may be treatable.

Table 21.1 summarizes the assisted reproductive technologies.

Key Concepts

1. Intrauterine insemination places donor sperm in a woman's reproductive tract.

2. A genetic and gestational surrogate mother is intrauterinely inseminated, becomes pregnant, then gives the baby to the father and his partner. A gestational surrogate mother gestates a baby conceived *in vitro* with gametes from a man and a woman who cannot carry a fetus.

3. In IVF, sperm and oocyte unite outside the body, and the resulting embryo is transferred to the uterus. Early embryos can be frozen.

4. In GIFT, sperm and oocytes are placed in a uterine tube past a blockage. In ZIFT, an IVF embryo is placed in a uterine tube.

5. In embryo adoption, a woman who has had intrauterine insemination has an early embryo washed out of her uterus and transferred to a woman who lacks oocytes.

6. PGD removes cells from early embryos and screens them for genetic or chromosomal abnormalities.

21.4 Extra Embryos

Sometimes assisted reproductive technologies leave "extra" oocytes, fertilized ova, or very early embryos. **Table 21.2** lists the possible fates of this biological material.

In the United States, nearly half a million embryos derived from IVF sit in freezers; some have been there for years. Most couples who donate embryos to others do so anonymously, with no intention of learning how their genetic offspring are raised. Scott and Glenda Lyons chose a different path when they learned that their attempt at IVF had yielded too many embryos.

In 2001, two of Glenda's eighteen embryos were transferred to her uterus, and developed into twins Samantha and Mitchell. Through a website where couples chat about fertility issues, Scott and Glenda met and selected Bruce and Susan Lindeman to receive fourteen remaining embryos. This second couple had tried IVF three times, with no luck. The Lyons's frozen embryos were shipped cross-country to a clinic where two were implanted in Susan's uterus. In July 2003, Chase and Jack Lindeman were born—genetic siblings of Samantha and Mitchell Lyons. But there were still embryos left. The Lyons allowed the Lindemans to send twelve embryos to a third couple, who used two to have twin daughters in August 2004. They are biological siblings of Samantha and Mitchell Lyons and Chase and Jack Lindeman (**figure 21.5**).

Another alternative to disposing of fertilized ova and embryos is to donate them for use in research. The results of experiments sometimes challenge long-held ideas, indicating that we still have much to learn about early human prenatal development. This was the case for a study from Royal Victoria Hospital in Montreal. Researchers examined the chromosomes of sperm from a man with XXY syndrome. Many of the sperm would be expected to have an extra X chromosome, due to nondisjunction (see figure 13.12), which could lead to a preponderance of XXX and XXY offspring. Surprisingly, only 3.9 percent of the man's *sperm* had extra chromosomes, but five out of ten of his spare *embryos* had an abnormal X, Y, or chromosome 18. That is, even though most of the man's sperm were normal, his embryos weren't. The source of reproductive problems in XXY syndrome, therefore, might not be in the sperm, but in early embryos—a finding that was previously unknown and not expected, and was only learned because of observing early human embryos.

In another study, Australian researchers followed the fates of single blastomeres that had too many or too few chromosomes. They wanted to see whether the abnormal cells preferentially ended up in the inner cell mass, which develops into the embryo, or the trophectoderm, which becomes extraembryonic membranes. The study showed that cells with extra or missing chromosomes become part of the inner cell mass much more frequently than expected by chance. This finding indicates that the ability of a blastomere sampled for PGD to predict health may depend on whether it is fated to be part of the inner cell mass.

Using fertilized ova or embryos designated for discard in research is controversial. Without regulations on privately funded research, ethically questionable experiments can happen. For example, researchers reported at a conference that they had mixed human cells from male embryos with cells from female

Table 21.2	Fates of Frozen Embryos

1. Store indefinitely.

2. Store and destroy after a set time.

3. Donate for embryonic stem cell derivation and research.

4. Thaw later for use by biological parents.

5. Thaw later for use by other parents.

6. Discard.

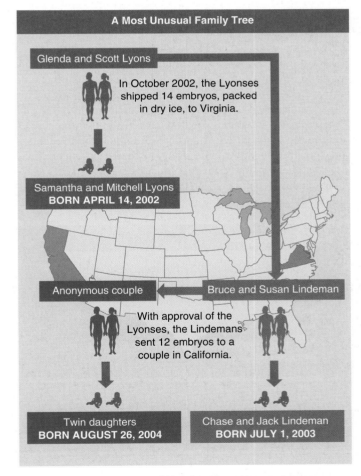

Figure 21.5 Using extra embryos. Six children resulted from Glenda and Scott Lyons's embryos. The Lyonses had a boy and a girl, then donated embryos to the Lindemans, who had twin boys. Finally, a couple in California used the Lyons's embryos to have twin daughters. The Lyonses and Lindemans have become friends.

embryos, to see if the normal male cells could "save" the female cells with a mutation. Sex was chosen as a marker because the Y chromosome is easy to detect, but the idea of human embryos with mixed sex parts caused a public outcry.

In vitro fertilized ova not chosen for reproduction have been the major source of donations for research. However, an experimental ART called **polar body biopsy** may reduce this supply by increasing the success of IVF. Polar body biopsy is based on Mendel's first law, the segregation of alleles. In the technique, if a polar body resulting from the first meiotic division in a woman who is a carrier of an X-linked disorder has the mutant allele, then the oocyte to which it clings is inferred to lack that allele. Oocytes that pass this test can be fertilized *in vitro* and the resulting embryo implanted. Polar body biopsy is possible because the polar body is attached to the much larger oocyte. A large pipette is used to hold the two cells in place, and a smaller pipette is used to separate the polar body. Then, DNA probes and FISH are used to look at genes and chromosomes in the polar body and infer the genotype of the oocyte (**figure 21.6**). Polar body biopsy followed by PGD is quite effective in avoiding conceptions with chromosome abnormalities or certain single-gene disorders.

ARTs introduce ownership and parentage issues (**table 21.3**). Another controversy is that human genome information is providing more traits to track and perhaps control in coming generations. When we can routinely scan the human genome in gametes, fertilized ova, or early embryos, who will decide which traits are worth living with, and which aren't?

ARTs operate on molecules and cells, but with repercussions for individuals and families. Ultimately, if it becomes widespread enough, these interventions into reproduction may affect the gene pool. Let us hope that regulations will evolve along with the technologies to assure that they are applied sensibly and humanely.

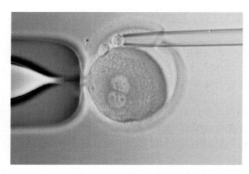

Figure 21.6 Polar body biopsy. The fact that an oocyte shares a woman's divided genetic material with a much smaller polar body allows the screening of oocytes for use in IVF.

Table 21.3	Assisted Reproductive Disasters

1. A physician in California used his own sperm to perform intrauterine insemination on 15 patients, telling them that he had used sperm from anonymous donors.

2. A plane crash killed the wealthy parents of two early embryos stored at −320°F (−195°C) in a hospital in Melbourne, Australia. Adult children of the couple were asked to share their estate with two 8-celled siblings-to-be.

3. Several couples in Chicago planning to marry discovered that they were half-siblings. Their mothers had been inseminated with sperm from the same donor.

4. Two Rhode Island couples sued a fertility clinic for misplacing several embryos.

5. Several couples in California sued a fertility clinic for implanting their oocytes or embryos in other women without donor consent. One woman requested partial custody of the resulting children if her oocytes were taken, and full custody if her embryos were used, even though the children were of school age and she had never met them.

6. A man sued his ex-wife for possession of their frozen fertilized ova. He won, and donated them for research. She had wanted to be pregnant.

7. The night before *in vitro* fertilized embryos were to be implanted in a 40-year-old woman's uterus after she and her husband had spent four years trying to conceive, the man changed his mind, and wanted the embryos destroyed.

Key Concepts

1. IVF produces extra fertilized ova and early embryos that may be implanted, frozen, donated, discarded, or used in research.

2. Embryos in research add to our knowledge of early human development.

3. Polar body biopsy enables physicians to select out defective oocytes.

Summary

21.1 Savior Siblings and More

1. **Assisted reproductive technologies** replace what is missing in reproduction, using laboratory procedures and people other than the infertile couple.

21.2 Infertility and Subfertility

2. **Infertility** is the inability to conceive a child after a year of unprotected intercourse. Subfertile individuals or couples manufacture gametes, but take longer than usual to conceive.

3. Causes of infertility in the male include low sperm count, a malfunctioning immune system, a varicose vein in the scrotum, structural sperm defects, drug exposure, vasectomy reversal, and abnormal hormone levels. Mutation may impair fertility.

4. Causes of infertility in the female include absent or irregular ovulation, blocked uterine tubes, an inhospitable or malshaped uterus, antisperm secretions, or lack of sperm-attracting biochemicals. Early pregnancy loss due to abnormal chromosome number is more common in older women and may appear to be infertility.

21.3 Assisted Reproductive Technologies

5. In **intrauterine insemination,** donor sperm are introduced into a woman's reproductive tract in a clinical setting.

6. A gestational and genetic surrogate mother provides her oocyte. Then intrauterine insemination is performed with sperm from a man whose partner cannot conceive or carry a fetus. The surrogate also provides her uterus for 9 months. A gestational surrogate mother receives an *in vitro* **fertilized** ovum that belongs genetically to the couple who ask her to carry it.

7. In IVF, oocytes and sperm meet in a dish, fertilized ova divide a few times, and embryos are placed in the woman's body, circumventing blocked tubes or malfunctioning sperm.

Intracytoplasmic sperm injection introduces immature or nonmotile sperm into oocytes.

8. Embryos can be frozen and thawed and then complete development when placed in a woman's uterus.

9. **GIFT** introduces oocytes and sperm into a uterine tube past a blockage; fertilization occurs in the woman's body. **ZIFT** places an early embryo in a uterine tube.

10. Oocytes can be frozen and stored. In embryo adoption, a woman undergoes intrauterine insemination. A week later, the embryo is washed out of her uterus and introduced into the reproductive tract of the woman whose partner donated the sperm.

11. Seven-celled embryos can develop normally if a blastomere is removed at the 8-cell stage and cleared for abnormal chromosomes or genes. This is **preimplantation genetic diagnosis**.

21.4 Extra Embryos

12. Extra fertilized ova and early embryos generated in IVF are used, donated to couples, stored, donated for research, or discarded. They enable researchers to study aspects of early human development that they could not investigate in other ways.

13. **Polar body biopsy** enables physicians to perform genetic tests on polar bodies and to infer the genotype of the accompanying oocyte.

www.mhhe.com/lewisgenetics10

Answers to all end-of-chapter questions can be found at **www.mhhe.com/lewisgenetics10.** You will also find additional practice quizzes, animations, videos, and vocabulary flashcards to help you master the material in this chapter.

Review Questions

1. Which assisted reproductive technologies might help the following couples? (More than one answer may fit some situations.)

    a. A woman is born without a uterus, but manufactures healthy oocytes.
    b. A man has cancer treatments that damage his sperm.
    c. A genetic test reveals that a woman will develop Huntington disease. She wants a child, but does not want to pass on the disease.
    d. Two women wish to have and raise a child together.
    e. A man and woman are each carriers of sickle cell disease. They do not want to have an affected child or terminate a pregnancy to avoid the birth of an affected child.
    f. A woman's uterine tubes are scarred and blocked.
    g. A young woman must have radiation to treat ovarian cancer, but wishes to have a child.

2. Why are men typically tested for infertility before women?

3. What are some of the causes of infertility among older women?

4. Cite a situation in which both man and woman contribute to subfertility.

5. How does ZIFT differ from GIFT? How does it differ from IVF?

6. Explain how preimplantation genetic diagnosis is similar to and different from CVS and amniocentesis, described in chapter 13.

7. How do each of the following assisted reproductive technologies deviate from the normal biological process?

    a. *in vitro* fertilization
    b. GIFT
    c. embryo adoption
    d. gestational-only surrogacy
    e. intrauterine insemination
    f. cytoplasmic donation

8. Explain how PGD works, and list two events in early prenatal development that might explain cases where the PGD result is inaccurate.

9. Why is it much easier to freeze and revive early embryos than oocytes?

10. Describe a scenario in which each of the following technologies is abused:

a. surrogate mother
b. gamete donation
c. IVF
d. PGD

11. Explain how polar body biopsy is based on Mendel's first law.

Applied Questions

1. How are the twiblings Violet and Kieran described in the chapter opener related to each other genetically?

2. Neil Patrick Harris is a TV actor who had twins with his partner, David Burtka. They used a surrogate mother, who carried two embryos that had been fertilized *in vitro*, one with Neil's sperm and the other with David's sperm. In terms of genetics, how closely are the babies, a boy and a girl, related to each other? (The fathers do not know who fathered which child.)

3. Some assisted reproductive technologies were invented to help people who could not have children for medical reasons, or to avoid conceiving a child with a genetic disease known to be in the family. With time, as the technologies became more familiar, in the United States, people with economic means began to use them for other reasons, such as a celebrity who does not wish to lose her shape during pregnancy. Remembering that the U.S. government does not regulate ARTs, do you think that any measures should be instituted to select candidates for ARTs? How can these technologies be made more affordable?

4. A woman in New Jersey applied for Social Security benefits for twins born by IVF after their father had died. She was denied, sued, lost, and appealed, and the appeals court upheld the denial, stating that she must show proof that the children were dependents at the time of the father's death. Do you think that the children should receive benefits? State a reason for your opinion.

5. Singer Sir Elton John and his partner David Furnish recently had a baby. Sir Elton wanted his sperm to be used because he thinks that his songwriting talent is inherited. Which ART did they use to have their son Zachary?

6. A man reads his medical chart and discovers that the results of his sperm analysis indicate that 22 percent of his sperm are shaped abnormally. He wonders why the physician said he had normal fertility if so many sperm are abnormally shaped. Has the doctor made an error?

7. An embryo bank in Texas offers IVF leftover embryos, which would otherwise remain in the deep freeze or be discarded, to people wanting to have children, for $2,500 each. The bank circumvents bioethical concerns by claiming that it sells a service, not an embryo. People in favor of the bank claim that purchasing an embryo is not different from paying for sperm or eggs, or an adopted child. Those who object to the bank claim that it makes an embryo a commodity.

a. Do you think that an embryo bank is a good idea, or is it unethical?
b. Whose rights are involved in the operation of the embryo bank?
c. Who should be liable if a child that develops from the embryo has a medical problem?
d. Is the bank elitist because the cost is so high?

8. A newspaper columnist wrote that frozen human embryos are "microscopic Americans." Former president George W. Bush called them "unique and genetically complete, like every other human being." A stem cell researcher referred to embryonic stem cells as "like any other cell in an adult, no different from the skin cells you rub off with a towel after a shower."

a. What is your opinion of the status of an 8-cell human embryo?
b. Does the status of a cell from an 8-cell human embryo in culture depend upon whether that cell is tested and discarded, or part of the embryo is implanted in a woman?
c. Do you think that there is any harm in an influential person, such as a journalist, politician, or researcher, stating the status of an embryo?
d. How might the following individuals respond to or feel about these definitions?
    i. A woman who ended a pregnancy, for whatever reason
    ii. A couple who have had a spontaneous abortion
    iii. A stem cell researcher
    iv. A person with a disease that one day may be treated with stem cells
    v. A couple who have tried to conceive for a decade

9. At the same time that 62- and 63-year-old women gave birth, actors Tony Randall and Anthony Quinn became fathers at ages 77 and 78—and didn't receive nearly as much criticism as the women. Do you think this is an unfair double standard, or a fair criticism based on valid biological information?

10. Many people spend thousands of dollars pursuing pregnancy. What is an alternative solution to their quest for parenthood?

11. An Oregon man anonymously donated sperm that were used to conceive a child. The man later claimed, and won, rights to visit his child. Is this situation for the man more analogous to a genetic and gestational surrogate mother, or an oocyte donor who wishes to see the child she helped to bring into existence?

12. Big Tom is a bull with valuable genetic traits. His sperm are used to conceive 1,000 calves. Mist, a dairy cow with exceptional milk output, has many oocytes removed, fertilized *in vitro,* and implanted into surrogate mothers. With their help, Mist becomes the genetic mother of 100 calves—far more than she could give birth to naturally. Which two reproductive technologies performed on humans are based on these two agricultural examples?

13. State who the genetic parents are and who the gestational mother is in each of the following cases:

    a. A man was exposed to unknown burning chemicals and received several vaccines during the first Gulf war, and abused drugs for several years before and after that. Now he wants to become a father, but he is concerned that past exposures to toxins have damaged his sperm. His wife undergoes intrauterine insemination with sperm from the husband's brother, who has led a calmer and healthier life.

    b. A 26-year-old woman has her uterus removed because of cancer. However, her ovaries are intact and her oocytes are healthy. She has oocytes removed and fertilized *in vitro* with her husband's sperm. Two resulting embryos are implanted into the uterus of the woman's best friend.

    c. Max and Tina had a child by IVF and froze three extra embryos. Two are thawed years later and implanted into the uterus of Tina's sister, Karen. Karen's uterus is healthy, but she has ovarian cysts that often prevent her from ovulating.

    d. Forty-year-old Christopher wanted children, but not a partner. He donated sperm, which were used for intrauterine insemination of an Indiana mother of one. The woman carried the resulting fetus to term for a fee, and gave birth to a daughter, Kelsey.

    e. Two men want to raise a child together. They go to a fertility clinic, have their sperm collected, mixed, and used to inseminate a friend. Nine months later she turns the baby over to them.

14. Delaying childbirth until after age 35 is associated with certain physical risks, yet an older woman is often more mature and financially secure. Many women delay childbirth so that they can establish careers. Suggest societal changes, perhaps using a reproductive technology, that would allow women to more easily have children and careers.

15. An IVF attempt yields twelve more embryos than the couple who conceived them can use. What could they do with the extras?

16. What do you think children born of an assisted reproductive technology should be told about their origins?

17. An IVF program in India offers preimplantation genetic diagnosis to help couples who already have a daughter to conceive a son. The reasoning is that because having a male heir is of such great importance in this society, offering PGD can enable couples to avoid aborting second and subsequent female pregnancies. Do you agree or disagree that PGD should be used for sex selection in this sociological context?

18. ICSI can help men who have small deletions in their Y chromosomes that stop sperm from maturing to become parents, but this passes on their infertility. Suggest an ART that they could use to prevent male infertility.

19. Novelist Jodi Picoult wrote about a savior sibling in *My Sister's Keeper.* She referred to preimplantation genetic diagnosis as "genetic engineering." Is this correct? Why or why not?

Web Activities

1. Invent a situation for a couple trying to have a baby with IVF and use http://www.ivfpredict.com to estimate their chances of success.

2. The U.S. government bans use of federal funds to create human embryos for research purposes, but does not regulate the human reproductive technology field at all. Consult websites to learn about regulations of stem cell research and ARTs in other nations.

3. Go to the Centers for Disease Control and Prevention website. Click on ART Trends, and use the information to answer the following questions.

    a. Since 1996, to what extent has the use of ARTs in the United States increased?

    b. Which is more successfully implanted into an infertile woman's uterus, a fresh or frozen donor oocyte?

    c. Which is more successfully implanted into an infertile woman's uterus, a donated oocyte or one of her own?

    d. What are two factors that could complicate data collection on ART success rates?

4. A company called Extend Fertility provides oocyte freezing services, telling women to "set your own biological clock." The home page states, "Today's women lead rich and busy lives—obtaining advanced degrees, pursuing successful careers, and taking better care of ourselves. As a result of this progress, many of us choose to have children later than our mothers did."

    Look at this or another egg-freezing website. Discuss how it might be viewed by the following individuals:

    a. A 73-year-old father of a healthy baby

    b. A 26-year-old woman, married with no children but who wants them, facing 6 months of chemotherapy

    c. An orphaned 10-year-old in Thailand

    d. A healthy 28-year-old woman in the United States who wants to earn degrees in medicine, law, and business before becoming pregnant

    e. A young mother in Mexico who is giving her son up for adoption because she cannot afford to raise him

5. Read the posts on surrogatemother.com and describe a match between a potential surrogate and a couple or individual wishing to use her services.

Case Studies and Research Results

1. Doola is 32 years old and is trying to decide if she and her husband are ready for parenthood when she learns that her 48-year-old mother has Alzheimer disease. The mother's physician tells Doola that because of the early onset, the Alzheimer disease could be inherited through a susceptibility gene. Doola is tested and indeed has the same dominant allele. She wants to have a child right away, so that she can enjoy many years as a mother. Her husband David feels that it wouldn't be fair to have a child knowing that Alzheimer disease likely lies in Doola's future.

   a. Who do you agree with, and why?

   b. David is also concerned that Doola could pass on her Alzheimer gene variant to a child. Which technology might help them avoid this?

   c. Is Doola correct in assuming that she is destined to develop Alzheimer disease?

2. Natallie Evans had to have her ovaries removed at a young age because they were precancerous, so she and her partner had IVF and froze their embryos for use at a later time. Under British law, both partners must consent for the continued storage of frozen embryos. Evans and her partner split, and he revoked his consent. She sued for the right to use the embryos. She told the court, "I am pleased to have the opportunity to ask the court to save my embryos and let me use them to have the child I so desperately want."

   What information should the court consider in deciding this case? Whose rights do you think should be paramount?

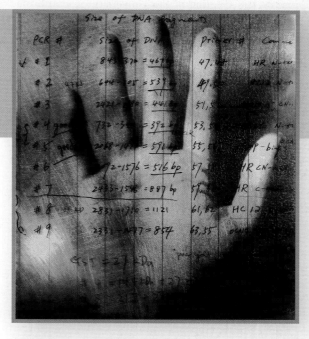

What can we learn from our genome sequences?

Genomics

Learning Outcomes

22.1 From Genetics to Genomics

1. Explain how linkage studies led to the idea to sequence the human genome.

2. Distinguish between the two approaches used to sequence the human genome.

22.2 DNA Sequencing and Genome Synthesis

3. Describe the Sanger method of DNA sequencing.

4. Describe how newer methods of DNA sequencing work.

5. Define *synthetic genome*.

22.3 Ways of Looking at Genomes

6. List questions that comparative genomics can address.

7. Discuss what the analysis of a representative 1 percent of the human genome revealed.

8. Define *human microbiome*.

22.4 Personal Genome Sequencing

9. What types of information can personal genome sequencing provide?

10. What are limitations of the utility of personal genome sequencing?

The Big Picture: Just over a decade ago we saw the first human genome sequences. Today the cost has plummeted to the point that personal genome sequencing is possible. The question is now not could we, but should we?

100,000 Genomes and Counting

Genome researchers like numbers. They spent the 1990s anticipating the sequencing of the 3.2 billion bits of information that comprise a human genome. Even before that was done, the focus had already turned to identifying the points where human genomes vary, which led to the 3.5 million variant bases of the HapMap project. Meanwhile, as technology improved, the feasibility of routinely sequencing human genomes grew.

The first few people to have their genomes sequenced were celebrities and/or millionaires. Craig Venter, first to the finish line in sequencing the human genome, was also first to have his genome sequence published, in late 2007. Six months later came the genome sequence of James Watson. Then ten prominent people joined the Personal Genome Project ("PGP-10") led by Harvard University geneticist George Church, and soon thousands had signed up. The goal: developing a new approach to preventing and treating disease.

The "1,000 Genomes Project" assessed 2,500 genomes from 27 populations to identify 99 percent of human genetic diversity. It tracked genetic variants that affect more than 1 percent of any population, including 15 million SNPs, a million copy number variants, and 20,000 chromosome alterations. The project has enabled researchers to distinguish genetic variants that affect health from those that do not.

22.1 From Genetics to Genomics

Genetics is a young science, genomics younger still. As one field has evolved into another, milestones have come at oddly regular intervals. A century after Gregor Mendel announced and published his findings, the genetic code was deciphered; a century after his laws were rediscovered, the human genome was sequenced.

Geneticist H. Winkler coined the term *genome* in 1920. A hybrid of "gene" and "chromosome," genome then denoted a complete set of chromosomes and genes. The modern definition refers to all the DNA in a haploid set of chromosomes. The term *genomics,* credited to T. H. Roderick in 1986, indicates the study of genomes. Thoughts of sequencing genomes echoed through much of the twentieth century, as researchers described the units of inheritance from different perspectives.

Beginnings in Linkage Studies

Sequencing the human genome unofficially began in the 1980s with deciphering signposts and developing shortcuts. Many of the initial steps and tools grew from existing technology. Linkage maps and studies of rare families that associated chromosomal aberrations with syndromes enabled researchers to assign some genes to their chromosomes. Then automated DNA sequencing took genetic analysis to a new level—information.

The evolution of increasingly detailed genetic maps is similar to zooming in on a geographical satellite map (**figure 22.1**). A cytogenetic map is like a map of California within a map of the United States, highlighting only the largest cities. A linkage map is like a map that depicts the smaller cities and large towns, and a physical map is similar to a geographical map indicating all towns. Finally, a sequence map is the equivalent of a Google map showing specific buildings.

Before the human genome was sequenced, researchers matched single genes to specific diseases using an approach called positional cloning. The technique began with examining a particular phenotype corresponding to a Mendelian disorder in large families, then identifying known parts of the genome that only affected relatives shared. Such sequences served as markers of the presumably tightly linked disease-causing gene. Throughout the 1980s and 1990s, positional cloning experiments discovered the genes behind Duchenne muscular dystrophy, cystic fibrosis, Huntington disease, and many others. So slow was the process that it took a decade to go from discovery of a genetic marker for Huntington disease to discovery of the gene.

Sequencing the Human Genome

The idea to sequence the human genome occurred to many researchers at about the same time, but with different goals (see the Technology Timeline on page 425). It was first brought up at a meeting held by the Department of Energy (DOE) in 1984 to discuss the long-term population genetic effects of exposure to low-level radiation. In 1985, researchers meeting at the University of California, Santa Cruz, called for an institute to sequence the human genome, because sequencing of viral genomes had shown that it could be done. The next year, virologist Renato Dulbecco proposed that the key to understanding the origin of cancer lay in knowing the human genome sequence. Later that year, scientists packed a room at the Cold Spring Harbor Laboratory on New York's Long Island to discuss the feasibility of a project to sequence the human genome. At first those against the project outnumbered those for it 5 to 1. The major fear was the shifting of goals of life science research from inquiry-based experimentation to amassing huge amounts of data.

A furious debate ensued. Detractors claimed that the project would be more gruntwork than a creative intellectual endeavor, comparing it to conquering Mt. Everest just because it is there. Practical benefits would be far in the future. Some researchers feared that such a "big science" project would divert government funds from basic research and AIDS. Finally, the

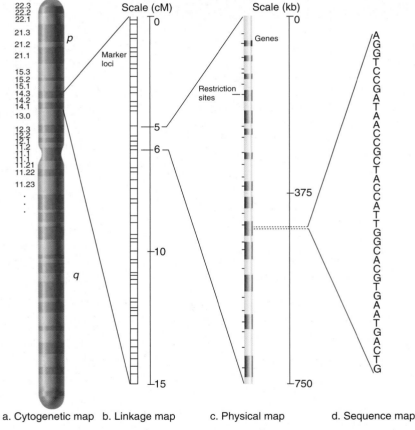

a. Cytogenetic map   b. Linkage map   c. Physical map   d. Sequence map

Figure 22.1 **Different levels of genetic maps are like zooming up the magnification on a geographical map.** **(a)** A cytogenetic map, based on associations between chromosome aberrations and syndromes, can distinguish DNA sequences that are at least 5,000 kilobases (kb) apart. **(b)** A linkage map derived from recombination data distinguishes genes hundreds of kb apart. **(c)** A physical map constructed from overlapped DNA pieces distinguishes genes tens of kb apart. **(d)** A sequence map depicts the order of nucleotide bases.

Technology Timeline

EVOLUTION OF GENOME PROJECTS

1985–1988   Idea to sequence human genome suggested at several scientific meetings.

1988   Congress authorizes the Department of Energy and the National Institutes of Health to fund the human genome project.

1989   Researchers at Stanford and Duke Universities invent DNA microarrays.

1990   Human genome project officially begins.

1991   Expressed sequence tag (EST) technology identifies protein-encoding DNA sequences.

1992   First DNA microarrays available.

1993   Need to automate DNA sequencing recognized.

1994   U.S. and French researchers publish preliminary map of 6,000 genetic markers, one every 1 million bases along the chromosomes.

1995   Emphasis shifts from gene mapping to sequencing. First genome sequenced: *Hemophilus influenzae*.

1996   Resolution to make all data public and updated daily at GenBank website. First eukaryote genome sequenced—yeast.

1998   Public Consortium releases preliminary map of pieces covering 98 percent of human genome. Millions of sequences in GenBank. Directions for DNA microarrays posted on Internet. First multicellular organism's genome sequenced: roundworm.

1999   Rate of filing of new sequences in GenBank triples. Public Consortium and two private companies race to complete sequencing. First human chromosome sequenced (22).

2000   Completion of first draft human genome sequence announced. Microarray technology flourishes. First plant and fungal genomes sequenced.

2001   Two versions of draft human genome sequence published.

2003   Finished version of human genome sequence announced to coincide with fiftieth anniversary of discovery of DNA structure. Entire protein-encoding part of human genome available on DNA microarrays.

2004   Final version of human genome sequence published.

2005   Annotation of human genome sequence continues gradually. Number of species with sequenced genomes soars.

2007   Detailed analysis of 1 percent of the human genome reveals that most of it is transcribed. First individual human genome sequenced. Human Microbiome Project begins.

2008   First genome synthesized: *Mycoplasma genitalium*.

2010   Several human genomes sequenced. First synthetic genome supports a bacterial cell. By now, 4,000 bacterial and viral and 250 eukaryotic species' genomes have been sequenced.

2012   1,000 Genomes Project completes cataloging of human genetic diversity.

?   Personal human genome sequencing is routine.

National Academy of Sciences convened a committee representing both sides to debate the feasibility, risks, and benefits of the project. The naysayers were swayed to the other side. In 1988, Congress authorized the National Institutes of Health (NIH) and the DOE to fund the $3 billion, 15-year human genome project, which began in 1990 with James Watson at the helm. The project set aside 3 percent of its budget for the Ethical, Legal and Social Implications (ELSI) Research Program. It has helped ensure that genetic information is not used to discriminate. Eventually, an international consortium as well as a private company, Celera Genomics, sequenced the human genome. They worked separately, but the efforts are referred to here as "the human genome project."

A series of technological improvements sped the genome project. In 1991, a shortcut called expressed sequence tag (EST) technology enabled researchers to quickly pick out genes most likely to be implicated in disease. This was a foreshadowing for

Random fragments:	AGTCCT CTAG AGCTA
	CTACT TAGAGT CCTAGC
Alignment:	CTAG
	TAGAGT
	AGTCCT
	CCTAGC
	AGCTA
	CTACT
Sequence:	CTAGAGTCCTAGCTACT

Figure 22.2 Deriving a DNA sequence. Automated DNA sequencers first determine the sequences of short pieces of DNA, or sometimes of just the ends of short pieces. Then algorithms search for overlaps. By overlapping the pieces, the software derives the overall DNA sequence.

future efforts to focus on the **exome,** which is the part of the genome that encodes protein. ESTs are cDNAs (see figure 19.5) made from the mRNAs in a cell type that is abnormal in a particular illness. Also in 1991, researchers began using DNA microarrays to display short DNA molecules. This technology became important in DNA sequencing (tiling arrays) as well as in assessing gene expression (expression arrays).

Computer algorithms eased the assembly of many short pieces of DNA with overlapping end sequences into longer sequences (**figure 22.2**). (The next section discusses the actual DNA sequencing.) When the project began, researchers cut several genomes' worth of DNA into overlapping pieces of about 40,000 bases (40 kilobases), then randomly cut the pieces into small fragments. The greater the number of overlaps, the more complete the final assembled sequence. The sites of overlap had to be unique sequences, found in only one place in the genome. Overlaps of repeated sequences found in several places in the genome could lead to more than one derived overall sequence—a little like searching a document for the word "that" versus searching for an unusual word, such as "dandelion." Searching for "dandelion" is more likely to lead to a specific part of a document, whereas "that" may occur in several places—just like repeats in a genome.

The use of unique sequences is why the human genome project did not uncover copy number variants. For example, the sequence CTACTACTA would appear only as CTA. Researchers did not at first appreciate the fact that repeats are a different form of information and source of variation than DNA base sequences. A balance was necessary between using DNA pieces large enough to be unique, but not so large that the sequencing would take a very long time.

Two general approaches were used to build the long DNA sequences to initially derive the sequence of the human genome (**figure 22.3**). The "clone-by-clone" technique the U.S. government-funded group used aligned DNA pieces one chromosome at a time. The "whole-genome shotgun" approach Celera Genomics used shattered the entire genome, then used an algorithm to identify and align overlaps in a continuous sequence. The task can be compared to cutting the binding off a large book, throwing it into the air, and reassembling the dispersed pages in order. A "clone-by-clone" dismantling of the book would divide it into bound chapters. The whole-genome shotgun approach would free every page. Whole-genome shotgunning is faster, but it misses some sections (particularly repeats) that the clone-by-clone method detects.

Technical advances continued. In 1995, DNA sequencing became automated, and software was developed that could rapidly locate the unique sequence overlaps among many small pieces of DNA and assemble them, eliminating the need to gather large guidepost pieces. In 1999, the race to sequence the human genome became intensely competitive. The battling factions finally called a truce. On June 26, 2000, Craig Venter from Celera Genomics and Francis Collins, representing the International Consortium (and now director of the NIH), flanked President Clinton in the White House rose garden to unveil the "first draft" of the human genome sequence. The milestone capped a decade-long project involving thousands of researchers, culminating a century of discovery. The historic June 26 date came about because it was the only opening on the White House calendar! In other words, the work was monumental; its announcement, staged. **Figure 22.4** is an overview of genome sequencing.

Looking back a decade, was sequencing the human genome valuable, or was it indeed like climbing Mt. Everest just because it is there? Availability of the sequence has accelerated gene discovery beyond the slow disease-by-disease approach of the pre-genome era, but gene expression profiling (see chapter 11) and genome-wide association studies (see chapter 7 and others) have also impacted health care and revealed human diversity. In addition, the early focus on determining "the" number of human genes seems misplaced in hindsight. Just as learning the words of a language is not the same as reading stories in that language, it is the interactions among genes that are important, not their number. Even if we do know a genome sequence, epigenetic changes that reflect response to the environment in gene expression may be of more immediate utility, in terms of relevant health information.

Knowing the human genome sequence changed our view of our genetic instruction manual. The human genome has

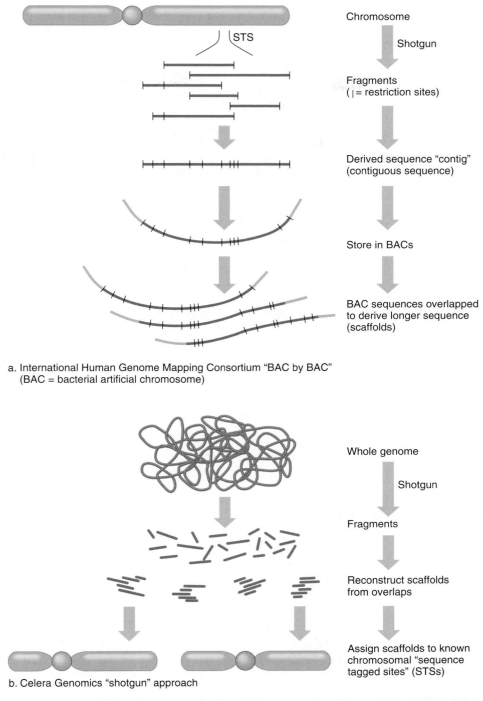

Chromosome

Shotgun

Fragments
(| = restriction sites)

Derived sequence "contig"
(contiguous sequence)

Store in BACs

BAC sequences overlapped
to derive longer sequence
(scaffolds)

a. International Human Genome Mapping Consortium "BAC by BAC"
   (BAC = bacterial artificial chromosome)

Whole genome

Shotgun

Fragments

Reconstruct scaffolds
from overlaps

Assign scaffolds to known
chromosomal "sequence
tagged sites" (STSs)

b. Celera Genomics "shotgun" approach

Figure 22.3 **Two routes to sequencing the human genome. (a)** The International Consortium began with known chromosomal sites and overlapped large pieces, called contigs, that were reconstructed from many small, overlapping pieces. "STS" stands for "sequence tagged site," which refers to specific known parts of chromosomes. A BAC is a cloning vector that uses bacterial DNA. **(b)** Celera Genomics shotgunned several copies of a genome into small pieces, overlapped them to form scaffolds, and then assigned scaffolds to known chromosomal sites. They used some Consortium data.

one-fifth the total number of predicted protein-encoding genes. The overestimate largely reflects the "genes in pieces" architecture in which one exon can be part of more than one gene. The exome was once the focus of human genetics; today we know it is a minuscule part of the total information.

Now that human genomes are being sequenced regularly, researchers are constructing several new types of maps. These depictions display patterns of gene expression, distributions of SNPs, methylation sites, protein folding, and networks of gene interactions. The "diseasome" of figure 1.8 shows one of

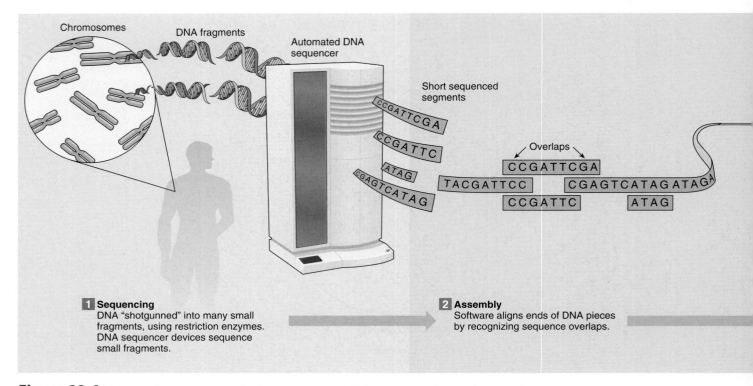

Figure 22.4 Sequencing genomes. The first-generation, whole-genome "shotgun" approach to genome sequencing overlapped DNA pieces cut from several copies of a genome, then assembled the overall sequence. Second-and third-generation techniques use microfluidics and nanomaterials to sequence DNA.

these new ways of looking at genetic information. All of these approaches suggest novel ways of treating disease.

Key Concepts

1. Human genome sequencing was built on linkage and cytogenetic information from decades of work.

2. Positional cloning located specific disease-causing genes in families.

3. The idea to sequence the human genome emerged in the mid-1980s with several goals. The project officially began in 1990.

4. Genome sequencing cuts several copies of a genome, sequences the pieces, then uses algorithms to overlap the pieces.

5. Clone-by-clone sequencing assembled chromosomes individually. Whole-genome shotgunning shattered the entire genome and rebuilt it.

22.2 DNA Sequencing and Genome Synthesis

Several new ways to sequence DNA are available today. One of the original methods for sequencing DNA, invented in 1977 by Frederick Sanger, is still used. The Sanger method generates a series of DNA fragments of identical sequence that are complementary to the DNA sequence of interest. These fragments differ in length from each other by one end base, as follows:

Sequence of interest: T A C G C A G T A C
Complementary sequence:
Series of fragments:

A T G C G T C A T G
T G C G T C A T G
G C G T C A T G
C G T C A T G
G T C A T G
T C A T G
C A T G
A T G
T G
G

The Sanger technique, also called chain termination, is used with the polymerase chain reaction (see figure 19.2) to copy a sequence of DNA multiple times, incorporating chemically altered bases corresponding to and substituting for one of the four types of bases. The result is a collection of partial sequences from which the end bases reveal the sequence. **Figure 22.5** shows how DNA sequence data derived from the Sanger method appear in scientific papers, and **figure 22.6** shows how to read the sequence from the end bases.

Newer approaches to DNA sequencing use a microfluidics environment, which is a small, fluid-filled chamber. One

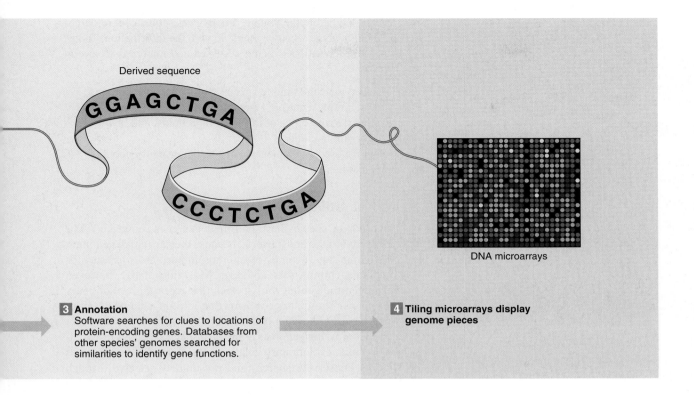

Derived sequence

GGAGCTGA
CCCTCTGA

DNA microarrays

3 Annotation
Software searches for clues to locations of protein-encoding genes. Databases from other species' genomes searched for similarities to identify gene functions.

4 Tiling microarrays display genome pieces

CTNGCTTTGGAGAAAGGCTCCATTGNCAATCAAGACACACA
CTatGCTTTGGAGAAAGGCTCCATTGgCAATCAAGACACACA

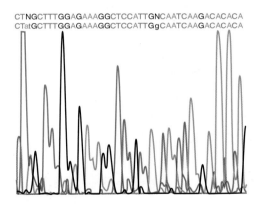

Figure 22.5 **DNA sequence data.** In automated first-generation DNA sequencing, a readout of sequenced DNA is a series of wavelengths that represent the terminal DNA base labeled with a fluorescent molecule.

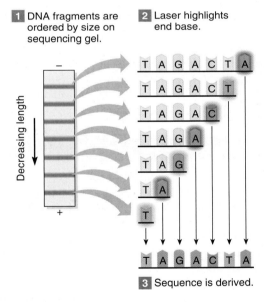

1 DNA fragments are ordered by size on sequencing gel.

2 Laser highlights end base.

Decreasing length

T A G A C T A
T A G A C T
T A G A C
T A G A
T A G
T A
T

T A G A C T A

3 Sequence is derived.

Figure 22.6 **Reading a DNA sequence.** A computer algorithm detects and records the end base from a series of size-ordered DNA fragments.

technique sequences many 100-base-long single-stranded DNA pieces that are attached to tiny beads in a water-oil mixture. A laser reads off fluorescently tagged bases that are added, according to the complementary base pair rules, as individual nucleotides stream past the strands. The method, called 454 sequencing, was invented by a father seeking a way to quickly sequence his newborn son's genome. The technique can sequence 20 million bases in about 4.5 hours.

Even newer approaches use nanomaterials to distinguish how each of the four nucleotide bases disrupts an electrical field as a DNA molecule passes through nanopores at about 1,000 bases per second. One such material is a one-atom-thick sheet of carbon called graphene that is strong, very thin, and conducts electricity. DNA is threaded through nanopores in it. Each of the four DNA bases disrupts the electrical field in a slightly different, but detectable way (**figure 22.7**). The changes are used to deduce the base sequence of the DNA molecule.

If it is possible to *sequence* a genome, it should also be possible to *synthesize* a genome. Researchers at the J. Craig Venter Institute created a "synthetic genome" for the 582,970 bases of *Mycoplasma genitalium,* a tiny bacterium. The

Figure 22.7 Graphene is a nanomaterial used to sequence DNA. A single DNA molecule passes through an aperture, changing an electrical field differently for each of the four types of bases.

Source: © Institute of Physics (the "Institute") and IOP Publishing.

researchers worked with 5,000- to 7,000-base-long "cassettes" of the known sequence, synthesizing sets of pieces 100 bases long that overlapped at the ends. They then delivered the pieces into other cells (*E. coli* or yeast), using vectors called bacterial artificial chromosomes (BACs). Natural recombination joined the pieces, building first a quarter of the genome, then half, then three-quarters, and finally the entire genome.

Key Concepts

1. In the Sanger method of DNA sequencing, complementary copies of an unknown DNA sequence are cut into different-sized pieces differing from each other by an end base. The pieces are overlapped by size and the labeled end bases read off.
2. Newer techniques use short DNA sequences and nanomaterials.
3. Researchers have built the first genome.

22.3 Ways of Looking at Genomes

If Wikipedia were transcribed into old-fashioned printed volumes, the information would fill many books. So it is with a human genome. Just as you wouldn't read all of Wikipedia to learn about toenail fungus or opera, not all of the 3.2 billion bases of a human genome hold information that is useful to a particular person in a particular circumstance.

Long before the human genome was sequenced, researchers thought about what we might do with the information. Today, genome information is being accessed and analyzed at several levels:

- between and among species, for views on evolution;
- in a representative 1 percent to discover how a genome works, the types of information it holds, and how it is organized; and
- in subsets of sequence, such as the protein-encoding exomes.

Comparative Genomics

Hundreds of species have had their genomes sequenced. The first were viruses and bacteria, because they are small genomes. Next came the genomes of animals important to us, such as mice, rats, chimps, cats, and dogs. Most informative, however, have been the genomes of species that represent evolutionary crossroads. These are organisms that introduced a new trait or were the last to have an old one.

Comparing genomes of modern species enables researchers to infer evolutionary relationships from DNA sequences that are conserved (shared) and presumably selected through time. **Figure 22.8** shows one way of displaying short sequence similarities, called a pictogram. DNA sequences from different species are aligned, and the bases at different points indicated. A large letter A, C, T, or G indicates, for example, that all species examined have the same base at that site. A polymorphic site, in contrast, has different bases for different species.

Comparative genomics uses conserved sequences to identify biologically important genome regions, assuming that persistence means evolutionary success. But there are exceptions—conserved sequences with no apparent function in humans. Either we haven't discovered the functions, or

Human
Mouse
Rat
Chicken

a. Not highly conserved

b. Highly conserved

Figure 22.8 A pictogram indicates conservation of the DNA sequence. These pictograms are for short sequences in corresponding regions of the human, mouse, rat, and chicken genomes. A large letter means that all four species have the same base at that site. If four letters appear in one column, then the species differ. Pictogram **(a)** is not highly conserved; **(b)** is.

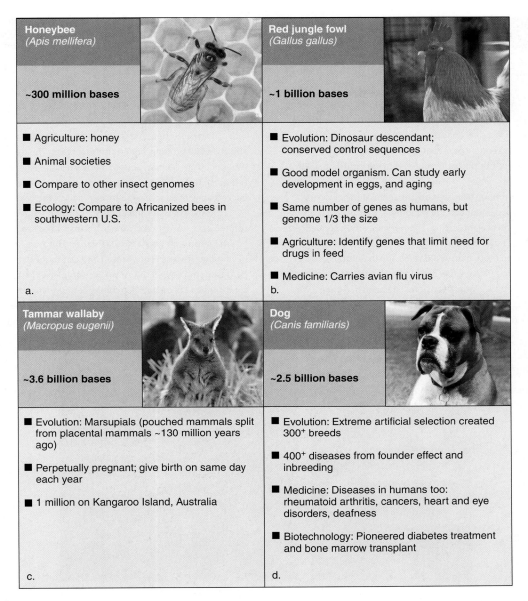

Figure 22.9 A sampling of animal genomes.

genomes include "raw material" for future functions. About 6 percent of the human genome sequence is highly conserved. **Figure 22.9** presents a few organisms whose genomes have been sequenced. Following are examples of the types of information inferred from conserved DNA sequences. **Reading 22.1** is a closer look at another important genome—the one that gives us chocolate.

The Minimum Gene Set Required for Life

The smallest microorganism known to be able to reproduce is *Mycoplasma genitalium*. It infects cabbage, citrus fruit, corn, broccoli, honeybees, and spiders, and causes respiratory illness in chickens, pigs, cows, and humans. Researchers call its tiny genome the "near-minimal set of genes for independent life."

Of 480 protein-encoding genes, 265 to 350 are essential. Considering how *Mycoplasma* uses its genes reveals the fundamental challenges of being alive. This was the first organism to have its genome synthesized.

Fundamental Distinctions Among the Three Domains of Life

Methanococcus jannaschii is a microorganism that lives at the bottom of 2,600-meter-tall "white smoker" chimneys in the Pacific Ocean, at high temperature and pressure and without oxygen. As archaea, these cells lack nuclei, yet replicate DNA and synthesize proteins in ways similar to multicellular organisms. The genome sequence confirmed that this organism represents a third form of life.

Sequencing the *Cacao* Genome

The race was on. The two teams of scientists projected that they would sequence the entire genome in 5 years. The competition culminated faster than expected in a dual announcement of success and a flurry of media coverage. Two sequences meant more information. The genome of interest wasn't ours, but that of *Theobroma cacao*—the source of chocolate.

The candy company Mars was on one team, Hershey on the other. The Mars group sequenced the genome of the cacao variety used in most chocolate, called Matina. The Hershey group sequenced an elite type called Criollo that descends from the strain that the Mayans domesticated more than 3,000 years ago in Central America. Today nearly 7 million farmers grow cacao, mostly in West Africa but also in Asia, South America, and elsewhere. The bean in **figure 1** was grown in Cuba. The trees grow in the shade, and the beans, which are the source of chocolate, produce various organic compounds—oils,

terpenes, and flavonoids—that contribute to the taste and texture of chocolate.

The cacao genome is a mere 420 million bases that encode about 36,000 proteins on ten chromosomes. The two varieties sequenced are homozygous at many sites in the genome, which eased sequencing efforts. The researchers used the 454 sequencing method, backed up with Sanger sequencing to fill in the gaps.

Researchers are analyzing the genome sequences for genes that control yield, resistance to fungal and insect pests, flavonoid synthesis for health reasons (they are antioxidants), and of course taste and texture. Having the genome sequence at hand can suggest new ways to increase yield while improving the crop. For example, 84 genes control and contribute to cocoa butter, a component used to make chocolate, that contributes much of its flavor. For those who enjoy chocolate-covered strawberries, the wild strawberry genome has been sequenced, too.

a.   b.

Figure 1 Researchers have sequenced the genome of *Theobroma cacao*. The fruits **(a)** provide one of life's pleasures—chocolate **(b)**.

The Simplest Organism with a Nucleus

The yeast *Saccharomyces cerevisiae* is single-celled with only about 6,000 genes, but a third of them have counterparts among mammals, including at least seventy genes implicated in human diseases. Understanding what a gene does in yeast can provide clues to how it affects human health, such as counterparts of mutations in cell cycle control genes in yeast causing cancer in humans.

The Basic Blueprints of an Animal

The genome of the tiny, transparent, 959-celled nematode worm *Caenorhabditis elegans* is packed with information on what it takes to be an animal. The worm's signal transduction

pathways, cytoskeleton, immune system, apoptotic pathways, and brain proteins are very similar to our own. Curiously, the fruit fly (*Drosophila melanogaster*) genome has 13,601 genes, fewer than the 18,425 in the much simpler worm. Of 289 disease-causing genes in humans, 177 have counterparts in *Drosophila*. The fly is a model for testing new treatments.

Life on Land

Before 450 million years ago, life was confined to the seas, where it was abundant and diverse. Algae, microorganisms, and jawless fishes shared the depths. The first organisms to colonize land were the mosses, and for this reason, the genome of the modern moss *Physcomitrella paten* was sequenced.

Mosses lack stems and leaves and have only a few cell types. They dominated landscapes until plants that had seeds and vessels evolved.

By the time animals ventured onto land, plants had already taken root. Sea residents whose descendants were probably among those first land dwellers were the lobe-finned fishes, which have fleshy, strong fins that could have evolved into limbs. Two types of lobe-finned fishes persist today and resemble their fossilized forms—the lungfishes and two species of coelacanths. Because the lungfish genome is huge, researchers analyzed the smaller coelacanth genome. Once thought to be extinct, coelacanths today live in the Indian Ocean. Information in the coelacanth genome may reveal the traits necessary for the evolution of the tetrapods—vertebrates with four limbs.

From Birds to Mammals

The sequencing of the chicken genome (figure 22.9*b*) marked a number of milestones. It was the first agricultural animal, the first bird, and, as such, the first direct descendant of dinosaurs. The genome of the red jungle fowl *Gallus gallus* is remarkably like our own, minus many repeats, but its genome organization is intriguing. Like other birds, fishes, and reptiles, but not mammals, the chicken genome is distributed among very large macrochromosomes and tiny microchromosomes. Repeats may have been responsible for the larger sizes of mammalian chromosomes.

From Chimps to Humans

Most comparisons of the human genome to those of other species seek similarities. Comparisons of our genome to that of the chimpanzee, however, seek genetic *differences*, which may reflect what makes humans unique (see Reading 16.1). The human and chimp genomes differ by 1.2 percent, equaling about 40 million DNA base substitutions. Within those differences, as well as copy number distinctions, may lie the answers to compelling medical questions. Why do only humans get malaria and Alzheimer disease? Why is HIV infection deadlier than the chimp version, SIV?

Probing Parts of the Human Exome

One way of examining the human genome is to randomly sample a small portion of it, and catalog what the DNA sequences do, like reading parts of a book. The ENCODE (Encyclopedia of DNA Elements) project did this for a representative 1 percent of the human genome. The project wasn't just sequencing, but also cataloged gene expression patterns in 92 cell types. ENCODE considered DNA sequences found among the chromosomes, from known, well-studied genes as well as uncharted territory. The result: a "parts list" of 30 million of our 3 billion DNA bases. Overall the project revealed that much more of the genome controls the protein-encoding genes than had been suspected, and that most of the genome is transcribed. Such control sequences are found both next to the genes they control as well as elsewhere in the genome.

The 1 percent of the genome analyzed in ENCODE is one way of looking at the genome. Another is to focus on the exome, which consists of DNA sequences that encode the parts of genes that are transcribed into RNA and then translated into protein.

The exome accounts for about 1.5 percent of the human genome and includes about 180,000 exons. Section 4.5 and the opener to chapter 10 illustrate applications of whole exome sequencing, which now costs less than $1,000 using a single DNA microarray chip. Sequencing the exome reveals new mutations when a child has a gene variant not in the genome of either parent. Analysis of human exomes so far confirms and extends the observation that very little of our genome encodes protein. The exome is not very variable, and most gene variants do not affect the phenotype because they are recessive or are harmless SNPs.

Many researchers study subsets of exomes—that is, classes of genes whose encoded proteins are functionally related. The lipidome, for example, includes genes whose protein products participate in the synthesis or breakdown of lipids (fats). The secretome includes proteins that are secreted from the cell. "Omes" already mentioned include the diseasome, metagenome, and microbiome (see chapter 1), the transcriptome (see chapter 11), and the pharmacogenome (see chapter 20). "Omics" are becoming more and more specific. For example, the "kinome" includes all of the genes that encode a type of enzyme called a kinase.

Yet another way to organize information in and about the human genome is to compile catalogs of the mutations that cause specific disorders in "locus-specific databases." A global effort called the Human Variome Project is bringing together the hundreds of existing databases so that researchers can be aware of each other's work. Many of the databases represent the efforts of disease-specific organizations run by parents of children who have inherited disorders.

It will be interesting to see how genome-level views and locus-specific databases come together. The many genome-wide association studies (GWAS) that take sweeping looks at variation will help to bridge the gap. Still another approach to genomics is to consider DNA within us that is not *of* us.

The Human Microbiome

The human genome isn't the only one in our bodies. About 90 percent of the genomes in us are from the microorganisms that live on our skin, in the twists of our intestines, inside our noses and along our respiratory passages, and in more private areas. More bacteria live on the palm of the hand than there are people on the planet! We are unaware of most of these microbial inhabitants, but others help us to digest fiber, metabolize certain drugs, and contribute to body odor. For example, bacteria determine whether or not acetaminophen is effective, and how well we extract energy from carbohydrates, putting on weight.

All these others in and on our bodies form the human **microbiome**. The first study of the microbiome, described in chapter 1, catalogued the contents of filled diapers for the first year of life for 14 infants. The Human Microbiome Project is now surveying five body parts: the skin, mouth, nasal passages, digestive system, and urogenital tract (**figure 22.10**).

Analysis of the skin microbiome illustrates how useful this information can be. The skin microbiome is complex, differs markedly among individuals, yet stays fairly consistent

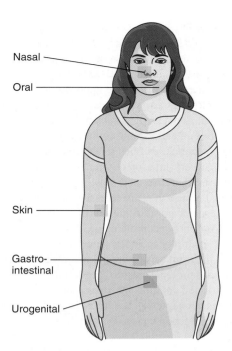

Nasal

Oral

Skin

Gastro-
intestinal

Urogenital

Figure 22.10 **The Human Microbiome Project is sequencing DNA from viruses and single-celled organisms from five body parts.**

in individuals over time. Wash an area of skin, and the types of bacteria that repopulate it will be very much like those that were scrubbed away. Because a person's "skin bacterial communities" are transferred when touching an object, microbiome information provides a new type of forensic tool. Even monozygotic (identical) twins have distinctive skin microbiomes, as does each finger in an individual. A tap of a computer key, brush of clothing, or grab of a weapon can leave a microbiome fingerprint that lasts 2 weeks.

Researchers are sampling skin microbiomes from other body parts too. It is not surprising that a perpetually damp armpit houses different types of bacteria than a scaly, dry leg. One study compared those sites and skin from the back of the head, inside the nostril, the bottom of a foot, and the area between the buttocks. Another study compared microbiomes for circumcised and uncircumcised penises. The differences explain why an uncircumcised organ is more likely to spread infection to a partner.

Key Concepts

1. Comparative genomics infers evolutionary relationships from conserved DNA sequences.
2. The ENCODE project is taking an in-depth look at 1 percent of the human genome.
3. The exome is a collection of the exons in the human genome. Only a small proportion of its variation affects the phenotype. "Omics" are subsets of exomes.
4. Locus-specific databases compile mutations in single genes. Researchers study both entire genomes and single genes.

22.4 Personal Genome Sequencing

Two decades ago, when "the human genome project" was just getting under way, probably the most important word, in hindsight, was "the." Today, with increasing focus on how we differ genetically from each other, the age of personal genome sequencing is here. Genome sequencing can provide a canvas on which other types of information can be painted, to give a fuller picture of how our bodies function and malfunction. **Reading 22.2** discusses the first human genomes sequenced.

Although genome sequencing is important for investigating our ancestry and our diversity today, it will perhaps be most practically valuable in health care. For this reason, a 40-year-old Stanford University engineer and co-inventor of a third-generation DNA sequencing device laid his genetic self bare in the pages of *The Lancet*, the oldest medical journal. He thereby joined the ranks of scientists who experimented on themselves, such as Jonas Salk, who took his own polio vaccine. The question: *What could Stephen Quake's genome sequence tell him that would be useful?*

Quake's quest began where most genetic studies begin: a family history (**figure 22.11a**). The 19-year-old son of his first cousin had died of a heart condition, and other relatives had heart and blood vessel troubles. Yet Stephen Quake himself is quite healthy—he exercises, eats well, doesn't smoke, and his conventional disease risk factors tend to be the ones he cannot control, such as male sex and age. The sequencing of Stephen Quake's genome used third-generation sequencing plus Sanger sequencing, and then was scanned for 2.6 million SNPs and 752 copy number variants. The analysis focused on four types of genetic information:

- mutations in known genes;
- gene variants that affect drug response;
- gene variants associated with complex traits and diseases; and
- novel mutations.

To provide context to the genome sequencing, Dr. Quake also had standard blood tests and tests based on his family history of cardiovascular disease (an electrocardiogram to assess heart function, an echocardiogram to assess heart structure, and a battery of exercise exams such as stress tests).

What did Dr. Quake learn about himself from his genome sequence? He indeed has gene variants—some common, some novel—that raise his risk of developing heart and blood vessel disease. He is a carrier of hereditary hemochromatosis (see section 20.2), epidermolysis bullosa (a condition in which the skin layers separate), and biotinidase deficiency (see Reading 2.1). More important than confirming what he already knew from family history and identifying recessive disorders that will not affect his health was the drug information. Taking a statin drug would keep his cholesterol levels down and not cause the side effect of muscle damage. Based on his blood work a statin didn't seem necessary, but his family history and genome sequence convinced his personal physician otherwise, and so the statin drug may save his life. He also learned that should his inherited

The First Three Human Genome Sequences

The first human genomes to be sequenced were actually composites of different individuals. The first two genomes from specific individuals to be sequenced, of genome research pioneers Craig Venter and James Watson, yielded few medical surprises. Instead, they showed that we had greatly underestimated genetic variation by focusing only on the DNA sequence. The numbers of copies of short sequences—copy number variants, or CNVs—contribute significantly to genetic variation.

"Back in 2001, we thought we differed from chimps by 1.27% of our genomes. Now we know that we differ from each other by as much as 1 to 3%. If we count all the differences, we are about 5 to 6% from the chimp. In the way we put sequences in public databases, we lost the insertions and deletions," said Venter to an American Society of Human Genetics meeting after he'd had a year to think about what his personal genome sequencing had revealed. He already knew much of it from his family history and personal experience.

Venter has gene variants associated with increased risk of Alzheimer disease and cardiovascular disease. He has alleles for dry earwax, blue eyes, lactose intolerance, a preference for activities in the evening, and a tendency toward antisocial behavior, novelty seeking, and substance abuse. He metabolizes caffeine fast, which he also knew. "I can have two double lattes and wash it down with a Red Bull and not be affected by it," he said.

James Watson, according to his genome sequence, carries a dozen rare recessive disorders that would affect glycogen storage, vision, and DNA repair if homozygous, and he is at elevated risk for twenty other disorders. Science journals deemed Watson's results "of thin clinical value" and yielding "few biological insights." Watson and Venter differ in inherited drug responses, supporting the value of pharmacogenetics/genomics (discussed in chapter 20). Said Venter, "You probably wouldn't suspect this based on our appearance—we are both bald, white scientists."

The third person to have his genome sequenced was called, simply, "YH." He is Han Chinese, an East Asian population that accounts for 30 percent of modern humanity. He has no inherited diseases in his family, but his genome includes 116 gene variants that cause recessive disorders, as well as many risk alleles. He shares with Craig Venter a tendency to tobacco addiction and high-risk alleles for Alzheimer disease.

An overall comparison of the first three genome sequences of individuals provides a peek at our variation. Each man has about 1.2 million SNPs, but a unique collection. Each has only .20 to .23 percent of SNPs that are nonsynonymous, meaning that they alter an encoded amino acid, and the men share only 37 percent of these more meaningful SNPs. The math indicates, therefore, that about .07 percent of our SNPs may affect our phenotypes.

propensity for developing blood clots come true, the drug clopidogrel (Plavix) would not work, and he would need to take low doses of the blood thinner warfarin (Coumadin) (see section 20.2). Should he develop diabetes, as his genome sequence suggests, the drug metformin (Glucophage) would likely not be effective. Figure 20.10b shows one way that the medical journal displayed the medically relevant parts of Stephen Quake's personal genome sequence.

The sequencing of Dr. Quake's genome took 4 weeks and $50,000. Sequencing the first two human genomes took a decade and cost about as many dollars as there are DNA bases. By the time you read this, the cost will surely be lower than $1,000. This availability raises questions of access. Who will pay for genome sequencing? How can the 3,000 genetic counselors and 1,200 clinical geneticists in the United States handle the many thousands of people who may want their genomes sequenced and interpreted? On the other hand, will the day come when societal pressure drives people to have their genomes sequenced, much as nearly everyone now has e-mail? Could governments make genome sequencing compulsory, and could the practice introduce new ways to discriminate against people? We have come a very long way from comparing genome sequencing to conquering Mt. Everest.

The state of the science of genome sequencing raises other questions. Much of the information is difficult to interpret, if not meaningless. The Mendelian disorders from which the field of human genetics began are very rare. Most of our ills result from differing degrees of input from variants of many genes, all against the ever-changing backdrop of the environment. The meaning of particular gene variants will change as we identify more gene variants. A sequence of DNA that is associated with elevated heart disease risk in 2012 may not indicate elevated risk after a protective gene variant is discovered in 2015. How will people react to finding DNA sequences "of unknown significance"? Whatever happens, personal genome sequencing will be humbling, for none of us has a "perfect" genome.

What will the coming flood of genetic information ultimately mean? Will it tell us where we came from more than family lore and documents? Will physicians consult strings of A, C, T, and G to determine how best to treat their patients, or will signs, symptoms, family history, and a patient's observations turn out to be more valuable types of information? Only time will tell.

I hope that this book has offered you glimpses of the future and stimulated you to think about the choices that genetic technology will present. For continuing coverage of human genetics, see my blog (Genetic Linkage) at www.rickilewis.com. Let me know your thoughts!

Ricki Lewis
rickilewis54@gmail.com

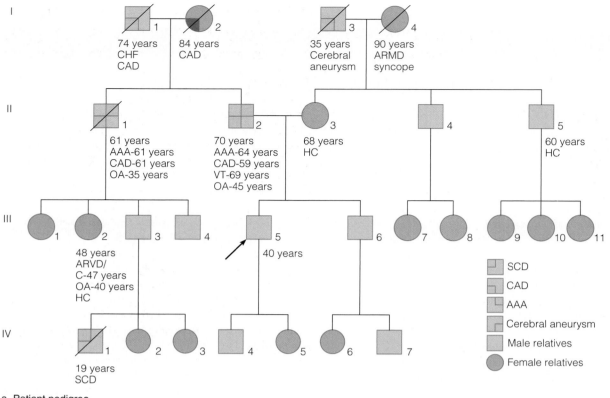

I

74 years
CHF
CAD

84 years
CAD

35 years
Cerebral
aneurysm

90 years
ARMD
syncope

II

61 years
AAA-61 years
CAD-61 years
OA-35 years

70 years
AAA-64 years
CAD-59 years
VT-69 years
OA-45 years

68 years
HC

60 years
HC

III

48 years
ARVD/
C-47 years
OA-40 years
HC

40 years

SCD

CAD

AAA

Cerebral aneurysm

Male relatives

Female relatives

IV

19 years
SCD

a. Patient pedigree

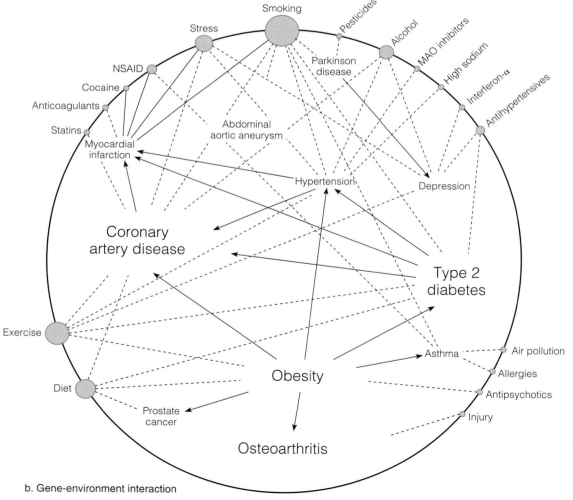

b. Gene-environment interaction

Figure 22.11 Genetic analysis old and new. (a) A pedigree is still a valuable way to see family history at-a-glance. The arrow shows the patient. Diagonal lines show deceased relatives. Years are age at death or diagnosis. A sudden death in a teenaged relative (individual IV1) compelled bioengineer Stephen Quake to sequence his genome. The abbreviations stand for diseases (AAA = abdominal aortic aneurysm, ARMD = age-related macular degeneration, ARVD/C = arrhythmogenic right ventricular dysplasia or cardiomyopathy, CAD = coronary artery disease, CHF = congestive heart failure, HC = hypercholesterolemia, OA = osteoarthritis, SCD = sudden cardiac death (presumed), VT = paroxysmal ventricular tachycardia). (b) This diagram relates environmental factors (by dotted lines) to the health problems for which an individual has elevated genetic risk. Font size reflects degree of risk.

Summary

22.1 From Genetics to Genomics

1. Genetic maps have increased in detail and resolution, from cytogenetic and linkage maps to physical and sequence maps.

2. Positional cloning discovered individual genes by beginning with a phenotype and gradually identifying a causative gene, localizing it to part of a chromosome.

3. The human genome project began in 1990 under the direction of the DOE and NIH. Technological advances sped the sequencing.

4. Several copies of a genome are cut and the pieces sequenced, overlapped, and aligned to derive the continuous sequence. For the human genome, the International Consortium used a chromosome-by-chromosome approach and Celera Genomics used whole-genome shotgunning.

5. Since the sequencing of the human genome, other types of genome information are recognized as being equally or more important.

22.2 DNA Sequencing and Genome Synthesis

6. In the Sanger chain termination method of DNA sequencing, DNA fragments differing in size and with substituted end bases are aligned, and the sequence read off from them.

7. Newer sequencing methods use microfluidics and nanomaterials.

8. Researchers synthesized the genome of a small bacterium.

22.3 Ways of Looking at Genomes

9. Identifying conserved regions among genomes of different species reveals some genes with vital functions.

10. Analysis of a representative 1 percent of the human genome showed that nearly all DNA is transcribed and that gene function is more complex than was thought.

11. The **exome** is the collection of exons in the human genome. Very little of the variation in the exome affects the phenotype. Other "omes" consider subsets of the exome.

22.4 Personal Genome Sequencing

12. An individual human genome sequence can confirm what is known from family history, detect disease-associated recessive alleles, and predict drug responses.

13. Calculating medical risks from genome information is difficult because knowing the genome sequence does not reveal degree of risk that particular DNA sequence variants confer, account for gene interactions, or consider environmental influences.

www.mhhe.com/lewisgenetics10

Answers to all end-of-chapter questions can be found at **www.mhhe.com/lewisgenetics10.** You will also find additional practice quizzes, animations, videos, and vocabulary flashcards to help you master the material in this chapter.

Review Questions

1. Distinguish between the exome and the genome.

2. How did family linkage patterns and chromosomal aberrations lay the groundwork for sequencing the human genome?

3. Describe how the four levels of genetic maps differ, and what new types of maps depict.

4. Explain how positional cloning was used to identify disease-causing genes.

5. Why was there initial disagreement over whether the human genome should be sequenced?

6. Why must several copies of a genome be cut to sequence it using the Sanger method?

7. How is graphene being used in DNA sequencing?

8. Do you think that placing a synthetic genome in a bacterium that can then divide is creating life? Give a reason for your answer.

9. From the 1960s until the human genome was sequenced, researchers thought that most of the human genome encodes protein. Then, when they discovered the true relative size of the exome, they hypothesized that most of the genome is not transcribed. What has the human genome sequence revealed about the proportion of the information that encodes protein, and about how much of the genome is transcribed?

10. Explain how conserved DNA sequences hold information about evolution.

11. List four types of information derived from comparing genomes of different species.

12. If you have your genome sequenced, what will you do with the information?

13. Why were Craig Venter and James Watson not upset at the disease-related gene variants found in their genomes?

Applied Questions

1. When the idea of sequencing the human genome was first discussed, some researchers thought that amassing large amounts of information was not creative. Discuss one way that genome sequencing has turned out to be more complicated and/or more creative than anticipated.

2. What lessons learned from the human genome project can be used to guide ethical use of personal genome information?

3. Suggest a species you believe should have its genome sequenced, and what information you think the sequence might reveal or questions it might answer.

4. Restriction enzymes break a sequence of DNA bases into the following pieces:

   T T A A T A T C G
   C G T T A A T A T C G C T A G
   G C T T C G T T
   A A T A T C G C T A G C T G C A
   C T T C G T
   T A G C T G C A
   G T T A A T A T C G C T A G C T G C A

   Reconstruct the original sequence.

5. In medical practice, an incidentaloma is a test result that a physician wasn't looking for, like finding a cancerous tumor on an X ray of a broken bone. Discuss how personal genome sequencing could provide "too much information" in the form of incidentalomas.

6. Name a gene variant that you would want to know is in your genome, one that you wouldn't, and one that you consider to be frivolous.

7. One newly identified human gene has counterparts (homologs) in bacteria, yeast, roundworms, mustard weed, fruit flies, mice, and chimpanzees. A second gene has homologs in fruit flies, mice, and chimpanzees only. What does this information suggest about the functions of these two human genes with respect to each other?

8. Headlines about sequencing genomes of such unusual organisms as sea squirts and pufferfish often serve as material for comedians. Why, scientifically, is it important to sequence the genomes of a variety of organisms?

9. What do you think is the most useful type of information that can be revealed in a personal genome sequence?

10. What can we learn from the genome sequences of adopted individuals?

11. Genome sequencing of a fetus's DNA reveals two recessive disease-causing mutations in a particular gene. What further information can distinguish whether the individual is a compound heterozygote who will have the disease or is a carrier?

12. Explain why the bioinformatics of cataloging the human microbiome is much more complex than sequencing a human genome.

13. Discuss how an old tool such as a pedigree is still useful in interpreting whole-genome or exome sequences.

14. What can comparing the genome sequences of children and parents reveal?

15. Stephen Quake had his genome sequenced because his cousin's young son died of sudden heart failure. From the pedigree in figure 22.11a determine the proportion of the genome that Quake shares with his young relative.

Web Activities

1. Invent an "omics." Consult Omics.org for hints.

2. Go to the Personal Genome Project website (www.personalgenomes.org). Would you participate? What would you gain, and what might you sacrifice?

3. Go to the Ethical, Legal and Social Implications (ELSI) Research Program website (http://www.genome.gov/10001618). Discuss one societal concern arising from genomics, and how it might affect you.

4. Go to the website for The Institute for Genomic Research (http://www.tigr.org/). List five species whose genomes have been sequenced and the diseases that they cause in humans.

5. Go to www.medomics.com. What is a diagnostic genome?

6. Go to www.1000genomes.org and update the coverage of the 1,000 Genomes Project in the opener to this chapter.

Forensics Focus

1. This book has discussed many types of genetic information, from the SNPs that represent single bases, to genome-wide patterns of SNPs and copy number variants, to single-gene mutations, to sequenced exomes, to sequenced genomes. Select one of these and discuss how it would provide more detailed information than the traditional approach of identifying individuals on the basis of comparing a few dozen markers.

2. Wrote one researcher, "The collective genomes of our microbial symbionts may be more personally identifying than our own human genomes." Explain how this statement led to development of a new forensic tool.

Case Studies and Research Results

1. After the tsunami that devastated Japan in 2011, many organisms never before seen washed up on shore, thrown from the deep sea and some quite bizarre in appearance. Researchers collected specimens and sequenced DNA to try to classify the animals. Consider the following 8-base sequence that is similar among the species:

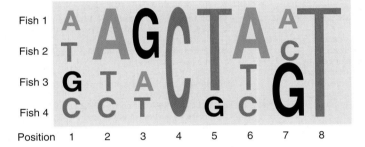

Fish 1	A	A	G	C	T	A	A	T
Fish 2	T	A	G	C	T	A	C	T
Fish 3	G	T	A	C	T	T	T	G
Fish 4	C	C	T	C	G	C	C	G
Position	1	2	3	4	5	6	7	8

   a. Write the DNA sequences for the two most closely related fishes.
   b. Which position(s) in the sequence are highly conserved?
   c. Which position(s) in the sequence are the least conserved?
   d. Which site is probably not essential, and how do you know this?
   e. A coelacanth has C T A C T G G T for this section of the genome. Which of the mystery fishes is the coelacanth's closest relative?

2. Explain how the fact that many genes were homozygous in the two varieties of *Cacao* whose genomes were sequenced eased the effort.

3. The 1,000 Genomes Project extends the HapMap project, which identified the 3.5 million SNPs that have served as the basis of most GWAS. Of the 15 million SNPs cataloged in the 1,000 Genomes Project (see the chapter opener), half are newly known. Researchers are reviewing old research reports based on HapMap data and adding new SNPs from the 1,000 Genomes Project. Adding new data to perform a statistical analysis on old data is a technique called imputation. Do you think that this is a valid or useful approach? Provide a reason for your answer.

4. Researchers sequenced the genome of a Sumatran orangutan named Susie, as well as parts of the genomes of ten others, five from Sumatra and five from Borneo. These islands are the only two places where orangutans live. Orangutans once lived throughout Southeast Asia. The animals were pushed onto their islands due to encroaching human activity—mining, logging, and construction of oil-palm plantations. Their genomes are very different from each other, and each species is more than twice as variable as the human genome. The sequencing was done by the whole-genome shotgun method, using six copies of the genome. Comparison to the genomes of other primates showed that the orangutans split from the lineage leading to humans 12 to 16 million years ago, and the two orangutan species split from each other about 400,000 years ago.

   a. Explain how Susie's genome was sequenced.
   b. What do you think the future holds for the orangutans?
   c. How closely related are we to orangutans compared to apes? (See chapter 16.)
   d. The orangutan genome shows signs of positive selection for several proteins that are parts of metabolic pathways that function abnormally in certain neurodegenerative diseases of humans. Suggest how this finding can be used.

Glossary

Pronunciations are provided for technical terms.

KEY ə = eh

- ‾ = long vowel sound
- ˘ = short vowel sound
- ″ = heavy accent
- ′ = light accent
- ^ = aw

A

acrocentric chromosome (ăk′rō sĕn′trĭk krōmə sōm) A chromosome in which the centromere is near one end.

adaptive immunity (ĭ-myōō′nĭ-tē) A slow, specific immune response following exposure to a foreign antigen.

adenine (ăd′en-ēn′) One of two purine nitrogenous bases in DNA and RNA.

affected sibling pair study A gene identification approach that looks for gene variants that siblings with a particular condition or trait share but that siblings who do not share the trait do not share.

allelic disorders (ə-lēl′ik) Different diseases caused by mutations in the same gene.

allele (ə-lēl′) An alternate (variant) form of a gene.

alternate splicing Building different proteins by combining exons of a gene in different ways.

amino (ə-mē′nō) **acid** A small organic molecule that is a protein building block.

amniocentesis (ăm′nē-ō-sĕn-tē′sĭs) A test that examines fetal chromosomes and biochemicals in amniotic fluid.

anaphase (ănə-fāz′) Stage of mitosis when the centromeres of replicated chromosomes part.

aneuploid (ăn′yū-ploid′) A cell with one or more extra or missing chromosomes.

angiogenesis (an″je-o-jen′ə-sis) Extension of blood vessels.

antibody (ăn′tē-bŏd′ē) A multisubunit protein, produced by B cells, that binds a specific foreign antigen, alerting the immune system or destroying the antigen.

anticodon (ăntē-kō′dŏn) A three-base sequence on one loop of a transfer RNA molecule that is complementary to an mRNA codon, and joins the appropriate amino acid and its mRNA.

antigen (ăn′tē-jən) A molecule that elicits an immune response.

antigen binding sites (ăn′tē-jən) Specialized ends of antibody chains.

antigen (ăn′tē-jən) **-presenting cell** A cell displaying a foreign antigen.

antiparallel The head-to-tail position of the entwined chains of the DNA double helix.

apoptosis (āpō-tō′sis) A form of cell death that is a normal part of growth and development.

assisted reproductive technologies Procedures that replace a gamete or the uterus to help people with fertility problems have children.

autoantibodies (ô′tō-ăn′tē-bŏdēz) Antibodies that attack the body's own cells.

autoimmunity (ô′tō-ĭ-myōō′nĭ-tē) An immune attack against one's own body.

autosomal (ôtə-sōməl) **dominant** The mode of inheritance in which one autosomal allele causes a phenotype. Such a trait can affect males and females and does not skip generations.

autosomal (ôtə-sōməl) **recessive** The mode of inheritance in which two autosomal alleles are required to cause a phenotype. Such a trait can affect males and females and can skip generations.

autosome (ôtə-sōm) A chromosome that does not have a gene that determines sex.

B

balanced polymorphism (pŏl′ē-môr′fizəm) Maintenance of a harmful recessive allele in a population because the heterozygote has a reproductive advantage.

base excision repair Removal of up to five contiguous DNA nucleotides to correct oxidative damage.

B cell A type of lymphocyte that secretes antibody proteins in response to nonself antigens displayed on other immune system cells.

bioethics A field that addresses personal issues that arise in applying medical technology and genetic information.

bioremediation Use of plants or microorganisms to detoxify environmental pollutants.

biotechnology The alteration of cells or biochemicals with a specific application.

blastocyst (blăs′tō-sĭst′) A hollow ball of cells descended from a fertilized ovum.

blastomere (blăstō′-mēr′) A cell of a blastocyst.

C

cancer (kăn′sər) A group of disorders resulting from loss of cell cycle control.

cancer stem cells Stem cells that divide and yield cancer cells and abnormal specialized cells.

carbohydrate (kar″bo-hī′-drăt) An organic compound that consists of carbon, hydrogen, and oxygen in a 1:2:1 ratio. Includes sugars and starches.

carcinogen (kar-sĭnə-jən) A substance that causes cancer.

case-control study An epidemiological method in which people with a particular condition are compared to individuals as much like them as possible, but without the disease.

cDNA library A collection of DNA molecules reverse transcribed from the mRNAs in a particular cell type.

cell (sel) The fundamental unit of life.

cell (sel) **cycle** A cycle of events describing a cell's preparation for division and division itself.

cellular adhesion A precise series of interactions among the proteins that connect cells.

cellular immune (ĭ-myōōn′) **response** T cells release cytokines to stimulate and coordinate an immune response.

centriole (sĕn′-trē-ohl) A structure in cells that organizes microtubules into the mitotic spindle.

centromere (sĕn′-trō-mîr) The largest constriction in a chromosome, located at a specific site in each chromosome type.

centrosome (sĕn′-trō-sōm) A structure built of centrioles and proteins that organizes microtubules into a spindle during cell division.

chaperone protein A protein that binds a polypeptide and guides folding.

chorionic villus (kôrē-ŏn′ĭk vĭl-us) **sampling** (CVS) A prenatal diagnostic technique that analyzes chromosomes in chorionic villus cells, which, like the fetus, descend from the fertilized ovum.

chromatid (krō′ mə-tĭd) A single, very long DNA molecule and its associated proteins, forming a longitudinal half of a replicated chromosome.

chromatin (krō′ mə-tĭn) DNA and its associated proteins.

chromatin (krō′ mə-tĭn) **remodeling** Adding or removing chemical groups to or from histones, which can alter gene expression.

chromosome (krō′ mə-sōm′) A highly wound continuous molecule of DNA and the proteins wrapped around it.

chromosome microarray analysis A technique that detects small copy number variants used with amniocentesis or chorionic villus sampling.

cleavage (klēvĭj) A series of rapid mitotic cell divisions after fertilization.

clines (klīnz) Allele frequencies that change from one geographical area to another.

cloning vector A piece of DNA used to transfer DNA from a cell of one organism into the cell of another.

coding strand The side of the DNA double helix for a particular gene from which RNA is not transcribed.

codominant A heterozygote in which both alleles are fully expressed.

codon (kō′dŏn) A continuous triplet of mRNA that specifies a particular amino acid.

coefficient of relatedness The proportion of genes that two people related in a certain way share.

cohort study An investigation that follows a large group of individuals over time while recording or assessing a health measure.

collectins (ko-lek′tinz) Immune system molecules that detect viruses, bacteria, and yeasts.

comparative genomics (jə-nō′mĭks) Identifying conserved DNA sequences among genomes of different species.

comparative genomic hybridization (CGH) A technique using fluorescent labels to detect copy number variants.

complement Plasma proteins that have a variety of immune functions.

complementary base pairs The pairs of DNA bases that hydrogen bond; adenine bonds to thymine and guanine to cytosine.

complementary DNA (cDNA) A DNA molecule that is the complement of an mRNA, copied using reverse transcriptase.

compound heterozygote An individual with two different mutations in the same gene.

concordance (kən-kôr′dens) A measure indicating the degree to which a trait is inherited; percentage of twin pairs in which both members express a trait.

conditional mutation (myōō-tā′shən) A genotype that is expressed only under certain environmental conditions.

conformation The three-dimensional shape of a molecule.

consanguinity (kŏnsăn-gwĭn′ĭ-tē) Blood relatives having children together.

copy number variant A DNA sequence present in different numbers of copies in different individuals; repeats.

critical period The time during prenatal development when a structure is sensitive to damage from a mutation or an environmental intervention.

crossing over An event during prophase I when homologs exchange parts.

cytogenetics (sītō-jə-nĕt′ĭks) Matching phenotypes to detectable chromosomal abnormalities.

cytokine (sītō-kīn′) A biochemical that a T cell secretes that controls immune function.

cytokinesis (sī-tō-kin-ē′-sis) Division of the cytoplasm and its contents.

cytoplasm (sī′-tō-plăzm) Cellular contents other than organelles.

cytosine (sī′-tō-sēn) One of the two pyrimidine nitrogenous bases in DNA and RNA.

cytoskeleton (sī-tō-skĕl′ĭ-tn) A framework of protein tubules and rods that supports the cell and gives it a distinctive form.

D

dedifferentiated A cell less specialized than the cell it descends from. A characteristic of a cancer cell.

deletion mutation (myōō-ta′-shən) A missing sequence of DNA or part of a chromosome.

deoxyribonucleic (dē-ŏksē-rībō-nōō-klā′ĭk) **acid** (DNA) The genetic material; the biochemical that forms genes.

deoxyribose (dē-ŏksē-rī′bōs) 5-carbon sugar in a DNA nucleotide.

differentiation Cell specialization, reflecting differential gene expression.

dihybrid cross Breeding individuals heterozygous for two traits.

diploid (dip′ loid) A cell containing two sets of chromosomes.

dizygotic (dīzī-gŏt′ĭk) **(DZ) twins** Twins that originate as two fertilized ova; fraternal twins.

DNA *See* **deoxyribonucleic acid.**

DNA damage response DNA repair.

DNA microarray *See* **microarray.**

DNA polymerase (pə-lim′ər-ās) (DNAP) An enzyme that adds new bases to replicating DNA and corrects mismatched base pairs.

DNA probe A labeled short sequence of DNA that binds its complement in a biological sample.

DNA profiling A biotechnology that detects differences in the number of copies of certain DNA repeats among individuals. Used to rule out or establish identity.

DNA replication Construction of a new DNA double helix using the information in parental strands as a template.

dominant A gene variant expressed when present in one copy.

duplication An extra copy of a DNA sequence, usually caused by misaligned pairing in meiosis.

E

ectoderm (ĕktō-dûrm) The outermost primary germ layer.

embryo (ĕm′brē-ō′) In humans, prenatal development until the end of the eighth week, when all basic structures are present.

embryonic (ĕmbrē-ŏn′ĭk) **stem (ES) cell** A cell, derived in laboratory culture from inner cell mass cells of very early embryos, that can self-renew and differentiate as any cell type.

empiric risk Probability that a trait will recur based upon its incidence in a population.

endoderm (ĕn′dō-dûrm) The innermost primary germ layer of the primordial embryo.

endoplasmic reticulum (ĕndō-plăzmi k rə-tik′u-ləm) (ER) An organelle consisting of a labyrinth of membranous tubules on which proteins, lipids, and sugars are synthesized.

enzyme (ĕnzīm) A type of protein that speeds the rate of a specific biochemical reaction.

epigenetic (ĕpē-jə-nĕt′ĭk) A layer of information placed on a gene that is a modification other than a change in DNA sequence, such as methylation.

epistasis (ĕpē-stā-sis) A gene masking the expression of another.

epitope (ep′ĭ-tōp) Part of an antigen that an antibody binds.

equational division The second meiotic division, producing four cells from two.

euchromatin (yōō-krō′mə-tin′) Parts of chromosomes that do not stain and that contain active genes.

eugenics (yōō-jĕn′ĭks) The control of individual reproductive choices to achieve a societal goal.

eukaryotic cell (yōō-kar′ē-ŏt′ĭk sel) A complex cell containing organelles, including a nucleus.

euploid (yōō′-ploid) A somatic cell with the normal number of chromosomes for the species.

ex vivo **gene therapy** Patching of mutant genes in cells growing in the laboratory and then introducing the cells into a patient.

excision repair Enzyme-catalyzed removal of pyrimidine dimers in DNA.

exome (x-ōm) The part of the genome that encodes protein.

exon (x-on) A part of a gene that encodes amino acids.

expressivity Degree of severity of a phenotype.

F

fetus (fē′təs) The prenatal human after the eighth week of development, when structures grow and specialize.

founder effect A type of genetic drift in which a few individuals found a new settlement, perpetuating a subset of alleles from the original population.

frameshift mutation (myōō-tā′shən) A mutation that alters a gene's reading frame.

fusion protein A protein that forms from translation of transcripts from two genes.

G

G₀ An offshoot of the cell cycle in which the cell remains specialized but does not replicate its DNA or divide.

G₁ The stage of the cell cycle following mitosis in which the cell resumes synthesis of proteins, lipids, and carbohydrates.

G₂ The stage of the cell cycle following S phase but before mitosis, when certain proteins are synthesized.

gamete (găm′ēt) A sex cell.

gamete intrafallopian (găm′ēt intrə-fə-lōpē-ən) **transfer** (GIFT) An infertility treatment in which sperm and oocytes are placed in a woman's uterine tube.

gastrula (găstrə-lə) A three-layered embryo.

gene (jēn) A sequence of DNA that instructs a cell to produce a particular protein.

gene expression Transcription of a gene's DNA into RNA.

gene expression profiling Use of DNA microarrays to detect the types and amounts of cDNAs reverse transcribed from the mRNAs in a particular cell source.

gene pool All the genes in a population.

gene therapy Replacing a malfunctioning gene to correct an abnormality.

genetic (jə-nĕt′ĭk) **code** The correspondence between specific mRNA triplets and the amino acids they specify.

genetic counselor A medical specialist who calculates risk of recurrence of inherited disorders in families, applying the laws of inheritance to pedigrees and interpreting genetic test results.

genetic determinism Attributing a trait to a gene or genes.

genetic drift Changes in allele frequencies in small groups reproductively separated from a larger population.

genetic heterogeneity A phenotype that can be caused by variants of any of several genes.

genetic load The collection of deleterious recessive alleles in a population.

genetic marker DNA sequences near genes of interest that are co-inherited unless separated by a crossover. SNPs and copy number variants are used as markers.

genetics The study of inherited variation.

genome (jē′nōm) The complete set of genetic instructions in the cells of a particular type of organism.

genome-wide association study (GWAS) A case-control study in which millions of variants (single nucleotide polymorphisms or copy number variants) that form haplotypes are compared between people with a condition and unaffected individuals.

genomic (jē nō′m ĭ k) **imprinting** Differing of the phenotype depending upon which parent transmits a particular allele.

genomic (jē nōm ĭ k) **library** A collection of DNA pieces representing the genome of an individual, including introns.

genomics (jē nōm ĭ ks) The study of the functions and interactions of many genes or other DNA sequences, or comparing genomes.

genotype (jē n′ə- tīp) The allele combinations in an individual that cause particular traits or disorders.

genotypic (jēn′ə- tīp′ĭk) **ratio** The ratio of genotype classes expected in the progeny of a particular cross.

germline gene therapy Genetic alterations of gametes or fertilized ova, which perpetuate the change throughout the organism and transmit it to future generations.

germline mutation A mutation in every cell in an individual because it was present in the fertilized ovum.

Golgi (gōl′jē) **apparatus** An organelle, consisting of flattened, membranous sacs, that packages secretion components.

gonads (gō′-nadz) Paired structures in the reproductive system where sperm or oocytes are manufactured.

growth factor A protein that stimulates mitosis.

guanine (gwa′ nēn) One of the two purine nitrogenous bases in DNA and RNA.

H

haploid (hăp′ loid) A cell with one set of chromosomes.

haplogroup (hăp′ lō-groōp) In DNA ancestry testing, a specific set of markers on the Y or mitochondrial chromosome.

haplotype (hăp′ lō tip) A series of known DNA sequences or single nucleotide polymorphisms linked on a chromosome.

Hardy-Weinberg equilibrium An idealized state in which allele frequencies in a population do not change from generation to generation.

heavy chains The two longer polypeptide chains of an antibody subunit.

hemizygous (hĕm′ ē-zī′ gəs) The sex that has half as many X-linked genes as the other; a human male.

heredity Transmission of inherited traits from generation to generation.

heritability An estimate of the proportion of phenotypic variation in a group due to genes.

heterochromatin (hĕtə-rō-krō′mətĭn) Dark-staining chromosome parts that have few protein-encoding genes.

heterogametic (hĕt′ə-rō-gə-mē′tĭk) **sex** The sex with two different sex chromosomes; a human male.

heteroplasmy (hĕt′ə-rō-plăz-mē) Mitochondria in the same cell having different alleles of a particular gene.

heterozygous (hĕtə-rō-zī′gəs) Having two different alleles of a gene.

histone (his′tōn) A type of protein around which DNA entwines.

hominins (hŏm′ə-nĭnz) Animals ancestral to humans only.

hominoids (hŏmə-noĭdz) Animals ancestral to apes and humans only.

homogametic (hō′mō-gə-mē′tĭk) **sex** The sex with identical types of sex chromosomes; the human female.

homologous (hō-mŏl′ə-gəs) **pairs** Chromosomes with the same gene sequence.

homozygosity (hōmō-zī-gəs′-ĭ-tē) **mapping** An approach to gene discovery that correlates stretches of homozygous DNA base sequence in the genomes of related individuals to certain traits or disorders.

homozygous (hōmō-zī′ gəs) Having two identical alleles of a gene.

hormone (hor′ mōn) A biochemical produced in a gland and carried in the blood to a target organ, where it exerts an effect.

human leukocyte antigen (lōōkə·sīt′ ăn′tĭ-jən) (HLA) **complex** Genes closely linked on the short arm of chromosome 6 that encode cell surface proteins important in immune system function.

humoral (yōō′ mər-əl) **immune response** Process in which B cells secrete antibodies into the bloodstream.

I

idiotype (ĭd′ē-o-tīp) Part of an antibody molecule that binds an antigen.

incidence The number of new cases of a disease during a certain time in a particular population.

incomplete dominance A heterozygote intermediate in phenotype between either homozygote.

independent assortment The random arrangement of homologous chromosome pairs, in terms of maternal or paternal origin, down the center of a cell in metaphase I. Inheritance of a gene on one chromosome does not influence inheritance of a gene on a different chromosome. (Mendel's second law)

induced pluripotent stem (iPS) cells Somatic cells that are genetically reprogrammed to differentiate into another cell type.

infertility The inability to conceive a child after a year of unprotected intercourse.

inflammation Part of the innate immune response that causes an infected or injured area to swell with fluid, turn red, and attract phagocytes.

innate immunity (ĭ-myōō′nĭ-tē) Components of immune response that are present at birth and do not require exposure to an environmental stimulus.

inner cell mass A clump of cells on the inside of the blastocyst that will continue developing into an embryo. Source of embryonic stem cells.

insertional translocation A rare type of translocation in which a part of one chromosome is part of a nonhomologous chromosome.

insertion mutation (myōō-tā′-shən) A mutation that adds DNA bases.

interferon (in″tər-fēr′on) A type of cytokine.

interleukin (in″tər-loo′kin) A type of cytokine.

intermediate filament A type of cytoskeletal component made of different proteins in different cell types.

interphase (in′tər-fāz′) Stage when a cell is not dividing.

intracytoplasmic (in′trə-sītō-plăzmĭk) **sperm injection** (ICSI) An infertility treatment that injects a sperm cell nucleus into an oocyte, to overcome lack of sperm motility.

intrauterine (in′trə-yōō′tər-in) **insemination** An infertility treatment that places donor sperm in the cervix or uterus.

intron (in trŏn) Part of a gene that is transcribed but is excised from the mRNA before translation into protein.

in vitro (in vē′trō) **fertilization** (IVF) Placing oocytes and sperm in a laboratory dish with appropriate biochemicals so that fertilization occurs, then, after a few cell divisions, transferring the embryos to a woman's uterus.

in vivo gene therapy Introduction of vectors carrying therapeutic human genes directly into the body part where they will act.

K

karyotype (kărē-ō-tīp) A size-order chromosome chart.

L

law of independent assortment *See* **independent assortment.**

law of segregation *See* **segregation.**

lethal allele (ə-lēl′) An allele that causes death before reproductive maturity or halts prenatal development.

ligand (lī′gənd) A molecule that binds to a receptor.

ligase (lī′gās) An enzyme that catalyzes the formation of covalent bonds in the sugar-phosphate backbone of a nucleic acid.

light chains The two shorter polypeptide chains of an antibody subunit.

linkage Genes on the same chromosome.

linkage disequilibrium Extremely tight linkage between DNA sequences.

linkage maps Maps that show gene order on chromosomes, determined from crossover frequencies between pairs of genes.

lipid (lĭpĭd) A type of organic molecule that has more carbon and hydrogen atoms than oxygen atoms. Includes fats and oils.

lysosome (lī′sō-sōm) A saclike organelle containing enzymes that degrade debris.

M

macroevolution (măk′rō-ĕv′ə-lōōshən) Genetic change sufficient to form a new species.

major histocompatibility (histō-kəm-pătə-bĭlĭ-tē) **complex** (MHC) A gene cluster, on chromosome 6 in humans, that includes many genes that encode components of the immune system.

manifesting heterozygote (hĕt′ə-rō-zīgōt) A female carrier of an X-linked recessive gene who expresses the phenotype because the normal allele is inactivated in some tissues.

meiosis (mī-ō′sĭs) Cell division that halves the number of chromosomes to form haploid gametes.

memory cells B or T cell descendants that carry out a secondary immune response.

mesoderm (mĕz-ō-dûrm) The middle primary germ layer.

messenger RNA (mRNA) A molecule of RNA complementary in sequence to the template strand of a gene that specifies a protein product.

metacentric chromosome (mĕtə-sĕn′trĭk krōmə sōm) A chromosome with the centromere approximately in the center.

metaphase (mĕtə-fāz) The stage of mitosis when chromosomes align along the center of the cell.

metastasis (mĕtə-stā′-sĭs) Spread of cancer from its site of origin to other parts of the body.

microarray A set of target genes embedded in a glass chip, to which labeled cDNAs from a sample bind and fluoresce. Microarrays show patterns of gene expression.

microevolution Change of allele frequency in a population.

microfilament A solid rod of actin protein that forms part of the cytoskeleton.

microRNA A 21- or 22-base-long RNA that binds to certain mRNAs, blocking their translation into protein.

microtubule (mīkrō-tōōbyōōl) A hollow structure built of tubulin protein that forms part of the cytoskeleton.

mismatch repair Proofreading of DNA for misalignment of short, repeated segments.

missense (mis′sĕns) A single base change mutation that alters an amino acid.

mitochondrion (mītō-kŏn′drē-ən) An organelle consisting of a double membrane that houses enzymes that catalyze reactions that extract energy from nutrients.

mitosis (mī-tōsĭs) Division of somatic (nonsex) cells.

mode of inheritance The pattern in which a gene variant passes from generation to generation. It may be dominant or recessive, autosomal or X- or Y-linked.

molecular evolution Changes in protein and DNA sequences over time used to estimate how recently species diverged from a shared ancestor.

monohybrid (mŏn′ō-hībrĭd) **cross** A cross of two individuals who are heterozygous for a single trait.

monosomy (mŏn′ō-sō′mē) A human cell with 45 (one missing) chromosomes.

monozygotic (mŏnō-zī-gŏt′ĭk) (MZ) **twins** Twins that originate as a single fertilized ovum; identical twins.

morula (môr′ yə-lə) The very early prenatal stage that resembles a mulberry.

multifactorial trait A trait or illness determined by several genes and the environment.

mutagen (myōō′tə-jən) A substance that changes, adds, or deletes a DNA base.

mutant (myōōt′nt) An allele that differs from the normal or most common allele in a population that alters the phenotype.

mutation (myōō-tā′shən) A change in a protein-encoding gene that affects the phenotype and occurs in less than 1 percent of a population.

N

natural selection Differential survival and reproduction of individuals with particular phenotypes in particular environments, which may alter allele frequencies in subsequent generations.

neural (nōōr′əl) **tube** A structure in the embryo that develops into the brain and spinal cord.

neurexin A protein in the presynaptic membrane of a neuron that uses the neurotransmitter glutamate and is involved in autism.

neuroglia Several types of cells in the nervous system that support neurons.

neuron (nōrin′) A nerve cell.

neuroligin A protein in the postsynaptic membrane of a neuron that uses the neurotransmitter glutamate and is involved in autism.

neurotransmitter A molecule that transmits messages in the nervous system.

nitrogenous (nī-trŏj′ə-nəs) **base** A nitrogen-containing base that is part of a nucleotide.

nondisjunction (nŏndis-jŭngk′shən) The unequal partitioning of chromosomes into gametes during meiosis.

nonsense mutation (myōōtā′shən) A point mutation that changes an amino-acid-coding codon into a stop codon, prematurely terminating synthesis of the encoded protein.

nonsense suppression Action of a drug that enables protein synthesis to ignore a nonsense mutation (stop codon not at the end of a gene).

nonsynonymous codon (kō′don) A codon that encodes a different amino acid from another codon.

nucleic (nōō-klē′ĭk) **acid** DNA or RNA.

nucleolus (nōō-klē′ə-ləs) A structure in the nucleus where ribosomes are assembled from ribosomal RNA and protein.

nucleosome (nōō′-klē-ō-sōm) A unit of chromatin structure.

nucleotide (nōō-klē-ō-tīd) The building block of a nucleic acid, consisting of a phosphate group, a nitrogenous base, and a 5-carbon sugar.

nucleotide (nōō-klē-ō-tīd) **excision repair** Replacement of up to 30 nucleotides to correct DNA damage of several types.

nucleus (nōō-klē-əs) A large, membrane-bounded region of a eukaryotic cell that houses DNA.

O

oncogene (ŏn′kə-jēn) A gene that normally controls the cell cycle, but causes cancer when overexpressed.

oocyte (ō′ə-sīt) The female gamete (sex cell).

oogenesis (ōə-jĕn′i-sĭs) Oocyte development.

organelle (ōr′gə-nĕl′) A specialized structure in a eukaryotic cell that carries out a specific function.

ovaries (o′və-rēz) The female gonads.

P

paracentric (para sĕn′-trĭk) **inversion** An inverted chromosome that does not include the centromere.

pedigree A chart of symbols connected by lines that depict the genetic relationships and transmission of inherited traits in related individuals.

penetrance Percentage of individuals with a genotype who have an associated phenotype.

pericentric (pər-ē sĕn-trik) **inversion** An inverted chromosome that includes the centromere.

peroxisome (pə-rŏk′si-sōm) An organelle consisting of a double membrane that houses enzymes with various functions.

pharmacogenetics (farm a kō jə-nĕt-i ks) Testing for single gene variants that affect drug metabolism.

pharmacogenomics (farm a kō jə-nōm-i ks) Testing for variants of many genes or gene expression profiles that affect drug metabolism.

phenocopy (fē′ nō-kŏp′ē) An environmentally caused trait that occurs in a familial pattern, mimicking inheritance.

phenotype (fē′ nō-tīp) The expression of a gene in traits or symptoms.

plasma cell A cell descended from a B cell that produces abundant antibodies of a single type.

plasma membrane (plăz′mə mĕm′brān) The selective barrier around a cell, consisting of proteins, glycolipids, glycoproteins, and lipid rafts on or in a phospholipid bilayer.

plasmid (plăz′ mi d) A small circle of double-stranded DNA found in some bacteria. Used as a vector in recombinant DNA technology.

pleiotropic (plēə-trōpĭ k) A single-gene disorder with several symptoms. Different symptom subsets may occur in different individuals.

point mutation (myōō-tā′ shən) A single base change in DNA.

polar body A product of female meiosis that contains little cytoplasm and does not continue to develop into an oocyte.

polar body biopsy (bī′ ŏp sē) A genetic test performed on a polar body to infer the genotype of the attached oocyte.

polygenic (pŏlē-jēn′ ik) **traits** Traits determined by more than one gene.

polymerase (pŏlə′-mə-rās) **chain reaction** (PCR) A nucleic acid amplification technique in which a DNA sequence is replicated in a test tube to rapidly produce many copies.

polymorphism (pŏlē-môr′ fiz əm) A DNA base or sequence at a certain chromosomal locus that varies in at least 1 percent of individuals in a population.

polyploid (pŏl′ē-ploid) A cell with one or more extra sets of chromosomes.

population A group of interbreeding individuals.

population bottleneck Decrease in allele diversity resulting from an event that kills many members of a population, followed by restoration of population numbers.

population genetics (jə-nĕt′iks) The study of allele frequencies in different groups of individuals.

population study Comparison of disease incidence in different groups of people.

preimplantation genetic (jə-nĕt′ik) **diagnosis** (PGD) Removing a cell from an 8-celled embryo and testing it for a mutation to deduce the genotype of the embryo.

prevalence The number of cases of a disease in a population at a particular time.

primary germ layers The three layers of an embryo.

primary immune response Immune system's response to initial encounter with a nonself antigen.

primary (1°) structure The amino acid sequence of a protein.

progenitor cell A cell whose descendants can follow any of several developmental pathways, but not all.

prokaryotic cell (prō-kārē-ŏt′ik sĕl) A cell that does not have a nucleus or other organelles. One of the three domains of life. Bacteria.

promoter A control sequence near the start of a gene.

pronuclei (prō-nōō′klē ī) DNA packets in the fertilized ovum.

prophase (prō′fāz) The first stage of mitosis or meiosis, when chromatin condenses.

prospective study A study that follows two or more groups.

proteasome (prō-tē-ə-sōm) A multiprotein structure in a cell shaped like a barrel through which misfolded proteins pass and are refolded or dismantled.

protein A type of macromolecule that is the direct product of genetic information; a chain of amino acids.

proteome (prō′tē-ōm) The set of proteins a cell produces.

proteomics (prōtē-ō′ miks) Study of the proteins produced in a particular cell type under particular conditions.

proto-oncogene (prōtō-ŏn′kə-jēn) A gene that normally controls the cell cycle, but when overexpressed causes cancer.

pseudogene (sōō′ dō jēn) A gene that does not encode protein, but whose sequence very closely resembles that of a coding gene.

Punnett square A diagram used to follow parental gene contributions to offspring.

purine (pyōō r′ēn) A DNA base with a two-ring structure; adenine and guanine are purines.

pyrimidine (pi-rim′i-dēn) A DNA base with a single-ring structure; cytosine, thymine, and uracil are pyrimidines.

Q

quantitative trait loci Genes that determine polygenic traits.

quaternary (4°) structure A protein that has more than one polypeptide subunit.

R

reading frame The grouping of DNA base triplets encoding an amino acid sequence.

receptor A structure on a cell that binds a specific molecule.

recessive An allele whose expression is masked by another allele.

reciprocal translocation A chromosome aberration in which two nonhomologous chromosomes exchange parts, conserving genetic balance but rearranging genes.

recombinant (rē-kŏm′bə-nənt) A series of alleles on a chromosome that differs from the series of either parent.

recombinant (rē-kŏm′bə-nənt) **DNA technology** Transferring genes between species.

reduction division The first meiotic division, which halves the chromosome number.

replication fork Locally opened portion of a replicating DNA double helix.

ribonucleic acid (RNA) (rī bō-nōō-klē′ik) A nucleic acid whose bases are A, C, U, and G.

ribose (rī′bōs) A 5-carbon sugar in RNA.

ribosomal (rī′bōs-ō′məl) **RNA** (rRNA) RNA that, with proteins, comprises ribosomes.

ribosome (rī′bō sōm) An organelle consisting of RNA and protein that is a scaffold for protein synthesis.

risk factor A characteristic or experience associated with increased likelihood of developing a particular medical condition.

RNA interference Introduction of a small interfering RNA molecule that binds to and prevents translation of a specific mRNA.

RNA polymerase (RNAP) (pŏl′ə-mə-rās) An enzyme that adds RNA nucleotides to a growing RNA chain.

Robertsonian (Răb-ərt - sō′-nē-ən) **translocation** A chromosome aberration in which two short arms of nonhomologous chromosomes break and the long arms fuse, forming one unusual, large chromosome.

run of homozygosity Regions of the genome in which contiguous SNPs (single nucleotide polymorphisms) are homozygous, indicating a shared ancestor with another person with the same pattern.

S

S phase The stage of interphase when DNA replicates.

secondary immune response Immune system activation in response to a second or subsequent encounter with a pathogen.

secondary (2°) structure Folds in a polypeptide caused by attractions between amino acids close together in the primary structure.

segregation The distribution of alleles of a gene into separate gametes during meiosis. (Mendel's first law)

self-renewal Defining property of a stem cell; the ability to yield a daughter cell like itself.

semiconservative replication DNA synthesis along each half of the double helix.

sex chromosome (krŏ'mə-sōm) A chromosome containing genes that specify sex.

sex-influenced trait Phenotype caused when an allele is recessive in one sex but dominant in the other.

sex-limited trait A trait that affects a structure or function present in only one sex.

sex ratio Number of males divided by number of females multiplied by 1,000 for people of a certain age in a population.

short tandem repeats (STRs) Repeats of 2 to 10 DNA bases that are compared in DNA profiling.

signal transduction A series of biochemical reactions and interactions that pass information from outside a cell to inside, triggering a response.

single nucleotide polymorphism (nōōklē̄o-tĭd pŏlē-môr' fiz'əm) (SNP) Single base sites that differ among individuals. A SNP is present in at least 1 percent of a population.

somatic cell (sō-măt'ĭk sĕl) A nonsex cell, with 23 pairs of chromosomes in humans.

somatic (sō-măt'ĭk) **gene therapy** Genetic alteration of a specific cell type, not transmitted to future generations.

somatic mutation (sō-măt'ĭk myōō-tā'shən) A genetic change in a nonsex cell.

spermatogenesis (spər-măt'ə-jĕn'ĭ-sis) Sperm cell differentiation.

spermatogonium (sper''mah-to-gō' ne-um) An undifferentiated cell in a seminiferous tubule that can give rise to a sperm cell in meiosis.

spermatozoon (spər-măt'ə-zō'ŏn) (sperm) A mature male reproductive cell (meiotic product).

spindle A structure composed of microtubules that pulls sets of chromosomes apart in a dividing cell.

spontaneous mutation (myōō-tā'sheən) A genetic change that results from mispairing when the replication machinery encounters a base in its rare tautomeric form.

SRY **gene** The sex-determining region of the Y. If the *SRY* gene is activated, the gonad develops into a testis; if not, an ovary forms under direction of other genes.

stem cells Cells that give rise to other stem cells, as well as to cells that differentiate.

submetacentric chromosome (sŭb mĕt-ə-sĕn'trĭk krŏ'mə-sōm) A chromosome in which the centromere establishes a long arm and a short arm.

sugar-phosphate backbone The "rails" of a DNA double helix, consisting of alternating deoxyribose and phosphate groups, oriented opposite one another.

synapse The space between two neurons that a neurotransmitter must cross to transmit a message.

synonymous codons (kō d ŏnz) DNA triplets that specify the same amino acid.

synteny (sin'tə-nē) Correspondence of genes on the same chromosome in several species.

T

tandem duplication A duplicated DNA sequence next to the original sequence.

T cell A type of lymphocyte that produces cytokines and coordinates the immune response.

telomerase (tə-lŏm'ə-rās) An enzyme, including a sequence of RNA, that adds DNA to chromosome tips.

telomere (tĕl'ə-mîr) A chromosome tip.

telophase (tĕlə-fāz) The stage of mitosis or meiosis when daughter cells separate.

template strand The DNA strand carrying the information to be transcribed.

teratogen (tə-rāt'ə-jən) A substance that causes a birth defect.

tertiary (3°) structure Folds in a polypeptide caused by interactions between amino acids and water. This draws together amino acids that are far apart in the primary structure.

testes (tes'tēz) The male gonads.

thymine (thī'mēn) One of the two pyrimidine bases in DNA.

transcription Manufacturing RNA from DNA.

transcription factor A protein that activates the transcription of certain genes.

transfer RNA (tRNA) A type of RNA that connects mRNA to amino acids during protein synthesis.

transgenic organism (trăns-jĕn'ĭk) An individual with a genetic modification in every cell.

transition A point mutation altering a purine to a purine or a pyrimidine to a pyrimidine.

translation Assembly of an amino acid chain according to the sequence of base triplets in a molecule of mRNA.

translocation Exchange between nonhomologous chromosomes.

translocation carrier An individual with exchanged chromosomes but no signs or symptoms. The person has the usual amount of genetic material, but it is rearranged.

transposon (trăns-pōzŏn) A gene or DNA segment that moves to another chromosome.

transversion A point mutation altering a purine to a pyrimidine or vice versa.

trisomy (trī sō'mē) A human cell with 47 chromosomes (one extra).

tumor suppressor gene (tōōmər səprĕs'ər jēn) A recessive gene whose normal function is to limit the number of divisions a cell undergoes.

U

uniparental disomy (yû-ni-pə'rent-əl dī sō mē) Inheriting two copies of the same gene from one parent.

uracil (yōōr'ə-sil) One of the four types of bases in RNA; a pyrimidine.

V

vaccine (vak-sē'n) An inactive or partial form of a pathogen that stimulates antibody production.

variable number of tandem repeats (VNTRs) Repeats of 10 to 80 DNA bases that are compared in DNA profiles.

vesicles (ves-ə-kulz) Bubble-like membrane-bounded organelles that participate in secretion.

virus (vī rəs) An infectious particle built of nucleic acid in a protein coat.

W

wild type The most common phenotype in a population for a particular gene.

X

X inactivation The inactivation of one X chromosome in each cell of a female mammal, occurring early in embryonic development.

X-linked Genes on an X chromosome.

X-Y homologs (hŏm'ə-lôgz) Y-linked genes that are similar to genes on the X chromosome.

Y

Y-linked Genes on a Y chromosome.

Z

zygote (zī'gōt) A prenatal human from the fertilized ovum stage until formation of the primordial embryo, at about two weeks.

zygote intrafallopian transfer (zī' gōt in'trə-fə-lō' pē-ən) (ZIFT) An assisted reproductive technology in which an ovum fertilized *in vitro* is placed in a woman's uterine tube.

Credits

Line Art and Text

Chapter 3
Page 54, Text Art 3.1: © Tribune Media Services, Inc. All Rights Reserved. Reprinted with permission.

Chapter 4
Reading 4.2 Figure 1: Source: Cystic Fibrosis Foundation. Patient Registry 2006 Annual Report.

Chapter 6
In Their Own Words, Figure 1: Courtesy Professor Jennifer A. Marshall-Graves, Australian National University.; In Their Own Words, Figure 3: Courtesy David Page, Massachusetts Institute of Technology, Howard Hughes Medical Institute Investigator.

Chapter 7
Figure 7.2: Data and reprint from Gordon Mendenhall, Thomas Mertens, and Jon Hendrix, "Fingerprint Ridge Count" in *The American Biology Teacher*, vol. 51, no. 4, April 1989, pp. 204–206. AMERICAN BIOLOGY TEACHER by MENDENHALL, MERTENS, AND HENDRIX. Copyright 1989 by NATIONAL ASSOCIATION OF BIOLOGY TEACHERS in the format Textbook via Copyright Clearance Center.; Figure 7.8: Originally published in the 4 May issue of *The New Yorker* Magazine, p. 43. © Charles Addams. With permission Tee and Charles Addams Foundation.; Figure 7.12: Graph from www.unitedhealthfoundation.org. Reprinted by permission.

Chapter 8
Table 8.1: Source: Psychiatric Genomics, Inc., Gaithersburg, MD. The information was collated from the Surgeon General's 1999 Report on Mental Health.; Figure 8.8: © Robert Gilliam.; Figure 8.10: Copyright 2009 National Academy of Sciences, U.S.A. All rights reserved. Bernard Crespi, Philip Stead, Michael Elliot, *Comparative genomics of autism & schizophrenia*, Proceedings of the National Academy of Sciences, http://www.pnas.org/content/107/suppl.1/1736. full.pdf+html?sid=5f4e02ab-8bbb-47c7-be77-1adb47a213c0.

Chapter 13
Figure 13.7: From *Color Atlas of Genetics* by Eberhard Passage, p. 401. Copyright © 2001. Reprinted by permission of Thieme Medical Publishers, Inc.

Chapter 14
Figure 14.10: Copyright 2008 National Academy of Sciences. All rights reserved. http://www.koshland-science-museum.org.

Chapter 16
Figure 16.7: Reprinted by permission from Macmillan Publishers Ltd: *Nature*. Krause et al. "The complete mitochondrial DNA genome of an unknown hominin from southern Siberia." *Nature* 464:894–7, April 8, 2010. Figure 3, page 897. Copyright 2010.; Figure 16.8: Reprinted with permission of the artist, Karol Schauer, and the State Office for Heritage Management and Archaeology Saxony-Anhalt.; Reading 16.2 Figure 1: Reprinted from *The American Journal of Human Genetics*, Vol. 79;2, Hua Liu, Frank Prugnolle, Andrea Manica and François Balloux, A Geographically Explicit Genetic Model of Worldwide Human-Settlement History, Copyright 2006, with permission of Elsevier.

Chapter 18
Figure 18.9: Reprinted by permission from Macmillan Publishers Ltd: *Nature,* Neurobiology: At the root of brain cancer, 432:281–282, copyright 2004.; Figure 18.12a: Reprinted by permission from Macmillan Publishers Ltd: *Nature,* Tissue repair and stem cell renewal in carcinogenesis, 432:324–331, copyright 2004.; Reading 18.1 Figure 2: Adapted from "Drug therapy: Imatinib mesylate—A new oral targeted therapy" by Savage & Antman: *New England Journal of Medicine* 346:683–693. Copyright © 2002 Massachusetts Medical Society. All rights reserved. Reprinted by permission.

Chapter 22
Figure 22.11a: Reprinted from *The Lancet,* vol. 375, Ashley et al., "Clinical Assessment Incorporating a Personal Genome," p. 1527, Copyright 2010, with permission from Elsevier.; Figure 22.11b: Reprinted from *The Lancet,* vol. 375, Ashley et al., "Clinical Assessment Incorporating a Personal Genome," p. 1532, Copyright 2010, with permission from Elsevier.

Photo Credits

About the Author
Page V: Courtesy Wendy Josephs.

Chapter 1
Opener: © Comstock Images/Jupiter Images RF; 1.1: © Royalty-Free/Corbis; 1.2: © CNRI/Photo Researchers; 1.4(top left): © Royalty-Free/Corbis; 1.4(top right): © PhotoDisc/Getty RF; 1.4(center left): © Byrappa Venkatesh, IMCB, Singapore; 1.4(center): © PhotoDisc/Getty RF; 1.4(center right): © PhotoDisc/Getty RF; 1.4(bottom left): © David M. Phillips/Photo Researchers; 1.4(bottom right): © Dr. Stanley Flegler/Visuals Unlimited; 1.5a: © Lester Bergman/Project Masters, Inc.; 1.5b: © Steve Mason/Getty RF; 1.6: © AP Images/Leslie Close; 1.7: © Jay Sand; 1.9: © Alexis Rockman, 2000. Courtesy Leo Koenig Inc.

Chapter 2
Opener: © The McGraw-Hill Companies, Inc./Al Telser, photographer; 2.1: Photo courtesy the Muscular Dystrophy Association; 2.2: © Manfred Kage/Photolibrary; p. 21: © SPL/Photo Researchers; 2.3(Top): © David M. Phillips/The Population Council/Science Source/Photo Researchers; 2.3(Bottom left): © K.R. Porter/Photo Researchers; 2.3(Bottom Right): © EM Research Services, Newcastle University; 2.6: © Prof. P. Motta & T. Naguro/SPL/Photo Researchers; 2.7: © Bill Longcore/Photo Researchers; 2.10: © Visuals Unlimited; p. 29: Courtesy, Lori Sames; 2.12: © P. Motta/SPL/Photo Researchers; 2.13b: © Bart's Medical Library/Phototake; 2.15b: © From Dr. A.T. Sumner, "Mammalian Chromosomes from Prophase to Telophase," Chromosoma, 100:410–418, 1991. @Springer-Verlag; 2.16(all): © Ed Reschke; 2.18: From L. Chong, et al. 1995. "A Human Telomeric Protein." *Science*, 270:1663–1667. @1995 American Association for the Advancement of Science. Photo courtesy, Dr. Titia DeLange; 2.19(left): © David McCarthy/Photo Researchers; 2.19(right): © Peter Skinner/Photo Researchers.

Chapter 3
3.9: Illustration by Nicolaas Hartsoeker, from Essai de Dioptrique, 1695; 3.10: © Prof. P.M. Motta/Univ. "La Sapienza", Rome/Photo Researchers; 3.13b: © Brand X Pictures RF; 3.14(left): © Petit Format/Nestle/Science Source/Photo Researchers; 3.14(center): © P.M. Motta & J. Van Blerkom/SPL/Photo Researchers; 3.14(right): © Petit Format/Nestle/Science Source/Photo Researchers; 3.17: Courtesy, Brittany and Abby Hensel; 3.18a: © Petit Format/Nestle/Photo Researchers; 3.18b: © Petit Format/SPL/Photo Researchers; 3.21: © Ingram Publishing/Alamy RF; 3.22: Courtesy, The Progeria Research Foundation; p. 65: © Mitch Wojnarowicz/The Image Works.

Chapter 4
Opener, p. 77: Courtesy, Cystic Fibrosis Foundation; 4.15: © Arie Kievit/Hollandse Hoogte/Redux; 4.18: © Andersen Ross/Getty RF

Chapter 5
Opener: Courtesy, Jennifer Pletcher; 5.1a: © Porterfield-Chickering/Photo Researchers; 5.2: From Genest, Jacques, Jr., Lavoie, Marc-Andre. August 12, 1999. "Images in Clinical Medicine." New England Journal of Medicine, pp 490. © 1999, Massachusetts Medical Society. All Rights Reserved; 5.7: Library of Congress, Print & Photographs Division [LC-USZ62-95719]; p. 97: © The McGraw-Hill Companies; 5.12: © Stock Montage.

Chapter 6

6.1: © Biophoto Associates/Photo Researchers; p. 114: © Dr. Walter Just; 6.5: Courtesy, Dr. Mark A. Crowe; 6.7b: Courtesy, Richard Alan Lewis M.D., M.S., Baylor College of Medicine; 6.8a: From J.M. Cantu, et al. 1984. Human Genetics, 66:66–70. © Springer-Verlag, Gmbh & Co. KG. Photo courtesy of Pragna I. Patel, Ph.D/Baylor College of Medicine; 6.10: © Animal Attraction/OS50/Getty RF; 6.11: "Reprinted from Stephen R.F. Twigg, et al., "Mutations of ephrin-B1 (EFNB1), a marker of tissue boundary formation, cause craniofrontonasal syndrome," PNAS 2004 101, pg. 8653. Image courtesy of Stephen Twigg and Andrew Wilkie" © 2004 National Academy of Sciences, U.S.A; 6.12: Designed by Mark Sherman. Provided by Arthur Riggs and Craig Cooney; p. 125: Courtesy, International Rett Syndrome Foundation; 6.14b: © Carla D. Kipper; 6.14c: Courtesy Roxanne De Leon and Angelman Syndrome Foundation.

Chapter 7

Opener: © Royalty-Free/Corbis; 7.1: © Brand X Pictures; p. 133: © The McGraw-Hill Companies, Inc./Al Telser, photographer; 7.3a: From Albert & Blakeslee, Corn and Man, *Journal of Heredity,* 1914, Vol. 5, pg. 51. By permission of Oxford University Press; 7.3b: © Peter Morenus/University of Connecticut at Storrs; 7.4b: © Jamie Hanson/Newspix; 7.5: Courtesy, The Smile Train, Mark Atkinson, photographer.

Chapter 8

Opener: © Royalty-Free/Corbis; 8.2: © BananaStock/Punchstock; 8.3: © Stanford University Center for Narcolepsy; 8.6: © PhotoDisc/Getty RF.

Chapter 9

Opener: © Dr. Gopal Murti/Photo Researchers; 9.4a: © Science Source/Photo Researchers; 9.4b: From "The Double Helix" by James D. Watson, 1968, Atheneum Press, NY. Courtesy Cold Spring Harbor Laboratory Archives; 9.5: © A. Barrington Brown/Photo Researchers; 9.9b: © M.C. Escher's "Drawing Hands" © 2007 The M.C. Escher Company-Holland. All rights reserved. www.mcescher.com; 9.13(top): © Ada Olins/Biological Photo Service; 9.13(bottom): © Science VU/Visuals Unlimited.

Chapter 10

Opener: © Ingram Publishing/Superstock RF; p. 195: © The Nobel Foundation, 1976.

Chapter 11

Opener: Courtesy, Bomber Command Museum of Canada; 11.4a: © Petit Format/Photo Researchers; 11.4b: © PhotoDisc/Getty RF; 11.10a: © Bristol Biomed Image Archive, University of Bristol. Image by Dr. John Eveson.

Chapter 12

Figure 12.1: © AP Images/Mike Wintroath; 12.2: © The McGraw-Hill Companies; 12.3: © Image Source/Getty RF; 12.4: © Science Photo Library/Photo Researchers; 12.8: © Brand X Pictures/Getty RF; p. 222: Courtesy, Rebekah Lieberman; p. 224(a): © Dr. Christine Harrison/Visuals Unlimited; p. 224(b): From R. Simensen, R. Curtis Rogers, "Fragile X Syndrome," American Family Physician, 39:186 May 1989. @ American Academy of Family Physicians; 12.12a: © Stockdisc/Punchstock RF; 12.12b: Bain, B.J., Diagnosis from the Blood Smear, New England Journal of Medicine, 2005; 353:498–507. © 2005 Massachusetts Medical Society. All rights reserved. Image courtesy Barbara J. Bain, Professor of Diagnostic Haematology, Imperial College, London; 12.16: © Kenneth Greer/Visuals Unlimited.

Chapter 13

Figure 13.1: © Science VU/Visuals Unlimited; 13.3: Courtesy, National Human Genome Research Institute; 13.6: © GE Medical Systems; 13.8a: Drawing by Walther Flemming; 13.8b: © CNRI/Photo Researchers; 13.9: © Courtesy Genzyme Corporation; 13.11: CNRI/Photo Researchers; 13.13: © Stockbyte/Veer RF; 13.14b: Courtesy, Allison Bradley; 13.16: Courtesy, Kathy Naylor; 13.19b: Courtesy Lawrence Livermore National Laboratory; 13.23: Courtesy, Ring Chromosome 20 Foundation.

Chapter 14

Opener: © Royalty-Free/Corbis; 14.1: © Comstock/Punchstock; 14.9: Image courtesy Esther N. Signer; 14.11: © PhotoDisc/Getty RF; 14.12(left): © Allan Baxter/Getty Images RF; 14.12(center top): © Image Source/Alamy RF; 14.12(center bottom): © Image Source/Punchstock RF; 14.12(right): © Chinafotopress/Getty Images.

Chapter 15

Opener: © Digital Vision/Getty RF; 15.2: © Stapleton Collection/Corbis; 15.7: Dr. Victor McKusick/Johns Hopkins University School of Medicine; 15.11: © Goodshoot/Alamy RF; 15.12: © Deanne Fitzmaurice/K&D Photography, Inc.; 15.13: © The McGraw-Hill Companies, Inc./Barry Barker, photographer; p. 289: © Centers for Disease Control and Prevention/Janice Carr; p. 295: Courtesy, Marie Deatherage; 15.18: Courtesy of the Harry H. Laughlin Papers Collection, Pickler Memorial Library, Truman State University. Image courtesy of Dolan DNA Learning Center.

Chapter 16

Opener: © Royalty-Free/Corbis; 16.3: Michael Hagelberg/Arizona State University Research Publications; 16.4: © Volker Steger/Nordstar-4 Million Years of Man/SPL/Photo Researchers; 16.5: Homo sapiens idaltu reconstruction © 2002 Jay H. Matternes; 16.6: © Joe McNally/Joe McNally Photography; 16.9: © Burt Silverman/Silverman Studios; 16.10a: © Peter Johnson/Corbis; 16.11a, c: Courtesy, James H. Asher, Jr.; 16.11b: © Trajano Paiva/Alamy; p. 313: © Imgram Publishing; 16.13: From F.R. Goodman and P.J. Scambler. "Human HOX Gene Mutations", Clinical Genetics, 59, Jan. 2001, page 2, 1A. By permission of John Wiley and Sons; 16.14: © Yann Arthus-Bertrand/Corbis.

Chapter 17

Opener: © CDC/C. Goldsmith, P. Feorino, E.L. Palmer, W.R. McManus; 17.1: © CDC/Janice Haney Carr; p. 327: © Royalty-Free/Corbis; 17.2: © Manfred Kage/Photolibrary; 17.5: © Biology Media/Photo Researchers; 17.11: © Bettmann/Corbis; 17.14: © Royalty-Free/Corbis; p. 338: © Courtesy, Dr. Maureen Mayes; 17.15(left): © Loisjoy Thurstun/Bubbles Photolibrary/Alamy; 17.15(right): © Pixtal/age fotostock; 17.16: © Science VU/Visuals Unlimited.

Chapter 18

Opener: © Dr. Ken Greer/Visuals Unlimited; 18.1a: © The McGraw-Hill Companies, Inc.; 18.1b: © Nancy Kedersha/Photo Researchers; 18.2b: © Steve Gschmeissner/SPL/Getty Images; 18.12b: © Centers for Disease Control ; p. 358: Courtesy, Erin Zammett-Ruddy. Image © 2005 Basil Childers; 18.14a: © The McGraw-Hill Companies, Inc./photographer; 18.15(top): © PhotoDisc/Getty RF; 18.15(bottom): © PhotoDisc/Getty RF; 18.16: S.A. Armstrong, et al. "MLL translocations specify a distinct gene expression profile that distinguishes a unique leukemia." Nature Genetics, Vol. 30, Fig. 5, p. 41–47, January 2002.

Chapter 19

Opener: Courtesy Barry Palevitz; 19.1: © Eye of Science/Photo Researchers; 19.4(left): © Maximilian Stock Ltd./SPL/Photo Researchers; 19.4(right): © Eric Kamp/Photolibrary; p. 380: © Image Source/Getty RF; 19.6: Courtesy MRC Harwell; 19.7: © Coast Guard photo by Petty Officer 2nd Class John D. Miller.

Chapter 20

Opener: Courtesy Ilyce Randell; 20.1: © Mediacolor's/Alamy; 20.3: © Charles Lewis/The Buffalo News; 20.6: © MedicalRF.com/Getty Images; 20.7: © Anna Powers; 20.8: © Courtesy Paul and Migdalia Gelsinger. Photo: Arizona Daily Star; 20.11(both): Courtesy, Wendy Josephs.

Chapter 21

Figure 21.1: © Keri Pickett/World Picture Network; p. 409: © Tony Brain/SPL/Photo Researchers; 21.3: © CNRI/Phototake; 21.4c: Integra. Photo courtesy of Ronald Carson, The Reproductive Science Center of Boston; 21.6: Courtesy, Dr. Anver Kuliev.

Chapter 22

Opener: © Don Carstens/Artville; 22.9a: Courtesy National Human Genome Research Institute; 22.9b: © IT Stock/Punchstock; 22.9c: © Photo courtesy of the state of Victoria (Australia), Department of Innovation, Industry and Regional Development; 22.9d: Courtesy, The Broad Institute of M.I.T. and Harvard University; p. 432(a): © Author's Image/Punchstock; p. 432(b): © C Squared Studios/Getty Images.

Index

Note: Information in figures and tables is denoted by *f* and *t*.

B

bacteria, weight and, 146
bacterial resistance, 289
balanced polymorphism, 290–292
banking
    oocyte, 413–414
    of stem cells, 40
Bardet-Biedl syndrome, 28
Barr body, 122, 123f, 248
base excision repair, 228–229
base pairs, complementary, 169–170, 171f
Basque people, 282, 309
Bateson, William, 71, 101
B cells, 328, 329f, 330–332, 334t
Beaudet, Arthur, 255
Becker muscular dystrophy, 216t, 221
Beckwith-Wiedemann syndrome, 126
behavior, genes and, 150–151
benign tumors, 349
"Berlin patient," 325
beta globin, 200, 212–213
beta radiation, 219
beta thalassemia, 213
binding proteins, 174f, 175
Binet, Alfred, 153
biobanks, population, 273
bioethics, 2
Biological Weapons Convention, 345
biopsy, polar body, 418
bioremediation, 381
biotechnology, 12–13, 372
bioterrorism, 345
biotinidase deficiency, 21
bioweapons, 344–345
bipolar disorder, 140t, 151t, 157
bird flu, 344
birth defects, 60–62
births, multiple, 57–58
bladder
    female, 46f
    male, 45f
blastocyst, 53, 55f, 56t
blastomeres, 53
blood fluidity, 133
blood groups, 326
blood plasma, 201
blood pressure
    concordance of, in twins, 140t
    heritability of, 138t
blood types, 92–93, 92f, 103, 279
Bloom syndrome, 285t
"blue person disease," 226
BMI. *see* body mass index (BMI)
body mass index (BMI)
    definition of, 144
    genetic control of, 145t
    heritability of, 138t
body weight, 144–146
Bombay phenotype, 93
bone marrow, 329f
bone marrow transplant, 342
bonobo, 238
bottlenecks, population, 284–285
Bouchard, Thomas, 140
brain size, 313
BRCA1 gene, 284, 361–362, 367, 373
breast cancer, 284, 285t, 358, 361–362, 367

Brown, Timothy, 325
Buck, Carrie, 296, 296f
bulbourethral gland, 45f
Bulgaria, 280–281
"burning man syndrome," 27
"bystander effect," 219

C

Caenorhabditis elegans, 432
CAH. *see* congenital adrenal hyperplasia (CAH)
calcium channel, 216t
CALHM1 gene, 97
CAMs. *see* cellular adhesion molecules (CAMs)
Canavan disease, 285t, 388, 400
cancer
    angiogenesis and, 353
    breast, 284, 285t, 358, 361–362, 367
    characteristics of cells in, 352–354, 354t
    colon, 229, 363–364
    definition of, 349
    development of, 352f
    diagnosis, 366–367
    environmental causes of, 364–366
    fusion proteins and, 358
    gene expression and, 357–358
    genes contributing to, 362–364
    gene therapy and, 397
    genome, 364
    in history, 349
    inherited vs. sporadic, 351
    invasiveness of, 353
    lung, 132, 155–156, 353f
    metastasis of, 349
    microRNAs and, 356–362
    mutation and, 218
    oncogenes and, 357–360
    origins of cells in, 354, 355f
    p53 gene and, 360
    prostate, 202
    skin, 360
    stem cells, 354
    stomach, 360–361
    telomerase and, 350–351
    thyroid, 218, 348
    treatment of, 366–367
    tumor suppressors and, 360–362
    vegetables and, 365, 365f
cannibalism, 195–196, 292
Caravan gene, 402
carbohydrases, 180t
carbohydrates
    in cells, 20
    inborn errors of metabolism and, 21
    in plasma membrane, 26f
carcinogens, 218
cardiovascular health, 133
caretaker genes, 363
Caroline Islands, 284
CARTaGENE, 273
case-control study, 142, 366
casein, 180t
cataplexy, 152
cats
    calico, 123
    common chromosomes with, 315t
Caucasians, cystic fibrosis in, 264t
CCR5 delta 32 mutation, 325

CCR5 protein, 212
CCR5 receptor, 335
cDNA library. *see* complementary DNA (cDNA) library
cell(s), 20, 22f
    adhesion of, 36, 354t
    antigen-presenting, 327
    cancerous, 352–354
    carbohydrates in, 20
    chemical constituents of, 20
    components of, 19–30
    cycle, 31–34, 350–351
    cytoskeleton in, 26f, 28–30
    death of, 34–35
    diploid, 19, 80f
    division of, 30–34
    enzymes in, 20
    eukaryotic, 19f
    genes and, 2
    haploid, 19, 80f
    interaction of, 19, 35–36
    lineages, 36–37
    lipids in, 20, 23f, 24
    microtubules in, 22f, 28
    in mitosis, 31–32
    nucleic acids in, 20
    organelles in, 19, 20–25, 25t
    overview of, 19
    plasma membrane in, 20, 22f, 23f, 24f, 25–27
    premature aging of, 54
    prokaryotic, 19, 19f
    proteins in, 20
    secretion in, 20–24, 23f
    signal transduction in, 35
    somatic, 19
    specialization of, 38f
    stem, 5, 18, 19, 36–40
    surfaces, importance of, 326–328
cellular adhesion molecules (CAMs), 36
cellular immune response, 329, 332–333
CENP-A. *see* centromere protein A (CENP-A)
centenarians, 65–66
centimorgans, 103
centrioles, 22f, 31
centromere protein A (CENP-A), 236
centromeres, 31, 207t, 208, 236–237
centrosomes, 31
cerebellar ataxia, 412
cervix, 46f
CFS. *see* chronic fatigue syndrome (CFS)
chains
    heavy, 331
    light, 331
chain switching, 200–201
Chamorro people, 309
chaperone proteins, 191–192
Chargaff, Erwin, 167t
Chase, Martha, 166, 167t
Chernobyl disaster, 218
chimpanzee, pygmy, 238
China
    eugenics in, 296
    genomic medicine in, 15t
    sex ratios in, 116
    vaccine technology in, 340
chloride channels, 27
chondrodysplasia, 214t
chorion, 56f, 57f

growth hormone, 379*t*
GTP. *see* guanosine triphosphate (GTP)
guanine, 3, 167, 169*f*
guanosine triphosphate (GTP), 188
Guthrie test, 393
gypsies, 280

H

Haldane, J. B. S., 173
hands, in Rett syndrome, 125
haplogroups, 317
haploid cells, 19, 47*f*, 80*f*
haplotypes, 105, 106*f*
Hardy, Godfrey Harold, 262
Hardy-Weinberg equation, 262–263
Hardy-Weinberg equilibrium, 262–265
HD. *see* Huntington disease (HD)
health care
   copy number variants in, 225f
   genetics and, 10–12
   proteomics and, 202f
   race and, 136
   stem cells in, 37–40
heart health, 133
heavy chains, 331
height, 134–135, 138*t*
"HeLa" cells, 352
helicase, 174*f*, 175
helper T cells, 329*f*, 334*t*
heme, 94
Hemings, Sally, 9, 9*f*
hemizygous, 116
hemochromatosis, hereditary, 394
hemoglobin, 180*t*, 200–201, 200*f*
hemoglobin C, 226
hemoglobin M, 226
hemoglobin S, 226
hemolytic anemia, 227, 337*t*
hemophilia A, 215*t*, 223, 265
hemophilia B, 119–120, 119*f*, 217
Hensel, Abigail and Britney, 58
hepatic lipase, 8
Herceptin, 395
HERC2 gene, 74
Hereditary Genius (Galton), 153
hereditary hemochromatosis (HH), 394
hereditary nonpolyposis colon cancer (HNPCC), 229
hereditary spherocytosis, 29–30
HER2 gene, 367, 395
heritability, 137–139
hermaphroditism, 112
herpes simplex, as teratogen, 62
Herrick, James, 213
Hershey, Alfred, 166, 167*t*
HERVs. *see* human endogenous retroviruses (HERVs)
heterochromatin, 236
heterogametic sex, 111–112
heterogeneity, genetic, 95–96, 98*t*
heteroplasmy, 99–100
heterozygote, manifesting, 123
heterozygous, 71
HGPRT enzyme, 124
HH. *see* hereditary hemochromatosis (HH)
Hillenbrand, Laura, 149
Hill People of New Guinea, 308
Hippocrates, 349
Hirschsprung disease, 216*t*

Hispanic Americans, cystic fibrosis in, 264*t*
histocompatibility complex, major, 326–328
histone complex, 203, 204*f*
histones, 171, 172*f*
history, 9–10
HIV. *see* human immunodeficiency virus (HIV)
HLAs. *see* human leukocyte antigens (HLAs)
HMGA2 gene, 142
HNPCC. *see* hereditary nonpolyposis colon cancer (HNPCC)
Hobbits, 301
Hodge, Nancy, 267
Holland, 199
Holocaust survivors, 272
homeobox proteins, 314–315
homeotics, 55
hominins, 302–305
hominoids, 302–305
Homo, 305–306
homocystine metabolism, 133*t*
Homo erectus, 306
homogametic sex, 111
Homo habilis, 306
homologous pairs, 46
homoplasmy, 100
homosexuality, 113, 115
homozygosity mapping, 142–143
homozygous, 71
Hope for Trisomy 13 + 18, 244*t*
Hopi Indians, 279
hormone, definition of, 34
HOX genes, 314–315
Huli people, 309
human endogenous retroviruses (HERVs), 207
Human Genome Project, 2, 424–426
human growth hormone, 379*t*
human immunodeficiency virus (HIV)
   AIDS and, 334–336
   drugs for, 336t
   evolution of, 289–290
   stem cell transplants for, 325
   T cells and, 334–335
   tuberculosis and, 288
human immunodeficiency virus (HIV), CCR5 protein and, 212
human leukocyte antigens (HLAs), 327
human microbiome, 433–434
Human Microbiome Project, 13, 433–434
human origins, 302–309
humans, modern, 308
humoral immune response, 329, 330–332
Hunter syndrome, 123
Huntingtin, 215*t*
Huntington disease (HD), 10, 69, 76, 82–83, 193, 193*t*, 215*t*, 424
Hutchinson-Gilford progeria syndrome, 64, 64*t*
hypercholesterolemia, familial, 21, 91–92, 92*f*, 215*t*, 221
hypertension, 140*t*
hypertrichosis, congenital generalized, 120, 121*f*
hypocretin, 152

I

ICF syndrome, 204*t*
ichthyosis, 117, 117*f*
ICSI. *see* intracytoplasmic sperm injection (ICSI)
Idaltu Man, 305–306
identity

   DNA profiling and, 271–272
   establishment of, with genetics, 9–10
   in mixtures of DNA, 272
   sexual, 115t
ideogram, 242–243
idiotypes, 331
immune deficiencies
   inherited, 333–334
immune system
   abnormal responses of, 333–339
   adaptive, 329–333
   allergies and, 337–339
   alterations of, 340–342
   autoimmunity and, 336–337
   inflammation in, 328
   innate, 328–329
   lymphatics in, 328
   major histocompatibility complex and, 326–328
   physical barriers in, 328–329
   pregnancy and, 338
   rejection by, 342
   transplants and, 342–343
   vaccines and, 340–341
   viruses and, 327
immunoglobulin types, 332*t*
immunotherapy, 341
implantation, 53–54
implants, stem cells and, 39–40
imprinting, genomic, 124–127
imprinting disorders, 126–127
inborn errors of metabolism, 21, 240
Incas, 344
incidence, definition of, 137
incomplete dominance, 91–92, 92*f*, 98*t*
incontinentia pigmenti (IP), 120
indels, 311
independent assortment, 48, 79, 80*f*, 103, 104*f*
India
   genomic medicine in, 15t
   sex ratios in, 116
indigenous peoples, 309–310
indigo, 378–379
Indonesia, 317*t*
induced mutation, 218–219
induced pluripotent stem (iPS) cells, 37
infertility, 407–410
infertility testing, 410
infidelity testing, 175–176
inflammation, 328
influenza, schizophrenia and, 158
information maximization, 205–206
Ingram, V. M., 213
inheritance
   autosomal dominant, 70
   autosomal recessive, 70
   of linked genes, 101–102
   of longevity, 65
   in mitochondria, 98, 100f
   modes of, 70, 75–78
   recombinant, 101
   segregation and, 71–73
   on sex chromosomes, 116–124
   sex-influenced, 122
   sex-limited, 122
   single-gene, 73–79
   X-linked dominant, 120
   X-linked recessive, 117–120
inherited cancer, 351

inherited colon cancer, 229
inherited immune deficiencies, 333–334, 398
initiation codon, 189
initiation complex, 188–189, 189*f*
innate immunity, 328–329
inner cell mass, 53–54, 55*f*
Innocence Project, 260
insemination, intrauterine, 410–412
insertion mutations, 220*t*, 221
insomnia, fatal familial, 194
insulin, 180*t*, 378, 379*t*
integrins, 36
intelligence
   definition of, 153
   heritability of, 138t, 154t
   success and, 154f
intelligence quotient (IQ) test, 153–154
interference, RNA, 382–384
interferons, 329, 333*t*, 341, 379*t*
interleukin-2, 379*t*
interleukins, 329, 333*t*
intermediate filaments, 28
interphase, 31
intersex, 112
intolerance, lactose, 278
intracellular digestion, 24–25
intracytoplasmic sperm injection (ICSI),
   412–413, 415
intrafallopian transfer, 413
intrauterine growth retardation (IUGR), 63
intrauterine insemination (IUI), 410–412, 416*t*
introns, 185, 207*t*
Inuit people, 309
invasiveness
   of cancer cells, 353
   of gene therapy, 397f
inversion, 243*t*, 253–254, 255*t*
in vitro fertilization (IVF), 412–413, 415, 416*t*
in vivo gene therapy, 397
ion channels, faulty, 27
ionizing radiation, 219
IP. *see* incontinentia pigmenti (IP)
iPS. *see* induced pluripotent stem (iPS) cells
IQ test. *see* intelligence quotient (IQ) test
Irons, Ernest, 212
isoagglutinin, 92
isochromosome, 243*t*, 254–255
isograft, 342
isolation, of genes, 376–377
isotretinoin, 61
IUGR. *see* intrauterine growth retardation (IUGR)
IUI. *see* intrauterine insemination (IUI)
IVF. *see in vitro* fertilization (IVF)

J

Jacobs syndrome, 250
Japan, 218
Jefferson, Thomas, 9
Jeffreys, Alec, 266, 272, 273
Jenner, Edward, 340
Joubert syndrome, 313
Jumping Frenchmen of Maine syndrome, 74

K

Kalahari people, 309
Kallmann syndrome, 121
karyotypes
   chromosomes and, 237–238

definition of, 5
evolution and, 238
Kelly, Hunter, 392
keratin, 180*t*
kinases, 180*t*
King George III, 94
Kinsearch Registry, 9
Klinefelter syndrome, 242, 249
knockouts, from gene targeting, 384
Köhler, George, 341
Komi people, 309
Kozma, Chahira, 231
Krabbe disease, 392
Kurds, 309
kuru, 195–196, 195*f*, 292

L

labia majora, 46*f*
labia minora, 46*f*
Lacks, Henrietta, 352
lactase deficiency, 21
lactase persistence, 278
lactocytes, 24
lactose, 24
lactose intolerance, 278
lactose tolerance, 278
lamin A gene, 215
Lancaster County, Pennsylvania, 283
"laughing disease," 195–196
law of segregation, 71, 72*f*, 73*t*, 98
LCA. *see* Leber congenital amaurosis (LCA)
LD. *see* linkage disequilibrium (LD)
Lebanon, 317*t*
Leber congenital amaurosis (LCA), 89, 95
Leber's congenital amaurosis II, 401
Lee, Pearl, 213
legislation, on genetic information, 12
Leigh syndrome, 99
Lemba people, 10, 10*f*
lens crystallins, 95
LEOPARD syndrome, 18
leptin, 144–145, 145*t*
leptin receptor, 145*t*
leptin transporter, 145*t*
Lesch-Nyhan syndrome, 21, 123–124
lethal alleles, 90, 98*t*
leukemia
   acute lymphoblastic, 366
   acute promyelocytic, 358–359
   Gleevec for, 358–359
   mixed lineage, 204
leukocyte adhesion, 133*t*
leukotreine A4 hydrolase, 136
Levan, Albert, 241
levels of organization, 6*f*
Levene, Phoebus, 167*t*
Lewis blood group, 326*t*
Lewy body dementia, 193*t*
library
   complementary DNA, 377
   genomic, 377
Lieberman, Rebekah, 222
ligand, definition of, 26
ligase, 174*f*, 175
light chains, 331
limbic system, 155
Lindeman, Chase, 417
Lindeman, Jack, 417

lineages, cell, 36–37
linkage, 101–106
   crossing over and, 102f
   definition of, 101
   discovery of, 101–102
   independent assortment and, 103, 104f
   maps, 102–103
   studies, 424
linkage disequilibrium (LD), 105
lip, cleft, 137*t*, 140*t*
lipases, 180*t*
"lipid rafts," 26–27
lipids
   in adrenoleukodystrophy, 25
   in cells, 20, 23f, 24
   inborn errors of metabolism and, 21
   in plasma membrane, 26, 26f
   in Tay-Sachs disease, 24
lipoprotein lipase, 133
liver, in gene therapy, 398*f*
load, genetic, 285
LOD scores, 105
longevity, 65–66
long-QT syndrome, 27
lung cancer, 132, 155–156, 353*f*
lungs, in gene therapy, 398*f*
lung surfactant protein, 379*t*
lupus, 336, 337*t*
lymphatics, 328
lymphocytes, 328
lymphoproliferative disease, X-linked, 334*t*
Lyon, Mary, 122
Lyons, Glenda, 417
Lyons, Scott, 417
lysosomal enzymes, 24*f*
lysosomal storage disease, 24, 395*t*
lysosome, 22*f*, 23*f*, 24–25, 24*f*, 25*t*

M

MAbs. *see* monoclonal antibodies (MAbs)
MacLeod, Colin, 166, 167*t*
macroevolution, 262
macrophages, 329*f*, 334*t*
mad cow disease, 196
major depressive disorder, 151*t*, 156–157
major histocompatibility complex (MHC),
   326–328
Malagasy, 309
malaria, 226*t*, 290–291
Malaysia, 317*t*
male infertility, 408
male reproductive system, 45
male sex chromosome aneuploids, 249–250
malformations of cortical development (MCD),
   179
malignant tumors, 349. *see also* cancer
manifesting heterozygote, 123
manure, pig, 371
Maori people, 309
maple syrup urine disease, 21
mapping, homozygosity, 142–143
maps, linkage, 102–103
Marfan syndrome, 95, 215, 215*t*, 216*t*
marijuana, 155
markers, genetic, 105, 315–316
Marshall-Graves, Jennifer A., 114
mass spectrometry, 393

N

NAA. *see* N-acetylaspartate (NAA)
N-acetylaspartate (NAA), 400
nail-patella syndrome, 104, 105*f*
narcolepsy, 152–153
Nash, Adam, 407
Nash, Jack, 407
Nash, John, Jr., 157
Nash, Lisa, 407
Nash, Molly, 407
Nathans, Jeremy, 118
Native Americans, 319–320
natural disaster victims, 271
natural killer cells, 334*t*
natural selection, 261, 286–292
Naura Island, 145
Nazis, 199
N-CAM. *see* neural cellular adhesion molecule (N-CAM)
ncRNAs. *see* noncoding RNAs
Neanderthals, 306, 307–308
Neel, James, 145
negative selection, 286–287
Nelmes, Sarah, 340
NEMO gene, 120
Neolithic age, 306*t*
nervous tissue
    overview of, 6*t*
neural cellular adhesion molecule (N-CAM), 154
neural stem cells, 179
neural tube, 58–59
neural tube defect (NTD), 58–59, 137, 295
neurexins, 159, 159*f*
neurofibromatosis type 1 (NF1), 35, 215*t*, 216
neurofibromin, 215*t*
neuroligins, 159, 159*f*
neuron, 150*f*
neuropathy, peripheral, 216*t*
neuropeptide Y, 145*t*
neurotransmission, 150
neutrophil immunodeficiency syndrome, 334*t*
newborn screening, genetic, 391*t*, 392–393
New World, population of, 319–320
NF1. *see* neurofibromatosis type 1 (NF1)
nicotine, 61, 155–156, 218
nicotinic receptor, 155
Niemann-Pick disease type A, 285*t*
Nile River, 281
Nirenberg, Marshall, 188
Nixon, Richard, 345
noncoding RNAs, 207–208
non-disjunction, 244
nonrandom mating, 261, 279–280
nonsense mutations, 220–221, 220*t*
nonsynonymous codons, 188
Northern Finland Birth Cohort, 134–135
NTD. *see* neural tube defect (NTD)
nuclear envelope, 22*f*
nuclear pore, 22*f*, 23*f*
nucleases, 180*t*
nucleic acids. *see also* deoxyribonucleic acid (DNA); ribonucleic acid (RNA)
    in cells, 20
    inborn errors of metabolism and, 21
nucleolus, 22*f*
nucleosomes, 171, 172*f*

nucleotide, 169, 169*f*
nucleotide excision repair, 228
nucleus, cell, 19, 22*f*, 23*f*, 25*t*
nutrients, as teratogens, 61–62
nutrigenetics testing, 394
nutrition, prenatal, 199

O

obesity, 144–146
obsessive compulsive disorder, 151*t*
OCA2 gene, 74
occupational hazards, 62
odor, of urine, 393
Ohno, Susumo, 114
Old Order Amish, 282
oncogenes, 216*t*, 350, 357–360
1,000 Genomes Project, 423
oocyte, 45
oocyte banking, 413–414
oocyte donation, 413–414
oocyte formation, 51–52
oocyte freezing, 416*t*
oogenesis, 51–52
oogonium, 51
opsin, 118
orangutans, 315*t*
orexin, 152
organ development, 201
organelles, 19, 20–25, 22*f*, 25*t*
organization, levels of, 6*f*
organogenesis, 58
orientation, sexual, 113, 115
origins, human, 302–309
ornithine transcarbamylase (OTC) deficiency, 398–400
Orrorin tugenensis, 302
osteoarthritis, 214*t*
osteogenesis imperfecta, 63, 95, 214, 214*t*
osteoporosis, 8
OTC deficiency. *see* ornithine transcarbamylase (OTC) deficiency
Ötzi people, 308
"out of Africa" hypothesis, 319
ovaries, 45, 46*f*, 55*f*
ovulation, irregular, 409

P

Page, David, 114
PAH. *see* phenylalanine hydrolase (PAH)
pain
    absent, 27
    extreme, 27
Painter, Theophilus, 241
Paleolithic age, 306*t*
paleontology, 302
pancreas, 201
panic disorder, 151*t*
paracentric inversion, 253, 254*f*
paralysis, sleep, 152
Parkinson disease, 193*t*
paroxysmal extreme pain disorder, 27
parsimony analysis, 315, 316*f*, 317, 319
Pasteur, Louis, 372
Patau syndrome, 247, 248
patenting DNA, 372–373
Pauling, Linus, 213

PCR. *see* polymerase chain reaction (PCR)
pedigrees
    analysis of, 82–84
    definition of, 5, 82
    mitochondrial, 98, 100, 101
penetrance, 93, 98*t*
penis, 45*f*
Pennington, Robert, 343
pericentric inversion, 254
period gene, 153
peripheral neuropathy, 216*t*
peroxisome, 22*f*, 24–25, 25*t*
Personal Genome Project, 423
personal genome sequencing, 434–435
PGD. *see* preimplantation genetic diagnosis (PGD)
p53 gene, 360
pharmacogenetic test, 394
pharmacogenomic test, 394
pharmacogenomic tests, 10*t*
pharmacological chaperone therapy, 395*t*, 396
phenocopies, 96, 98*t*
phenotype
    alleles and, 72
    definition of, 5
    frequencies, 261
    mutant, 72, 212
    wild type, 72
    X inactivation and, 122–124
phenylalanine, 193–194
phenylalanine hydrolase (PAH), 293
phenylketonuria (PKU), 90–91, 193–194, 193*t*, 261*t*, 291*t*, 292–294, 393
Philippines, 317*t*
Phipps, James, 340
phobias, 151*t*
phospholipids, membrane, 26, 26*f*
Physcomitrella paten, 432
physical barriers, in immune system, 328–329
pig grafts, 343
pig manure, 371
Pima Indians, 145
Pingelapese blindness, 284
Pingelapese people, 284
PKU. *see* phenylketonuria (PKU)
plasma, blood, 201
plasma cells, 329*f*, 330
plasma membrane, 20, 22*f*, 23*f*, 24*f*, 25–27
plasmid, 377
Plasmodium falciparum, 290
plastin 3 gene, 93
Plavix, 435
pleiotropy, 93–95, 94*f*, 98*t*
point mutations, 220–221
polar body, 51, 53*f*
polar body biopsy, 418
polyacrylamide, 266
polycystic kidney disease, 63
polydactyly, 8*f*, 93
polyendocrinopathy syndrome type 1, 337*t*
polygenic trait(s)
    definition of, 132
    fingerprints as, 134
    heart health as, 133
    height as, 134–135
    investigation of, 137–141
    quantitative trait loci and, 132